Index of Study Strategies

Elementary Algebra

Elementary Algebra

George Woodbury

College of the Sequoias

PEARSON

Addison
Wesley

Boston San Francisco New York
London Toronto Sydney Tokyo Singapore Madrid
Mexico City Munich Paris Cape Town Hong Kong Montreal

Editorial Director:	Christine Hoag
Editor in Chief:	Maureen O'Connor
Executive Editor:	Jennifer Crum
Executive Project Manager:	Kari Heen
Senior Project Editor:	Lauren Morse
Assistant Editor:	Antonio Arvelo
Production Manager:	Ron Hampton
Cover Designer:	Joyce Consentino Wells
Media Producers:	Sharon Tomasulo, Carl Cottrall, Ceci Fleming
Software Development:	Monica Salud, MathXL; Ted Hartman, TestGen
Senior Marketing Manager:	Michelle Renda
Marketing Coordinator:	Nathaniel Koven
Market Development Manager:	Dona Kenly
Senior Author Support/Technology Specialist:	Joseph K. Vetere
Senior Prepress Supervisor:	Caroline Fell
Senior Manufacturing Buyer:	Carol Melville
Senior Media Buyer:	Ginny Michaud
Composition/Production Coordination:	Pre-Press PMG
Cover photo:	© Keiji Watanabe/Getty Images

Library of Congress Cataloging-in-Publication Data

Woodbury, George 1967–
    Elementary algebra / George Woodbury.
        p. cm.
    ISBN-10: 0-321-16642-6   ISBN-13: 978-0-321-16642-5 (student)
    ISBN-10: 0-321-54204-5   ISBN-13: 978-0-321-54204-5 (preview edition)
    1. Algebra—Textbooks. I. Title.
QA152.3.W659 2008
512.9—dc22   2007060830

5 6 7 8 9 10—VHP—11

To Tina, Dylan, and Alycia
You make everything meaningful and worthwhile—
yesterday, today, and tomorrow.

Contents

Preface

George Woodbury's primary goal as a teacher is to empower his students to succeed in algebra and beyond. To successfully teach developmental algebra courses, instructors must find a way to reach two equally important goals: 1) fully prepare students for collegiate-level mathematics, and 2) build students' confidence in their mathematical abilities.

One of the ways George prepares his students for future mathematics courses is by introducing the fundamental algebraic concepts of graphing and functions early in *Elementary Algebra*. He then consistently and frequently incorporates these concepts throughout the text. George also understands that strong exercise sets that provide both volume and variety are critical to student success, so he has written extensive sets that incorporate not only a substantial amount of skill and drill, but also plenty of writing exercises to encourage critical thinking and creativity. The key to success for developmental students is often not just about the math—it's about making sure that students understand how to study and prepare for a math class. To help give students the confidence they require to succeed, George fully integrates targeted study strategies throughout the text and also provides further guidance via a *Math Study Skills Workbook*, written specifically for his texts by Alan Bass.

Success breeds further success. With the content, features, and ancillaries of *Elementary Algebra*, George Woodbury strives to help his students not simply to "make it through" to intermediate algebra or college algebra or precalculus, but to truly have the skills and conceptual understanding necessary for further success in those courses.

Practice Makes Perfect!

Examples   Based on his experiences in the classroom, George Woodbury has included an abundance of clearly and completely worked-out examples.

Quick Checks   The opportunity for practice shouldn't be designated only for the exercise sets. Nearly every example in this text is immediately followed by a *Quick Check* exercise, allowing students to practice what they have learned and assess their understanding of newly learned concepts. Answers to the *Quick Check* exercises are provided in the back of the book so students can independently check their understanding.

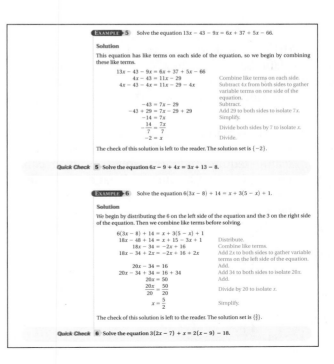

EXAMPLE 5   Solve the equation $13x - 43 - 9x = 6x + 37 + 5x - 66$.

Solution

This equation has like terms on each side of the equation, so we begin by combining these like terms.

$$13x - 43 - 9x = 6x + 37 + 5x - 66$$
$$4x - 43 = 11x - 29 \qquad \text{Combine like terms on each side.}$$
$$4x - 43 - 4x = 11x - 29 - 4x \qquad \text{Subtract } 4x \text{ from both sides to gather variable terms on one side of the equation.}$$
$$-43 = 7x - 29 \qquad \text{Subtract.}$$
$$-43 + 29 = 7x - 29 + 29 \qquad \text{Add 29 to both sides to isolate } 7x.$$
$$-14 = 7x \qquad \text{Simplify.}$$
$$\frac{-14}{7} = \frac{7x}{7} \qquad \text{Divide both sides by 7 to isolate } x.$$
$$-2 = x \qquad \text{Divide.}$$

The check of this solution is left to the reader. The solution set is $\{-2\}$.

Quick Check 5   Solve the equation $6x - 9 + 4x = 3x + 13 - 8$.

EXAMPLE 6   Solve the equation $6(3x - 8) + 14 = x + 3(5 - x) + 1$.

Solution

We begin by distributing the 6 on the left side of the equation and the 3 on the right side of the equation. Then we combine like terms before solving.

$$6(3x - 8) + 14 = x + 3(5 - x) + 1$$
$$18x - 48 + 14 = x + 15 - 3x + 1 \qquad \text{Distribute.}$$
$$18x - 34 = -2x + 16 \qquad \text{Combine like terms.}$$
$$18x - 34 + 2x = -2x + 16 + 2x \qquad \text{Add } 2x \text{ to both sides to gather variable terms on the left side of the equation.}$$
$$20x - 34 = 16 \qquad \text{Add.}$$
$$20x - 34 + 34 = 16 + 34 \qquad \text{Add 34 to both sides to isolate } 20x.$$
$$20x = 50 \qquad \text{Add.}$$
$$\frac{20x}{20} = \frac{50}{20} \qquad \text{Divide by 20 to isolate } x.$$
$$x = \frac{5}{2} \qquad \text{Simplify.}$$

The check of this solution is left to the reader. The solution set is $\{\frac{5}{2}\}$.

Quick Check 6   Solve the equation $3(2x - 7) + x = 2(x - 9) - 18$.

Exercises   Woodbury's text provides more exercises than most other texts, allowing students ample opportunity to develop their skills and increase their understanding. The exercise sets are filled with both traditional skill- and drill-type exercises as well as unique exercise types that require thoughtful and creative responses. In addition, a great deal of focus has been placed on reading and interpreting graphs, graphing equations, and understanding and working with functions.

Types of Exercises

- **Vocabulary Exercises:** Each exercise set starts out with a series of exercises that check student understanding of the basic vocabulary covered in the preceding section. (See pg. 94)

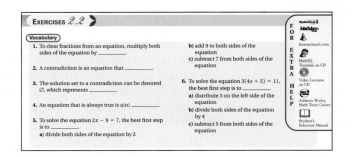

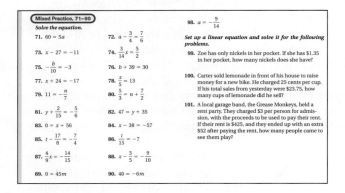

- **Mixed Practice Exercises:** Mixed Practice exercises are provided as appropriate throughout the book to give students an opportunity to practice multiple types of problems in one setting. These exercises require students to focus on determining the correct method to use to solve the problem and reduce their tendency to simply memorize steps to solving the problems for each objective. (See pg. 181)

- **Writing in Mathematics Exercises:** Asking students to explain their answer in written form is an important skill that often leads to a higher level of understanding as acknowledged by the *AMATYC Standards*. (See pg. 96)

- **Solutions Manual Exercises:** Solutions Manual exercises found within the Writing in Mathematics feature require students to solve a problem completely with step-by-step explanations as if they were writing their own solutions manual. (See pg. 122)

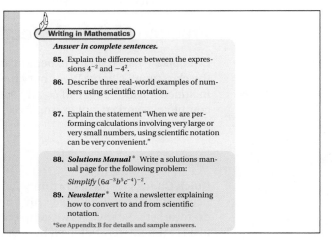

- **Newsletter Exercises:** As a way to encourage students to be creative in their mathematical writing, students are asked to explain a mathematical topic. The explanation should be in the form of a short, visually appealing article that might be published in a newsletter read by people who are interested in learning mathematics. (See pp. 122, 134, 152)

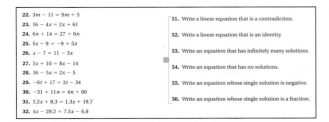

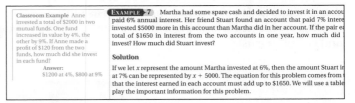

- **Quick Review Exercises:** Appearing once per chapter, Quick Review Exercises are a short selection of review exercises geared toward helping students maintain skills they have previously learned and preparing them for upcoming concepts. (See pg. 122)

- **Community Exercises:** Building a community of learners within the classroom can make for a much more dynamic classroom and provide a sense of trust and collaboration among students. The Community Exercises, indicated in the Annotated Instructor's Edition with the **CE** icon, require thought, understanding, and communication. These optional exercises are provided as appropriate throughout the text. (See pp. 132, 183, 192)

- **Stretch Your Thinking Exercises:** Appearing at the end of each chapter, Stretch Your Thinking exercises ask students to go the extra mile by taking what they have learned in the chapter and applying it to a problem that requires some additional thought. These exercises can be done individually or in groups. (See pg. 238)

- **Classroom Examples:** Having in-class practice problems at your fingertips is extremely helpful to both new and experienced instructors. These instructor examples, called Classroom Examples, are included in the margins of the AIE as in-class practice problems. (See pg. 76)

- **Worksheets for Classroom or Lab Practice** offer extra practice exercises for every section of the text with ample space for students to show their work. These lab- and classroom-friendly workbooks also list the learning objectives and key vocabulary terms for every text section, along with vocabulary practice problems.

Early-and-Often Approach to Graphing and Functions

Woodbury introduces the primary algebraic concepts of graphing and functions early in the text (Chapter 3) and then consistently incorporates them throughout the text, providing optimal opportunity for their use and review. Introducing functions and graphing early helps students become comfortable with reading and interpreting graphs and function notation. Working with these topics throughout the text establishes a basis for understanding that better prepares students for future math courses.

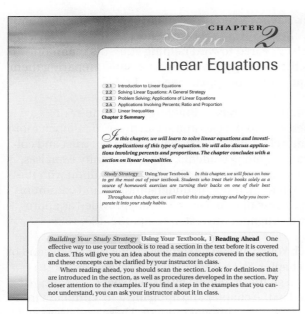

Building Your Study Strategies

Woodbury introduces a *Study Strategy* in each chapter opener. Each strategy is then revisited and expanded upon prior to each section's exercise set in *Building Your Study Strategy* boxes and then again at the end of the chapter. These helpful Study Strategies outline good study habits and ask students to apply these skills as they progress through the textbook. Study Strategy topics include Study Groups, Using Your Textbook Effectively, Test Taking, Overcoming Math Anxiety, and many more. (See pgs. 141, 149, 177)

The ***Math Study Skills Workbook***, created specifically to complement the Woodbury text, gives students suggestions and activities that will help to make them better mathematics students. Students are given further guidance on better note-taking, time-management, test-taking skills, and other strategies.

Additional Student Support

Applying Skills and Solving Problems

Problem solving is a skill that is required daily in the real world and in mathematics. Based on George Pólya's text, *How to Solve It*, George Woodbury presents a **six-step problem-solving strategy** in Chapter 2 that lays the foundation for solving applied problems. He then expands on this problem-solving strategy throughout the text by incorporating **hundreds of applied problems** on topics such as motion, geometry, and mixture problems. Interesting themes in the applied problems include investing and saving money, understanding sports statistics, landscaping, home ownership, and cell phone usage.

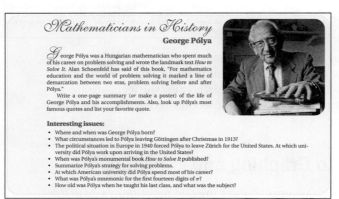

Mathematicians in History

Mathematicians in History activities provide a structured opportunity for students to learn about the rich and diverse history of mathematics. These short research projects, which ask students to investigate the life of a prominent mathematician, can be assigned as independent work or used as a collaborative-learning activity. (See pg. 139)

Using Your Calculator

The optional *Using Your Calculator* feature is presented throughout the text, giving students guided calculator instruction with screen shots, as appropriate, to complement the material being covered. (See pg. 89)

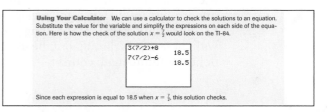

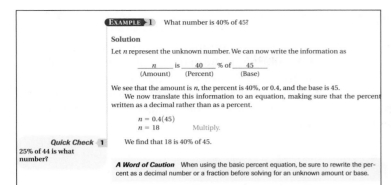

A Word of Caution

A Word of Caution tips, located throughout the text, help students avoid misconceptions by pointing out typical errors that students often make. (See pg. 114)

End-of-Chapter Content

Each chapter concludes with a **Chapter Summary,** a **summary** of the chapter's **Study Strategies, Chapter Review Exercises,** and a **Chapter Test.** Together, these are an excellent resource for extra practice and test preparation. Full solutions to highlighted chapter review exercises are provided in the back of the text as yet one more way for students to assess their understanding and check their work. A set of **Cumulative Review** exercises can be found after Chapters 3, 6, and 9. These exercises are strategically placed to help students review before midterm and final exams.

Pass the Test: Chapter Test Solutions on CD provides students with video footage of an instructor working through the complete solution to all the exercises in all chapter tests in the textbook. Additionally, the videos have optional English captioning. The CD also contains interactive vocabulary flashcards, a Spanish glossary, and videos with tips for time management.

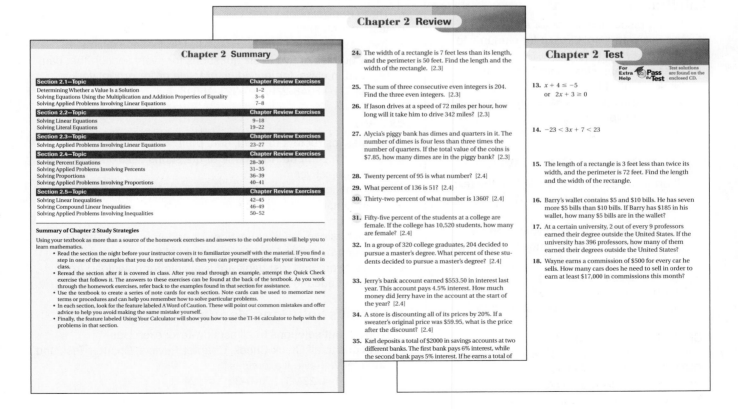

Overview of Supplements

The supplements available to students and instructors are designed to provide the extra support needed to help students be successful. As you can see from the list of supplements that follows, all areas of support are covered from tutoring help (Math Tutor Center) to guided solutions (video lectures and solutions manuals) to help in being a better math student (*Math Study Skills Workbook*). We hope that these additional supplements will help students master the skills, gain confidence in their mathematical abilities, and move onto the next course.

Student Supplements	Instructor Supplements

Math Study Skills Workbook

- Created specifically to complement the Woodbury text
- Helps to make better mathematics students through study suggestions and activities to be completed by the student
- Further guidance on better note-taking, time-management, test-taking skills, and other strategies
- Written by Alan Bass

ISBNs: 0-321-54257-6, 978-0-321-54257-1

Worksheets for Classroom or Lab Practice

- Extra practice exercises for every section of the text with ample space for students to show their work.
- These lab- and classroom-friendly workbooks also list the learning objectives and key vocabulary terms for every text section, along with vocabulary practice problems

ISBNs: 0-321-52310-5, 978-0-321-52310-5

Video Lectures on CD or DVD

- Short lectures for each section of the text presented by author George Woodbury and Mark Tom
- Complete set of digitized videos on CD-ROMs or DVDs for student use at home or on campus
- Ideal for distance learning or supplemental instruction
- Video lectures include optional English captions

CD ISBNs: 0-321-50652-9, 978-0-321-50652-8
DVD ISBNs: 0-321-54797-7, 978-0-321-54797-2

Annotated Instructor's Edition

- Contains answers to all exercises in the textbook and highlights special Community Exercises
- Provides Teaching Tips and Classroom Examples

ISBNs: 0-321-46757-4, 978-0-321-46757-7

Instructor and Adjunct Support Manual

- Includes resources to help both new and adjunct faculty with course preparation and classroom management by offering helpful teaching tips correlated to the sections of the text.

ISBNs: 0-321-46975-5, 978-0-321-46975-5

Printed Test Bank

- Three free-response tests per chapter and two multiple-choice tests per chapter
- Two free-response and two multiple-choice final exams
- Short quizzes of five to six questions for every section that can be used in class, for individual practice or for group work

ISBNs: 0-321-46548-2, 978-0-321-46548-1

TestGen

- Enables instructors to build, edit, print, and administer tests
- Features a computerized bank of questions developed to cover all text objectives
- Algorithmically based, allowing instructors to create multiple, but equivalent, versions of the same question or test with the click of a button
- Instructors can modify questions or add new questions
- Tests can be printed or administered online
- The software and test bank are available for download from Pearson Education's online catalog.

Instructor's Solutions Manual

- Worked-out solutions to all section-level exercises
- Solutions to all Quick Check, Chapter Review, Chapter Test, and Cumulative Review exercises

ISBNs: 0-321-47224-1, 978-0-321-47224-3

Student Supplements	**Instructor Supplements**

Pass the Test:
Chapter Test Solutions on CD

- Provides students with a video of an instructor working through all exercises from all chapter tests in the textbook.
- Includes interactive vocabulary flashcards, a Spanish glossary, and a video with tips for time management.
- Optional English captioning
- Automatically included in every new copy of the textbook.

Student's Solutions Manual

- Worked-out solutions for the odd-numbered section-level exercises
- Solutions to all problems in the Chapter Review, Chapter Test, and Cumulative Review exercises

ISBNs: 0-321-47442-2, 978-0-321-47442-1

MathXL Tutorials on CD

- Algorithmically generated practice exercises that correlate at the objective level to the content of the text
- Includes an example and a guided solution to accompany every exercise as well as video clips for selected exercises
- Recognizes student errors and provides appropriate feedback; generates printed summaries of student progress

ISBNs: 0-321-50668-5, 978-0-321-50668-9

The Math Tutor Center

The Math Tutor Center is staffed by qualified mathematics instructors who provide students with tutoring on examples and odd-numbered exercises from the textbook. Tutoring is available via toll-free telephone, toll-free fax, e-mail, or the Internet. White Board technology allows tutors and students to actually see problems worked while they "talk" in real time over the Internet during tutoring sessions.

www.mathtutorcenter.com

PowerPoint Slides

- Present key concepts and definitions and examples from the text
- Available in MyMathLab or can be downloaded from Pearson Education's online catalog

Active Learning Lecture Slides

- Multiple-choice questions available for each chapter of the text
- Formatted in PowerPoint and can be used with classroom response systems
- Available in MyMathLab or can be downloaded from Pearson Education's online catalog

Math Adjunct Support Center

The Math Adjunct Support Center is staffed by qualified mathematics instructors with over 50 years of combined experience at both the community college and university level. Assistance is provided for faculty in the following areas:

- Suggested syllabus consultation
- Tips on using materials packaged with the book
- Book-specific content assistance
- Teaching suggestions including advice on classroom strategies

For more information, visit
www.aw-bc.com/tutorcenter/math-adjunct.html

MathXL® www.mathxl.com

MathXL is a powerful online homework, tutorial, and assessment system that accompanies Pearson Education textbooks in mathematics or statistics. With MathXL, instructors can create, edit, and assign online homework and tests using algorithmically generated exercises correlated at the objective level to the textbook. They can also create and assign their own online exercises and import TestGen tests for added flexibility. All student work is tracked in MathXL's online gradebook. Students can take chapter tests in MathXL and receive personalized study plans based on their test results. The study plan diagnoses weaknesses and links students directly to tutorial exercises for the objectives they need to study and retest. Students can also access supplemental animations and video clips directly from selected exercises. MathXL is available to qualified adopters. For more information, visit our website at www.mathxl.com, or contact your Pearson Education sales representative.

MyMathLab® www.mymathlab.com

MyMathLab is a series of text-specific, easily customizable online courses for Pearson Education textbooks in mathematics and statistics. Powered by CourseCompass™ (Pearson Education's online teaching and learning environment) and MathXL® (our online homework, tutorial, and assessment system), MyMathLab gives instructors the tools they need to deliver all or a portion of their course online, whether students are in a lab setting or working from home. MyMathLab provides a rich and flexible set of course materials, featuring free-response exercises that are algorithmically generated for unlimited practice and mastery. Students can also use online tools, such as video lectures, animations, and a multimedia textbook, to independently improve their understanding and performance. Instructors can use MyMathLab's homework and test managers to select and assign online exercises correlated directly to the textbook, and they can also create and assign their own online exercises and import TestGen tests for added flexibility. MyMathLab's online gradebook—designed specifically for mathematics and statistics—automatically tracks students' homework and test results and gives the instructor control over how to calculate final grades. Instructors can also add offline (paper-and-pencil) grades to the gradebook. MyMathLab is available to qualified adopters. For more information, visit our website at www.mymathlab.com or contact your Pearson Education sales representative.

InterAct Math® Tutorial Web site: www.interactmath.com

Get practice and tutorial help online! This interactive tutorial Web site provides algorithmically generated practice exercises that correlate directly to the exercises in the textbook. Students can retry an exercise as many times as they like with new values each time for unlimited practice and mastery. Every exercise is accompanied by an interactive guided solution that provides helpful feedback for incorrect answers, and students can also view a worked-out sample problem that steps them through an exercise similar to the one they're working on.

Acknowledgments

Writing a textbook such as this is a monumental task, and I would like to take this opportunity to thank everyone who has helped me along the way. The following reviewers provided thoughtful suggestions and were instrumental in the development of *Elementary Algebra*.

Tim Allen	*TVI Community College*
Jan Archibald	*Ventura College*
Sonia Avetisian	*Orange Coast College*
David Casey	*Citrus College*
Kevin Cooper	*National American University*

Susi Curl	*McCook Community College*
Mary Deas	*Johnson County Community College*
Scott Dennison	*University of Wisconsin–Oshkosh*
Deborah Fries	*Wor-Wic Community College*
Emanuel Gerakios	*St. Petersburg College–Tarpon Springs Campus*
Pauline Hall	*Iowa State University*
Mary Beth Headlee	*Manatee Community College*
Nancy Johnson, Ph.D	*Manatee Community College*
Lonnie Larson	*Sacramento City College*
Mickey Levendusky	*Pima Community College–Downtown Campus*
Sandra Lofstock	*St. Petersburg College–Tarpon Springs Campus*
Carol Marinas	*Barry University*
Lauri McManus	*St. Louis Community College*
Jean Olsen	*Pikes Peak Community College*
Dorothy Polson	*Portland Community College*
Terri Seiver	*San Jacinto College–Central Campus*
Cindy Soderstrom	*Salt Lake Community College*
Timothy Thompson	*Oregon Institute of Technology*
Mary Lou Townsend	*Wor-Wic Community College*
Patrick Ward	*Illinois Central College*

I have truly enjoyed working with the team at Addison-Wesley. I do owe special thanks to my editor, Jenny Crum, as well as Lauren Morse, Antonio Arvelo, Ron Hampton, Jay Jenkins, Michelle Renda, Nathaniel Koven, and Sharon Tomasulo. Thanks are due to Greg Tobin and Maureen O'Connor for believing in my vision and taking a chance on me, and to Susan Winslow for getting this all started. Anne Scanlan-Rohrer is an outstanding developmental editor, and her hard work and suggestions have had a tremendous impact on this book. Stephanie Logan's assistance during the production process was invaluable, and Sarah Sponholz and Gary Williams also deserve credit for their help in accuracy checking the pages.

Thanks to Vineta Harper, Mark Tom, Don Rose, and Ross Rueger, my colleagues at College of the Sequoias, who have provided some great advice along the way and frequently listened to my ideas. I would also like to thank my students for keeping my fires burning. It truly is all about the students.

Most importantly, thanks to my wife Tina and our wonderful children Dylan and Alycia. They are truly my greatest blessing, and I love them more than words can say. The process of writing a textbook is long and difficult, and they have been supportive and understanding at every turn.

George Woodbury

Review of Real Numbers

This chapter reviews properties of real numbers and arithmetic that are necessary for success in algebra. The chapter also introduces several algebraic properties.

Study Strategy **Study Groups** *Throughout this book study strategies will help you learn and be successful in this course. In this chapter, we will focus on getting involved in a study group.*

Working with a study group is an excellent way to learn mathematics, improve your confidence and level of interest, and improve your performance on quizzes and tests. When working with a group, you will be able to work through questions about the material you are studying. Also, by being able to explain how to solve a particular type of problem to another person in your group, you will increase your ability to retain this knowledge.

We will revisit this study strategy throughout this chapter and help you incorporate it into your study habits. See the end of Section 1.1 for tips on how to get a study group started.

1.1

Integers, Opposites, and Absolute Value

Objectives

1 Graph whole numbers on a number line.
2 Determine which is the greater of two whole numbers.
3 Graph integers on a number line.
4 Find the opposite of an integer.
5 Determine which is the greater of two integers.
6 Find the absolute value of an integer.

A **set** is a collection of objects, such as the set consisting of the numbers 1, 4, 9, and 16. This set can be written as {1, 4, 9, 16}. The braces, {}, are used to denote a set, and the values listed inside are said to be **elements**, or members, of the set. A set with no elements is called the **empty set**. A **subset** of a set is a collection of some or all of the elements of the set. For example, {1, 9} is a subset of the set {1, 4, 9, 16}. A subset can also be an empty set.

Whole Numbers

Objective 1 Graph whole numbers on a number line. This text, for the most part, deals with the set of real numbers.

Real Numbers

A number is a **real number** if it can be written as a fraction or as a decimal number.

One subset of the set of real numbers is the set of natural numbers.

Natural Numbers

The set of **natural numbers** is the set {1, 2, 3, . . . }.

If we include the number 0 with the set of natural numbers, we have the set of **whole numbers**.

Whole Numbers

The set of **whole numbers** is the set {0, 1, 2, 3, . . . }. This set can be displayed on a number line as follows.

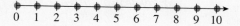

The arrow on the right-hand side of the number line indicates that the values continue to increase in this direction. There is no largest whole number, but we say that the values approach infinity (∞).

To graph any particular number on a number line, we place a point, or a dot, at that location on the number line.

EXAMPLE 1   Graph the number 6 on a number line.

Solution

To graph any number on a number line, we place a point at that number's location.

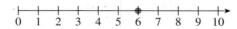

Quick Check 1
Graph the number 4 on
a number line.

Inequalities

Objective 2 **Determine which is the greater of two whole numbers.** When comparing two whole numbers a and b, we say that a is **greater than** b, denoted $a > b$, if the number a is to the right of the number b on the number line. The number a is **less than** b, denoted $a < b$, if a is to the left of b on the number line. The statements $a > b$ and $a < b$ are called **inequalities**.

EXAMPLE 2   Write the appropriate symbol, either $<$ or $>$, between the following:
6 _____ 4

Solution

Let's take a look at the two values graphed on a number line.

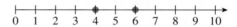

Since the number 6 is to the right of the number 4 on the number line, we say that 6 is greater than 4. So $6 > 4$.

EXAMPLE 3   Write the appropriate symbol, either $<$ or $>$, between the following:
2 _____ 5

Solution

Since 2 is to the left of 5 on the number line, $2 < 5$.

Quick Check 2
Write the appropriate
symbol, either $<$ or $>$,
between the following:

a) 8 _____ 3
b) 19 _____ 23

Integers

Objective 3 **Graph integers on a number line.** Another important subset of the real numbers is the set of integers.

Integers

The set of **integers** is the set $\{ \ldots, -3, -2, -1, 0, 1, 2, 3, \ldots \}$. We can display the set of integers on a number line as follows.

The arrow on the left side indicates that the values continue to decrease in this direction, and they are said to approach negative infinity $(-\infty)$.

Opposites

Objective **4** **Find the opposite of an integer.** The set of integers is the set of whole numbers together with the opposites of the natural numbers. The **opposite** of a number is a number on the other side of 0 on the number line and the same distance from 0 as that number. We denote the opposite of a real number a as $-a$. For example, -5 and 5 are opposites because they are both 5 units away from 0 and one is to the left of 0 while the other is to the right of 0.

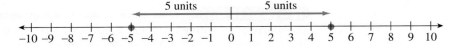

Numbers to the left of 0 on the number line are called **negative numbers**. Negative numbers represent a quantity less than 0. For example, if you have written checks that the balance in your checking account cannot cover, your balance would be a negative number. A temperature that is below 0° F, a golf score that is below par, and an elevation that is below sea level are other examples of quantities that can be represented by negative numbers.

EXAMPLE **4**   What is the opposite of 7?

Solution

The opposite of 7 is -7, since -7 is also 7 units away from 0 but is on the opposite side of 0.

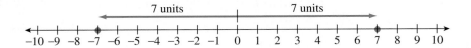

EXAMPLE **5**   What is the opposite of -6?

Solution

The opposite of -6 is 6.

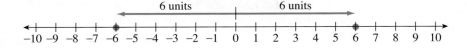

The opposite of 0 is 0 itself. Zero is the only number that is its own opposite.

Quick Check **3**
Find the opposite of the given integer.

a) -13 b) 8

Inequalities with Integers

Objective **5** **Determine which is the greater of two integers.** Inequalities for integers follow the same guidelines as they do for whole numbers. If we are given two integers a and b, the number that is greater is the number that is to the right on the number line.

EXAMPLE ▶ 6   Write the appropriate symbol, either < or >, between the following:
-3 ___ 5

Solution

Looking at the number line, we can see that -3 is to the left of 5, so $-3 < 5$.

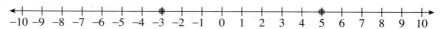

EXAMPLE ▶ 7   Write the appropriate symbol, either < or >, between the following:
-2 ___ -7

Solution

On the number line, -2 is to the right of -7, so $-2 > -7$.

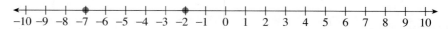

Quick Check ◀ **4**
Write the appropriate symbol, either < or >, between the following:

a) -14 ___ -11
b) 6 ___ -20

Absolute Values

Objective **6**   **Find the absolute value of an integer.**

Absolute Value

> The **absolute value** of a number a, denoted $|a|$, is the distance between a and 0 on the number line.

Distance cannot be negative, so the absolute value of a number a will always be 0 or higher.

EXAMPLE ▶ 8   Find the absolute value of 6.

Solution

The number 6 is 6 units away from 0 on the number line, so $|6| = 6$.

EXAMPLE ▶ 9   Find the absolute value of -4.

Solution

The number -4 is 4 units away from 0 on the number line, so $|-4| = 4$.

Quick Check ◀ **5**
Find the absolute value of -9.

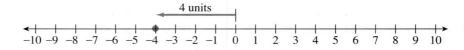

Building Your Study Strategy **Study Groups, 1 Who to Work With?** To form a study group, you must begin with the question, "Who do I want to work with?" Look for students who are serious about learning, who seem prepared for each class, who ask good questions during class. If a student does not appear to do the homework, or is constantly late for class, then this student may not be the best person to work with.

Look for students you feel you can get along with. You are about to spend a long period working with this group, sometimes under stressful conditions. Working with group members who get on your nerves will not help you to learn.

If you take advantage of tutorial services provided by your college, keep an eye out for classmates who do the same. There is a strong chance that classmates who use the tutoring center are serious about learning mathematics and earning good grades.

EXERCISES 1.1

Vocabulary

1. A set with no elements is called the _____.

2. A number m is _____ than another number n if it is located to the left of n on the number line.

3. The arrow on the right side of a number line indicates that the values approach _____.

4. $c > d$ if c is located to the _____ of d on the number line.

Graph the following whole numbers on a number line.

5. 7

6. 3

7. 6

8. 9

9. 2

10. 13

Write the appropriate symbol, either < or >, between the following whole numbers.

11. 3 ___ 13

12. 7 ___ 9

13. 8 ___ 6

14. 12 ___ 5

15. 45 ___ 27

16. 36 ___ 38

Graph the following integers on a number line.

17. -4

18. -7

19. 3

20. -9

21. -12

22. 6

Find the opposite of the following integers.

23. -7

24. 5

25. 14

26. -20

27. 0

28. -39

Write the appropriate symbol, either < or >, between the following integers.

29. -7 ___ -9

30. -5 ___ -2

31. -13 ___ -11

32. -8 ___ -14

33. -16 ___ 0

34. 5 ___ -3

35. -4 ___ 3

36. -105 ___ -129

Find the following absolute values.

37. $|-15|$

38. $|9|$

39. $|0|$

40. $|-6|$

41. $|25|$

42. $|-13|$

43. $-|7|$

44. $-|12|$

45. $-|-29|$

46. $-|-8|$

FOR EXTRA HELP

MyMathLab
MathXL
Interactmath.com
MathXL
Tutorials on CD
Video Lectures on CD
Tutor Center
Addison-Wesley Math Tutor Center
Student's Solutions Manual

Write the appropriate symbol, either < or >, between the following integers.

47. $|-7|$ ___ 4

48. $|-11|$ ___ -6

49. $|-14|$ ___ -14

50. $|8|$ ___ $|-19|$

51. $-|-24|$ ___ $|-47|$

52. $|-8|$ ___ $-|8|$

Identify whether the given number is a member of the following sets of numbers: A. natural numbers, B. whole numbers, C. integers, D. real numbers.

53. 8

54. -6

55. 0

56. 3.14

57. -9

58. 20

Find the missing number, if possible. There may be more than one number that works, so find as many as possible. There may be no number that works.

59. $|?| = 7$

60. $|?| = 14$

61. $|?| = 0$

62. $|?| = -3$

63. $|?| + 5 = 11$

64. $3 \cdot |?| + 8 = 20$

65. Use a number line to show that $|-5| = 5$.

66. Use a number line to show that the opposite of 8 is -8.

67. Use a number line to show that $-9 < -4$.

68. Use a number line to show that $3 > -6$.

Writing in Mathematics

Answer in complete sentences.

69. A fellow student tells you that to find the absolute value of any number, just make the number positive. Is this always true? Explain why or why not in your own words.

70. True or false: The opposite of the opposite of a number is the number itself.

71. If the opposite of a nonzero integer is equal to the absolute value of that integer, is the integer positive or negative? Explain your reasoning.

72. If an integer is less than its opposite, is the integer positive or negative? Explain your reasoning.

73. *Solutions Manual* * *A solutions manual page is an assignment in which you solve a problem as if you were creating a solutions manual for a textbook. Show and explain each step in solving the problem in such a way that a fellow student would be able to read and understand your work.*

Write a solutions manual page for the following problem:

Write the appropriate symbol, < or >, between the two integers. -18 ____ -22

74. *Newsletter* * *A newsletter assignment asks you to explain a certain mathematical topic. Your explanation should be in the form of a short, visually appealing article that might be published in a newsletter read by people who were interested in learning mathematics.*

Write a newsletter explaining how to determine the greater of two integers.

*See Appendix B for details and sample answers.

1.2

**Operations
with
Integers**

1 Add integers.
2 Subtract integers.
3 Multiply integers.
4 Divide integers.

Addition and Subtraction of Integers

Objective 1 **Add integers.** We can use the number line to help us learn how to add and subtract integers. Suppose we were trying to add the integers 3 and -7, which could be written as $3 + (-7)$. On a number line, we will start at 0 and move 3 units in the positive, or right, direction. Adding -7, tells us to move 7 units in the negative, or left, direction.

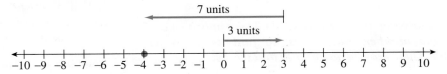

Since we end up at -4, this tells us that $3 + (-7) = -4$.

We can use a similar approach to verify an important property of opposites: the sum of two opposites is equal to 0.

Sum of Two Opposites

For any real number a, $a + (-a) = 0$.

Suppose that we wanted to add the opposites 4 and -4. Using the number line, we would begin at 0 and move 4 units to the right. We then move 4 units to the left, ending at 0. So, $4 + (-4) = 0$.

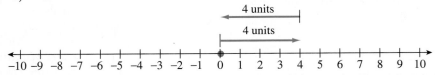

We could also see that $3 + (-7) = -4$ through the use of manipulatives, which are hands-on tools used to demonstrate mathematical properties. Suppose that we had a bag of green and red candies. We could let each piece of green candy represent a positive 1 and each piece of red candy represent a negative 1. To add $3 + (-7)$, we would begin by combining 3 green candies (positive 3) with 7 red candies (negative 7). Combining 1 red candy with 1 green candy has a net result of 0, as the sum of two opposites is equal to 0. So, each time we can make a pair of a green candy and a red candy, these two cancel each other's effect and can be discarded. After doing this, we would be left with 4 red candies. Our answer would be -4.

We now will examine another technique for finding the sum of a positive number and a negative number. In the sum $3 + (-7)$, the number 3 contributes to the sum in a positive fashion while the number –7 contributes to the sum in a negative fashion. The two numbers contribute to the sum in an opposite manner. We can think of the sum as the difference between these two contributions.

Adding a Positive Number and a Negative Number

1. **Take the absolute value of each number, and find the difference between these two absolute values.** This is the difference between the two numbers' contributions to the sum.
2. **The sign of the result is the same as the sign of the number that has the largest absolute value.**

For the sum $3 + (-7)$, we begin by taking the absolute value of each number: $|3| = 3$, $|-7| = 7$. The difference between the absolute values is 4. The sign of the sum is the same as the sign of the number that has the larger absolute value. In this case, –7 has the larger absolute value, so our result is negative. Therefore, $3 + (-7) = -4$.

EXAMPLE 1   Find the sum $12 + (-8)$.

Solution

$$|12| = 12; \quad |8| = 8$$ Find the absolute value of each number.
$$12 - 8 = 4$$ The difference between the absolute values is 4.
$$12 + (-8) = 4$$ Since the number with the larger absolute value is positive, the result is positive.

Quick Check 1 Find the sum $14 + (-6)$.

Notice that in the previous example $12 + (-8)$ is equivalent to $12 - 8$, which also equals 4. What the two expressions share in common is that there is one number (12) that contributes to the total in a positive fashion and a second number (8) that contributes to the total in a negative fashion.

EXAMPLE 2   Find the sum $3 + (-11)$.

Solution

Again we have one number (3) contributing to the total in a positive way and a second number (11) contributing in a negative way. The difference between their contributions is 8, and since the number making the larger contribution is negative, our result is -8.

Quick Check 2
Find the sum $4 + (-17)$.

$$3 + (-11) = -8$$

Note that $-11 + 3$ also equals -8. The rules for adding a positive integer and a negative integer still apply when the first number is negative and the second number is positive.

Adding Two (or More) Negative Numbers

1. Total the negative contributions of each number.
2. The sign of the result is negative.

EXAMPLE 3   Find the sum $-3 + (-7)$.

Solution

Both values contribute to the total in a negative fashion. Totaling the negative contributions of 3 and 7 gives us 10, and our result is negative because both of our numbers were negative.

$$-3 + (-7) = -10$$

Quick Check 3
Find the sum $-2 + (-9)$.

Objective 2 Subtract integers. To subtract a negative integer from another integer, we will use the following property:

Subtraction of Real Numbers

For any real numbers a and b, $a - b = a + (-b)$.

This property tells us that adding the opposite of b to a is the same as subtracting b from a. Suppose that we were subtracting a negative integer, such as in the example $-8 - (-19)$. The property for subtraction of real numbers tells us that subtracting -19 is the same as adding its opposite (19), so we would then convert this subtraction to $-8 + 19$. Remember that subtracting a negative number is equivalent to adding a positive number.

EXAMPLE 4   Subtract $6 - (-27)$.

Solution

$$6 - (-27) = 6 + 27 \quad \text{Subtracting } -27 \text{ is the same as adding 27.}$$
$$= 33 \quad \text{Add.}$$

Quick Check 4
Subtract $11 - (-7)$.

General Strategy for Adding/Subtracting Integers

- Rewrite "double signs." *Adding a negative number, $4 + (-5)$, can be rewritten as subtracting a positive number, $4 - 5$. Subtracting a negative number, $-2 - (-7)$, can be rewritten as adding a positive number, $-2 + 7$.*
- Look at each integer and determine whether it is contributing positively or negatively to the total.
- Add up any integers contributing positively to the total, if there are any, resulting in a single positive integer. In a similar fashion, add up all integers that are contributing to the total negatively, resulting in a single negative integer. Finish by finding the sum of these two integers.

Rather than being told to add or subtract, we may be asked to simplify a numerical expression. To **simplify** an expression, perform all arithmetic operations.

EXAMPLE 5   Simplify $17 - (-11) - 6 + (-13) - (-21) + 3$.

Solution

We begin by working on the *double signs*. This produces the following:

$$17 - (-11) - 6 + (-13) - (-21) + 3$$
$$= 17 + 11 - 6 - 13 + 21 + 3$$
$$= 52 - 19$$

$$= 33$$

Rewrite double signs. The four integers that contribute in a positive fashion (17, 11, 21, and 3) have a total of 52. The two integers that contribute in a negative fashion have a total of -19. Subtract.

Quick Check 5   Simplify $14 - 9 - (-22) - 6 + (-30) + 5$.

Using Your Calculator   When using your calculator, it is important to be able to distinguish between the subtraction key and the key for a negative number. On the TI-84, the subtraction key $\boxed{-}$ is listed above the addition key on the right side of the calculator, while the negative key $\boxed{(-)}$ is located to the left of the $\boxed{\text{ENTER}}$ key at the bottom of the calculator. Here are two different ways to simplify the expression from the previous example on the TI-84:

```
17-(-11)-6+(-13)
-(-21)+3
```

```
17+11-6-13+21+3
```

In either case, pressing the $\boxed{\text{ENTER}}$ key produces the result 33.

Multiplication and Division of Integers

Objective 3   Multiply integers.   The result obtained when multiplying two numbers is called the **product** of the two numbers. The numbers that are multiplied are called **factors**. When we multiply two positive integers, their product is also a positive integer. For example, the product of the two positive integers 4 and 7 is the positive integer 28. This can be written as $4 \cdot 7 = 28$. The product $4 \cdot 7$ can also be written as $4(7)$ or $(4)(7)$.

The product $4 \cdot 7$ is another way to represent the repeated addition of 7 four times.

$$4 \cdot 7 = 7 + 7 + 7 + 7$$
$$= 28$$

We will use this concept to show that the product of a positive integer and a negative integer is a negative integer. Suppose we want to multiply 4 by -7. We can rewrite this as -7 being added four times or $(-7) + (-7) + (-7) + (-7)$. From our work earlier in this section we know that this total is -28, so $4(-7) = -28$. Anytime we multiply a positive integer and a negative integer together the result is negative. So $(-7)(4)$ is also equal to -28.

Products of Integers

$$(\text{Positive}) \cdot (\text{Negative}) = \text{Negative}$$
$$(\text{Negative}) \cdot (\text{Positive}) = \text{Negative}$$

EXAMPLE 6   Multiply $5(-8)$.

Solution

We begin by multiplying 5 and 8, which equals 40. The next step is to determine the sign of our result. Whenever we multiply a positive integer by a negative integer the result is negative.

$$5(-8) = -40$$

Quick Check 6
Multiply $10(-6)$.

A Word of Caution   Note the difference between $5 - 8$ (a subtraction) and $5(-8)$ (a multiplication). A set of parentheses *without* a sign in front of them is used to imply multiplication.

Product	Result
$(3)(-5)$	-15
$(2)(-5)$	-10
$(1)(-5)$	-5
$(0)(-5)$	0
$(-1)(-5)$	?
$(-2)(-5)$	?

The product of two negative integers is a positive integer. Let's try to understand why this is true by considering the example $(-2)(-5)$. Examine the table to the left, which shows the products of some integers and -5.

Notice the pattern in the table. Each time the integer multiplied by -5 decreases by 1, the product increases by 5. As we go from $0(-5)$ to $(-1)(-5)$, the product should increase by 5. So $(-1)(-5) = 5$ and by the same reasoning $(-2)(-5) = 10$.

Product of Two Negative Integers

$$(\text{Negative}) \cdot (\text{Negative}) = \text{Positive}$$

EXAMPLE 7   Multiply $(-9)(-8)$.

Solution

We begin by multiplying 9 and 8, which equals 72. The next step is to determine the sign of our result. Whenever we multiply a negative integer by a negative integer the result is positive.

$$(-9)(-8) = 72.$$

Quick Check 7
Multiply $(-7)(-9)$.

Products of Integers

- If a product contains an *odd number* of negative factors, then the result is negative.
- If a product contains an *even number* of negative factors, then the result is positive.

The main idea behind this principle is that every two negative factors multiply to be positive. If there are three negative factors, the product of the first two is a positive number. Multiplying this positive product by the third negative factor produces a negative product.

[handwritten: 10 · 3 = -30]

EXAMPLE ▶ 8   Multiply $7(-2)(-5)(-3)$.

Solution

Since there are three negative factors, our product will be negative.

$$7(-2)(-5)(-3) = -210$$

Quick Check **8** Multiply $-4(-10)(5)(-2)$.

[handwritten: 40 · 5 = 200   = -400]

Using Your Calculator   Here is what the screen should look like when using the TI-84 to multiply the expression in the previous example:

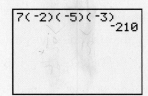

Notice that parentheses can be used to indicate multiplication without using the ⊠ key.

Before continuing on to division, we should mention multiplication by 0. Any real number multiplied by 0 will be 0; this is the multiplication property of 0.

Multiplication by 0

For any real number *x*,

$$0 \cdot x = 0$$
$$x \cdot 0 = 0.$$

Objective 4   Divide integers. When dividing one number called the **dividend** by another number called the **divisor**, the result obtained is called the **quotient** of the two numbers:

The statement "6 divided by 3 is equal to 2" is true because the product of the quotient and the divisor, $2 \cdot 3$, is equal to the dividend 6.

$$6 \div 3 = 2 \longleftrightarrow 2 \cdot 3 = 6$$

When we divide two integers that have the same sign (both positive or both negative), the quotient is positive. When we divide two integers that have different signs (one negative, one positive), the quotient is negative. Note that this is consistent with our rules for multiplication.

Quotients of Integers

(Positive) ÷ (Positive) = Positive   (Positive) ÷ (Negative) = Negative

(Negative) ÷ (Negative) = Positive   (Negative) ÷ (Positive) = Negative

EXAMPLE 9   Divide $(-54) \div (-6)$.

Solution

When we divide a negative number by another negative number, the result is positive.

$(-54) \div (-6) = 9$ Note that $-6 \cdot 9 = -54$.

EXAMPLE 10   Divide $(-33) \div 11$.

Solution

When we divide a negative number by a positive number, the result is negative.

Quick Check **9**

Divide $72 \div (-8)$.

$(-33) \div 11 = -3$

Whenever 0 is divided by any integer (except 0), the quotient is 0. For example, $0 \div 16 = 0$. We can check that this quotient is correct by multiplying the quotient by the divisor. Since $0 \cdot 16 = 0$, the quotient is correct.

Division by Zero

Whenever an integer is divided by 0, the quotient is said to be **undefined**.

We use the word **undefined** to state that an operation cannot be performed or is meaningless. For example, $41 \div 0$ is undefined. Suppose there was a real number a for which $41 \div 0 = a$. In that case, the product $a \cdot 0$ would be equal to 41. Since we know that the product of 0 and any real number is equal to 0, there does not exist such a number a.

Building Your Study Skills Study Groups, 2 **When to Meet**   Once you have formed a study group, determine where and when to meet. It is a good idea to meet at least twice a week, for at least an hour per session. Consider a location where quiet discussion is allowed, such as the library or tutorial center.

Clearly, the group must meet at times when each person is available. Some groups like to meet during the hour before class, using the study group as a way to prepare for class. Other groups prefer to meet during the hour after class, allowing them to go over material while it is fresh in their minds. Another suggestion is to meet at a time when your instructor is holding office hours. This way, if the entire group is struggling with the same question, one or more members can visit the instructor for assistance.

EXERCISES 1.2

FOR EXTRA HELP

MyMathLab

MathXL

Interactmath.com

MathXL
Tutorials on CD

Video Lectures
on CD

Tutor Center

Addison-Wesley
Math Tutor Center

Student's
Solutions Manual

Vocabulary

1. When finding the sum of a positive integer and a negative integer, the sign of the result is determined by _____.

2. The sum of two negative integers is a _____ integer.

3. Subtracting a negative integer can be rewritten as adding _____.

4. The product of a positive integer and a negative integer is a _____ integer.

5. The product of a negative integer and a negative integer is a _____ integer.

6. If a product contains a(n) _____ number of negative integers, the product is negative.

7. In a division problem, the number we divide by is called the _____.

8. Division by 0 results in a quotient that is _____.

Add.

9. $3 + (-5)$
10. $4 + (-2)$
11. $7 + (-25)$
12. $64 + (-105)$
13. $(-4) + 5$
14. $-9 + 2$
15. $-14 + 22$
16. $(-35) + 50$
17. $-5 + (-6)$
18. $-9 + (-9)$

Subtract.

19. $8 - 6$
20. $13 - 9$
21. $5 - 11$
22. $4 - 12$
23. $(-5) - 3$
24. $(-9) - 6$
25. $-7 - 14$
26. $-29 - 17$
27. $15 - (-12)$
28. $41 - (-35)$

Simplify.

29. $8 + 13 - 6$
30. $7 - 15 - 4$
31. $-9 + 7 - 4$
32. $-5 - 8 + 23$
33. $6 - (-16) + 5$
34. $18 - 21 - (-62)$
35. $4 + (-15) - 13 - (-25)$
36. $-13 + (-12) - (-1) - 29$

37. A mother with $30 in her purse paid $22 for her family to get into a movie. How much money did she have remaining?

38. A student had $60 in his checking account prior to writing an $85 check to the bookstore for books and supplies. What is his account's new balance?

39. The temperature at 6 A.M. in Fargo, North Dakota, was $-8°$ C. By 3 P.M., the temperature had risen by $12°$ C. What was the temperature at 3 P.M.?

40. If a golfer completes a round at 3 strokes under par, her score is denoted -3. A professional golfer had rounds of -4, -2, 3, and -6 in a recent tournament. What was her total score for this tournament?

41. Dylan drove from a town located 400 feet below sea level to another town located 1750 feet above sea level. What was the change in elevation traveling from one town to another?

42. After withdrawing $80 from her bank using an ATM card, Alycia had $374 remaining in her savings account. How much money did Alycia have in the account prior to withdrawing the money?

Multiply.

43. $8(-6)$ **44.** $-3(5)$

45. $-4 \cdot 9$ **46.** $-7(-8)$

47. $-11(-13)$ **48.** $-6 \cdot 15$

49. $82(-1)$ **50.** $-1 \cdot 19$

51. $-6 \cdot 0$ **52.** $0(-240)$

53. $-6(-3)(5)$ **54.** $-2(-4)(-8)$

55. $5 \cdot 3(-2)(-6)$ **56.** $-7 \cdot 2(-7)(-2)$

Divide, if possible.

57. $36 \div (-6)$ **58.** $54 \div (-9)$

59. $-63 \div 7$ **60.** $-48 \div 12$

61. $-21 \div (-3)$ **62.** $-100 \div (-5)$

63. $126 \div (-9)$ **64.** $-420 \div 14$

65. $0 \div (-13)$ **66.** $0 \div 11$

67. $29 \div 0$ **68.** $-15 \div 0$

(**Mixed Practice, 69–80**)

Simplify.

69. $-11(-12)$

70. $126 \div (-6)$

71. $5 - 13$

72. $5(-13)$

73. $17 - (-11) - 49$

74. $8(-7)(-6)$

75. $-432 \div 3$

76. $-5 \cdot 17$

77. $9(-24)$

78. $9 + (-24)$

79. $5 \cdot 3(-17)(-29)(0)$

80. $-16 + (-11) - 42 - (-58)$

81. Kevin owns 8 baseball cards, each of which is worth $12. What is the total worth of these cards?

82. An Internet startup company had an operating loss of $10,000 last month. If the four partners decide to divide up the loss equally among themselves, what is the loss for each person?

83. Tina owns 400 shares of a stock that dropped in value by $3 per share last month. She also owns 500 shares of a stock that went up by $2 per share last month. What is Tina's net income on these two stocks for last month?

84. Mario took over as the CEO for a company that lost $20 million dollars in the year 2000. The company lost three times as much in 2001. The company went on to lose $13 million more in 2002 than it had lost in 2001. How much money did Mario's company lose in 2002?

85. When a certain integer is added to -11, the result is 15. What is that integer?

86. Fourteen less than a certain integer is -23. What is that integer?

87. When a certain integer is divided by -7, the result is 12. What is that integer?

88. When a certain integer is multiplied by -3, the result is -417. What is that integer?

True or False (If false, give an example that shows why the statement is false.)

89. The sum of two integers is always an integer.

90. The difference of two integers is always an integer.

91. The sum of two whole numbers is always a whole number.

92. The difference of two whole numbers is always a whole number.

93. Use a number line to demonstrate that $3 + 6 = 9$.

94. Use a number line to demonstrate that $5 + (-2) = 3$.

95. Use a number line to demonstrate that $-7 + 3 = -4$.

96. Use a number line to demonstrate that $-4 - 5 = -9$.

Writing in Mathematics

Explain each of the following in your own words.

97. Explain why subtracting a negative integer from another integer is the same as adding the opposite of that integer to it. Use the example $11 - (-5)$ in your explanation.

98. Explain why a positive integer times a negative integer produces a negative integer.

99. Explain why a negative integer times another negative integer produces a positive integer.

100. Explain why division by 0 is undefined.

101. *Solutions Manual* * Write a solutions manual page for the following problem:

Simplify.
$16 - (-7) + (-19) - 42 + 6 - 37$

102. *Newsletter* * Write a newsletter explaining how to add and subtract integers.

103. *Newsletter* * Write a newsletter explaining how to multiply and divide integers.

*See Appendix B for details and sample answers.

1.3
Fractions

Objectives

1. **Find the factor set of a natural number.**
2. **Determine whether a natural number is prime.**
3. **Find the prime factorization of a natural number.**
4. **Simplify a fraction to lowest terms.**
5. **Change an improper fraction to a mixed number.**
6. **Change a mixed number to an improper fraction.**

Factors

Objective 1   Find the factor set of a natural number. To factor a natural number, we express it as the product of two natural numbers. For example, one way to factor 12 is to rewrite it as 3 · 4. In this example, 3 and 4 are said to be factors of 12. The collection of all factors of a natural number is called its **factor set**. The factor set of 12 can be written as {1, 2, 3, 4, 6, 12}, since 1 · 12 = 12, 2 · 6 = 12, and 3 · 4 = 12.

EXAMPLE 1   Write the factor set for 18.

Solution

Quick Check 1
Write the factor set for 36.

Since 18 can be factored as 1 · 18, 2 · 9, and 3 · 6, its factor set is {1, 2, 3, 6, 9, 18}.

Prime Numbers

Objective 2   Determine whether a natural number is prime.

Prime Numbers

A natural number is **prime** if it is greater than 1 and its only two factors are 1 and itself.

For instance, the number 13 is prime because its only two factors are 1 and 13. The first 10 prime numbers are 2, 3, 5, 7, 11, 13, 17, 19, 23, and 29. The number 8 is not prime because it has factors other than 1 and 8, namely, 2 and 4. A natural number greater than 1 that is not prime is called a **composite** number. The number 1 is considered to be neither prime nor composite.

EXAMPLE 2   Determine whether the following numbers are prime or composite:
a) 26   **b)** 37

Solution

a) The factor set for 26 is {1, 2, 13, 26}. Since 26 has factors other than 1 and itself, it is a composite number.

b) Since the number 37 has no factors other than 1 and itself, 37 is a prime number.

Quick Check **2** Determine whether the following numbers are prime or composite.

    a) 57　　　　**b)** 47　　　　**c)** 48

Prime Factorization

Objective **3** **Find the prime factorization of a natural number.** When we rewrite a natural number as a product of prime factors, we obtain the **prime factorization** of the number. The prime factorization of 12 is $2 \cdot 2 \cdot 3$, because 2 and 3 are prime numbers and $2 \cdot 2 \cdot 3 = 12$. A **factor tree** is a useful tool for finding the prime factorization of a number. Here is an example of a factor tree for 72.

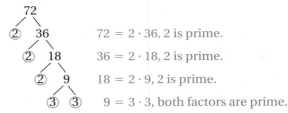

$72 = 2 \cdot 36$, 2 is prime.

$36 = 2 \cdot 18$, 2 is prime.

$18 = 2 \cdot 9$, 2 is prime.

$9 = 3 \cdot 3$, both factors are prime.

The prime factorization of 72 is $2 \cdot 2 \cdot 2 \cdot 3 \cdot 3$. It should be noted that we could have begun by rewriting 72 as $8 \cdot 9$, and then factored these two numbers. The process for creating a factor tree for a natural number is not unique, although the prime factorization for the number is unique.

EXAMPLE **3** Find the prime factorization for 51.

Solution

Quick Check **3**
Find the prime factorization for 63.

This number has only two factors: 3 and 17. The prime factorization is $3 \cdot 17$.

Fractions

Objective **4** **Simplify a fraction to lowest terms.**

Rational Numbers

A **rational number** is a real number that can be written as the quotient (or ratio) of two integers, the second of which is not zero. An **irrational number** is a real number that cannot be written in this way, such as the number π.

Rational numbers are often expressed using fraction notation such as $\frac{3}{7}$. Whole numbers, such as 7, can be written as a fraction whose denominator is 1, such as $\frac{7}{1}$. The number on the top of the fraction is called the **numerator**, and the number on the bottom of the fraction is called the **denominator**.

$$\frac{\text{numerator}}{\text{denominator}}$$

If the numerator and denominator do not have any common factors other than 1, then the fraction is said to be in **lowest terms**.

    To simplify a fraction to lowest terms, we can begin by finding the prime factorization of both the numerator and denominator. Then, divide both the numerator and denominator by their common factors.

EXAMPLE 4   Simplify $\frac{18}{30}$ to lowest terms.

Solution

$$\frac{18}{30} = \frac{2 \cdot 3 \cdot 3}{2 \cdot 3 \cdot 5}$$

Find the prime factorization of the numerator and denominator. $18 = 2 \cdot 3 \cdot 3, 30 = 2 \cdot 3 \cdot 5$.

$$= \frac{\overset{1}{\cancel{2}} \cdot \overset{1}{\cancel{3}} \cdot 3}{\underset{1}{\cancel{2}} \cdot \underset{1}{\cancel{3}} \cdot 5}$$

Divide out common factors.

$$= \frac{3}{5}$$

Simplify.

EXAMPLE 5   Simplify $\frac{4}{24}$ to lowest terms.

Solution

$$\frac{4}{24} = \frac{2 \cdot 2}{2 \cdot 2 \cdot 2 \cdot 3}$$

Find the prime factorization of the numerator and denominator.

$$\frac{\overset{1}{\cancel{2}} \cdot \overset{1}{\cancel{2}}}{\underset{1}{\cancel{2}} \cdot \underset{1}{\cancel{2}} \cdot 2 \cdot 3}$$

Divide out common factors.

$$= \frac{1}{6}$$

Simplify.

Quick Check **4**
Simplify to lowest terms:

a) $\frac{45}{210}$ b) $\frac{24}{384}$

A Word of Caution   It is customary to leave off the denominator of a fraction if it is equal to 1. For example, rather than writing $\frac{19}{1}$, we usually write 19. However, we cannot omit a numerator that is equal to 1. In the previous example, it would have been a mistake if we had written 6 instead of $\frac{1}{6}$.

Mixed Numbers and Improper Fractions

Objective 5 **Change an improper fraction to a mixed number.** An **improper fraction** is a fraction whose numerator is greater than or equal to its denominator, such as $\frac{7}{4}, \frac{400}{150}, \frac{8}{8}$, and $\frac{35}{7}$. (In contrast, a proper fraction's numerator is smaller than its denominator.) An improper fraction is often converted to a **mixed number**, which is the sum of a whole number and a fraction. For example, the improper fraction $\frac{14}{3}$ can be represented by the mixed number $4\frac{2}{3}$, which is equivalent to $4 + \frac{2}{3}$.

To convert an improper fraction to a mixed number, begin by dividing the denominator into the numerator. The quotient is the whole number portion of the mixed number, and the remainder becomes the numerator of the fractional part.

Denominator of fractional portion of mixed number → $3\overline{)14}$ ← Whole-number portion of mixed number

-12

2 ← Numerator of fractional portion of mixed number

EXAMPLE 6   Convert the improper fraction $\frac{71}{9}$ to a mixed number.

Solution

We begin by dividing 9 into 71, which divides in 7 times with a remainder of 8.

$$9\overline{)71}$$
$$\,7$$
$$-63$$
$$8$$

Quick Check 5
Convert the improper fraction $\frac{121}{13}$ to a mixed number.

The mixed number for $\frac{71}{9}$ is $7\frac{8}{9}$.

Objective 6   Change a mixed number to an improper fraction. Often we have to convert a mixed number such as $2\frac{7}{15}$ into an improper fraction before proceeding with arithmetic operations.

Rewriting a Mixed Number as an Improper Fraction

- Multiply the whole number part of the mixed number by the denominator of the fractional part of the mixed number.
- Add this product to the numerator of the fractional part of the mixed number.
- The sum is the numerator of the improper fraction. The denominator stays the same.

$2\frac{7}{15}$ ← Add product to numerator

Multiply

$$2\frac{7}{15} = \frac{37}{15}$$

EXAMPLE 7   Convert the mixed number $5\frac{4}{7}$ to an improper fraction.

Solution

Quick Check 6
Convert the mixed number $8\frac{1}{6}$ to an improper fraction.

We begin by multiplying $5 \cdot 7 = 35$. We add this product to 4 to produce a numerator of 39.

$$5\frac{4}{7} = \frac{39}{7}$$

Building Your Study Strategy Study Groups, 3   **Where to Meet**

- Some study groups prefer to meet off campus in the evening. One good place to meet would be at a coffee shop with tables large enough for everyone to work on, provided that the surrounding noise is not too distracting. The food court at a local mall would work as well.
- Some groups take advantage of study rooms at public libraries.
- Other groups like to meet at each other's homes. This typically provides a comfortable, relaxing atmosphere to work in. If your group is invited to meet at someone's house, try to be a perfect guest. Help to set up and clean up if you can, or bring beverages and healthy snacks.

EXERCISES *1.3*

Vocabulary

1. The collection of all the factors of a natural number is called its _____.

2. A natural number greater than 1 is _____ if its only factors are 1 and itself.

3. A natural number greater than 1 that is not prime is called a _____ number.

4. Define the prime factorization of a natural number.

5. The numerator of a fraction is _____.

6. The denominator of a fraction is _____.

7. A fraction is in lowest terms if _____.

8. A fraction whose numerator is less than its denominator is called a(n) _____ fraction.

9. A fraction whose numerator is greater than or equal to its denominator is called a(n) _____ fraction.

10. An improper fraction can be rewritten as either a whole number or as a _____.

11. Is 7 a factor of 247?

12. Is 13 a factor of 273?

13. Is 6 a factor of 4836?

14. Is 9 a factor of 32,057?

15. Is 15 a factor of 2835?

16. Is 103 a factor of 1754?

Write the factor set for the following numbers.

17. 27

18. 15

19. 20

20. 16

21. 31

22. 48

23. 60

24. 21

25. 49

26. 25

27. 69

28. 103

29. 221

30. 209

FOR EXTRA HELP

MyMathLab

MathXL

Interactmath.com

MathXL Tutorials on CD

Video Lectures on CD

Addison-Wesley Math Tutor Center

Student's Solutions Manual

Write the prime factorization of the following numbers. (If the number is prime, state this.)

31. 18   **32.** 20

33. 42   **34.** 28

35. 33   **36.** 50

37. 27   **38.** 32

39. 125   **40.** 49   **41.** 29

42. 38   **43.** 99   **44.** 90

45. 31   **46.** 91   **47.** 119

48. 53   **49.** 360

50. 126   **51.** 104

52. 109

Simplify the following fractions to lowest terms.

53. $\dfrac{10}{16}$ **54.** $\dfrac{30}{36}$ **55.** $\dfrac{9}{45}$

56. $\dfrac{38}{2}$ **57.** $\dfrac{126}{294}$ **58.** $\dfrac{60}{84}$

59. $\dfrac{27}{64}$ **60.** $\dfrac{66}{154}$ **61.** $\dfrac{160}{176}$

62. $\dfrac{56}{45}$

Convert the following mixed numbers to improper fractions.

63. $3\dfrac{4}{5}$ **64.** $7\dfrac{2}{9}$

65. $2\dfrac{16}{17}$ **66.** $6\dfrac{5}{14}$

67. $22\dfrac{13}{19}$ **68.** $37\dfrac{6}{25}$

Convert the following improper fractions to whole numbers or mixed numbers.

69. $\dfrac{39}{5}$ **70.** $\dfrac{56}{8}$

71. $\dfrac{101}{7}$ **72.** $\dfrac{12}{4}$

73. $\dfrac{61}{11}$ **74.** $\dfrac{85}{13}$

75. List four fractions that are equivalent to $\frac{3}{4}$.

76. List four fractions that are equivalent to $1\frac{2}{3}$.

77. List four whole numbers that have at least three different prime factors.

78. List four whole numbers greater than 100 that are prime.

Writing in Mathematics

Answer in complete sentences.

79. Describe a real-world situation involving fractions. Describe a real-world situation involving mixed numbers.

80. Describe a situation in which you should convert an improper fraction to a mixed number.

81. *Solutions Manual** Write a solutions manual page for the following problem:

Find the prime factorization for 840.

82. *Newsletter** Write a newsletter explaining how to find the prime factorization of a natural number.

83. *Newsletter** Write a newsletter explaining how to simplify a fraction to lowest terms.

*See Appendix B for details and sample answers.

1.4
Operations with Fractions

Objectives

1 Multiply fractions and mixed numbers.
2 Divide fractions and mixed numbers.
3 Add and subtract fractions and mixed numbers with the same denominator.
4 Find the least common multiple (LCM) of two natural numbers.
5 Add and subtract fractions and mixed numbers with different denominators.

Multiplying Fractions

Objective 1 Multiply fractions and mixed numbers. To multiply fractions, we multiply the numerators together and multiply the denominators together. When multiplying fractions, we may simplify any individual fraction, as well as divide out a common factor from a numerator and a different denominator. Dividing out a common factor in this fashion is often referred to as **cross-canceling**.

EXAMPLE 1 Multiply $\frac{4}{11} \cdot \frac{5}{6}$.

Solution

The first numerator (4) and the second denominator (6) have a common factor of 2 that we can eliminate through division.

$$\frac{4}{11} \cdot \frac{5}{6} = \frac{\overset{2}{\cancel{4}}}{11} \cdot \frac{5}{\underset{3}{\cancel{6}}} \qquad \text{Divide out the common factor 2.}$$

$$= \frac{2}{11} \cdot \frac{5}{3} \qquad \text{Simplify.}$$

$$= \frac{10}{33} \qquad \text{Multiply the two numerators and the two denominators.}$$

Quick Check 1
Multiply $\frac{10}{63} \cdot \frac{9}{16}$.

EXAMPLE 2 Multiply $3\frac{1}{7} \cdot \frac{14}{55}$.

Solution

When multiplying a mixed number by another number, we should convert the mixed number to an improper fraction before proceeding.

$$3\frac{1}{7} \cdot \frac{14}{55} = \frac{22}{7} \cdot \frac{14}{55} \qquad \text{Convert } 3\frac{1}{7} \text{ to the improper fraction } \frac{22}{7}.$$

$$= \frac{\overset{2}{\cancel{22}}}{\underset{1}{\cancel{7}}} \cdot \frac{\overset{2}{\cancel{14}}}{\underset{5}{\cancel{55}}} \qquad \text{Divide out common factors of 11 and 7.}$$

Quick Check 2
Multiply $2\frac{2}{3} \cdot 8\frac{5}{8}$.

$$= \frac{4}{5} \qquad \text{Multiply.}$$

Dividing Fractions

Objective 2 Divide fractions and mixed numbers.

Reciprocal

When we invert a fraction, such as $\frac{3}{5}$ to $\frac{5}{3}$, the resulting fraction is called the **reciprocal** of the original fraction.

Consider the fraction $\frac{a}{b}$, where a and b are nonzero real numbers. The reciprocal of this fraction is $\frac{b}{a}$. Notice that if we multiply a fraction by its reciprocal, such as $\frac{a}{b} \cdot \frac{b}{a}$, then the result is 1. This property will be important in Chapter 2.

Reciprocal Property

For any nonzero real numbers a and b, $\frac{a}{b} \cdot \frac{b}{a} = 1$.

To divide a number by a fraction, we invert the divisor and then multiply.

EXAMPLE 3   Divide $\frac{16}{25} \div \frac{22}{15}$.

Solution

$$\frac{16}{25} \div \frac{22}{15} = \frac{16}{25} \cdot \frac{15}{22}$$ Invert the divisor and multiply.

$$= \frac{\overset{8}{16}}{\underset{5}{25}} \cdot \frac{\overset{3}{15}}{\underset{11}{22}}$$ Divide out the common factors 2 and 5.

$$= \frac{24}{55}$$ Multiply.

Quick Check 3
Divide $\frac{12}{25} \div \frac{63}{10}$.

A Word of Caution   When dividing a number by a fraction, we must invert the divisor (not the dividend) before dividing out a common factor from a numerator and a denominator.

When performing a division involving a mixed number, begin by rewriting the mixed number as an improper fraction.

EXAMPLE 4   Divide $2\frac{5}{8} \div 3\frac{3}{10}$.

Solution

We begin by rewriting each mixed number as an improper fraction.

$$2\frac{5}{8} \div 3\frac{3}{10} = \frac{21}{8} \div \frac{33}{10}$$ Rewrite each mixed number as an improper fraction.

$$= \frac{21}{8} \cdot \frac{10}{33}$$ Invert the divisor and multiply.

$$= \frac{\overset{7}{\cancel{21}}}{\underset{4}{\cancel{8}}} \cdot \frac{\overset{5}{\cancel{10}}}{\underset{11}{\cancel{33}}} \qquad \text{Divide out common factors.}$$

$$= \frac{35}{44} \qquad \text{Multiply.}$$

Quick Check **4**
Divide $\frac{20}{21} \div 2\frac{2}{3}$.

Adding and Subtracting Fractions

Objective **3** **Add and subtract fractions and mixed numbers with the same denominator.** To add or subtract fractions that have the same denominator, we add or subtract the numerators, placing the result over the common denominator. Be sure to simplify the result to lowest terms.

EXAMPLE **5**   Add $\frac{1}{12} + \frac{5}{12}$.

Solution

Quick Check **5**
Add $\frac{7}{16} + \frac{3}{16}$.

$$\frac{1}{12} + \frac{5}{12} = \frac{6}{12} \qquad \text{Since the denominators are the same, add the numerators and place the sum over 12.}$$

$$= \frac{1}{2} \qquad \text{Simplify to lowest terms.}$$

EXAMPLE **6**   Subtract $\frac{3}{8} - \frac{9}{8}$.

Solution

The two denominators are both the same (8), so we subtract the numerators. When we subtract $3 - 9$ the result is -6. Although we may leave the negative sign in the numerator, it is often brought out in front of the fraction itself.

$$\frac{3}{8} - \frac{9}{8} = -\frac{6}{8} \qquad \text{Subtract the numerators.}$$

$$= -\frac{3}{4} \qquad \text{Simplify to lowest terms.}$$

Quick Check **6**
Subtract $\frac{17}{20} - \frac{5}{20}$.

When performing an addition involving a mixed number, we can begin by rewriting the mixed number as an improper fraction.

EXAMPLE **7**   Add $3\frac{5}{12} + 2\frac{11}{12}$.

Solution

We begin by rewriting each mixed number as an improper fraction.

$$3\frac{5}{12} + 2\frac{11}{12} = \frac{41}{12} + \frac{35}{12} \qquad \text{Rewrite } 3\frac{5}{12} \text{ and } 2\frac{11}{12} \text{ as improper fractions.}$$

$$= \frac{76}{12} \qquad \text{Add the numerators.}$$

$$= \frac{19}{3} \qquad \text{Simplify to lowest terms.}$$

$$= 6\frac{1}{3} \qquad \text{Rewrite as a mixed number.}$$

It is not necessary to rewrite the result as a mixed number, but this is often done when performing arithmetic operations on mixed numbers.

Quick Check **7**
Add $6\frac{1}{8} + 5\frac{5}{8}$.

Objective **4** **Find the least common multiple (LCM) of two natural numbers.**
Two fractions are said to be **equivalent fractions** if they have the same numerical value and can both be simplified to be the same fraction when simplifying to lowest terms. In order to add or subtract two fractions with different denominators, we must first convert them to equivalent fractions with the same denominator. To do this, we find the **least common multiple (LCM)** of the two denominators. This is the smallest number that is a multiple of both denominators. For example, the LCM of 4 and 6 is 12 because 12 is the smallest multiple of both 4 and 6.

To find the LCM for two numbers, we can begin by factoring them into their prime factorizations.

Finding the LCM of Two or More Numbers

- Find the prime factorization of each number.
- Find the common factors of the numbers.
- Multiply the common factors by the remaining factors of the numbers.

EXAMPLE **8**   Find the LCM of 24 and 30.

Solution

We begin with the prime factorizations of 24 and 30.

$$24 = 2 \cdot 2 \cdot 2 \cdot 3$$
$$30 = 2 \cdot 3 \cdot 5$$

$$24 = ②\cdot 2 \cdot 2 \cdot ③$$
$$30 = ②\cdot③\cdot 5$$

The common factors are 2 and 3. Additional factors are a pair of 2's, as well as a 5. So to find the LCM, we multiply the common factors (2 and 3) by the additional factors (2, 2, and 5).

$$2 \cdot 3 \cdot 2 \cdot 2 \cdot 5 = 120$$

Quick Check **8**
Find the least common multiple of 18 and 42.

The least common multiple of 24 and 30 is 120.

Another technique for finding the LCM for two numbers is to start listing the multiples of the larger number until we find a multiple that is also a multiple of the smaller number. For example, the first few multiples of 6 are

$$6\text{: } 6, 12, 18, 24, 30, \ldots$$

The first multiple listed that is also a multiple of 4 is 12, so the LCM of 4 and 6 is 12.

Objective 5 **Add and subtract fractions and mixed numbers with different denominators.** When adding or subtracting two fractions that do not have the same denominator, we first find a common denominator by finding the LCM of the two denominators. We then convert each fraction to an equivalent fraction whose denominator is that common denominator. Once we rewrite the two fractions so they have the same denominator, we can then add (or subtract) as done previously in this section.

Adding or Subtracting Fractions with Different Denominators

- Find the LCM of the denominators.
- Rewrite each fraction as an equivalent fraction whose denominator is the LCM of the original denominators.
- Add or subtract the numerators, placing the result over the common denominator.
- Simplify to lowest terms, if possible.

EXAMPLE 9 Add $\frac{5}{12} + \frac{9}{14}$.

Solution

The prime factorization of 12 is $2 \cdot 2 \cdot 3$, and the prime factorization of 14 is $2 \cdot 7$. The two denominators have a common factor of 2. If we multiply this common factor by the other factors of these two numbers, 2, 3, and 7, we see that the LCM of these two denominators is 84. We begin by rewriting each fraction as an equivalent fraction whose denominator is 84. We will multiply the first fraction by $\frac{7}{7}$ and the second fraction by $\frac{6}{6}$. Since $\frac{7}{7}$ and $\frac{6}{6}$ are both equal to 1, we are not changing the value of either fraction.

$$\frac{5}{12} + \frac{9}{14} = \frac{5}{12} \cdot \frac{7}{7} + \frac{9}{14} \cdot \frac{6}{6}$$

Multiply the first fraction's numerator and denominator by 7. Multiply the second fraction's numerator and denominator by 6.

$$= \frac{35}{84} + \frac{54}{84}$$

Multiply.

$$= \frac{89}{84}$$

Add.

Quick Check **9**

Add $\frac{4}{9} + \frac{2}{15}$.

This fraction is already in lowest terms.

EXAMPLE 10 Simplify $\frac{5}{6} - \frac{3}{8} + \frac{2}{9}$.

Solution

Here is the prime factorization for each denominator:

$$6 = 2 \cdot 3 \qquad 8 = 2 \cdot 2 \cdot 2 \qquad 9 = 3 \cdot 3$$

The LCM for these three denominators is 72. We now convert each fraction to an equivalent fraction whose denominator is 72.

$$\frac{5}{6} - \frac{3}{8} + \frac{2}{9} = \frac{5}{6} \cdot \frac{12}{12} - \frac{3}{8} \cdot \frac{9}{9} + \frac{2}{9} \cdot \frac{8}{8}$$

Multiply the first fraction by $\frac{12}{12}$, the second fraction by $\frac{9}{9}$, and the third fraction by $\frac{8}{8}$.

$$= \frac{60}{72} - \frac{27}{72} + \frac{16}{72}$$ Multiply.

$$= \frac{49}{72}$$ Simplify. $60 - 27 + 16 = 49$.

Quick Check 10
Simplify $\frac{3}{4} + \frac{2}{3} - \frac{7}{12}$.

This fraction is in lowest terms.

When performing a subtraction involving a mixed number, we can begin by rewriting the mixed number as an improper fraction.

EXAMPLE 11   Subtract $4\frac{1}{3} - \frac{3}{4}$.

Solution

We begin by rewriting $4\frac{1}{3}$ as an improper fraction.

$$4\frac{1}{3} - \frac{3}{4} = \frac{13}{3} - \frac{3}{4}$$ Rewrite $4\frac{1}{3}$ as an improper fraction.

$$= \frac{13}{3} \cdot \frac{4}{4} - \frac{3}{4} \cdot \frac{3}{3}$$ The LCM of the denominators is 12. Multiply the first fraction by $\frac{4}{4}$. Multiply the second fraction by $\frac{3}{3}$.

$$= \frac{52}{12} - \frac{9}{12}$$ Multiply.

$$= \frac{43}{12}$$ Subtract.

$$= 3\frac{7}{12}$$ Rewrite as a mixed number.

Quick Check 11
Subtract $5\frac{1}{5} - 3\frac{5}{6}$.

Building Your Study Strategy **Study Groups, 4 Going Over Homework**
A study group can go over homework assignments together. It is important that each group member work on the assignment before arriving at the study session. If you struggled with a problem, or couldn't do it at all, ask for help or suggestions from your group members.

If there was a problem that you seem to understand better than your group, be sure to share your knowledge; explaining how to do a certain problem increases your chances of retaining that knowledge until the exam and beyond.

At some point, try several even-numbered problems from the text that are representative of the problems on the homework assignment. Check your solutions with other members of your group.

At the end of each session, quickly review what the group has accomplished. Finish the meeting by planning your next session, including the time and location of the meeting and what you plan to work on at that meeting.

EXERCISES 1.4

1. To multiply fractions, we _____.

2. Before multiplying by a mixed number, we convert it to a(n) _____.

3. When we invert a fraction, the result is called its _____.

4. Explain how to divide by a fraction.

5. In order to add or subtract fractions they must have the same _____.

6. The smallest number that is a multiple of two numbers is called their _____.

7. A board is cut into two pieces which measure $4\frac{1}{6}$ feet and $3\frac{3}{8}$ feet, respectively. Which operation will give the length of the original board?

a) $4\frac{1}{6} + 3\frac{3}{8}$ b) $4\frac{1}{6} - 3\frac{3}{8}$

c) $4\frac{1}{6} \cdot 3\frac{3}{8}$ d) $4\frac{1}{6} \div 3\frac{3}{8}$

8. Ralph has 65 ounces of perfume that he wants to pour into bottles that each hold $1\frac{5}{8}$ ounces. Which operation will give the number of $1\frac{5}{8}$ ounce bottles that Ralph will need?

a) $65 + 1\frac{5}{8}$ b) $65 - 1\frac{5}{8}$

c) $65 \cdot 1\frac{5}{8}$ d) $65 \div 1\frac{5}{8}$

Multiply. Your answer should be in lowest terms.

9. $\frac{5}{8} \cdot \frac{6}{25}$ 10. $\frac{3}{4} \cdot \frac{18}{23}$

11. $-\frac{9}{10} \cdot \frac{14}{15}$ 12. $\frac{21}{10} \cdot \frac{30}{49}$

13. $4\frac{2}{7} \cdot \frac{14}{25}$

14. $-3\frac{5}{9} \cdot 2\frac{1}{6}$

15. $5 \cdot 6\frac{3}{10}$

16. $8 \cdot \frac{7}{12}$

17. $\frac{2}{3} \cdot \frac{8}{9}$

18. $-\frac{12}{35}\left(-\frac{14}{99}\right)$

Divide. Your answer should be in lowest terms.

19. $\frac{3}{4} \div \frac{9}{16}$ 20. $\frac{2}{5} \div \frac{7}{15}$

21. $-\frac{1}{9} \div \frac{4}{3}$ 22. $\frac{5}{6} \div \frac{13}{14}$

23. $\frac{17}{40} \div \frac{1}{2}$ 24. $\frac{7}{30} \div \left(-\frac{1}{5}\right)$

25. $\frac{4}{11} \div 3\frac{1}{5}$ 26. $3\frac{3}{8} \div 6$

27. $-2\frac{4}{5} \div 6\frac{2}{3}$ 28. $7\frac{1}{9} \div 13\frac{3}{8}$

Add or subtract.

29. $\frac{7}{15} + \frac{4}{15}$ 30. $\frac{1}{8} + \frac{5}{8}$

31. $\frac{5}{9} + \frac{4}{9}$ 32. $\frac{3}{10} + \frac{9}{10}$

33. $\frac{2}{5} - \frac{4}{5}$ 34. $\frac{17}{18} - \frac{5}{18}$

35. $\frac{9}{16} - \frac{3}{16}$ 36. $\frac{13}{42} - \frac{29}{42}$

Find the LCM of the given numbers.

37. 12, 18 38. 6, 8
39. 10, 14 40. 9, 15
41. 9, 36 42. 9, 10
43. 4, 6, 9 44. 8, 10, 12

Simplify. Your answer should be in lowest terms.

45. $\dfrac{1}{2} + \dfrac{1}{3}$ **46.** $\dfrac{2}{3} + \dfrac{1}{5}$

47. $\dfrac{4}{5} + \dfrac{3}{4}$ **48.** $\dfrac{4}{7} + \dfrac{1}{4}$

49. $\dfrac{7}{10} + \dfrac{5}{8}$ **50.** $\dfrac{3}{4} + \dfrac{5}{6}$

51. $\dfrac{21}{63} + \dfrac{1}{5}$ **52.** $\dfrac{1}{4} + \dfrac{60}{80}$

53. $6\dfrac{1}{5} + 5$ **54.** $3 + 8\dfrac{3}{7}$

55. $6\dfrac{2}{3} + 5\dfrac{1}{6}$ **56.** $11\dfrac{4}{9} + 5\dfrac{1}{3}$

57. $\dfrac{5}{6} - \dfrac{1}{3}$ **58.** $\dfrac{7}{12} - \dfrac{3}{4}$

59. $\dfrac{2}{3} - \dfrac{7}{15}$ **60.** $\dfrac{1}{2} - \dfrac{7}{9}$

61. $\dfrac{3}{4} - \dfrac{2}{7}$ **62.** $\dfrac{5}{8} - \dfrac{5}{6}$

63. $\dfrac{11}{20} - \dfrac{5}{12}$ **64.** $\dfrac{5}{7} - \dfrac{3}{14}$

65. $7\dfrac{1}{2} - 3\dfrac{1}{4}$ **66.** $12\dfrac{2}{3} - 6\dfrac{2}{5}$

67. $12\dfrac{3}{10} - 9$ **68.** $6 - 4\dfrac{3}{4}$

69. $-\dfrac{5}{9} - \dfrac{7}{12}$ **70.** $-\dfrac{9}{10} - \dfrac{11}{14}$

71. $-\dfrac{9}{16} + \dfrac{5}{24}$ **72.** $-\dfrac{3}{8} + \dfrac{13}{24}$

73. $\dfrac{6}{7} - \left(-\dfrac{8}{15}\right)$ **74.** $\dfrac{1}{12} - \left(-\dfrac{19}{30}\right)$

75. $-\dfrac{4}{15} + \left(-\dfrac{13}{18}\right)$ **76.** $-\dfrac{17}{24} + \left(-\dfrac{25}{42}\right)$

77. $\dfrac{3}{16} + \dfrac{9}{20} - \dfrac{11}{12}$ **78.** $\dfrac{10}{21} - \dfrac{13}{18} + \dfrac{8}{15}$

(**Mixed Practice, 79–96**)

Simplify.

79. $\dfrac{8}{9} \cdot \dfrac{3}{5}$ **80.** $\dfrac{19}{20} - \dfrac{7}{20}$

81. $\dfrac{5}{12} + \dfrac{11}{12}$ **82.** $\dfrac{3}{4} + \dfrac{7}{10}$

83. $\dfrac{7}{30} \div \dfrac{35}{48}$ **84.** $\dfrac{12}{35} \cdot \dfrac{14}{27}$

85. $\dfrac{1}{6} - \dfrac{7}{8}$ **86.** $\dfrac{7}{24} - \dfrac{29}{40}$

87. $\dfrac{19}{30} + \dfrac{11}{18}$ **88.** $3\dfrac{1}{5} \cdot 4\dfrac{3}{8}$

89. $13 \div \dfrac{1}{8}$ **90.** $\dfrac{3}{5} - \dfrac{2}{3} - \dfrac{7}{10}$

91. $3\dfrac{4}{7} + 6\dfrac{3}{5} - 8$ **92.** $12\dfrac{1}{3} + 7\dfrac{1}{6} - 5\dfrac{1}{2}$

93. $\dfrac{15}{56} + \left(-\dfrac{16}{21}\right)$ **94.** $\dfrac{7}{12} - \left(-\dfrac{23}{30}\right)$

95. $-\dfrac{9}{13} + \dfrac{19}{36}$ **96.** $-\dfrac{3}{8} - \dfrac{81}{100}$

Find the missing number.

97. $\dfrac{11}{24} + \dfrac{5}{?} = \dfrac{13}{12}$ **98.** $\dfrac{?}{10} - \dfrac{1}{3} = \dfrac{1}{6}$

99. $\dfrac{10}{21} \cdot \dfrac{?}{75} = \dfrac{4}{45}$ **100.** $\dfrac{11}{40} + \dfrac{?}{40} = \dfrac{9}{10}$

101. Bruce is fixing a special dinner for his girlfriend. The three recipes he is preparing call for $\frac{1}{2}$ cup, $\frac{3}{4}$ cup, and $\frac{1}{3}$ cup of olive oil, respectively. In total, how much olive oil does Bruce need to make these three recipes?

102. Sue gave birth to twins. One of the babies weighed $4\frac{7}{8}$ pounds at birth, and the other baby weighed $5\frac{1}{4}$ pounds. Find the total weight of the twins at birth.

103. A chemist has $\frac{23}{40}$ fluid ounce of a solution. If she needs $\frac{1}{8}$ fluid ounce of the solution for an experiment, how much of the solution will remain?

104. A popular weed spray concentrate recommends using $1\frac{1}{4}$ tablespoons of concentrate for each quart of water. How much concentrate needs to be mixed with 6 quarts of water?

105. A board that is $4\frac{1}{5}$ feet long needs to be cut into 6 pieces of equal length. How long will each piece be?

106. Ross makes a batch of hot sauce that will be poured into bottles that hold $5\frac{3}{4}$ fluid ounces. If Ross has 115 fluid ounces of hot sauce, how many bottles can he fill?

107. A craftsman is making a rectangular picture frame. Two of the sides will each be $\frac{5}{6}$ of a foot long, while the other two sides will each be $\frac{2}{3}$ of a foot long. If the craftsman has one board that is $2\frac{3}{4}$ feet long, is this enough to make the picture frame? Explain why or why not.

108. A pancake recipe calls for $1\frac{1}{3}$ cups of whole-wheat flour to make 12 pancakes. How much flour would be needed in order to make 48 pancakes?

109. A pancake recipe calls for $1\frac{1}{3}$ cups of whole-wheat flour to make 12 pancakes. How much flour would be needed to make 6 pancakes?

110. Explain, using your own words, the difference between dividing a number in half and dividing a number by one half.

111. Draw a picture to show that $\frac{2}{7} + \frac{4}{7} = \frac{6}{7}$. (Hint: The fraction $\frac{2}{7}$ can be represented as

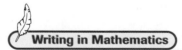
.)

112. Draw a picture to show that $\frac{1}{8} + \frac{5}{8} = \frac{3}{4}$.

113. Draw a picture to show that $\frac{5}{9} - \frac{2}{9} = \frac{1}{3}$.

114. List a pair of numbers whose LCM is 84.

Writing in Mathematics

Answer in complete sentences.

115. Explain why we cannot divide a common factor of 2 from the numbers 4 and 6 in the expression $\frac{4}{7} \cdot \frac{6}{11}$.

116. Explain why it would be a bad idea to rewrite fractions with a common denominator before multiplying them.

117. ***Solutions Manual*** * Write a solutions manual page for the following problem:

Simplify. $\frac{2}{9} - \frac{3}{2} + \frac{5}{6}$

118. ***Newsletter*** * Write a newsletter explaining how to add and subtract fractions.

See Appendix B for details and sample answers.

1 Perform arithmetic operations with decimals.
2 Rewrite a fraction as a decimal number.
3 Rewrite a decimal number as a fraction.
4 Rewrite a fraction as a percent.
5 Rewrite a decimal as a percent.
6 Rewrite a percent as a fraction.
7 Rewrite a percent as a decimal.

1.5 Decimals and Percents

Decimals

Rational numbers can also be represented using **decimal notation**. The decimal 0.23 is equivalent to the fraction $\frac{23}{100}$ or twenty-three hundredths. The digit 2 is said to be in the tenths place, and the digit 3 is in the hundredths place. Here is a chart showing several place values for decimals:

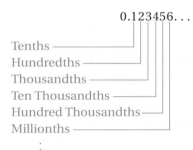

$$0.123456\ldots$$

Tenths
Hundredths
Thousandths
Ten Thousandths
Hundred Thousandths
Millionths

Objective 1 **Perform arithmetic operations with decimals.** Here is a brief summary of arithmetic operations using decimals:

- To add or subtract two decimal numbers, align the decimal points and add or subtract as you would with integers.

$$3.96 + 12.072$$

$$\begin{array}{r} 3.96 \\ + 12.072 \\ \hline 16.032 \end{array}$$

If the two decimals do not have the same number of decimal places, we can extend the number with fewer places by adding 0's to the end of that number. For instance, 3.96 can be written as 3.960 in the preceding example.

- To multiply two decimal numbers, multiply them as you would integers. The total number of decimal places in the two factors tell us how many decimal places are in the product.

$$-2.09 \cdot 3.1$$

In this example, the two factors have a total of three decimal places, so our product must have three decimal places. We multiply these two numbers as if they were 209 and 31 and then insert the decimal point in the appropriate place, leaving three digits to the right of the decimal point.

$$
\begin{array}{r}
-209 \\
\times \quad 31 \\
\hline
-6479
\end{array}
\qquad
\begin{array}{r}
-2.09 \\
\times \quad 3.1 \\
\hline
-6.479
\end{array}
$$

3 places

- To divide two decimal numbers, we move the decimal point in the divisor to the right so that it becomes an integer. We must then move the decimal point in the other number (dividend) to the right by the same number of spaces. The decimal point in the answer will be aligned with this new location of the decimal point in the dividend.

$$8.24 \div 0.4$$

$$0.4\overline{)8.24}$$ We begin by moving *each* decimal point one place to the right.

$$0.4\overline{)8.24}$$

$$\begin{array}{r} 20.6 \\ 4\overline{)82.4} \end{array}$$ Then we perform the division.

Rewriting Fractions as Decimals and Decimals as Fractions

Objective 2 Rewrite a fraction as a decimal number. To rewrite any fraction as a decimal, we divide its numerator by its denominator. The fraction line is simply another way to write "÷".

EXAMPLE 1   Rewrite the fraction $\frac{5}{8}$ as a decimal.

Solution

To rewrite this fraction as a decimal, we will divide 5 by 8. Since 8 does not divide into 5, we will place a decimal point after the 5 and add a 0 after it.

$$8\overline{)5.0}$$

We can now begin the division, adding 0's to the end of the dividend until there is no remainder.

$$
\begin{array}{r}
0.625 \\
8\overline{)5.000} \\
\underline{-4\,8} \\
20 \\
\underline{-16} \\
40 \\
\underline{-40} \\
0
\end{array}
$$

Quick Check 1

Rewrite the fraction $\frac{3}{4}$ as a decimal.

$$\frac{5}{8} = 0.625$$

EXAMPLE 2   Rewrite the fraction $\frac{23}{30}$ as a decimal.

Solution

When we divide 23 by 30, our result is a decimal that does not terminate ($0.76666\ldots$). The pattern continues repeating the digit 6 forever. This is an example of a **repeating decimal**. We may place a bar over the repeating digit(s) to denote a repeating decimal.

$$\frac{23}{30} = 0.7\overline{6}$$

Quick Check 2
Rewrite the fraction $\frac{5}{18}$ as a decimal.

Objective 3   Rewrite a decimal number as a fraction. Suppose we wanted to rewrite a decimal number such as 0.48 as a fraction. This decimal is read as "forty-eight hundredths" and is equivalent to the fraction $\frac{48}{100}$. Simplifying this fraction shows us that $0.48 = \frac{12}{25}$.

EXAMPLE 3   Rewrite the decimal 0.8 as a fraction in lowest terms.

Solution

The decimal ends in the tenths place, so we will begin with a fraction whose denominator is 10.

$$0.8 = \frac{8}{10} \qquad \text{Remove the decimal point to find the numerator.}$$

$$= \frac{4}{5} \qquad \text{Simplify to lowest terms.}$$

Quick Check 3
Rewrite the decimal 0.04 as a fraction in lowest terms.

Using Your Calculator   To rewrite a decimal as a fraction using the TI-84, we type the decimal, press the [MATH] key, and select option 1. The following screens show how to rewrite the decimal 0.8 as a fraction using the TI-84 as in Example 3:

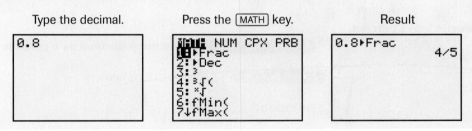

| Type the decimal. | Press the [MATH] key. | Result |

EXAMPLE 4   Rewrite the decimal 0.164 as a fraction in lowest terms.

Solution

This decimal ends in the thousandths place, so we start with a fraction of $\frac{164}{1000}$.

Quick Check 4
Rewrite the decimal 0.425 as a fraction in lowest terms.

$$0.164 = \frac{164}{1000} \qquad \text{Rewrite as a fraction whose denominator is 1000.}$$

$$= \frac{41}{250} \qquad \text{Simplify to lowest terms.}$$

Percents

Objective 4 Rewrite a fraction as a percent. Percents (%) are used to represent numbers as parts of one hundred. One percent, which can be written as 1%, is equivalent to 1 part of 100 or $\frac{1}{100}$ or 0.01. The fraction $\frac{27}{100}$ is equivalent to 27%. Percents, decimals, and fractions are all ways to write a rational number. We will learn to convert back and forth between percents and fractions, as well as between percents and decimals.

Rewriting Fractions and Decimals as Percents

To rewrite a fraction or a decimal as a percent, we multiply it by 100%.

EXAMPLE 5 Rewrite $\frac{2}{5}$ as a percent.

Solution

We begin by multiplying by 100% and simplifying.

$$\frac{2}{\underset{1}{\cancel{5}}} \cdot \overset{20}{\cancel{100}}\% = 40\%$$

Quick Check 5
Rewrite $\frac{7}{10}$ as a percent.

When rewriting a fraction as a percent, it may be helpful to multiply by $\frac{100}{1}\%$.

EXAMPLE 6 Rewrite $\frac{3}{8}$ as a percent.

Solution

Again, we multiply by 100%. Occasionally, as in this example, we will end up with an improper fraction, which can be changed to a mixed number.

$$\frac{3}{\underset{2}{\cancel{8}}} \cdot \overset{25}{\cancel{100}}\% = \frac{75}{2}\%, \text{ which can be rewritten as } 37\frac{1}{2}\%.$$

Quick Check 6
Rewrite $\frac{21}{40}$ as a percent.

Objective 5 Rewrite a decimal as a percent.

EXAMPLE 7 Rewrite 0.3 as a percent.

Solution

When we multiply a decimal by 100, the result is the same as moving the decimal point two places to the right.

$$0.30.$$

$$0.3 \cdot 100\% = 30\%$$

Quick Check 7
Rewrite 0.42 as a percent.

Rewriting Percents as Fractions and Decimals

Objective 6 Rewrite a percent as a fraction. To rewrite a percent as a fraction or a decimal, we can divide it by 100 and omit the percent sign. When rewriting a percent as a fraction, we may choose to multiply by $\frac{1}{100}$ rather than dividing by 100.

EXAMPLE ▸8   Rewrite 44% as a fraction.

Solution

We will begin by multiplying by $\frac{1}{100}$ and omitting the percent sign.

$$\overset{11}{\cancel{44}} \cdot \frac{1}{\underset{25}{\cancel{100}}} = \frac{11}{25}$$

Quick Check **8** Rewrite 35% as a fraction.

EXAMPLE ▸9   Rewrite $16\frac{2}{3}\%$ as a fraction.

Solution

To begin, we rewrite the mixed number $16\frac{2}{3}$ as an improper fraction $\left(\frac{50}{3}\right)$ and then simplify.

$$\frac{\overset{1}{\cancel{50}}}{3} \cdot \frac{1}{\underset{2}{\cancel{100}}} = \frac{1}{6}$$

Quick Check **9** Rewrite $11\frac{2}{3}\%$ as a fraction.

Objective 7 **Rewrite a percent as a decimal.** In the next examples, we will rewrite percents as decimals, rather than as fractions.

EXAMPLE ▸10   Rewrite 56% as a decimal.

Solution

We start by dropping the percent sign and dividing by 100. Keep in mind that dividing a decimal number by 100 is the same as moving the decimal point two places to the left.

$$.56,$$

$$56 \div 100 = 0.56$$

Quick Check **10** Rewrite 8% as a decimal.

EXAMPLE ▸11   Rewrite 143% as a decimal.

Solution

When a percent is greater than 100%, its equivalent decimal must be greater than 1.

$$143 \div 100 = 1.43$$

Quick Check **11** Rewrite 240% as a decimal.

Building Your Study Strategy Study Groups, 5 **Three Questions** Another way to structure a group study session is to have each member bring a list of three questions to the meeting. The questions could be about specific homework problems or they could be about topics or procedures that have been covered in class. Begin by having each person read one of their questions. Once the questions have been read, the group should attempt to come up with answers that each member understands. Repeat this process until all questions have been answered. If the group cannot answer a question, ask your instructor at the beginning of the next class or during office hours.

EXERCISES 1.5

Vocabulary

1. The first place to the right of a decimal point is the _____ place.

2. The third place to the right of a decimal point is the _____ place.

3. To rewrite a fraction as a decimal _____.

4. To rewrite a fraction as a percent _____.

5. To rewrite a percent as a fraction _____.

6. Percents are used to represent numbers as parts of _____.

Simplify the following decimal expressions.

7. $4.23 + 3.62$

8. $13.89 - 2.54$

9. $-7(5.2)$

10. $69.54 \div 6$

11. $8.4 - 3.7$

12. $-7.9 + (-4.5)$

13. $13.568 \div 0.4$

14. $3.6(4.7)$

15. $-2.2 \cdot 3.65$

16. $6.2 - 15.9$

17. $13.47 - (-21.562)$

18. $5.283 \div 0.25$

19. $-6.3(3.9)(-2.25)$

20. $-4.84 \div (-0.016)$

21. $37.278 + 56.722$

22. $109.309 - 27.46 - 52.3716$

Rewrite the following fractions as decimal numbers.

23. $\frac{3}{5}$

24. $\frac{7}{10}$

25. $-\frac{7}{8}$

26. $\frac{33}{4}$

27. $\frac{16}{25}$

28. $-\frac{9}{16}$

29. $24\frac{29}{50}$

30. $7\frac{3}{20}$

Rewrite the following decimal numbers as fractions in lowest terms.

31. 0.2

32. 0.5

33. 0.44

34. 0.35

35. −0.75   **36.** −0.68

37. 0.375   **38.** 0.204

39. The normal body temperature for humans is 98.6° F. If Melody has a temperature that is 2.8° F above normal, what is her temperature?

40. Paul spent the following amounts on gifts for his wife's birthday: $32.95, $16.99, $47.50, $12.37, and $285. How much did Paul spend on gifts for his wife?

41. The balance of Jessica's checking account is $293.42. If she writes a check for $47.47, what will her new balance be?

42. The S&P 500 Index is one method used to measure the performance of the stock market. Over a three-day period, the S&P 500 lost 7.26 points, gained 3.17 points, and lost 12.69 points. If the index was at 905.36 before these three days, what was the index at after these three days?

43. An office manager bought 12 cases of paper. If each case cost $21.47, what was the total cost for the 12 cases?

44. Jean gives Chris a $20 bill and tells him to go to the grocery store and buy as many hot dogs as he can. If each package of hot dogs costs $2.65, how many packages can Chris buy? How much change will Chris have?

Rewrite as percents.

45. $\dfrac{1}{2}$ **46.** $\dfrac{1}{4}$

47. $\dfrac{3}{4}$ **48.** $\dfrac{3}{5}$

49. $\dfrac{4}{5}$ **50.** $\dfrac{5}{8}$

51. $\dfrac{7}{8}$ **52.** $\dfrac{11}{12}$

53. $\dfrac{27}{4}$ **54.** $\dfrac{12}{5}$

55. 0.4   **56.** 0.6

57. 0.15   **58.** 0.87

59. 0.09   **60.** 0.03

61. 3.2   **62.** 2.75

Rewrite as fractions.

63. 53%   **64.** 71%

65. 84%   **66.** 80%

67. 7%   **68.** 2%

69. $11\dfrac{1}{9}\%$ **70.** $18\dfrac{2}{11}\%$

71. 520%   **72.** 275%

Rewrite as decimals.

73. 32%   **74.** 57%

75. 16%   **76.** 29%

77. 4%   **78.** 1%

79. 0.3%   **80.** 61.3%

81. 400%   **82.** 320%

83. Find three fractions that are equivalent to 0.375.

84. Find three fractions that are equivalent to 0.4.

Complete the following table.

	Fraction	Decimal	Percent
85.		0.2	
86.	$\dfrac{7}{40}$		
87.			32%
88.			45%
89.	$\dfrac{13}{8}$		
90.		0.64	

Writing in Mathematics

Answer in complete sentences.

91. Until recently, stock prices at the New York Stock Exchange were reported as fractions. Now prices are reported as decimals. Do you feel that this was a good idea? Explain why.

92. Describe a real-world application involving decimals.

93. *Solutions Manual** Write a solutions manual page for the following problem:

Josh's checking account had a balance of $287.29. After making a deposit of $384.16, Josh wrote checks for $109.50, $302.23, and $147.86. What is his new balance?

94. *Newsletter** Write a newsletter explaining how to convert a fraction to a percent and a percent to a fraction.

*See Appendix B for details and sample answers.

1.6

Representing Data Graphically

Objectives
1 **Create a pie chart.**
2 **Create a bar graph.**
3 **Display data over time using a line graph.**

In today's data-driven society, we constantly see graphs presenting **data** or information on the television news, as well as in newspapers and magazines. Not only is it important to be able to read and interpret graphs, we must be able to create them as well. This section focuses on the creation and interpretation of pie charts, bar graphs, and line graphs that display data over a period of time.

Pie Charts

Objective 1 **Create a pie chart.** A **pie chart** is a circle used to display how a set of data can be divided into categories by percent. The entire circle represents 100%, or the total. Each category is displayed as a wedge of the circle, showing the percentage of the total that the category represents. Here is an example of a pie chart, showing the market share of leading pizza makers:

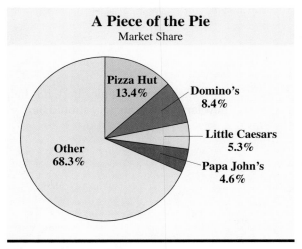

A Piece of the Pie
Market Share

Pizza Hut 13.4%
Domino's 8.4%
Little Caesars 5.3%
Papa John's 4.6%
Other 68.3%

(*Source: Advertising Age*)

From the pie chart, we can see that Pizza Hut (13.4%) had the highest market share of any one chain. We also see that the category labeled as Other, which includes independent pizzerias and some minor chains, has a majority of the market share (68.3%).

EXAMPLE 1 A survey of 200 registered voters regarding an upcoming bond measure showed that 90 voters were in favor of the bond, 72 were opposed, and the rest were undecided. Create a pie chart that shows the percent of these voters who were in favor of the bond, opposed to it, or undecided.

Solution

We begin by finding what percent of the total is represented by each group. Since 90 of the 200 voters were in favor of the bond, this category is $\frac{90}{200}$ of the total. To rewrite this fraction as a percent, we multiply by 100%.

$$\frac{90}{200} \cdot 100\% = 45\% \qquad \text{Multiply by 100\% and simplify.}$$

45% of these voters were in favor of the bond. To find the percent of these voters who opposed the bond, we need to rewrite the fraction $\frac{72}{200}$ as a percent, which is 36%.

   Since the three categories must total 100%, we know that the percent of these voters who were undecided must be 19% because $100 - 45 - 36 = 19$. Here is the pie chart:

Quick Check 1
150 students were asked whether there was adequate parking on their campus. Thirty-three of the students said yes, 87 said no, and the remaining 30 were unsure. Create a pie chart showing the percent of each response.

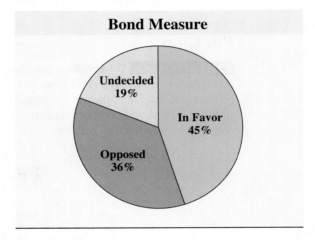

When creating a pie chart, it is helpful to use the following template for determining how large each category should be.

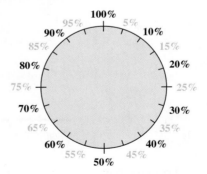

Bar Graphs

Objective 2 Create a bar graph. A **bar graph** is another graph that can be used to compare categories. To construct a bar graph, each category must be associated with a number. A bar is drawn above each category's name. The height of the bar is determined by the number associated with the category, while the width of all bars is the same. For example, the following graph shows the number of drive-in theaters in five different states:

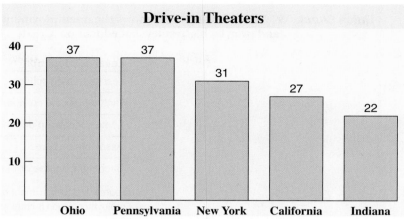

Drive-in Theaters

(*Source:* United Drive-In Theatre Owners Association and AP)

EXAMPLE 2   The following table shows the average tuition and fees in 2006–07 for three different types of colleges and universities:

Type	Average Tuition and Fees
Private Four-Year	$22,218
Public Four-Year	$5836
Public Two-Year	$2272

Create a bar graph for this set of data. (*Source:* Annual Survey of Colleges, The College Board)

Solution

We will use a scale of $4000 for the average tuition and fees.

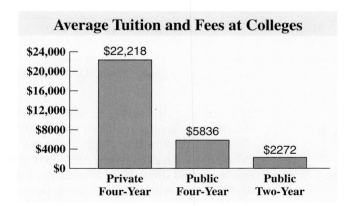

Average Tuition and Fees at Colleges

Quick Check **2** The following table shows the unemployment rates in 2006 for persons 25 years old and over, by highest level of education. Create a bar graph for this set of data.

Level of Education	Percent Unemployed
Less than high school	7.6
High school, no college	4.7
Some college, no degree	4.2
Associate's degree	3.3
Bachelor's degree or higher	2.3

(*Source:* U.S. Department of Labor, Bureau of Labor Statistics, Office of Employment and Unemployment Statistics, Current Population Survey, 2005)

Line Graphs

Objective **3** **Display data over time using a line graph.** The final graph introduced in this section is a **line graph**, which is used to show how a numerical quantity changes over time. The following line graph shows how the percentage of public school classrooms with Internet access changed during the years 1996–2001:

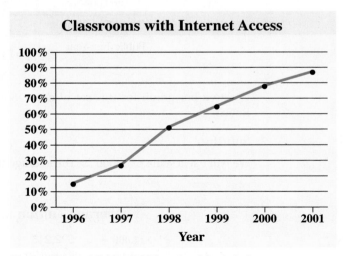

(*Source:* National Center for Educational Statistics)

By examining this graph, we can see that the percentage of public school classrooms with Internet access increased rapidly during this period.

To construct a line graph, we begin with a horizontal number line that is used to show time, such as the years 1996–2001 in the previous graph. We also draw a vertical line that is used to show the numerical quantity we are investigating, such as the percentage of classrooms with Internet access. For each period, we place a point at the appropriate height, and connect adjacent points with lines.

EXAMPLE 3   The median U.S. household income grew from approximately $8000 in 1969 to approximately $44,000 in 2004. Create a line graph for this set of data. *(Source: U.S. Census Bureau)*

Year	1969	1974	1979	1984	1989	1994	1999	2004
Median Salary ($ Thousands)	8	11	16	22	29	32	41	44

Solution

The horizontal line used for showing the year needs to be labeled from 1969 to 2004, in steps of 5 years. The vertical line used for showing the median income must extend at least to 44, so we will use a scale of 5 and label the line up to 45. The point for the year 1969 needs to be placed at a height of 8. We plot the remainder of the points, and connect them with lines.

Quick Check **3**
Here is the amount of money raised for charity by runners and walkers for the years from 2002 to 2005. Create a line graph for this set of data. (*Source: USA Track & Field*)

Year	Money Raised (Millions)
2002	520
2003	560
2004	575
2005	656

Building Your Study Strategy **Study Groups, 6   Keeping in Touch**   It is important for group members to share phone numbers and e-mail addresses. This will allow you to contact other group members if you misplace the homework assignment for that day. You can also contact others in your group if you have to miss class. This allows you to find out what was covered in class and which problems were assigned for homework. The person you contact can also give you advice on certain problems.

Some members of study groups agree to call each other if they are having trouble with homework problems. Calling a group member for this purpose should be a last resort. Be sure that you have used all available resources (examples in the text, notes, etc.) and have given the problem your fullest effort. Otherwise, you may end up calling for help on all of the problems.

Vocabulary

1. A(n) _____ is a circle used to display how a set of data can be divided into categories by percent.

2. When creating a pie chart, we can find the percent of the total represented by a group by dividing the number in that group by _____.

3. A bar graph is a graph used to _____.

4. A _____ is a graph used to display how a numerical quantity changes over time.

For Exercises 5–14, construct a pie chart to represent the given set of data.

5. In 2000–01, 79% of public school teachers were women and 21% were men. (*Source:* National Center for Education Statistics)

6. In the 2004 presidential election, 64% of voting-age citizens voted while 36% did not. (*Source:* U.S. Census Bureau)

7. A 2005 study revealed that 36% of all Americans 25 years and older held a college degree and the remaining 64% did not. (*Source:* U.S. Census Bureau, Current Population Survey)

8. Thirty-one percent of fourth-grade public school students in the United States performed at or above the "proficient" level in the 2003 National Assessment of Educational Progress math test, while 69% performed below this level. (*Source:* National Center for Education Statistics)

9. A number of business executives were asked how often they check in while on vacation. Here are the results.

Occasionally	Daily	Constantly	Never
49%	15%	6%	30%

(*Source:* Interland.com 2005 Business Barometer)

10. Here is a table showing how long Americans planned to take to pay off their holiday debt:

Will not have any debt	57%
Less than a month	10%
One month to 5 months	22%
More than 5 months	7%
Not sure	4%

(*Source:* Maritz Poll survey)

11. Four hundred moviegoers were asked whether they felt that they should pay less to attend a movie if the theater runs ads on the screen. Of the 400 respondents, 284 said yes, 40 said no, and 76 said they were unsure. (*Source:* Based on an InsightExpress survey)

12. A sample of 1400 workers who were asked whether they would not mind using electronic fingerprint or hand-recognition technology to record their time and attendance at work yielded the following results:

Would not mind	Privacy invasion	Not sure
728	280	392

(*Source:* American Payroll Association survey)

13. A sample of 400 high school freshmen who were asked when they planned to attend college produced the following results:

Right after graduation	256
After working for at least 2 years	48
Never	68
Unsure	28

14. A community college math department is offering 15 prealgebra classes, 21 elementary algebra classes, and 24 intermediate algebra classes this semester.

For Exercises 15–16, construct two pie charts to represent the given sets of data and illustrate the difference between them.

15. In 2003–04, 55% of first-time college freshmen were females. In the same year, 43% of college faculty members were females. (*Source:* National Center for Educational Statistics)

16. In 2003–04, 47% of the bachelor's degrees awarded in mathematics were earned by females. In the same year, 28% of the doctoral degrees awarded in mathematics were earned by females. (*Source:* National Center for Educational Statistics)

For Exercises 17–22, construct a bar chart to represent the given set of data.

17. A survey of 1009 adults in September 2006 asked if the price of gasoline caused any financial hardship in their household.

Response	Frequency
Severe hardship	121
Moderate hardship	414
No hardship	464
Unsure	10

(*Source:* Opinion Research Corporation)

18. Citizens of four industrialized countries were asked in May 2006 to state a fair price for a gallon of gasoline. Here are the average responses for each country:

Country	Cost/Gallon
USA	$1.99
Canada	$2.72
Germany	$4.87
United Kingdom	$5.16

(*Source:* Ipsos Public Affairs poll)

19. Here are the percentages of public school eighth-graders in certain cities who scored at or above "proficient" on national assessments of reading skills in 2003:

City	% Proficient
Atlanta	11%
Boston	22%
Chicago	15%
San Diego	20%
Washington, DC	10%

(*Source:* National Center for Educational Statistics)

20. Here are the percentages of public school eighth-graders in certain cities who scored at or above "proficient" on national assessments of math skills in 2003:

City	% Proficient
Atlanta	6%
Boston	17%
Chicago	9%
San Diego	18%
Washington, DC	6%

(*Source:* National Center for Educational Statistics)

21. Here is a list of the companies that were granted the most patents in 2005:

Company	Number of Patents
IBM	2941
Canon	1828
Hewlett-Packard	1797
Matsushita Electric	1688
Samsung	1641

(*Source:* U.S. Department of Commerce)

22. In 2006, some Americans were asked about how much they planned to spend on holiday gifts. Here are the average results, rounded to the nearest $5:

Gifts for …	Amount
Family	$450
Self	$100
Friends	$85
Coworkers	$20
Others	$45

(*Source:* National Retail Federation)

For Exercises 23–30, create a line graph for the given set of data.

23. Here are government estimates for the number of hours U.S. adults spent listening to recorded music, for the years 1999–2004.

Year	1999	2000	2001	2002	2003	2004
Hours	281	258	229	200	184	185

(*Source:* U.S. Census Bureau)

24. The percentage of U.S. households that have at least one mutual fund has grown steadily since the early 1980s.

1980	1984	1988	1992	1996	2000	2004
6%	12%	24%	27%	37%	49%	48%

(*Source:* Investment Company Institute)

27. The following table shows that the average size of a new home, in square feet, has grown over time:

Year	1950	1960	1970	1980	1990	2000
Average Size	1100	1400	1500	1740	2080	2270

(*Source:* National Association of Home Builders)

25. Here are the number of worldwide spam messages sent daily, in billions, for selected years.

1999	2000	2001	2002	2003	2004
1.0	2.3	4.0	5.6	7.3	23.0

(*Source:* IDC)

28. Here are the number of U.S. cell phone subscribers, in millions, for selected years.

1998	1999	2000	2001	2002	2003	2004	2005
69	86	109	128	141	159	182	208

(*Source:* Cellular Telecommunications & Internet Associates)

26. Here are the number of mobile homes in the United States since 1960, in millions:

1960	1970	1980	1990	2000
0.8	2.1	4.7	7.4	8.8

(*Source:* U.S. Bureau of the Census)

29. The number of obese people having weight-loss surgery, primarily gastric bypass surgery, has increased dramatically. (*Source:* American Society for Bariatric Surgery)

1997	2002	2003	2004	2005
23,100	63,100	103,200	140,000	171,000

30. Here is the total value of online retail sales in the United States, in billions, for the years 1998–2005. (*Source:* U.S. Census Bureau)

Year	1998	1999	2000	2001	2002	2003	2004	2005
Sales	$6	$12	$24	$31	$41	$54	$67	$81

31. Buy a bag of plain M&M candies, pour out the first 50 candies and create a pie chart for the colors of these 50 candies.

32. In a public location on campus, record the gender of the first 200 students you see. Create a pie chart to display your results.

Writing in Mathematics

Answer in complete sentences.

33. Explain why it is important for each bar in a bar graph to have the same width.

34. Explain why the following bar graph is misleading.

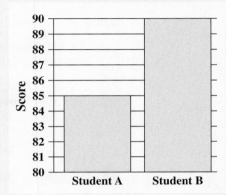

Score on Math Exam

35. *Solutions Manual** Write a solutions manual page for the following problem:

Create a line graph for the following data.

College Enrollment Rates of High School Graduates					
Year	1964	1974	1984	1994	2004
Percent	48%	47%	55%	62%	67%

(*Source:* National Center for Educational Statistics)

36. *Newsletter** Write a newsletter explaining how to create pie charts.

*See Appendix B for details and sample answers.

QUICK REVIEW EXERCISES

Section 1.6

Simplify.

1. $-19 + 12$

2. $(-4)(-5)(-6)$

3. $-9 + 21 - (-6) - 38$

4. $\dfrac{4}{9} \div \left(-\dfrac{14}{165}\right)$

1.7

Exponents and Order of Operations

Objectives

1. Simplify exponents.
2. Use the order of operations to simplify arithmetic expressions.

Exponents

Objective 1 Simplify exponents. The same number used repeatedly as a factor can be represented using **exponential notation**. For example, $3 \cdot 3 \cdot 3 \cdot 3 \cdot 3$ can be written as 3^5.

Base, Exponent

For the expression 3^5, the number being multiplied (3) is called the **base**. The **exponent** (5) tells us how many times that the base is used as a factor.

We read 3^5 as "*three raised to the fifth power*" or simply as "*three to the fifth power*." When raising a base to the second power, we usually say that the base is being **squared**. When raising a base to the third power, we usually say that the base is being **cubed**. Exponents of 4 or higher do not have a special name.

EXAMPLE 1   Simplify 2^7.

Solution

In this example, 2 is a factor seven times.

$$2^7 = 2 \cdot 2 \cdot 2 \cdot 2 \cdot 2 \cdot 2 \cdot 2 \qquad \text{Write 2 as a factor seven times.}$$
$$= 128 \qquad \text{Multiply.}$$

Quick Check **1**
Simplify 4^3.

Using Your Calculator   Many calculators have a key that can be used to simplify expressions with exponents. When using the TI-84, we use the $\boxed{\wedge}$ key. Other calculators may have a key that is labeled $\boxed{y^x}$ or $\boxed{x^y}$. Here is the screen that you should see when using the TI-84 to simplify the expression in Example 1:

```
2^7
            128
```

EXAMPLE 2   Simplify $\left(\frac{2}{3}\right)^3$.

Solution

When the base is a fraction, the same rules apply. We use $\frac{2}{3}$ as a factor three times.

$$\left(\frac{2}{3}\right)^3 = \frac{2}{3} \cdot \frac{2}{3} \cdot \frac{2}{3} \qquad \text{Write as a product.}$$
$$= \frac{8}{27} \qquad \text{Multiply.}$$

Quick Check **2**
Simplify $\left(\frac{1}{8}\right)^4$.

Consider the expression $(-3)^2$. The base is -3; so we multiply $(-3) \cdot (-3)$, and the result is positive 9. However, in the expression -2^4, the negative sign is not included in a set of parentheses with the 2. So the base of this expression is 2 and not -2. We will use 2 as a factor 4 times and then take the opposite of our result: $-2^4 = -(2 \cdot 2 \cdot 2 \cdot 2) = -16$.

Order of Operations

Objective 2 **Use the order of operations to simplify arithmetic expressions.**
Suppose we were asked to simplify the expression $2 + 4 \cdot 3$. We could obtain two different results, depending on whether we performed the addition or the multiplication first. Performing the addition first would give us $6 \cdot 3 = 18$, while performing the multiplication first would give us $2 + 12 = 14$. Only one result is correct. The **order of operations agreement** is a standard order in which arithmetic operations are performed, ensuring a single correct answer.

Order of Operations

1. **Remove grouping symbols.** We begin by simplifying all expressions within parentheses, brackets, and absolute value bars. We also perform any operations in the numerator or denominator of a fraction. This is done by following steps 2–4, presented next.
2. **Perform any operations involving exponents.** After all grouping symbols have been removed from the expression, we simplify any exponential expressions.
3. **Multiply and divide.** These two operations have equal priority. We perform multiplications or divisions in the order they appear from left to right.
4. **Add and subtract.** At this point, the only remaining operations should be additions and subtractions. Again, these operations are of equal priority, and we perform them in the order that they appear from left to right. We can also use the strategy we used in totaling integers from Section 1.2.

Considering this information, we find that the correct result when we simplify $2 + 4 \cdot 3$ is 14, because the multiplication takes precedence over the addition.

$$2 + 4 \cdot 3 = 2 + 12 \qquad \text{Multiply } 4 \cdot 3.$$
$$= 14 \qquad \text{Add.}$$

EXAMPLE 3   Simplify $4 + 3 \cdot 5 - 2^6$.

Solution

In this example, the operation with the highest priority is 2^6, since simplifying exponents takes precedence over addition or multiplication. Be sure to note that the base is 2, and not -2, since the negative sign is not grouped with 2 inside a set of parentheses.

Quick Check 3

Simplify
$-2 \cdot 7 + 11 - 3^4.$

$$4 + 3 \cdot 5 - 2^6 = 4 + 3 \cdot 5 - 64 \qquad \text{Raise 2 to the 6}^{\text{th}} \text{ power.}$$
$$= 4 + 15 - 64 \qquad \text{Multiply } 3 \cdot 5.$$
$$= -45 \qquad \text{Simplify.}$$

EXAMPLE 4   Simplify $(-5)^2 - 4(-2)(6)$.

Solution

We begin by squaring negative 5, which equals 25.

$$
\begin{aligned}
(-5)^2 - 4(-2)(6) &= 25 - 4(-2)(6) & &\text{Square } -5. \\
&= 25 - (-48) & &\text{Multiply } 4(-2)(6). \\
&= 25 + 48 & &\text{Write as a sum, eliminating the} \\
& & &\text{double signs.} \\
&= 73
\end{aligned}
$$

Quick Check 4
Simplify
$9^2 - 4(-2)(-10)$.

A Word of Caution   When we square a negative number, such as $(-5)^2$ in the previous example, the result is a positive number.

EXAMPLE 5   Simplify $8 \div 2 + 3(7 - 4 \cdot 5)$.

Solution

The first step is to simplify the expression inside the set of parentheses. Here the multiplication takes precedence over the subtraction. Once we have simplified the expression inside the parentheses, we proceed to multiply and divide. We finish by subtracting.

$$
\begin{aligned}
8 \div 2 + 3(7 - 4 \cdot 5) &= 8 \div 2 + 3(7 - 20) & &\text{Multiply } 4 \cdot 5. \\
&= 8 \div 2 + 3(-13) & &\text{Subtract } 7 - 20. \\
&= 4 + 3(-13) & &\text{Divide } 8 \div 2. \\
&= 4 - 39 & &\text{Multiply } 3(-13). \\
&= -35 & &\text{Subtract.}
\end{aligned}
$$

Using Your Calculator   When using your calculator to simplify an expression using the order of operations, you may want to perform one operation at a time. However, if you are careful to enter all the parentheses, you can enter the entire expression at one time. Here is how we can simplify the expression in Example 5 using the TI-84:

```
8/2+3(7-4*5)
                -35
```

Quick Check 5
Simplify
$20 \div 5 \cdot 10(3 \cdot 6 - 9)$.

Occasionally, an expression will have one set of grouping symbols inside another set, such as the expression $3[7 - 4(9 - 3)] + 5 \cdot 4$. We call this **nesting** grouping symbols. We begin by simplifying the innermost set of grouping symbols and work our way out from there.

EXAMPLE ▸ **6**   Simplify $3[7 - 4(9 - 3)] + 5 \cdot 4$.

Solution

We begin by simplifying the expression inside the set of parentheses. Once that has been done, we simplify the expression inside the square brackets.

$$3[7 - 4(9 - 3)] + 5 \cdot 4$$

$= 3[7 - 4 \cdot 6] + 5 \cdot 4$	Subtract $9 - 3$, to simplify the expression inside the parentheses. Now we turn our attention to simplifying the expression inside the square brackets.
$= 3[7 - 24] + 5 \cdot 4$	Multiply $4 \cdot 6$.
$= 3[-17] + 5 \cdot 4$	Subtract $7 - 24$.
$= -51 + 20$	Multiply $3[-17]$ and $5 \cdot 4$.
$= -31$	Simplify.

Quick Check ◂ **6**
Simplify
$4[3^2 + 5(2 - 8)]$.

Building Your Study Strategy Study Groups, 7 **Productive Group Members**   For any group or team to be effective, it must be made up of members who are committed to giving their fullest effort to success. Here are some pointers for being a productive study group member:

- **Arrive at each session fully prepared.** Be sure that you have completed all required homework assignments. Bring a list of any questions that you have.
- **Stay focused during study sessions.** Do not spend your time socializing. It is all right to be friendly and begin each session with small talk, but be sure to shift the focus to math after a few minutes.
- **Be open minded.** During study sessions, you may be told that you are wrong. Keep in mind that the goal is to learn mathematics, not to be always correct initially.
- **Consider the feelings of others.** When a person has made a mistake or does not know how to solve a particular problem, be supportive and encouraging.
- **Know when to speak and when to listen.** A study group works best when it is a collaborative body. One person should not be asking all the questions, and one person should not be answering all the questions.

EXERCISES *1.7*

F
O
R

E
X
T
R
A

H
E
L
P

Interactmath.com

MathXL
Tutorials on CD

Video Lectures
on CD

Tutor
Center

Addison-Wesley
Math Tutor Center

Student's
Solutions Manual

Vocabulary

1. In exponential notation, the factor being multiplied repeatedly is called the _____.

2. In exponential notation, the exponent tells us _____.

3. When a base is _____, it is raised to the second power.

4. When a base is _____, it is raised to the third power.

5. Symbols such as ( ) and [ ] are called _____ symbols.

6. When simplifying arithmetic expressions, once all grouping symbols have been removed, we perform any operations involving _____.

Rewrite the given expression using exponential notation.

7. $2 \cdot 2 \cdot 2$

8. $9 \cdot 9 \cdot 9 \cdot 9 \cdot 9$

9. $\dfrac{2}{9} \cdot \dfrac{2}{9} \cdot \dfrac{2}{9} \cdot \dfrac{2}{9} \cdot \dfrac{2}{9} \cdot \dfrac{2}{9}$

10. $\dfrac{1}{7} \cdot \dfrac{1}{7} \cdot \dfrac{1}{7}$

11. $(-2)(-2)(-2)(-2)$

12. $(-6)(-6)(-6)(-6)(-6)$

13. $-3 \cdot 3 \cdot 3 \cdot 3 \cdot 3$

14. $-\dfrac{3}{5} \cdot \dfrac{3}{5} \cdot \dfrac{3}{5} \cdot \dfrac{3}{5}$

15. *"five to the third power"*

16. *"seven squared"*

Simplify the given expression.

17. 2^5

18. 5^2

19. 7^4

20. 6^3

21. 10^3

22. 10^4

23. 1^{723}

24. 0^{2364}

25. $\left(\dfrac{3}{4}\right)^3$

26. $\left(\dfrac{1}{5}\right)^5$

27. 0.2^5

28. 0.5^4

29. $(-3)^4$

30. -3^4

31. -2^7

32. $(-2)^7$

33. $2^3 \cdot 3^2$

34. $4^2 \cdot 5^3$

35. $-3^2 \cdot (-4)^3$

36. $-\left(\dfrac{2}{3}\right)^2 \cdot \left(\dfrac{3}{5}\right)^3$

Simplify the given expression.

37. $1 + 2 \cdot 3$

38. $5 - 4 \cdot 3$

39. $12 - 8 \div 4$

40. $6 + 2 \cdot 11 - 27$

41. $4 + 6^2 \div 3$

42. $7 \cdot 6 - 3 \cdot 13$

43. $3.2 + 2.8(-6.3)$

44. $(3.2 + 2.8) - 6.3$

45. $17.1 - 8.58 \div 3.9$

46. $4.7 \cdot 13.9 - 3.6^2$

47. $(-3)^2 - 4(-5)(-4)$

48. $3^2 + 4^2$

49. $(3 + 4)^2$

50. $5(7 - 2)$

51. $-4 - 5(7 - 3 \cdot 6)$

52. $3 - 6(5^2 - 4 + 2 \cdot 7)$

53. $\dfrac{1}{2} + \dfrac{1}{2} \cdot \dfrac{4}{7}$

54. $\dfrac{3}{5} \cdot \dfrac{2}{3} + \dfrac{15}{7} \cdot \dfrac{21}{20}$

55. $\dfrac{2}{9} \div \dfrac{5}{3} \cdot \dfrac{33}{50}$

56. $4 \cdot \dfrac{3}{8} - \left(\dfrac{7}{6}\right)^2$

57. $\dfrac{1}{9}\left(\dfrac{19}{28}-\dfrac{1}{4}\right)$

58. $\dfrac{8}{25}\div\left(\dfrac{4}{15}-\dfrac{1}{3}+\dfrac{3}{5}\right)\cdot\dfrac{7}{18}$

59. $\dfrac{2^2-9}{3^3-7}$

60. $\dfrac{3+5\cdot7-4\cdot2}{1+2\cdot17}$

61. $\dfrac{(3+5)\cdot6-8}{1+3^2+2}$

62. $\dfrac{10^3+9^3}{1^3+12^3}$

63. $2-3[5-(3+6)]$

64. $6[14-7(3-5)-5^2]$

65. $5^2+|9-2\cdot7|$

66. $3\cdot|7\cdot2-4\cdot8|-6\cdot5$

67. $9-|3^2+2^3-10\cdot9|$

68. $|-4^2+19|-|(-7)^2+5(10)|$

69. $-21(|4-3\cdot9|+|8(13-4)|)$

70. $(6\cdot5+40\div20)(|8^2-2\cdot17|-|10^2-2\cdot51|)$

Construct an "order-of-operations problem" of your own that involves at least four numbers and produces the given result.

71. 41

72. 0

73. -13

74. -19

The size of a computer's memory is measured by the number of bytes that it can store. The following table lists the number of bytes in commonly used storage units.

1 kilobyte (KB)	2^{10} bytes
1 megabyte (MB)	2^{20} bytes
1 gigabyte (GB)	2^{30} bytes
1 terabyte (TB)	2^{40} bytes

Calculate the number of bytes in each of the following.

75. 1 kilobyte

76. 1 megabyte

77. 1 gigabyte

78. 1 terabyte

Find the missing number.

79. $2^? = 64$

80. $7^? = 16,807$

81. $?^4 = 1296$

82. $?^5 = -243$

83. $\left(\dfrac{3}{?}\right)^4 = \dfrac{81}{625}$

84. $\left(\dfrac{7}{4}\right)^? = \dfrac{2401}{256}$

Use arithmetic operation signs such as $+$, $-$, $\cdot$, and $\div$ between the values to produce the desired results. (You may use parentheses as well.)

85. $3\quad 5\quad 9\quad 8 = 14$

86. $4\quad 7\quad 2\quad 6 = 14$

87. $3\quad 7\quad 9\quad 2 = 33$

88. $8\quad 2\quad 3\quad 14\quad 4\quad 7 = 2$

Writing in Mathematics

Answer in complete sentences.

89. Explain the difference between $(-2)^6$ and -2^6.

90. Explain how to use the order of operations to simplify expressions.

91. *Solutions Manual* * Write a solutions manual page for the following problem:

Simplify. $3-4(20\div4-6\cdot8)-5^2$

92. *Newsletter* * Write a newsletter explaining how to use the order of operations to simplify an expression.

See Appendix B for details and sample answers.

Objectives

1 **Build variable expressions.**
2 **Evaluate algebraic expressions.**
3 **Understand and use the commutative and associative properties of real numbers.**
4 **Use the distributive property.**
5 **Identify terms and their coefficients.**
6 **Simplify variable expressions.**

Variables

The number of students absent from a particular English class changes from day to day, as does the closing price for a share of Microsoft stock and the daily high temperature in Providence, Rhode Island. Quantities that change, or vary, are often represented by variables. A **variable** is a letter or symbol that is used to represent a quantity that either changes or has an unknown value.

Variable Expressions

Objective 1 **Build variable expressions.** Suppose a buffet restaurant charges $7 per person to eat. To determine the bill for a family to eat in the restaurant, we multiply the number of people by $7. The number of people can change from family to family, so we can represent this quantity by a variable such as x. The bill for a family with x people can be written as $7 \cdot x$, or simply $7x$. This expression for the bill, $7x$, is known as a variable expression. A **variable expression** is a combination of one or more variables with numbers or arithmetic operations. When the operation is multiplication such as $7 \cdot x$, we often omit the multiplication dot. Here are some other examples of variable expressions:

$$3a + 5, \qquad x^2 + 3x - 10, \qquad \frac{y + 5}{y - 3}$$

EXAMPLE 1   Write an algebraic expression for "the sum of a number and 17."

Quick Check 1
Write an algebraic expression for "9 more than a number."

Solution

We must choose a variable to represent the unknown number. Let x represent the number. The expression is $x + 17$.

Other phrases to look for that suggest addition are *plus, increased by, more than,* and *total.*

EXAMPLE 2   Write an algebraic expression for "five less than a number."

Solution

Let x represent the number. The expression is $x - 5$.

Quick Check 2
Write an algebraic expression for "a number decreased by 25."

Be careful with the order of subtraction when the phrase *less than* is used. *Five less than a number* tells us that we need to subtract 5 from that number. A common error is to write the subtraction in the opposite order.

Other phrases that suggest subtraction are *difference, minus,* and *decreased by.*

A Word of Caution Five less than a number is written as $x - 5$, not $5 - x$.

EXAMPLE 3 Write an algebraic expression for "the product of 3 and two different numbers."

Quick Check 3
Write an algebraic expression for "twice a number."

Solution

Since we are building an expression involving two different unknown numbers, we need to introduce two variables. Let x and y represent the two numbers. The expression is $3xy$.

Other phrases that suggest multiplication are *times, multiplied by, of,* and *twice.*

EXAMPLE 4 Four friends decide to rent a fishing boat for the day. If all 4 friends decide to split evenly the cost of renting the boat, write a variable expression for the amount each friend would pay.

Quick Check 4
Write an algebraic expression for "the quotient of a number and 20."

Solution

Let c represent the cost of the boat. The expression is $c \div 4$ or $\frac{c}{4}$.

Other phrases that suggest division are *quotient, divided by,* and *ratio.*

Here are some different phrases that translate to $x + 5$, $x - 10$, $8x$, and $\frac{x}{6}$:

$x + 5$	The sum of a number and 5
	A number plus 5
	A number increased by 5
	5 more than a number
	The total of a number and 5
$x - 10$	10 less than a number
	The difference of a number and 10
	A number minus 10
	A number decreased by 10
$8x$	The product of 8 and a number
	8 times a number
	8 multiplied by a number
$\frac{x}{6}$	The quotient of a number and 6
	A number divided by 6
	The ratio of a number and 6

Evaluating Variable Expressions

Objective 2 Evaluate algebraic expressions. We will often have to evaluate variable expressions for particular values of variables. To do this, we substitute the appropriate numerical value for each variable and then simplify the resulting expression using the order of operations.

EXAMPLE 5   Evaluate $2x - 7$ for $x = 6$.

Solution

The first step when evaluating a variable expression is to rewrite the expression, replacing each variable by a set of parentheses. For example, rewrite $2x - 7$ as $2(\ ) - 7$. Then we can substitute the appropriate value for each variable and simplify.

$$
\begin{aligned}
& 2x - 7 \\
& 2(6) - 7 && \text{Substitute 6 for } x. \\
&= 12 - 7 && \text{Multiply.} \\
&= 5 && \text{Subtract.}
\end{aligned}
$$

Quick Check **5**
Evaluate $5x + 2$ for $x = 11$.

The expression $2x - 7$ is equal to 5 for $x = 6$.

EXAMPLE 6   Evaluate $x^2 - 5x + 6$ for $x = 3$.

Solution

$$
\begin{aligned}
& x^2 - 5x + 6 \\
& (3)^2 - 5(3) + 6 && \text{Substitute 3 for } x. \\
&= 9 - 5(3) + 6 && \text{Square 3.} \\
&= 9 - 15 + 6 && \text{Multiply.} \\
&= 0 && \text{Simplify.}
\end{aligned}
$$

Quick Check **6**
Evaluate $x^2 - 13x - 40$ for $x = -8$.

EXAMPLE 7   Evaluate $b^2 - 4ac$ for $a = 3$, $b = -2$, and $c = -10$.

Solution

$$
\begin{aligned}
& b^2 - 4ac \\
& (-2)^2 - 4(3)(-10) && \text{Substitute 3 for } a, -2 \text{ for } b, \text{ and } -10 \text{ for } c. \\
&= 4 - 4(3)(-10) && \text{Square negative 2.} \\
&= 4 - (-120) && \text{Multiply.} \\
&= 4 + 120 && \text{Rewrite without double signs.} \\
&= 124 && \text{Total.}
\end{aligned}
$$

Quick Check **7**
Evaluate $b^2 - 4ac$ for $a = -1$, $b = 5$, and $c = 18$.

Properties of Real Numbers

Objective 3   Understand and use the commutative and associative properties of real numbers. We will now examine three properties of real numbers. The first is the **commutative property**, which is used for addition and multiplication.

Commutative Property

For all real numbers a and b,

$$a + b = b + a \quad \text{and} \quad a \cdot b = b \cdot a.$$

This property states that changing the order of the numbers in a sum or in a product does not change the result. For example, $3 + 9 = 9 + 3$ and $7 \cdot 8 = 8 \cdot 7$. Note that this property does not work for subtraction or for division. Changing the order of the numbers in a subtraction or a division will generally change the result.

The second property that we will cover is the **associative property**, which also holds for addition and multiplication.

Associative Property

For all real numbers a, b, and c,

$$(a + b) + c = a + (b + c) \qquad \text{and} \qquad (ab)c = a(bc).$$

This property states that changing the grouping of the numbers in a sum or in a product does not change the result. Check for yourself; does $(2 + 7) + 3$ equal $2 + (7 + 3)$? The first expression simplifies to $9 + 3$, while the second becomes $2 + 10$, both of which equal 12.

EXAMPLE 8 Simplify $5(12x)$ by applying the associative property.

Solution

By the associative property we know that $5(12x) = (5 \cdot 12)x$. We can now multiply $5 \cdot 12$ to produce a result of $60x$.

Quick Check **8**
Simplify $8(3a)$ by applying the associative property.

EXAMPLE 9 Simplify $(3x + 7) + 15$ by applying the associative property.

Solution

By the associative property we know that $(3x + 7) + 15 = 3x + (7 + 15)$. This allows us to add 7 and 15. The simplified expression is $3x + 22$.

Quick Check **9**
Simplify $(9x + 20) + 14$ by applying the associative property.

Objective 4 Use the distributive property. The third property of real numbers is the **distributive property**.

Distributive Property

For all real numbers a, b, and c,

$$a(b + c) = ab + ac.$$

This property tells us that we can distribute the factor outside of the parentheses to each number being added in the parentheses, perform the multiplications, and then add. (This property also holds true when the operation inside the parentheses is subtraction.) Consider the expression $3(5 + 4)$. The order of operations tells us that this is equal to $3 \cdot 9$, or 27. Here is how we simplify the expression using the distributive property:

$$
\begin{aligned}
3(5 + 4) &= 3 \cdot 5 + 3 \cdot 4 &&\text{Apply the distributive property.}\\
&= 15 + 12 &&\text{Multiply } 3 \cdot 5 \text{ and } 3 \cdot 4.\\
&= 27 &&\text{Add.}
\end{aligned}
$$

Either way we get the same result.

EXAMPLE 10   Simplify $2(4 + 3x)$ using the distributive property.

Solution

Quick Check 10
Simplify $7(5x - 4)$ using the distributive property.

$$2(4 + 3x) = 2 \cdot 4 + 2 \cdot 3x \qquad \text{Distribute the 2.}$$
$$= 8 + 6x \qquad \text{Multiply. Recall from the example regarding the associative property that } 2 \cdot 3x \text{ equals } 6x.$$

A Word of Caution   The expression $8 + 6x$ is not equal to $14x$.

EXAMPLE 11   Simplify $7(2 + 3a - 4b)$ by using the distributive property.

Quick Check 11
Simplify $12(x - 2y + 3z)$ using the distributive property.

Solution

When there are more than two terms inside the parentheses, we distribute the factor to each term. Also, it is a good idea to be able to distribute the factor and multiply mentally.

$$7(2 + 3a - 4b) = 14 + 21a - 28b$$

EXAMPLE 12   Simplify $-4(2x - 5)$ using the distributive property.

Solution

When the factor outside the parentheses is negative, the negative number must be distributed to each term inside the parentheses. This will change the sign of each term inside the parentheses.

$$-4(2x - 5) = (-4) \cdot 2x - (-4) \cdot 5 \qquad \text{Distribute the } -4.$$
$$= -8x - (-20) \qquad \text{Multiply.}$$
$$= -8x + 20 \qquad \text{Rewrite without double signs.}$$

Quick Check 12
Simplify $-6(4x + 11)$ using the distributive property.

Simplifying Variable Expressions

Objective 5 Identify terms and their coefficients. In an algebraic expression, a **term** is either a number, a variable, or a product of a number and variables. Terms in an algebraic expression are separated by addition or subtraction. In the expression $7x - 5y + 3$, there are three terms: $7x$, $-5y$, and 3. The numerical factor of a term is its **coefficient**. The coefficients of these terms are 7, -5, and 3.

EXAMPLE 13   For the expression $-5x + y + 3xy - 19$, determine the number of terms, list them, and state the coefficient for each term.

Solution

This expression has four terms: $-5x$, y, $3xy$, and -19.

What is the coefficient for the second term? Although it does not appear to have a coefficient, its coefficient is 1. This is because y is the same as $1 \cdot y$. So the four coefficients are -5, 1, 3, and -19.

Quick Check 13 For the expression $x^3 - x^2 + 23x - 59$, determine the number of terms, list them, and state the coefficient for each term.

Objective 6 **Simplify variable expressions.** Two terms that have the same variable factors with the same exponents, or that are both constants, are called **like terms**. Consider the expression $9x + 8y + 6x - 3y + 8z - 5$. There are two sets of like terms in this expression: $9x$ and $6x$, as well as $8y$ and $-3y$. There are no like terms for $8z$, since no other term has z as its sole variable factor. Similarly, there are no like terms for the constant term -5.

Suppose we had to simplify the expression $8y - 3y$. This expression could be rewritten as $(8 - 3)y$ using the distributive property, or simply as $5y$.

Combining Like Terms

When simplifying variable expressions, we can combine like terms into a single term by adding or subtracting the coefficients of the like terms.

EXAMPLE 14   Simplify $4x + 11x$ by combining like terms.

Solution

These two terms are like terms, since they both have the same variable factors. We can simply add the two coefficients together to produce the expression $15x$.

$$4x + 11x = 15x$$

Quick Check 14
Simplify $3x + 7y + y - 5x$ by combining like terms.

A general strategy for simplifying algebraic expressions is to begin by performing any distributive multiplications. After all multiplications have been performed, all that remains should be addition and subtraction. At this point, we can combine any like terms.

EXAMPLE 15   Simplify $8(3x - 5) - 4x + 7$.

Solution

The first step is to use the distributive property by distributing the 8 to each term inside the parentheses. Then we will be able to combine like terms.

Quick Check 15
Simplify $5(2x + 3) + 9x - 8$.

$$8(3x - 5) - 4x + 7 = 24x - 40 - 4x + 7 \quad \text{Distribute the 8.}$$
$$= 20x - 33 \quad \text{Combine like terms.}$$

EXAMPLE 16   Simplify $5(9 - 7x) - 10(3x + 4)$.

Solution

We must be sure to distribute the *negative* 10 into the second set of parentheses.

$$5(9 - 7x) - 10(3x + 4) = 45 - 35x - 30x - 40 \quad$$

Distribute the 5 to each term in the first set of parentheses and distribute the -10 to each term in the second set of parentheses.
Combine like terms.

$$= -65x + 5$$

Quick Check 16

Simplify
$3(2x - 7) - 8(x - 9)$.

Usually we write our simplified result with variable terms preceding constant terms, but it would also be correct to write $5 - 65x$, because of the commutative property.

A Word of Caution   If a factor in front of a set of parentheses is negative or has a subtraction sign in front of it, we must distribute the negative sign along with the factor.

Building Your Study Strategy   Study Groups, 8   **Dealing with Unproductive Group Members**   Occasionally, a group will have a member who is not productive or is even disruptive. You should not allow one person to prevent your group from being successful. If you have a group member who is not contributing to the group in a positive way, try talking to that person one on one, discreetly. Once the person knows how important this study group is in helping you to learn mathematics, that person may change his or her ways and start being an effective group member.

Some groups prefer to sit down as a group with the nonproductive or disruptive member and air their grievances. Always be honest with the person, but try not to be offensive. Try to find solutions that would be acceptable to the group and the group member in question. Seek advice from your instructor. Undoubtedly, your instructor has seen a group in a similar situation and may have valuable advice for you.

If you have worked on improving the situation but have seen no improvement, then it may be time to remove the person from the group. Try to be professional about it rather than adversarial. Remember, you will continue to see this person in class each day, and it would be best to separate on good terms.

EXERCISES *1.8*

F
O
R

E
X
T
R
A

H
E
L
P

MyMathLab

Math XL

Interactmath.com

MathXL
Tutorials on CD

Video Lectures
on CD

Tutor
Center

Addison-Wesley
Math Tutor Center

Student's
Solutions Manual

Vocabulary

1. A(n) _____ is a letter or symbol used to represent a quantity that either changes or has an unknown value.

2. A combination of one or more variables with numbers and/or arithmetic operations is a(n) _____.

3. To _____ a variable expression, substitute the appropriate numerical value for each variable and then simplify the resulting expression using the order of operations.

4. For any real numbers a and b, the _____ property states that $a + b = b + a$ and $a \cdot b = b \cdot a$.

5. For any real numbers a, b, and c, the _____ property states that $(a + b) + c = a + (b + c)$ and $(a \cdot b) \cdot c = a \cdot (b \cdot c)$.

6. For any real numbers a, b, and c, the _____ property states that $a(b + c) = ab + ac$.

7. A(n) _____ is either a number, a variable, or a product of numbers and variables.

8. Give the definition for a coefficient.

9. Two terms that have the same variable factors, or that are both constant terms, are called _____.

10. To combine two like terms, add their _____.

Build a variable expression for the following phrases.

11. The sum of a number and 11

12. A number decreased by 9

13. Five times a number

14. A number divided by 14

15. Thirteen minus a number

16. Twice a number

17. Four less than three times a number

18. Five times a number, increased by 17

19. The sum of two different numbers

20. One number divided by another

21. Seven times the difference of two numbers

22. Half the sum of a number and 25

23. A college charges \$325 per credit for tuition. If we let c represent the number of credits that a student is taking, build a variable expression for the student's tuition.

24. Thomas Magnum, a private investigator from the 1980s TV show *Magnum, P. I.,* charged \$200 per day to take a case. If we let d represent the number of days it took Magnum to crack the case, build a variable expression for the amount of money he would charge for the case.

25. A professional baseball player is appearing at a baseball card convention. The promoter agreed to pay the player a flat fee of \$25,000 plus \$22 per autograph signed. If we let a represent the number of autographs signed, build a variable expression for the amount of money the player will be paid.

26. Jim Rockford, a private investigator from the 1970s TV show *The Rockford Files,* charged \$200 per day plus expenses to take a case. If we let d represent the number of days that Rockford worked on a case, and if he had \$425 in expenses, build a variable expression for the amount of money he would charge for the case.

Evaluate the following algebraic expressions under the given conditions.

27. $4x - 9$ for $x = 2$ **28.** $7 + 6x$ for $x = 5$

29. $3a - 2b$ for $a = 2$ and $b = -7$

30. $-5w + 4z$ for $w = -1$ and $z = 3$

31. $3(x - 5)$ for $x = 7$

32. $(x + 3)(2x - 5)$ for $x = -2$

33. $x^2 + 9x + 18$ for $x = 5$

34. $a^2 - 7a - 30$ for $a = 3$

35. $y^2 + 4y - 17$ for $y = -3$

36. $x^2 - 5x - 9$ for $x = -4$

37. $m^2 - 9$ for $m = 3$ **38.** $5 - x^2$ for $x = -2$

39. $x^7 + 5x^6 + 303x^3 - 754$ for $x = 0$

40. $a^3 - 4a - 13$ for $a = -7$

41. $b^2 - 4ac$ for $a = 2$, $b = 5$ and $c = -3$

42. $b^2 - 4ac$ for $a = -5$, $b = 7$ and $c = -4$

43. $b^2 - 4ac$ for $a = -1$, $b = -2$ and $c = 10$

44. $b^2 - 4ac$ for $a = 15$, $b = -6$ and $c = 9$

45. $5(x + h) - 17$ for $x = -3$ and $h = 0.01$

46. $-3(x + h) + 4$ for $x = -6$ and $h = 0.001$

47. $(x + h)^2 - 5(x + h) - 14$ for $x = -3$ and $h = 0.1$

48. $(x + h)^2 + 4(x + h) - 28$ for $x = 6$ and $h = 1$

49. The commutative property works for addition but does not work in general for subtraction.

a) Give an example of two numbers a and b such that $a - b \neq b - a$.

b) Can you find two numbers a and b such that $a - b = b - a$?

50. The commutative property works for multiplication but does not work in general for division.

a) Give an example of two numbers a and b such that $\frac{a}{b} \neq \frac{b}{a}$.

b) Can you find two numbers a and b such that $\frac{a}{b} = \frac{b}{a}$?

Simplify, where possible.

51. $3(x - 9)$

52. $4(2x + 5)$

53. $5(3 - 7x)$

54. $12(5x - 7)$

55. $7x + 9x$

56. $14x + 3x$

57. $3x - 8x$

58. $22a - 15a$

59. $3x - 7 + 4x + 11$

60. $15x + 32 - 19x - 57$

61. $32 - 49a + 87a - 95$

62. $52 + 32a - 17a + 45$

63. $4a + 5b$

64. $13x - 12y + 357$

65. $5x - 3y - 7x - 19y$

66. $15m - 11n - 6m + 22n$

67. $3(2x - 5) + 4x + 7$

68. $8 - 19k + 6(3k - 5)$

69. $-2(4x - 9) - 11$

70. $3(2a + 11b) - 13a$

71. $6y - 5(3y - 17)$

72. $5 - 9(3 - 4x)$

73. $3(4z - 7) + 9(2z + 3)$

74. $-6(5x - 9) - 7(13 - 12x)$

75. $-2(5a + 4b - 13c) + 3b$

76. $-7(-2x + 3y - 17z) - 5(3x - 11)$

For the following expressions,
a) determine the number of terms;
b) write down each term; and
c) write down the coefficient for each.

Be sure to simplify each expression before answering.

77. $5x^3 + 3x^2 - 7x - 15$

78. $-3a^2 - 7a - 10$

79. $3x - 17$

80. $9x^4 - 10x^3 + 13x^2 - 17x + 329$

81. $5(3x^2 - 7x + 11) - 6x$

82. $4(3a - 5b - 7c - 11) - 2(6b - 5c)$

83. $5(-7a - 3b + 5c) - 3(6b - 9c - 23)$

84. $2(-4x + 7y) - (3x + 9y)$

85. List 4 different algebraic expressions that simplify to $12x - 15$.

86. List 4 different algebraic expressions that evaluate to equal 16 for $x = 2$.

Writing in Mathematics

Answer in complete sentences.

87. Give an example of a real-world situation that can be described by the variable expression $60x$. Explain why the expression fits your situation.

88. Give an example of a real-world situation that can be described by the variable expression $20x + 35$. Explain why the expression fits your situation.

89. *Solutions Manual* * Write a solutions manual page for the following problem:

Simplify. $4(2x - 5) - 3(3x - 8) - 7x$

90. *Newsletter* * Write a newsletter explaining how to evaluate expressions.

*See Appendix B for details and sample answers.

Chapter 1 Summary

Section 1.1—Topic	Chapter Review Exercises
Determining Inequalities Involving Integers	1–2
Finding Absolute Values	3–4
Finding Opposites of Integers	5–6

Section 1.2—Topic	Chapter Review Exercises
Finding Sums and Differences of Integers	7–12
Finding Products and Quotients of Integers	13–14
Solving Applied Problems Involving Integers	55–57

Section 1.3—Topic	Chapter Review Exercises
Finding the Factor Set of a Number	15–16
Finding the Prime Factorization of a Number	17–18
Simplifying a Fraction to Lowest Terms	19–22
Rewriting Mixed Numbers as Improper Fractions	23–24
Rewriting Improper Fractions as Mixed Numbers	25–26

Section 1.4—Topic	Chapter Review Exercises
Multiplying and Dividing Fractions	27–30, 35, 37
Adding and Subtracting Fractions	31–34, 36, 38
Solving Applied Problems Involving Fractions	58

Section 1.5—Topic	Chapter Review Exercises
Performing Arithmetic Operations with Decimals	39–42
Rewriting Fractions as Decimals	43–44
Rewriting Decimals as Fractions	45–46
Converting Fractions and Decimals to Percents	47–50
Converting Percents to Fractions and Decimals	51–54

Section 1.6—Topic	Chapter Review Exercises
Creating a Pie Chart	59
Creating a Bar Chart	60
Creating a Line Graph	61

Section 1.7—Topic	Chapter Review Exercises
Simplifying Expressions Involving Exponents	62–65
Simplifying Expressions Using the Order of Operations	66–71

Section 1.8—Topic	Chapter Review Exercises
Building Variable Expressions	72–77
Evaluating Variable Expressions	78–83
Simplifying Variable Expressions	84–89
Identifying Terms and Their Coefficients	90–91

Summary of Chapter 1 Study Strategies

- Creating a study group is one of the best ways to help you learn mathematics and improve your performance on quizzes and exams.
- Sometimes, a concept will be easier for you to understand if it is explained to you by a peer. One of your group members may have some insight into a particular topic, and that person's explanation may be just the trick to turn on the "lightbulb" in your head.
- When you explain a certain topic to a fellow student, or show a student how to solve a particular problem, you increase your chances of retaining this knowledge. Because you are forced to choose your words in such a way that the other student will completely understand, you have demonstrated that you truly understand.
- You will also find it helpful to have a support group: students who can lean on each other when times are bad. By being supportive of each other, you are increasing the chances that you will all learn and be successful in your class.

Write the appropriate symbol, either < or >, between the following integers. [1.1]

1. -10 ___ -7 **2.** 3 ___ -12

Find the following absolute values. [1.1]

3. $|8|$ **4.** $|-13|$

Find the opposite of the following integers. [1.1]

5. -6 **6.** 12

Add or subtract. [1.2]

7. $9 + (-16)$ **8.** $-10 + 7$

9. $-22 - 19$ **10.** $4 - (-23)$

Simplify. [1.2]

11. $3 - 16 - 24$ **12.** $8 - (-19) - 7$

Multiply or divide. [1.2]

13. $9(-6)$ **14.** $-144 \div (-9)$

Write the factor set for the following numbers. [1.3]

15. 42

16. 108

Write the prime factorization of the following numbers. (If prime, state this.) [1.3]

17. 32 **18.** 60

Simplify the following fractions to lowest terms. [1.3]

19. $\dfrac{24}{42}$ **20.** $\dfrac{9}{72}$ **21.** $\dfrac{40}{234}$ **22.** $\dfrac{98}{7}$

Rewrite the following mixed numbers as improper fractions. [1.3]

23. $5\dfrac{2}{3}$ **24.** $12\dfrac{7}{25}$

Rewrite the following improper fractions as mixed numbers. [1.3]

25. $\dfrac{38}{10}$ **26.** $\dfrac{55}{6}$

Multiply or divide. [1.4]

27. $\dfrac{9}{16} \cdot \dfrac{28}{75}$ **28.** $5\dfrac{1}{2} \cdot \dfrac{14}{33}$

29. $\dfrac{8}{15} \div \dfrac{20}{21}$ **30.** $3\dfrac{1}{5} \div 7\dfrac{1}{2}$

Add or subtract. [1.4]

31. $\dfrac{5}{8} + \dfrac{7}{12}$ **32.** $\dfrac{13}{20} + \dfrac{5}{6}$

33. $\dfrac{11}{18} - \dfrac{4}{9}$ **34.** $\dfrac{7}{36} - \dfrac{13}{42}$

Simplify. [1.4]

35. $\dfrac{30}{49} \cdot \dfrac{35}{66}$ **36.** $\dfrac{7}{16} - \dfrac{11}{48}$

37. $\dfrac{9}{40} \div \dfrac{39}{35}$ **38.** $\dfrac{7}{20} + \dfrac{1}{15}$

Simplify the following decimal expressions. [1.5]

39. $8.7 + 3.92$ **40.** $24.308 - 15.49$

41. $8.4 \cdot 3.6$ **42.** $40.92 \div 4.65$

Rewrite the following fractions as decimal numbers. [1.5]

43. $\dfrac{8}{25}$ **44.** $\dfrac{15}{16}$

Rewrite the following decimal numbers as fractions in lowest terms. [1.5]

45. 0.75 **46.** 0.28

Rewrite as a percent. [1.5]

47. $\dfrac{2}{5}$ **48.** $\dfrac{7}{25}$

49. 0.9 **50.** 0.45

Rewrite as a fraction. [1.5]

51. 30% **52.** 55%

Rewrite as a decimal. [1.5]

53. 90% **54.** 4%

55. Jeff bought a book at a yard sale for \$23. If he had \$40 prior to buying the book, how much money does Jeff have now? [1.2]

56. Gray had \$78 in his checking account prior to writing a \$125 check to the bookstore for books and supplies. What is his account's new balance? [1.2]

57. Three investors plan to start a new company. If start-up costs are \$37,800, how much will each person have to invest? [1.2]

Worked-out solutions to Review Exercises marked with ◯ *can be found on page AN–3.*

58. If one recipe calls for $1\frac{1}{2}$ cups of flour and a second recipe calls for $2\frac{2}{3}$ cups of flour, how much flour is needed to make both recipes? [1.4]

59. 42% of entering freshmen at two-year colleges enrolled in at least one remedial course in fall 2005, while 58% did not. Create a pie chart to display this set of data.

60. The following table shows how many World Cups different nations have won in soccer. Create a bar graph for this set of data. (*Source:* FIFA)

Country	Number
Uruguay	2
Italy	4
W. Germany	3
Brazil	5
England	1
Argentina	2
France	1

61. The median age at first marriage has been increasing for U.S. women. Create a line graph for this set of data. (*Source:* Census Bureau)

Year	Median Age at First Marriage
1960	20.3
1970	20.8
1980	22.0
1990	23.9
2000	25.1

Simplify the given expression. **[1.7]**

62. 4^3

63. $\left(\frac{2}{5}\right)^3$

64. -2^6

65. $3^5 \cdot 5^2$

Simplify the given expression. **[1.7]**

66. $5 + 8 \cdot 4$

67. $25 - 15 \div 5$

68. $3 + 13 \cdot 5 - 20$

69. $(3 + 13) \cdot 5 - 20$

70. $54 - 27 \div 3^2$

71. $\frac{3}{4} + \frac{1}{4} \cdot \frac{12}{25}$

Build a variable expression for the following phrases. **[1.8]**

72. The sum of a number and 14

73. A number decreased by 20

74. Eight less than twice a number

75. Nine more than 6 times a number

76. A coffeehouse charges $3.55 for a cup of coffee. If we let c represent the number of cups of coffee that a coffeehouse sells on a particular day, build a variable expression for the revenue from coffee sales. [1.8]

77. A rental company rents moving vans for $20, plus $0.15 per mile. If we let m represent the number of miles, build a variable expression for the cost to rent a moving van from this rental company. [1.8]

Evaluate the following algebraic expressions under the given conditions. [1.8]

78. $3x + 17$ for $x = 9$ **79.** $9 - 8x$ for $x = -2$

80. $10a - 4b$ for $a = 2$ and $b = -9$

81. $(8x - 9)(2x - 11)$ for $x = 4$

82. $x^2 - 7x - 30$ for $x = -3$

83. $b^2 - 4ac$ for $a = -1$, $b = -8$ and $c = 5$

Simplify. [1.8]

84. $5(x + 7)$

85. $6x + 21x$

86. $8x - 25 - 3x + 17$

87. $8y - 6(4y - 21)$

88. $15 - 23k + 7(4k - 9)$

89. $-8(2x + 25) - (103 - 19x)$

For the following expressions,
a) determine the number of terms;
b) write down each term; and
c) write down the coefficient for each.

90. $x^3 - 4x^2 - 10x + 41$

91. $-x^2 + 5x - 30$

Write the appropriate symbol, either < or >, between the following integers.

1. -15 ___ -18

Find the following absolute value.

2. $|-17|$

Simplify.

3. $7 + (-13)$

4. $-7(-9)$

5. Write the factor set for 45.

6. Write the prime factorization of 108. (If the number is prime, state this.)

7. Simplify $\dfrac{60}{84}$ to lowest terms.

8. Rewrite $\dfrac{67}{18}$ as a mixed number.

Simplify.

9. $\dfrac{11}{63} \cdot \dfrac{15}{44}$

10. $\dfrac{2}{9} \div \dfrac{8}{21}$

11. $\dfrac{3}{5} + \dfrac{11}{12}$

12. $\dfrac{5}{24} - \dfrac{4}{9}$

13. Simplify $8.05(2.27)$.

14. Rewrite 0.36 as a fraction. Your answer should be in lowest terms.

15. Eleanor bought 13 computers for $499 each. What was the total cost for the 13 computers?

16. After a deposit of $407.83, the balance in Lindsay's checking account was $1203.34. What was the balance before the deposit?

17. Rewrite 72% as a fraction.

18. Rewrite 6% as a decimal.

19. At a certain college, 4400 of the students are female and 3600 of the students are male. Create a pie chart to display this set of data.

Simplify the given expression.

20. $16 - 8 \cdot 5$

21. $-9 + 4 \cdot 13 - 6 \cdot 3$

22. $\dfrac{4^2 + 3^2}{3 + 4 \cdot 13}$

23. Build a variable expression for the phrase *seven less than four times a number.*

24. A landscaper charges $50, plus $20 per hour, for yard maintenance. If we let h represent the number of hours spent working on a particular yard, build a variable expression for the landscaper's charge.

Evaluate the following algebraic expressions under the given conditions.

25. $16 - 5x$ for $x = -9$

26. $x^2 + 6x - 17$ for $x = -8$

Simplify.

27. $5(2x - 13)$

28. $7y - 8(2y - 30)$

Mathematicians in History
Srinivasa Ramanujan

Srinivasa Ramanujan was a self-taught Indian mathematician, viewed by many to be one of the greatest mathematical geniuses in history. As a student, he became so totally immersed in his work with mathematics that he ignored his other subjects and failed his college exams. Ramanujan's life is chronicled in the biography *The Man Who Knew Infinity: A Life of the Genius Ramanujan.*

Write a one-page summary (or make a poster) of the life of Srinivasa Ramanujan and his accomplishments.

Interesting issues:

- Where and when was Srinivasa Ramanujan born?
- At age 16, Ramanujan borrowed a mathematics book that strongly influenced his life as a mathematician. What was the name of the book?
- Ramanujan got married on July 14, 1909. The marriage was arranged by his mother. How old was his bride at the time?
- What jobs did Ramanujan hold in India?
- Which renowned mathematician invited Ramanujan to England in 1914?
- Ramanujan's health in England was poor. What was the cause of his poor health?
- What is the significance of the taxicab number 1729?
- What were the circumstances that led to Ramanujan's death, and what was his age when he died?

Smoothie Superb makes an orange juice and strawberry-banana smoothie called the Berry Ana. The store charges $3.78 for a regular size and gives the customer one free nutritional boost. For each additional nutritional boost, they charge 43 cents. Let b represent the number of nutritional boosts in a drink.

a) Build a variable expression for the total cost of a Berry Ana smoothie with b nutritional boosts.

b) What would be the total cost for a customer to order a Berry Ana smoothie with two nutritional boosts?

During December, Smoothie Superb makes a specialty drink called the Cold Terminator. The Cold Terminator is the exact same smoothie as the Berry Ana, but contains four nutritional boosts. A regular-size Cold Terminator sells for $4.99.

c) If a customer wanted to order a Berry Ana smoothie with four nutritional boosts, would it be more cost effective to buy a Cold Terminator? Explain your answer.

d) Set up an inequality statement (using either $<$ or $>$) comparing the costs of the Cold Terminator and a Berry Ana with four boosts.

Linear Equations

*I*n this chapter, we will learn to solve linear equations and investigate applications of this type of equation. We will also discuss applications involving percents and proportions. The chapter concludes with a section on linear inequalities.

Study Strategy   **Using Your Textbook**   *In this chapter, we will focus on how to get the most out of your textbook. Students who treat their books solely as a source of homework exercises are turning their backs on one of their best resources.*

Throughout this chapter, we will revisit this study strategy and help you incorporate it into your study habits.

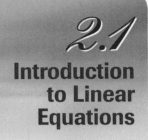

2.1

Introduction to Linear Equations

Linear Equations

Objective 1 Identify linear equations. An **equation** is a mathematical statement of equality between two expressions. It is a statement that asserts that the value of the expression on the left side of the equation is equal to the value of the expression on the right side. Here are a few examples of equations:

$$2x = 8 \qquad x + 17 = 20 \qquad 3x - 8 = 2x + 6 \qquad 5(2x - 9) + 3 = 7(3x + 16)$$

All of these are examples of linear equations. A **linear equation** in one variable has a single variable, and the exponent for that variable is 1. For example, if the variable in a linear equation is x, then the equation cannot have terms containing x^2 or x^3. The variable in a linear equation cannot appear in a denominator either. Here are some examples of equations that are not linear equations:

$$x^2 - 5x - 6 = 0 \qquad \text{Variable is squared.}$$
$$m^3 - m^2 + 7m = 7 \qquad \text{Variable has exponents greater than 1.}$$
$$\frac{5x + 3}{x - 2} = -9 \qquad \text{Variable is in denominator.}$$

Solutions of Equations

Objective 2 Determine whether a value is a solution of an equation. A **solution** of an equation is a value that, when substituted for the variable in the equation, produces a true statement, such as $5 = 5$.

EXAMPLE 1   Is $x = 3$ a solution of $9x - 7 = 20$?

Solution

To check whether a particular value is a solution of an equation, we substitute that value for the variable in the equation. If, after simplifying both sides of the equation, we have a true mathematical statement, then the value is a solution.

$$9x - 7 = 20$$
$$9(3) - 7 = 20 \qquad \text{Substitute 3 for } x.$$
$$27 - 7 = 20 \qquad \text{Multiply.}$$
$$20 = 20 \qquad \text{Subtract.}$$

Since 20 is equal to 20, we know that $x = 3$ is a solution.

EXAMPLE 2   Is $x = -2$ a solution of $3 - 4x = -5$?

Solution

Again, we substitute for x and simplify both sides of the equation.

$$3 - 4x = -5$$
$$3 - 4(-2) = -5 \qquad \text{Substitute } -2 \text{ for } x.$$
$$3 + 8 = -5 \qquad \text{Multiply.}$$
$$11 = -5 \qquad \text{Add.}$$

This statement, $11 = -5$, is not true because 11 is not equal to -5. Therefore, $x = -2$ is not a solution of this equation.

Quick Check **1**
Is $x = -7$ a solution of $4x + 23 = -5$?

The set of all solutions of an equation is called its **solution set**. The process of finding an equation's solution set is called **solving the equation**. When we find all of the solutions to an equation, we write these values using set notation inside braces { }.

When solving a linear equation, our goal is to convert it to an equivalent equation that has the variable isolated on one side with a number on the other side; for example, $x = 3$ or $-5 = y$. The value that is on the opposite side of the equation from the variable after it has been isolated is the solution of the equation.

Multiplication Property of Equality

Objective 3 **Solve linear equations using the multiplication property of equality.** Our first tool for solving linear equations is the **multiplication property of equality**. It tells us that for any equation, if we multiply both sides of the equation by the same nonzero number, then both sides remain equal to each other.

Multiplication Property of Equality

For any algebraic expressions A and B, and any nonzero number n,

if $A = B$ then $n \cdot A = n \cdot B$.

Think of a scale that holds two weights in balance. If we double the amount of weight on each side of the scale, will the scale still be balanced? Of course it will.

We can think of an equation as a scale, and the expressions on each side as the weights that are balanced. Multiplying both sides by the same nonzero number leaves both sides still balanced and equal to each other.

Why is it so important that the number we multiply both sides of the equation by not be 0? If we multiply both sides of an equation by 0, the resulting equation will be $0 = 0$. Since this equation no longer contains a variable, we will not be able to isolate the variable on one side of the equation.

EXAMPLE 3   Solve $\frac{x}{3} = 4$ by using the multiplication property of equality.

Solution

The goal when solving this linear equation is to isolate the variable x on one side of the equation. We begin by multiplying both sides of this equation by 3. The expression $\frac{x}{3}$ is equivalent to $\frac{1}{3}x$, so when we multiply by 3, we are multiplying by the reciprocal of $\frac{1}{3}$. The product of reciprocals is equal to 1, so the resulting expression on the left side of the equation is $1x$, or x.

$$\frac{x}{3} = 4$$

$$3 \cdot \frac{x}{3} = 3 \cdot 4 \qquad \text{Multiply both sides by 3.}$$

$$\overset{1}{\cancel{3}} \cdot \frac{x}{\underset{1}{\cancel{3}}} = 3 \cdot 4 \qquad \text{Divide out common factors.}$$

$$x = 12 \qquad \text{Multiply.}$$

The solution that we have found is $x = 12$. Before moving on, we must check this value to be sure that we have made no mistakes and that it is actually a solution.

Check

$$\frac{x}{3} = 4$$

$$\frac{12}{3} = 4 \qquad \text{Substitute 12 for } x.$$

$$4 = 4 \qquad \text{Divide.}$$

Quick Check **2**
Solve $\frac{x}{8} = -2$ by using the multiplication property of equality.

This is a true statement, so $x = 12$ is a solution and the solution set is $\{12\}$.

The multiplication property of equality also allows us to divide both sides of an equation by the same nonzero number without affecting the equality of the two sides. This is because dividing both sides of an equation by a nonzero number, n, is equivalent to multiplying both sides of the equation by the reciprocal of the number, $\frac{1}{n}$.

EXAMPLE 4   Solve $5y = 40$ by using the multiplication property of equality.

Solution

We begin by dividing both sides of the equation by 5, which will isolate the variable y.

$$5y = 40$$

$$\frac{5y}{5} = \frac{40}{5} \qquad \text{Divide both sides by 5.}$$

$$\frac{\overset{1}{\cancel{5}}y}{\underset{1}{\cancel{5}}} = \frac{40}{5} \qquad \text{Divide out common factors on the left side.}$$

$$y = 8 \qquad \text{Simplify.}$$

Again, we check our solution before writing it in solution set notation.

Check

$$5y = 40$$
$$5(8) = 40 \qquad \text{Substitute 8 for } y.$$
$$40 = 40 \qquad \text{Multiply.}$$

$y = 8$ is indeed a solution and the solution set is $\{8\}$.

Quick Check   3 Solve $4a = 56$ by using the multiplication property of equality.

If the coefficient of the variable term is negative, we must divide both sides of the equation by that negative number.

EXAMPLE 5   Solve $-7n = -56$ by using the multiplication property of equality.

Solution

In this example, the coefficient of the variable term is -7. To solve this equation, we need to divide both sides by -7.

$$-7n = -56$$
$$\frac{-7n}{-7} = \frac{-56}{-7} \qquad \text{Divide both sides by } -7.$$
$$n = 8 \qquad \text{Simplify.}$$

The check of this solution is left to the reader. The solution set is $\{8\}$.

Quick Check   4 Solve $-9a = 144$ by using the multiplication property of equality.

A Word of Caution   When dividing both sides of an equation by a negative number, keep in mind that this will change the sign of the number on the other side of the equation.

Consider the equation $-x = 16$. The coefficient of the variable term is -1. To solve this equation we can either multiply both sides of the equation by -1, or divide both sides by -1. Either way, the solution is $x = -16$. We could also have solved the equation by inspection by reading the equation $-x = 16$ as *"the opposite of x is 16."* If the opposite of x is 16, then we know that x must be equal to -16.

We will now learn how to solve an equation in which the coefficient of the variable term is a fraction.

EXAMPLE 6   Solve $\frac{3}{8}a = -\frac{5}{2}$ by using the multiplication property of equality.

Solution

In this example, we have a variable multiplied by a fraction. In such a case, we can multiply both sides of the equation by the reciprocal of the fraction. When we multiply a fraction by its reciprocal, the result is 1. This will leave the variable isolated.

$$\frac{3}{8}a = -\frac{5}{2}$$

$$\overset{1}{\underset{3}{\cancel{8}}} \cdot \overset{1}{\underset{1}{\frac{\cancel{3}}{\cancel{8}}}} a = \overset{4}{\underset{3}{\frac{\cancel{8}}{3}}}\left(-\frac{5}{\underset{1}{\cancel{2}}}\right) \qquad \text{Multiply by the reciprocal of the fraction } \tfrac{3}{8} \text{ and divide out common factors.}$$

$$a = -\frac{20}{3} \qquad \text{Simplify.}$$

Quick Check 5
Solve $\frac{9}{16}x = \frac{21}{8}$ using the multiplication property of equality.

The check of this solution is left to the reader. The solution set is $\left\{-\frac{20}{3}\right\}$.

Addition Property of Equality

Objective 4 Solve linear equations using the addition property of equality.
The **addition property of equality** tells us that we can add the same number to both sides of an equation, or subtract the same number from both sides of an equation, without affecting the equality of the two sides.

Addition Property of Equality

> For any algebraic expressions A and B, and any number n,
>
> $$\text{if} \quad A = B \quad \text{then} \quad A + n = B + n$$
> $$\text{and} \quad A - n = B - n.$$

This property helps us to solve equations in which we have a number either added to or subtracted from a variable on one side of an equation.

EXAMPLE 7   Solve $x + 4 = 11$ using the addition property of equality.

Solution

In this example, the number 4 is being added to the variable x. To isolate the variable, we use the addition property of equality to subtract 4 from both sides of the equation.

$$x + 4 = 11$$
$$x + 4 - 4 = 11 - 4 \qquad \text{Subtract 4 from both sides.}$$
$$x = 7 \qquad \text{Simplify.}$$

To check this solution, we will substitute 7 for x in the original equation.

Check
$$x + 4 = 11$$
$$7 + 4 = 11 \qquad \text{Substitute 7 for } x.$$
$$11 = 11 \qquad \text{Add.}$$

We have a true statement, so the solution set is $\{7\}$.

Quick Check **6** Solve $a + 22 = -8$ using the addition property of equality.

EXAMPLE 8 Solve $13 = y - 9$ using the addition property of equality.

Solution

When a value is subtracted from a variable, we isolate the variable by adding that value to both sides of the equation. In this example, we will add 9 to both sides.

$$13 = y - 9$$
$$13 + 9 = y - 9 + 9 \qquad \text{Add 9 to both sides.}$$
$$22 = y \qquad \text{Add.}$$

Check
$$13 = y - 9$$
$$13 = 22 - 9 \qquad \text{Substitute 22 for } y.$$
$$13 = 13 \qquad \text{Subtract.}$$

This is a true statement, so the solution set is $\{22\}$.

Quick Check **7** Solve $x - 28 = -13$ using the addition property of equality.

EXAMPLE 9 Solve $m + \frac{5}{2} = \frac{2}{3}$ using the addition property of equality.

Solution

We begin by subtracting $\frac{5}{2}$ from both sides of the equation.

$$m + \frac{5}{2} = \frac{2}{3}$$

$$m + \frac{5}{2} - \frac{5}{2} = \frac{2}{3} - \frac{5}{2} \qquad \text{Subtract } \frac{5}{2} \text{ from both sides.}$$

$$m = \frac{4}{6} - \frac{15}{6} \qquad \text{Simplify the left side. On the right side, rewrite the fractions as equivalent fractions with a common denominator of 6.}$$

$$m = -\frac{11}{6} \qquad \text{Subtract.}$$

The check of this solution is left to the reader. The solution set is $\left\{-\frac{11}{6}\right\}$.

Quick Check **8** Solve $x - \frac{7}{4} = \frac{5}{6}$ using the addition property of equality.

In the next section, we will learn how to solve equations requiring us to use both the multiplication and addition properties of equality.

Applications

Objective 5 **Solve applied problems using the multiplication property of equality or the addition property of equality.** We finish this section with an example of an applied problem that can be solved with a linear equation.

EXAMPLE 10   Admission to the county fair is $8 per person, so the admission price for a group of x people can be represented by $8x$. If a Cub Scout group paid a total of $208 for admission to the county fair, how many people were in the group?

Solution

We can express this relationship in an equation by using the idea that the total cost of admission is equal to $208. Since the total cost for x people can be represented by $8x$, our equation is $8x = 208$.

$$8x = 208$$
$$\frac{8x}{8} = \frac{208}{8} \qquad \text{Divide both sides by 8.}$$
$$x = 26 \qquad \text{Divide.}$$

You can verify that this value checks as a solution. Whenever we work on an applied problem, we should present our solution as a complete sentence. There were 26 people in the Cub Scout group at the county fair.

Quick Check 9 Josh spent a total of $26.50 to take a date to the movies. This left him with only $38.50 in his pocket. How much money did he have with him before going to the movies?

Building Your Study Strategy Using Your Textbook, 1 **Reading Ahead**   One effective way to use your textbook is to read a section in the text before it is covered in class. This will give you an idea about the main concepts covered in the section, and these concepts can be clarified by your instructor in class.

When reading ahead, you should scan the section. Look for definitions that are introduced in the section, as well as procedures developed in the section. Pay closer attention to the examples. If you find a step in the examples that you cannot understand, you can ask your instructor about it in class.

EXERCISES 2.1

F O R   E X T R A   H E L P

MyMathLab

MathXL

Interactmath.com

MathXL Tutorials on CD

Video Lectures on CD

Tutor Center

Addison-Wesley Math Tutor Center

Student's Solutions Manual

Vocabulary

1. A(n) _____ is a mathematical statement of equality between two expressions.

2. A(n) _____ of an equation is a value that, when substituted for the variable in the equation, produces a true statement.

3. Define a solution set of an equation.

4. State the multiplication property of equality.

5. State the addition property of equality.

6. Freebird's Pizza charges $12 for a pizza. If the bill for an office pizza party is $168, which equation can be used to determine how many pizzas were ordered?
    a) $x + 12 = 168$ b) $x - 12 = 168$
    c) $12x = 168$ d) $\dfrac{x}{12} = 168$

Is the given equation a linear equation? If not, explain why not.

7. $5x^2 - 7x = 3x + 8$

8. $4x - 9 = 17$

9. $3x - 5(2x + 3) = 8 - x$

10. $\dfrac{7}{x^2} - \dfrac{5}{x} - 13 = 0$

11. $y = 5$

12. $x^4 - 1 = 0$

13. $\dfrac{3x}{11} - \dfrac{5}{4} = \dfrac{2x}{7}$

14. $x \cdot 4 - 1 = 0$

15. $x + \dfrac{3}{x} + 18 = 0$

16. $3(4x - 9) + 7(2x + 5) = 15$

Check to determine whether the given value is a solution of the equation.

17. $x = 7$, $5x - 9 = 26$

18. $x = 3$, $2x - 11 = x + 2$

19. $a = 8$, $3 - 2a = a + 2a - 11$

20. $m = 4$, $15 - 8m = 3m - 29$

21. $z = \dfrac{1}{4}$, $\dfrac{2}{3}z + \dfrac{11}{6} = 2$

22. $t = \dfrac{5}{3}$, $\dfrac{1}{10}t + \dfrac{1}{3} = \dfrac{1}{2}$

23. $m = 3.4$, $3m - 2 = 2m + 0.4$

24. $a = -2.5$, $4a - 6 = 9 + 10a$

Solve using the multiplication property of equality.

25. $3x = 54$ 26. $14a = -56$

27. $4y = -40$ 28. $13x = 91$

29. $6b = 10$ 30. $8z = -42$

31. $5a = 0$ 32. $0 = 12x$

33. $-5t = 35$ 34. $-11x = -44$

35. $-2x = -28$ 36. $-9h = 54$

37. $-t = 45$ 38. $-y = -31$

39. $\dfrac{x}{3} = 7$ 40. $\dfrac{x}{4} = 3$

41. $-\dfrac{t}{8} = 12$ 42. $-\dfrac{g}{13} = -7$

43. $\dfrac{2}{3} = \dfrac{x}{12}$ 44. $\dfrac{3}{8} = \dfrac{x}{4}$

45. $-\dfrac{2}{5}x = 4$ 46. $-\dfrac{5}{6}x = 15$

47. $3.2x = 6.944$

48. $-4.7x = 42.5538$

Solve using the addition property of equality.

49. $a + 7 = 10$

50. $b + 5 = 13$

51. $x + 8 = 4$

52. $x + 1 = -15$

53. $x + 13 = 30$

54. $n - 19 = 11$

55. $a + 3.2 = 5.7$

56. $x - 14.9 = -21.2$

57. $12 = x + 3$

58. $4 = m + 10$

59. $b - 7 = 13$

60. $a - 9 = 99$

61. $t - 7 = -4$

62. $n - 3 = -18$

63. $-4 + x = 19$

64. $-15 + x = -7$

65. $x + 9 = 0$

66. $b - 17 = 0$

67. $a + \dfrac{5}{12} = \dfrac{5}{9}$

68. $x - \dfrac{7}{10} = \dfrac{4}{15}$

69. $a + 5 + 6 = 7$

70. $x + 4 - 9 = -3$

Mixed Practice, 71–90

Solve the equation.

71. $60 = 5a$

72. $a - \dfrac{3}{4} = \dfrac{7}{6}$

73. $x - 27 = -11$

74. $\dfrac{3}{14}x = \dfrac{5}{2}$

75. $-\dfrac{b}{10} = -3$

76. $b + 39 = 30$

77. $x + 24 = -17$

78. $\dfrac{x}{5} = 13$

79. $11 = -\dfrac{n}{7}$

80. $\dfrac{5}{3} = n + \dfrac{7}{2}$

81. $y + \dfrac{2}{15} = -\dfrac{5}{6}$

82. $47 = y + 35$

83. $0 = x + 56$

84. $x - 38 = -57$

85. $t - \dfrac{17}{8} = -\dfrac{7}{4}$

86. $\dfrac{t}{15} = -7$

87. $\dfrac{4}{9}x = -\dfrac{14}{15}$

88. $x - \dfrac{3}{5} = -\dfrac{9}{10}$

89. $0 = 45m$

90. $40 = -6m$

The values given are solutions of a linear equation. Give an equation that can be solved with the multiplication property of equality and that has the given solution.

91. $x = 7$

92. $x = -13$

93. $n = \dfrac{5}{2}$

94. $m = -\dfrac{3}{10}$

The values given are solutions of a linear equation. Give an equation that can be solved with the addition property of equality and that has the given solution.

95. $x = -6$

96. $x = 14$

97. $b = \dfrac{1}{6}$

98. $a = -\dfrac{9}{14}$

Set up a linear equation and solve it for the following problems.

99. Zoe has only nickels in her pocket. If she has $1.35 in her pocket, how many nickels does she have?

100. Carter sold lemonade in front of his house to raise money for a new bike. He charged 25 cents per cup. If his total sales from yesterday were $23.75, how many cups of lemonade did he sell?

101. A local garage band, the Grease Monkeys, held a rent party. They charged $3 per person for admission, with the proceeds to be used to pay their rent. If their rent is $425, and they ended up with an extra $52 after paying the rent, how many people came to see them play?

102. Ross organized a tour of a local winery. Attendees paid Ross $12 to go on the tour, plus another $5 for lunch. If Ross collected $493, how many people came on the tour?

103. An insurance company hired 8 new employees. This brought their total to 174 employees. How many employees did the company have before these 8 people were hired?

104. Geena scored 17 points higher than Jared on the last math exam. If Geena's score was 85, find Jared's score.

105. As a cold front was moving in, the temperature in Visalia dropped by 19° F in a two-hour period. If the temperature dropped to 37° F, what was the temperature before the cold front moved in?

106. Stephanie switched to a vegan diet and lost 13 pounds in three months. If her new weight is 134 pounds, what was her weight prior to switching to a vegan diet?

Writing in Mathematics

Answer in complete sentences.

107. Explain why the equation $0x = 15$ cannot be solved.

108. Explain why we cannot multiply both sides of an equation by 0 when applying the multiplication property of equality.

109. Find a value for x such that the expression $x + 21$ is less than -39.

110. Find a value of x such that the expression $-8x$ is greater than 95.

111. *Solutions Manual** Write a solutions manual page for the following problem:

Solve. $x - 8 = -17$

112. *Newsletter** Write a newsletter explaining how to solve linear equations using the addition and multiplication properties of equality.

See Appendix B for details and sample answers.

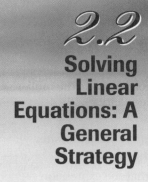

2.2

Solving Linear Equations: A General Strategy

1 Solve linear equations using both the multiplication property of equality and the addition property of equality.
2 Solve linear equations containing fractions.
3 Solve linear equations using the five-step general strategy.
4 Identify linear equations with no solution.
5 Identify linear equations with infinitely many solutions.
6 Solve literal equations for a specified variable.

In the previous section, we solved equations that required only one operation to isolate the variable. In this section, we will learn how to solve equations requiring the use of both the multiplication and addition properties of equality.

Solving Linear Equations

Objective **1** **Solve linear equations using both the multiplication property of equality and the addition property of equality.** Suppose we needed to solve the equation $4x - 7 = 17$. Should we divide both sides by 4 first? Should we add 7 to both sides first? We refer to the order of operations. This tells us that in the expression $4x - 7$, we first multiply 4 by x and then subtract 7 from the result. To isolate the variable x, we undo these operations in the opposite order. We will first add 7 to both sides to undo the subtraction, and then we will divide both sides by 4 to undo the multiplication.

<table>
<tr><td align="center">**Solution**</td><td align="center">**Check**</td></tr>
<tr><td align="center">$4x - 7 = 17$</td><td align="center">$4(6) - 7 = 17$</td></tr>
<tr><td align="center">$4x - 7 + 7 = 17 + 7$</td><td align="center">$24 - 7 = 17$</td></tr>
<tr><td align="center">$4x = 24$</td><td align="center">$17 = 17$</td></tr>
<tr><td align="center">$\dfrac{4x}{4} = \dfrac{24}{4}$</td><td></td></tr>
<tr><td align="center">$x = 6$</td><td></td></tr>
</table>

Since our solution $x = 6$ checks, the solution set is $\{6\}$.

EXAMPLE 1   Solve the equation $3x + 41 = 8$.

Solution

To solve this equation, we begin by subtracting 41 from both sides. This will isolate the term $3x$. We can then divide both sides of the equation by 3 to isolate the variable x.

$$3x + 41 = 8$$
$$3x + 41 - 41 = 8 - 41 \qquad \text{Subtract 41 from both sides to isolate } 3x.$$
$$3x = -33 \qquad \text{Subtract.}$$
$$\frac{3x}{3} = -\frac{33}{3} \qquad \text{Divide both sides by 3 to isolate } x.$$
$$x = -11 \qquad \text{Simplify.}$$

Now we check our solution.

$$3x + 41 = 8$$
$$3(-11) + 41 = 8 \qquad \text{Substitute } -11 \text{ for } x.$$
$$-33 + 41 = 8 \qquad \text{Multiply.}$$
$$8 = 8 \qquad \text{Simplify.}$$

Quick Check 1
Solve the equation
$5x - 2 = 33$.

Since $x = -11$ produced a true statement, the solution set is $\{-11\}$.

EXAMPLE 2 Solve the equation $-8x - 19 = 13$.

Solution

$$-8x - 19 = 13$$
$$-8x - 19 + 19 = 13 + 19 \qquad \text{Add 19 to both sides to isolate } -8x.$$
$$-8x = 32 \qquad \text{Add.}$$
$$\frac{-8x}{-8} = \frac{32}{-8} \qquad \text{Divide by } -8 \text{ to isolate } x.$$
$$x = -4 \qquad \text{Simplify.}$$

Now we check our solution.

$$-8x - 19 = 13$$
$$-8(-4) - 19 = 13 \qquad \text{Substitute } -4 \text{ for } x.$$
$$32 - 19 = 13 \qquad \text{Multiply.}$$
$$13 = 13 \qquad \text{Simplify.}$$

Quick Check 2
Solve the equation
$6 - 4x = 38$.

The solution set is $\{-4\}$.

Solving Linear Equations Containing Fractions

Objective 2 Solve linear equations containing fractions. If an equation contains fractions, it can be helpful to convert it to an equivalent equation that does not contain fractions before solving it. This can be done by multiplying both sides of the equation by the LCM of the denominators.

EXAMPLE 3 Solve the equation $\frac{2}{3}x - 5 = \frac{3}{4}$.

Solution

This equation contains two fractions, and the denominators are 3 and 4. The LCM of these two denominators is 12, so we will begin by multiplying both sides of the equation by 12.

$$\frac{2}{3}x - 5 = \frac{3}{4}$$
$$12\left(\frac{2}{3}x - 5\right) = 12\left(\frac{3}{4}\right) \qquad \text{Multiply both sides by the LCM 12.}$$
$$\overset{4}{\cancel{12}} \cdot \frac{2}{\underset{1}{3}}x - 12 \cdot 5 = \overset{3}{\cancel{12}} \cdot \frac{3}{\underset{1}{4}} \qquad \text{Distribute and divide out common factors.}$$

$$8x - 60 = 9 \qquad \text{Multiply.}$$
$$8x - 60 + 60 = 9 + 60 \qquad \text{Add 60 to both sides to isolate } 8x.$$
$$8x = 69 \qquad \text{Add.}$$
$$\frac{8x}{8} = \frac{69}{8} \qquad \text{Divide by 8 to isolate } x.$$
$$x = \frac{69}{8} \qquad \text{Simplify.}$$

Now we check our solution.

$$\frac{2}{3}x - 5 = \frac{3}{4}$$

$$\frac{2}{3}\left(\frac{69}{8}\right) - 5 = \frac{3}{4} \qquad \text{Substitute } \frac{69}{8} \text{ for } x.$$

$$\frac{\overset{1}{\cancel{2}}}{\underset{1}{\cancel{3}}}\left(\frac{\overset{23}{\cancel{69}}}{\underset{4}{\cancel{8}}}\right) - 5 = \frac{3}{4} \qquad \text{Divide out common factors.}$$

$$\frac{23}{4} - 5 = \frac{3}{4} \qquad \text{Simplify.}$$

$$\frac{23}{4} - \frac{20}{4} = \frac{3}{4} \qquad \text{Rewrite 5 as } \frac{20}{4}.$$

$$\frac{3}{4} = \frac{3}{4} \qquad \text{Simplify.}$$

Quick Check **3** The solution set is $\left\{\frac{69}{8}\right\}$.

Solve the equation
$\frac{2}{7}x + \frac{1}{2} = \frac{4}{3}$.

A Word of Caution   When multiplying both sides of an equation by the LCM of the denominators, be sure to multiply each term by the LCM, including any terms that do not contain fractions.

A General Strategy for Solving Linear Equations

Objective **3**   **Solve linear equations using the five-step general strategy.** We will now examine a process that can be used to solve any linear equation. This process works not only for types of equations we have already learned to solve but also for more complicated equations, such as $7x + 4 = 3x - 20$ and $5(2x - 9) + 3x = 7(3 - 8x)$.

Solving Linear Equations

1. **Simplify each side of the equation completely.**
   - Use the distributive property to clear any parentheses.
   - If there are fractions in the equation, multiply both sides of the equation by the LCM of the denominators to clear the fractions from the equation.
   - Combine any like terms that are on the same side of the equation. After you have completed this step, the equation should contain at most one variable term and at most one constant term on each side.
2. **Collect all variable terms on one side of the equation.** If there is a variable term on each side of the equation, use the addition property of equality to place both variable terms on the same side of the equation.

continued

3. **Collect all constant terms on the other side of the equation.** If there is a constant term on each side of the equation, use the addition property of equality to isolate the variable term.
4. **Divide both sides of the equation by the coefficient of the variable term.** At this point our equation should be of the form $ax = b$. We use the multiplication property of equality to find our solution.
5. **Check your solution.** Check that the value creates a true equation when substituted for the variable in the equation.

In the next example, we will solve equations that have variable terms and constant terms on both sides of the equation.

EXAMPLE 4   Solve the equation $3x + 8 = 7x - 6$.

Solution

In this equation there are no parentheses or fractions to clear, and there are no like terms to be combined. We begin by gathering the variable terms on one side of the equation.

$$3x + 8 = 7x - 6$$
$$3x + 8 - 3x = 7x - 6 - 3x \qquad \text{Subtract } 3x \text{ from both sides to gather the variable terms on the right side of the equation.}$$
$$8 = 4x - 6 \qquad \text{Subtract.}$$
$$8 + 6 = 4x - 6 + 6 \qquad \text{Add 6 to both sides to isolate } 4x.$$
$$14 = 4x \qquad \text{Add.}$$
$$\frac{14}{4} = \frac{4x}{4} \qquad \text{Divide both sides by 4 to isolate } x.$$
$$\frac{7}{2} = x \qquad \text{Simplify.}$$

Quick Check 4
Solve the equation
$6x + 19 = 3x - 8$.

The check of this solution is left to the reader. The solution set is $\{\frac{7}{2}\}$.

Using Your Calculator   We can use a calculator to check the solutions to an equation. Substitute the value for the variable and simplify the expressions on each side of the equation. Here is how the check of the solution $x = \frac{7}{2}$ would look on the TI-84.

```
3(7/2)+8
            18.5
7(7/2)-6
            18.5
```

Since each expression is equal to 18.5 when $x = \frac{7}{2}$, this solution checks.

In the next example, we will solve an equation containing like terms on the same side of the equation.

EXAMPLE 5 Solve the equation $13x - 43 - 9x = 6x + 37 + 5x - 66$.

Solution

This equation has like terms on each side of the equation, so we begin by combining these like terms.

$$13x - 43 - 9x = 6x + 37 + 5x - 66$$
$$4x - 43 = 11x - 29 \qquad \text{Combine like terms on each side.}$$
$$4x - 43 - 4x = 11x - 29 - 4x \qquad \text{Subtract } 4x \text{ from both sides to gather variable terms on one side of the equation.}$$
$$-43 = 7x - 29 \qquad \text{Subtract.}$$
$$-43 + 29 = 7x - 29 + 29 \qquad \text{Add 29 to both sides to isolate } 7x.$$
$$-14 = 7x \qquad \text{Simplify.}$$
$$-\frac{14}{7} = \frac{7x}{7} \qquad \text{Divide both sides by 7 to isolate } x.$$
$$-2 = x \qquad \text{Divide.}$$

The check of this solution is left to the reader. The solution set is $\{-2\}$.

Quick Check 5 Solve the equation $6x - 9 + 4x = 3x + 13 - 8$.

EXAMPLE 6 Solve the equation $6(3x - 8) + 14 = x + 3(5 - x) + 1$.

Solution

We begin by distributing the 6 on the left side of the equation and the 3 on the right side of the equation. Then we combine like terms before solving.

$$6(3x - 8) + 14 = x + 3(5 - x) + 1$$
$$18x - 48 + 14 = x + 15 - 3x + 1 \qquad \text{Distribute.}$$
$$18x - 34 = -2x + 16 \qquad \text{Combine like terms.}$$
$$18x - 34 + 2x = -2x + 16 + 2x \qquad \text{Add } 2x \text{ to both sides to gather variable terms on the left side of the equation.}$$
$$20x - 34 = 16 \qquad \text{Add.}$$
$$20x - 34 + 34 = 16 + 34 \qquad \text{Add 34 to both sides to isolate } 20x.$$
$$20x = 50 \qquad \text{Add.}$$
$$\frac{20x}{20} = \frac{50}{20} \qquad \text{Divide by 20 to isolate } x.$$
$$x = \frac{5}{2} \qquad \text{Simplify.}$$

The check of this solution is left to the reader. The solution set is $\left\{\frac{5}{2}\right\}$.

Quick Check 6 Solve the equation $3(2x - 7) + x = 2(x - 9) - 18$.

EXAMPLE ▶7   Solve the equation $8x - 7(3x - 5) = 6(4 - x) + x - 7$.

Solution

When applying the distributive property, we must be careful to distribute -7 (not just 7) on the left side. This changes the signs of the terms inside the parentheses.

$$8x - 7(3x - 5) = 6(4 - x) + x - 7$$

$8x - 21x + 35 = 24 - 6x + x - 7$ Distribute. (Be careful with signs.)

$-13x + 35 = -5x + 17$ Combine like terms.

$-13x + 35 + 13x = -5x + 17 + 13x$ Add $13x$ to both sides to gather variable terms on the right side of the equation.

$35 = 8x + 17$ Simplify.

$35 - 17 = 8x + 17 - 17$ Subtract 17 from both sides to isolate $8x$.

$18 = 8x$ Subtract.

$\dfrac{18}{8} = \dfrac{8x}{8}$ Divide both sides by 8 to isolate x.

$\dfrac{9}{4} = x$ Simplify.

The check of this solution is left to the reader. The solution set is $\left\{\frac{9}{4}\right\}$.

Quick Check   7 Solve the equation $5 - 4(3x - 8) + 7x = 8 - (x - 13)$.

In the next example, we will clear the equation of fractions before solving.

EXAMPLE ▶8   Solve the equation $\frac{3}{5}x - \frac{2}{3} = \frac{3}{4}x + \frac{5}{6} + 2x$.

Solution

We begin by finding the LCM of the four denominators (5, 3, 4, and 6), which is 60. We then multiply both sides of the equation by 60 to clear the equation of fractions.

$$\frac{3}{5}x - \frac{2}{3} = \frac{3}{4}x + \frac{5}{6} + 2x$$

$60\left(\dfrac{3}{5}x - \dfrac{2}{3}\right) = 60\left(\dfrac{3}{4}x + \dfrac{5}{6} + 2x\right)$ Multiply both sides by LCM (60).

$\overset{12}{\cancel{60}} \cdot \dfrac{3}{\cancel{5}_1}x - \overset{20}{\cancel{60}} \cdot \dfrac{2}{\cancel{3}_1} = \overset{15}{\cancel{60}} \cdot \dfrac{3}{\cancel{4}_1}x + \overset{10}{\cancel{60}} \cdot \dfrac{5}{\cancel{6}_1} + 60 \cdot 2x$ Distribute and divide out common factors.

$36x - 40 = 45x + 50 + 120x$ Multiply.

$36x - 40 = 165x + 50$ Combine like terms.

$36x - 40 - 36x = 165x + 50 - 36x$ Subtract $36x$ from both sides to gather variable terms on the right side of the equation.

$-40 = 129x + 50$ Subtract.

$-40 - 50 = 129x + 50 - 50$ Subtract 50 from both sides to isolate $129x$.

$$-90 = 129x \qquad \text{Simplify.}$$
$$-\frac{90}{129} = \frac{129x}{129} \qquad \text{Divide both sides by 129 to isolate } x.$$
$$-\frac{30}{43} = x \qquad \text{Simplify.}$$

Quick Check 8
Solve the equation
$\frac{1}{10}x - \frac{2}{5} = \frac{1}{20}x - \frac{7}{10}$.

The check of this solution is left to the reader. The solution set is $\left\{-\frac{30}{43}\right\}$.

Contradictions and Identities

Objective 4 Identify linear equations with no solution. In each equation that we have solved to this point, there has been exactly one solution. This will not always be the case. We now turn our attention to two special types of equations: contradictions and identities.

A **contradiction** is an equation that will always be false, regardless of the value that we substitute for the variable. A contradiction has no solution, so its solution set is the empty set { }. The empty set is also known as the **null set** and is denoted by the symbol ∅.

EXAMPLE 9   Solve the equation $5x + 5 = 5x + 6$.

Solution

We begin solving this equation by gathering both variable terms on the same side of the equation. This can be done by subtracting $5x$ from both sides of the equation.

$$5x + 5 = 5x + 6$$
$$5x + 5 - 5x = 5x + 6 - 5x \qquad \text{Subtract } 5x \text{ from both sides.}$$
$$5 = 6 \qquad \text{Simplify.}$$

Quick Check 9
Solve the equation
$3x - 4 = 3x + 4$.

We are left with an equation that is a false statement, since 5 can never be equal to 6. This equation is a contradiction and has no solutions. Its solution set is ∅.

Objective 5 Identify linear equations with infinitely many solutions. An **identity** is an equation that is always true. If we substitute any real number for the variable in an identity, it will produce a true statement. The solution set for an identity is the set of all real numbers. We will denote the set of all real numbers as ℝ. An identity has infinitely many solutions rather than a single solution.

EXAMPLE 10   Solve the equation $3x + 8 = 3x + 8$.

Solution

We will begin to solve this equation by attempting to gather all variable terms on one side of the equation.

$$3x + 8 = 3x + 8$$
$$3x + 8 - 3x = 3x + 8 - 3x \qquad \text{Subtract } 3x \text{ from both sides.}$$
$$8 = 8$$

Quick Check 10
Solve the equation
$5a + 4 = 4 + 5a$.

Since 8 is always equal to 8, regardless of the value of x, this equation is an identity. Its solution set is the set of all real numbers ℝ.

Literal Equations

Objective 6 **Solve literal equations for a specified variable.** A literal equation is an equation that contains two or more variables.

Perimeter of a Rectangle

> The **perimeter** of a rectangle is a measure of the distance around the rectangle. The formula for the perimeter (P) of a rectangle with length L and width W is $P = 2L + 2W$.

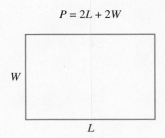

$$P = 2L + 2W$$

The equation $P = 2L + 2W$ is a literal equation with three variables. Often we will be asked to solve literal equations for one variable in terms of the other variables in the equation. In the perimeter example, if we solved for the width W in terms of the length L and the perimeter P, then we would have a formula for the width of a rectangle if we knew its length and perimeter.

We solve literal equations by isolating the specified variable. We will use the same general strategy that we used for solving linear equations. We treat the other variables in the equation as if they were constants.

EXAMPLE 11   Solve the literal equation $P = 2L + 2W$ (perimeter of a rectangle) for W.

Solution

We want to gather all terms containing the variable W on one side of the equation and gather all other terms on the other side. This can be done by subtracting $2L$ from both sides.

$$P = 2L + 2W$$
$$P - 2L = 2L + 2W - 2L \qquad \text{Subtract } 2L \text{ to isolate } 2W.$$
$$P - 2L = 2W \qquad \text{Subtract.}$$
$$\frac{P - 2L}{2} = \frac{2W}{2} \qquad \text{Divide by 2 to isolate } W.$$
$$\frac{P - 2L}{2} = W$$

Quick Check 11
Solve the literal equation
$x + 2y = 5$ for y.

We usually rewrite the equation so that the variable we solved for appears on the left side: $W = \frac{P - 2L}{2}$.

EXAMPLE 12   Solve the literal equation $A = \frac{1}{2}bh$ (area of a triangle) for h.

Solution

Since this equation has a fraction, we can begin by multiplying both sides by 2.

$$A = \frac{1}{2}bh$$

$$2 \cdot A = 2 \cdot \frac{1}{2}bh \qquad \text{Multiply both sides by 2 to clear fractions.}$$

$$2 \cdot A = \overset{1}{\cancel{2}} \cdot \frac{1}{\underset{1}{\cancel{2}}}bh \qquad \text{Divide out common factors.}$$

$$2A = bh \qquad \text{Multiply.}$$

$$\frac{2A}{b} = \frac{bh}{b} \qquad \text{Divide both sides by } b \text{ to isolate } h.$$

$$\frac{2A}{b} = h \quad \text{or} \quad h = \frac{2A}{b}.$$

Quick Check 12
Solve the literal equation
$\frac{1}{5}xy = z$ for x.

Some geometry formulas contain the symbol π, such as $C = 2\pi r$ and $A = \pi r^2$. The symbol π is the Greek letter pi and is used to represent an irrational number that is approximately equal to 3.14. When solving literal equations containing π, do not replace it by its approximate value.

> **Building Your Study Strategy** Using Your Textbook, 2 **Quick Check Exercises**
> The Quick Check exercises following most examples in this text are designed to be similar to the examples of the book. After reading through, and possibly reworking an example, try the corresponding Quick Check exercise. After checking your answer in the back of the book, you will be able to determine whether you are ready to move forward in the section or whether you need to practice more.

EXERCISES 2.2

Vocabulary

1. To clear fractions from an equation, multiply both sides of the equation by _____.

2. A contradiction is an equation that _____.

3. The solution set to a contradiction can be denoted ∅, which represents _____.

4. An equation that is always true is a(n) _____.

5. To solve the equation $2x - 9 = 7$, the best first step is to _____.
   a) divide both sides of the equation by 2

   b) add 9 to both sides of the equation
   c) subtract 7 from both sides of the equation

6. To solve the equation $3(4x + 5) = 11$, the best first step is to _____.
   a) distribute 3 on the left side of the equation
   b) divide both sides of the equation by 4
   c) subtract 5 from both sides of the equation

Solve.

7. $3x + 16 = 7$ **8.** $2x - 9 = 15$

9. $5 - 4x = 3$ **10.** $16 - 7x = -12$

11. $6 = 2a + 11$ **12.** $18 = 8 - 6b$

13. $9x + 24 = 24$

14. $-4x + 30 = -34$

15. $16.2x - 43.8 = 48.54$

16. $-9.5x - 72.35 = 130$

17. $\dfrac{1}{4}x - \dfrac{1}{3} = \dfrac{5}{12}$

18. $\dfrac{3}{11}x + \dfrac{5}{2} = 8$

19. $\dfrac{x}{12} - \dfrac{11}{6} = \dfrac{5}{4}$

20. $\dfrac{2}{5} - \dfrac{3}{8}x = -\dfrac{11}{10}$

21. $7a + 11 = 5a - 9$

22. $3m - 11 = 9m + 5$

23. $16 - 4x = 2x + 61$

24. $6n + 14 = 27 + 6n$

25. $5x - 9 = -9 + 5x$

26. $x - 7 = 11 - 3x$

27. $3x + 10 = 8x - 14$

28. $16 - 5x = 2x - 5$

29. $-6t + 17 = 3t - 34$

30. $-31 + 11n = 4n + 60$

31. $3.2x + 8.3 = 1.3x + 19.7$

32. $4x - 29.2 = 7.5x - 6.8$

33. $\dfrac{1}{2}x - 3 = \dfrac{11}{5} - \dfrac{3}{4}x$

34. $\dfrac{3}{5} - \dfrac{5}{6}x = 2x + \dfrac{4}{3}$

35. $3x + 8 + x = 2x - 9 + 13$

36. $3x + 14 + 8x - 90 = 4x + 11 + x + 27$

37. $27x - 16 - 6x = 14 + 21x - 30$

38. $-5n + 11 - 4n + 9 = n - 45$

39. $9n - 17 - 5n - 32 = 23 + 4n + 5$

40. $5t + 72 + 8t = 17t - 28 + 11t$

41. $2(2x - 5) - 7 = x - 12$

42. $10 + 4(3x + 1) = 5x - 7 - 2x$

43. $4(2x - 3) - 5x = 3(x + 4)$

44. $2(5x - 3) + 4(x - 8) = 11$

45. $13(3x + 4) - 7(2x - 5) = 5x - 13$

46. $3(2x - 8) - 5(15 - 6x) = 9(4x - 11)$

Find a linear equation that has the given solution. (Answers may vary.)

47. $x = 9$

48. $x = 6$

49. $a = \dfrac{1}{9}$

50. $b = \dfrac{5}{7}$

51. Write a linear equation that is a contradiction.

52. Write a linear equation that is an identity.

53. Write an equation that has infinitely many solutions.

54. Write an equation that has no solutions.

55. Write an equation whose single solution is negative.

56. Write an equation whose single solution is a fraction.

Solve the following literal equations for the specified variable.

57. $5x + y = -2$ for y

58. $-6x + y = 9$ for y

59. $7x + 2y = 4$ for y

60. $9x + 4y = 20$ for y

61. $6x - 3y = 13$ for y

62. $12x - 5y = -18$ for y

63. $P = a + b + c$ for b

64. $P = a + b + 2c$ for c

65. $d = r \cdot t$ for t

66. $d = r \cdot t$ for r

67. $C = 2\pi r$ for r

68. $S = 2\pi rh$ for r

Temperatures are usually reported using either the Celsius scale or the Fahrenheit scale. To convert a Celsius temperature to a Fahrenheit temperature, we multiply the Celsius temperature by $\frac{9}{5}$ and add 32 to the result.

69. If the temperature outside is 68° F, find the Celsius temperature.

70. If the normal body temperature for a person is 98.6° F, find the Celsius equivalent.

71. A number is doubled and then added to 7. If the result is 715, find the number.

72. If three-fourths of a number is added to 11, the result is 20. Find the number.

73. It costs $200 to rent a booth at a craft fair. Tina wants to sell homemade kites at the fair. It costs Tina $4 in material to make each kite.

    a) If she has $520 available to buy material and pay expenses, how many kites could she make to sell at the craft fair after paying the $200 rental charge?

    b) If she is able to sell all of the kites, how much would she need to charge for each kite in order to break even?

    c) How much should she charge for each kite in order to make a profit of $500?

74. A churro is a Mexican dessert pastry. Dan is able to buy churros for $0.75 each, which he plans to sell at a high school baseball game. Dan has to donate $50 to the high school team in order to be allowed to sell the churros at the game.

    a) If Dan has $110 to invest in this venture, how many churros will he be able to buy after paying $50 to the team?

    b) If Dan charges $2 for each churro, how many must he sell in order to break even?

    c) If Dan charges $2 for each churro and he is able to sell all of his churros, what will his profit be?

75. Find the missing value such that $x = 3$ is a solution of $5x - ? = 11$.

76. Find the missing value such that $x = -2$ is a solution of $3(2x - 5) + ? = 5(3 - x)$.

77. Find the missing value such that $x = -\frac{3}{2}$ is a solution of $2x + 9 = 6x + ?$.

78. Find the missing value such that $x = \frac{2}{5}$ is a solution of $4(2x + 7) = 7(3x + ?)$.

Writing in Mathematics

Answer in complete sentences.

79. Solve the linear equation $14 = 2x - 9$, showing all steps. Next to this, show all of the steps necessary to solve the literal equation $y = mx + b$ for x. Write a paragraph explaining how these two processes are similar and how they are different.

80. Write a word problem whose solution can be found from the equation $4.75x + 225 = 795$. Explain how the equation fits your problem.

81. *Solutions Manual** Write a solutions manual page for the following problem:

    Solve. $3(2x + 9) + 4x = 5(x - 6) + 17$

82. *Newsletter** Write a newsletter explaining the steps for solving linear equations.

*See Appendix B for details and sample answers.

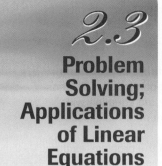

2.3
Problem Solving; Applications of Linear Equations

Objectives

Objectives

1 Understand the six steps for solving applied problems.

2 Solve problems involving an unknown number.

3 Solve problems involving geometric formulas.

4 Solve problems involving consecutive integers.

5 Solve problems involving motion.

6 Solve other applied problems.

Introduction to Problem Solving

Although the ability to perform mathematical computations and abstract algebraic manipulations is valuable, one of the most important skills developed in math classes is the skill of problem solving. Every day we are faced with making important decisions and predictions, and the thought process required in decision-making is quite similar to the process of solving mathematical problems.

When faced with a problem to solve in the real world, we first take inventory of the facts that we know. We also determine exactly what it is that we are trying to figure out. Then we develop a plan for taking what we already know and using it to help us figure out how to solve the problem. Once we have solved the problem, we reflect on the route that we took to solve it in order to make sure that that route was a logical way to solve the problem. After examining our solution for correctness and practicality, we often finish by presenting our solution to others for their consideration, information, or approval.

An example of one such real-world problem would be determining how early to leave for school on the first day of classes. Suppose your first class is at 9 A.M. The problem to solve is figuring out what time to leave your house so that you will not be late. Now think about the facts that you know: It's normally a 20-minute drive to school, and the classroom is a 10-minute walk from the parking lot. Add information specific to the first day of classes: Traffic near the school will be more congested than normal, and it will take longer to find a parking space in the parking lot. Based on your past experience, you figure an extra 15 minutes for traffic and another 10 minutes for parking. Adding up all of these times tells us that you need 55 minutes to get to your class. Adding an extra 15 minutes, just to be sure, you decide that you need to leave home by 7:50 A.M.

Objective 1 **Understand the six steps for solving applied problems.** In the previous example, we identified the problem to be solved, gathered our facts, and used them to find a solution to the problem. Solving applied math problems will require that we follow a similar procedure. George Pólya, this chapter's Mathematician in History, was a leader in the study of problem solving. Here is a general plan for solving applied math problems, based on the work of Pólya's text *How to Solve It*.

Solving Applied Problems

1. **Read the problem.** This step is often overlooked, but misreading or misinterpreting the problem essentially guarantees an incorrect solution. Read the problem once quickly to get a rough idea of what is going on, and then read it more carefully a second time to gather all of the important information.

continued

2. **List all of the important information.** Identify all known quantities presented in the problem, and determine which quantities we need to find. Creating a table to hold this information or a drawing to represent the problem are good ideas.
3. **Assign a variable to the unknown quantity.** If there is more than one unknown quantity, express each one in terms of the same variable, if possible.
4. **Find an equation relating the known values to the variable expressions representing the unknown quantities.** Sometimes, this equation can be translated directly from the wording of the problem. Other times the equation will depend on general facts that are known outside the statement of the problem, such as geometry formulas.
5. **Solve the equation.** Solving the equation is not the end of the problem, as the value of the variable is often not what we were originally looking for. If we were asked for the length and width of a rectangle and we reply "x equals 4," we have not answered the question. Check your solution to the equation.
6. **Present the solution.** Take the value of the variable from the solution to the equation and use it to figure out all unknown quantities. Check these values to be sure that they make sense in the context of the problem. For example, the length of a rectangle cannot be negative. Finally, present your solution in a complete sentence, using the proper units.

We will now put this strategy to use. All of the applied problems in this section will lead to linear equations.

Objective 2 **Solve problems involving an unknown number.**

EXAMPLE 1 Seven more than twice a number is 39. Find the number.

Solution

This is not a very exciting problem, but it provides us with a good opportunity to practice setting up applied problems.

There is one unknown in this problem. Let's use n to represent the *unknown number*.

Unknown
Number: n

Now we translate the sentence "Seven more than twice a number is 39" into an equation. In this case, twice a number is $2n$, so seven more than twice a number is $2n + 7$:

$$\underbrace{\text{Seven more than twice a number}}_{2n + 7} \underbrace{\text{is}}_{=} \underbrace{39}_{39}$$

Now we solve the equation.

$$2n + 7 = 39$$
$$2n = 32 \qquad \text{Subtract 7 from both sides.}$$
$$n = 16 \qquad \text{Divide both sides by 2.}$$

The solution of the equation is $n = 16$. Now we refer back to the table containing our unknown quantity. Since our unknown number was n, the solution to the problem is the same as the solution of the equation. The number is 16.

It is a good idea to check our solution. Twice our number is 32, and seven more than that is 39.

Quick Check ◀ **1**

Three less than four times a number is 29. Find the number.

Geometry Problems

Objective **3** **Solve problems involving geometric formulas.** The next example involves the perimeter of a rectangle, which is a measure of the distance around the outside of the rectangle. The perimeter P of a rectangle whose length is L and whose width is W is given by the formula $P = 2L + 2W$.

EXAMPLE ▶ **2**   Mark's vegetable garden is in the shape of a rectangle, and he can enclose it with 96 feet of fencing. If the length of his garden is 16 feet more than 3 times the width, find the dimensions of his garden.

Solution

The unknown quantities are the length and the width. Since the length is given in terms of the width, it is a good idea to pick a variable to represent the width. We can then write the length in terms of this variable. We will let w represent the width of the rectangle. The length is 16 feet more than 3 times the width, so we can express the length as $3w + 16$. One piece of information that we are given is that the perimeter is 96 feet. This information can be recorded in the following table or drawing:

Unknowns	Perimeter: 96 feet
Length: $3w + 16$	
Width: w	Width w
Known	Length $3w + 16$
Perimeter: 96 feet	

We start with the formula for perimeter and substitute the appropriate values and expressions.

$$P = 2L + 2W$$

$96 = 2(3w + 16) + 2w$	Substitute 96 for P, $3w + 16$ for L, and w for W.
$96 = 6w + 32 + 2w$	Distribute.
$96 = 8w + 32$	Combine like terms.
$64 = 8w$	Subtract 32 from both sides.
$8 = w$	Divide both sides by 8.

We now take this solution and substitute 8 for w in the expressions for length and width, which can be found in the table where we listed our unknowns:

Length: $3w + 16 = 3(8) + 16 = 40$
Width: $w = 8$

The length of Mark's garden is 40 feet, and the width is 8 feet. It checks that the perimeter of this rectangle is 96 feet.

Quick Check   **2** ▸ A rectangular living room has a perimeter of 80 feet. The length of the room is 8 feet less than twice the width of the room. Find the dimensions of the room.

Here is a summary of useful formulas and definitions from geometry:

Geometry Formulas and Definitions

- Perimeter of a triangle with sides s_1, s_2, and s_3: $P = s_1 + s_2 + s_3$

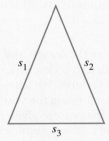

- Perimeter of a square with side s: $P = 4s$

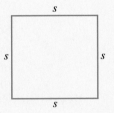

- Perimeter of a rectangle with length L and width W: $P = 2L + 2W$

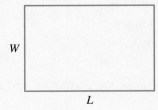

- **Circumference** of a circle with radius r: $C = 2\pi r$

- Area of a triangle with base b and height h: $A = \dfrac{1}{2}bh$

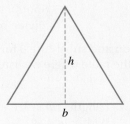

- Area of a square with side s: $A = s^2$

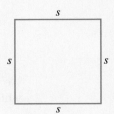

- Area of a rectangle with length L and width W: $A = LW$

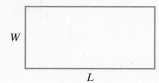

W

L

- Area of a circle with radius r: $A = \pi r^2$

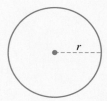

r

- An **equilateral triangle** is a triangle that has three equal sides.

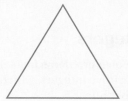

- An **isosceles triangle** is a triangle that has at least two equal sides.

EXAMPLE 3   An isosceles triangle has a perimeter of 30 cm. The third side is 3 cm shorter than each of the sides that have equal lengths. Find the lengths of the three sides.

Solution

In this problem, there are three unknowns, which are the lengths of the three sides. Because the triangle is an isosceles triangle, the first two sides have equal lengths. We can represent the length of each side by the variable x. The third side is 3 cm shorter than the other two sides, and its length can be represented by $x - 3$. Here is a summary of this information, together with the known perimeter:

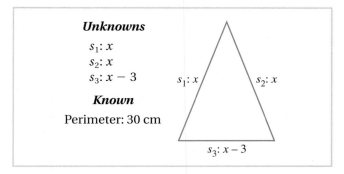

Unknowns

s_1: x
s_2: x
s_3: $x - 3$

Known

Perimeter: 30 cm

s_1: x s_2: x

s_3: $x - 3$

We now set up the equation and solve for x.

$$P = s_1 + s_2 + s_3$$
$$30 = x + x + (x - 3) \qquad \text{Substitute 30 for } P, x \text{ for } s_1 \text{ and } s_2, \text{ and } x - 3 \text{ for } s_3.$$
$$30 = 3x - 3 \qquad \text{Combine like terms.}$$
$$33 = 3x \qquad \text{Add 3 to both sides.}$$
$$11 = x \qquad \text{Divide both sides by 3.}$$

Looking back to the table, we substitute 11 for x.

$$s_1: x = 11$$
$$s_2: x = 11$$
$$s_3: x - 3 = 11 - 3 = 8$$

The three sides are 11 cm, 11 cm, and 8 cm. The perimeter of this triangle checks to be 30 cm.

Quick Check **3** The perimeter of a triangle is 112 inches. The longest side is 8 inches longer than 3 times the shortest side. The other side of the triangle is 20 inches longer than twice the shortest side. Find the lengths of the 3 sides.

Consecutive Integers

Objective 4 Solve problems involving consecutive integers. The following example involves consecutive integers, which are integers that are 1 unit apart from each other on the number line.

EXAMPLE 4   The sum of three consecutive integers is 87. Find them.

Solution

In this problem, the three consecutive integers are our three unknowns. We will let x represent the first integer. Since consecutive integers are 1 unit apart from each other, we can let $x + 1$ represent the second integer. Adding another 1, we can let $x + 2$ represent the third integer. Here is a table of the unknowns:

> ***Unknowns***
> First: x
> Second: $x + 1$
> Third: $x + 2$

We know that the sum of these three integers is 87, which leads to our equation.

$$
\begin{aligned}
x + (x + 1) + (x + 2) &= 87 \\
3x + 3 &= 87 \qquad \text{Combine like terms.}\\
3x &= 84 \qquad \text{Subtract 3 from both sides.}\\
x &= 28 \qquad \text{Divide both sides by 3.}
\end{aligned}
$$

We now substitute this solution to find our three unknowns.

> First: $x = 28$
> Second: $x + 1 = 28 + 1 = 29$
> Third: $x + 2 = 28 + 2 = 30$

Quick Check **4**
The sum of three consecutive integers is 213. Find them.

The three integers are 28, 29, and 30. The three integers do add up to 87.

Suppose that a problem involved consecutive *odd* integers rather than consecutive integers. Consider any string of consecutive odd integers, such as 15, 17, 19, 21, 23, ... How far apart are these consecutive odd integers? Each odd integer is 2 away from the previous one. When working with consecutive odd integers, or consecutive even integers for that matter, we will let x represent the first integer and then will add 2 to find the next consecutive integer of that type. Here is a table showing the pattern of unknowns for consecutive integers, consecutive odd integers, and consecutive even integers:

Consecutive Integers	Consecutive Odd Integers	Consecutive Even Integers
First: x	First: x	First: x
Second: $x + 1$	Second: $x + 2$	Second: $x + 2$
Third: $x + 2$	Third: $x + 4$	Third: $x + 4$
⋮	⋮	⋮

Motion Problems

Objective 5   **Solve problems involving motion.**

Distance Formula

> When an object such as a car moves at a constant rate of speed, r, for an amount of time, t, then the distance traveled, d, is given by the formula $d = r \cdot t$.

Suppose that a person drove 238 miles in 3.5 hours. To determine the car's rate of speed for this trip, we can substitute 238 for d and 3.5 for t in the equation $d = r \cdot t$.

$$d = r \cdot t$$
$$238 = r \cdot 3.5 \qquad \text{Substitute 238 for } d \text{ and 3.5 for } t.$$
$$\frac{238}{3.5} = \frac{r \cdot 3.5}{3.5} \qquad \text{Divide both sides by 3.5.}$$
$$68 = r \qquad \text{Simplify.}$$

The car's rate of speed for this trip was 68 mph.

EXAMPLE 5   Tina drove her car at a rate of 60 mph from her home to Rochester. Her mother Linda made the same trip at a rate of 80 mph, and it took her 2 hours less than Tina to make the trip. How far is it from Tina's home to Rochester?

Solution

For this problem, we can begin by setting up a table showing the relevant information. We will let t represent the time it took for Tina to make the trip. Since it took Linda 2 hours less to make the trip, we can represent her time by $t - 2$. We multiply each person's rate of speed by the time she traveled to find the distance she traveled.

	Rate	Time	Distance ($d = r \cdot t$)
Tina	60	t	$60t$
Linda	80	$t - 2$	$80(t - 2)$

Since we know that both trips were exactly the same distance, we get our equation by setting the expression for Tina's distance equal to the expression for Linda's distance, and we then solve for t.

$$60t = 80(t - 2)$$
$$60t = 80t - 160 \qquad \text{Distribute.}$$
$$-20t = -160 \qquad \text{Subtract } 80t \text{ from both sides.}$$
$$t = 8 \qquad \text{Divide both sides by } -20.$$

At this point, we must be careful to answer the appropriate question. We were asked to find the distance that was traveled. We can substitute 8 for t in either of the expressions for distance. Using the expression for the distance traveled by Tina, $60t = 60(8) = 480$. Using the expression for the distance traveled by Linda, $80(t - 2)$, produces the same result. It is 480 miles from Tina's home to Rochester.

Quick Check **5** Jake drove at a rate of 85 mph for a certain amount of time. His brother Elwood then took over as the driver. Elwood drove at a rate of 90 mph for 3 hours longer than Jake had driven. If the two brothers drove a total of 970 miles, how long did Elwood drive?

Other Problems

Objective 6 **Solve other applied problems.**

EXAMPLE 6   One number is 5 more than twice another number. If the sum of the two numbers is 80, find the two numbers.

Solution

In this problem, there are two unknown numbers. We will let one of the numbers be x. Since we know that one number is 5 more than twice the other number, we can represent the other number as $2x + 5$.

Unknowns
First: x
Second: $2x + 5$

Finally, we know that the sum of these two numbers is 80, which leads to the equation $x + 2x + 5 = 80$.

$$x + 2x + 5 = 80$$
$$3x + 5 = 80 \qquad \text{Combine like terms.}$$
$$3x = 75 \qquad \text{Subtract 5 from both sides.}$$
$$x = 25 \qquad \text{Divide both sides by 3.}$$

We now substitute this solution to find the two unknown numbers:

> First: $x = 25$
>
> Second: $2x + 5 = 2(25) + 5 = 50 + 5 = 55$

The two numbers are 25 and 55. The sum of these two numbers is 80.

Quick Check **6** A number is one more than five times another number. If the sum of the two numbers is 103, find the two numbers.

We conclude this section with a problem involving coins.

EXAMPLE 7 Ernie has nickels and dimes in his pocket. He has 5 more dimes than nickels. If Ernie has $2.30 in his pocket, how many nickels and dimes does he have?

Solution

We know the amount of money Ernie has in nickels and dimes totals $2.30. Since the number of dimes is given in terms of the number of nickels, we will let n represent the number of nickels. Since we know that Ernie has 5 more dimes than nickels, we can represent the number of dimes by $n + 5$. A table can be quite helpful in organizing our information, as well as in finding our equation.

	Number of Coins	Value of Coin	Amount of Money
Nickels	n	0.05	$0.05n$
Dimes	$n + 5$	0.10	$0.10(n + 5)$

The amount of money that Ernie has in nickels and dimes is equal to $2.30.

$$
\begin{aligned}
0.05n + 0.10(n + 5) &= 2.30 \\
0.05n + 0.10n + 0.50 &= 2.30 && \text{Distribute and multiply.} \\
0.15n + 0.50 &= 2.30 && \text{Combine like terms.} \\
0.15n &= 1.80 && \text{Subtract 0.50 from both sides.} \\
n &= 12 && \text{Divide both sides by 0.15.}
\end{aligned}
$$

The number of nickels is 12. The number of dimes $(n + 5)$ is $12 + 5$ or 17. Ernie has 12 nickels and 17 dimes. It is left to the reader to verify that these coins are worth a total of $2.30.

Quick Check **7** Bert has a pocketful of nickels and quarters in his pocket. He has 13 more nickels than quarters. If Bert has $3.35 in his pocket, how many nickels and quarters does he have?

Building Your Study Strategy Using Your Textbook, 3 **Supplementing Notes and Using Examples** When you are rewriting your classroom notes, your textbook can be helpful in supplementing your notes. If your instructor gave you a set of definitions in class, look up the corresponding definitions in the textbook. You can also look for examples similar to the ones provided by your instructor during class.

The textbook should also be kept available while you are working on your homework. If you are stuck on a particular problem, look through the section for an example that may give you an idea about how to proceed.

EXERCISES 2.3

Vocabulary

1. Which equation can be used to solve "7 less than twice a number is 23?"

a) $2(n - 7) = 23$ **b)** $2n - 7 = 23$
c) $7 - 2n = 23$

2. State the formula for the perimeter of a rectangle.

3. If a rectangle has a perimeter of 80 feet and its length is 8 feet less that 3 times its width, which equation can be used to find the dimensions of the rectangle?

a) $2(W - 8) + 2(3W) = 80$
b) $2(3W - 8) + 2W = 80$
c) $2(L) + 2(3L - 8) = 80$

4. Consecutive integers are integers that are _____ unit(s) apart on a number line.

5. If we let x represent the first of three consecutive integers, then we can represent the other two integers by _____ and _____.

6. If we let x represent the first of three consecutive even integers, then we can represent the other two integers by _____ and _____.

7. If we let x represent the first of three consecutive odd integers, then we can represent the other two integers by _____ and _____.

8. Charles has 7 fewer $5 bills than $10 bills. The total value of these bills is $265. Which equation can be used to determine how many of each type of bill Charles has?

a) $5(n - 7) + 10n = 265$
b) $5n - 7 + 10n = 265$
c) $5n + 10(n - 7) = 265$
d) $5n + 10n - 7 = 265$

9. Five more than twice a number is 97. Find the number.

10. Eighteen more than 7 times a number is 109. Find the number.

11. Eight less than 5 times a number is 717. Find the number.

12. Nine less than 3 times a number is 42. Find the number.

13. When 3.2 is added to 6 times a number, the result is 56.6. Find the number.

14. When 17.2 is subtracted from twice a number, the result is 11.4. Find the number.

15. The width of a rectangle is 5 meters less than its length, and the perimeter is 26 meters. Find the length and the width of the rectangle.

16. A rectangle has a perimeter of 62 feet. If its length is 9 feet longer than its width, find the dimensions of the rectangle.

17. A schoolyard sandbox is in the shape of a rectangle. It has 54 feet of boards around the border that enclose the sandbox. If the length of the sandbox is 3 feet less than twice the width, find the dimensions of the sandbox.

18. A rectangular sheet of paper has a perimeter of 39 inches. The length of the paper is 6 inches less than twice its width. Find the dimensions of the sheet of paper.

19. A rectangle has a perimeter of 130 cm. If the length of the rectangle is 4 times its width, find the dimensions of the rectangle.

20. A homeowner has built a rectangular shuffleboard court in her backyard. The length of the court is 7 times as long as the width, and the perimeter is 96 feet. Find the dimensions of the court.

21. A square has a perimeter of 220 feet. What is the length of one of its sides?

22. The bases in a Major League Baseball infield are set out in the shape of a square. When Manny Ramirez hits a home run, he has to run 360 feet to make it all the way around the bases. How far is the distance from home plate to first base?

23. An equilateral triangle has a perimeter of 96 cm. Find the length of each side of this triangle.

24. The third side of an isosceles triangle is 2 inches shorter than each of the other two sides. If the perimeter of this triangle is 67 inches, find the length of each side.

25. One side of a triangle is 4 inches longer than the side that is the base of the triangle. The third side is 7 inches longer than the base. If the perimeter is 26 inches, find the length of each side.

26. Two sides of a triangle are 3 feet and 6 feet longer than the third side of the triangle, respectively. If the perimeter of the triangle is 36 feet, find the length of each side.

27. A circle has a circumference of 69.08 inches. Use $\pi \approx 3.14$ to find the radius of the circle.

28. A circle has a circumference of 48π inches. Find its radius.

Two positive angles are said to be **complementary** if their measures add up to 90°.

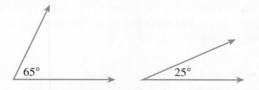

29. Angles A and B are complementary angles. If the measure of angle A is 30° more than the measure of angle B, find the measures of the two angles.

30. Angles A and B are complementary angles. Angle A is 15° less than angle B. Find the measures of the two angles.

31. An angle is 10° more than 3 times its complementary angle. Find the measures of the two angles.

32. An angle is 6° less than twice its complementary angle. Find the measures of the two angles.

Two positive angles are said to be **supplementary** if their measures add up to 180°.

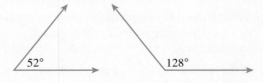

33. An angle is 12° more than 3 times its supplementary angle. Find the measures of the two angles.

34. An angle is 39° less than twice its supplementary angle. Find the measures of the two angles.

The measures of the three angles inside a triangle total 180°.

35. One angle in a triangle is 10° more than the smallest angle in the triangle, while the other angle is 20° more than the smallest angle. Find the measures of the three angles.

36. A triangle contains three angles: *A*, *B*, and *C*. Angle *B* is twice angle *A*, and angle *C* is 20° more than angle *A*. Find the measures of the three angles.

37. If the perimeter of a rectangle is 104 inches and the length of the rectangle is 6 inches less than its width, find the area of the rectangle.

38. If the perimeter of a rectangle is 76 inches and the width of the rectangle is 4 inches more than its length, find the area of the rectangle.

39. A rectangle has a perimeter of 46 inches. The width of the rectangle is 7 inches shorter than twice its length. If a square's side is as long as the width of this rectangle, find the perimeter of the square.

40. A rectangular pen with a perimeter of 200 feet is divided up into four square pens by building three fences inside the rectangular pen as follows:

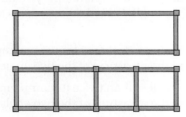

   a) What are the original dimensions of the rectangular pen?

   b) What is the total length of the three sections of fencing that were added to the inside of the rectangular pen?

   c) If the additional fencing cost $2.85 per foot, find the cost of the fencing required to make the four rectangular pens.

41. The sum of two consecutive integers is 305. Find the two integers.

42. The sum of three consecutive integers is 186. Find the three integers.

43. If the sum of three consecutive integers is 237, what is the sum of the two smaller integers?

44. If the sum of five consecutive integers is 755, what is the product of the smallest integer and the largest integer?

45. The sum of three consecutive odd integers is 225. Find the three odd integers.

46. The sum of three consecutive even integers is 96. Find the three even integers.

47. The sum of four consecutive odd integers is 360. Find the largest of the four odd integers.

48. The sum of five odd integers is 575. Find the average of these five odd integers.

49. Arthur has a book open and the two page numbers that he is looking at have a sum of 277. What are the page numbers of the pages he is looking at?

50. Tyler's dance recitals are held on three consecutive days in the month of June. The three dates add up to 72. What is the date of the first recital?

51. The smallest of three consecutive integers is 18 less than the sum of the two larger integers. Find the three integers.

52. There are three consecutive odd integers, and the sum of the smaller two integers is 49 less than 3 times the larger integer. Find the three odd integers.

53. A. J. drove his race car at a rate of 125 miles per hour for 4 hours. How far did he drive?

54. Angela made the 300-mile drive to Las Vegas in 6 hours. What was her average rate for the trip?

55. If Mario drives at a rate of 68 miles per hour, how long will it take him to drive 374 miles?

56. A bullet train travels 240 kilometers per hour. How far can it travel in 45 minutes?

57. Janet swam a 50-meter race in 28 seconds. What was her speed in meters per second?

58. In 1927, Charles Lindbergh flew his airplane, *Spirit of St. Louis,* 3500 miles from New York City to Paris in 33.5 hours. What was his speed for the trip, to the nearest tenth of a mile per hour?

59. Susan averaged 80 miles per hour on the way to Phoenix to make a sales call. On the way home, she averaged 70 miles per hour and it took her 1 hour longer to drive home than it did to drive to Phoenix.

  a) What was the total driving time for Susan's trip?

  b) How far does Susan live from Phoenix?

60. Geoffrey drove 60 miles per hour on the way to visit his parents. On the way back to school he drove 75 miles per hour and it took him 1 hour less time than it did to drive to his parents' house.

  a) What was the total driving time for Geoffrey's trip?

  b) How far is Geoffrey's school from his parents' house?

61. Don started driving east at 6 A.M. at a rate of 75 miles per hour. If Dennis leaves the same place heading east at 8 A.M. at a rate of 80 miles per hour, how long will it take him to catch up to Don?

62. At 7 A.M., a train left the station, heading directly north at 50 miles per hour. At 9 A.M., a bus left the same station heading directly south at 60 miles per hour. At what time will the train and the bus be 650 miles apart?

63. One number is 17 more than another. If the sum of the two numbers is 71, find the two numbers.

64. One number is 20 more than another. If the sum of the two numbers is 106, find the two numbers.

65. One number is 8 less than another number. If the sum of the two numbers is 96, find the two numbers.

66. One number is 27 less than another number. If the sum of the two numbers is 153, find the two numbers.

67. One number is 5 more than another number. If the smaller number is doubled and added to 3 times the larger number, the sum is 80. Find the two numbers.

68. One number is 11 more than another number. If 3 times the smaller number is added to 5 times the larger number, the sum is 191. Find the two numbers.

69. After saving quarters for a month, Larry has $13.75 in quarters. How many quarters does Larry have?

70. A roll of nickels is worth $2. How many nickels are in a roll?

71. Chris has a jar with dimes and quarters in it. The jar has 7 more quarters in it than it has dimes. If the total value of the coins is $5.60, how many quarters are in the jar?

72. Kay's change purse has nickels and dimes in it. The number of nickels is 5 more than twice the number of dimes. If the total value of the coins is $4.05, how many nickels are in the purse?

73. The film department holds a fund-raiser by showing the movie π. Admission was \$4 for students and \$7 for nonstudents. The number of students who attended was 10 more than 4 times the number of nonstudents who attended. If the department raised \$500, how many students attended the movie?

74. A movie theater charges \$4.50 for children to see a matinee and \$6.75 for adults. At today's matinee there were 20 more children than adults, and the total receipts were \$405. How many children were at today's matinee?

Writing in Mathematics

Answer in complete sentences.

75. Write a word problem involving a rectangle whose length is 50 feet and whose width is 20 feet. Explain how you created your problem.

76. Write a word problem whose solution is "There were 70 children and 40 adults in attendance." Explain how you created your problem.

77. Write a word problem for the given table and equation. Explain how you created your problem.
Equation: $55t = 70(t - 3)$

	Rate	Time	Distance $(d = r \cdot t)$
Steve	55	t	$55t$
Ross	70	$t - 3$	$70(t - 3)$

78. Write a word problem for the given table and equation. Explain how you created your problem.
Equation: $18(t - 2) = 5t + 9$

	Rate	Time	Distance $(d = r \cdot t)$
?	5	t	$5t$
?	18	$t - 2$	$18(t - 2)$

79. *Solutions Manual** Write a solutions manual page for the following problem:

The width of a rectangle is 3 inches less than twice its length. If the perimeter of the rectangle is 36 inches, find its length and width.

80. *Newsletter** Write a newsletter explaining how to solve consecutive integer problems.

*See Appendix B for details and sample answers.

2.4

Applications Involving Percents; Ratio and Proportion

Objectives

1 Use the basic percent equation to find an unknown amount, base, or percent.
2 Solve applied problems using the basic percent equation.
3 Solve applied problems involving percent increase or percent decrease.
4 Solve problems involving interest.
5 Solve mixture problems.
6 Solve for variables in proportions.
7 Solve applied problems involving proportions.

The Basic Percent Equation

Objective 1 Use the basic percent equation to find an unknown amount, base, or percent. We know that 40 is one-half of 80. Since the fraction $\frac{1}{2}$ is the same as 50%, we could also say that 40 is 50% of 80. In this example, the number 40 is referred to as the **amount** and the number 80 is referred to as the **base**. The basic equation relating these two quantities reflects that the amount is a percentage of the base, or

$$\text{Amount} = \text{Percent} \cdot \text{Base}$$

In this equation, it is important to express the percent as a decimal or a fraction rather than as a percent. Obviously, it is crucial to identify the amount, the percent, and the base correctly. It is a good idea to write the information in the following form:

$$\underset{\text{(Amount)}}{\underline{\hspace{2cm}}} \text{ is } \underset{\text{(Percent)}}{\underline{\hspace{2cm}}} \text{ % of } \underset{\text{(Base)}}{\underline{\hspace{2cm}}}$$

EXAMPLE 1   What number is 40% of 45?

Solution

Let n represent the unknown number. We can now write the information as

$$\underset{\text{(Amount)}}{\underline{\quad n \quad}} \text{ is } \underset{\text{(Percent)}}{\underline{\quad 40 \quad}} \text{ % of } \underset{\text{(Base)}}{\underline{\quad 45 \quad}}$$

We see that the amount is n, the percent is 40%, or 0.4, and the base is 45.

We now translate this information to an equation, making sure that the percent is written as a decimal rather than as a percent.

$$n = 0.4(45)$$
$$n = 18 \qquad \text{Multiply.}$$

We find that 18 is 40% of 45.

Quick Check 1
25% of 44 is what number?

> **A Word of Caution**   When using the basic percent equation, be sure to rewrite the percent as a decimal number or a fraction before solving for an unknown amount or base.

EXAMPLE ▸ 2 Eight percent of what number is 12?

Solution

Letting n represent the unknown number, we can rewrite this information as

$$\underbrace{12}_{\text{(Amount)}} \text{ is } \underbrace{8}_{\text{(Percent)}} \text{ \% of } \underbrace{n}_{\text{(Base)}}$$

The amount is 12, the percent is 8%, and the base is unknown. We must be sure to write the percent as a decimal before working with the equation.

$$12 = 0.08n$$
$$\frac{12}{0.08} = \frac{0.08n}{0.08} \qquad \text{Divide both sides by 0.08.}$$
$$150 = n \qquad \text{Simplify.}$$

Quick Check 2
39 is 60% of what number?

Eight percent of 150 is 12.

EXAMPLE ▸ 3 What percent of 56 is 35?

Solution

In this problem, the percent is unknown. If we let p represent the percent, then we can write this information as

$$\underbrace{35}_{\text{(Amount)}} \text{ is } \underbrace{p}_{\text{(Percent)}} \text{ \% of } \underbrace{56}_{\text{(Base)}}$$

We can now translate directly to the basic percent equation and solve it.

$$35 = p \cdot 56$$
$$\frac{35}{56} = \frac{p \cdot 56}{56} \qquad \text{Divide both sides by 56.}$$
$$\frac{5}{8} = p \qquad \text{Simplify.}$$

When the percent is unknown, we must convert our result from a fraction or a decimal to a percent by multiplying by 100%.

$$p = \frac{5}{8} \cdot 100\% \qquad \text{Multiply by 100\%.}$$
$$p = \frac{5}{\overset{}{\underset{2}{8}}} \cdot \overset{25}{100}\% \qquad \text{Divide out common factors.}$$
$$p = \frac{125}{2}\% \qquad \text{Simplify.}$$
$$p = 62\frac{1}{2}\% \qquad \text{Write as a mixed number.}$$

Quick Check 3
56 is what percent of 80?

35 is $62\frac{1}{2}\%$ of 56.

A Word of Caution After using the basic percent equation to solve for an unknown percent, be sure to rewrite the solution as a percent by multiplying by 100%.

When using the basic percent equation to solve for an unknown percent, keep in mind that the base is not necessarily the larger of the two given numbers. For example, if we are trying to determine what percent of 80 is 200, the base is the smaller number (80).

Applications

Objective 2 Solve applied problems using the basic percent equation. We will now use our basic percent equation to solve applied problems. We will continue to use the strategy for solving applied problems developed in Section 2.3.

EXAMPLE 4   There are 7107 female students at a certain community college. If 60% of all the students at the college are female, what is the total enrollment of the college?

Solution

We know that 60% of the students are female, so the number of female students (7107) is 60% of the total enrollment. If we let n represent the unknown total enrollment, then our information can be written as

$$\underline{\quad 7107 \quad} \text{ is } \underline{\quad 60 \quad} \text{ % of } \underline{\quad n \quad}$$
$$\text{(Amount)} \qquad \text{(Percent)} \qquad \text{(Base)}$$

We translate this sentence to an equation and solve.

$$7107 = 0.6n$$
$$11{,}845 = n \qquad \text{Divide both sides by 0.6.}$$

Quick Check 4
Of the 500 children who attend a particular elementary school, 28% buy their lunch at school. How many children buy their lunch at this school?

There are 11,845 students at the college.

Percent Increase and Percent Decrease

Objective 3 Solve applied problems involving percent increase or percent decrease. **Percent increase** is a measure of how much a quantity has increased from its original value. The amount of increase is the amount in the basic percent equation, while the original value is the base. **Percent decrease** is a similar measure of how much a quantity has decreased from its original value.

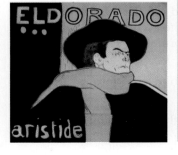

EXAMPLE 5   An art collector bought a lithograph for $2500. After three years, the lithograph was valued at $3800. Find the percent increase in the value of the lithograph over the three years.

Solution

The amount of increase in value is $1300, because $3800 - 2500 = 1300$. We need to find out what percent of the original value is the increase in value. If we let p represent the unknown percent, then we can write our information as

$$\underline{\quad 1300 \quad} \text{ is } \underline{\quad p \quad} \text{ % of } \underline{\quad 2500 \quad}$$
$$\text{(Amount)} \qquad \text{(Percent)} \qquad \text{(Base)}$$

This translates to the equation $1300 = p \cdot 2500$, which we can solve.

$$1300 = p \cdot 2500$$
$$0.52 = p \qquad \text{Divide both sides by 2500.}$$

Converting this decimal result to a percent, we see that the lithograph increased in value by 52%.

Quick Check 5
Lee bought a new car for $30,000 three years ago, and now it is worth $13,800. Find the percent decrease in the value of this car.

A Word of Caution When solving problems involving percent increase or percent decrease, the base is always the *original amount*.

Percents are involved in a great number of business applications. One such problem is determining the sale price of an item after a percent discount has been applied.

EXAMPLE 6 A department store is having a "20% off" sale, so all prices have been reduced by 20% of the original price. If a robe was originally priced at $37.50, what is the sale price after the 20% discount?

Solution

We begin by finding the amount by which the original price has been discounted. Let d represent the amount of the discount. Since the amount of the discount is 20% of the original price ($37.50), we can write our information as

$$\underset{\text{(Amount)}}{\underline{\quad d \quad}} \text{ is } \underset{\text{(Percent)}}{\underline{\quad 20 \quad}} \% \text{ of } \underset{\text{(Base)}}{\underline{\quad \$37.50 \quad}}.$$

This translates to the equation $d = 0.2(37.50)$, which we solve for d.

$$d = 0.2(37.50)$$
$$d = 7.50 \qquad \text{Multiply.}$$

Quick Check 6
A department store bought a shipment of CD players for $42 wholesale. If the store marks the CD players up by 45%, what is the selling price of a CD player?

The discount is $7.50, so the sale price is $37.50 − $7.50 or $30.00.

Interest

Objective 4 **Solve problems involving interest.** **Interest** is a fee paid by a borrower for the privilege of borrowing a sum of money. Banks also pay interest to customers who deposit money in their bank. Simple interest is calculated as a percentage of the **principal**, which is the amount borrowed or deposited. If r is the annual interest rate, then the simple interest (I) owed on a principal (P) is given by the formula $I = P \cdot r \cdot t$, where t is time in years.

If an investor deposited $3000 in an account that paid 4% annual interest for one year, we could determine how much money would be in the account at the end of one year by substituting 3000 for P, 0.04 for r, and 1 for t in the formula $I = P \cdot r \cdot t$.

$$I = P \cdot r \cdot t$$
$$I = 3000(0.04)(1) \qquad \text{Substitute 3000 for } P, 0.04 \text{ for } r, \text{ and 1 for } t.$$
$$I = 120 \qquad \text{Multiply.}$$

The interest earned in one year is $120. There is $3000 plus the $120 in interest, or $3120, in the account after 1 year.

EXAMPLE 7   Martha had some spare cash and decided to invest it in an account that paid 6% annual interest. Her friend Stuart found an account that paid 7% interest, and invested $5000 more in this account than Martha did in her account. If the pair earned a total of $1650 in interest from the two accounts in one year, how much did Martha invest? How much did Stuart invest?

Solution

If we let x represent the amount Martha invested at 6%, then the amount Stuart invested at 7% can be represented by $x + 5000$. The equation for this problem comes from the fact that the interest earned in each account must add up to $1650. We will use a table to display the important information for this problem.

Quick Check **7**

Mark invested money in two accounts. One account paid 3% annual interest, and the other account paid 5% annual interest. He invested $2000 more in the account that paid 5% interest than in the account that paid 3% interest. If Mark earned a total of $500 in interest from the two accounts in one year, how much did he invest in each account?

Account	Principal (P)	Rate (r)	Time (t)	Interest $I = P \cdot r \cdot t$
6%	x	0.06	1	$0.06x$
7%	$x + 5000$	0.07	1	$0.07(x + 5000)$
Total				1650

The interest from the first account is $0.06x$, and the interest from the second account is $0.07(x + 5000)$, so their sum must be $1650. Essentially, we can read this equation in the final column in the table.

$$
\begin{aligned}
0.06x + 0.07(x + 5000) &= 1650 \\
0.06x + 0.07x + 350 &= 1650 && \text{Distribute.} \\
0.13x + 350 &= 1650 && \text{Combine like terms.} \\
0.13x &= 1300 && \text{Subtract 350 from both sides.} \\
x &= 10{,}000 && \text{Divide both sides by 0.13.}
\end{aligned}
$$

Since $x = 10{,}000$, the amount Martha invested at 6% interest (x) is $10,000 and the amount Stuart invested at 7% interest $(x + 5000)$ is $15,000. The reader may verify that Martha earned $600 in interest and Stuart earned $1050 in interest, which totals $1650.

EXAMPLE 8   Donald loaned a total of $13,500 to two borrowers, Carolyn and George. Carolyn paid 8% annual interest, while George had to pay 12%. If Donald earned a total of $1320 in interest from the two borrowers in one year, how much did Carolyn borrow? How much did George borrow?

Solution

If we let x represent the amount that Carolyn borrowed, then the amount George borrowed can be represented by $13,500 - x$. (By subtracting the amount that Carolyn borrowed from $13,500, we are left with the amount that George borrowed.) The equation for this problem comes from the fact that the interest earned by Donald must add up to $1320. Again, we will use a table to display the important information for this problem.

Borrower	Principal (P)	Rate (r)	Time (t)	Interest $I = P \cdot r$
Carolyn	x	0.08	1	$0.08x$
George	$13,500 - x$	0.12	1	$0.12(13,500 - x)$
Total	13,500			1320

The interest paid by Carolyn is $0.08x$ and the interest paid by George is $0.12(13,500 - x)$, so their sum must be $1320. Again, we can read this equation in the final column in the table.

$$0.08x + 0.12(13,500 - x) = 1320$$
$$0.08x + 1620 - 0.12x = 1320 \qquad \text{Distribute.}$$
$$-0.04x + 1620 = 1320 \qquad \text{Combine like terms.}$$
$$-0.04x = -300 \qquad \text{Subtract 1620 from both sides.}$$
$$x = 7500 \qquad \text{Divide both sides by } -0.04.$$

Since $x = 7500$, the amount Carolyn borrowed is $7500 and the amount George borrowed $(13,500 - x)$ is $6000. The reader may verify that Carolyn paid $600 in interest and George paid $720 in interest, which totals $1320.

Quick Check **8**

Patsy invested a total of $1900 in two accounts. One account paid 4% annual interest, and the other account paid 5% annual interest. If Patsy earned a total of $86.50 in interest from the two accounts, how much did she invest in each account?

Mixture Problems

Objective 5 **Solve mixture problems.** The next example is a **mixture problem**. In this problem, we will be mixing two solutions that have different concentrations of alcohol to produce a mixture whose concentration of alcohol is somewhere in between the two individual solutions. We will determine how much of each solution should be used in order to produce a mixture of the desired specifications.

If two quantities total to some number V, and we let x represent one of the quantities, then the other quantity can be expressed as $V - x$. For example, if two numbers add up to 75 and we let x represent one of the other numbers, then the other number can be expressed as $75 - x$.

EXAMPLE 9   A chemist has two solutions. The first is 20% alcohol, and the second is 30% alcohol. How many milliliters (ml) of each solution should she use if she wants to make 40 ml of a solution that is 24% alcohol?

Solution

There are two unknown quantities in this problem: the amount of the 20% alcohol solution that she should use and the amount of the 30% alcohol solution that she should use. We will let x represent the amount of the 20% alcohol solution in ml. We need to express the amount of the 30% solution in terms of x as well. Since the two quantities must add up to 40 ml, the amount of the 30% solution can be represented by $40 - x$.

The equation that we will use to solve this problem is based on the amount of alcohol in each solution, as well as the amount of alcohol in the mixture. When we add the amount of alcohol that comes from the 20% solution to the amount of alcohol that comes from the 30% solution, it should equal the amount of alcohol in the mixture.

Since we want to end up with 40 ml of a solution that is 24% alcohol, this solution should contain $0.24(40) = 9.6$ ml of alcohol. The following table, which is quite similar to the one used in the previous example involving interest, presents all of the information:

Solution	Amount of Solution (ml)	% Alcohol	Amount of Alcohol (ml)
Solution 1 (20%)	x	0.2	$0.2x$
Solution 2 (30%)	$40 - x$	0.3	$0.3(40 - x)$
Mixture	40	0.24	$0.24(40) = 9.6$

Alcohol from solution 1 + Alcohol from solution 2 = Alcohol in mixture

$$0.2x \qquad + \qquad 0.3(40 - x) \qquad = \qquad 9.6$$

Now we solve the equation for x.

$$\begin{aligned}
0.2x + 0.3(40 - x) &= 9.6 \\
0.2x + 12 - 0.3x &= 9.6 & &\text{Distribute.} \\
-0.1x + 12 &= 9.6 & &\text{Combine like terms.} \\
-0.1x &= -2.4 & &\text{Subtract 12 from both sides.} \\
x &= 24 & &\text{Divide both sides by } -0.1.
\end{aligned}$$

Since x represented the amount of the 20% alcohol solution that the chemist should use, she should use 24 ml of the 20% solution. Subtracting this amount from 40 ml tells us that she should use 16 ml of the 30% alcohol solution.

> *Quick Check* 9
> **Marie had one solution that was 30% alcohol and a second solution that was 42% alcohol. How many milliliters of each solution should be mixed to make 80 milliliters of a solution that is 39% alcohol?**

Ratio and Proportion

A **ratio** is a comparison of two quantities using division and is usually written as a fraction. Suppose a first-grade student has 20 minutes to eat her lunch and this is followed by a 30-minute recess. The ratio of the time for eating to the time for recess is $\frac{20}{30}$, which can be simplified to $\frac{2}{3}$. This means that for every 2 minutes of eating time, there are 3 minutes of recess time.

Objective 6 Solve for variables in proportions. Ratios are important tools for solving some applied problems involving proportions. A **proportion** is a statement of equality between two ratios. In other words, a proportion is an equation in which two ratios are equal to each other. Here are a few examples of proportions:

$$\frac{5}{8} = \frac{n}{56} \qquad \frac{8}{11} = \frac{5}{n} \qquad \frac{9}{2} = \frac{4x + 5}{x}$$

The **cross products** of a proportion are the two products obtained when we multiply the numerator of one fraction by the denominator of the other fraction. The two cross products for the proportion $\frac{a}{b} = \frac{c}{d}$ are $a \cdot d$ and $b \cdot c$.

When we solve a proportion, we are looking for the value of the variable that produces a true statement. We will use the fact that the cross products of a proportion are equal if the proportion is indeed true.

$$\text{If } \frac{a}{b} = \frac{c}{d}, \text{ then } a \cdot d = b \cdot c.$$

Consider the two equal fractions $\frac{3}{6}$ and $\frac{5}{10}$. If we write these two fractions in the form of an equation $\frac{3}{6} = \frac{5}{10}$, the two cross products ($3 \cdot 10$ and $6 \cdot 5$) are both equal to 30. This holds true for any proportion.

To solve a proportion, we will multiply each numerator by the denominator on the other side of the equation, and we will set these products equal to each other. We then solve the resulting equation for the variable in the problem. This process is called **cross multiplying** and will be demonstrated in the next example.

EXAMPLE 10 Solve $\frac{3}{4} = \frac{n}{68}$.

Solution

We begin by cross multiplying. We multiply the numerator on the left by the denominator on the right $(3 \cdot 68)$, and we multiply the denominator on the left by the numerator on the right $(4 \cdot n)$.

Then we write an equation that states that the two products are equal to each other, and we solve this equation.

$$\frac{3}{4} = \frac{n}{68}$$

$3 \cdot 68 = 4 \cdot n$ Cross multiply.

$204 = 4n$ Multiply.

$51 = n$ Divide both sides by 4.

Quick Check 10
Solve $\frac{3}{17} = \frac{n}{187}$.

The solution set is $\{51\}$.

A Word of Caution When solving a proportion, we cross multiply. *Do not "cross cancel."*

Applications of Proportions

Objective 7 Solve applied problems involving proportions. Now we turn our attention to applied problems involving proportions.

EXAMPLE 11 Studies show that one out of every nine people is left-handed. In a sample of 216 people, how many would be left handed according to these studies?

Solution

We begin by setting up a ratio based on the known information. Since there is one person out of every nine who is left-handed, the ratio would be $\frac{1}{9}$. Notice that the numerator represents the number of left-handed people while the denominator represents the number of all people. When we set up a proportion it is crucial to keep this ordering consistent. This given ratio will be the left side of the proportion. To set up the right side of the proportion, we need to write a second ratio relating the unknown quantity to the given quantity. The unknown quantity would be the number of left-handed people in the group of 216 people. Let n represent the number of left-handed people. The proportion would then be $\frac{1}{9} = \frac{n}{216}$. Again, be very careful that the ratio on the left side $\left(\frac{\text{left -handed}}{\text{total}}\right)$ is consistent with the ratio on the right side. Thus,

$$\frac{1}{9} = \frac{n}{216}$$

$216 = 9n$ Cross multiply.

$24 = n$ Divide both sides by 9.

Quick Check 11
Studies show that 4 out of 5 dentists recommend sugarless gum for their patients that chew gum. In a group of 85 dentists, how many would recommend sugarless gum to their patients according to these studies?

According to these studies, there should be 24 left-handed people in the group of 216 people.

Building Your Study Strategy Using Your Textbook, 4 **Creating Notecards**
Your textbook can be used to create a series of note cards for studying new terms
and procedures or difficult problems. Each time you find a new term in the textbook, write the term on one side of a note card and its definition on the other
side. Collect all of these cards and cycle through them, reading the term and trying to recite the definition by memory. Check your definition on the other side of
the card. The same thing can be done for each new procedure introduced in the
textbook.

If there is a particular type of problem you are struggling with, find an example that has been worked out in the text. Write the problem on the front of the card
and write the complete solution on the back of the card. Collect all of the cards
containing difficult problems and select one at random. Try to solve the problem
and then check the solution given on the back of the card. Repeat this for the
other cards.

Some students only have trouble recalling what the first step is for solving
particular problems. If you are in this group, you can write a series of problems
on the front of some note cards and the first step on the back of the cards. You
can then cycle through these cards in the same way. It is easy to carry a set of note
cards with you. Anytime you have 5 to 15 free minutes, cycle through them.

EXERCISES 2.4

Vocabulary

1. State the basic percent equation.

2. In a percent problem, the _____ is a
percentage of the base.

3. Percent increase is a measure of how much a
quantity has increased from its _____.

4. Simple interest is calculated as a percentage of the
_____.

5. State the formula for calculating simple interest.

6. A(n) _____ is a comparison of two quantities
using division and is usually written as a fraction.

7. A(n) _____ is a statement of equality between
two ratios.

8. If $\frac{a}{b} = \frac{c}{d}$, then $a \cdot d =$ _____.

9. Fourteen percent of 50 is what
number?

10. What percent of 80 is 24?

11. Forty percent of what number is 56?

12. Seventeen is 4% of what number?

13. What percent of 190 is 133?

14. What number is 29% of 62?

15. What is $37\frac{1}{2}\%$ of 104?

16. Fifteen is what percent of 36?

17. What is 240% of 68?

18. What percent of 24 is 108?

19. Fifty-five is 125% of what number?

20. What is 600% of 53?

21. Thirty percent of the M&M's in a bowl are brown. If
there are 280 M&M's in the bowl, how many are
brown?

22. A doctor helped 320 women deliver their babies last
year. Eight of these women had twins. What percent
of the doctor's patients had twins?

23. Forty percent of the registered voters in a certain precinct are registered as Democrats. If 410 of the registered voters in the precinct are registered as Democrats, how many registered voters are there in the precinct?

24. An online retailer adds a 15% charge to all items for shipping and handling. If a shipping-and-handling fee of $96 is added to the cost of a computer, what was the original price of the computer?

25. A certain brand of rum is 40% alcohol. How many milliliters of alcohol are there in a 750-milliliter bottle of rum?

26. A certificate of deposit (CD) pays 3.08% interest. Find the amount of interest earned in one year on a $5000 CD.

27. Dylan bought a baseball card for $11.50. The card increased in value by 60% in one year. How much was the card worth after one year?

28. Kristy bought shares of a stock valued at $48.24. If the stock price soared to $66.33, find the percent increase in the value of the stock.

29. The price of a gallon of milk increased by $0.26. This represented a price increase of 8%. What was the original price of the milk?

30. A bookstore charges its customers 32% over the wholesale cost of the book. How much does the bookstore charge for a book that has a wholesale price of $64?

31. A store is discounting all of its prices by 15%. If a sweater's original price was $24.95, what is the price after the discount?

32. The average score for a class on test 1 was 88. The average score for test 2 was 77. What was the percent decrease in the average score from test 1 to test 2?

33. A community college has 1600 parking spots on campus. When it builds its new library, the college will lose 120 of its parking spots. What will the percent decrease in the number of available parking spots be?

34. A community college had its budget slashed by $3,101,000. This represents a cut of 7% of its total budget from last year. What was last year's budget?

35. Last year, the average SAT math score for students at a certain high school was 500. This year the average score decreased by 20%. By what percent do scores need to increase next year to bring the average score back up to 500? (*Hint:* The answer is not 20%.)

36. During the year 2006, an auto plant produced 25% fewer cars than it had in the year 2005. Production fell by another 20% during 2007. By what percent will production have to increase in 2008 to make the same number of cars as the plant produced in 2005?

37. Janet invested her savings in two accounts. One of the accounts paid 2% interest, and the other paid 5%. Janet put twice as much money in the account that paid 5% as she put in the account that paid 2%. If she earned a total of $84 in interest from the two accounts in one year, find the amount invested in each account.

38. One of Jaleel's mutual funds made a 10% profit last year, while the other mutual fund made an 8% profit. Jaleel had invested $1400 more in the fund that made an 8% profit than he did in the fund that made a 10% profit. If he made $45 more from the 8% fund than he did from the 10% fund, how much was invested in each fund?

39. Kamiran deposits a total of $6500 in savings accounts at two different banks. The first bank pays 4% interest, while the second bank pays 5% interest. If he earns a total of $300 in interest in one year, how much was deposited at each bank?

40. Dianne was given $35,000 when she retired. She invested some at 7% interest and the rest at 9% interest. If she earned $2910 in interest in one year, how much was invested in each account?

41. A mechanic has two antifreeze solutions. One is 70% antifreeze and the other is 40% antifreeze. How much of each solution should be mixed to make 60 gallons of a solution that is 52% antifreeze?

42. A chemist has two saline solutions. One is 2% salt and the other is 6% salt. How much of each solution should be mixed to make 2 liters of a saline solution that is 3% salt?

43. A dairyman has some milk that is 5% butterfat, as well as some lowfat milk that is 2% butterfat. How much of each needs to be mixed together to make 1000 gallons of milk that is 3.2% butterfat?

44. Patti has two solutions. One is 27% acid and the other is 39% acid. How much of each solution should be mixed together to make 9 liters of a solution that is 29% acid?

45. Paula has 400 milliliters of a solution that is 80% alcohol. She plans to dilute it so that it is only 50% alcohol. How much water does she have to add in order to do this?

46. A bartender has rum that is 40% alcohol. How much rum and how much cola need to be mixed together to make 5 liters of rum-and-cola that is 16% alcohol?

Solve the proportion.

47. $\dfrac{2}{3} = \dfrac{n}{48}$

48. $\dfrac{5}{8} = \dfrac{n}{32}$

49. $\dfrac{20}{n} = \dfrac{45}{81}$

50. $\dfrac{22}{30} = \dfrac{55}{n}$

51. $\dfrac{2}{5} = \dfrac{n}{126}$

52. $\dfrac{3}{8} = \dfrac{n}{200}$

53. $\dfrac{7}{6} = \dfrac{n}{75}$

54. $\dfrac{3}{100} = \dfrac{n}{412}$

55. $\dfrac{3}{4} = \dfrac{n + 12}{28}$

56. $\dfrac{5}{9} = \dfrac{3n - 2}{18}$

57. $\dfrac{2n + 15}{18} = \dfrac{5n + 13}{24}$

58. $\dfrac{n + 10}{40} = \dfrac{1 - n}{48}$

Set up a proportion and solve it for the following problems.

59. At a certain community college, three out of every five students are female. If the college has 9200 students, how many are female?

60. A day care center has a policy that there will be at least 4 staff members present for every 18 children. How many staff members are necessary to accommodate 63 children?

61. The directions for a powdered plant food state that 3 tablespoons of the food need to be mixed with 2 gallons of water. How many tablespoons need to be mixed with 15 gallons of water?

62. Eight out of every nine people are right-handed. If a factory has 352 right-handed workers, how many left-handed workers are there at the factory?

63. If 5 out of every 6 teachers in a city meet the minimum qualifications to be teaching and the city has 864 teachers, how many do not meet the minimum qualifications?

64. A survey showed that 7 out of every 10 registered voters in a county are in favor of a bond measure to raise money for a new college campus. If there are 140,000 registered voters in the county, how many are not in favor of the bond measure?

65. Two hundred people were asked about their habits when traveling internationally. The following pie chart shows the responses by category:

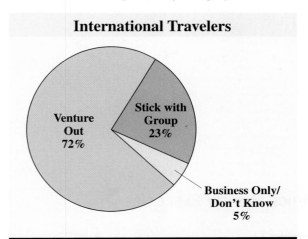

International Travelers

Venture Out 72%

Stick with Group 23%

Business Only/ Don't Know 5%

(*Source:* Harris Interactive Survey for InterContinental Hotels & Resorts)

Approximately how many of these people venture out on their own?

66. In 2002, 5524 people were killed in work-related incidents in the United States. The following pie chart shows the deaths by category:

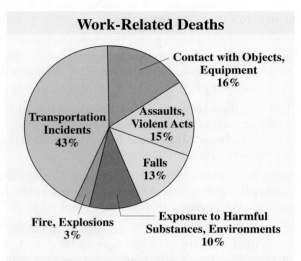

Work-Related Deaths

Contact with Objects, Equipment 16%

Assaults, Violent Acts 15%

Transportation Incidents 43%

Falls 13%

Fire, Explosions 3%

Exposure to Harmful Substances, Environments 10%

(*Source:* Bureau of Labor Statistics, Census of Fatal Occupational Injuries)

Approximately how many deaths were due to transportation incidents?

Writing in Mathematics

Answer in complete sentences.

67. Write a word problem for the following table and equation, and explain how you created your problem.

Principal (P)	Rate (r)	Time (t)	Interest $I = P \cdot r \cdot t$
x	0.03	1	$0.03x$
$x + 10{,}000$	0.05	1	$0.05(x + 10{,}000)$
Total			2100

Equation: $0.03x + 0.05(x + 10{,}000) = 2100$

68. Write a word problem for the following table and equation, and explain how you created your problem.

Solution	Amount of Solution (ml)	% Alcohol	Amount of Alcohol (ml)
Solution 1	x	0.38	$0.38x$
Solution 2	$60 - x$	0.5	$0.5(60 - x)$
Mixture	60	0.4	24

Equation: $0.38x + 0.5(60 - x) = 24$

69. *Solutions Manual** Write a solutions manual page for the following problem:

Approximately 59% of all master's degrees awarded in 2004 were earned by female students. If approximately 560,000 students earned master's degrees, how many were female?

70. *Newsletter** Write a newsletter explaining how to solve percent increase problems.

*See Appendix B for details and sample answers.

QUICK REVIEW EXERCISES

Section 2.4

Solve.

1. $4x + 17 = 41$

2. $5x - 19 = 3x - 51$

3. $2x + 2(x + 8) = 68$

4. $x + 3(x + 2) = 2(x + 4) + 30$

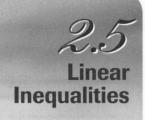

2.5

Linear Inequalities

Objectives

1 Present the solutions of an inequality on a number line.
2 Present the solutions of an inequality using interval notation.
3 Solve linear inequalities.
4 Solve compound inequalities involving the union of two linear inequalities.
5 Solve compound inequalities involving the intersection of two linear inequalities.
6 Solve applied problems using linear inequalities.

Suppose that you went to a doctor and she told you that your temperature was normal. This would mean that your temperature was 98.6° F. However, if the doctor told you that you had a fever, could you tell what your temperature was? No, all you would know is that it was above 98.6° F. If we let the variable t represent your temperature, this relationship could be written as $t > 98.6$. The expression $t > 98.6$ is an example of an inequality. An inequality is a mathematical statement comparing two quantities using the symbols $<$ (*less than*) or $>$ (*greater than*). It states that one quantity is smaller than (or larger than) the other quantity.

Two other symbols that may be used in an inequality are $\leq$ and $\geq$. The symbol $\leq$ is read as *"less than or equal to,"* and the symbol $\geq$ is read as *"greater than or equal to."* The inequality $x \leq 7$ is used to represent all real numbers that are either less than 7 or equal to 7. Inequalities involving the symbols $\leq$ or $\geq$ are often called **weak inequalities**, while inequalities involving the symbols $<$ or $>$ are called **strict inequalities** because they involve numbers that are strictly less than (or greater than) a given value.

Presenting Solutions of Inequalities Using a Number Line

Objective **1** Present the solutions of an inequality on a number line.

Linear Inequality

A **linear inequality** is an inequality containing linear expressions.

A linear inequality often has infinitely many solutions, and it is not possible to list every solution. Consider the inequality $x < 3$. There are infinitely many solutions of this inequality, but the one thing that all of the solutions share is that they are all to the left of 3 on the number line. Values located to the right of 3 on the number line are greater than 3. All shaded numbers to the left are solutions of the inequality and can be represented on a number line as follows:

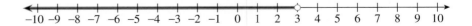

An **endpoint** of an inequality is a point on a number line that separates values that are solutions from values that are not. The open circle at $x = 3$ tells us that the endpoint is not included in the solution, because 3 is not less than 3. If the inequality were $x \leq 3$, then we would fill in the circle at $x = 3$.

Here are three more inequalities, with their solutions graphed on a number line:

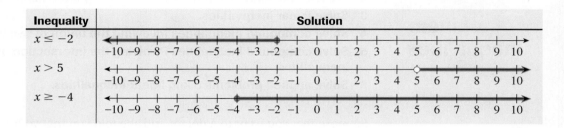

Inequality	Solution
$x \leq -2$	
$x > 5$	
$x \geq -4$	

Saying that a is less than b is equivalent to saying that b is greater than a. In other words, the inequality $a < b$ is equivalent to the inequality $b > a$. This is an important piece of information, as we will want to rewrite an inequality so that the variable is on the left side of the inequality before graphing it on a number line.

EXAMPLE ▸ 1   Graph the solutions of the inequality $1 > x$ on a number line.

Solution

Before we graph the solutions, we should rewrite the inequality so that the variable is on the left side in order to avoid confusion about which direction should be shaded on the number line. Saying that 1 is greater than x is the same as saying that x is less than 1, or $x < 1$. The values that are less than 1 are to the left of 1 on the number line.

Quick Check ◂ 1
Graph the solutions of the inequality $8 < x$ on a number line.

Interval Notation

Objective 2 **Present the solutions of an inequality using interval notation.**
Another way to present the solutions of an inequality is by using **interval notation**. A range of values on a number line, such as the solutions of the inequality $x \geq 4$, is called an **interval**. Inequalities typically have one or more intervals as their solutions. Interval notation presents an interval that is a solution by listing its left and right endpoints. We use parentheses around the endpoints if the endpoints are not included as solutions, and we use brackets if the endpoints are included as solutions.

When an interval continues on indefinitely to the right on a number line, we will use the symbol ∞ (infinity) in the place of the right endpoint and follow it with a parenthesis. Let's look again at the number line associated with the solutions of the inequality $x \geq -4$.

The solutions begin at -4 and include any number that is greater than -4. The interval is bounded by -4 on the left side and extends without bound on the right side, which can be written in interval notation as $[-4, \infty)$. The endpoint -4 is included in the interval, so we write a square bracket in front of it, because it is a solution, while we will always write a parenthesis after ∞, as it is not an actual number that ends the interval at that point.

If the inequality had been $x > -4$, instead of $x \geq -4$, then we would have written a parenthesis rather than a square bracket before -4. In other words, the solutions of the inequality $x > -4$ can be expressed in interval notation as $(-4, \infty)$.

For intervals that continue indefinitely to the left on a number line, we will use $-\infty$ in the place of the left endpoint.

Here is a summary of different inequalities with the solutions presented on a number line and in interval notation:

Inequality	Number Line	Interval Notation
$x > 4$		$(4, \infty)$
$x \geq 4$		$[4, \infty)$
$x < 4$		$(-\infty, 4)$
$x \leq 4$		$(-\infty, 4]$

Solving Linear Inequalities

Objective 3   Solve linear inequalities. Solving a linear inequality, such as $7x - 11 \leq -32$, is very similar to solving a linear equation. In fact, there is only one difference between the two procedures:

> Whenever we multiply both sides of an inequality by a negative number or divide both sides by a negative number, the direction of the inequality changes.

Why is this? Consider the inequality $3 < 5$, which is a true statement. If we multiply each side by -2, is the left side $(-2 \cdot 3 = -6)$ still less than the right side $(-2 \cdot 5 = -10)$? No, $-6 > -10$ because -6 is located to the right of -10 on the number line. Changing the direction of the inequality after multiplying (or dividing) by a negative number produces an inequality that is still a true statement.

EXAMPLE 2   Solve the inequality $7x - 11 \leq -32$. Present your solutions on a number line and using interval notation.

Solution

Think about the steps that you would take to solve the equation $7x - 11 = -32$. The same steps will be followed in solving this inequality.

$$7x - 11 \le -32$$
$$7x \le -21 \quad \text{Add 11 to both sides.}$$
$$x \le -3 \quad \text{Divide both sides by 7.}$$

Now we present our solution on a number line.

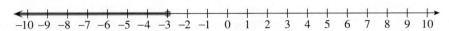

This can be expressed in interval notation as $(-\infty, -3]$.

Quick Check **2** Solve the inequality $4x + 3 < 31$. **Present your solutions on a number line and using interval notation.**

EXAMPLE **3**   Solve the inequality $-2x - 5 > 9$. Present your solutions on a number line and using interval notation.

Solution

We begin by solving the inequality.

$$-2x - 5 > 9$$
$$-2x > 14 \quad \text{Add 5 to both sides.}$$
$$\frac{-2x}{-2} < \frac{14}{-2} \quad \text{Divide both sides by } -2 \text{ to isolate } x. \text{ Notice that the direction of the inequality must change because we are dividing by a negative number.}$$
$$x < -7 \quad \text{Simplify.}$$

Now we present our solutions.

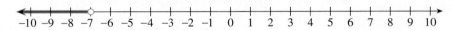

In interval notation this is $(-\infty, -7)$.

Quick Check **3** Solve the inequality $7 - 4x \le -13$. **Present your solutions on a number line and using interval notation.**

EXAMPLE **4**   Solve the inequality $5x - 12 > 8 - 3x$. Present your solutions on a number line and using interval notation.

Solution

We need to collect all variable terms on one side of the inequality and all constant terms on the other side.

$$5x - 12 > 8 - 3x$$
$$8x - 12 > 8 \quad \text{Add } 3x \text{ to both sides.}$$
$$8x > 20 \quad \text{Add 12 to both sides.}$$
$$x > \frac{5}{2} \quad \text{Divide both sides by 8 and simplify.}$$

Here is our solution on a number line:

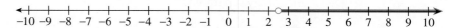

In interval notation, this is written as $\left(\frac{5}{2}, \infty\right)$.

Quick Check **4** Solve the inequality $3x + 8 < x + 2$. Present your solutions on a number line and using interval notation.

EXAMPLE ▶ 5   Solve the inequality $3(7 - 2x) + 6x \leq 5x - 4$. Present your solutions on a number line and using interval notation.

Solution

As with equations, we begin by simplifying each side completely.

$$3(7 - 2x) + 6x \leq 5x - 4$$
$$21 - 6x + 6x \leq 5x - 4 \qquad \text{Distribute.}$$
$$21 \leq 5x - 4 \qquad \text{Combine like terms.}$$
$$25 \leq 5x \qquad \text{Add 4 to both sides.}$$
$$5 \leq x \qquad \text{Divide both sides by 5.}$$

We can rewrite this solution as $x \geq 5$, which will help us when we display our solutions on a number line.

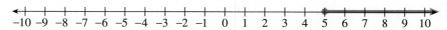

This can be expressed in interval notation as $[5, \infty)$.

Quick Check **5** Solve the inequality $(11x - 3) - (2x - 13) \geq 4(2x + 1)$. Present your solutions on a number line and using interval notation.

Compound Inequalities

Objective **4** **Solve compound inequalities involving the union of two linear inequalities.** A **compound inequality** is made up of two or more individual inequalities. One type of compound inequality involves the word "or." In this type of inequality, we are looking for real numbers that are solutions of one inequality or the other. For example, the solutions of the compound inequality $x < 2$ or $x > 4$ are real numbers that are either less than 2 or greater than 4. The solutions of this compound inequality can be displayed on a number line as follows:

To express the solutions using interval notation, we write both intervals with the symbol for **union** ($\cup$) between them. The interval notation for $x < 2$ or $x > 4$ is $(-\infty, 2) \cup (4, \infty)$.

To solve a compound inequality involving "or," we simply solve each individual inequality separately.

EXAMPLE ▸6   Solve $9x + 20 < -34$ or $5 + 4x \geq 19$.

Solution

We must solve each inequality separately. We begin with $9x + 20 < -34$.

$$9x + 20 < -34$$
$$9x < -54 \qquad \text{Subtract 20 from both sides.}$$
$$x < -6 \qquad \text{Divide both sides by 9.}$$

Now we solve the second inequality, $5 + 4x \geq 19$.

$$5 + 4x \geq 19$$
$$4x \geq 14 \qquad \text{Subtract 5 from both sides.}$$
$$x \geq \frac{7}{2} \qquad \text{Divide both sides by 4 and simplify.}$$

We combine our solutions on a single number line.

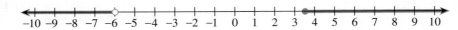

Our solutions can be expressed in interval notation as $(-\infty, -6) \cup [\frac{7}{2}, \infty)$.

Quick Check   **6**   Solve $3x - 2 \leq 10$ or $4x - 13 \geq 15$.

Objective 5 **Solve compound inequalities involving the intersection of two linear inequalities.** Another type of compound inequality involves the word "and," and the solutions of this compound inequality must be solutions of each individual inequality. For example, the solutions of the inequality $x > -5$ and $x < 1$ are real numbers that are, at the same time, greater than -5 and less than 1. This type of inequality is generally written in the condensed form $-5 < x < 1$. You can think of the solutions as being between -5 and 1.

To solve a compound inequality of this type, we need to isolate the variable in the middle part of the inequality between two real numbers. A key aspect of our approach will be to work on all three parts of the inequality at once.

EXAMPLE ▸7   Solve $-5 \leq 2x - 1 < 9$.

Solution

We are trying to isolate x in the middle of this inequality. The first step will be to add 1 to all three parts of this inequality, after which we will divide by 2.

$$-5 \leq 2x - 1 < 9$$
$$-4 \leq 2x < 10 \qquad \text{Add 1 to each part of the inequality.}$$
$$-2 \leq x < 5 \qquad \text{Divide each part of the inequality by 2.}$$

The solutions are presented on the following number line:

$$\xleftarrow{\qquad} \begin{array}{ccccccccccccccccccccc} & & & & & & & & & & & & & & & & & & & \\ -10 & -9 & -8 & -7 & -6 & -5 & -4 & -3 & -2 & -1 & 0 & 1 & 2 & 3 & 4 & 5 & 6 & 7 & 8 & 9 & 10 \end{array} \xrightarrow{\qquad}$$

These solutions can be expressed in interval notation as $[-2, 5)$.

Quick Check   7 Solve $-11 < 4x + 1 < 7$.

Applications

Objective 6   Solve applied problems using linear inequalities. Many applied problems involve inequalities rather than equations. Here are some key phrases and their translations into inequalities:

Key Phrases for Inequalities

x is greater than a x is more than a x is higher than a x is above a	$x > a$
x is at least a x is a or higher	$x \geq a$
x is less than a x is lower than a x is below a	$x < a$
x is at most a x is a or lower	$x \leq a$
x is between a and b, exclusive x is more than a but less than b	$a < x < b$
x is between a and b, inclusive x is at least a but no more than b	$a \leq x \leq b$

EXAMPLE 8   A sign next to an amusement park ride says that you must be at least 48″ tall in order to get on the ride. Set up an inequality that shows the heights of people who are able to get on the ride.

Solution

Let h represent the height of a person. If a person must be at least 48″ tall, then his or her height must be 48″ or above. In other words, the person's height must be greater than or equal to 48″. The inequality is $h \geq 48$.

A Word of Caution   The phrase "at least" means "greater than or equal to" ($\geq$) and does not mean "less than."

EXAMPLE ▸ 9   An instructor tells her students that they need a score of 70 or higher out of a possible 100 on the final exam to pass. Set up an inequality that shows the scores that are not passing.

Solution

Let *s* represent a student's score. Since a score of 70 or higher will pass, any score lower than 70 will not pass. This inequality can be written as $s < 70$. If we happen to know that the lowest possible score is 0, we could also write the inequality as $0 \leq s < 70$.

Quick Check   8 ▸ The Brainiac Club has a bylaw that states a person must have an IQ of at least 130 in order to be admitted to the club. Set up an inequality that shows the IQs of people who are unable to join the club.

We will now set up and solve applied problems involving inequalities.

EXAMPLE ▸ 10   If a student averages 90 or higher on the five tests given in a math class, then the student will earn a grade of A. Sean's scores on the first four tests are 82, 87, 93, and 92. What score on the fifth test will give Sean an A?

Solution

To find the average of five test scores, we add the five scores and divide by 5. Let *x* represent the score of the fifth test. The average can then be expressed as $\frac{82 + 87 + 93 + 92 + x}{5}$. We are interested in which scores on the fifth test give an average that is 90 or higher.

$$\frac{82 + 87 + 93 + 92 + x}{5} \geq 90$$

$$\frac{354 + x}{5} \geq 90 \qquad \text{Simplify the numerator.}$$

$$5 \cdot \frac{354 + x}{5} \geq 5 \cdot 90 \qquad \text{Multiply both sides by 5 to clear the fraction.}$$

$$354 + x \geq 450 \qquad \text{Simplify.}$$

$$x \geq 96 \qquad \text{Subtract 354 from both sides.}$$

Sean must score at least 96 on the fifth test to earn an A.

Quick Check   9 ▸ If a student averages 70 or lower on the four tests given in a math class, then the student will fail the class. Bobby's scores on the first three tests are 74, 78, and 80. What scores on the fourth test will result in Bobby failing the class?

Building Your Study Strategy **Using Your Textbook, 5 A Word of Caution/ Using Your Calculator** Two features of this textbook that you may find helpful are labeled A Word of Caution and Using Your Calculator. Each Word of Caution box warns you about common errors for certain problems, and tells you how to avoid repeating the same mistake. After a section is covered in class, look through the section for this feature before attempting the homework exercises.

Using Your Calculator shows how you can use the Texas Instruments TI-84 calculator to help with selected examples in the text. If you own one of these calculators, look for this feature before beginning your homework. If you own a different calculator, you should be able to interpret the directions for your calculator.

EXERCISES 2.5

Vocabulary

1. A(n) _____ is an inequality containing linear expressions.

2. When graphing an inequality, we use a(n) _____ circle to indicate an endpoint that is not included as a solution.

3. We can express the solutions of an inequality using either a number line or _____ notation.

4. When we are solving a linear inequality, we change the direction of the inequality whenever _____.

5. An inequality that is composed of two or more individual inequalities is called a(n) _____ inequality.

6. Which inequality can be associated with the statement *The person's height is at least 80 inches?*
a) $x \leq 80$ **b)** $x < 80$ **c)** $x \geq 80$ **d)** $x > 80$

Graph each inequality on a number line, and write it in interval notation.

7. $x < 3$

8. $x > -6$

9. $x \geq -1$

10. $x \leq 13$

11. $-2 < x < 8$

12. $3 < x \leq 12$

13. $x > \dfrac{9}{2}$

14. $x \geq -\dfrac{7}{3}$

15. $x \leq 2.4$

16. $x > 5.5$

17. $x > 8$ or $x \leq 2$

18. $x < -5$ or $x > 0$

19. $x < \dfrac{3}{2}$ or $x > \dfrac{13}{4}$

20. $-5 \leq x \leq 5$

Write an inequality associated with the given graph or interval notation.

21.

```
 <---+---+---+---+---+---+---+---+---+---+---+---o===+===+===+===+===>
    -10 -9 -8 -7 -6 -5 -4 -3 -2 -1  0  1  2  3  4  5  6  7  8  9 10
```

22.

```
 <===+===+===+===+===+===+===+===•---+---+---+---+---+---+---+---+--->
    -10 -9 -8 -7 -6 -5 -4 -3 -2 -1  0  1  2  3  4  5  6  7  8  9 10
```

23.

```
 <---+---+---+---+---+---•===+===+===+===+===+===+===+===•---+---+--->
    -10 -9 -8 -7 -6 -5 -4 -3 -2 -1  0  1  2  3  4  5  6  7  8  9 10
```

24.

```
 <===+===+===+===+===+===o---+---+---+---+---+---+---+---o===+===+===>
    -10 -9 -8 -7 -6 -5 -4 -3 -2 -1  0  1  2  3  4  5  6  7  8  9 10
```

25. $(-\infty, 2) \cup (9, \infty)$

26. $(-3, -2)$ **27.** $(-\infty, -4)$

28. $[-8, \infty)$

Solve. Graph your solution on a number line, and express it in interval notation.

29. $x + 9 < 5$

30. $x - 7 > 3$

31. $2x > 12$

32. $4x \geq -20$

33. $-5x > 15$

34. $-8x < -24$

35. $3x \geq -10.5$

36. $2x \leq 9.4$

37. $3x + 2 < 14$

38. $2x - 7 > 11$

39. $5 + 13x \leq 31$

40. $-3 + 8x \geq -37$

41. $-2x + 9 \leq 29$

42. $7 - 4x > 19$

43. $8 < 3x - 13$

44. $1 \geq 5x + 16$

45. $\dfrac{2}{3}x > \dfrac{10}{9}$

46. $\dfrac{5}{9}x - \dfrac{4}{3} < 7$

47. $7(x + 4) \leq -21$

48. $3(x - 5) > 18$

49. $4(2x - 7) + 10 < 38$

50. $5(x - 3) - 1 \geq -31$

51. $5x + 13 > 2x - 11$

52. $-2x + 51 < 9 - 8x$

53. $5(x + 3) + 2 > 3(x + 4) - 6$

54. $5(2x - 3) - 3 < 2(2x + 9) - 4$

55. $2x < -8$ or $x + 7 > 5$

56. $x - 3 \leq -5$ or $6x > -6$

57. $9x - 7 < -25$ or $3x + 11 > 25$

58. $4x - 9 < -17$ or $4x - 9 > 17$

59. $-2x + 17 < 5$ or $10 - x > 7$

60. $-4x + 13 \leq 9$ or $6 - 5x \geq 21$

61. $\dfrac{1}{4}x + \dfrac{7}{24} \leq \dfrac{2}{3}$ or $\dfrac{1}{4}x + \dfrac{9}{20} \geq \dfrac{6}{5}$

62. $\dfrac{1}{8}x - \dfrac{1}{5} < \dfrac{1}{12}x - \dfrac{3}{40}$ or $\dfrac{1}{2}x - \dfrac{3}{20} > \dfrac{1}{4}x + \dfrac{11}{10}$

63. $-4 < x - 3 < 1$

64. $2 \leq x + 7 \leq 5$

65. $-12 \leq 8x \leq 18$

66. $-6 < 2x < 0$

67. $-2 < 5x - 7 < 28$

68. $18 \leq 4x - 10 \leq 50$

69. $6 < 3x + 15 \leq 33$

70. $-12 \leq 4x - 18 < 7$

71. $\dfrac{1}{5} \leq \dfrac{1}{2}x - \dfrac{1}{3} \leq \dfrac{7}{4}$

72. $-\dfrac{3}{10} < \dfrac{1}{5}x - \dfrac{1}{2} < \dfrac{3}{5}$

73. Explain, in your own words, why the inequality $8 < 9x + 7 < 3$ has no solutions.

74. Explain, in your own words, why solving the compound inequality $x + 7 < 9$ or $x + 7 < 13$ is the same as solving the inequality $x + 7 < 13$ only.

75. Adrian needs to score at least 2 goals in the final game of the season to set a new scoring record. Write an inequality that shows the number of goals that will set a new record.

76. If Eddie sells fewer than seven cars this week, he will be fired. Write an inequality that shows the number of cars that will make Eddie unemployed.

77. A certain type of shrub is tolerant to 30° F, which means that it can survive at temperatures down to 30° F. Write an inequality that shows the temperatures at which the shrub cannot survive.

78. If Chad scores below 40% on the final exam, he will not pass the class. Write an inequality that shows the scores at which Chad will pass the class.

79. The manager of a winery tells the owner that he has lost at least 12,500 grapevines due to Pierce's disease, spread by an insect called a glassy-winged sharp-shooter. Write an inequality for the number of grapevines that have been lost.

80. A community college must have 11,200 or more students this semester in order to be eligible for growth funding from the state. Write an inequality that shows the number of students that makes the college eligible for growth funding.

81. John is paid to get signatures on petitions. He gets paid 10 cents for each signature. How many signatures does he need to gather today in order to earn at least $30?

82. Carlo attends a charity wine-tasting festival. There is a $10 admission fee. In addition, there is a $4 charge for each variety tasted. If Carlo brings $40 with him, how many varieties can he taste?

83. Angelica is going out of town on business. Her company will reimburse her up to $85 for a rental car. She rents a car for $19.95 plus 7 cents per mile. How many miles can Angelica drive without going over the amount that her company will reimburse?

84. An elementary school's booster club is holding a fund-raiser by selling cookie dough. Each tub of cookie dough sells for $11, and the booster club gets to keep half of the proceeds. If the booster club wants to raise at least $12,000, how many tubs of cookie dough do they need to sell?

85. Charles told his son Harry that he should always give at least 15% of his earnings to charity. If Harry earned $37,500 last year, how much should he have given to charity?

86. Karina is looking for a formal dress to wear to a ball, and she has $400 to spend. If the store charges 7% sales tax on each dress, what price range should Karina be considering?

87. Students in a real-estate class must have an average score of at least 80% on their exams in order to pass. If Jacqui has scored 92, 93, 85, and 96 on the first four exams, what scores on the fifth exam would allow her to pass the course? (Assume the highest possible score is 100.)

88. Students with an average test score below 70 after the third test will be sent an Early Alert warning. Robert scored 62 and 59 on the first two tests. What scores on the third exam will save Robert from receiving an Early Alert warning? (Assume the highest possible score is 100.)

Writing in Mathematics

Answer in complete sentences.

89. Explain, in your own words, why we must change the direction of an inequality when we divide both sides of that inequality by a negative number.

90. *Solutions Manual*** Write a solutions manual page for the following problem:

    Solve. $-10 < 3x + 2 \le 29$

91. *Newsletter** Write a newsletter explaining how to solve noncompound linear inequalities.

*See Appendix B for details and sample answers.

Chapter 2 Summary

Section 2.1—Topic	Chapter Review Exercises
Determining Whether a Value Is a Solution	1–2
Solving Equations Using the Multiplication and Addition Properties of Equality	3–6
Solving Applied Problems Involving Linear Equations	7–8

Section 2.2—Topic	Chapter Review Exercises
Solving Linear Equations	9–18
Solving Literal Equations	19–22

Section 2.3—Topic	Chapter Review Exercises
Solving Applied Problems Involving Linear Equations	23–27

Section 2.4—Topic	Chapter Review Exercises
Solving Percent Equations	28–30
Solving Applied Problems Involving Percents	31–35
Solving Proportions	36–39
Solving Applied Problems Involving Proportions	40–41

Section 2.5—Topic	Chapter Review Exercises
Solving Linear Inequalities	42–45
Solving Compound Linear Inequalities	46–49
Solving Applied Problems Involving Inequalities	50–52

Summary of Chapter 2 Study Strategies

Using your textbook as more than a source of the homework exercises and answers to the odd problems will help you to learn mathematics.

- Read the section the night before your instructor covers it to familiarize yourself with the material. If you find a step in one of the examples that you do not understand, then you can prepare questions for your instructor in class.
- Reread the section after it is covered in class. After you read through an example, attempt the Quick Check exercise that follows it. The answers to these exercises can be found at the back of the textbook. As you work through the homework exercises, refer back to the examples found in that section for assistance.
- Use the textbook to create a series of note cards for each section. Note cards can be used to memorize new terms or procedures and can help you remember how to solve particular problems.
- In each section, look for the feature labeled A Word of Caution. These will point out common mistakes and offer advice to help you avoid making the same mistake yourself.
- Finally, the feature labeled Using Your Calculator will show you how to use the TI-84 calculator to help with the problems in that section.

Determine whether the given value is a solution of the equation. [2.1]

1. $x = 7, 4x + 9 = 19$

2. $a = -9, (4a - 15) - (2a + 7) = -2(11 - a)$

Solve using the multiplication property of equality. [2.1]

3. $2x = 32$ **4.** $-9x = 54$

Solve using the addition property of equality. [2.1]

5. $x - 3 = -9$ **6.** $a + 14 = 8$

For each of the following problems, set up a linear equation and solve it. [2.1]

7. A local club charged an $8 cover charge to see a "Battle of the Bands." After paying the winning band a prize of $250, the club made a profit of $694. How many people came to see the "Battle of the Bands?"

8. Thanks to proper diet and exercise, Randy's cholesterol dropped by 23 points in the last two months. If his new cholesterol level is 189 points, what was his cholesterol level two months ago?

Solve. [2.2]

9. $5x - 8 = 27$ **10.** $-11 = 3a + 16$

11. $-4x + 15 = 7$ **12.** $51 - 2x = -21$

13. $\dfrac{x}{6} - \dfrac{5}{12} = \dfrac{5}{4}$ **14.** $\dfrac{1}{5}x - \dfrac{1}{2} = \dfrac{5}{4}$

15. $4x - 19 = 5x + 14$

16. $2m + 23 = 17 - 8m$

17. $2n - 11 + 3n + 4 = 7n - 39$

18. $3(2x - 7) - (x - 9) = 2x - 33$

Solve the following literal equations for the specified variable. [2.2]

19. $7x + y = 8$ for y

20. $15x + 7y = 30$ for y

21. $C = \pi \cdot d$ for d

22. $P = 3a + b + 2c$ for a

23. One number is 26 more than another number. If the sum of the two numbers is 98, find the two numbers. [2.3]

Worked-out solutions to Review Exercises marked with ⬤ can be found on page AN–5.

24. The width of a rectangle is 7 feet less than its length, and the perimeter is 50 feet. Find the length and the width of the rectangle. [2.3]

25. The sum of three consecutive even integers is 204. Find the three even integers. [2.3]

26. If Jason drives at a speed of 72 miles per hour, how long will it take him to drive 342 miles? [2.3]

27. Alycia's piggy bank has dimes and quarters in it. The number of dimes is four less than three times the number of quarters. If the total value of the coins is $7.85, how many dimes are in the piggy bank? [2.3]

28. Twenty percent of 95 is what number? [2.4]

29. What percent of 136 is 51? [2.4]

30. Thirty-two percent of what number is 1360? [2.4]

31. Fifty-five percent of the students at a college are female. If the college has 10,520 students, how many are female? [2.4]

32. In a group of 320 college graduates, 204 decided to pursue a master's degree. What percent of these students decided to pursue a master's degree? [2.4]

33. Jerry's bank account earned $553.50 in interest last year. This account pays 4.5% interest. How much money did Jerry have in the account at the start of the year? [2.4]

34. A store is discounting all of its prices by 20%. If a sweater's original price was $59.95, what is the price after the discount? [2.4]

35. Karl deposits a total of $2000 in savings accounts at two different banks. The first bank pays 6% interest, while the second bank pays 5% interest. If he earns a total of $117 in interest in the first year, how much did he deposit at each bank? [2.4]

Solve the proportion. [2.4]

36. $\dfrac{n}{6} = \dfrac{18}{27}$ **37.** $\dfrac{4}{9} = \dfrac{n}{72}$

38. $\dfrac{32}{13} = \dfrac{192}{n}$ **39.** $\dfrac{38}{n} = \dfrac{57}{75}$

Set up a proportion and solve it to solve the following problems. [2.4]

40. At a certain community college, 3 out of every 7 students plan to transfer to a four-year university. If the college has 14,952 students, how many plan to transfer to a four-year university?

41. Six tablespoons of a concentrated weed killer must be mixed with 1 gallon of water before spraying. How many tablespoons of the weed killer need to be mixed with $2\frac{1}{2}$ gallons of water?

Solve. Graph your solution on a number line, and express it in interval notation. [2.5]

42. $x + 5 < 2$

43. $x - 3 > -8$

44. $2x - 7 \geq 3$

45. $3x + 4 \leq -8$

46. $3x < -18$ or $x + 3 > 10$

47. $x + 5 \leq -3$ or $4x \geq -12$

48. $-21 \leq 3x - 6 \leq 24$

49. $-11 < 2x - 9 < 11$

50. Erin needs to make at least seven sales this week to qualify for a bonus. Write an inequality that shows the number of sales that will qualify Erin to receive a bonus. [2.5]

51. A high school yearbook staff is selling advertisements for $45. If the yearbook staff needs to raise at least $3000 in advertisement revenue, how many advertisements do they need to sell? [2.5]

52. If Steve's test average is at least 70, but lower than 80, then his final grade will be a C. His scores on the first four tests were 58, 72, 66, and 75. What scores on the fifth exam will give him a C for the class? (Assume that the highest possible score is 100.) [2.5]

Chapter 2 Test

For Extra Help | Pass the Test | Test solutions are found on the enclosed CD.

1. Solve $-8x = 56$ using the multiplication property of equality.

2. Solve $x + 17 = -22$ using the addition property of equality.

Solve.

3. $7x - 20 = 36$

4. $72 = 30 - 8a$

5. $3x + 34 = 5x - 21$

6. $3(5x - 16) - 9x = x - 63$

7. $\dfrac{1}{8}x - \dfrac{2}{3} = \dfrac{5}{6}$

8. $\dfrac{n}{100} = \dfrac{22}{40}$

9. Solve the literal equation $4x + 3y = 28$ for y.

10. What number is 16% of 175?

11. What percent of 660 is 99?

Solve. Graph your solution on a number line, and express it in interval notation.

12. $6x + 8 \leq -7$

13. $x + 4 \leq -5$
    or $2x + 3 \geq 0$

14. $-23 < 3x + 7 < 23$

15. The length of a rectangle is 3 feet less than twice its width, and the perimeter is 72 feet. Find the length and the width of the rectangle.

16. Barry's wallet contains $5 and $10 bills. He has seven more $5 bills than $10 bills. If Barry has $185 in his wallet, how many $5 bills are in the wallet?

17. At a certain university, 2 out of every 9 professors earned their degree outside the United States. If the university has 396 professors, how many of them earned their degrees outside the United States?

18. Wayne earns a commission of $500 for every car he sells. How many cars does he need to sell in order to earn at least $17,000 in commissions this month?

Mathematicians in History
George Pólya

$\mathcal{G}$eorge Pólya was a Hungarian mathematician who spent much of his career on problem solving and wrote the landmark text *How to Solve It*. Alan Schoenfeld has said of this book, "For mathematics education and the world of problem solving it marked a line of demarcation between two eras, problem solving before and after Pólya."

Write a one-page summary (*or* make a poster) of the life of George Pólya and his accomplishments. Also, look up Pólya's most famous quotes and list your favorite quote.

Interesting issues:

- Where and when was George Pólya born?
- What circumstances led to Pólya leaving Göttingen after Christmas in 1913?
- The political situation in Europe in 1940 forced Pólya to leave Zürich for the United States. At which university did Pólya work upon arriving in the United States?
- When was Pólya's monumental book *How to Solve It* published?
- Summarize Pólya's strategy for solving problems.
- At which American university did Pólya spend most of his career?
- What was Pólya's mnemonic for the first fourteen digits of π?
- How old was Pólya when he taught his last class, and what was the subject?

Step 1: Form groups of no more than four students. Each group should have a regular-sized bag of plain M&M's.

Step 2: Sort the M&M's into the six colors. Count and record the number of M&M's you have of each color. Write these values in the "Number" column of the following table:

Color	Number	Variable Representation	Observed Percent	Expected Percent	Expected Number
Brown		n		13%	
Yellow				14%	
Red				13%	
Blue				24%	
Orange				20%	
Green				16%	

Step 3: Create a variable representation for the number you have of each color. Let the number of brown M&M's be n and use this variable in the creation of the representations for the other colors. For example, if you have 13 brown M&M's and 20 yellow M&M's, you could represent the number of yellow M&M's by $n + 7$ or $2n - 6$. Be creative in how you describe the number of yellow, red, blue, orange, and green M&M's.

Step 4: Using the variable representations of all six colors, create an equation that relates these variables to your total number of M&M's.

Step 5: Once all the groups have created equations, pass yours on to another group and have them solve it. Did they get the correct numbers for all six colors? If not, first check to make sure that your variable representations are correct. If you believe your representations are correct, have the other group check their solving process. If the group still cannot solve your equation correctly, have your instructor check your representations and the other groups' process of solving the equation to see where the error may have occurred. If the group found solving your equation an easy process, see if you can create more elaborate variable representations to try and stump them.

Step 6: Each bag of plain M&M's is supposed to have a certain percentage of each color present. (See the "Expected Percent" column for those percentages.) Using the numbers you counted, calculate the percentage of each color that is actually present in your bag of M&M's and list these values in the "Observed Percent" column. Remember that the total of your "Observed Percent" column should be very close to 100%. Are your percentages close to the expected percents?

Step 7: Calculate the expected number of each color of M&M's that you should have found in your bag and write these values in the "Expected Number" column. How would you calculate this number? Discuss possible processes with your group. Here's a hint: It has to do with the expected percentages the candy company claims to have in each bag and your total number of M&M's.

Step 8: Submit your completed chart and equation(s) to your instructor for review.

Graphing Linear Equations

In this chapter, we will examine linear equations in two variables. Equations involving two variables relate two unknown quantities to each other, such as a person's height and weight, or the number of items sold by a company and the profit that the company made. The primary focus is on graphing linear equations, which is the technique that we use to display the solutions to an equation in two variables.

Study Strategy **Making the Best Use of Your Resources** *In this chapter, we will focus on how to get the most out of your available resources. What works best for one student may not work for another student. Consider all of the different resources presented in this chapter, and use the ones you feel could help you.*

Throughout this chapter, we will revisit this study strategy and help you incorporate it into your study habits.

3.1

The Rectangular Coordinate System; Equations in Two Variables

Objectives

1. Determine whether an ordered pair is a solution of an equation in two variables.
2. Plot ordered pairs on a rectangular coordinate plane.
3. Complete ordered pairs for a linear equation in two variables.
4. Graph linear equations in two variables by plotting points.

Equations in Two Variables and Their Solutions; Ordered Pairs

Objective 1 Determine whether an ordered pair is a solution of an equation in two variables. In this section, we will begin to examine equations in two variables, which relate two different quantities to each other. Here are some examples of equations in two variables:

$$y = 4x + 7 \qquad 2a + 3b = -6$$
$$y = 3x^2 - 2x - 9 \qquad t = \frac{4n + 9}{3n - 7}$$

As in the previous chapter, we will be looking for solutions of these equations. A solution of an equation in two variables is a pair of values that, when substituted for the two variables, produce a true statement. Consider the equation $3x + 4y = 10$. One solution of this equation is $x = 2$ and $y = 1$. We see that this is a solution by substituting 2 for x and 1 for y into the equation:

$$3x + 4y = 10$$
$$3(2) + 4(1) = 10 \qquad \text{Substitute 2 for } x \text{ and 1 for } y.$$
$$6 + 4 = 10 \qquad \text{Multiply.}$$
$$10 = 10 \qquad \text{Add.}$$

Since this produces a true statement, our pair of values together form a solution. Solutions to this type of equation are written as **ordered pairs** (x, y). In our example, the ordered pair $(2, 1)$ is a solution. We call it an ordered pair because the two values are listed in the specific order of x first and y second. The values are often referred to as **coordinates.**

EXAMPLE 1 Is the ordered pair $(-3, 2)$ a solution of the equation $y = 3x + 11$?

Solution

To determine whether this ordered pair is a solution, we need to substitute -3 for x and 2 for y:

$$y = 3x + 11$$
$$(2) = 3(-3) + 11 \qquad \text{Substitute } -3 \text{ for } x \text{ and 2 for } y.$$
$$2 = -9 + 11 \qquad \text{Multiply.}$$
$$2 = 2 \qquad \text{Simplify.}$$

Since this is a true statement, the ordered pair $(-3, 2)$ is a solution of the equation $y = 3x + 11$.

EXAMPLE 2   Is $(4, 0)$ a solution of the equation $y - 2x = 8$?

Solution

Again, we check to see whether this ordered pair is a solution by substituting the given values for x and y. We must be careful to substitute 4 for x and 0 for y, not vice versa:

$$y - 2x = 8$$
$$(0) - 2(4) = 8 \qquad \text{Substitute 4 for } x \text{ and 0 for } y.$$
$$0 - 8 = 8 \qquad \text{Multiply.}$$
$$-8 = 8 \qquad \text{Subtract.}$$

Since this is a false statement, $(4, 0)$ is not a solution of the equation $y - 2x = 8$.

Quick Check 1

Is $(0, -6)$ a solution of the equation $3x + 2y = -12$?

The Rectangular Coordinate Plane

Objective 2 **Plot ordered pairs on a rectangular coordinate plane.** Ordered pairs can be displayed graphically using the **rectangular coordinate plane.** The rectangular coordinate plane is also known as the **Cartesian plane,** named after the famous French philosopher and mathematician René Descartes.

The rectangular coordinate plane is made up of two number lines, known as **axes.** The axes are drawn at right angles to each other, intersecting at 0 on each axis. The accompanying graph shows how it looks.

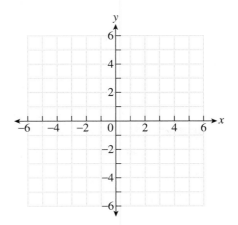

x-axis, y-axis, origin

> The **horizontal axis,** or **x-axis,** is used to show the first value of an ordered pair. The **vertical axis,** or **y-axis,** is used to show the second value of an ordered pair. The point where the two axes cross is called the **origin** and represents the ordered pair $(0, 0)$.

Positive values of x are found to the right of the origin, and negative values of x are found to the left of the origin. Positive values of y are found above the origin, and negative values of y are found below the origin.

We can use the rectangular coordinate plane to represent ordered pairs that are solutions of an equation. To plot an ordered pair (x, y) on a rectangular coordinate plane, we begin at the origin. The x-coordinate tells us how far to the left or right of the origin to move. The y-coordinate then tells us how far up or down to move from there.

Suppose we wanted to plot the ordered pair $(-3, 2)$ on a rectangular coordinate plane. The x-coordinate is -3, and that tells us to move three units to the left of the origin. The y-coordinate is 2, so we move up two units to find the location of our ordered pair, and we place a point at this location. We will often refer to an ordered pair as a **point** when we plot it on a graph.

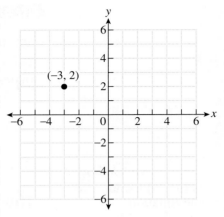

One final note about the rectangular coordinate plane is that the two axes divide the plane into four sections called **quadrants.** The four quadrants are labeled I, II, III, and IV. A point that lies on one of the axes is not considered to be in one of the quadrants.

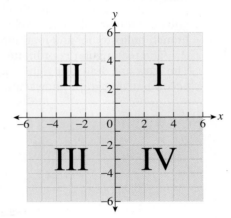

EXAMPLE 3   Plot the ordered pairs $(5, 1)$, $(4, -2)$, $(-2, -4)$, and $(-4, 0)$ on a rectangular coordinate plane.

Quick Check 2

Plot the ordered pairs $(3, 8)$, $(5, -4)$, $(-3, 6)$, and $(0, 7)$ on a rectangular coordinate plane.

Solution

The four ordered pairs are plotted on the following graph:

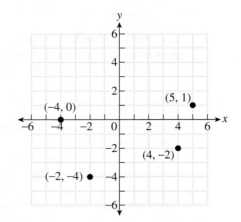

Choosing the appropriate scale for a graph is an important skill. To plot the ordered pair $\left(\frac{7}{2}, -4\right)$ on a rectangular coordinate plane, we could use a scale of $\frac{1}{2}$ on the axes. A rectangular plane with axes showing values ranging from -10 to 10 is not large enough to plot the ordered pair $(-45, 25)$. A good choice would be to label the axes showing values that range from -50 to 50. Since this is a wide range of values, it makes sense to increase the scale. Since each coordinate is a multiple of 5, we will use a scale of 5 units.

It is important to be able to read the coordinates of an ordered pair from a graph.

EXAMPLE 4   Find the coordinates of points *A* through *G* on the following graph:

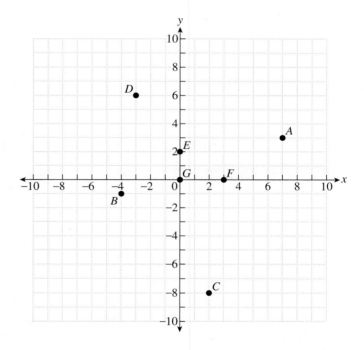

Solution

When determining coordinates for a point on a rectangular coordinate plane, we begin by finding the *x*-coordinate. We do this by determining how far to the left or right of the origin the point is. If the point is to the right, then the *x*-coordinate is positive. If the point is to the left, then the *x*-coordinate is negative. If the point is neither to the right nor to the left of the origin, then the *x*-coordinate is 0.

We next determine the *y*-coordinate of the point by measuring how far above or below the origin the point is. If the point is above the origin, then the *y*-coordinate is positive. If the point is below the origin, then the *y*-coordinate is negative. If the point is neither above nor below the origin, then the *y*-coordinate is 0.

Here is a table containing the coordinates of each point:

Point	A	B	C	D	E	F	G
Coordinates	(7, 3)	(−4, −1)	(2, −8)	(−3, 6)	(0, 2)	(3, 0)	(0, 0)

Quick Check **3** Find the coordinates of points *A* through *E* on the following graph:

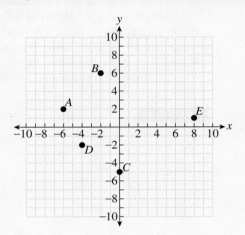

Linear Equations in Two Variables

Objective 3 Complete ordered pairs for a linear equation in two variables.

Standard Form of a Linear Equation

A linear equation in two variables is an equation that can be written in the form $Ax + By = C$, where A, B, and C are real numbers and A and B are not both 0. This is called the **standard form** of a linear equation in two variables.

To find an ordered pair that is a solution of a linear equation, we can select an arbitrary value for one of the variables, substitute this value into the equation, and solve for the remaining variable. The next example gives us practice with finding solutions of linear equations in two variables.

EXAMPLE 5 Find the missing coordinate such that the ordered pair $(2, y)$ is a solution of the equation $y = 5x - 4$.

Solution

Quick Check **4**
Find the missing coordinate such that the ordered pair $(3, y)$ is a solution of the equation $y = -3x + 5$.

We are given an *x*-coordinate of 2, so we substitute this for *x* in the equation.

$y = 5(2) - 4$ Substitute 2 for *x* and solve for *y*.
$y = 10 - 4$ Multiply.
$y = 6$ Subtract.

The ordered pair $(2, 6)$ is a solution of this equation.

The solutions of any linear equation in two variables follow a pattern. We will now begin to explore this pattern.

EXAMPLE 6   Complete the following table of ordered pairs with solutions of the linear equation $x + y = 5$:

x	3		0
y		−2	

Plot the solutions on a rectangular coordinate plane.

Solution

The first ordered pair has an x-coordinate of 3. Substituting this for x gives us the equation $(3) + y = 5$, which has a solution of $y = 2$. $(3, 2)$ is a solution.

The second ordered pair has a y-coordinate of −2. Substituting for y, this produces the equation $x + (−2) = 5$. The solution of this equation is $x = 7$, so $(7, −2)$ is a solution.

The third ordered pair has a x-coordinate of 0. When we substitute this for x, we get the equation $(0) + y = 5$. This equation has a solution of $y = 5$, so $(0, 5)$ is a solution. The completed table is

x	3	7	0
y	2	−2	5

Quick Check 5
Complete the table of ordered pairs with solutions of the linear equation $x − y = 6$:

x	4		−1
y		3	

Plot the solutions on a rectangular coordinate plane.

When we plot these three points on a rectangular coordinate plane, it looks like the following graph. Notice that the three points appear to fall on a straight line.

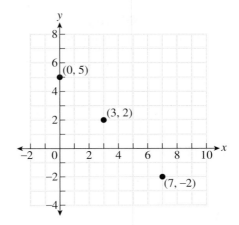

Graphing Linear Equations by Plotting Points

Objective 4   Graph linear equations in two variables by plotting points.
When we plot the points for each set of solutions, we should notice that the three points appear to be in line with each other. In each case, we can draw a straight line that passes through each of the three points. In addition to the three points on the graph, every

point that lies on the line is a solution of the equation. Every ordered pair that is a solution must lie on this line.

We can sketch a graph of the solutions of a linear equation by first plotting ordered pairs that are solutions and then drawing the straight line through these points. The line we sketch continues indefinitely in both directions. Although only two points are needed to determine a line, we will find three points. The third point is used as a check to make sure that the first two points are accurate. We will pick three arbitrary values for one of the variables and find the other coordinates.

EXAMPLE 7 Graph the linear equation $y = 3x - 4$.

Solution

When the equation has y isolated on one side of the equals sign, it is a good idea to select values for x and find the corresponding y-values. We find three points on the line by letting $x = 0$, $x = 1$, and $x = 2$. We did not need to use 0, 1, and 2 for x; they were chosen arbitrarily. If you start with other values, your ordered pairs will still be on the same line. The following table shows the calculations:

$x = 0$	$x = 1$	$x = 2$
$y = 3(0) - 4$	$y = 3(1) - 4$	$y = 3(2) - 4$
$y = 0 - 4$	$y = 3 - 4$	$y = 6 - 4$
$y = -4$	$y = -1$	$y = 2$

Our three ordered pairs are $(0, -4)$, $(1, -1)$, and $(2, 2)$. We plot these three points on a rectangular coordinate plane and draw a line through them.

Quick Check 6
Graph the linear equation $y = 2x + 1$.

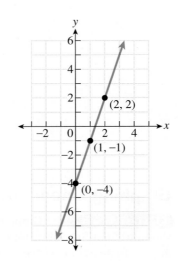

Building Your Study Strategy **Using Your Resources, 1   Your Instructor**   One of the most important resources available to you is your instructor. Take advantage of your instructor's experience and knowledge. If you have a question during the lecture, be sure to ask. The question that you have is probably on the minds of several other students as well.

Learn where your instructor's office is located and when office hours are held. Visiting your instructor during office hours is a great way to get one-on-one help. There are some questions that are difficult for the instructor to answer completely during the class period that are more easily handled in your instructor's office. Your instructor may also be able to give you extra problems to practice with. In addition to answering your questions during class, some instructors are available for questions either right before or right after class.

EXERCISES *3.1*

Vocabulary

1. The standard form of a linear equation in two variables is _____.

2. Solutions to an equation in two variables are written as a(n) _____.

3. The _____ is used to show the first coordinate of an ordered pair on a rectangular coordinate plane.

4. The _____ is used to show the second coordinate of an ordered pair on a rectangular coordinate plane.

5. The point at $(0, 0)$ is called the _____.

6. A(n) _____ on a rectangular coordinate plane is used to show the solutions of a linear equation in two variables.

Is the ordered pair a solution of the given equation?

7. $(9, 3)$, $2x + y = 21$

8. $(5, 4)$, $3x + 2y = 23$

9. $(1, -4)$, $x + 2y = 7$

10. $(3, -3)$, $4x - 3y = 21$

11. $(-6, 2)$, $5x - 4y = -38$

12. $(-4, 3)$, $6x - 5y = 9$

13. $(3, 2)$, $-2x + 5y = 4$

14. $(5, -2)$, $-x + 7y = -9$

15. $(-3, 0)$, $-4x + y = 12$

16. $(0, 4)$, $2x - 5y = -20$

Plot the points on a rectangular coordinate plane.

17. $(5, 2)$ $(3, 4)$ $(2, -6)$ $(-4, 1)$

18. $(-3, -5)\,(-5, 3)\,(4, -3)\,(-2, -1)$

Find the coordinates of the labeled points A, B, C, and D.

23.

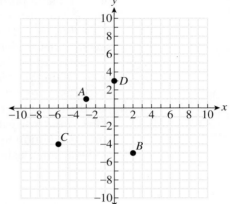

19. $(0, 4)\,(-2, 0)\,(-4, 2)$
$(2, 4)$

20. $(-4, 0)\,(-5, 0)\,(0, -3)$
$(0, 2)$

21. 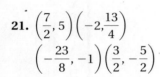 $\left(\dfrac{7}{2}, 5\right)\left(-2, \dfrac{13}{4}\right)$
$\left(-\dfrac{23}{8}, -1\right)\left(\dfrac{3}{2}, -\dfrac{5}{2}\right)$

24.

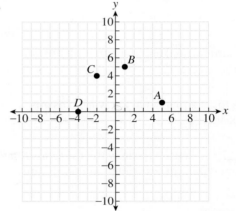

22. $(-70, 30)\,(20, -90)$
$(45, 25)\,(-50, -68)$

25.

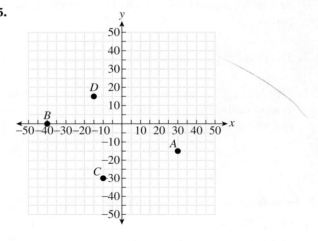

26.

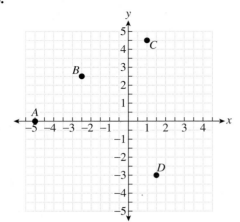

In which quadrant (I, II, III, or IV) is the given point?

27. $(5, -1)$

28. $(-2, 8)$

29. $(3, 4)$

30. $(-6, -4)$

31. $(2, 5)$

32. $(-3, -3)$

33. $\left(\dfrac{1}{2}, -\dfrac{3}{5}\right)$

34. $(-700, 1000)$

Using the given coordinate, find the other coordinate that makes the ordered pair a solution of the equation.

35. $3x - 2y = 8$, $(2, y)$

36. $x + 4y = 7$, $(x, 3)$

37. $-2x + 5y = -6$, $(x, -4)$

38. $5x + y = 3$, $(3, y)$

39. $-x - 3y = -4$, $(10, y)$

40. $-3x - 2y = 5$, $(x, -4)$

41. $7x - 11y = -22$, $(0, y)$

42. $4x + 3y = -20$, $(x, 0)$

43. $3x + \dfrac{1}{2}y = -2$, $(x, 8)$

44. $\dfrac{1}{5}x + 3y = 20$, $(10, y)$

Complete each table with ordered pairs that are solutions.

45. $2x + y = 7$

x	y
1	
	1
0	

46. $x + 3y = 8$

x	y
2	
	3
0	

47. $5x - 3y = -12$

x	y
-3	
0	
3	

48. $-2x + 4y = -16$

x	y
-2	
0	
2	

49. $y = 4x - 9$

x	y
0	
1	
4	

50. $y = -3x + 2$

x	y
-1	
0	
3	

51. $y = \dfrac{2}{3}x - 2$

x	y
-3	
0	
6	

52. $y = \dfrac{7}{5}x + 1$

x	y
-5	
0	
5	

53. Find three equations that have the ordered pair $(5, 2)$ as a solution.

54. Find three equations that have the ordered pair $(3, -4)$ as a solution.

55. Find three equations that have the ordered pair $(0, 6)$ as a solution.

56. Find three equations that have the ordered pair $(-8, 0)$ as a solution.

Find three ordered pairs that are solutions of the given linear equation, and use them to graph the equation.

57. $y = 4x + 1$

58. $y = 2x - 3$

59. $y = -2x + 7$

60. $y = -3x + 6$

61. $y = 3x$

62. $y = -5x$

63. $y = \dfrac{1}{4}x + 3$

64. $y = \dfrac{1}{2}x - 3$

65. $y = -\dfrac{3}{5}x + 6$

66. $y = -\dfrac{5}{2}x + 4$

67. $3x + y = 9$

68. $2x + 3y = 6$

Writing in Mathematics

Answer in complete sentences.

69. Explain, in your own words, why the order of the coordinates in an ordered pair is important. Is there a difference between plotting the ordered pair (a, b) and the ordered pair (b, a)?

70. If the ordered pair (a, b) is in quadrant II, in which quadrant is (b, a)? In which quadrant is $(-a, b)$? Explain how you determined your answers.

71. *Solutions Manual** Write a solutions manual page for the following problem:

Find three ordered pairs that are solutions to the linear equation $y = -2x + 8$, and use them to graph the equation.

72. *Newsletter** Write a newsletter explaining how to plot ordered pairs on a rectangular coordinate plane.

**See Appendix B for details and sample answers.*

3.2

Graphing Linear Equations and Their Intercepts

Objectives

1. Find the *x*- and *y*-intercepts of a line from its graph.
2. Find the *x*- and *y*-intercepts of a line from its equation.
3. Graph linear equations using their intercepts.
4. Graph linear equations that pass through the origin.
5. Graph horizontal lines.
6. Graph vertical lines.
7. Interpret the graph of an applied linear equation.

In the previous section we learned that we can use a rectangular coordinate plane to display the solutions of an equation in two variables. We also learned that the solutions of a linear equation in two variables were all on a straight line. In this section, we will learn to graph the line associated with a linear equation in two variables in a more systematic fashion.

Intercepts

Objective 1   Find the *x*- and *y*-intercepts of a line from its graph.

A point at which a graph crosses the *x*-axis is called an **x-intercept**, and a point at which a graph crosses the *y*-axis is called a **y-intercept.**

EXAMPLE 1   Find the coordinates of any *x*-intercepts and *y*-intercepts.

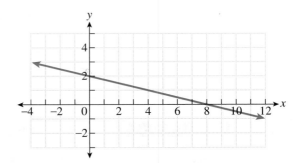

Solution

An *x*-intercept is a point at which the graph crosses the *x*-axis. We see that this graph crosses the *x*-axis at the point $(8, 0)$, so the *x*-intercept is the point $(8, 0)$.

This graph crosses the *y*-axis at the point $(0, 2)$, so its *y*-intercept is $(0, 2)$.

Quick Check   **1** Find the coordinates of any *x*-intercepts and *y*-intercepts.

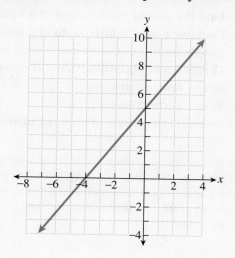

Some lines do not have both an *x*- and *y*-intercept.

This horizontal line does not cross the *x*-axis, so it does not have an *x*-intercept. Its *y*-intercept is the point $(0, -6)$.

The *x*-intercept of this vertical line is the point $(4, 0)$. This graph does not cross the *y*-axis, so it does not have a *y*-intercept.

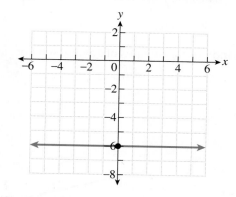

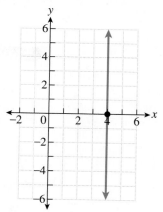

Objective 2 **Find the *x*- and *y*-intercepts of a line from its equation.** Any point that lies on the *x*-axis has a *y*-coordinate of 0. Similarly, any point that lies on the *y*-axis has an *x*-coordinate of 0. We can use these facts to find the intercepts of an equation's graph algebraically. To find an *x*-intercept, we substitute 0 for *y* and solve for *x*. To find a *y*-intercept, we substitute 0 for *x* and solve for *y*.

Finding Intercepts

To Find an *x*-Intercept
Substitute 0 for *y*, and solve the resulting equation for *x*.

To Find a *y*-Intercept
Substitute 0 for *x*, and solve the resulting equation for *y*.

EXAMPLE 2   Find the x- and y-intercepts for the equation $5x - 2y = 10$.

Solution

Let's begin with the x-intercept by substituting 0 for y and solving for x.

$$5x - 2y = 10$$
$$5x - 2(0) = 10 \qquad \text{Substitute 0 for } y.$$
$$5x = 10 \qquad \text{Simplify the left side of the equation.}$$
$$x = 2 \qquad \text{Divide both sides by 5.}$$

The x-intercept is at $(2, 0)$. To find the y-intercept, we substitute 0 for x and solve for y.

$$5x - 2y = 10$$
$$5(0) - 2y = 10 \qquad \text{Substitute 0 for } x.$$
$$-2y = 10 \qquad \text{Simplify the left side of the equation.}$$
$$y = -5 \qquad \text{Divide both sides by } -2.$$

The y-intercept is $(0, -5)$.

> **Quick Check 2**
> Find the x- and y-intercepts for the equation $-3x + 4y = -12$.

EXAMPLE 3   Find the x- and y-intercepts for the equation $y = -3x - 7$.

Solution

We will begin by finding the x-intercept. To do this, we substitute 0 for y and solve for x.

$$y = -3x - 7$$
$$0 = -3x - 7 \qquad \text{Substitute 0 for } y.$$
$$3x = -7 \qquad \text{Add } 3x \text{ to both sides.}$$
$$x = -\frac{7}{3} \qquad \text{Divide both sides by 3.}$$

The x-intercept is $\left(-\frac{7}{3}, 0\right)$.
To find the y-intercept, we substitute 0 for x and solve for y.

$$y = -3x - 7$$
$$y = -3(0) - 7 \qquad \text{Substitute 0 for } x.$$
$$y = -7 \qquad \text{Simplify.}$$

> **Quick Check 3**
> Find the x- and y-intercepts for the equation $y = 2x - 9$.

The y-intercept is at $(0, -7)$.

Graphing a Linear Equation Using Its Intercepts

Objective 3 Graph linear equations using their intercepts. When graphing a linear equation, we will begin by finding any intercepts. This will often give us two points for our graph. Although only two points are necessary to graph a straight line, we will plot at least three points to be sure that our first two points are accurate. In other words, we will use a third point as a "check" for the first two points we find. We find this third point by selecting a value for x, substituting it into the equation, and solving for y.

EXAMPLE 4   Graph $3x + y = 6$. Label any intercepts.

Solution

We begin by finding the x- and y-intercepts.

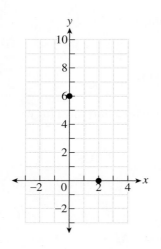

x-intercept		**y-intercept**	
$3x + (0) = 6$	Substitute 0 for y.	$3(0) + y = 6$	Substitute 0 for x.
$3x = 6$	Simplify.	$y = 6$	Simplify.
$x = 2$	Divide both sides by 3.		

The x-intercept is $(2, 0)$. The y-intercept is $(0, 6)$.

Before looking for a third point, let's plot the two intercepts on a graph. This can help us select a value to substitute for x when finding a third point. It is a good idea to select a value for x that is somewhat near the x-coordinates of the other two points. For example, $x = 1$ would be a good choice for this example. If we choose a value of x that is too far away from the x-coordinates of the other two points, then we may have to change the scale of the axes or extend them to show this third point.

$3(1) + y = 6$	Substitute 1 for x.
$3 + y = 6$	Simplify the left side of the equation.
$y = 3$	Subtract 3.

The point $(1, 3)$ is our third point. Keep in mind that the third point can vary with our choice for x, but the three points should still be on the same line. After plotting the third point, we finish by drawing a straight line that passes through all three points.

Quick Check **4**
Graph $2x - 3y = 12$.
Label any intercepts.

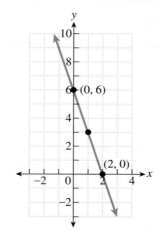

EXAMPLE 5   Graph $y = 2x + 5$. Label any intercepts.

Solution

We will begin by finding the x- and y-intercepts.

x-intercept ($y = 0$)	**y-intercept** ($x = 0$)
$0 = 2x + 5$	$y = 2(0) + 5$
$-5 = 2x$	$y = 5$
$-\dfrac{5}{2} = x$	$(0, 5)$
$\left(-\dfrac{5}{2}, 0\right)$	

Choosing $x = 1$ to find our third point is a good choice: It is close to the two intercepts, and it will be easy to substitute 1 for x in the equation.

$$y = 2(1) + 5 \qquad \text{Substitute 1 for } x \text{ in the equation.}$$
$$y = 7 \qquad \text{Simplify.}$$

Our third point is $(1, 7)$. Here is the graph:

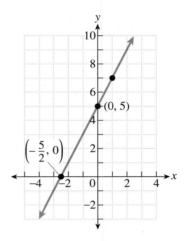

Quick Check **5**

Graph $y = -3x - 9$.
Label any intercepts.

Using Your Calculator   We can use the TI-84 to graph a line. To enter the equation of the line that we graphed in Example 5, press the $\boxed{Y=}$ key. Enter 2x + 5 next to Y_1, using the key labeled $\boxed{X,T,\Theta,n}$ to type the variable x. Now press the key labeled $\boxed{GRAPH}$ to graph the line.

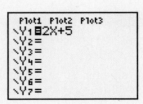

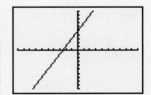

Graphing a Line by Using Its Intercepts

- Find the x-intercept by substituting 0 for y and solving for x.
- Find the y-intercept by substituting 0 for x and solving for y.
- Select a value of x, substitute it into the equation, and solve for y to find the coordinates of a third point on the line.
- Graph the line that passes through these three points.

Graphing Linear Equations That Pass through the Origin

Objective **4** **Graph linear equations that pass through the origin.** Some lines do not have two intercepts. For example, a line that passes through the origin has its x- and y-intercepts at the same point. In this case, we will need to find two additional points.

EXAMPLE 6   Graph $2x - 8y = 0$. Label any intercepts.

Solution

Let's start by finding the x-intercept.

$$2x - 8(0) = 0 \qquad \text{Substitute 0 for } y.$$
$$2x = 0 \qquad \text{Simplify.}$$
$$x = 0 \qquad \text{Divide both sides by 2.}$$

The x-intercept is at the origin $(0, 0)$. This point is also the y-intercept, which you can verify by substituting 0 for x and solving for y. We will need to find two more points before we graph the line. Suppose that we choose $x = 1$. Then we have

$$2(1) - 8y = 0 \qquad \text{Substitute 1 for } x.$$
$$2 - 8y = 0 \qquad \text{Simplify.}$$
$$2 = 8y \qquad \text{Add } 8y.$$
$$\frac{1}{4} = y \qquad \text{Divide both sides by 8 and simplify.}$$

The point $\left(1, \frac{1}{4}\right)$ is on the line. However, it may not be easy for us to put this point on the graph because of its fractional y-coordinate. We can try to avoid this problem by solving the equation for y before choosing a value for x.

$$2x - 8y = 0$$
$$2x = 8y \qquad \text{Add } 8y \text{ to both sides.}$$
$$\frac{1}{4}x = y \quad \text{or} \quad y = \frac{1}{4}x. \qquad \text{Divide both sides by 8 and simplify.}$$

If we choose a value for x that is a multiple of 4, then we will have a y-coordinate that is an integer. We will use $x = 4$ and $x = -4$, although there are many other choices that would work.

$x = 4$	$x = -4$
$y = \frac{1}{4}(4)$	$y = \frac{1}{4}(-4)$
$y = 1$	$y = -1$
$(4, 1)$	$(-4, -1)$

Here is the graph.

Graph $y = 3x$. Label any intercepts.

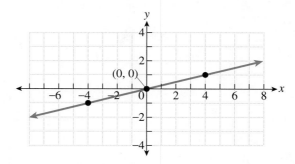

Graphing a Line Passing through the Origin

- Determine that the x- and y-intercepts are both at the origin.
- Find a second point on the line by selecting a value of x and substituting it into the equation. Solve the equation for y.
- Find a third point on the line by selecting another value of x and substituting it into the equation. Solve the equation for y.
- Graph the line that passes through these three points.

Horizontal Lines

Objective 5 **Graph horizontal lines.** Can we draw a line that does not cross the x-axis? Yes; here's an example.

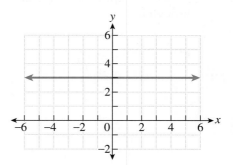

What is the equation of the line in this graph? Notice that each point has the same y-coordinate (3). The equation for the line is $y = 3$. The equation for a horizontal line is always of the form $y = b$, where b is a real number. To graph the line $y = b$, plot the y-intercept $(0, b)$ on the y-axis and draw a horizontal line through this point.

EXAMPLE 7   Graph $y = -2$. Label any intercepts.

Solution

This graph will be a horizontal line, as its equation is of the form $y = b$. If an equation does not have a term containing the variable x, then its graph will be a horizontal line.

We begin by plotting the point $(0, -2)$, which is the y-intercept. We then draw a horizontal line through this point.

Quick Check **7**
Graph $y = 7$. Label any intercepts.

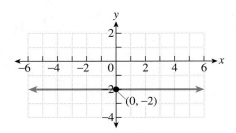

Vertical Lines

Objective **6** **Graph vertical lines.** Vertical lines have equations of the form $x = a$, where a is a real number. Just as the equation of a horizontal line does not have any terms containing the variable x, the equation of a vertical line does not have any terms containing the variable y. We begin to graph an equation of this form by plotting its x-intercept at $(a, 0)$. We then draw a vertical line through this point.

EXAMPLE **8**   Graph $x = 6$. Label any intercepts.

Solution

The x-intercept for this line is at $(6, 0)$. We plot this point and then draw a vertical line through it.

Quick Check **8**
Graph $x = -15$. Label any intercepts.

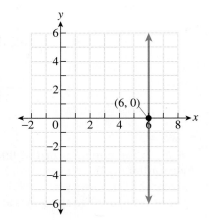

Horizontal Lines
Equation: $y = b$
- Plot the y-intercept at $(0, b)$.
- Graph the horizontal line passing through this point.

Vertical Lines
Equation: $x = a$
- Plot the x-intercept at $(a, 0)$.
- Graph the vertical line passing through this point.

Applications of the Graphs of Linear Equations

Objective 7 **Interpret the graph of an applied linear equation.** The intercepts of a graph have important interpretations within the context of an applied problem. The y-intercept often gives us an initial condition, such as the amount of money owed on a loan, or the fixed costs to attend college before adding in the number of units that a student is taking. The x-intercept often shows us a break-even point, such as when a business goes from a loss to a profit.

EXAMPLE 9   Chris borrowed $3600 from his friend Mack, promising to pay him $50 per month until he had paid off the loan. The amount of money (y) that Chris owes Mack after x months have passed is given by the equation $y = 3600 - 50x$.

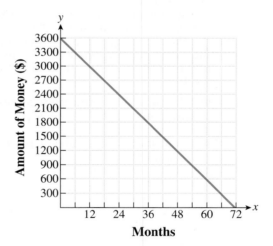

Months

Use the graph of this equation to answer the following questions (since the number of months cannot be negative, the graph begins at $x = 0$):

a) What does the y-intercept represent in this situation?

Solution

The y-intercept is at $(0, 3600)$. It tells us that after borrowing the money from Mack, Chris owes Mack $3600.

b) What does the x-intercept represent in this situation?

Solution

The x-intercept is at $(72, 0)$. It tells us that Chris will pay off his debt in 72 months.

Using Your Calculator   When graphing a line using the TI-84, we will occasionally need to resize the window. In Example 9, the *x*-intercept is at the point $(72, 0)$ and the *y*-intercept is at $(0, 3600)$. Our graph should show these important points. After entering $3600 - 50x$ for Y_1, press the [WINDOW] key. Fill in the screen as follows, and press the [GRAPH] key to display the graph:

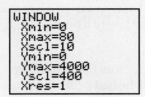

```
WINDOW
  Xmin=0
  Xmax=80
  Xscl=10
  Ymin=0
  Ymax=4000
  Yscl=400
  Xres=1
```

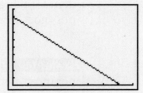

Quick Check   9   Nancy invests $10,000 in a new business and expects to make a $2000 profit each month. Nancy's net profit (*y*) after *x* months have passed is given by the equation $y = 2000x - 10,000$.

Use the graph of this equation to answer the following questions (since the number of months cannot be negative, the graph begins at $x = 0$):

a) What does the *y*-intercept represent in this situation?

b) What does the *x*-intercept represent in this situation?

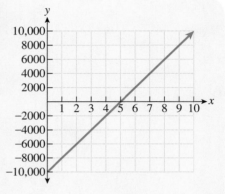

Building Your Study Strategy   Using Your Resources, 2   **Tutors**   Most college campuses have a tutorial center that provides free tutoring in math and other subjects. Some tutorial centers offer walk-in tutoring, others schedule appointments for one-on-one or group tutoring, and some combine these approaches.

Walk-in tutoring allows you to come in whenever you want to and ask questions as they arise. One-on-one tutoring is designed for students who will require more assistance than simple questions and answers, such as if you need someone to help explain the topic to you. If you do not completely understand a tutor's answer or directions, do not be afraid to ask for further explanation.

While it is a good idea to ask a tutor if you are doing a problem correctly or to give you an idea of a starting point for a problem, it is not a good idea to ask a tutor to do a problem for you. We learn by doing, not by watching. It may look easy while the tutor is working a problem, but this will not help you to do the next problem.

EXERCISES 3.2

F O R E X T R A H E L P

MyMathLab
MathXP

Interactmath.com

MathXL
Tutorials on CD
Video Lectures
on CD
Tutor Center
Addison-Wesley
Math Tutor Center
Student's
Solutions Manual

Vocabulary

1. A point at which a graph crosses the *x*-axis is called a(n) _____.

2. A point at which a graph crosses the *y*-axis is called a(n) _____.

3. A(n) _____ line has no *y*-intercept.

4. A(n) _____ line has no *x*-intercept.

5. State the procedure for finding an *x*-intercept.

6. State the procedure for finding an *y*-intercept.

Find the x- and y-intercepts from the graph.

7.

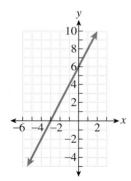

8.

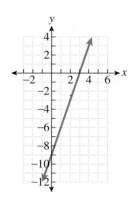

9.

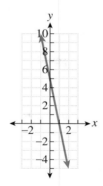

10.

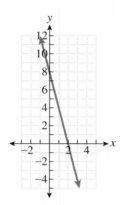

11.

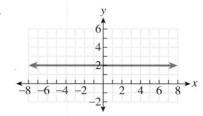

12.

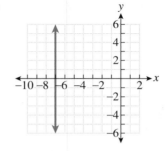

Find the x- and y-intercepts.

13. $x + y = 6$

14. $x + y = -3$

15. $x - y = -8$

16. $x - y = 4$

17. $3x + y = 9$

18. $2x + y = -10$

19. $-3x + 5y = 45$

20. $-2x - 6y = 18$

21. $-4x + 6y = 14$

22. $5x + 8y = -60$

23. $\frac{6}{5}x + \frac{3}{10}y = 10$

24. $\frac{1}{3}x + \frac{1}{2}y = 2$

25. $\frac{3}{5}x + \frac{1}{2}y = 3$

26. $\frac{2}{9}x - \frac{1}{12}y = \frac{2}{3}$

27. $y = -x + 8$

28. $y = x - 5$

29. $y = 3x - 12$

30. $y = -2x + 16$

31. $y = 3x$

32. $y = -2x$

33. $y = \frac{3}{5}x - 2$

34. $y = \frac{4}{7}x - 4$

35. $y = 2$

36. $x = 10$

37. Find three equations that have an x-intercept at $(4, 0)$.

38. Find three equations that have an x-intercept at $(-7, 0)$.

39. Find three equations that have a y-intercept at $(0, -10)$.

40. Find three equations that have a y-intercept at $(0, 2)$.

Find the intercepts, and then graph the line.

41. $x + y = 4$

42. $x + y = -2$

43. $2x - 5y = 10$

44. $-4x + 2y = 8$

45. $-3x + 4y = -24$

46. $6x + 5y = -30$ **47.** $3x + y = 9$ **51.** $y = x - 4$

52. $y = x + 6$

48. $x + 2y = 8$

53. $y = 5x$ **54.** $y = -4x$

49. $-4x + 7y = 14$

50. $3x - 4y = 18$

55. $y = -2x + 7$ **56.** $y = 4x - 2$

57. $y = \dfrac{7}{2}x - 7$

58. $y = -\dfrac{1}{3}x + 2$

59. $y = 4$

60. $y = -6$

61. $x = 3$

62. $x = -8$

63. A lawyer bought a new copy machine for her office in the year 2007, paying $12,000. For tax purposes, the copy machine depreciates at $1500 per year. If we let x represent the number of years after 2007, then the value of the copier (y) after x years is given by the equation $y = 12{,}000 - 1500x$.

   a) Find the y-intercept of this equation. Explain, in your own words, what this intercept signifies.

   b) Find the x-intercept of this equation. Explain, in your own words, what this intercept signifies.

   c) In 2010, what will be the value of the copy machine?

64. Jerry was ordered to perform 400 hours of community service as a reading tutor. Jerry spends eight hours each Saturday teaching people to read. The equation that tells the number of hours (y) that remain on the sentence after x Saturdays is $y = 400 - 8x$.

   a) Find the y-intercept of this equation. Explain, in your own words, what this intercept signifies.

b) Find the *x*-intercept of this equation. Explain, in your own words, what this intercept signifies.

c) After 12 Saturdays, how many hours of community service does Jerry have remaining?

65. At a country club, members pay an annual fee of $5000. In addition, they must pay $75 for each round of golf that they play. The equation that gives the cost (*y*) to play *x* rounds of golf per year is $y = 5000 + 75x$.

a) Find the *y*-intercept of this equation. Explain, in your own words, what this intercept signifies.

b) In your own words, explain why this equation has no *x*-intercept within the context of this problem.

c) If a person plays one round of golf per week, how much will she pay to the country club in one year?

66. In addition to paying $24 per unit, a community college student pays a registration fee of $100 per semester. The equation that gives the cost (*y*) to take *x* units is $y = 24x + 100$.

a) Find the *y*-intercept of this equation. Within the context of this problem, is this cost possible? Explain why or why not.

b) In your own words, explain why this equation has no *x*-intercept within the context of this problem.

c) A full load for a student is 12 units. How much will a full-time student pay per semester?

(**Writing in Mathematics**)

Answer in complete sentences.

67. *Solutions Manual** * Write a solutions manual page for the following problem:

Find the x- and y-intercepts, and then graph the line. $3x - 4y = 18$

68. *Newsletter** * Write a newsletter explaining how to graph a line using its intercepts.

*See Appendix B for details and sample answers.

QUICK REVIEW EXERCISES

Section 3.2

Solve for y.

1. $2x + y = 8$

2. $4x - y = 6$

3. $-8x + 2y = 10$

4. $4x + 6y = 12$

3.3

Slope of a Line

1 **Understand the slope of a line.**
2 **Find the slope of a line from its graph.**
3 **Find the slope of a line passing through two points using the slope formula.**
4 **Find the slopes of horizontal and vertical lines.**
5 **Find the slope and *y*-intercept of a line from its equation.**
6 **Find the equation of a line given its slope and *y*-intercept.**
7 **Graph a line using its slope and *y*-intercept.**
8 **Interpret the slope and *y*-intercept in real-world applications.**

Slope of a Line

Objective 1 **Understand the slope of a line.** Here are four different lines that pass through the point $(0, 2)$:

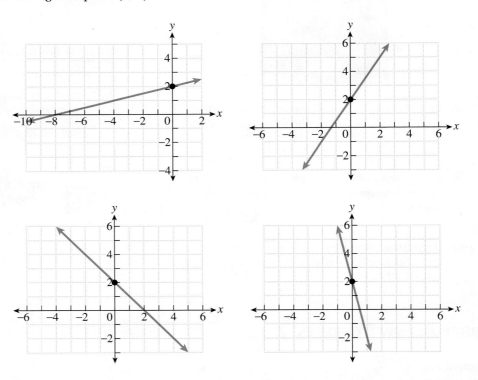

Notice that some of the lines are rising to the right, while others are falling to the right. Also, some are rising or falling more steeply than the others. The characteristic that distinguishes these lines is their slope.

Slope

The **slope** of a line is a measure of how steeply a line rises or falls as it moves to the right. We use the letter *m* to represent the slope of a line.

If a line rises as it moves to the right, then its slope is positive. If a line falls as it moves to the right, then its slope is negative.

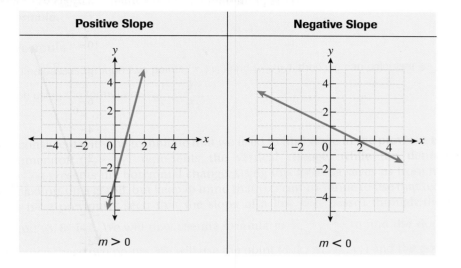

| Positive Slope | Negative Slope |

$m > 0$ $m < 0$

To find the slope of a line, we begin by selecting two points that are on the line. As we move from the point on the left to the point on the right, we measure how much the line rises or falls. The slope is equal to this vertical distance divided by the distance traveled from left to right. This is often referred to as "rise over run," or as the change in y divided by the change in x. Sometimes the change in y is written as Δy, and the change in x is written as Δx. The Greek letter Δ (delta) is often used to represent change in a quantity. The slope of a line is represented as

$$m = \frac{\text{rise}}{\text{run}} \text{ or } m = \frac{\Delta y}{\Delta x}.$$

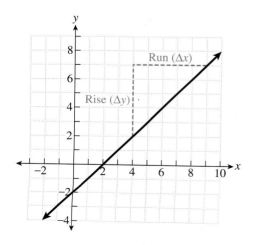

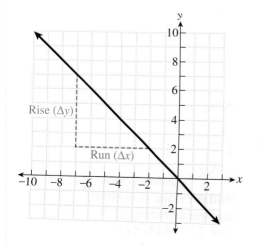

A Word of Caution   It does not matter which point is labeled as (x_1, y_1) and which point is labeled as (x_2, y_2). It *is* important that the order in which the y-coordinates are subtracted in the numerator is also the order in which the x-coordinates are subtracted in the denominator.

EXAMPLE ▶ 3   Use the formula $m = \dfrac{y_2 - y_1}{x_2 - x_1}$ to find the slope of the line that passes through the two points $(-5, -2)$ and $(-1, 4)$.

Solution

In this example, we learn to apply the formula to points that have negative x- or y-coordinates.

$$m = \frac{4 - (-2)}{-1 - (-5)} \qquad \text{Substitute into the formula.}$$

$$= \frac{4 + 2}{-1 + 5} \qquad \text{Eliminate ``double signs.''}$$

$$= \frac{6}{4} \qquad \text{Simplify the numerator and denominator.}$$

$$= \frac{3}{2} \qquad \text{Simplify.}$$

The slope of this line is $\frac{3}{2}$. The line rises by 3 units for every 2 units that it moves to the right.

Quick Check 3   Use the formula $m = \dfrac{y_2 - y_1}{x_2 - x_1}$ to find the slope of the line that passes through the two points $(-3, 5)$ and $(1, -5)$.

Horizontal and Vertical Lines

Objective 4   **Find the slopes of horizontal and vertical lines.** Let's look at the line that passes through the points $(1, 3)$ and $(4, 3)$.

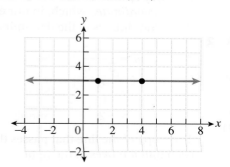

This is a horizontal line. What is its slope?

$$m = \frac{3 - 3}{4 - 1}$$
$$= \frac{0}{3}$$
$$= 0$$

The slope of this line is 0. The same is true for any horizontal line. The vertical change between any two points on a horizontal line is always equal to 0, and when we divide 0 by any nonzero number the result is equal to 0. Thus, the slope of any horizontal line is equal to 0.

Here is a brief summary of the properties of horizontal lines.

Horizontal Lines

> Equation: $y = b$, where b is a real number
> y-intercept: $(0, b)$
> Slope: $m = 0$

Look at the following vertical line:

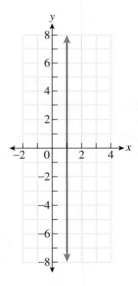

Notice that each point on this line has an x-coordinate of $x = 1$. We can attempt to find the slope of this line by selecting any two points on the line, such as $(1, 2)$ and $(1, 5)$, and using the slope formula.

$$m = \frac{5 - 2}{1 - 1}$$
$$= \frac{3}{0}$$

The fraction $\frac{3}{0}$ is undefined. So the slope of a vertical line is said to be undefined.

Here is a brief summary of the properties of vertical lines.

Vertical Lines

> Equation: $x = a$, where a is a real number
> x-intercept: $(a, 0)$
> Slope: Undefined

EXAMPLE 4   Find the slope, if it exists, of the line $x = -2$.

Solution

Quick Check 4
Find the slope, if it
exists, of the line $x = 5$.

This line is a vertical line. The slope of a vertical line is undefined, so this line has undefined slope.

EXAMPLE 5   Find the slope, if it exists, of the line $y = 4$.

Solution

This line is a horizontal line. The slope of any horizontal line is 0, so the slope of this line is 0.

Quick Check 5
Find the slope, if it
exists, of the line
$y = -2$.

> *A Word of Caution*   Avoid stating that a line has "no slope." This is a vague phrase; some may take it to mean that the slope is equal to 0, while others may interpret it as meaning that the slope is undefined.

Slope–Intercept Form of a Line

Objective 5   Find the slope and y-intercept of a line from its equation. If we solve a given equation for y so that it is in the form $y = mx + b$, this is the **slope–intercept form** of a line. The number being multiplied by x is the slope m, while b represents the y-coordinate of the y-intercept.

Slope–Intercept Form of a Line

> $y = mx + b$
> m: Slope of the line
> b: y-coordinate of the y-intercept

This means that if we were graphing the line $y = 2x + 5$, then the line would have a slope of 2 and a y-intercept of $(0, 5)$. Let's verify that this is true, beginning with the slope. Every time we increase the value of x by 1, y will increase by 2 because x is being multiplied by 2. If y increases by 2 (vertical change) every time x increases by 1 (horizontal change), this means that the slope of the line is equal to 2. As for the y-intercept, we can verify that this is true by substituting 0 for x in the equation.

$$y = 2x + 5$$
$$y = 2(0) + 5$$
$$y = 5$$

This shows that the y-intercept is at $(0, 5)$. Repeating this for the general form $y = mx + b$, we see that the y-coordinate of the y-intercept is b.

EXAMPLE 6   Find the slope and y-intercept for the line $y = -3x + 7$.

Solution

Quick Check 6
Find the slope and the y-intercept of the line $y = \frac{5}{2}x - 6$.

Since this equation is already solved for y, we can read the slope and the y-intercept directly from the equation. The slope is -3, which is the coefficient of the term containing x in the equation. The y-intercept is $(0, 7)$, since 7 is the constant in the equation.

EXAMPLE 7   Find the slope and y-intercept for the line $2x + 2y = 11$.

Solution

We begin by solving for y.

$$2x + 2y = 11$$
$$2y = -2x + 11 \qquad \text{Subtract } 2x \text{ from both sides.}$$
$$\frac{2y}{2} = \frac{-2x}{2} + \frac{11}{2} \qquad \text{Divide each term by 2.}$$
$$y = -x + \frac{11}{2} \qquad \text{Simplify.}$$

Quick Check 7
Find the slope and the y-intercept of the line $6x + 2y = 10$.

The slope of this line is -1. When we see $-x$ in the equation, we need to remember that this is the same as $-1x$. The y-intercept is $(0, \frac{11}{2})$.

A Word of Caution   We cannot determine the slope and y-intercept of a line from its equation unless the equation has been solved for y first.

Objective 6   Find the equation of a line given its slope and y-intercept.

EXAMPLE 8   Find the equation of a line that has a slope of 2 and a y-intercept of $(0, -15)$.

Solution

Quick Check 8
Find the equation of a line that has a slope of $\frac{1}{2}$ and a y-intercept of $(0, -\frac{4}{7})$.

We will use the slope–intercept form ($y = mx + b$) to help us find the equation of this line. Since the slope is 2, we can replace m by 2. Also, since the y-intercept is $(0, -15)$, we can replace b with -15. The equation of this line is $y = 2x - 15$.

Graphing a Line by Using Its Slope and y-Intercept

Objective 7   Graph a line by using its slope and y-intercept.   Once we have an equation in slope–intercept form, we can use this information to graph the line. We can begin by plotting the y-intercept as our first point. We can then use the slope of the line to find a second point.

EXAMPLE 9   Graph the line $y = -4x + 8$ using its slope and y-intercept.

Solution

We will start at the y-intercept, which is $(0, 8)$. Since the slope is -4, this tells us that the line moves down four units as it moves one unit to the right. This would give us a second point at $(1, 4)$. Here is the graph:

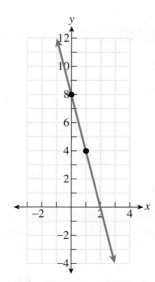

Quick Check **9**

Graph the line
$y = 2x + 6$ using its
slope and y-intercept.

EXAMPLE 10   Graph the line $y = \frac{1}{2}x - 3$ using its slope and y-intercept.

Solution

We will start by plotting the y-intercept at $(0, -3)$. The slope is $\frac{1}{2}$, which tells us that the line rises by one unit as it moves two units to the right. This will give us a second point at $(2, -2)$.

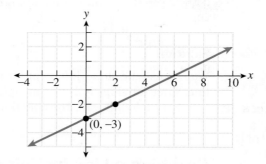

Quick Check 10
Graph the line
$y = -\dfrac{2}{3}x + 4$ using its
slope and y-intercept.

Graphing a Line by Using Its Slope and y-Intercept

- Plot the y-intercept $(0, b)$.
- Use the slope m to find another point on the line.
- Graph the line that passes through these 2 points.

We have two techniques for graphing lines that are not horizontal or vertical lines. We can graph a line either by finding its x- and y-intercepts or by using the y-intercept and the slope of the line. There are certain times when one technique is more efficient to apply than the other, and knowing which technique to use for a particular linear equation can save time and prevent errors. In general, if the equation is already solved for y, such as $y = 3x + 7$ or $y = -4x + 5$, then graphing the line using the y-intercept and the slope is a good choice. If the equation is not already solved for y, such as $2x + 3y = 12$ or $5x - 2y = -10$, consider finding both the x- and y-intercepts first and then drawing the line that passes through these two points.

Applications

Objective 8 Interpret the slope and y-intercept in real–world applications.
The concept of slope becomes much more important to us when we apply it to real-world situations.

EXAMPLE 11   The number of Americans (y), in millions, who have cellular phones can be approximated by the equation $y = 13x + 77$, where x is the number of years after 1999. Interpret the slope and y-intercept of this line. (*Source:* Cellular Telecommunications Industry Association)

Solution

The slope of this line is 13, and this tells us that each year we can expect an additional 13 million Americans to have cellular phones. (The number of Americans who have cellular phones is increasing because the slope is positive.) The y-intercept at $(0, 77)$ tells us that approximately 77 million Americans had cellular phones in 1999.

Quick Check 11 The number of women (y) accepted to medical school in a given year can be approximated by the equation $y = 218x + 7485$, where x is the number of years after 1997. Interpret the slope and y-intercept of this line. (*Source:* Association of American Medical Colleges)

Building Your Study Strategy Using Your Resources, 3 **Study Skills Courses**
Some colleges run short-term study skills courses or seminars, and these courses can be quite helpful to students who are learning mathematics. These courses cover everything from note taking and test taking to overcoming math anxiety. The tips and suggestions that you pick up can only help to improve your understanding of mathematics. To find out whether there are any such courses or seminars at your school, ask an academic counselor.

EXERCISES 3.3

F O R E X T R A H E L P

MyMathLab
MathXL
Interactmath.com
MathXL
Tutorials on CD
Video Lectures on CD
Tutor Center
Addison-Wesley
Math Tutor Center
Student's
Solutions Manual

Vocabulary

1. The _____ of a line is a measure of how steeply a line rises or falls as it moves to the right.

2. A line that rises from left to right has _____ slope.

3. A line that falls from left to right has _____ slope.

4. State the formula for the slope of a line that passes through two points.

5. A(n) _____ line has a slope of 0.

6. The slope of a(n) _____ line is undefined.

7. The _____ form of the equation of a line is $y = mx + b$.

8. In order to determine the slope of a line from its equation, the equation must be solved for _____.

Determine whether the given line has a positive or negative slope.

9.

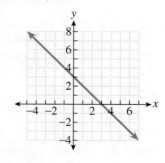

10.

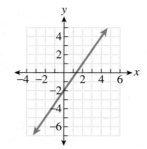

11.

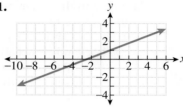

12.

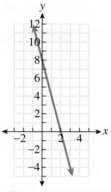

Find the slope of the given line. If the slope is undefined, state this.

13.

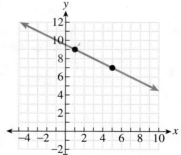

14.

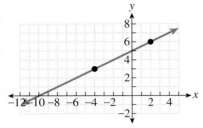

15.

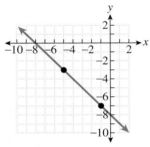

16.

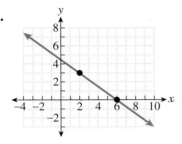

17.

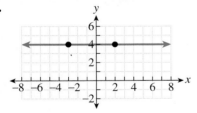

18.

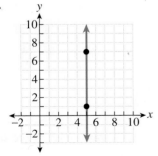

Find the slope of a line that passes through the given two points. If the slope is undefined, state this.

19. $(3, 5)$ and $(4, 7)$

20. $(2, 6)$ and $(3, 1)$

21. $(-2, 3)$ and $(-5, -6)$

22. $(5, -4)$ and $(-1, 5)$

23. $(3, 6)$ and $(-4, 6)$

24. $(-2, -6)$ and $(2, 6)$

25. $(0, 4)$ and $(0, 7)$

26. $(0, 0)$ and $(4, 7)$

Graph. Label any intercepts and determine the slope of the line. If the slope is undefined, state this.

27. $y = 4$

28. $x = -2$

29. $x = 0$

30. $y = -\dfrac{17}{5}$

31. $x = 8$

32. $y = 0$

Determine the equation of the given line, as well as the slope of the line. If the slope is undefined, state this.

33.

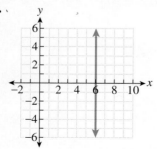

34.

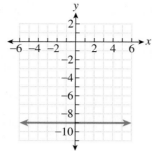

35.

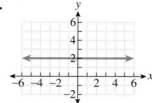

36.

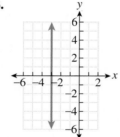

Find the slope and the y-intercept of the given line.

37. $y = 6x - 7$

38. $y = 4x + 11$

39. $y = -2x + 3$

40. $y = -5x - 8$

41. $6x + 4y = -10$

42. $8x - 6y = 12$

43. $x - 5y = 8$

44. $x + 4y = 14$

Find the equation of a line with the given slope and y-intercept.

45. Slope -2, y-intercept $(0, 5)$

46. Slope 4, y-intercept $(0, 3)$

47. Slope 3, y-intercept $(0, -6)$

48. Slope -5, y-intercept $(0, -2)$

57. $y = 4x$

58. $y = -2x$

49. Slope 0, y-intercept $(0, -4)$

50. Slope 0, y-intercept $(0, 1)$

Graph using the slope and y-intercept.

51. $y = 3x + 6$

52. $y = 2x - 8$

59. $y = \dfrac{7}{2}x - 7$

60. $y = -\dfrac{5}{4}x + 10$

53. $y = -2x - 6$

54. $y = -5x + 10$

(Mixed Practice, 61–82)

Graph using the most efficient technique.

61. $y = -x + 2$

55. $y = x - 3$

56. $y = -x - 1$

62. $y = -\dfrac{1}{4}x + 2$

63. $y = -2x + 8$

64. $y = -3x + 7$

71. $y = \dfrac{4}{5}x$

72. $x = \dfrac{9}{2}$

65. $y = 6x - 9$

66. $y = -2$

73. $y = 4$

74. $y = x + 7$

75. $x + 4y = 6$

76. $-10x + 4y = 10$

67. $y = -x - 5$

68. $y = -5x$

77. $y = \dfrac{3}{4}x + 3$

78. $x = -3$

69. $3x - 2y = -12$

70. $2x + 5y = 15$

79. $3x + y = 1$

80. $y = \dfrac{2}{5}x + 1$

Graph the two lines and find the coordinates of the point of intersection.

83. $y = 3x - 6$ and $y = -x - 2$

81. $5x - y = -5$

82. $-8x + 2y = -10$

84. $5x - 4y = 20$ and $y = \dfrac{1}{2}x + 1$

85. Over a 10-year period from 1990 to 2000, the value of a baseball card increased by \$40. If we let x represent the number of years after 1990, the value (y) of the card is given by the equation $y = 4x + 50$.

    a) Find the slope of the equation. In your own words, explain what this slope signifies.

    b) Find the y-intercept of the equation. In your own words, explain what this y-intercept signifies.

    c) Use this equation to predict the value of the card in the year 2015.

86. The number of nursing students in college (y) at a time x years after 1999 is approximated by the equation $y = -3600x + 60{,}000$. (Based on data from the years 1995–1999. *Source:* American Association of Colleges of Nursing)

a) Find the slope of the equation. In your own words, explain what this slope signifies.

b) Find the y-intercept of the equation. In your own words, explain what this y-intercept signifies.

c) Use this equation to predict the number of nursing students in the year 2025. Does this number seem possible? Explain why or why not.

87. On a 1000-foot stretch of highway through the mountains, the road rises by a total of 50 feet. Find the slope of the road.

88. George walked 1 mile (5280 feet) on a treadmill. The grade was set to 4%, which means that the slope of the treadmill is 4% or 0.04. Over the course of his workout, how many feet did George climb (vertically)?

Writing in Mathematics

Answer in complete sentences.

89. In general, the larger the animal, the slower its heart rate. Would an equation relating heart rate (y) to an animal's weight (x) have a positive or negative slope? Explain why.

90. The governor of a state predicts unemployment rates will increase for the next five years. Would an equation relating unemployment rate (y) to the number of years after the statement was made (x) have a positive or negative slope? Explain why.

91. Consider the equations $y = -\frac{3}{8}x + 7$ and $20x - 8y = 320$. For each equation, do you feel that graphing by finding the x- and y-intercepts or by using the slope and y-intercept is the most efficient way to graph the equation? Explain your choices.

92. *Solutions Manual*[*] Write a solutions manual page for the following problem:

Find the slope of the line that passes through the points $(-2, 6)$ *and* $(4, -8)$.

93. *Newsletter*[*] Write a newsletter explaining how to graph a line using its slope and y-intercept.

*See Appendix B for details and sample answers.

3.4
Linear Functions

Objectives

1. Define *function, domain,* and *range.*
2. Evaluate functions.
3. Graph linear functions.
4. Interpret the graph of a linear function.
5. Determine the domain and range of a function from its graph.

A coffeehouse sells coffee for $2 per cup. We know that it would cost $4 to buy 2 cups, $6 for 3 cups, $8 for 4 cups, and so on. We also know that the general formula for the cost of x cups in dollars is $2 \cdot x$. The cost depends on the number of cups bought. We say that the cost is a function of the number of cups bought.

Functions; Domain and Range

Objective 1   Define *function, domain,* and *range.* A **relation** is a rule that takes an input value from one set and assigns a particular output value from another set to it. A relation for which each input value is assigned one and only one output value is called a function. In the example about coffee, the input value is the number of cups of coffee bought, and the output value is the cost. For each number of cups bought, there is only one possible cost. The rule for determining the cost is to multiply the number of cups by $2.

Cups	1	2	3	4	. . .	x	. . .
	↓	↓	↓	↓	. . .	↓	
Cost	$2	$4	$6	$8		$2 \cdot x$	

The set of input values for a function is called the **domain** of the function. The domain for the example about coffee is the set of natural numbers $\{1, 2, 3, \ldots\}$. The set of output values for a function is called the **range** of the function. The range of the function in the coffee example is $\{2, 4, 6, \ldots\}$.

Function, Domain, and Range

A **function** is a rule that takes an input value and assigns one and only one output value to it. The **domain** of the function is the set of input values, and the **range** is the set of output values.

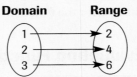

If an input value can be associated with more than one output value, then the relation is not a function. For example, a relation that took a month of the year as its input and listed the people at your school who were born in that month as its output would not be a function because each month has more than one person born in that month.

EXAMPLE 1 A mail-order company is selling holiday ornaments for $4 each. There is an additional $4.95 charge for shipping and handling per order. Find the function for the total cost of an order, as well as the domain and range of the function.

Solution

To determine the cost of an order, we begin by multiplying the number of ornaments by $4. To this we still need to add $4.95 for shipping and handling.

Function: If n is the number of ornaments ordered, then the cost is $4n + 4.95$.
Domain: Set of possible number of ornaments ordered $\{1, 2, 3, \ldots\}$
Range: Set of possible costs of the orders $\{\$8.95, \$12.95, \$16.95, \ldots\}$

Quick Check **1** A rental agency rents small moving trucks for $19.95 plus $0.15 per mile. Find the function for the total cost to rent a truck, as well as the domain and range of the function.

Function Notation

The perimeter of a square with side x can be found using the formula $P = 4x$. This is a function that can also be expressed using function notation as $P(x) = 4x$. **Function notation** is a way to present the output value of a function for the input x. The notation on the left side, $P(x)$, tells us the name of the function, P, as well as the input variable, x. $P(x)$ tells us that P is a function of x, and is read as "P of x." The parentheses on the left side are used to identify the input variable and are not used to indicate multiplication. We used P as the name of the function (P for perimeter), but we could have used any letter that we wanted. The letters f and g are frequently used for function names. The expression on the right side, $4x$, is the formula for the function.

$$P(x) = 4x$$
$\uparrow$

The variable inside the parentheses is the input variable for the function.

$$P(x) = 4x$$
$\uparrow$

$P(x)$ is the output value of the function P when the input value is x.

$$P(x) = 4x$$
$\uparrow$

The expression on the right side of the equation is the formula for this function.

Evaluating Functions

Objective 2 **Evaluate functions.** Suppose we wanted to find the perimeter of a square with a side of 3 inches. We are looking to evaluate our perimeter function, $P(x) = 4x$, for an input of 3, or, in other words, $P(3)$. When we find the output value of a function for a particular value of x, this is called **evaluating** the function. To evaluate a function for a particular value of the variable, we substitute that value for the variable in the function's formula, and then we simplify the resulting expression. To evaluate $P(3)$, we substitute 3 for x in the formula and simplify the resulting expression.

$$P(x) = 4x$$
$$P(3) = 4(3)$$
$$= 12$$

Since $P(3) = 12$, the perimeter is 12 inches.

EXAMPLE 2 Let $f(x) = x + 7$. Find $f(4)$.

Solution

We need to replace x in the function's formula by 4 and simplify the resulting expression.

$$f(4) = 4 + 7 \quad \text{Substitute 4 for } x.$$
$$= 11 \quad \text{Add.}$$

$f(4) = 11$. This means that when the input is $x = 4$, the output of the function is 11.

EXAMPLE 3 Let $g(x) = 3x - 8$. Find $g(-2)$.

Solution

In this example, we need to replace x by -2. As our functions become more complicated, it is a good idea to use parentheses when we substitute our input value.

$$g(-2) = 3(-2) - 8 \quad \text{Substitute } -2 \text{ for } x.$$
$$= -6 - 8 \quad \text{Multiply.}$$
$$= -14 \quad \text{Simplify.}$$

Quick Check 2
Let $g(x) = 2x + 9$. Find $g(-6)$.

EXAMPLE 4 Let $g(x) = 7x + 11$. Find $g(0)$.

Solution

Often, we will evaluate a function at $x = 0$. In this example, the term $7x$ is equal to 0 when $x = 0$, leaving the function equal to 11.

$$g(0) = 7(0) + 11 \quad \text{Substitute 0 for } x.$$
$$= 11 \quad \text{Simplify.}$$

Quick Check 3
Let $g(x) = 9x - 23$. Find $g(0)$.

EXAMPLE 5 Let $g(x) = 8 - 3x$. Find $g(a + 3)$.

Solution

In this example, we are substituting a variable expression for x in the function. After we replace x by $a + 3$, we will need to simplify the resulting variable expression.

$$g(a + 3) = 8 - 3(a + 3) \quad \text{Replace } x \text{ by } a + 3.$$
$$= 8 - 3a - 9 \quad \text{Distribute } -3.$$
$$= -3a - 1 \quad \text{Combine like terms.}$$

Quick Check 4
Let $g(x) = 7x - 12$. Find $g(a + 8)$.

Linear Functions and Their Graphs

Objective 3 **Graph linear functions.**

A **linear function** is a function of the form $f(x) = mx + b$, where m and b are real numbers.

Some examples of linear functions are $f(x) = x - 9$, $f(x) = 5x$, $f(x) = 3x + 11$, and $f(x) = 6$. We now turn our attention towards graphing linear functions. We graph any function $f(x)$ by plotting points of the form $(x, f(x))$. The output value of the function $f(x)$ is treated as the variable y was when we were graphing linear equations in two variables. When we graph a function $f(x)$, the vertical axis is used to represent the output values of the function.

We can begin to graph a linear function by finding the y-intercept. As with a linear equation that is in slope–intercept form, the y-intercept for the graph of a linear function $f(x) = mx + b$ is the point $(0, b)$. For example, the y-intercept for the graph of the function $f(x) = 5x - 8$ is the point $(0, -8)$. In general, to find the y-intercept of any function $f(x)$, we can find $f(0)$. After plotting the y-intercept, we can then use the slope m to find other points.

EXAMPLE 6   Graph the linear function $f(x) = \dfrac{1}{2}x + 2$.

Solution

Quick Check 5
Graph the linear function
$f(x) = \dfrac{3}{4}x - 6.$

We can start with the y-intercept, which is $(0, 2)$. The slope of the line is $\frac{1}{2}$, so beginning at the point $(0, 2)$, we move up one unit and two units to the right. This leads to a second point at $(2, 3)$.

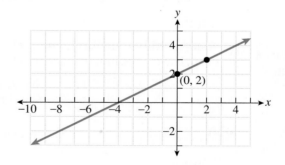

EXAMPLE 7   Graph the linear function $f(x) = -3$.

Solution

This function is known as a **constant function.** The function is constantly equal to -3, regardless of the input value x. Its graph is a horizontal line with a y-intercept at $(0, -3)$.

Quick Check **6**
**Graph the linear
function $f(x) = 4$.**

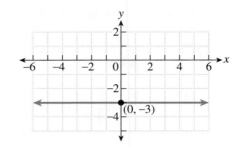

Interpreting the Graph of a Linear Function

Objective 4 **Interpret the graph of a linear function.** Here is the graph of a function $f(x)$:

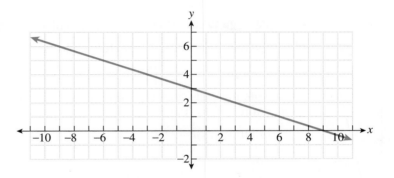

The ability to read and interpret a graph is an important skill. This line has an x-intercept at the point $(9, 0)$, so $f(9) = 0$. The y-intercept is at the point $(0, 3)$, so $f(0) = 3$.

Suppose that we wanted to find $f(3)$ for this particular function. We can do so by finding a point on the line that has an x-coordinate of 3. The y-coordinate of this point is $f(3)$. In this case, $f(3) = 2$.

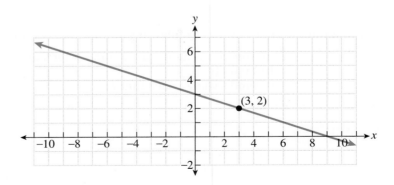

We can also use this graph to solve the equation $f(x) = 6$. Look for the point on the graph that has a y-coordinate of 6. The x-coordinate of this point is -9, so $x = -9$ is the solution of the equation $f(x) = 6$.

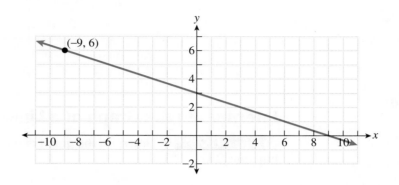

EXAMPLE ▶ 8 Consider the graph of a function $f(x)$ to the left.

a) Find the x-intercept and the y-intercept.

Solution

Note the scale on this graph, measured in units equal to $\frac{1}{2}$. The x-intercept is $(\frac{1}{2}, 0)$. The y-intercept is $(0, -2)$.

b) Find $f(1\frac{1}{2})$.

Solution

We are looking for a point on the line that has an x-coordinate of $1\frac{1}{2}$. The point is $(1\frac{1}{2}, 4)$, so $f(1\frac{1}{2}) = 4$.

c) Find a value x such that $f(x) = 2$.

Solution

We are looking for a point on the line that has a y-coordinate of 2, and this point is $(1, 2)$. The value that satisfies the equation $f(x) = 2$ is $x = 1$.

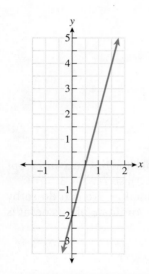

Quick Check **7** Consider the following graph of a function:

a) Find the x-intercept and the y-intercept.
b) Find $f(6)$.
c) Find a value x such that $f(x) = 6$.

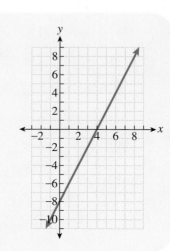

Objective **5**   Determine the domain and range of a function from its graph.
The domain and range of a function can also be read from a graph. Recall that the
domain of a function is the set of all input values. This corresponds to all of the
x-coordinates of the points on the graph. The domain is the interval of values on the
x-axis for which the graph exists. We read the domain from left to right on the graph.

The domain of a linear function is the set of all real numbers, which can be written in
interval notation as $(-\infty, \infty)$. The graphs of linear functions continue on to the left and to
the right. Look at the following three graphs of linear functions.

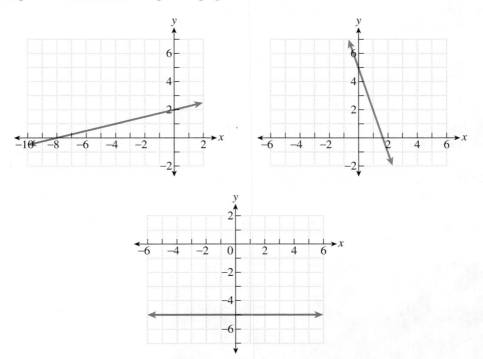

Each line continues to the left as well as to the right. This tells us that the graph exists for
all values of *x* in the interval $(-\infty, \infty)$.

The range of a function can be read vertically from a graph. The range goes from the
lowest point on the graph to the highest point. All linear functions have $(-\infty, \infty)$ as their
range, except constant functions.

Since a constant function of the form $f(x) = c$, where *c* is a real number, has only
one possible output value, its range would be $\{c\}$. For example, the range of the constant
function $f(x) = 3$ is $\{3\}$.

Building Your Study Strategy **Using Your Resources, 4 Student's Solutions
Manual**   The Student's Solutions Manual for a mathematics textbook can be a
valuable resource when used properly but it can be detrimental to learning if
used improperly. You should refer to the solutions manual to check your work or
to see where you went wrong with your solution. The detailed solution can be
helpful in this case.

Some students do their homework with the solutions manual sitting open in
front of them, looking at what they should write for the next line. If you do noth-
ing except essentially copy the solutions manual onto your own paper, you will
most likely have tremendous difficulty on the next exam.

EXERCISES 3.4

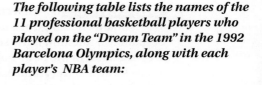

MyMathLab

MathXL

Interactmath.com

MathXL
Tutorials on CD

Video Lectures
on CD

Tutor Center

Addison-Wesley
Math Tutor Center

Student's
Solutions Manual

F O R   E X T R A   H E L P

Vocabulary

1. A(n) _____ is a relation that takes one input value and assigns one and only one output value to it.

2. The _____ of a function is the set of all possible input values.

3. The _____ of a function is the set of all possible output values.

4. Substituting a value for the function's variable and simplifying the resulting expression is called _____ the function.

5. A(n) _____ function is a function of the form $f(x) = mx + b$.

6. The _____ of a linear function is always $(-\infty, \infty)$.

The following table lists the names of the 11 professional basketball players who played on the "Dream Team" in the 1992 Barcelona Olympics, along with each player's NBA team:

Player	NBA Team
Patrick Ewing	New York Knicks
David Robinson	San Antonio Spurs
Karl Malone	Utah Jazz
Charles Barkley	Philadelphia 76ers
Larry Bird	Boston Celtics
Scottie Pippen	Chicago Bulls
Chris Mullin	Golden State Warriors
Michael Jordan	Chicago Bulls
Clyde Drexler	Portland Trail Blazers
Magic Johnson	Los Angeles Lakers
John Stockton	Utah Jazz

7. Would a relation that took a player's name from the "Dream Team" as an input and listed his NBA team as an output be a function? Explain why or why not.

8. Could a function be defined in the opposite direction, with the name of the NBA team as the input and the "Dream Team" player's name as the output? Explain why or why not.

9. Would a relation that took a person as an input and listed that person's mother as an output be a function? Why or why not?

10. Could a function be defined with a woman as an input and her child as an output? Why or why not?

For Exercises 11–14, determine whether a function exists with
a) set A as the input and set B as the output, and
b) set B as the input and set A as the output.

11. High Temperatures on December 16

Set A City	Set B High Temp.
Boston, MA	39° F
Orlando, FL	75° F
Providence, RI	39° F
Rochester, NY	26° F
Visalia, CA	57° F

12.

Set A Person	Set B Last 4 Digits of SSN
Jenny Crum	1234
Maureen O'Connor	5283
Dona Kenly	6405
Susan Winslow	5555
Lauren Morse	9200

13.

Set A Person	Set B Birthday
Greg Erb	December 15
Karen Guardino	December 17
Jolene Lehr	October 15
Siméon Poisson	June 21
Lindsay Skay	May 28
Sharon Smith	June 21

14.

Set A Football Player	Set B Uniform Number
Jeff Garcia	5
Garrison Hearst	20
Terrell Owens	81
Bryant Young	97
Julian Peterson	98

For the given set of ordered pairs, determine whether a function could be defined for which the input would be an x-coordinate and the output would be the corresponding y-coordinate. If a function cannot be defined in this manner, explain why.

15. $\{(-2, 4), (-1, 1), (0, 0), (1, 1), (2, 4), (3, 9)\}$

16. $\{(2, -2), (1, -1), (0, 0), (1, 1), (2, 2)\}$

17. $\{(5, 3), (2, 7), (-4, -6), (5, -2), (0, 4)\}$

18. $\{(-6, 3), (-2, 3), (1, 3), (5, 3), (11, 3)\}$

19. $\{(2, -5), (2, -1), (2, 0), (2, 3), (2, 5)\}$

20. $\{(1, 1), (2, 2), (3, 3), (4, 4), (5, 5)\}$

21. A Celsius temperature can be converted to a Fahrenheit temperature by multiplying it by $\frac{9}{5}$ and then adding 32.

a) Create a function $F(x)$ that converts a Celsius temperature x to a Fahrenheit temperature.

b) Use the function $F(x)$ from part a) to convert the following Celsius temperatures to Fahrenheit temperatures.

0° C   100° C   30° C   −10° C   −40° C

22. To convert a Fahrenheit temperature to a Celsius temperature we first subtract 32 from it, and then multiply that difference by $\frac{5}{9}$. Create a function $C(x)$ that converts a Fahrenheit temperature x into a Celsius temperature.

23. A college student takes a summer job selling newspaper subscriptions door-to-door. She is paid $36 for a 4-hour shift. She also earns $7 for each subscription sold.

   a) Create a function $f(x)$ for the amount she earns on a night when she sells x subscriptions.

   b) Use the function $f(x)$ to determine how much she earns on a night that she sells 12 subscriptions.

24. An elementary school holds a carnival. Attendees are charged $8 for dinner, plus 25¢ for each game played.

   a) Create a function $f(x)$ for the amount that a person pays if he eats dinner and plays x games.

   b) Use the function $f(x)$ to determine how much a person pays if he eats dinner and plays 45 games.

25. Create a linear function whose graph has a slope of 4 and a y-intercept at $(0, 3)$.

26. Create a linear function whose graph has a slope of 5 and a y-intercept at $(0, -9)$.

27. Create a linear function whose graph has a slope of -3 and a y-intercept at $(0, -4)$.

28. Create a linear function whose graph has a slope of $-\frac{1}{2}$ and a y-intercept at $(0, \frac{2}{3})$.

29. Create a linear function whose graph has a slope of 0 and a y-intercept at $(0, 6)$.

30. Create a linear function whose graph has a slope of 0 and a y-intercept at $(0, 0)$.

Evaluate the given function.

31. $g(x) = x + 2$, $g(-5)$

32. $h(x) = -5x$, $h(4)$

33. $f(x) = 2x - 3$, $f(5)$

34. $f(x) = 4x + 1$, $f(-2)$

35. $f(x) = -7x + 11$, $f(-3)$

36. $f(x) = 15 - 4x$, $f(4)$

37. $f(x) = 9x - 25$, $f(0)$

38. $f(x) = 6x + 13$, $f(0)$

39. $g(x) = 3x - 1$, $g\left(\dfrac{2}{3}\right)$

40. $g(x) = 5x + 7$, $g\left(\dfrac{9}{5}\right)$

41. $f(x) = 3x + 4$, $f(a)$

42. $f(x) = 2x - 3$, $f(b)$

43. $f(x) = 7x - 2$, $f(a + 3)$

44. $f(x) = 5x + 9$, $f(a - 7)$

45. $f(x) = 16 - 3x$, $f(2a - 5)$

46. $f(x) = 5 - 6x$, $f(3a - 4)$

47. $f(x) = 6x + 4$, $f(x + h)$

48. $f(x) = 3x + 11$, $f(x + h)$

Graph the linear function.

49. $f(x) = 6x - 6$ **50.** $f(x) = 2x + 6$

51. $f(x) = -3x + 3$ **52.** $f(x) = -x - 8$

53. $f(x) = \dfrac{4}{3}x + 4$ **54.** $f(x) = \dfrac{2}{5}x - 2$

55. $f(x) = 7x$

56. $f(x) = -9x$

61. Refer to the graph of the function $f(x)$.

   a) Find the x-intercept.

   b) Find the y-intercept.

   c) Find $f(5)$.

   d) Find a value a such that $f(a) = -8$.

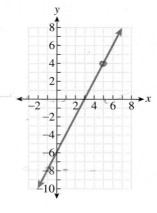

57. $f(x) = \dfrac{8}{3}x$

58. $f(x) = -\dfrac{3}{7}x$

62. Refer to the graph of the function $g(x)$.

   a) What is the x-intercept?

   b) What is the y-intercept?

   c) Find $g(-4)$.

   d) Find a value a such that $g(a) = -2$.

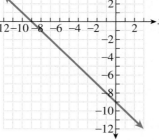

59. $f(x) = 4$

63. Refer to the graph of the function $f(x)$.

   a) What is the x-intercept?

   b) What is the y-intercept?

   c) Find $f(6)$.

   d) Find a value a such that $f(a) = 9$.

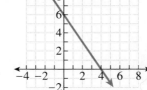

60. $f(x) = -6$

64. Refer to the graph of the funciton $f(x)$.

  a) What is the x-intercept?

  b) What is the y-intercept?

  c) Find $f(-1)$.

  d) Find a value a such that $f(a) = -8$.

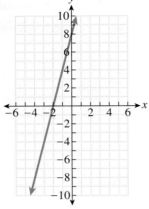

67.

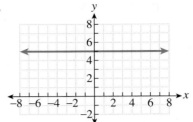

68.

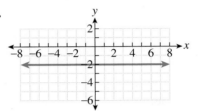

Find the domain and range of the given function.

65.

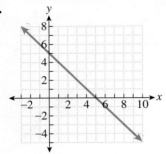

66.

Writing in Mathematics

Answer in complete sentences.

69. Give an example of two sets A and B and a rule that is a function from set A to set B. Explain why your rule meets the definition of a function. Give another rule that would not be a function from set B to set A. Explain why your rule does not meet the definition of a function.

70. *Solutions Manual** Write a solutions manual page for the following problem:

  Graph the linear function $f(x) = \frac{2}{3}x - 4$.

71. *Newsletter** Write a newsletter explaining how to graph linear functions.

*See Appendix B for details and sample answers.

QUICK REVIEW EXERCISES

Section 3.4

Find the slope of the given line.

  1. $6x + 2y = 17$

  2. $5x - 4y = -16$

  3. $x - 5y = 10$

  4. $14x + 10y = 30$

1 Determine whether two lines are parallel.
2 Determine whether two lines are perpendicular.

Parallel Lines

Objective 1 Determine whether two lines are parallel. In this section, we will examine the relationship between two lines. Consider the following pair of lines.

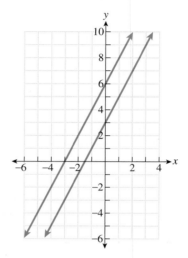

We notice that these two lines do not intersect. The lines have the same slope and are called **parallel lines.**

Parallel Lines

- Two nonvertical lines are **parallel** if they have the same slope. In other words, if we denote the slope of one line as m_1 and the slope of the other line as m_2, the two lines are parallel if $m_1 = m_2$.
- If two lines are both vertical lines, they are parallel.

Following are some examples of lines that are parallel:

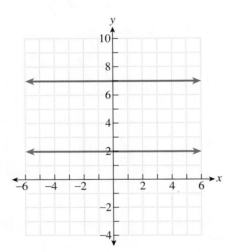

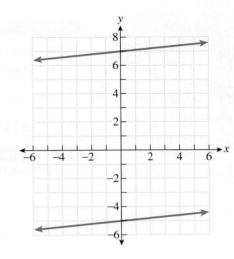

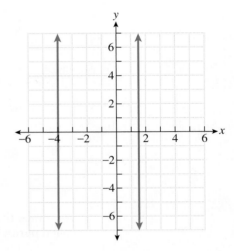

EXAMPLE ▸1   Are the two lines $y = 2x - 5$ and $y = 3x + 1$ parallel?

Quick Check ◂1   **Solution**

Are the two lines $y = 4x + 3$ and $y = -4x$ parallel?

These two equations are in slope–intercept form, so we can see that the slope of the first line is 2 and the slope of the second line is 3. Since the slopes are not equal, these two lines are not parallel.

EXAMPLE ▸2   Are the two lines $6x + 3y = -9$ and $y = -2x + 2$ parallel?

Solution

To find the slope of the first line, we solve the equation for y:

$$6x + 3y = -9$$
$$3y = -6x - 9 \qquad \text{Subtract } 6x \text{ from both sides.}$$
$$y = -2x - 3 \qquad \text{Divide by 3.}$$

The slope of the first line is -2. Since the second line is already in slope–intercept form, we see that the slope of that line is also -2. Since the two slopes are equal, the two lines are parallel.

Quick Check **2** Are the two lines $8x + 6y = 36$ and $y = -\dfrac{4}{3}x + 5$ parallel?

Any two horizontal lines, such as $y = 4$ and $y = 1$, are parallel to each other, since their slopes are both 0. Any two vertical lines, such as $x = 2$ and $x = -1$, are parallel to each other as well, by definition.

EXAMPLE ▸ **3**   Find the slope of a line that is parallel to the line $4x + 3y = 8$.

Solution

For a line to be parallel to $4x + 3y = 8$, it must have the same slope as this line. To find the slope of this line, we solve the equation for y.

$$4x + 3y = 8$$
$$3y = -4x + 8 \qquad \text{Subtract } 4x \text{ from both sides.}$$
$$y = -\frac{4}{3}x + \frac{8}{3} \qquad \text{Divide by 3.}$$

Quick Check **3**
Find the slope of a line that is parallel to the line $2x - 9y = 36$.

The slope of the line $4x + 3y = 8$ is $-\frac{4}{3}$. A line that is parallel to $4x + 3y = 8$ has slope $-\frac{4}{3}$.

Perpendicular Lines

Objective **2**   Determine whether two lines are perpendicular.

Two distinct lines that are not parallel intersect at one point. **Perpendicular lines** are one special type of intersecting lines. Here is an example of two lines that are perpendicular. Perpendicular lines intersect at right angles. Notice that one of these lines has a positive slope and the other line has a negative slope.

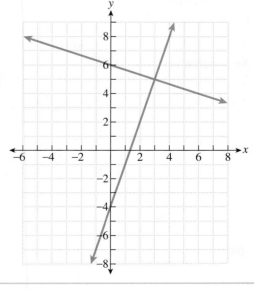

Perpendicular Lines

- Two nonvertical lines are **perpendicular** if their slopes are **negative reciprocals.** In other words, if we denote the slope of one line as m_1 and the slope of the other line as m_2, the two lines are perpendicular if $m_1 = -\dfrac{1}{m_2}$. (This is equivalent to saying that two lines are perpendicular if the product of their slopes is -1.)

- A vertical line is perpendicular to a horizontal line.

Two numbers are **negative reciprocals** if they are reciprocals that have opposite signs, such as 5 and $-\frac{1}{5}$, -2 and $\frac{1}{2}$, and $-\frac{3}{10}$ and $\frac{10}{3}$.

EXAMPLE 4   Are the two lines $y = 4x + 5$ and $y = -4x + 5$ perpendicular?

Solution

The slopes of these two lines are 4 and -4, respectively. While the signs of these two slopes are opposite, the slopes are not reciprocals. Therefore, the two lines are not perpendicular.

Quick Check 4
Are the two lines
$y = x + 7$ and
$y = -x + 3$
perpendicular?

EXAMPLE 5   Are the two lines $y = 3x + 7$ and $2x + 6y = 5$ perpendicular?

Solution

The slope of the first line is 3. To find the slope of the second line, we solve the equation for y.

$$2x + 6y = 5$$
$$6y = -2x + 5 \qquad \text{Subtract } 2x.$$
$$y = -\frac{2}{6}x + \frac{5}{6} \qquad \text{Divide by 6.}$$
$$y = -\frac{1}{3}x + \frac{5}{6} \qquad \text{Simplify.}$$

Quick Check 5
Are the two lines
$y = \dfrac{2}{3}x + 5$ and
$6x + 4y = 12$

perpendicular?

The slope of the second line is $-\frac{1}{3}$. The two slopes are negative reciprocals, so the lines are perpendicular.

Any vertical line, such as $x = 2$, is perpendicular to any horizontal line, such as $y = -2$.

EXAMPLE 6   Find the slope of a line that is perpendicular to the line $5x + 2y = 8$.

Solution

We begin by finding the slope of this line. This can be done by solving the equation $5x + 2y = 8$ for y.

$$5x + 2y = 8$$
$$2y = -5x + 8 \qquad \text{Subtract } 5x \text{ from both sides.}$$
$$y = -\frac{5}{2}x + \frac{8}{2} \qquad \text{Divide by 2.}$$
$$y = -\frac{5}{2}x + 4 \qquad \text{Simplify.}$$

The slope of this line is $-\frac{5}{2}$. To find the slope of a line perpendicular to this line, we take the reciprocal of this slope and change its sign from negative to positive. A line is perpendicular to $5x + 2y = 8$ if it has a slope of $\frac{2}{5}$.

Quick Check　**6**
Find the slope of a line that is perpendicular to the line $7x - 3y = 21$.

Determining Whether Two Lines Are Parallel, Perpendicular, or Neither

To summarize, nonvertical lines are parallel if and only if they have the same slope and are perpendicular if and only if their slopes are negative reciprocals. Horizontal lines are parallel to other horizontal lines, and vertical lines are parallel to other vertical lines. Finally, a horizontal line and a vertical line are perpendicular to each other. These concepts are summarized in the following table:

If . . .	the lines are parallel if . . .	the lines are perpendicular if . . .
two nonvertical lines have slopes m_1 and m_2	$m_1 = m_2$	$m_1 = -\dfrac{1}{m_2}$
one of the two lines is horizontal	the other line is horizontal	the other line is vertical
one of the two lines is vertical	the other line is vertical	the other line is horizontal

EXAMPLE　7　Are the two lines $y = 6x + 2$ and $y = \frac{1}{6}x - 3$ parallel, perpendicular, or neither?

Quick Check　**7**
Are the two lines $y = 2x - 5$ and $y = -2x$ parallel, perpendicular, or neither?

Solution

The slope of the first line is 6 and the slope of the second line is $\frac{1}{6}$. The slopes are not equal, so the lines are not parallel.

The slopes are not negative reciprocals, so the lines are not perpendicular either.

The two lines are neither parallel nor perpendicular.

EXAMPLE　8　Are the two lines $-3x + y = 7$ and $-6x + 2y = -4$ parallel, perpendicular, or neither?

Solution

We find the slope of each line by solving each equation for y.

$$-3x + y = 7$$
$$y = 3x + 7 \qquad \text{Add } 3x \text{ to both sides.}$$

The slope of the first line is 3. Now we find the slope of the second line.

Quick Check　**8**
Are the two lines $y = 4x + 3$ and $8x - 2y = 12$ parallel, perpendicular, or neither?

$$-6x + 2y = -4$$
$$2y = 6x - 4 \qquad \text{Add } 6x \text{ to both sides.}$$
$$y = 3x - 2 \qquad \text{Divide by 2.}$$

The slope of the second line is also 3.

Since the two slopes are equal, the lines are parallel.

EXAMPLE 9   Are the two lines $10x + 2y = 0$ and $x - 5y = 3$ parallel, perpendicular, or neither?

Solution

We begin by finding the slope of each line.

$$10x + 2y = 0$$
$$2y = -10x \qquad \text{Subtract } 10x \text{ from both sides.}$$
$$y = -5x \qquad \text{Divide by 2.}$$

The slope of the first line is -5. Now we find the slope of the second line.

$$x - 5y = 3$$
$$-5y = -x + 3 \qquad \text{Subtract } x \text{ from both sides.}$$
$$y = \frac{1}{5}x - \frac{3}{5} \qquad \text{Divide by } -5.$$

The slope of the second line is $\frac{1}{5}$. Since the two slopes are negative reciprocals, the lines are perpendicular.

Quick Check 9
Are the two lines
$y = -\frac{2}{5}x + 9$ and
$5x - 2y = -8$ parallel, perpendicular, or neither?

EXAMPLE 10   Find the equation of the line that is perpendicular to $y = \frac{1}{3}x + 2$ and has a y-intercept at $(0, 6)$.

Quick Check 10
Find the equation of the line that is parallel to $6x + 4y = 21$ and has a y-intercept at $(0, -7)$.

Solution

The slope of the line $y = \frac{1}{3}x + 2$ is $\frac{1}{3}$. Since we are looking for the equation of a line that is perpendicular to $y = \frac{1}{3}x + 2$, the slope of this line is -3. In addition to knowing the slope of the line, we know that the y-coordinate of the y-intercept is 6. We can now find the equation of this line using the slope–intercept form $(y = mx + b)$. The equation of the line is $y = -3x + 6$.

Building Your Study Strategy Using Your Resources, 5 **Classmates**   Some frequently overlooked resources are the students who are taking the class with you.

- If you have a quick question about a particular problem, a classmate can help you.
- If your notes are incomplete for a certain day, a classmate may allow you to use his or her notes to fill in the holes in your own notes.
- If you are forced to miss class, a quick call to a classmate can help you find out what was covered in class and what the homework assignment is.
- You can form a study group with your fellow classmates.
- A classmate can also help to motivate you when you are feeling low.

EXERCISES 3.5

Vocabulary

1. Two nonvertical lines are _____ if they have the same slope.

2. Two nonvertical lines are _____ if their slopes are negative reciprocals.

3. A vertical line is parallel to a(n) _____ line.

4. A vertical line is perpendicular to a(n) _____ line.

5. Two _____ lines do not intersect.

6. Two _____ lines intersect at right angles.

Are the two given lines parallel?

7. $y = 3x + 5, y = 3x - 2$

8. $y = 5x - 7, y = -5x + 3$

9. $4x + 2y = 9, 3y = 6x + 7$

10. $x + 3y = -4, 3x + 9y = 8$

11. $y = 6, y = -6$

12. $x = 2, x = 7$

Are the two given lines perpendicular?

13. $y = 4x, y = \frac{1}{4}x - 3$

14. $y = -\frac{3}{2}x + 2, y = \frac{2}{3}x + 1$

15. $15x + 3y = 11, x - 5y = -4$

16. $x + y = 6, x - y = -3$

17. $y = -7, y = \frac{1}{7}$

18. $x = 3, y = 4$

Are the two given lines parallel, perpendicular, or neither?

19. $y = 6x - 2, y = -6x + 5$

20. $y = 7x - 9, y = \frac{1}{7}x + 3$

21. $y = 4x + 3, y = -\frac{1}{4}x - 6$

22. $y = 8x - 16, y = 8x - 1$

23. $10x + 2y = 7, y = -5x + 12$

24. $9x + 3y = 15, 4x - 12y = 8$

25. $y = 4, y = \frac{1}{4}$

26. $x - 4y = 11, 8y = 2x + 3$

27. $30x + 10y = 17, 12x - 4y = 9$

28. $x = 3, x = 0$

29. $x = 5, y = -2$

30. $x + y = 16, y = x - 7$

Are the lines associated with the given functions parallel, perpendicular, or neither?

31. $f(x) = 7x - 5, g(x) = 7x + 3$

32. $f(x) = -2x - 5, g(x) = -\frac{1}{2}x + 4$

33. $f(x) = 6, g(x) = 6x + 4$

34. $f(x) = 4x + 9, g(x) = -\frac{1}{4}x + 3$

35. $f(x) = x + 7, g(x) = x + 2$

36. $f(x) = 5, g(x) = -5$

Find the slope of a line that is parallel to the given line. If the slope is undefined, state this.

37. $y = -6x + 7$

38. $y = 3x + 2$

39. $y = 5x$

40. $y = -3$

41. $12x + 4y = 8$

42. $2x + 6y = -7$

Find the slope of a line that is perpendicular to the given line. If the slope is undefined, state this.

43. $y = \dfrac{1}{4}x + 1$

44. $y = -3x + 14$

45. $y = \dfrac{3}{5}x - 6$

46. $14x - 6y = 17$

47. $15x + 24y = 39$

48. $x = -63$

49. Find the slope of a line that is parallel to the line $Ax + By = C$. (A, B, and C are real numbers, $A \neq 0$, and $B \neq 0$.)

50. Find the slope of a line that is perpendicular to the line $Ax + By = C$. (A, B, and C are real numbers, $A \neq 0$, and $B \neq 0$.)

51. Is the line that passes through $(3, 7)$ and $(5, -1)$ parallel to the line $y = -4x + 11$?

52. Is the line that passes through $(-6, 2)$ and $(4, 8)$ parallel to the line $-3x + 5y = 15$?

53. Is the line that passes through $(4, 7)$ and $(-1, 9)$ perpendicular to the line $-10x + 4y = 8$?

54. Is the line that passes through $(-3, -8)$ and $(2, 7)$ perpendicular to the line that passes through $(-5, 3)$ and $(7, 7)$?

55. Find the equation of a line that is parallel to $y = 2x + 9$ and has a y-intercept at $(0, -6)$. Graph the line and label any intercepts.

56. Find the equation of a line that is parallel to $3x + 2y = 7$ and has a y-intercept at $(0, 9)$. Graph the line and label any intercepts.

57. Find the equation of a line that is parallel to $y = 13$ and has a y-intercept at $(0, -4)$. Graph the line and label any intercepts.

58. Find the equation of a line that is perpendicular to $y = -3x + 12$ and has a y-intercept at $(0, 3)$. Graph the line and label any intercepts.

59. Find the equation of a line that is perpendicular to $5x - 3y = -9$ and has a y-intercept at $(0, -3)$. Graph the line and label any intercepts.

60. Find the equation of a line that is perpendicular to $x = 3$ and has a y-intercept at $(0, 7)$. Graph the line and label any intercepts.

Writing in Mathematics

Answer in complete sentences.

61. Describe three real-world examples of parallel lines.

62. Describe three real-world examples of perpendicular lines.

63. Explain the process for determining whether two lines of the form $Ax + By = C$ are parallel.

64. Explain the process for determining whether two lines of the form $Ax + By = C$ are perpendicular.

65. *Solutions Manual* * Write a solutions manual page for the following problem:

Are the two given lines parallel, perpendicular, or neither? $6x - 2y = 10, 3x + 9y = 21$

66. *Newsletter* * Write a newsletter explaining how to determine whether two lines are parallel or perpendicular.

*See Appendix B for details and sample answers.

QUICK REVIEW EXERCISES

Section 3.5

Find the slope of a line that passes through the given two points. If the slope is undefined, state this.

1. $(5, 7)$ and $(8, 1)$

2. $(-4, 2)$ and $(2, 10)$

3. $(-9, -5)$ and $(-7, 5)$

4. $(6, 0)$ and $(-2, -6)$

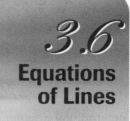

3.6
Equations of Lines

Objectives

1 Find the equation of a line using the point–slope form.

2 Find the equation of a line given two points on the line.

3 Find a linear equation to describe real data.

4 Find the equation of a parallel or perpendicular line.

We are already familiar with the slope–intercept form of the equation of a line: $y = mx + b$. When an equation is written in this form, we know both the slope of the line (m) and the y-coordinate of the y-intercept (b). This form is convenient for graphing lines. In this section, we will look at another form of the equation of a line. We will also learn how to find the equation of a line if we know the slope of a line and any point through which the line passes.

Point–Slope Form of the Equation of a Line

Objective 1 Find the equation of a line using the point–slope form. In Section 3.3, we learned that if we know the slope of a line and its y-intercept, we can write the equation of the line using the slope–intercept form of a line $y = mx + b$, where m is the slope of the line and b is the y-coordinate of the y-intercept. If we know the slope of a line and the coordinates of any point on that line, not just the y-intercept, we can write the equation of the line using the point–slope form of an equation.

Point–Slope Form of the Equation of a Line

> The **point–slope form** of the equation of a line with slope m that passes through the point (x_1, y_1) is
> $$y - y_1 = m(x - x_1).$$

This form can be derived directly from the slope formula $m = \dfrac{y_2 - y_1}{x_2 - x_1}$. If we let (x, y) represent an arbitrary point on the line, this formula becomes $m = \dfrac{y - y_1}{x - x_1}$. Multiplying both sides of that equation by $(x - x_1)$ produces the point–slope form of the equation of a line.

If a line has slope 2 and passes through the point $(3, 1)$, we find its equation by substituting 2 for m, 3 for x_1, and 1 for y_1. Its equation is $y - 1 = 2(x - 3)$. This equation can be left in that form, or converted to slope–intercept form or standard form.

EXAMPLE 1 Find the equation for a line with slope 4 that passes through the point $(-2, 5)$. Write the equation in slope–intercept form.

Solution

We will substitute 4 for m, -2 for x_1, and 5 for y_1 in the point–slope form. We will then solve the equation for y in order to write the equation in slope–intercept form.

$5x - 2y = 10 \quad (\overset{x}{2}, \overset{y}{5}) \quad y = mx + b$

24

10

13

11

14

12

15

$$y - 5 = 4(x - (-2)) \qquad \text{Substitute into point–slope form.}$$
$$y - 5 = 4(x + 2) \qquad \text{Eliminate double signs.}$$
$$y - 5 = 4x + 8 \qquad \text{Distribute.}$$
$$y = 4x + 13 \qquad \text{Add 5 to isolate } y.$$

The equation for a line with slope 4 that passes through $(-2, 5)$ is $y = 4x + 13$. Below is the graph of the line, showing that it passes through the point $(-2, 5)$.

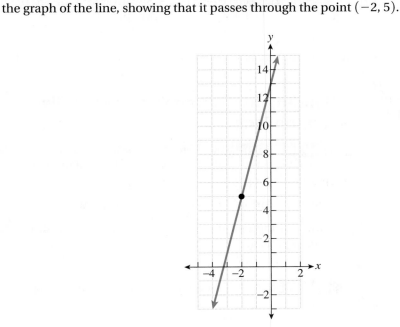

Quick Check 1

Find the equation for a line with slope 2 that passes through the point $(1, 5)$. Write the equation in slope–intercept form.

EXAMPLE 2 Find the equation for a line with slope -1 and that passes through the point $(4, -3)$. Write the equation in slope–intercept form.

Solution

We begin by substituting -1 for m, 4 for x_1, and -3 for y_1 in the point–slope form. After substituting, we solve the equation for y.

$$y - (-3) = -1(x - 4) \qquad \text{Substitute into point–slope form.}$$
$$y + 3 = -1(x - 4) \qquad \text{Eliminate double signs.}$$
$$y + 3 = -x + 4 \qquad \text{Distribute } -1.$$
$$y = -x + 1 \qquad \text{Subtract 3 to isolate } y.$$

Quick Check 2

Find the equation for a line with slope -3 and that passes through the point $(4, -6)$. Write the equation in slope–intercept form.

The equation for a line with slope -1 that passes through $(4, -3)$ is $y = -x + 1$.

In the previous example, we converted the equation from point–slope form to slope–intercept form. This is a good idea in general. We use the point–slope form because it is a convenient form for finding the equation of a line if we know its slope and the coordinates of a point on the line. We convert the equation to slope–intercept form because it is easier to graph a line when the equation is in this form.

EXAMPLE 3 The line associated with the linear function $f(x)$ has a slope of 2. If $f(-3) = -7$, find the function $f(x)$.

Solution

A linear function is of the form $f(x) = mx + b$. In this example, we know that $m = 2$, so $f(x) = 2x + b$. We will now use the fact that $f(-3) = -7$ to find b:

$$f(-3) = -7$$
$$2(-3) + b = -7 \qquad \text{Substitute } -3 \text{ for } x \text{ in the function } f(x).$$
$$-6 + b = -7 \qquad \text{Multiply.}$$
$$b = -1 \qquad \text{Add 6.}$$

Quick Check 3

The line associated with the linear function $f(x)$ has a slope of -4. If $f(-2) = 13$, find the function $f(x)$.

Now that we have found b, we know that $f(x) = 2x - 1$.

Finding the Equation of a Line Given Two Points on the Line

Objective 2   Find the equation of a line given two points on the line. Another use of the point–slope form is to find the equation of a line that passes through two given points. We begin by finding the slope of the line passing through those two points using the slope formula $m = \dfrac{y_2 - y_1}{x_2 - x_1}$. Then we use the point–slope form with this slope and either of the two points we were given. We finish by rewriting the equation in slope–intercept form.

EXAMPLE ▸ 4   Find the equation of a line that passes through the two points $(-1, 6)$ and $(2, 0)$.

Solution

We begin by finding the slope of the line that passes through these two points.

$$m = \frac{0 - 6}{2 - (-1)} \qquad \text{Substitute into the formula } m = \frac{y_2 - y_1}{x_2 - x_1}.$$
$$= \frac{-6}{3} \qquad \text{Simplify numerator and denominator.}$$
$$= -2 \qquad \text{Simplify.}$$

We can now substitute into the point–slope form using either of the two points with $m = -2$. We will use $(2, 0)$:

$$y - 0 = -2(x - 2) \qquad \text{Substitute in point–slope form.}$$
$$y = -2x + 4 \qquad \text{Distribute.}$$

Quick Check 4

Find the equation of a line that passes through the two points $(-2, 7)$ and $(6, -5)$.

The equation of the line that passes through these two points is $y = -2x + 4$. If we had used the point $(-1, 6)$ instead of $(2, 0)$, our result would have been the same.

At the right is the graph of the line, passing through the points $(-1, 6)$ and $(2, 0)$.

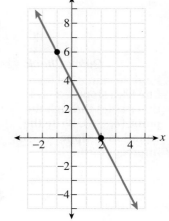

EXAMPLE 5   Find the equation of a line that passes through the two points $(8, 2)$ and $(8, 9)$.

Solution

We begin by attempting to find the slope of the line that passes through these two points.

$$m = \frac{9 - 2}{8 - 8} \qquad \text{Substitute into the slope formula.}$$

$$= \frac{7}{0} \qquad \text{Simplify the numerator and denominator.}$$

The slope is undefined, so this line is a vertical line. (We could have discovered this by plotting these two points on a graph.) The equation for this line is $x = 8$.

Quick Check **5**
Find the equation of a line that passes through the two points $(-6, 4)$ and $(3, 4)$.

Finding a Linear Equation to Describe Linear Data

Objective 3   Find a linear equation to describe real data.

EXAMPLE 6   In 2003, there were approximately 6.2 million U.S. households with a net worth of at least \$1 million. By 2005, that number had increased to 8.9 million U.S. households. Find a linear equation that describes the number (y) of U.S. millionaire households (in millions) x years after 2003. (*Source:* TNS Financial Services, Affluent Market Research Program)

Solution

Since x represents the number of years after 2003, then $x = 0$ for the year 2003 and $x = 2$ for the year 2005. This tells us that two points on the line are $(0, 6.2)$ and $(2, 8.9)$. We begin by calculating the slope of the line.

$$m = \frac{8.9 - 6.2}{2 - 0} \qquad \text{Substitute into the slope formula.}$$

$$= \frac{2.7}{2} \qquad \text{Simplify the numerator and the denominator.}$$

$$= 1.35 \qquad \text{Divide.}$$

The slope is 1.35, which tells us that the number of U.S. millionaires increases by 1.35 million per year. In this example, we know that the y-intercept is $(0, 6.2)$, so we can write the equation directly in slope–intercept form. The equation is $y = 1.35x + 6.2$. (If we did not know the y-intercept of the line, we would find the equation by substituting into the point–slope form.)

Quick Check **6** In 2000, approximately 84.6 million people traveled to a theme or amusement park. By 2002, this number had increased to approximately 92.4 million people. Find a linear equation that tells the number (y) of people, in millions, who traveled to a theme or amusement park x years after 2000. (*Source: Domestic Travel Market Report, Travel Industry Association of America*)

Finding the Equation of a Parallel or Perpendicular Line

Objective **4** **Find the equation of a parallel or perpendicular line.** To find the equation of a line, we must know the slope of the line and the coordinates of a point on the line. In the previous examples, either we were given the slope or we calculated the slope using two points which were on the line. Sometimes, the slope of the line is given in terms of another line. We could be given the equation of a line either parallel or perpendicular to the line for which we are trying to find the equation.

EXAMPLE 7 Find the equation of a line that is parallel to the line $y = -\frac{3}{4}x + 15$ and that passes through $(-8, 5)$.

Solution

Since the line is parallel to $y = -\frac{3}{4}x + 15$, its slope must be $-\frac{3}{4}$. We will substitute this slope, along with the point $(-8, 5)$, into the point–slope form to find the equation of this line:

$$y - 5 = -\frac{3}{4}(x - (-8)) \qquad \text{Substitute into point–slope form.}$$

$$y - 5 = -\frac{3}{4}(x + 8) \qquad \text{Eliminate double signs.}$$

$$y - 5 = -\frac{3}{4}x - \frac{3}{\overset{1}{\cancel{4}}} \cdot \overset{2}{\cancel{8}} \qquad \text{Distribute and divide out common factors.}$$

$$y - 5 = -\frac{3}{4}x - 6 \qquad \text{Simplify.}$$

$$y = -\frac{3}{4}x - 1 \qquad \text{Add 5.}$$

Quick Check **7**
Find the equation of a line that is parallel to the line $y = \frac{2}{5}x - 9$ and that passes through $(5, 4)$.

The equation of the line parallel to $y = -\frac{3}{4}x + 15$ and passing through $(-8, 5)$ is $y = -\frac{3}{4}x - 1$.

EXAMPLE 8 Find the equation of a line that is perpendicular to the line $y = -\frac{1}{2}x + 7$ and that passes through $(5, 3)$.

Solution

Quick Check **8**
Find the equation of a line that is perpendicular to the line $9x - 12y = 4$ and that passes through $(6, -2)$.

The slope of the line $y = -\frac{1}{2}x + 7$ is $-\frac{1}{2}$, so the slope of a perpendicular line must be the negative reciprocal of $-\frac{1}{2}$, or 2. We now can substitute into the point–slope form:

$$y - 3 = 2(x - 5) \qquad \text{Substitute into point–slope form.}$$
$$y - 3 = 2x - 10 \qquad \text{Distribute.}$$
$$y = 2x - 7 \qquad \text{Add 3.}$$

The equation of the line is $y = 2x - 7$.

If we are given . . .	We find the equation by . . .
The slope and the y-intercept	Substituting m and b into the slope–intercept form $y = mx + b$
The slope and a point on the line	Substituting m and the coordinates of the point into the point–slope form $y - y_1 = m(x - x_1)$
Two points on the line	Calculating m using the slope formula $m = \dfrac{y_2 - y_1}{x_2 - x_1}$ and then substituting the slope and the coordinates of one of the points into the point–slope form $y - y_1 = m(x - x_1)$
A point on the line and the equation of a parallel line	Substituting the slope of that line and the coordinates of the point into the point–slope form $y - y_1 = m(x - x_1)$
A point on the line and the equation of a perpendicular line	Substituting the negative reciprocal of the slope of that line and the coordinates of the point into the point–slope form $y - y_1 = m(x - x_1)$

Keep in mind that if the slope of the line is undefined, then the line is vertical and its equation is of the form $x = a$, where a is the x-coordinate of the given point. We do not use the point–slope form or the slope–intercept form to find the equation of a vertical line.

Building Your Study Strategy Using Your Resources, 6   **Internet Resources**
Consider using the Internet as a resource that will allow you to conduct further research on topics you are learning. There are a great number of Web sites dedicated to mathematics. These sites have alternative explanations, examples, and practice problems. If you find a Web site that helps you with one particular topic, check that site when you are researching another topic. Keep in mind that anyone can create a Web site, so there is no guarantee that the information is correct. If you have any questions, be sure to ask your instructor.

EXERCISES *3.6* ❯

F O R   E X T R A   H E L P

MyMathLab

Math*XP*

Interactmath.com

MathXL
Tutorials on CD

Video Lectures
on CD

Tutor Center
Addison-Wesley
Math Tutor Center

Student's
Solutions Manual

Vocabulary

1. The point–slope form of the equation of a line with slope m that passes through the point (x_1, y_1) is _____.

2. State the procedure for finding the equation of a line that passes through the points (x_1, y_1) and (x_2, y_2).

Write the following equations in slope–intercept form.

3. $4x + 2y = 8$

4. $-3x + 6y = 9$

5. $x - 5y = 10$

6. $-12x - 16y = 10$

7. $7x - y = 3$

8. $6x + 4y = 0$

Find the equation of a line with the given slope and y-intercept.

9. Slope -3, y-intercept $(0, 5)$

10. Slope 2, y-intercept $(0, -4)$

11. Slope $\frac{2}{3}$, y-intercept $(0, -3)$

12. Slope 0, y-intercept $(0, 9)$

13. Slope 5, y-intercept $(0, 8)$

14. Slope $-\frac{3}{5}$, y-intercept $(0, 2)$

Find the slope–intercept form of the equation of a line with the given slope that passes through the given point. Graph the line.

15. Slope 1, through $(3, 2)$

16. Slope -2, through $(1, 6)$

17. Slope -3, through $(-2, 9)$

18. Slope -1, through $(-6, -3)$

19. Slope $\frac{3}{2}$, through $(-6, -9)$

20. Slope $\dfrac{2}{5}$, through $(-5, -4)$

28. $(4, 6), (8, 10)$

21. Slope 0, through $(8, 5)$

22. Undefined slope, through $(8, 5)$

29. $(-3, -3), (1, 9)$ **30.** $(-6, 9), (-2, -3)$

23. Find a linear function $f(x)$ with slope -2 such that $f(-4) = 23$.

24. Find a linear function $f(x)$ with slope 5 such that $f(3) = 12$.

25. Find a linear function $f(x)$ with slope $\dfrac{3}{5}$ such that $f(-15) = -17$.

26. Find a linear function $f(x)$ with slope 0 such that $f(316) = 228$.

31. $(-10, -12), (6, 20)$ **32.** $(-7, 4), (2, 4)$

Find the slope–intercept form of the equation of a line that passes through the given points. Graph the line.

27. $(2, -3), (7, 2)$

33. $(-2, -9), (-2, -3)$ **34.** $(8, -4), (8, 2)$

For Exercises 35–40, find the slope–intercept form of the equation of a line whose x-intercept and y-intercept are given.

	x-intercept	**y-intercept**
35.	$(6, 0)$	$(0, 2)$
36.	$(-9, 0)$	$(0, 3)$
37.	$(4, 0)$	$(0, -4)$
38.	$(-10, 0)$	$(0, -2)$
39.	$(6, 0)$	$(0, -6)$
40.	$(10, 0)$	$(0, 4)$

41. Jamie sells newspaper subscriptions door-to-door to help pay her tuition. She is paid a certain salary each night, plus a commission on each sale she makes. On Monday, she sold 3 subscriptions and was paid $66. On Tuesday she sold 8 subscriptions and was paid $116.

a) Find a linear equation that calculates Jamie's nightly pay on a night that she sells x subscriptions.

b) How much will Jamie be paid on a night that she makes no sales?

42. Luis contracted with a landscaper to install a brick patio in his backyard. The landscaper charged $10,000. Luis paid the landscaper a deposit the first month and agreed to make a fixed monthly payment until the balance was paid in full. After three months, Luis owed $6800. After seven months, Luis still owed $5200.

a) Find a linear equation for Luis's balance (y) after x months.

b) How much was the deposit that Luis paid?

c) How many months will it take to pay off the entire balance?

43. Members of a country club pay an annual membership fee. In addition, they pay a greens fee each time they play a round of golf. Last year, Dave played golf 37 times and paid the country club a total of $6790 (membership fee and greens fees). Last year, Vijay played golf 15 times and paid the country club $5250.

a) Find a linear equation for the amount owed (y) by a member who plays x rounds of golf in a year.

b) What is the annual membership fee at this country club?

c) What is the cost for a single round of golf?

44. A computer repairman charges a service fee in addition to an hourly rate to fix a computer. To fix Frank's computer, the repairman took two hours and charged a total of $225 (service fee plus hourly rate for two hours of work). Bundy's computer was in deeper trouble, and took six hours to fix. The charge to Bundy was $505.

   a) Find a linear equation for the charge (y) for a repair that takes x hours to perform.

   b) How much is the repairman's service fee?

   c) How much is the repairman's hourly rate?

Find the slope–intercept form of the equation of a line that is parallel to the given line and that passes through the given point.

45. Parallel to $y = -2x + 13$, through $(-6, -3)$

46. Parallel to $y = 4x - 11$, through $(3, 8)$

47. Parallel to $-3x + y = 9$, through $(4, -6)$

48. Parallel to $10x + 2y = 40$, through $(-2, 9)$

49. Parallel to $y = 5$, through $(2, 7)$

50. Parallel to $x = -5$, through $(-6, -4)$

Find the slope–intercept form of the equation of a line that is perpendicular to the given line and that passes through the given point.

51. Perpendicular to $y = 3x - 7$, through $(9, 4)$

52. Perpendicular to $y = -\dfrac{1}{2}x + 5$, through $(1, -3)$

53. Perpendicular to $4x + 5y = 9$, through $(-4, 2)$

54. Perpendicular to $3x - 2y = 8$, through $(-6, -7)$

55. Perpendicular to $y = 3$, through $(-5, -3)$

56. Perpendicular to $x = -8$, through $(6, 1)$

Find the slope–intercept form of the equation of a line that is parallel to the graphed line and that passes through the point plotted on the graph.

57.

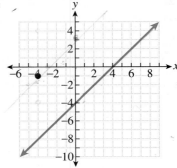

58.

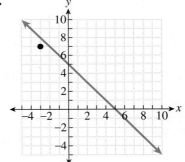

Find the slope–intercept form of the equation of a line that is perpendicular to the graphed line and that passes through the point plotted on the graph.

59.

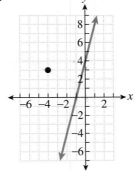

60.

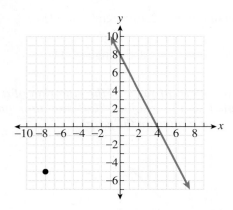

Writing in Mathematics

Answer in complete sentences.

61. Explain why the equation of a vertical line has the form $x = a$ and the equation of a horizontal line has the form $y = b$.

62. Explain why a horizontal line has a slope of 0 and a vertical line has undefined slope.

63. *Solutions Manual** Write a solutions manual page for the following problem:

Find the equation of the line with slope $m = -\frac{4}{3}$ that passes through the point $(6, -7)$.

64. *Newsletter** Write a newsletter explaining how to find the equation of a line that passes through two given points.

*See Appendix B for details and sample answers.

QUICK REVIEW EXERCISES

Section 3.6

Graph. Label any x- and y-intercepts.

1. $5x - 2y = 10$

2. $y = -3x - 8$

3. $y = \frac{2}{3}x - 4$

4. $y = -8$

3.7

Linear Inequalities

1 Determine whether an ordered pair is a solution of a linear inequality in two variables.

2 Graph a linear inequality in two variables.

3 Graph a linear inequality involving a horizontal or vertical line.

4 Graph linear inequalities associated with applied problems.

In this section, we will learn how to solve **linear inequalities** in two variables. Here are some examples of linear inequalities in two variables.

$$2x + 3y \le 6 \qquad 5x - 4y \ge -8 \qquad -x + 9y < -18 \qquad -3x - 4y > 7$$

Solutions of Linear Inequalities in Two Variables

Objective 1 Determine whether an ordered pair is a solution of a linear inequality in two variables.

Solutions of Linear Inequalities

A solution of a linear inequality in two variables is an ordered pair (x, y) such that when the coordinates are substituted into the inequality, a true statement results.

For example, consider the linear inequality $2x + 3y \le 6$. The ordered pair $(2, 0)$ is a solution because, when we substitute these coordinates into the inequality, it produces the following result:

$$2x + 3y \le 6$$
$$2(2) + 3(0) \le 6$$
$$4 + 0 \le 6$$
$$4 \le 6$$

The last inequality is a true statement, so $(2, 0)$ is a solution. The ordered pair $(3, 4)$ is not a solution, because $2(3) + 3(4)$ is not less than or equal to 6. Any ordered pair (x, y) for which $2x + 3y$ evaluates to be less than or equal to 6 is a solution of this inequality, and there are infinitely many solutions to this inequality. We will display our solutions on a graph.

Graphing Linear Inequalities in Two Variables

Objective 2 Graph a linear inequality in two variables. Ordered pairs that are solutions of the inequality $2x + 3y \le 6$ are one of two types. Ordered pairs (x, y) for which $2x + 3y = 6$ are solutions, and ordered pairs (x, y) for which $2x + 3y < 6$ are also solutions. Points satisfying $2x + 3y = 6$ lie on a line, so we begin by graphing this line. Since the equation is in standard form, a quick way to graph this line is by finding its x- and y-intercepts.

x-intercept ($y = 0$)	**y-intercept ($x = 0$)**
$2x + 3(0) = 6$	$2(0) + 3y = 6$
$2x + 0 = 6$	$0 + 3y = 6$
$2x = 6$	$3y = 6$
$x = 3$	$y = 2$
$(3, 0)$	$(0, 2)$

Here is the graph of the line:

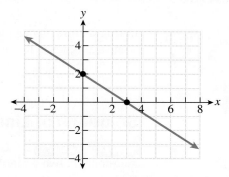

This line divides the plane into two half-planes and is the dividing line between ordered pairs for which $2x + 3y < 6$ and ordered pairs for which $2x + 3y > 6$. To finish graphing our solutions, we must determine which half-plane contains the ordered pairs for which $2x + 3y < 6$. To do this we will use a **test point,** which is a point that is not on the graph of the line, whose coordinates are used for determining which half-plane contains the solutions of the inequality. We substitute the test point's coordinates into the original inequality, and if the resulting inequality is true, this point and all other points on the same side of the line are solutions, and we will shade that half-plane. If the resulting inequality is false, then the solutions are on the other side of the line, and we shade that half-plane instead. A wise choice for the test point is the origin $(0, 0)$ if it is not on the line that has been graphed, since its coordinates are easy to work with when substituting into the inequality. Since $(0, 0)$ is not on the line, we will use it as a test point.

$$\text{Test Point: } (0, 0)$$
$$2(0) + 3(0) \leq 6$$
$$0 + 0 \leq 6$$
$$0 \leq 6$$

Since the last line is a true inequality, $(0, 0)$ is a solution, and we will shade the half-plane containing this point.

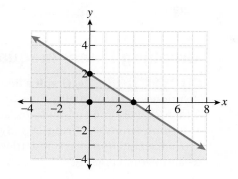

EXAMPLE 1   Graph $y \geq 4x + 3$.

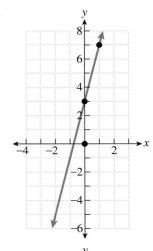

Solution

We begin by graphing the line $y = 4x + 3$. This equation is in slope–intercept form, so we can graph it by plotting its y-intercept $(0, 3)$ and using the slope (up four units, one unit to the right) to find other points on the line.

Since the line does not pass through the origin, we can use $(0, 0)$ as a test point.

$$\text{Test Point: } (0, 0)$$
$$0 + 4(0) \geq 3$$
$$0 + 0 \geq 3$$
$$0 \geq 3$$

The last inequality is false, so the solutions are in the half-plane that does not contain the origin. The graph of the inequality is shown at the right.

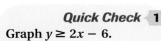

Quick Check 1
Graph $y \geq 2x - 6$.

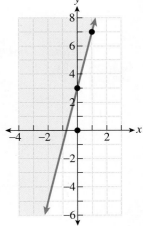

The first two inequalities we graphed were **weak linear inequalities,** which are inequalities involving the symbols $\leq$ or $\geq$. We now turn our attention to **strict linear inequalities,** which involve the symbols $<$ or $>$. An example of a strict linear inequality would be $x - 5y < 5$. Ordered pairs for which $x - 5y = 5$ are not solutions to this inequality, so the points on the line are not included as solutions. We denote this on the graph by graphing the line as a dashed or broken line. We still pick a test point and shade the appropriate half-plane the same way we did in the previous examples.

EXAMPLE 2   Graph $x - 5y < 5$.

Solution

We start by graphing the line $x - 5y = 5$ as a dashed line. This equation is in standard form, so we can graph the line by finding its intercepts.

x-intercept $(y = 0)$	y-intercept $(x = 0)$
$x - 5(0) = 5$	$0 - 5y = 5$
$x - 0 = 5$	$0 - 5y = 5$
$x = 5$	$-5y = 5$
	$y = -1$
$(5, 0)$	$(0, -1)$

At the right is the graph of the line, with an x-intercept at $(5, 0)$ and a y-intercept at $(0, -1)$. Notice that the line is a dashed line.

Since the line does not pass through the origin, we will use $(0, 0)$ as a test point.

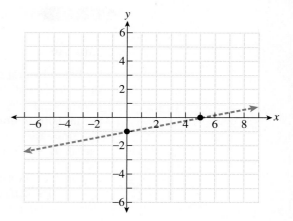

$$\text{Test Point: } (0, 0)$$
$$0 - 5(0) < 5$$
$$0 - 0 < 5$$
$$0 < 5$$

This inequality is true, so the origin is a solution. We shade the half-plane containing the origin.

Quick Check **2**
Graph $x + 2y < 6$.

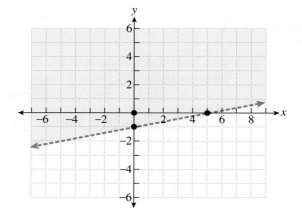

EXAMPLE 3   Graph $y > 3x$.

Solution

This is a strict inequality, so we begin by graphing the line $y = 3x$ as a dashed line.

The equation is in slope–intercept form, with a slope of 3 and a y-intercept at $(0, 0)$. After plotting the y-intercept, we can use the slope to find a second point at $(1, 3)$. The graph of the line is shown at the right.

Since the line passes through the origin, we cannot use $(0, 0)$ as a test point. We will try to choose a point that is clearly not on the line, such as $(4, 0)$, which is to the right of the line.

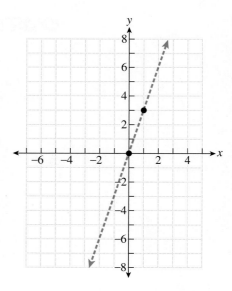

Quick Check 3

Graph $y \leq -\dfrac{4}{3}x.$

Test Point: $(4, 0)$
$0 > 3(4)$
$0 > 12$

This inequality is false, so we will shade the half-plane that does not contain the test point $(4, 0)$. At the right is the graph.

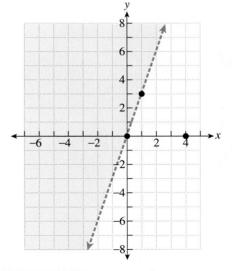

Graphing Linear Inequalities Involving Horizontal or Vertical Lines

Objective 3 **Graph a linear inequality involving a horizontal or vertical line.**
A linear inequality involving a horizontal or vertical line has only one variable but can still be graphed on a plane rather than on a number line. After graphing the related line, we find that it is not necessary to use a test point. Instead, we can use reasoning to determine where to shade. However, we may continue to use test points if we choose.

To graph the inequality $x > 5$, we begin by graphing the vertical line $x = 5$, using a dashed line. The values of x that are greater than 5 are to the right of this line, so we shade the half-plane to the right of $x = 5$. If we had used the origin as a test point, the resulting inequality $(0 > 5)$ would be false, so we would shade the half-plane that does not contain the origin, producing the same graph.

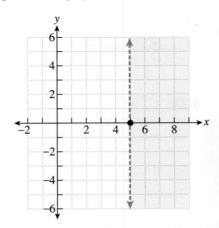

To graph the inequality $y \leq -2$, we begin by graphing the horizontal line $y = -2$, using a solid line. The values of y that are less than -2 are below this line, so we shade the half-plane below the line.

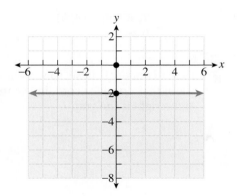

Here is a brief summary of how to graph linear inequalities in two variables:

Graphing Linear Inequalities in Two Variables

- Graph the line related to the inequality, which is found by replacing the inequality symbol with an equals sign.
- If the inequality symbol includes equality ($\leq$ or $\geq$), graph the line as a solid line.
- If the inequality symbol does not include equality ($<$ or $>$), graph the line as a dashed or broken line.
- Select a test point that is not on the line. Use the origin if possible. Substitute the coordinates of the point into the original inequality. If the resulting inequality is true, then shade the region on the side of the line that contains the test point. If the resulting inequality is false, shade the region on the side of the line that does not contain the test point.

Applications

Objective 4 Graph linear inequalities associated with applied problems. We turn our attention to an application problem to end the section.

EXAMPLE 4 A movie theater has 120 seats. The number of adults and children that can be admitted cannot exceed the number of seats. Set up and graph the appropriate inequality.

Solution

There are two unknowns in this problem: the number of adults and the number of children. We will let x represent the number of adults and y represent the number of children. Since the total number of adults and children cannot exceed 120, we know that $x + y \leq 120$. (We also know that $x \geq 0$ and $y \geq 0$, since we cannot have a negative number of children or adults. Therefore, we are restricted to the first quadrant, the positive x-axis and the positive y-axis.)

To graph this inequality, we begin by graphing $x + y = 120$ as a solid line. Since this equation is in standard form, we graph the line using its x-intercept $(120, 0)$ and its y-intercept $(0, 120)$.

The origin is not on this line, so we can use $(0, 0)$ as a test point.

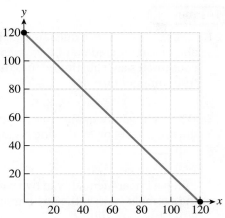

$x + y \leq 120$

$0 + 0 \leq 120$ Substitute 0 for x and 0 for y.

$0 \leq 120$ True.

This is a true statement, so we will shade on the side of the line that contains the origin.

Quick Check 4
To make a fruit salad, Irv needs a total of at least 60 pieces of fruit. If Irv decides to buy only apples and pears, set up and graph the appropriate inequality.

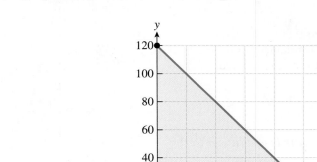

Building Your Study Strategy Using Your Resources, 7 **MyMathLab**   This textbook has an online resource called MyMathLab.com. At this site, you will be able to access video clips of lectures for each section in the book. These are useful if there is a particular topic you are struggling with or if you were forced to miss a day of class. You can also find practice tutorial exercises that provide feedback. Based on your performance, MyMathLab.com will generate a study plan for you. At this Web site, you will also have access to sample tests, the Student's Solutions Manual, and other supplementary information.

Vocabulary

1. A(n) _____ to a linear inequality in two variables is an ordered pair (x, y) such that when the coordinates are substituted into the inequality, a true statement results.

2. When a linear inequality in two variables involves the symbols $\leq$ or $\geq$, the line is graphed as a(n) _____ line.

3. When a linear inequality in two variables involves the symbols $<$ or $>$, the line is graphed as a(n) _____ line.

4. A point that is not on the graph of the line and whose coordinates are used for determining which half-plane contains the solutions to the linear inequality is called a(n) _____.

Determine whether the ordered pair is a solution to the given linear inequality.

5. $5x + 3y \leq 22$

    a) $(0, 0)$ **b)** $(8, -4)$
    c) $(2, 4)$ **d)** $(-3, 9)$

6. $2x - 4y > 6$

    a) $(4, 1)$ **b)** $(2, -3)$
    c) $(7, 2)$ **d)** $(5, 1)$

7. $y < 6x - 11$

    a) $(5, 8)$ **b)** $(0, 0)$
    c) $(2, 1)$ **d)** $(-4, -13)$

8. $x - 7y \geq -3$

    a) $(4, 1)$ **b)** $(9, 2)$
    c) $(0, 0)$ **d)** $(-16, -2)$

9. $y < 7$

    a) $(0, 0)$ **b)** $(8, 6)$
    c) $(-3, 10)$ **d)** $(6, 7)$

10. $x \leq -2$

    a) $(0, 0)$ **b)** $(-1, 3)$
    c) $(-2, 19)$ **d)** $(-5, -4)$

Complete the solution of the linear inequality by shading the appropriate region.

11. $4x + y \geq 7$

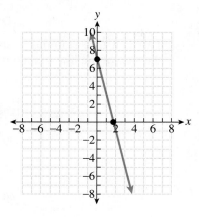

12. $-3x + 2y \leq -6$

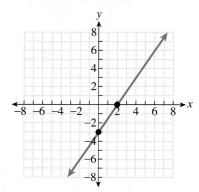

13. $2x + 8y < -4$

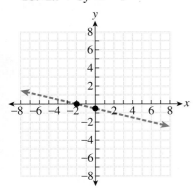

14. $10x - 2y > 0$

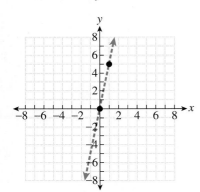

Which graph, A or B, represents the solution to the linear inequality? Explain how you chose your answer.

15. $3x + 4y \geq 24$

A)   B)

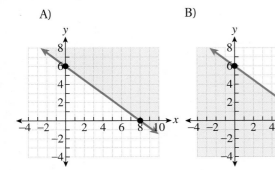

16. $y \leq \dfrac{1}{8}x$

A)   B)

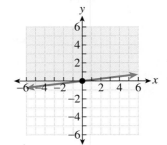   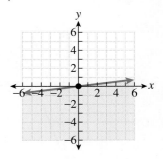

17. $3x - 2y < -18$

A)   B)

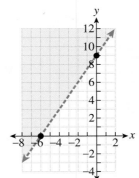   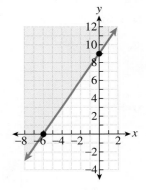

18. $5x - 4y > 20$

A)   B)

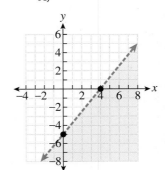   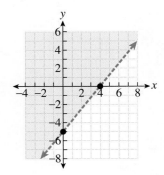

Determine the missing inequality sign ($<$, $>$, $\leq$, or $\geq$) for the linear inequality based on the given graph.

19. $8x - 3y$ _____ 24

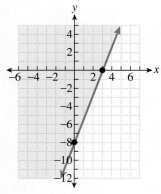

20. $3x + 6y$ _____ -27

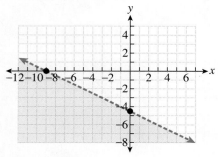

21. $-5x + 8y$ _____ 0

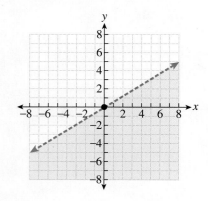

22. y _____ $4x - 6$

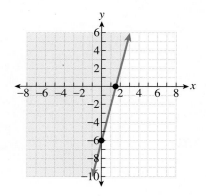

Graph the linear inequality.

23. $6x - 5y < -30$

24. $2x + 4y \leq 16$

25. $-x + 3y \geq 9$

26. $7x - 2y > -14$

31. $y \geq \dfrac{2}{3}x - 6$

32. $y > 4x - 8$

27. $-9x + 4y \leq 0$

28. $y < -\dfrac{1}{4}x$

33. $y \leq -2x - 4$

34. $y < -5$

29. $y > 3$

30. $x \geq 5$

35. $y > -\dfrac{3}{2}x - \dfrac{3}{2}$

36. $y \geq \dfrac{3}{5}x$

37. $x \le -2$

40. $y \ge -x + 15$

38. $y < x$

Determine the linear inequality associated with the given solution. (First, find the equation of the line. Then rewrite this equation as an inequality with the appropriate inequality sign: $<$, $>$, $\le$, or $\ge$.)

41.

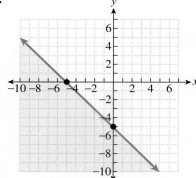

39. $y > 2x - 30$

42.

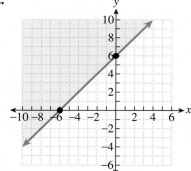

43.

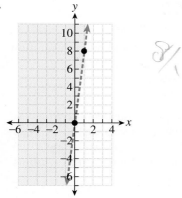

44.

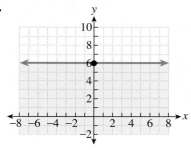

45. More than 50 people attended a charity basketball game. Some of the people were faculty members and the rest were students. Set up and graph an inequality involving the number of faculty members and the number of students in attendance.

46. An elevator has a warning posted inside the car that the maximum capacity is 1200 pounds. Suppose that the average weight of an adult is 160 pounds and the average weight of a child is 60 pounds.

   a) Set up and graph an inequality involving the number of adults and the number of children that can safely ride on the elevator.

   b) From the graph from part a), can two adults and 16 children ride in the elevator at the same time?

Find the region that contains ordered pairs that are solutions to both inequalities. First graph each inequality separately, and then shade the region that the two graphs have in common.

47. $x + y < 3$ and $y > x - 7$

48. $2x + 3y \leq 18$ and $2x + y \geq 8$

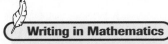

Writing in Mathematics

Answer in complete sentences.

49. True or false: every ordered pair is a solution of either $3x - 8y < 24$ or $3x - 8y > 24$. Explain your answer.

50. Explain why we used a dashed line when graphing inequalities involving the symbols $<$ and $>$ and why we use a solid line when graphing inequalities involving the symbols $\leq$ and $\geq$.

51. *Solutions Manual* * Write a solutions manual page for the following problem:

Graph the linear inequality $2x + 3y < 12$.

52. *Newsletter* * Write a newsletter explaining how to graph linear inequalities in two variables.

*See Appendix B for details and sample answers.

Chapter 3 Summary

Section 3.1—Topic	Chapter Review Exercises
Finding the Coordinates of a Point on a Graph	1
Determining the Quadrant a Point Is In	2–3
Determining Whether an Ordered Pair Is a Solution of an Equation	4–6

Section 3.2—Topic	Chapter Review Exercises
Finding the x- and y-Intercepts of a Line	7–10
Graphing a Line by Using Its Intercepts	11–16, 51–56

Section 3.3—Topic	Chapter Review Exercises
Finding the Slope of a Line Passing through Two Points	17–20
Finding the Slope and y-Intercept of a Line from Its Equation	21–26, 36–37
Graphing a Line by Using Its Slope and y-Intercept	27–32, 51–56

Section 3.4—Topic	Chapter Review Exercises
Solving Applications Involving Linear Functions	38
Evaluating Linear Functions	39–41
Interpreting the Graph of a Function	42

Section 3.5—Topic	Chapter Review Exercises
Determining Whether Two Lines Are Parallel, Perpendicular, or Neither	33–35

Section 3.6—Topic	Chapter Review Exercises
Finding the Equation of a Line Given Its Slope and a Point on the Line	43–45
Finding the Equation of a Line Given Two Points on the Line	46–48
Finding the Equation of a Line from Its Graph	49–50

Section 3.7—Topic	Chapter Review Exercises
Graphing Linear Inequalities	57–60

Summary of Chapter 3 Study Strategies

You have a great number of resources at your disposal to help you learn mathematics, and your success may depend on how well you use them.

- Your instructor is a valuable resource. He or she can answer your questions and provide advice. In addition to being available during the class period and office hours, some instructors take questions from students right before class begins or just after class ends.
- The tutorial center is another valuable resource. Keep in mind that the goal is for you to understand the material and to be able to solve the problems yourself.
- Many colleges offer short courses or seminars in study skills. Ask an academic counselor about these courses and whether they would be helpful.
- Some other resources discussed in this chapter were the Student's Solutions Manual, your classmates, and the Internet.
- Finally, the online supplement to this textbook at MyMathLab.com can be quite helpful. At this site you will find video clips, tutorial exercises, and much, much more.

Find the coordinates of the labeled points A, B, C, and D. [3.1]

1.

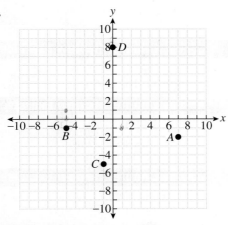

In which quadrant (I, II, III, or IV) is the given point located? [3.1]

2. $(-3, -3)$ **3.** $(-9, 4)$

Is the ordered pair a solution to the given equation? [3.1]

4. $(5, -2), 4x - 2y = 16$

5. $\left(\dfrac{7}{2}, \dfrac{3}{5}\right), 4x + 5y = 17$

6. $(-6, -4), x - 3y = 6$

Find the x-intercept and y-intercept, if possible. [3.2]

7. $2x - 8y = -40$

8. $5x - y = -15$

9. $3x - 7y = 0$

10. $4x - 7y = 14$

Find the x- and y-intercepts, and use them to graph the equation. [3.2]

11. $-3x + 2y = 12$

12. $x - 4y = -8$

13. $2x + 3y = 0$

14. $4x + 3y = -6$

15. $y = -\dfrac{3}{2}x + 6$ **16.** $y = 2x - 4$

Find the slope of the line that passes through the given points. [3.3]

17. $(-5, 3)$ and $(-3, 9)$

18. $(6, -2)$ and $(3, 10)$

19. $(3, -8)$ and $(8, -3)$

20. $(2, -7)$ and $(7, -7)$

Worked-out solutions to Review Exercises marked with 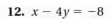 can be found on page AN–12.

Find the slope and the y-intercept of the given line. [3.3]

21. $y = -2x + 7$

22. $y = 4x - 6$

23. $y = -\dfrac{2}{3}x$

24. $4x - 2y = 9$

25. $5x + 3y = 18$

26. $y = -7$

Graph using the slope and y-intercept. [3.3]

27. $y = 3x - 6$ **28.** $y = -5x - 5$

29. $y = \dfrac{2}{5}x + 2$

30. $y = x$

31. $y = -\dfrac{7}{3}x + 3$ **32.** $y = \dfrac{1}{2}x + \dfrac{3}{2}$

Are the two given lines parallel, perpendicular, or neither? [3.5]

33. $y = 2x - 9$
$y = -2x + 9$

34. $4x + y = 11$
$y = \dfrac{1}{4}x + \dfrac{5}{2}$

35. $12x - 9y = 17$
$-8x + 6y = 10$

Find the equation of a line with the given slope and y-intercept. [3.3]

36. Slope -4, y-intercept $(0, -2)$

37. Slope $\dfrac{2}{5}$, y-intercept $(0, -6)$

38. Monique's vacation fund has $720 in it. She plans to add $50 to this fund each month. [3.4]

a) Create a function $f(x)$ for the amount of money in Monique's vacation fund after x months.

b) Use the function from part a) to determine how much money will be in the vacation fund after 11 months.

c) Use the function from part a) to determine how many months it will take Monique to save enough money for a trip to Hawaii, which will cost $2750.

Evaluate the given function. [3.4]

39. $f(x) = 9x + 7$, $f(-2)$

40. $g(x) = 3 - 8x$, $g(-5)$

41. $f(x) = 3x + 7$, $f(5a - 1)$

42. Consider the following graph of a function $f(x)$. [3.4]

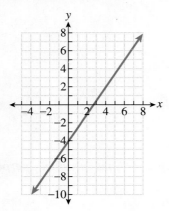

a) Find $f(-2)$.

b) Find all values x for which $f(x) = 5$.

c) Find the domain of $f(x)$.

d) Find the range of $f(x)$.

Find the slope–intercept form of the equation of a line with the given slope that passes through the given point. [3.6]

43. Slope 1, through $(4, -2)$

44. Slope -5, through $(1, 3)$

45. Slope $-\dfrac{3}{2}$, through $(-4, 9)$

Find the slope–intercept form of the equation of a line that passes through the two given points. [3.6]

46. $(-2, 1)$ and $(2, -7)$

47. $(-6, -5)$ and $(2, 7)$

48. $(5, 3)$ and $(-9, 3)$

Find the equation of the line that has been graphed. [3.6]

49.

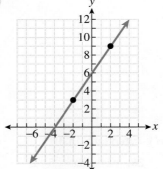

50.

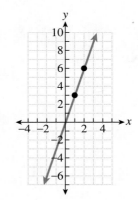

Graph. [3.2/3.3]

51. $y = 4x + 8$

52. $y = -\dfrac{1}{2}x + 3$

53. $y = -5x + 2$

54. $x = 5$

55. $3x - 4y = 24$

56. $y = -6$

Graph the inequality on a plane. [3.7]

57. $3x + y < -9$

58. $2x + 7y \geq 14$

59. $y \leq -4$

60. $y > \dfrac{3}{2}x + 5$

In which quadrant (I, II, III, or IV) is the given point located?

1. $(-3, 7)$

Is the ordered pair a solution to the given equation?

2. $(4, -2)$, $-3x - 7y = 2$

Find the x-intercept and y-intercept, if possible.

3. $5x - 6y = -15$

4. $y = -\dfrac{7}{2}x + 21$

Find the x- and y-intercepts, and use them to graph the equation.

5. $4x - y = 6$

6. $y = \dfrac{2}{3}x - 4$

Find the slope of the line that passes through the given points.

7. $(2, 8)$ and $(-3, 9)$

Find the slope and the y-intercept of the given line.

8. $y = 7x - 8$

9. $3x + 5y = 45$

Graph using the slope and y-intercept.

10. $y = -2x + 3$

11. $y = -\dfrac{6}{5}x + 5$

Are the two given lines parallel, perpendicular, or neither?

12. $4y = 8x - 11$

$2x - y = 44$

Find the equation of a line with the given slope and y-intercept.

13. Slope -6, y-intercept $(0, 5)$

Evaluate the given function.

14. $f(x) = 7 - 2x$, $f(-3)$

15. Find the slope–intercept form of the equation of a line with a slope of $-\frac{5}{2}$ that passes through the point $(-2, 9)$.

16. Find the slope–intercept form of the equation of a line that passes through the points $(-6, 4)$ and $(-1, -6)$.

Graph.

17. $y = -4x + 6$

18. $-4x + 5y = 20$

19. $y = 2$

20. Graph the inequality on a plane. $x - 6y < -3$

Mathematicians in History
René Descartes

René Descartes was an early 17th-century philosopher who made important contributions to mathematics. Cartesian geometry resulted from his application of algebra to geometry. The rectangular coordinate plane is also called the Cartesian plane in his honor. Descartes once said, "Mathematics is a more powerful instrument of knowledge than any other that has been bequeathed to us by human agency."

Write a one-page summary (*or* make a poster) of the life of René Descartes and his accomplishments. Also, look up Descartes's most famous quotes and list your favorite quote.

Interesting issues:

- Where and when was René Descartes born?
- How old was Descartes when he enrolled at the Jesuit College at La Fleche?
- Descartes's first major treatise on physics was *Le Monde, ou Traité de la Lumierè*. Why did he choose not to publish his results?
- Descartes's most famous quote, in Latin, is "Cogito, ergo sum." What is the English translation of this quote?
- How did a fly help Descartes come up with the idea for the rectangular coordinate system?
- Queen Christina of Sweden invited Descartes to Sweden in 1649, where he died shortly thereafter. Describe the circumstances that led to his death.

Step 1: Form groups of no more than 4 or 5 students. Write the members' names in the chart below. For the best results, try to form groups in which all members are of the same sex. Each group should have a meter stick, ruler, or some type of measuring device.

Step 2: Measure the length of each member's foot (without shoes). Then measure and record the height of each member. Record these values in the appropriate column of the following table:

Member's Name	Length of Foot	Height	Ordered Pair

Step 3: For each member, create an ordered pair in which the first coordinate is the length of the member's foot and the second coordinate is his or her height.

Step 4: Create an appropriate axis system with the foot length along the horizontal axis and the height along the vertical axis.

Step 5: Plot your ordered pairs in the axis system you created. Try to be as accurate as possible. Starting from left to right, label your points A, B, C, D, etc. Visually, does it look like these points form a straight line? Would a straight line be an appropriate model for your data? Use a straightedge to visually check whether your group was correct.

Step 6: Find the slope of the line that would connect point A and point B. Find the slope of the line that would connect point B and point C. Continue until you have connected all your points. Are these slopes the same? Should the slopes be the same if the points don't line up in a straight line? Have a group discussion and write down the group's thoughts on these questions.

Step 7: Using a straightedge, try to find a straight line that best fits your ordered pairs. Calculate the equation of this line and draw it on your plane.

Step 8: Along with your completed chart, turn in the linear equation you created that your group feels best fits your data.

Write the appropriate symbol, either < or >, between the following integers. [1.1]

1. -3 ____ -8

Simplify. [1.2]

2. $-9 - 25$ 3. $15(-11)$

4. $-234 \div (-18)$

Write the prime factorization of the following numbers. (If the number is prime, state this.) [1.3]

5. 72

Simplify. Your answer should be in lowest terms. [1.4]

6. $\dfrac{6}{35} \cdot \dfrac{5}{21}$ 7. $\dfrac{3}{4} + \dfrac{9}{10}$

Simplify the following decimal expressions. [1.5]

8. $9.8 - 3.72$ 9. 4.7×3.8

10. A group of 32 baseball fans decided to attend a game together. If each ticket cost $14, how much did the group spend on tickets? [1.2]

11. If one recipe calls for $1\frac{1}{3}$ cups of flour and a second recipe calls for $2\frac{3}{4}$ cups of flour, how much flour is needed to make both recipes? [1.4]

Simplify the given expression. [1.7]

12. $9 + 3 \cdot 4 - 2^6$

13. $7 + 2(9 - 4 \cdot 5) - 6(-5)$

Evaluate the following algebraic expressions under the given conditions. [1.8]

14. $5x - 8$ for $x = -4$

Simplify. [1.8]

15. $8x - 11 - 3x + 8$

16. $3(3x - 8) - 5(4x + 9)$

Solve. [2.1, 2.2]

17. $3x - 4 = -31$ 18. $-2x + 15 = 8$

19. $3x + 17 = 6x - 25$

20. $2(2x + 5) - 3(x - 8) = 20$

21. The width of a rectangle is 9 feet shorter than its length, and the perimeter is 70 feet. Find the length and the width of the rectangle. [2.3]

22. Celia has $5 and $10 bills in her purse, worth a total of $205. She has 8 more $5 bills than $10 bills. How many $5 bills does Celia have? [2.3]

23. Twenty percent of 125 is what number? [2.4]

24. What percent of 275 is 66? [2.4]

25. Monisha bought a new house for $220,000. The house increased in value by 13% in one year. How much was the house worth after one year? [2.4]

Solve the proportion. [2.4]

26. $\dfrac{n}{8} = \dfrac{27}{36}$

27. At a certain community college, five out of every nine students plan to transfer to a four-year university. If the college has 8307 students, how many plan to transfer to a 4-year university? [2.4]

Solve. Graph your solution on a number line, and express it in interval notation. [2.5]

28. $2x - 5 \geq 3x - 8$,

29. $-7 \leq 3x + 8 \leq 11$,

Find the x- and y-intercepts, and use them to graph the equation. [3.2]

30. $2x - y = 6$

31. $-3x + 4y = -24$

Find the slope of the line that passes through the given points. [3.3]

32. $(-3, -7)$ and $(3, 5)$

Find the slope and the y-intercept of the given line. [3.3]

33. $2x + 3y = 21$

Graph using the slope and y-intercept. Find the x-intercept and label it on the graph. [3.3]

34. $y = 2x - 3$

35. $y = \dfrac{3}{4}x - 3$

Are the two given lines parallel, perpendicular, or neither? [3.5]

36. $5x + 3y = 9$
$3x - 5y = -5$

37. Bruce is saving his money to buy a new guitar. So far he has saved $220 and he plans to save $40 each month. [3.4]

    a) Create a function $f(x)$ for the amount of money that Bruce has saved after x months.

    b) Use the function from part a) to determine how much money Bruce will have saved after 8 months.

    c) Use the function from part a) to determine how many months it will take until Bruce has saved enough to buy the guitar, which will cost $800.

Evaluate the given function. [3.4]

38. $f(x) = 24x + 55, f(14)$

39. $g(x) = -7x + 905, g(-32)$

Find the slope–intercept form of the equation of a line with the given slope that passes through the given point. Graph the line. [3.6]

40. Slope -2 through $(3, -4)$

Find the slope–intercept form of the equation of a line that passes through the two given points. Graph the line. [3.6]

41. $(2, 1)$ and $(4, -5)$

Graph the inequality on a plane. [3.7]

42. $3x - 2y > 12$

Systems of Equations

In this chapter, we will learn to set up and solve systems of equations. A system of equations is a set of two or more associated equations containing two or more variables. We will learn to solve systems of linear equations in two variables by three methods: graphing, the substitution method, and the addition method. We will also learn to set up systems of equations to solve applied problems. We will finish the chapter by learning to solve systems of linear inequalities in two variables.

Study Strategy **Doing Your Homework** *Doing a homework assignment should not be viewed as just some requirement. Homework exercises are assigned to help you to learn mathematics. In this chapter, we will discuss how to do your homework and get the most out of your effort.*

4.1
Systems of Linear Equations; Solving Systems by Graphing

Objectives

1. Determine whether an ordered pair is a solution of a system of equations.
2. Identify the solution of a system of linear equations from a graph.
3. Solve a system of linear equations graphically.
4. Identify systems of linear equations with no solution or infinitely many solutions.

Systems of Linear Equations and Their Solutions

A **system of linear equations** consists of two or more linear equations. Here are some examples:

$$5x + 4y = 20 \qquad\qquad y = \frac{3}{5}x \qquad\qquad x + y = 11$$
$$2x - y = 6 \qquad\qquad y = -4x + 3 \qquad\qquad y = 5$$

Objective 1 Determine whether an ordered pair is a solution of a system of equations. The equations in a system are examined together, and we are looking for the solution(s) that the equations have in common.

Solution of a System of Linear Equations

> A **solution of a system of linear equations** is an ordered pair (x, y) that is a solution of each equation in the system.

EXAMPLE 1   Is the ordered pair $(2, -3)$ a solution of the given system of equations?

$$5x + y = 7$$
$$2x - 3y = 13$$

Solution

To determine whether the ordered pair $(2, -3)$ is a solution, we need to substitute 2 for x and -3 for y in each equation. If the resulting equations are both true, then the ordered pair is a solution of the system. Otherwise, it is not.

$$5x + y = 7 \qquad\qquad 2x - 3y = 13$$
$$5(2) + (-3) = 7 \qquad\qquad 2(2) - 3(-3) = 13$$
$$10 - 3 = 7 \qquad\qquad 4 + 9 = 13$$
$$7 = 7 \qquad\qquad 13 = 13$$

Since $(2, -3)$ is a solution of each equation, it is a solution of the system of equations.

EXAMPLE 2   Is the ordered pair $(-5, -1)$ a solution of the given system of equations?

$$y = 2x + 9$$
$$y = 14 - 3x$$

Solution

Again, we begin by substituting -5 for x and -1 for y in each equation.

$$y = 2x + 9 \qquad\qquad y = 14 - 3x$$
$$(-1) = 2(-5) + 9 \qquad (-1) = 14 - 3(-5)$$
$$-1 = -10 + 9 \qquad\qquad -1 = 14 + 15$$
$$-1 = -1 \qquad\qquad\qquad -1 = 29$$

Although $(-5, -1)$ is a solution of the equation $y = 2x + 9$, it is not a solution of the equation $y = 14 - 3x$. Therefore $(-5, -1)$ is not a solution of the system of equations.

Quick Check 1
Is the ordered pair $(-6, 3)$ a solution of the given system of equations?

$$5x - 2y = -36$$
$$-x + 7y = 27$$

Objective **2**   **Identify the solution of a system of linear equations from a graph.** Systems of two linear equations can be solved graphically. Since an ordered pair that is a solution of a system of two equations must be a solution of each equation, we know that this ordered pair is on the graph of each equation. To solve a system of two linear equations, we graph each line and look for points of intersection. Often, this will lead to a single solution. Here are some graphical examples of systems of equations that have the ordered pair $(3, 1)$ as a solution:

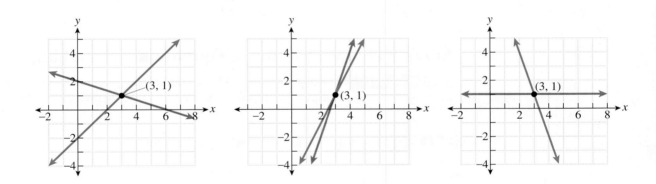

EXAMPLE **3**   Find the solution of the system of equations plotted on the following graph:

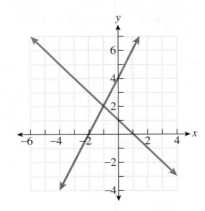

Solution

The two lines intersect at the point $(-1, 2)$, so the solution of the system of equations is $(-1, 2)$.

Quick Check **2** Find the solution of the system of equations plotted on the graph.

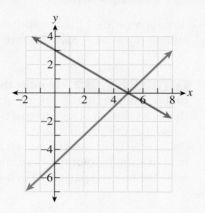

Solving Linear Systems of Equations by Graphing

Objective **3** **Solve a system of linear equations graphically.** Now we will solve a system of linear equations by graphing the lines and finding their point of intersection.

EXAMPLE **4**   Solve the following system by graphing:

$$y = 5x - 9$$
$$y = -\frac{7}{2}x + 8$$

Solution

Both equations are in slope–intercept form, so we will graph the lines by plotting the y-intercept and then using the slope of the line to find another point on the line.

The first line has a y-intercept at $(0, -9)$. From that point, we can find a second point by moving up five units and one unit to the right, since the slope is 5. The graph of the first line is shown at the top of the next page on the left.

The y-intercept of the second line is $(0, 8)$. Since the slope of this line is $-\frac{7}{2}$, we can move down seven units and two units to the right from this point to find a second point on this line. Both graphs are shown on the same set of axes at the top of the next page on the right.

Examining the graph, we can see that the two lines intersect at the point $(2, 1)$. The solution of the system of equations is the ordered pair $(2, 1)$. We could check that this ordered pair is a solution by substituting its coordinates into both equations. The check is left to the reader.

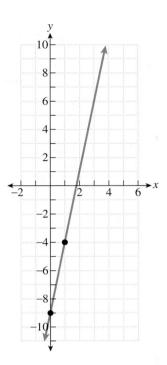

Quick Check 3

Solve the following
system by graphing:

$$y = 3x - 2$$
$$y = 5x - 6$$

Using Your Calculator   To solve a system of linear equations using the TI–84, we must graph each equation and then look for points of intersection. To graph the equations from Example 4, begin by pushing the [Y=] key. Next to Y_1 type $5x - 9$, and next to Y_2 type $-\frac{7}{2}x + 8$. Press the [GRAPH] key to display the graph.

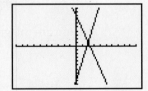

To find the points of intersection, push [2nd][TRACE] to access the CALC menu and select option **5: intersect.** You should see the following screen:

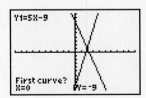

When prompted for the first curve, select Y_1 by pushing [ENTER]. When prompted for the second curve, select Y_2 by pushing [ENTER].

We must now make an initial guess for one of the solutions. In the next screen shot, the TI–84 asks us if we would like $(0, 8)$ to be our guess. Accept this guess by pressing ENTER. The solution is the point $(2, 1)$.

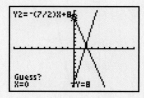   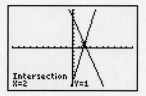

EXAMPLE 5   Solve the following system by graphing:

$$2x - 3y = 12$$
$$x + y = 1$$

Solution

Both equations are in general form, so we will graph them using their x- and y-intercepts.

2x − 3y = 12		x + y = 1	
x-intercept (y = 0)	**y-intercept (x = 0)**	**x-intercept (y = 0)**	**y-intercept (x = 0)**
$2x - 3(0) = 12$	$2(0) - 3y = 12$	$x + (0) = 1$	$(0) + y = 1$
$2x = 12$	$-3y = 12$	$x = 1$	$y = 1$
$x = 6$	$y = -4$		
$(6, 0)$	$(0, -4)$	$(1, 0)$	$(0, 1)$

Here are the graphs of each line on the same set of axes:

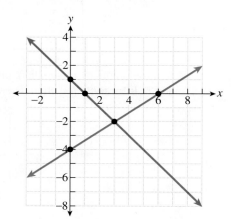

Quick Check 4

Solve the following system by graphing:

$$x + 5y = 5$$
$$x - y = -7$$

The two lines intersect at the point $(3, -2)$, so the solution of this set of equations is the ordered pair $(3, -2)$.

Objective 4 **Identify systems of linear equations with no solution or infinitely many solutions.** A system of two linear equations that has a single solution is called an **independent system.** Not every system of linear equations has a single solution. Suppose that the graphs of the two equations are parallel lines. Parallel lines do not intersect, so such a system has no solution. A system that has no solution is called an **inconsistent system.**

Occasionally, the two equations in a system will have identical graphs. In this case, every ordered pair that lies on the line is a solution, so there are infinitely many solutions of the system. Such a system is called a **dependent system.**

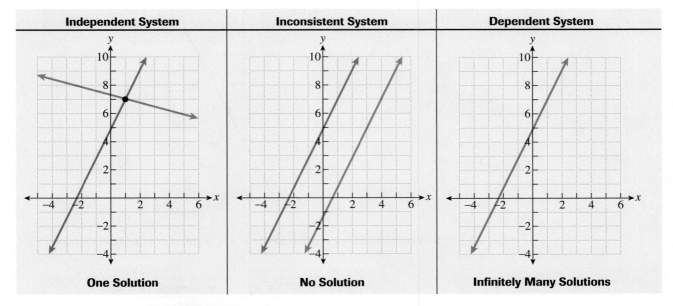

Independent System	Inconsistent System	Dependent System
One Solution	No Solution	Infinitely Many Solutions

EXAMPLE 6 Solve the following system by graphing:

$$y = -3x + 5$$
$$y = -3x - 3$$

Solution

Both equations are in slope–intercept form, so we can graph each line by plotting its y-intercept and then use the slope to find a second point on the line. The first line has a y-intercept at $(0, 5)$, and the slope is -3. The second line has a y-intercept at $(0, -3)$, and the slope is also -3. The graphs are at the right.

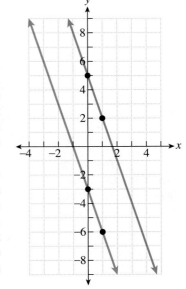

These two lines are parallel lines that do not intersect, so this system has no solution. The system is an inconsistent system. In general, if we know that the two lines have the same slope, but different y-intercepts, then we know that the system is an inconsistent system and has no solution. However, if the two lines have the same slope *and* the *same* y-intercept, then the system is a dependent system.

Quick Check 5
Solve the following system by graphing:

$$y = -4x + 6$$
$$12x + 3y = 15$$

EXAMPLE ▶ 7   Solve the following system by graphing:

$$y = 2x - 4$$
$$4x - 2y = 8$$

Solution

The first line is in slope–intercept form, so we can graph it by first plotting its y-intercept at $(0, -4)$. Using its slope of 2, we see that the line also passes through the point $(1, -2)$ as shown in the graph below.

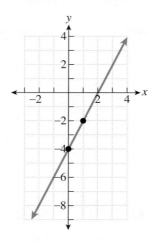

To graph the second line, we will find the x- and y-intercepts, as the equation is in general form.

x-intercept $(y = 0)$	y-intercept $(x = 0)$
$4x - 2(0) = 8$	$4(0) - 2y = 8$
$4x = 8$	$-2y = 8$
$x = 2$	$y = -4$
$(2, 0)$	$(0, -4)$

After plotting these intercepts on the graph, we see that the two lines are exactly the same. This system is a dependent system. We can display the solutions of a dependent system by showing the graph of either equation. Ordered pairs that are solutions of the system are of the form $(x, 2x - 4)$ for any real number x. To determine the form of these solutions, solve one of the equations for y and use that to replace y in the ordered pair (x, y). Notice that if we rewrite the equation $4x - 2y = 8$ in slope–intercept form, it is exactly the same as the first equation. Equations in a dependent linear system of equations are multiples of each other.

Quick Check 6

Solve the following system by graphing:

$$4x + 6y = 12$$
$$y = -\tfrac{2}{3}x + 2$$

Solving systems of equations by graphing does have a major drawback. Often, we will not be able to accurately determine the coordinates of the solution of an independent system. Consider the graph of the system

$$y = 2x - 6$$
$$y = \frac{2}{5}x + 1$$

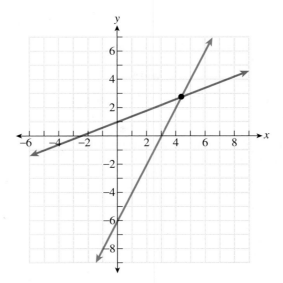

We can see that the x-coordinate of this solution is between 4 and 5 and that the y-coordinate of this solution is between 2 and 3. However, we cannot determine the exact coordinates of this solution from the graph. In the next two sections, we will develop algebraic techniques for finding solutions of linear systems of equations.

Building Your Study Strategy Doing Your Homework, 1 **Review First**   Before beginning any homework assignment, it is a good idea to review. Start by going over your notes from class to remind you of the types of problems covered by your instructor and of how your instructor solved them. Your notes may contain the steps for certain procedures and advice from your instructor about typical errors to avoid. After reviewing your notes, keep them handy for further reference as you proceed through the homework assignment.

You should then review the appropriate section in the text. Pay particular attention to the examples, as they show the solution to problems similar to the problems you are about to solve in the homework exercises. Look for the feature labeled A Word of Caution for advice on avoiding common errors. Finally, look for summaries of procedures introduced in the section. As you proceed through the homework assignment, refer back to the section as needed.

Vocabulary

1. A(n) _____ consists of two or more linear equations.

2. Give an example of a system of two linear equations in two unknowns.

3. An ordered pair (x, y) is a solution to a system of two linear equations if _____.

4. A system of two linear equations is a(n) _____ system if it has exactly one solution.

5. A system of two linear equations is a(n) _____ system if it has no solution.

6. A system of two linear equations is a(n) _____ system if it has infinitely many solutions.

Is the ordered pair a solution of the given system of equations?

7. $(3, 1)$, $\begin{aligned} 2x + y &= 7 \\ 3x - 2y &= 7 \end{aligned}$

8. $(0, 6)$, $\begin{aligned} 7x - 2y &= -12 \\ x + 5y &= 30 \end{aligned}$

9. $(-4, 5)$, $\begin{aligned} x + y &= 1 \\ x - 3y &= -11 \end{aligned}$

10. $(-2, -7)$, $\begin{aligned} 4x + y &= -15 \\ 6x - 2y &= 2 \end{aligned}$

11. $(-1, -3)$, $\begin{aligned} 4x - y &= -1 \\ -2x + 4y &= -10 \end{aligned}$

12. $(7, -8)$, $\begin{aligned} 9x - 5y &= 103 \\ x - y &= -1 \end{aligned}$

Find the solution of the system of equations on each graph. If there is no solution, state this.

13.

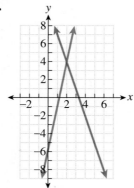

14.

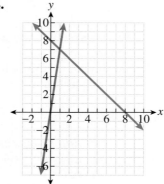

15.

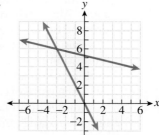

16.

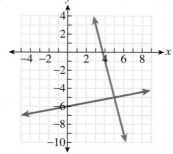

Solve the system by graphing. If the system is inconsistent and has no solution, state this. If the system is dependent, write the form of the solution for any real number x.

17. $y = 2x + 5$
$y = 4x + 1$

18. $y = -\dfrac{2}{3}x + 5$
$y = \dfrac{4}{3}x - 7$

19. $y = 5x - 3$
$x = 2$

20. $y = 3x$
$y = 3$

21. $x + 3y = 6$
$2x + y = -8$

22. $3x + 4y = -12$
$3x + 2y = 6$

23. $5x - 3y = 15$
$2x - y = 4$

24. $7x + 3y = 0$
$x + y = 4$

25. $4x - 5y = -20$
$-4x + 7y = 28$

26. $-2x + y = 10$
$2x + 3y = 6$

27. $y = 5x - 4$
$10x - 2y = 6$

28. $4x + 5y = 10$
$y = -\dfrac{3}{5}x + 3$

29. $9x - 2y = -18$
$y = \dfrac{1}{2}x - 7$

30. $y = -\dfrac{2}{7}x + 3$
$4x + 14y = 6$

31. Draw the graph of a linear equation that, along with the given line, forms an inconsistent system.

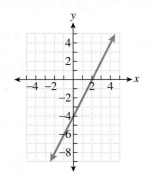

32. Draw the graph of a linear equation that, along with the given line, forms a dependent system.

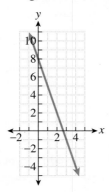

33. Draw the graph of a linear equation that, along with the given line, forms a system of equations whose single solution is the ordered pair $(5, 1)$.

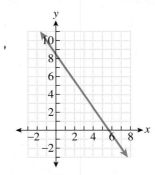

34. Draw the graphs of two linear equations that form an inconsistent system.

35. Draw the graphs of two linear equations that form a dependent system.

36. Draw the graphs of two linear equations that form a system of equations whose single solution is the ordered pair $(-3, 2)$.

Answer in complete sentences.

37. Explain why the solution of an independent system of equations is the ordered pair that is the point of intersection for the two lines.

38. When solving a system of equations by graphing, we may not be able to accurately determine the coordinates of the solution of an independent system. Explain why.

39. *Solutions Manual** Write a solutions manual page for the following problem:

*Solve the system of equations $y = -2x + 6$
by graphing. $x + 4y = -4$*

40. *Newsletter** Write a newsletter explaining how to identify inconsistent and dependent systems from a graph.

*See Appendix B for details and sample answers.

QUICK REVIEW EXERCISES

Section 4.1

Solve.

1. $8x + 3(2x - 5) = 27$

2. $7x - 5(3x + 9) = -13$

3. $2(y + 6) - 3y = 17$

4. $-3(6y + 1) + 7y = 74$

1 Solve systems of linear equations by using the substitution method.
2 Use the substitution method to identify inconsistent systems of equations.
3 Use the substitution method to identify dependent systems of equations.
4 Solve systems of equations without variable terms with a coefficient of 1 or -1 by the substitution method.
5 Solve applied problems by using the substitution method.

4.2
Solving Systems of Equations by Using the Substitution Method

Solving systems of equations by graphing each equation works well when the solution has integer coordinates. However, when the coordinates are not integers, or when the solution is outside the region that was graphed, exact solutions are difficult to find. In the next two sections, we will examine algebraic methods for solving systems of linear equations. The goal of each method is to combine the two equations with two variables into a single equation with only one variable. After solving for this variable, we will substitute the value of that variable into one of the original equations and solve for the other variable.

The Substitution Method

Objective 1 **Solve systems of linear equations by using the substitution method.** Consider the following system of equations:

$$y = x - 5$$
$$2x + 3y = 20$$

Looking at the first equation, $y = x - 5$, we see that y is equal to the expression $x - 5$. We can use this fact to replace y with $x - 5$ in the second equation $2x + 3y = 20$. We are substituting the expression $x - 5$ for y, since the two expressions are equivalent. The first example will solve this system by using the substitution method.

The Substitution Method

To solve a system of equations by using the substitution method, we solve one of the equations for one of the variables and then substitute that expression for the variable in the other equation.

EXAMPLE 1 Solve the following system by substitution:

$$y = x - 5$$
$$2x + 3y = 20$$

Solution

Since the first equation is solved for y in terms of x, we can substitute $x - 5$ for y in the second equation.

$$y = \boxed{x - 5}$$
$$2x + 3y = 20$$

This will produce a single equation with only one variable, x.

$$2x + 3y = 20$$
$$2x + 3(x - 5) = 20 \qquad \text{Substitute } x - 5 \text{ for } y.$$
$$2x + 3x - 15 = 20 \qquad \text{Distribute.}$$
$$5x - 15 = 20 \qquad \text{Combine like terms.}$$
$$5x = 35 \qquad \text{Add 15.}$$
$$x = 7 \qquad \text{Divide both sides by 5.}$$

We are now halfway to our solution. We know that the x-coordinate of the solution is 7, and we now need to find the y-coordinate.

A Word of Caution   The solution of a system of equations in two variables is an *ordered pair,* not a single value.

We can find the y-coordinate by substituting 7 for x in either of the original equations. It is easier to substitute the value into the equation $y = x - 5$, since y is already isolated in this equation. We could have chosen the other equation; it would produce the same solution. In general, it is a good idea to substitute the first value we find back into the equation that was used to make the original substitution.

$$y = x - 5$$
$$y = (7) - 5 \qquad \text{Substitute 7 for } x.$$
$$y = 2 \qquad \text{Subtract.}$$

The solution to this system is $(7, 2)$. This solution can be checked by substituting 7 for x and 2 for y in the equation $2x + 3y = 20$. The check is left to the reader.

Quick Check 1

Solve the following system by substitution:

$$y = x - 3$$
$$7x - 4y = 27$$

In the first example, one of the equations was already solved for y. To use the substitution method, we will often have to solve one of the equations for x or y. When choosing which equation to work with or which variable to solve for, look for an equation that has an x or y term with a coefficient of 1 or -1 ($x, -x, y$ or $-y$). When we find one of those terms, solving for that variable can be done easily through addition or subtraction.

EXAMPLE 2   Solve the following system by substitution:

$$x + 2y = 7$$
$$4x - 3y = -16$$

Solution

The first equation can easily be solved for x by subtracting $2y$ from both sides of the equation. This produces the equation $x = 7 - 2y$. We can now substitute the expression $7 - 2y$ for x in the second equation $4x - 3y = -16$.

A Word of Caution   Do not substitute $7 - 2y$ into the equation that it came from, $x + 2y = 7$, as the resulting equation will be $7 = 7$. Substitute it into the other equation in the system.

$$4x - 3y = -16$$
$$4(7 - 2y) - 3y = -16 \qquad \text{Substitute } 7 - 2y \text{ for } x.$$
$$28 - 8y - 3y = -16 \qquad \text{Distribute 4.}$$
$$28 - 11y = -16 \qquad \text{Combine like terms.}$$
$$-11y = -44 \qquad \text{Subtract 28 from both sides.}$$
$$y = 4 \qquad \text{Divide both sides by } -11.$$

Since the y-coordinate of the solution is 4, we can substitute this value for y into the equation $x = 7 - 2y$ in order to find the x-coordinate.

$$x = 7 - 2(4) \qquad \text{Substitute 4 for } y.$$
$$x = -1 \qquad \text{Simplify.}$$

The solution of this system of equations is $(-1, 4)$. We could check this solution by substituting -1 for x and 4 for y in the equation $4x - 3y = -16$. This is left to the reader. Although the first coordinate we solved for was y, keep in mind that the ordered pair must be written in the order (x, y).

Quick Check 2

Solve the following system by substitution:

$$x - y = -4$$
$$3x + 5y = 36$$

Inconsistent Systems

Objective 2   **Use the substitution method to identify inconsistent systems of equations.** Recall from the previous section that a linear system of equations with no solution is called an inconsistent system. The two lines associated with the equations of an inconsistent system are parallel lines. The next example shows how to determine that a system is an inconsistent system while using the substitution method.

EXAMPLE 3   Solve the following system using substitution:

$$x = 3y - 7$$
$$3x - 9y = 18$$

Solution

Since the first equation is solved for x, we will substitute $3y - 7$ for x in the second equation.

$$3x - 9y = 18$$
$$3(3y - 7) - 9y = 18 \qquad \text{Substitute } 3y - 7 \text{ for } x.$$
$$9y - 21 - 9y = 18 \qquad \text{Distribute.}$$
$$-21 = 18 \qquad \text{Combine like terms.}$$

The resulting equation is false, because -21 is not equal to 18. Since this equation can never be true, the system of equations has no solution ($\varnothing$) and is an inconsistent system.

Quick Check 3

Solve the following system by substitution:

$$-2x + y = 3$$
$$8x - 4y = 10$$

Dependent Systems

Objective 3   **Use the substitution method to identify dependent systems of equations.** A dependent system of linear equations has two equations with identical graphs and is a system with infinitely many solutions. When using the substitution method to solve a dependent system, we will obtain an identity, such as $3 = 3$, after making the substitution.

EXAMPLE ▶ 4   Solve the following system by substitution:

$$4x - y = 3$$
$$-8x + 2y = -6$$

Solution

We will solve the first equation for y.

$$4x - y = 3$$
$$-y = -4x + 3 \qquad \text{Subtract } 4x.$$
$$y = 4x - 3 \qquad \text{Divide both sides by } -1.$$

We now substitute $4x - 3$ for y in the second equation.

$$-8x + 2y = -6$$
$$-8x + 2(4x - 3) = -6 \qquad \text{Substitute } 4x - 3 \text{ for } y.$$
$$-8x + 8x - 6 = -6 \qquad \text{Distribute 2.}$$
$$-6 = -6 \qquad \text{Combine like terms.}$$

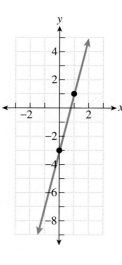

This equation is an identity that is true for all values of x. This system of equations is a dependent system with infinitely many solutions.

Since we have already shown that $y = 4x - 3$, the ordered pairs that are solutions have the form $(x, 4x - 3)$. The solutions are shown in the graph at the right, which is the graph of the line $y = 4x - 3$.

Quick Check **4**
Solve the following system by substitution:

$$x - 6y = 4$$
$$3x - 18y = 12$$

Objective **4**   **Solve systems of equations without variable terms with a coefficient of 1 or −1 by the substitution method.** All of the examples that have appeared so far in this section had at least one equation containing a variable term with a coefficient of 1 or −1. Such equations are easy to solve for one variable in terms of the other. In the next example, we will learn how to use the substitution method when this is not the case.

EXAMPLE ▶ 5   Solve the following system by substitution:

$$3x + 2y = -2$$
$$-2x + 7y = 43$$

Solution

There are no variable terms with a coefficient of 1 or −1 (x, $-x$, y, or $-y$). We will solve the first equation for y, although this selection is completely arbitrary. We could solve either equation for either variable and still find the same solution of the system.

$$3x + 2y = -2$$
$$2y = -3x - 2 \qquad \text{Subtract } 3x.$$
$$y = -\frac{3}{2}x - 1 \qquad \text{Divide both sides by 2.}$$

We now may substitute $-\frac{3}{2}x - 1$ for y in the second equation.

$$-2x + 7y = 43$$

$$-2x + 7\left(-\frac{3}{2}x - 1\right) = 43 \qquad \text{Substitute } -\frac{3}{2}x - 1 \text{ for } y.$$

$$-2x - \frac{21}{2}x - 7 = 43 \qquad \text{Distribute 7.}$$

$$2\left(-2x - \frac{21}{2}x - 7\right) = 2 \cdot 43 \qquad \begin{array}{l}\text{Multiply both sides by 2 to clear}\\ \text{fractions.}\end{array}$$

$$2(-2x) - \cancel{2}\left(\frac{21}{\cancel{2}_1}x\right) - 2 \cdot 7 = 2 \cdot 43 \qquad \begin{array}{l}\text{Distribute and divide out common}\\ \text{factors.}\end{array}$$

$$-4x - 21x - 14 = 86 \qquad \text{Multiply.}$$

$$-25x - 14 = 86 \qquad \text{Combine like terms.}$$

$$-25x = 100 \qquad \text{Add 14.}$$

$$x = -4 \qquad \text{Divide both sides by } -25.$$

We know that the x-coordinate of our solution is -4. We will substitute this value for x in the equation $y = -\frac{3}{2}x - 1$.

$$y = -\frac{3}{2}(-4) - 1 \qquad \text{Substitute } -4 \text{ for } x.$$

$$y = 5 \qquad \text{Simplify.}$$

Quick Check 5
Solve the following system by substitution:

$$5x - 3y = 17$$
$$2x + 4y = -14$$

The solution to this system is $(-4, 5)$.

Before turning our attention to an application of linear systems of equations, we present the following general strategy for solving these systems:

Solving a System of Equations by Using the Substitution Method

1. **Solve one of the equations for either variable.** If an equation has a variable term whose coefficient is 1 or -1, try to solve that equation for that variable.
2. **Substitute this expression for the variable in the other equation.** At this point, we will have an equation with only one variable.
3. **Solve this equation.** This gives us one coordinate of our ordered-pair solution.
4. **Substitute this value for the variable into the equation from step 1.** After simplifying, this will give us the other coordinate of our solution.
5. **Check the solution.**

Applications

Objective 5 **Solve applied problems by using the substitution method.** A college performed a play. There were 75 people in the audience. If admission was $6 for adults and $2 for children and the total box office receipts were $310, how many adults and how many children attended the play?

Solution

There are two unknowns in this problem: the number of adults and the number of children. As in Chapter 2, we start with a table of unknowns.

> ***Unknowns***
>
> Number of Adults: A
> Number of Children: C

Since there are two variables, we need a system of two equations. The first equation comes from the total attendance of 75 people. Since all of these people are either adults or children, the number of adults (A) plus the number of children (C) must equal 75. The first equation is $A + C = 75$.

The second equation comes from the amount of money paid. The amount of money paid by adults $(\$6A)$ plus the amount of money paid by children $(\$2C)$ must equal the total amount of money paid. The second equation is $6A + 2C = 310$. Here is the system of equations that we need to solve:

$$A + C = 75$$
$$6A + 2C = 310$$

Before we solve this system of equations, let's look at a table that contains all of the pertinent information for this problem.

	Number of Attendees	Cost per Ticket ($)	Money Paid ($)
Adults	A	6	$6A$
Children	C	2	$2C$
Total	75		310

Notice that the first equation in the system $(A + C = 75)$ can be found in the first column of the table (labeled Number of Attendees). The second equation in the system $(6A + 2C = 310)$ can be found in the last column in the table (labeled Money Paid ($)).

The first equation in this system can be solved for either A or C; we will solve it for C. This produces the equation $C = 75 - A$, so the expression $75 - A$ can be substituted for C in the second equation.

$$
\begin{aligned}
6A + 2C &= 310 & \\
6A + 2(75 - A) &= 310 & \text{Substitute } 75 - A \text{ for } C. \\
6A + 150 - 2A &= 310 & \text{Distribute.} \\
4A + 150 &= 310 & \text{Combine like terms.} \\
4A &= 160 & \text{Subtract 150.} \\
A &= 40 & \text{Divide both sides by 4.}
\end{aligned}
$$

There were 40 adults. We now substitute this value for A in the equation $C = 75 - A$.

$$
\begin{aligned}
C &= 75 - (40) & \text{Substitute 40 for } A. \\
C &= 35 & \text{Subtract.}
\end{aligned}
$$

There were 40 adults and 35 children at the play.

We should check our solution for accuracy. If there were 40 adults and 35 children, then that is a total of 75 people. Also, the 40 adults each paid $6, which is $240 dollars. The 35 children each paid $2, which is another $70. The total paid is $310, so our solution checks.

Quick Check **6** A small-business owner bought 15 new computers for his company. He paid $700 for each desktop computer and $1000 for each laptop computer. If he paid $11,700 for the computers, how many desktop computers and how many laptop computers did he buy?

> **Building Your Study Strategy** Doing Your Homework, 2 **Neat and Complete**
> Two important words to keep in mind when working on your homework exercises are *neat* and *complete*. When your homework is neat, it is easier to check your work, it helps an instructor spot an error if you need help with a particular problem, and it will be easier to understand when you review an old homework assignment prior to an exam.
>   Be as complete as you can when working on the homework exercises. By listing each step, you are increasing your chances of being able to remember all of the steps necessary to solve a similar problem on an exam or quiz. When you review an old homework assignment prior to an exam, you may have difficulty remembering how to solve a problem if you did not write down all of the necessary steps. If you make a mistake while working on a particular exercise, make note of the mistake that you made, to avoid making a similar mistake later.

EXERCISES 4.2

Vocabulary

1. To solve a system of equations by substitution, begin by solving _____.

2. Once one of the equations has been solved for one of the variables, that expression is _____ for that variable in the other equation.

3. If substitution results in an equation that is an identity, then the system is a(n) _____ system.

4. If substitution results in an equation that is a contradiction, then the system is a(n) _____ system.

Solve the system by substitution. If the system is inconsistent and has no solution, state this. If the system is dependent, write the form of the solution for any real number x.

5. $y = 3x - 5$
    $x + 3y = 15$

6. $x = 4y + 3$
    $2x + 5y = -7$

7. $x = 2 - 5y$
    $7x + 10y = 89$

8. $x = 1 - 4y$
    $4x + 2y = -17$

9. $y = 5x - 10$
    $-10x + 2y = 20$

MyMathLab

MathXP

F O R   E X T R A   H E L P

Interactmath.com

MathXL
Tutorials on CD

Video Lectures
on CD

Tutor Center

Addison-Wesley
Math Tutor Center

Student's
Solutions Manual

10. $y = 11 - 5x$
$-7x - 2y = 14$

11. $18x + 6y = -12$
$y = -3x - 2$

12. $x = 2y - 15$
$3x + 14y = 5$

13. $y = 2.2x - 6.8$
$3x + 2y = 16$

14. $x = 4.7y - 40.6$
$5x - 2y = -31$

15. $x = \frac{2}{3}y - 4$
$6x + 7y = 75$

16. $y = \frac{3}{4}x - \frac{7}{2}$
$5x - 4y = 2$

17. $x + 6y = 5$
$10x - 4y = 2$

18. $2x + y = 3$
$12x - 2y = -12$

19. $x + y = 11$
$3x + 5y = 29$

20. $4x + 2y = 34$
$x - y = -8$

21. $3x + 8y = 20$
$2x + y = 9$

22. $x + 3y = 14$
$3x + 9y = 28$

23. $5x - y = 18$
$4x - 3y = 10$

24. $5x + 6y = 41$
$-x + 2y = 11$

25. $6x + y = 21$
$-18x - 3y = -63$

26. $10x - y = -23$
$-20x + 2y = -46$

27. $x + 2.8y = 3.4$
$2x - 9y = -37$

28. $-3.4x + y = -30.8$
$2.1x + 3.2y = 57.2$

29. $\frac{3}{2}x + y = -7$
$-3x + 2y = 34$

30. $x + \frac{3}{8}y = 24$
$16x - 11y = -24$

31. $9x - 8y = 12$
$y = 3$

32. $1776x - 3y = 270$
$x = 0$

33. $4x + 2y = 10$
$5x - 4y = 6$

34. $9x + 3y = 36$
$7x - 6y = 78$

35. $3x + 5y = 14$
$6x - 5y = -32$

36. $9x + 4y = 23$
$5x - 2y = -2$

37. Give an equation that, along with the equation $y = 2x - 14$, forms an inconsistent system.

38. Give an equation that, along with the equation $y = 5x - 7$, forms a dependent system.

39. Give an equation that, along with the equation $y = 7x - 2$, forms a system whose only solution is $(-1, -9)$.

40. Make up two equations that form a system whose only solution is $(3, -2)$.

41. The college basketball team charges $5 admission to the general public for its games, while students at the college pay only $2. If the attendance for last night's game was 1200 people and the total receipts were $5250, how many of the people in attendance were students and how many were not students?

42. Kenny walks away from a blackjack table with a total of $62 in $1 and $5 chips. If Kenny has 26 chips, how many are $1 chips and how many are $5 chips?

43. A sorority set up a charity dinner to help raise money for a local children's support group. They sold chicken dinners for $8 and steak dinners for $10. If 142 people ate dinner at the fundraiser and the sorority raised $1326, how many people ordered chicken and how many people ordered steak?

44. Jeannie has some $10 bills and some $20 bills. If she has 273 bills worth a total of $4370, how many of the bills are $10 bills and how many are $20 bills?

45. To celebrate a productive week in his office, Devin stopped by a coffee shop on his way to work Friday. He bought some regular coffees for $1.75 each and some café mochas for $3.25 each. If he bought 18 drinks and spent $39, how many regular coffees and how many café mochas did he buy?

46. Laurel scored 47 points in last night's basketball game. Eight of her points were scored on free throws, which count as 1 point each, and the rest were scored on 2-point shots and 3-point shots. If the number of 2-point shots that Laurel made was 2 more than twice the number of 3-point shots that she made, how many 2-point shots did she make?

47. The median age of men, y, at their first marriage in a particular year can be approximated by the equation $y = 0.12x + 23.2$, where x represents the number of years after 1970 and y is in years. The median age of women, y, at their first marriage in a particular year can be approximated by the equation $y = 0.14x + 20.8$, where x represents the number of years after 1970 and y is in years. If the current trend continues, in what year will the median age of men at their first marriage equal the median age of women at their first marriage? (*Source:* U.S. Census Bureau)

48. The number of doctoral degrees conferred on women, y, in a particular year can be approximated by the equation $y = 473x + 17,811$, where x represents the number of years after 1995. The number of doctoral degrees conferred on men, y, in a particular year can be approximated by the equation $y = -423x + 26,841$, where x represents the number of years after 1995. In what year were the number of doctoral degrees conferred on women equal to the number of doctoral degrees conferred on men? (*Source:* U.S. Department of Education, National Center for Educational Statistics)

Writing in Mathematics

Answer in complete sentences.

49. Explain, in your own words, how to solve a system of linear equations using the substitution method.

50. When using the substitution method, how can you tell that a system of equations is inconsistent and has no solutions? How can you tell that a system of equations is dependent and has infinitely many solutions? Explain your answer.

51. ***Solutions Manual*** * Write a solutions manual page for the following problem:

Solve the system of equations $y = 3x - 7$
by the substitution method $2x - y = 2$

52. ***Newsletter*** * Write a newsletter explaining how to solve a system of two linear equations by using the substitution method.

*See Appendix B for details and sample answers.

4.3

Solving Systems of Equations by Using the Addition Method

1 Solve systems of equations by using the addition method.
2 Use the addition method to identify inconsistent and dependent systems of equations.
3 Solve systems of equations with coefficients that are fractions or decimals by using the addition method.
4 Solve applied problems by using the addition method.

The Addition Method

Objective 1   Solve systems of equations by using the addition method. In this section, we will examine the **addition method** for solving systems of linear equations.

The Addition Method

> The addition method is an algebraic alternative to the substitution method. The two equations in a system of linear equations will be combined into a single equation with a single variable by adding them together.

In the last section, we saw that solving a system of equations by substitution is fairly easy when one of the equations contains a variable term with a coefficient of 1 or -1. The substitution method can become quite tedious for solving a system when this is not the case.

As with the substitution method, the goal of the addition method is to combine the two given equations into a single equation with only one variable. This method is based on the addition property of equality from Chapter 2, which tells us that when we add the same number to both sides of an equation, the equation remains true. (For any real numbers a, b, and c, if $a = b$, then $a + c = b + c$.) This property can be extended to cover adding equal expressions to both sides of an equation: if $a = b$ and $c = d$, then $a + c = b + d$.

Consider the system

$$3x + 2y = 20$$
$$5x - 2y = -4$$

If we add the left side of the first equation ($3x + 2y$) to the left side of the second equation ($5x - 2y$), this will be equal to the total of the two numbers on the right side of the equations. Here is the result of this addition:

$$\begin{array}{r} 3x + 2y = 20 \\ \underline{5x - 2y = -4} \\ 8x = 16 \end{array}$$

Notice that when we added these two equations, the two terms containing y were opposites and summed to zero, leaving an equation containing only the variable x. We next solve this equation for x and substitute this value for x in either of the two original equations to find y. The first example will walk through this entire process.

EXAMPLE 1   Solve the following system by addition:

$$3x + 2y = 20$$
$$5x - 2y = -4$$

Solution

Since the two terms containing y are opposites, we can use addition to eliminate y.

$$3x + 2y = 20$$
$$\underline{5x - 2y = -4} \qquad \text{Add the two equations.}$$
$$8x = 16$$

$$x = 2 \qquad \text{Divide both sides by 8.}$$

The x-coordinate of our solution is 2. We now substitute this value for x in the first original equation. (We could have chosen the other equation, since it will produce exactly the same solution.)

$$3x + 2y = 20$$
$$3(2) + 2y = 20 \qquad \text{Substitute 2 for } x.$$
$$6 + 2y = 20 \qquad \text{Multiply.}$$
$$2y = 14 \qquad \text{Subtract 6.}$$
$$y = 7 \qquad \text{Divide both sides by 2.}$$

The solution of this system is the ordered pair $(2, 7)$. We could check this solution by substituting 2 for x and 7 for y in the equation $5x - 2y = -4$. This is left to the reader.

Quick Check 1
Solve the following
system by addition:

$$4x + 3y = 31$$
$$-4x + 5y = -23$$

If the two equations in a system do not contain a pair of opposite variable terms, we can still use the addition method. We must first multiply both sides of one or both equations by a constant(s) in such a way that two of the variable terms become opposites. Then we proceed as in the first example.

EXAMPLE 2   Solve the following system by addition:

$$2x - 3y = -16$$
$$-6x + 5y = 32$$

Solution

Adding these two equations at this point would not be helpful, as the sum would still contain both variables. If we multiply the first equation by 3, the terms containing x will be opposites. This allows us to proceed with the addition method.

$$2x - 3y = -16 \quad \xrightarrow{\text{Multiply by 3}} \quad 6x - 9y = -48$$
$$-6x + 5y = 32 \qquad\qquad\qquad\qquad -6x + 5y = 32$$

$$6x - 9y = -48$$
$$\underline{-6x + 5y = 32} \qquad \text{Add.}$$
$$-4y = -16$$

$$y = 4 \qquad \text{Divide both sides by } -4.$$

A Word of Caution   Be sure to multiply *both* sides of the equation by the same number. Do not just multiply the side containing the variable terms.

We now substitute 4 for y in the equation $2x - 3y = -16$ and solve for x.

$$2x - 3(4) = -16 \qquad \text{Substitute 4 for } y.$$
$$2x - 12 = -16 \qquad \text{Multiply.}$$
$$2x = -4 \qquad \text{Add 12.}$$
$$x = -2 \qquad \text{Divide both sides by 2.}$$

The solution of this system is $(-2, 4)$. (Be sure to write the x-coordinate first in the ordered pair, even though we solved for y first.)

Quick Check 2

Solve the following system by addition:

$$2x - 9y = 33$$
$$4x + 3y = 3$$

Before continuing, let's outline the basic strategy for using the addition method to solve systems of linear equations.

Using the Addition Method

1. **Write each equation in standard form ($Ax + By = C$).** For each equation, gather all variable terms on the left side and all constants on the right side.
2. **Multiply one or both equations by the appropriate constant(s).** The goal of this step is to make either the terms containing x or the terms containing y opposites.
3. **Add the two equations together.** At this point, we should have one equation containing a single variable.
4. **Solve the resulting equation.** This will give us one of the coordinates of the solution.
5. **Substitute this value for the appropriate variable in either of the original equations, and solve for the other variable.** Choose the equation that you feel will be easier to solve. When we solve this equation, we will find the other coordinate of the solution.
6. **Write the solution as an ordered pair.**
7. **Check your solution.**

EXAMPLE 3   Solve the following system by addition:

$$5x + 8y + 11 = 0$$
$$7x = 6y + 19$$

Solution

To use the addition method, we must first write each equation in standard form:

$$5x + 8y = -11$$
$$7x - 6y = 19$$

Now we must decide which variable to eliminate. A wise choice for this system is to eliminate y, as the two coefficients already have opposite signs. If we multiply the first equation by 3 and the second equation by 4, our terms containing y will be $24y$ and $-24y$.

$$5x + 8y = -11 \xrightarrow{\text{Multiply by 3}} 15x + 24y = -33$$
$$7x - 6y = 19 \xrightarrow{\text{Multiply by 4}} 28x - 24y = 76$$

$$
\begin{array}{rcl}
15x + 24y &=& -33 \\
\underline{28x - 24y} &=& \underline{76} \qquad \text{Add to eliminate } y. \\
43x &=& 43 \\
x &=& 1 \qquad \text{Divide both sides by 43.}
\end{array}
$$

The x-coordinate of our solution is 1. Substituting 1 for x in the equation $5x + 8y = -11$ will allow us to find the y-coordinate.

$$
\begin{array}{rcl}
5(1) + 8y &=& -11 \qquad \text{Substitute 1 for } x. \\
5 + 8y &=& -11 \qquad \text{Multiply.} \\
8y &=& -16 \qquad \text{Subtract 5.} \\
y &=& -2 \qquad \text{Divide both sides by 8.}
\end{array}
$$

Quick Check **3**

Solve the following system by addition.

$$3x + 2y = -26$$
$$5y = 2x + 11$$

The solution to this system is $(1, -2)$. The check is left to the reader.

Objective 2 Use the addition method to identify inconsistent and dependent systems of equations. Recall from the previous section that when solving an inconsistent system (no solution), we obtain an equation that is a contradiction such as $0 = -6$. When solving a dependent system (infinitely many solutions), we will end up with an equation that is an identity such as $7 = 7$. The same is true when applying the addition method.

EXAMPLE ▶ 4 Solve the following system by addition:

$$3x + 5y = 15$$
$$6x + 10y = 30$$

Solution

Suppose we chose to eliminate the variable x. We can do so by multiplying the first equation by -2. This produces the following system:

$$3x + 5y = 15 \xrightarrow{\text{Multiply by } -2} -6x - 10y = -30$$
$$6x + 10y = 30 \phantom{\xrightarrow{\text{Multiply by } -2}} 6x + 10y = 30$$

$$
\begin{array}{rcl}
-6x - 10y &=& -30 \\
\underline{6x + 10y} &=& \underline{30} \qquad \text{Add to eliminate } x. \\
0 &=& 0
\end{array}
$$

Because the resulting equation is an identity, this is a dependent system with infinitely many solutions. To determine the form of our solutions, we can solve the equation $3x + 5y = 15$ for y.

$$
\begin{array}{rcl}
3x + 5y &=& 15 \\
5y &=& -3x + 15 \qquad \text{Subtract } 3x \text{ to isolate } 5y. \\
y &=& -\dfrac{3}{5}x + 3 \qquad \text{Divide both sides by 5 and simplify.}
\end{array}
$$

The ordered pairs that are solutions have the form $(x, -\frac{3}{5}x + 3)$. We could also display the solutions by graphing the line $3x + 5y = 15$, or $y = -\frac{3}{5}x + 3$.

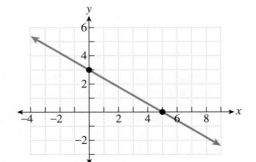

EXAMPLE 5   Solve the following system by addition:

$$6x - 3y = 7$$
$$-2x + y = -3$$

Solution

To eliminate the variable y, we can multiply the second equation by 3. The variable term containing y in the second equation would then be $3y$, which is the opposite of $-3y$ in the first equation.

$$6x - 3y = 7 \qquad\qquad\qquad\qquad 6x - 3y = 7$$
$$-2x + y = -3 \xrightarrow{\text{Multiply by 3}} -6x + 3y = -9$$

$$\begin{array}{r} 6x - 3y = 7 \\ -6x + 3y = -9 \\ \hline 0 = -2 \end{array} \qquad \text{Add to eliminate } y.$$

Since the resulting equation is false, this is an inconsistent system with no solution.

Systems of Equations with Coefficients That Are Fractions or Decimals

Objective 3 **Solve systems of equations with coefficients that are fractions or decimals by using the addition method.** If one or both equations in a system of linear equations contain fractions or decimals, we can clear the fractions or decimals first and then solve the resulting system.

EXAMPLE 6   Solve the following system by addition:

$$\frac{5}{4}x + \frac{1}{3}y = 3$$
$$\frac{3}{8}x - \frac{5}{6}y = \frac{13}{2}$$

Solution

The LCM of the denominators in the first equation is 12, so if we multiply both sides of the first equation by 12, we can clear the equation of fractions.

$$12\left(\frac{5}{4}x + \frac{1}{3}y\right) = 12 \cdot 3 \qquad \text{Multiply both sides by 12.}$$

$$\overset{3}{\cancel{12}} \cdot \frac{5}{\underset{1}{\cancel{4}}}x + \overset{4}{\cancel{12}} \cdot \frac{1}{\underset{1}{\cancel{3}}}y = 12 \cdot 3 \qquad \text{Distribute and divide out common factors.}$$

$$15x + 4y = 36 \qquad \text{Simplify.}$$

Multiplying both sides of the second equation by 24 will clear the fractions from that equation.

$$24\left(\frac{3}{8}x - \frac{5}{6}y\right) = 24 \cdot \frac{13}{2} \qquad \text{Multiply both sides by 24.}$$

$$\overset{3}{\cancel{24}} \cdot \frac{3}{\underset{1}{\cancel{8}}}x - \overset{4}{\cancel{24}} \cdot \frac{5}{\underset{1}{\cancel{6}}}y = \overset{12}{\cancel{24}} \cdot \frac{13}{\underset{1}{\cancel{2}}} \qquad \text{Distribute and divide out common factors.}$$

$$9x - 20y = 156 \qquad \text{Simplify.}$$

Here is the resulting system of equations:

$$15x + 4y = 36$$
$$9x - 20y = 156$$

Since the coefficient of the term containing y in the second equation is a multiple of the coefficient containing y in the first equation, we will eliminate y. This can be done by multiplying the first equation by 5.

$$
\begin{array}{l}
15x + 4y = 36 \\
9x - 20y = 156
\end{array}
\xrightarrow{\text{Multiply by 5}}
\begin{array}{l}
75x + 20y = 180 \\
9x - 20y = 156
\end{array}
$$

$$
\begin{array}{l}
75x + 20y = 180 \\
\underline{9x - 20y = 156} \qquad \text{Add to eliminate } y. \\
84x \qquad\quad = 336
\end{array}
$$

$$x = 4 \qquad \text{Divide both sides by 84.}$$

The x-coordinate of our solution is 4. To solve for y, we can substitute 4 for x in either of the original equations, or we can substitute into either of the equations that was created by clearing the fractions. We will use the equation $15x + 4y = 36$.

$$15(4) + 4y = 36 \qquad \text{Substitute 4 for } x.$$
$$60 + 4y = 36 \qquad \text{Multiply.}$$
$$4y = -24 \qquad \text{Subtract 60.}$$
$$y = -6 \qquad \text{Divide both sides by 4.}$$

Quick Check 6

Solve the following system by addition:

$$-\frac{2}{3}x + \frac{5}{2}y = 14$$
$$\frac{5}{12}x + \frac{3}{4}y = \frac{1}{2}$$

The solution of this system is the ordered pair $(4, -6)$.

If one or both equations in a system of linear equations contain decimals, we can clear the decimals first by multiplying by the appropriate power of 10 and then solve the resulting system.

EXAMPLE 7 Solve the system by addition.

$$0.3x + 2y = 7.5$$
$$0.04x + 0.07y = 0.41$$

Solution

If we multiply both sides of the first equation by 10, we can clear the equation of decimals.

$$10(0.3x + 2y) = 10(7.5) \quad \text{Multiply both sides by 10.}$$
$$3x + 20y = 75 \quad \text{Distribute and multiply.}$$

For the second equation, we multiply both sides by 100 to clear the decimals.

$$100(0.04x + 0.07y) = 100(0.41) \quad \text{Multiply both sides by 100.}$$
$$4x + 7y = 41 \quad \text{Distribute and multiply.}$$

Here is the resulting system of equations:

$$3x + 20y = 75$$
$$4x + 7y = 41$$

We eliminate x by multiplying the first equation by 4 and the second equation by -3.

$$3x + 20y = 75 \quad \xrightarrow{\text{Multiply by 4}} \quad 12x + 80y = 300$$
$$4x + 7y = 41 \quad \xrightarrow{\text{Multiply by } -3} \quad -12x - 21y = -123$$

$$
\begin{array}{r}
12x + 80y = 300 \\
-12x - 21y = -123 \\
\hline
59y = 177
\end{array}
\quad \text{Add to eliminate } x.
$$

$$y = 3 \quad \text{Divide both sides by 59.}$$

The y-coordinate of our solution is 3. To solve for x we substitute 3 for y in either of the original equations, or either of the equations that was created by clearing the decimals. The equation $3x + 20y = 75$ is a good choice, as it will be easier to solve for x in an equation that does not contain decimals.

$$3x + 20(3) = 75 \quad \text{Substitute 3 for } y.$$
$$3x + 60 = 75 \quad \text{Multiply.}$$
$$3x = 15 \quad \text{Subtract 60.}$$
$$x = 5 \quad \text{Divide both sides by 3.}$$

The solution to this system is the ordered pair $(5, 3)$.

Quick Check 7 Solve the system by addition.

$$0.03x + 0.05y = 29.5$$
$$x + y = 700$$

The substitution and addition methods for solving a system of equations will solve any system of linear equations. There are certain times when one method is easier to apply than the other, and knowing which method to choose for a particular system of equations can save time and prevent errors. In general, if one of the equations in the system is already solved for one of the variables, such as $y = 3x$ or $x = -4y + 5$, then using the substitution method is a good choice. The substitution method also works well when one of the equations contains a variable term with a coefficient of 1 or -1 (such as $x + 6y = 17$ or $7x - y = 14$). Otherwise, consider using the addition method.

Applications

Objective 4   Solve applied problems by using the addition method. We conclude this section with an application of solving systems of equations using the addition method.

EXAMPLE 8   Dylan has $3.20 in change in his pocket. Each coin is either a dime or a quarter. If Dylan has 17 coins in his pocket, how many are dimes and how many are quarters?

Solution

There are two unknowns in this problem: the number of dimes and the number of quarters. We will start with a table of unknowns.

> *Unknowns*
> Number of dimes: d
> Number of quarters: q

The first equation in our system comes from the fact that Dylan has 17 coins. Since all of these coins are either dimes or quarters, the number of dimes (d) plus the number of quarters (q) must equal 17. The first equation is $d + q = 17$.

The second equation comes from the amount of money Dylan has in his pocket. The amount of money in dimes ($0.10d$) plus the amount of money in quarters ($0.25q$) must equal the total amount of money he has. The second equation is $0.10d + 0.25q = 3.20$.

	Number of Coins	Value per Coin ($)	Money in Pocket ($)
Dimes	d	0.10	$0.10d$
Quarters	q	0.25	$0.25q$
Total	17		3.20

Here is the system of equations that we will need to solve:

$$\begin{aligned} d + \quad q &= 17 \\ 0.10d + 0.25q &= 3.20 \end{aligned}$$

Multiplying both sides of the second equation by 100 will clear the equation of decimals, producing the following system:

$$\begin{aligned} d + \quad q &= 17 \\ 10d + 25q &= 320 \end{aligned}$$

We can now eliminate d by multiplying the first equation by -10.

$$
\begin{array}{l}
d + q = 17 \\
10d + 25q = 320
\end{array}
\xrightarrow{\text{Multiply by } -10}
\begin{array}{l}
-10d - 10q = -170 \\
10d + 25q = 320
\end{array}
$$

$$
\begin{array}{rl}
-10d - 10q = & -170 \\
\underline{10d + 25q = 320} & \qquad \text{Add to eliminate } d. \\
15q = & 150 \\
q = & 10 \qquad \text{Divide both sides by 15.}
\end{array}
$$

Quick Check 8

Erica has \$5.35 in change in her pocket. Each coin is either a nickel or a dime. If Erica has 75 coins in her pocket, how many are nickels and how many are dimes?

There are 10 quarters in Dylan's pocket. To find the number of dimes that he has, we can substitute 10 for q in the equation $d + q = 17$ and solve for d.

$$
\begin{array}{ll}
d + (10) = 17 & \qquad \text{Substitute 10 for } q. \\
d = 7 & \qquad \text{Subtract 10.}
\end{array}
$$

Dylan has 7 dimes and 10 quarters in his pocket. You can verify that the total amount of money in his pocket is \$3.20.

Building Your Study Strategy Doing Your Homework, 3 **Difficult Problems**
Eventually, there will be a homework exercise that you cannot answer correctly. Rather than giving up, here are some options to consider:

- Review your class notes. There may be a similar problem that was discussed in class. If so, you can use the solution of this problem to help you figure out the homework exercise that you were unable to do.
- Review the related section in the text. You may be able to find a similar example, or there may be a Word of Caution warning you about typical errors on that type of problem.
- Call someone in your study group. A member of your study group may have already completed the problem and can help you figure out what to do.

If you still cannot solve the problem, move on and try the next problem. Be sure to ask your instructor the very next day about this problem.

Vocabulary

1. If an equation in a system of equations contains fractions, we can clear the fractions by multiplying both sides of the equation by _____.

2. If an equation in a system of equations contains decimals, we can clear the decimals by multiplying both sides of the equation by the appropriate power of _____.

3. If the addition method results in an equation that is an identity, then the system is a(n) _____ system.

4. If the addition method results in an equation that is a contradiction, then the system is a(n) _____ system.

Solve the system by addition. If the system is inconsistent and has no solution, state this. If the system is dependent, write the form of the solution for any real number x.

5. $4x + 3y = -17$
   $2x - 3y = 5$

6. $7x + 5y = 2$
   $-7x + 4y = -11$

7. $-2x + y = -7$
   $2x + 9y = 17$

8. $8x - 6y = -22$
   $5x + 6y = -43$

9. $3x + 2y = 20$
   $7x - 4y = 38$

10. $6x - 9y = -18$
    $-3x + 8y = 23$

11. $2x - 4y = 11$
    $-4x + 8y = -22$

12. $8x - 3y = 34$
    $-2x + 9y = 8$

13. $x + 4y = 11$
    $3x + 16y = 37$

14. $10x + 2y = 44$
    $3x + 8y = 65$

15. $12x + 9y = -69$
    $3x + 5y = -31$

16. $2x + 3y = 13$
    $10x + 15y = 26$

17. $8x + 7y = -5$
    $-x + 4y = -14$

18. $7x + 11y = -31$
    $4x - y = 26$

19. $3x + 4y = 26$
    $5x - 3y = 24$

20. $7x + 2y = -70$
    $6x - 5y = -13$

21. $-8x + 5y = -28$
    $3x + 8y = -29$

22. $5x - 5y = 45$
    $-4x + 4y = -36$

23. $4x - 3y = -6$
    $6x + 5y = 29$

24. $-7x + 4y = -25$
    $5x + 14y = 60$

25. $-6x + 9y = 17$
    $-4x + 6y = 12$

26. $3x - 11y = 43$
    $2x - 7y = 27$

27. $5x + 7y = 16$
    $4x + 8y = 14$

28. $4x + 6y = -20$
    $6x + 13y = -48$

29. $4y = 3x - 13$
    $6x + 5y = 52$

30. $7x = -18 - 4y$
    $3x - 2y = -30$

31. $5x = 3y + 21$
    $-5x + 4y = -23$

32. $5y = 2x + 13$
    $4x + 3y = 65$

33. $\dfrac{1}{2}x + \dfrac{2}{5}y = \dfrac{7}{5}$
    $\dfrac{1}{4}x - \dfrac{1}{6}y = \dfrac{1}{3}$

34. $\dfrac{1}{4}x + \dfrac{1}{2}y = \dfrac{11}{4}$
    $-\dfrac{1}{9}x + \dfrac{1}{3}y = \dfrac{4}{9}$

35. $\dfrac{2}{3}x + \dfrac{1}{6}y = 4$
    $\dfrac{5}{3}x - y = -7$

36. $\dfrac{1}{7}x + \dfrac{1}{28}y = \dfrac{1}{2}$
    $\dfrac{1}{3}x - \dfrac{2}{9}y = -\dfrac{23}{18}$

$$r + c = 148$$
$$(c + 42) + c = 148 \qquad \text{Substitute } c + 42 \text{ for } r.$$
$$2c + 42 = 148 \qquad \text{Combine like terms.}$$
$$2c = 106 \qquad \text{Subtract 42.}$$
$$c = 53 \qquad \text{Divide both sides by 2.}$$

Carolyn had 53 DVDs. To find out how many DVDs Ross had, we can substitute 53 for c in the equation $r = c + 42$.

$$r = (53) + 42 \qquad \text{Substitute 53 for } c.$$
$$r = 95 \qquad \text{Add.}$$

Quick Check 1

At a certain community college, there are 68 instructors who teach math or English. The number of English instructors is 12 more than the number of math instructors. How many English instructors are there?

Ross had 95 DVDs and Carolyn had 53 DVDs. Combined, this is a total of 148 DVDs.

The setup for the previous example can be used for any problem in which we know the sum and difference of two quantities. The same system of equations would be used to find two numbers whose difference is 42 and whose sum is 148.

Geometry Problems

Objective 2 **Solve geometry problems by using a system of equations.** In Section 2.3, we solved problems involving the perimeter of a rectangle. We will now use a system of two equations in two variables to solve the same type of problem.

EXAMPLE 2 The length of a rectangle is five feet more than twice its width. If the perimeter of the rectangle is 124 feet, find the length and the width of the rectangle.

Solution

The two unknowns are the length and the width of the rectangle. We will use the variables l and w for the length and width, respectively.

> ***Unknowns***
> Length: l
> Width: w

We are told that the length is 5 feet more than twice the width of the rectangle. We express this relationship in an equation as $l = 2w + 5$. The second equation in the system comes from the fact that the perimeter is 124 feet. Since the perimeter of a rectangle is equal to twice the length plus twice the width, our second equation can be written as $2l + 2w = 124$. Here is the system we are solving:

$$l = 2w + 5$$
$$2l + 2w = 124$$

We will use the substitution method to solve this system, as the first equation is already solved for l. We begin by substituting the expression $2w + 5$ for l in the equation $2l + 2w = 124$.

$$2(2w + 5) + 2w = 124 \qquad \text{Substitute } 2w + 5 \text{ for } l.$$
$$4w + 10 + 2w = 124 \qquad \text{Distribute.}$$
$$6w + 10 = 124 \qquad \text{Combine like terms.}$$
$$6w = 114 \qquad \text{Subtract 10.}$$
$$w = 19 \qquad \text{Divide both sides by 6.}$$

The width of the rectangle is 19 feet. To solve for the length, we substitute 19 for w in the equation $l = 2w + 5$.

$$l = 2(19) + 5 \qquad \text{Substitute 19 for } w.$$
$$l = 43 \qquad \text{Simplify.}$$

Quick Check **2**

The length of the rectangle is 43 feet, and the width is 19 feet.

A rectangular concrete slab is being poured for a new house. The perimeter of the slab is 230 feet and the length of the slab is 35 feet longer than the width. Find the dimensions of the concrete slab.

Interest Problems

Objective 3 Solve interest problems by using a system of equations. For certain problems, it's easier to set up a system of equations in two variables than to try to express both unknown quantities in terms of the same variable. For example, the mixture and interest problems covered in Section 2.4 are often easier to solve when working with a system of equations in two variables. We will begin with a problem involving interest.

EXAMPLE 3   Serena invested $8000 in two certificates of deposit (CDs). Some of the money was deposited in a CD that paid 7% annual interest, and the rest was deposited in a CD that paid 6% annual interest. If Serena earned $530 in interest in the first year, how much did she invest in each CD?

Solution

The two unknowns are the amount invested at 7% and the amount invested at 6%. We will let x represent the amount invested at 7% interest, and y represent the amount invested at 6% interest. To determine the interest earned, we multiply the principal (the amount of money invested) by the interest rate by the time in years. If the first account earns 7% interest, then we can represent the interest earned in one year by $0.07x$. In a similar fashion, we can represent the interest earned in one year from the investment at 6% interest as $0.06y$. The following table summarizes this information:

Account	Principal	Interest Rate	Time (years)	Interest Earned
CD 1	x	0.07	1	$0.07x$
CD 2	y	0.06	1	$0.06y$
Total	8000			530

The first equation in our system comes from the fact that the amount invested at 7% interest plus the amount invested at 6% is equal to the total amount invested. As an equation, this can be represented as $x + y = 8000$. This equation can be found in our table in the column labeled Principal. The second equation comes from the amount of interest earned and can be found in the column labeled Interest Earned. We know that Serena earned $0.07x$ in interest from the first CD and she earned $0.06y$ in interest from the second CD. Since the total amount of interest earned comes from these two CDs, the second equation in our system is $0.07x + 0.06y = 530$. Here is the system we must solve:

$$x + y = 8000$$
$$0.07x + 0.06y = 530$$

We will clear the second equation of decimals by multiplying both sides of the equation by 100.

$$x + \quad y = 8000 \qquad\qquad\qquad\qquad x + \quad y = \quad 8000$$
$$0.07x + 0.06y = \quad 530 \xrightarrow{\text{Multiply by 100}} 7x + 6y = 53{,}000$$

To solve this system of equations, we can use the addition method. (The substitution method would work just as well.) If we multiply the first equation by -6, then the variable terms containing y are opposites.

Quick Check 3

$$x + \quad y = \quad 8000 \xrightarrow{\text{Multiply by } -6} \quad -6x - 6y = -48{,}000$$
$$7x + 6y = 53{,}000 \qquad\qquad\qquad\qquad 7x + 6y = \quad 53{,}000$$

$$
\begin{array}{rl}
-6x - 6y = & -48{,}000 \\
7x + 6y = & 53{,}000 \qquad \text{\small Add to eliminate } y. \\
\hline
x \quad = & 5{,}000
\end{array}
$$

We know that Serena invested \$5000 at 7% interest. To find the amount invested at 6%, we substitute 5000 for x in the original equation $x + y = 8000$.

$$(5000) + y = 8000 \qquad \text{\small Substitute 5000 for } x.$$
$$y = 3000 \qquad\quad\; \text{\small Subtract 5000.}$$

Serena invested \$5000 at 7% interest and \$3000 at 6% interest.

Quick Check 3
Annika invested \$4200 in two certificates of deposit (CDs). Some of the money was deposited in a CD that paid 5% annual interest, and the rest was deposited in a CD that paid 4% annual interest. If Annika earned \$200 in interest in the first year, how much did she invest in each CD?

Mixture Problems

Objective 4 **Solve mixture problems by using a system of equations.**

EXAMPLE 4 Gunther works at a coffee shop that sells Kona coffee beans for \$40 per pound and Colombian coffee beans for \$15 per pound. The coffee shop also sells a blend, which is made up of Kona coffee and Colombian coffee, for \$20 per pound. If Gunther's boss tells him to make 30 pounds of this blend, how much Kona coffee and Colombian coffee must be mixed together?

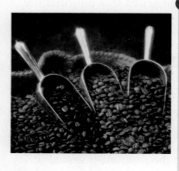

Solution

The two unknowns are the number of pounds of Kona coffee (k) and the number of pounds of Colombian coffee (c) that need to be mixed together to form the blend. Since Gunther needs to make 30 pounds of coffee with a value of \$20 per pound, the total value of this mixture can be found by multiplying 30 by 20. The total cost of this blend is \$600. The following table summarizes this information:

Coffee	Pounds	Cost per Pound	Total Cost
Kona	k	40	$40k$
Colombian	c	15	$15c$
Blend	30	20	600

The first equation in our system comes from the fact that the weight of the Kona coffee plus the weight of the Colombian coffee must equal 30 pounds. As an equation, this can be represented by $k + c = 30$. This equation can be found in our table in the column labeled Pounds. The second equation comes from the total cost of the coffee. We know that the total cost is $600, so the second equation in our system is $40k + 15c = 600$. This equation can be found in the column labeled Total Cost. Here is the system we must solve:

$$k + c = 30$$
$$40k + 15c = 600$$

Quick Check 4

Hannibal is making 48 pounds of a mixture of almonds and cashews that his nut shop will sell for $6 per pound. If the nut shop sells almonds for $5 per pound and cashews for $8 per pound, how many pounds of almonds and cashews should Hannibal mix together?

To solve this system of equations, we can use the addition method. If we multiply the first equation by -15, then the variable terms containing c are opposites.

$$
\begin{array}{l}
k + c = 30 \\
40k + 15c = 600
\end{array}
\xrightarrow{\text{Multiply by } -15}
\begin{array}{l}
-15k - 15c = -450 \\
40k + 15c = 600
\end{array}
$$

$$
\begin{array}{rl}
-15k - 15c = & -450 \\
\underline{40k + 15c = 600} & \qquad \text{Add to eliminate } c. \\
25k = 150 \\
k = 6 & \qquad \text{Divide both sides by 25.}
\end{array}
$$

We know that Gunther must use 6 pounds of Kona coffee. To determine how much Colombian coffee Gunther must use, we substitute 6 for k in the equation $k + c = 30$.

$$
\begin{array}{ll}
(6) + c = 30 & \qquad \text{Substitute 6 for } k. \\
c = 24 & \qquad \text{Subtract 6.}
\end{array}
$$

Gunther must use 6 pounds of Kona coffee and 24 pounds of Colombian coffee.

The previous example is a mixture problem in which two items of different values or costs are combined into a mixture with a specific cost. In other mixture problems, two different liquid solutions of different concentration levels are mixed together to form a single solution with a given strength, as in the next example.

EXAMPLE 5   Woody has one solution that is 28% alcohol and a second solution that is 46% alcohol. He wants to combine these solutions to make 45 liters of a solution that is 40% alcohol. How many liters of each original solution need to be used?

Solution

The two unknowns are the volume of the 28% alcohol solution and the volume of the 46% alcohol solution that need to be mixed together in order to make 45 liters of a solution that is 40% alcohol.

Solution	Volume of Solution	% Concentration of Alcohol	Volume of Alcohol
28%	x	0.28	$0.28x$
46%	y	0.46	$0.46y$
Mixture (40%)	45	0.40	18

The first equation in our system comes from the fact that the volume of the two solutions must equal 45 liters. As an equation, this can be represented by $x + y = 45$. This equation can be found in our table in the column labeled Volume of Solution. The second equation comes from the total volume of alcohol, under the column labeled Volume of Alcohol. We know that the total volume of alcohol is 18 liters (40% of the 45 liters in the mixture is alcohol), so the second equation in our system is $0.28x + 0.46y = 18$. Here is the system we must solve:

$$x + y = 45$$
$$0.28x + 0.46y = 18$$

We begin by clearing the second equation of decimals. This can be done by multiplying both sides of the second equation by 100.

$$\begin{array}{l} x + y = 45 \\ 0.28x + 0.46y = 18 \end{array} \xrightarrow{\text{Multiply by 100}} \begin{array}{l} x + y = 45 \\ 28x + 46y = 1800 \end{array}$$

To solve this system of equations, we can use the addition method. If we multiply the first equation by -28, then the variable terms containing x are opposites.

Quick Check 5

A chemist has one solution that is 60% acid and a second solution that is 50% acid. She wants to combine the solutions to make a new solution that is 54% acid. If she needs to make 400 milliliters of this new solution, how many milliliters of the two solutions should be mixed?

$$\begin{array}{l} x + y = 45 \\ 28x + 46y = 1800 \end{array} \xrightarrow{\text{Multiply by } -28} \begin{array}{l} -28x - 28y = -1260 \\ 28x + 46y = 1800 \end{array}$$

$$\begin{array}{r} -28x - 28y = -1260 \\ \underline{28x + 46y = 1800} \quad \text{Add to eliminate } x. \\ 18y = 540 \end{array}$$

$$y = 30 \quad \text{Divide both sides by 18.}$$

We know that Woody must use 30 liters of the solution that is 46% alcohol. To determine how much of the 28% alcohol solution Woody must use, we substitute 30 for y in the equation $x + y = 45$.

$$x + (30) = 45 \quad \text{Substitute 30 for } y.$$
$$x = 15 \quad \text{Subtract 30.}$$

Woody must use 15 liters of the 28% alcohol solution and 30 liters of the 46% alcohol solution.

Motion Problems

Objective 5 **Solve motion problems by using a system of equations.** We will now turn our attention to motion problems, which were introduced in Section 2.3. Recall that the equation for motion problems is $d = r \cdot t$, where r is the rate of speed, t is the time and *d* is the distance traveled. For example, if a car is traveling at a speed of 50 miles per hour for 3 hours, then the distance traveled is $50 \cdot 3$ or 150 miles.

EXAMPLE 6   Efrain is finishing up his training for the Shaver Lake Triathlon. Yesterday he focused on running and biking, spending a total of 5 hours on the two activities. Efrain runs at a speed of 10 miles per hour and rides his bike at a speed of 30 miles per hour. If Efrain covered a total of 120 miles yesterday, how much time did he spend running and how much time did he spend riding?

Solution

The two unknowns in this problem are the amount of time Efrain ran and the amount of time he rode his bike. We will let *x* represent the time that Efrain ran and *y* represent the amount of time that he rode his bike. All of our information is displayed in the following table:

	Rate	Time	Distance $(d = r \cdot t)$
Running	10	x	$10x$
Biking	30	y	$30y$
Total		5	120

We know that Efrain spent a total of 5 hours on these two activities, so $x + y = 5$. We also know that the total distance traveled was 120 miles. This leads to the equation $10x + 30y = 120$. Here is the system of equations we must solve:

$$x + y = 5$$
$$10x + 30y = 120$$

We can solve the first equation for x ($x = 5 - y$), so we will use the substitution method to solve this system of equations. We substitute the expression $5 - y$ for the variable x in the equation $10x + 30y = 120$.

Quick Check 6

On a trip to Las Vegas, Fong drove at an average speed of 70 miles per hour, except for a period when he was driving through a construction zone. In the construction zone, Fong's average speed was 40 miles per hour. If it took Fong 5 hours to drive 335 miles, how long was he in the construction zone?

$10(5 - y) + 30y = 120$	Substitute $5 - y$ for x.
$50 - 10y + 30y = 120$	Distribute.
$50 + 20y = 120$	Combine like terms.
$20y = 70$	Subtract 50.
$y = \dfrac{7}{2}$	Divide both sides by 20 and simplify.

Writing the fraction $\frac{7}{2}$ as a mixed number, we find that Efrain spent $3\frac{1}{2}$ hours riding his bike. To find out how long he was running, we substitute $3\frac{1}{2}$ for *y* in the equation $x = 5 - y$ and solve for *x*.

$x = 5 - 3\dfrac{1}{2}$	Substitute $3\frac{1}{2}$ for y.
$x = 1\dfrac{1}{2}$	Subtract.

Efrain spent $1\frac{1}{2}$ hours running and $3\frac{1}{2}$ hours riding his bike.

Suppose that a boat travels at a speed of 15 miles per hour in still water. The boat will travel faster than 15 miles per hour while heading downstream, because the speed of the current is pushing the boat forward as well. The boat will travel slower than 15 miles per hour while heading upstream, because the speed of the current is pushing the boat back. If the speed of the current is 3 miles per hour, this boat would travel 18 miles per hour downstream $(15 + 3)$ and 12 miles per hour upstream $(15 - 3)$. The speed of an airplane is affected in a similar fashion, depending on whether it is flying with or against the wind.

EXAMPLE 7 Juana's motorboat can take her from her campsite downstream to the nearest store, which is 27 miles away, in 1 hour. Returning upstream from the store to the campsite takes 3 hours. How fast is Juana's boat in still water, and what is the speed of the current?

Solution

The two unknowns in this problem are the speed of the boat in still water and the speed of the current. We will let b represent the speed of the boat in still water, and we will let c represent the speed of the current. The information for this problem can be summarized in the following table:

	Rate	Time	Distance $(d = r \cdot t)$
Downstream	$b + c$	1	$1(b + c) = b + c$
Upstream	$b - c$	3	$3(b - c) = 3b - 3c$

Since the distance in each direction is 27 miles, we know that both $b + c$ and $3b - 3c$ are equal to 27. Here is the system we need to solve:

$$b + c = 27$$
$$3b - 3c = 27$$

We will solve this system of equations by using the addition method. Multiplying both sides of the first equation by 3 will help us eliminate the variable c.

$$\begin{array}{l} b + c = 27 \\ 3b - 3c = 27 \end{array} \xrightarrow{\text{Multiply by 3}} \begin{array}{l} 3b + 3c = 81 \\ 3b - 3c = 27 \end{array}$$

$$\begin{array}{rl} 3b + 3c = & 81 \\ 3b - 3c = & 27 \\ \hline 6b \quad\;\; = & 108 \\ b = & 18 \end{array}$$ Add to eliminate c.

Divide both sides by 6.

The speed of the boat in still water is 18 miles per hour. To find the speed of the current, we substitute 18 for b in the original equation $b + c = 27$.

$$(18) + c = 27 \qquad \text{Substitute 18 for } b.$$
$$c = 9 \qquad \text{Subtract 18.}$$

The speed of the current is 9 miles per hour.

Quick Check 7
An airplane traveling with a tailwind can make a 2250-mile trip in $4\frac{1}{2}$ hours. However, traveling into the same wind would take 5 hours to fly 2250 miles. What is the speed of the plane in calm air, and what is the speed of the wind?

Building Your Study Strategy **Doing Your Homework, 4 Homework Diary**
Once you finish the last exercise of a homework assignment, it is not necessarily time to stop. Try summarizing what you have just accomplished in a homework diary. In this diary, you can list the types of problems you have solved, the types of problems you struggled with, and important notes about these problems. When you are ready to begin preparing for a quiz or exam, this diary will help you to decide where to focus your attention. If you are using note cards as a study aide, this would be a good time to create a set of study cards for this section. You now know which problems are more difficult for you, as well as which procedures you will need to know in the future.

Summarizing your homework efforts in these ways will help you to retain what you learned from the assignment.

EXERCISES 4.4

Vocabulary

1. State the formula for the perimeter of a rectangle.

2. State the formula for calculating simple interest.

3. State the formula for calculating distance.

4. A chemist is mixing a solution that is 35% alcohol with a second solution that is 55% alcohol to create 80 milliliters of a solution that is 40% alcohol. Which system of equations can be used to solve this problem?

a) $x + y = 40$
$0.35x + 0.55y = 80$

b) $x + y = 80$
$0.35x + 0.55y = 0.40$

c) $x + y = 80$
$0.35x + 0.55y = 0.40(80)$

5. A mathematics instructor is teaching an elementary algebra class and a statistics class. She has 17 more students in her statistics class than she has in her elementary algebra class. The two classes have 93 students in total. How many students does she have in each class?

6. A hardcover math book is 289 pages longer than a paperback math book. If the total number of pages in these two books is 1691, how many pages are there in each book?

7. Andre's father is 28 years older than he is. If we added Andre's age to his father's age, the total would be 98. How old is Andre? How old is his father?

8. Maggie's ma tells everyone that she is 14 years younger than she actually is. If we add her actual age to the age that she tells people she is, the total would be 122. What is the actual age of Maggie's ma (as reported by Bob Dylan)?

9. Two baseball players hit 79 home runs combined last season. The first player hit 7 more home runs than twice the number of home runs hit by the second player. How many home runs did each player hit?

10. A piece of wire 240 centimeters long was cut into two pieces. The longer of the two pieces is 25 centimeters longer than four times the length of the shorter piece. How long is each piece?

MyMathLab
MathXL
Interactmath.com
MathXL Tutorials on CD
Video Lectures on CD
Tutor Center
Addison-Wesley Math Tutor Center
Student's Solutions Manual

F O R E X T R A H E L P

40. A farmer has two types of feed that are 12% protein and 18% protein respectively. How many pounds of each need to be combined to make a mixture of 36 pounds of feed that is 13.5% protein?

41. A bartender mixes a batch of vodka (40% alcohol) and tonic (no alcohol). How much of each needs to be mixed to make 4 liters of vodka and tonic that is 22% alcohol?

42. An auto mechanic has an antifreeze solution that is 10% antifreeze. He needs to mix this solution with pure antifreeze to produce 5 liters of a solution that is 19% antifreeze. How much of each does he need to mix?

43. George and Tina drove from their house to Palm Springs for a vacation. During the first part of the trip, George drove at a speed of 70 miles per hour. They switched drivers and Tina drove the rest of the way at a speed of 85 miles per hour. It took them 6 hours to reach Palm Springs, which was 480 miles away. Find the length of time that George drove, as well as the length of time that Tina drove.

44. Matt was driving to a conference in Salt Lake City, Utah. During the first day of driving, Matt averaged 72 miles per hour. On the second day, he drove at an average speed of 80 miles per hour in order to arrive before sundown. If his total driving time for the 2 days was 15 hours and the distance traveled was 1148 miles, what was Matt's driving time for each day?

45. A racecar driver averaged 160 miles per hour over the first part of the Sequoia 500, but engine trouble dropped her average speed to 130 miles per hour for the rest of the race. If the 500-mile race took the driver 3.5 hours to complete, after what period did her engine trouble start?

46. A truck driver started for his destination at an average speed of 60 miles per hour. Once snow began to fall, he had to drop his speed to 40 miles per hour. It took the truck driver 9 hours to reach his destination, which was 425 miles away. How many hours after the driver started did the snow begin to fall?

47. A kayak can travel 18 miles downstream in 3 hours, while it would take 9 hours to make the same trip upstream. Find the speed of the kayak in still water, as well as the speed of the current.

48. A rowboat would take 3 hours to travel 6 miles upstream, while it would only take $\frac{3}{4}$ hour to travel 6 miles downstream. Find the speed of the rowboat in still water, as well as the speed of the current.

49. An airplane traveling with a tailwind can make a 3000-mile trip in 5 hours. However, traveling into the same wind would take 6 hours to fly 3000 miles. What is the speed of the plane in calm air, and what is the speed of the wind?

50. A small plane can fly 240 miles with a tailwind behind it in 3 hours. Flying the same distance into the same wind would take 4 hours. Find the speed of the plane with no wind.

Writing in Mathematics

51. Write a word problem whose solution is "The length of the rectangle is 15 m and the width is 8 m."

52. Write a word problem whose solution is "$5000 at 6% and $3000 at 5%."

53. Write a word problem whose solution is "80 ml of 40% acid solution and 10 ml of 58% acid solution."

54. Write a word problem whose solution is "Bill drove for 3 hours and Margie drove for 4 hours."

55. ***Solutions Manual**** Write a solutions manual page for the following problem:

> *A play was attended by 300 people. Admission was $8 for adults and $3 for children. If the total receipts for the play were $2150, how many children attended the play?*

56. ***Newsletter**** Write a newsletter explaining how to solve interest problems using a system of two equations.

**See Appendix B for details and sample answers.*

Quick Review Exercises

Section 4.4

Graph. Label any x- and y-intercepts.

1. $x - 4y = 6$

2. $y = 2x - 5$

3. $y = -\frac{1}{4}x - 2$

4. $y = 3$

Objectives

1 Solve a system of linear inequalities by graphing.
2 Solve a system of linear inequalities with more than two inequalities.

Just as we had systems of linear equations, we can also have a **system of linear inequalities**. A system of linear inequalities is made up of two or more linear inequalities. As with systems of equations, an ordered pair is a solution to a system of linear inequalities if it is a solution to each linear inequality in the system.

Solving a System of Linear Inequalities by Graphing

Objective 1 Solve a system of linear inequalities by graphing. To find the solution set to a system of linear inequalities, we begin by graphing each inequality separately. If the solutions to the individual inequalities intersect, the region of intersection is the solution set to the system of inequalities.

EXAMPLE 1   Graph the following system of inequalities:

$$y \geq 2x - 5$$
$$y < 3x - 7$$

Solution

We begin by graphing the inequality $y \geq 2x - 5$. To do this, we graph the line $y = 2x - 5$ as a solid line. Recall that we use a solid line when the points on the line are solutions to the inequality. Since this equation is in slope–intercept form, an efficient way to graph it is by plotting the y-intercept at $(0, -5)$ and then using the slope of 2 to find additional points on the line.

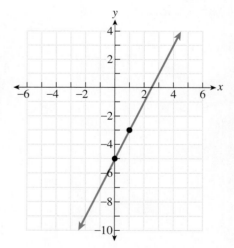

Since the origin is not on this line, we can use $(0, 0)$ as a test point. Substitute 0 for x and 0 for y in the inequality $y \geq 2x - 5$.

$$0 \geq 2(0) - 5$$
$$0 \geq 0 - 5$$
$$0 \geq -5$$

Substituting these coordinates into the inequality $y \geq 2x - 5$ produces a true statement, so we shade the half-plane containing $(0, 0)$ in the graph on the right.

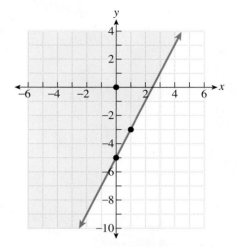

We now turn our attention to the second inequality in the system. To graph the inequality $y < 3x - 7$, we use a dashed line to graph the equation $y = 3x - 7$. Recall that the dashed line is used to signify that no point on the line is a solution to the inequality. Again, this equation is in slope–intercept form, so we graph the line by plotting the y-intercept at $(0, -7)$ and using the slope of 3 to find additional points in the graph on the right.

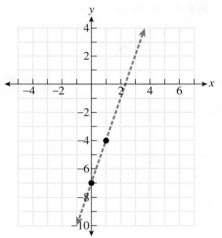

We can use the origin as a test point for this inequality as well. Substituting 0 for x and 0 for y produces a false statement.

$$0 < 3(0) - 7$$
$$0 < 0 - 7$$
$$0 < -7$$

We shade the half-plane on the opposite side of the line from the origin in the graph on the right.

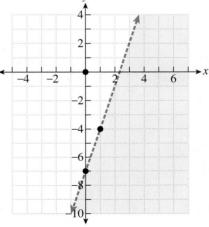

The solution of the system of linear inequalities is the region where the two solutions intersect.

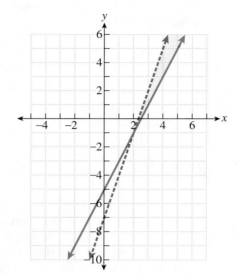

Quick Check 1
Graph the following system of inequalities:

$$y > x + 3$$
$$2x + 3y < 6$$

EXAMPLE 2 Graph the following system of inequalities:

$$y \leq -2x + 4$$
$$y \geq 3$$

Solution

We begin by graphing the line $y = -2x + 4$ using a solid line. This equation is in slope–intercept form, so we begin by plotting the y-intercept $(0, 4)$, and then we use the slope of -2 to find additional points on the line. Now we determine which half-plane to shade. Since the line does not pass through the origin, we can choose $(0, 0)$ as a test point.

$$y \leq -2x + 4$$
$$0 \leq -2(0) + 4$$
$$0 \leq 4$$

Substituting this ordered pair into the inequality produces a true statement. Thus, we shade the half-plane containing the origin in the graph on the right.

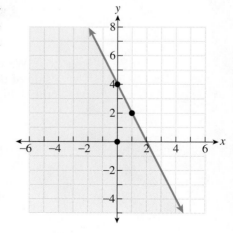

Now we turn our attention to the inequality $y \geq 3$. We begin by graphing the line $y = 3$ using a solid line. This line is a horizontal line with a y-intercept at $(0, 3)$. Again, we may choose the origin as a test point. Substituting the ordered pair $(0, 0)$ produces a false statement. We shade the half-plane above the line $y = 3$ that does not contain the origin. The graph is shown on the right.

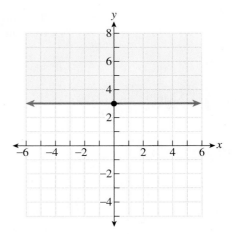

The solution of the system of equations is the region where the two shaded areas intersect in the graph on the right.

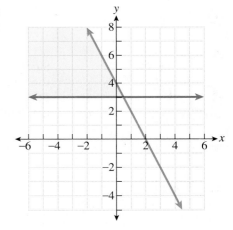

Quick Check **2**
Graph the following system of inequalities:

$$-3x + 4y \leq 8$$
$$y \geq 0$$

Solving a System of Linear Inequalities with More than Two Inequalities

Objective **2** **Solve a system of linear inequalities with more than two inequalities.** The next example involves a system of three linear inequalities. Regardless of the number of inequalities in a system, we still find the solution in the same manner. Graph each inequality individually, and then find the region where the shaded regions all intersect.

EXAMPLE **3**   Graph the following system of inequalities:

$$3x + 4y > 12$$
$$x < 6$$
$$y > 4$$

Solution

We begin by graphing the line $3x + 4y = 12$ with a dashed line. This is the line associated with the first inequality. Since the equation is in standard form, we can graph it by finding its x- and y-intercepts. The x-intercept for this line can be found by substituting 0 for y in the equation.

$3x + 4y = 12$	
x-intercept ($y = 0$)	**y-intercept ($x = 0$)**
$3x + 4(0) = 12$	$3(0) + 4y = 12$
$3x = 12$	$4y = 12$
$x = 4$	$y = 3$
$(4, 0)$	$(0, 3)$

The x-intercept is at $(4, 0)$. The y-intercept of the line is at $(0, 3)$. Since the line does not pass through the origin, we can use $(0, 0)$ as a test point. When we substitute 0 for x and 0 for y in the inequality $3x + 4y > 12$, we obtain a false statement.

$$3(0) + 4(0) > 12$$
$$0 > 12$$

Since this statement is not true, we shade the half-plane on the side of the line that does not contain the point $(0, 0)$ in the graph on the right.

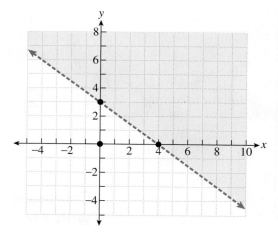

To graph the second inequality, we begin with the graph of the vertical line $x = 6$. This line is also dashed. The solutions of this inequality are to the left of this vertical line, as shown in the graph on the right.

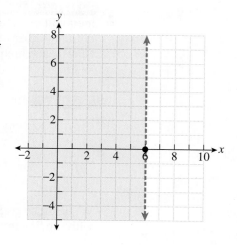

For the third inequality, we graph the dashed horizontal line $y = 4$. The solutions of this inequality are above the horizontal line $y = 4$, as shown in the graph on the right.

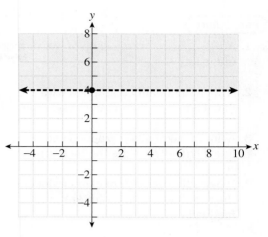

Graph the following system of inequalities:

$$x + y \le 8$$
$$x \ge 1$$
$$y \ge -1$$

We now display the solutions to the system of inequalities by finding the region of intersection for these three inequalities, shown in the graph on the right.

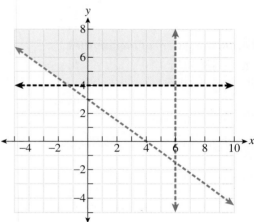

Building Your Study Strategy Doing Your Homework, 5 **Homework Schedule**
Try to establish a regular schedule for doing your homework. Some students find it advantageous to work on their homework as soon as possible after class, while the material is still fresh in their minds. You may be able to find other students in your class who can work with you in the library. If you have to wait until the evening to do your homework, try to complete your assignment before you get tired. If you have more than one subject to work on, begin with math.

It is crucial that you be able to complete an assignment before the next class session. Since math skills build on previous concepts, if you do not understand the material one day, then you will have difficulty understanding the next day's material.

Vocabulary

1. A(n)_____ is made up of two or more linear inequalities.

2. A linear inequality is graphed with a _____ line if it involves either of the symbols $<$ or $>$.

3. A linear inequality is graphed with a _____ line if it involves either of the symbols $\leq$ or $\geq$.

4. When solving a system of linear inequalities, shade the region where the solutions of the individual inequalities _____.

Graph the system of inequalities.

5. $y \geq 3x - 4$
   $y \leq -x + 5$

6. $y \geq 2x + 7$
   $y < x + 3$

7. $y \leq 5x - 4$
   $y > -2x - 6$

8. $y < 4x + 6$
   $y < 3$

9. $y < 5x + 5$
   $y > 5x - 5$

10. $y \geq \frac{1}{2}x + 3$
    $y < -4x$

11. $x + 4y < 8$
$3x - 2y < 6$

12. $5x + 2y \leq 10$
$-3x + y > 5$

17. $y \geq 3x$
$y > 3$

13. $-7x + 3y > 21$
$4x + 9y \leq 36$

18. $x < -4$
$2x + 6y > 12$

14. $8x + y \geq -8$
$3x + 4y < -12$

19. $y > -\dfrac{3}{2}x + 6$
$4x - 3y \geq -6$

15. $y \geq 5$
$5x + 2y < 0$

16. $-4x + y < 6$
$4x - y < 2$

20. $x < 4$
$y \geq -7$

21. $x > 1$
$8x - 4y > 12$

22. $3x + 10y > 15$
$\quad y \leq -\dfrac{5}{2}x + 7$

23. $y < 2x$
$\quad y < 6$
$\quad x > 2$

28. Write a system of linear inequalities whose solution is the entire third quadrant.

29. Write a system of linear inequalities that has no solution.

30. Is it possible to write a system of linear inequalities whose solution set is the entire rectangular coordinate plane? If so, write the system; if not, explain why in your own words.

31. Find the area of the region defined by the following system of linear inequalities:

$x \geq 2$
$y \geq 1$
$y \leq -x + 9$

24. $y > x - 4$
$\quad y < -2x + 5$
$\quad y > -2$

25. $3x - 2y \leq -10$
$\quad y < -4x - 6$
$\quad -x + 3y < 12$

32. Write a system of linear inequalities whose solution set is shown on the following graph:

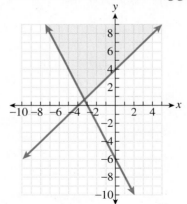

26. $x \geq -5$
$\quad x \leq 4$
$\quad y < \dfrac{1}{5}x + 2$

27. Write a system of linear inequalities whose solution is the entire first quadrant, as shown in the graph to the right:

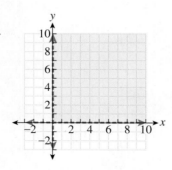

Writing in Mathematics

33. ***Solutions Manual*** * Write a solutions manual page for the following problem:

Solve the system of linear $y < 3x + 5$
equalities. $2x + 3y \geq 6$

34. ***Newsletter*** * Write a newsletter explaining how to solve a system of linear inequalities.

See Appendix B for details and sample answers.

Section 4.1—Topic	Chapter Review Exercises
Solving Systems of Linear Equations by Graphing	1–8

Section 4.2—Topic	Chapter Review Exercises
Solving Systems of Linear Equations by Using the Substitution Method	9–13

Section 4.3—Topic	Chapter Review Exercises
Solving Systems of Linear Equations by Using the Addition Method	14–18
Solving Systems of Linear Equations by Any Method	19–30

Section 4.4—Topic	Chapter Review Exercises
Solving Applications Using Systems of Linear Equations	31–40

Section 4.5—Topic	Chapter Review Exercises
Solving Systems of Linear Inequalities	41–44

Summary of Chapter 4 Study Strategies

A homework assignment should not be viewed as a chore to be completed but, rather, as an opportunity to increase your understanding of mathematics. Here is a summary of the ideas presented in this chapter:

- Review your notes and the text before beginning a homework assignment.
- Be neat and complete when doing a homework assignment. *If your work is neat, it will be easier to spot errors. When you are reviewing for an exam, your homework will be easier to review if it is neat. Do not skip any steps, even if you feel that showing them is not necessary.*
- Strategies for homework problems you cannot answer:
  - *Look in your notes and the text for similar problems or suggestions about this type of problem.*
  - *Call another student in the class for help.*
  - *Get help from a tutor.*
  - *Ask your instructor for help at the first opportunity.*
- Complete a homework session by summarizing your work. *Prepare notes on difficult problems or concepts. These hints will be helpful when you begin to review for an exam.*
- Keep up to date with your homework assignments. *Falling behind will make it difficult for you to learn the new material presented in class.*

Find the solution of the system of equations shown on each graph. If there is no solution, state this. [4.1]

1.

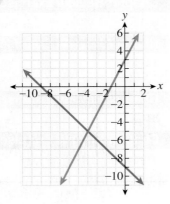

2.

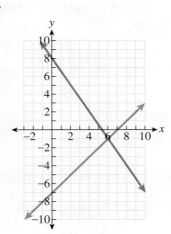

3.

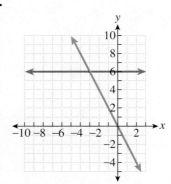

4.

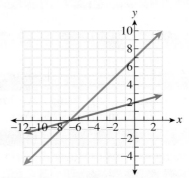

Solve the system by graphing. If the system is inconsistent and has no solution, state this. If the system is dependent, write the form of the solution for any real number x. [4.1]

5. $y = x - 5$
$y = -\dfrac{1}{2}x + 7$

6. $y = \dfrac{1}{3}x - 4$
$y = -x$

7. $y = -2x + 8$
$y = \dfrac{1}{2}x + 3$

8. $2x - y = -8$
$-3x + 2y = 10$

Solve the system by substitution. If the system is inconsistent and has no solution, state this. If the system is dependent, write the form of the solution for any real number x. [4.2]

9. $x = 2y + 1$
$2x - 5y = 4$

10. $y = -x - 3$
$3x + 4y = -5$

11. $4x + y = -5$
$5x - 3y = 15$

12. $-4x + 2y = 18$
$2x - y = 3$

13. $y = \dfrac{2}{3}x - 5$
$3x - 4y = 23$

Solve the system by addition. If the system is inconsistent and has no solution, state this. If the system is dependent, write the form of the solution for any real number x. [4.3]

14. $3x + 2y = 15$
$5x - 2y = 17$

15. $6x + 7y = -66$
$-3x + 4y = -27$

16. $2x - y = 4$
$-4x + 2y = -8$

Worked-out solutions to Review Exercises marked with ⬤ can be found on page AN–17.

17. $y = 4x + 5$
$5x - 2y = -4$

18. $\dfrac{1}{2}x + \dfrac{2}{3}y = 2$

$3x - \dfrac{1}{2}y = -15$

Solve the system by the method of your choice. If the system is inconsistent and has no solution, state this. If the system is dependent, write the form of the solution for any real number x. [4.2, 4.3]

19. $4x - 3y = 20$
$-x + 6y = -26$

20. $x + y = 4$
$-5x + 4y = 43$

21. $y = 2x - 1$
$3x - 7y = 29$

22. $8x - 9y = 67$
$x = 5$

23. $y = 3x - 2$
$-9x + 3y = -6$

24. $x + y = -4$
$4x - 6y = -1$

25. $3x - 2y = 12$
$y = \dfrac{4}{5}x - 6$

26. $x + 2y = -5$
$\dfrac{1}{3}x - \dfrac{1}{4}y = 13$

27. $-\dfrac{1}{2}x + 2y = 3$
$2x - 8y = -9$

28. $7x - 5y = -46$
$6x + 7y = 17$

29. $y = 0.3x + 2.4$
$3x - 2y = 7.2$

30. $x + y = 8000$
$0.04x + 0.05y = 340$

31. The sum of two numbers is 87. One number is 18 more than twice the other number. Find the two numbers. [4.4]

32. Hal has 284 more baseball cards in his collection than his brother Mac does. Together, they have 1870 baseball cards. How many cards does Hal have? [4.4]

33. A hockey team charges $10 for tickets to their games. On Wednesday nights, the team has a Ladies' Night promotion, and women of any age are admitted for half-price. If the total attendance on the last Ladies' Night was 1764 fans and the team collected $13,470, how many women were at the game? [4.4]

34. A movie theater charges $7.50 to see a movie, but senior citizens are admitted for only $4. If 206 people paid a total of $1384 to see a movie, how many were senior citizens? [4.4]

35. Jackie's home is on a rectangular lot. The fence that runs around the entire yard is 1320 feet long. If the length of the yard is 140 feet more than the width, find the dimensions of the yard. [4.4]

36. Jiana has a front door that lets cold drafts in. The height of the door is 4 inches more than twice its width. If she uses 224 inches of weather stripping around the entire door, how tall is her door? [4.4]

37. Juan deposits a total of $4000 in two different certificates of deposit (CDs). One CD pays 5% annual interest, while the other pays 4.25% annual interest. If Juan earned $179 interest during the first year, how much did he put in each CD? [4.4]

38. Marcia bought shares in two mutual funds, investing a total of $2500. One fund's shares went up by 20%, while the other fund went up in value by 8%. This was a total profit of $320 for Marcia. How much did she invest in each mutual fund? [4.4]

39. One type of cattle feed is 4% protein, while a second type is 13% protein. How many pounds of each type must be mixed together to make 45 pounds of a mixture that is 10% protein? [4.4]

40. A kayak can travel 12 miles upstream in 6 hours. It only takes $1\frac{1}{2}$ hours for the kayak to travel the same distance downstream. Find the speed of the kayak in still water. [4.4]

Solve the system of inequalities. **[4.5]**

41. $y \le 2x + 7$
$y > x - 5$

43. $y \le -4x + 7$
$y \ge \dfrac{3}{2}x$

42. $3x + 4y < 24$
$5x + y \ge 10$

44. $y > \dfrac{1}{2}x - 3$
$y > -2$

For Extra Help Pass the Test Test solutions are found on the enclosed CD.

1. Find the solution of the system of equations shown on the graph. If there is no solution, state this.

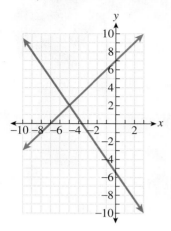

2. Solve the system that follows by graphing. If the system is inconsistent and has no solution, state this. If the system is dependent, write the form of the solution for any real number x.

$$y = -2x + 3$$
$$y = -3x + 7$$

3. Solve the system that follows by substitution. If the system is inconsistent and has no solution, state this. If the system is dependent, write the form of the solution for any real number x.

$$x + 4y = -9$$
$$2x - 5y = 21$$

4. Solve the system that follows by addition. If the system is inconsistent and has no solution, state this. If the system is dependent, write the form of the solution for any real number x.

$$5x + 2y = -22$$
$$2x - 5y = -3$$

Solve the given system by the method of your choice. If the system is inconsistent and has no solution, state this. If the system is dependent, write the form of the solution for any real number x.

5. $-x + 4y = 11$
   $3x - 4y = -17$

6. $4x + 2y = 9$
   $y = -2x + 18$

7. $x + y = 11$
   $5x - 2y = 13$

8. $7x - 11y = -47$
   $9x + 2y = 101$

9. $13x + 6y = 52$
   $x = 7y + 4$

10. $4x - 6y = 18$
    $-10x + 15y = -45$

11. Last season, a baseball player had 123 more walks than he had home runs. If we add the number of times he walked to the number of times he hit a home run, the total would be 237. How many walks did the player have?

12. Admission to an amusement park is $20, but children under 8 years old are admitted for $5. If there were 13,900 people at the amusement park yesterday and they paid a total of $197,000, how many children under 8 years old were at the park?

13. Steve's vegetable garden is in the shape of a rectangle, and it requires 66 feet of fencing to surround it. The length is 1 foot more than three times the width. Find the dimensions of Steve's garden.

14. Emma bought shares in two mutual funds, investing a total of $32,000. One fund's shares went up by 7%, while the other fund went up in value by 3%. This was a total profit of $1760 for Emma. How much did she invest in each mutual fund?

15. Solve the system of inequalities.

$$y < \frac{1}{2}x + 4$$
$$5x + 2y > -16$$

Mathematicians in History
Hypatia of Alexandria

Hypatia of Alexandria was one of the first women known to make major contributions to the field of mathematics. She was also active in the fields of astronomy and philosophy. One quote that has been attributed to her is, "Reserve your right to think, for even to think wrongly is better than not to think at all."

Write a one-page summary (*or* make a poster) of the life of Hypatia of Alexandria and her accomplishments.

Interesting issues:

- Where and when was Hypatia born?
- Who was Hypatia's father, and what fields was he involved in?
- Hypatia lectured on the philosophy of Neoplatonism. What are the principal ideas of Neoplatonism?
- What was Hypatia's relationship to Cyril and Orestes?
- In the year A.D. 415, Hypatia was murdered. Describe the circumstances of her death.

Stretch Your Thinking ❭ Chapter 4

Have you ever wondered how math problems are created? Well, today you will be able to explore the answer to that question by creating your very own math application. You have studied various applications of systems of equations (Section 4.4), and now it's your turn to create an application problem that can be solved by using systems of equations.

The first step is to decide upon the context of your problem. You need to have two unknowns for which you will be solving. Will you be solving for the measure of two angles? Will you be looking for the quantities of two types of coffee beans being mixed together? Will you be looking for the dimensions of a rectangular object? Try to be as creative as possible when thinking about the context.

Once you have decided upon the context, you will need to decide on the values of your two unknowns. You will then need to create two distinct equations that relate these values to each other. Think about the equations you saw in Section 4.4 and about how you can make your own equations.

After you have your two equations, solve the system by using one of the methods you learned in this chapter. Did you arrive at the correct values of your unknowns? Have a friend solve your system, and see if he or she was able to correctly solve it. You may need to talk to your instructor if you are having difficulties in making sure that your equations correctly relate your unknowns to each other.

Once you have finalized your equations, you need a well-formed paragraph that describes the context of your problem and gives your equations in mathematical language. Again, refer to Section 4.4 for examples of wording and paragraph structure. After you feel that your paragraph correctly states the problem you intended, switch with someone in class and see if he or she can solve your application.

Turn in your paragraph, the two equations, and your solution to your instructor. Have fun!

Exponents and Polynomials

The expressions and equations we have examined to this point in the text have been linear. In this chapter, we begin to examine expressions and functions that are nonlinear, including polynomials such as $-16t^2 + 32t + 128$ or $x^5 - 3x^4 + 8x^2 - 9x + 7$. After learning some basic definitions, we will learn to add, subtract, multiply, and divide polynomials. We will introduce properties for working with expressions containing exponents that will make using these operations easier. We will also cover scientific notation, which is used to represent very large or very small numbers and to perform arithmetic calculations with them.

Study Strategy **Test Taking** *To be successful in a math class, understanding the material is important, but you must also be a good test taker. In this chapter, we will discuss the test-taking skills necessary for success in a math class.*

5.1

Exponents

Objectives

1 Use the product rule for exponents.
2 Use the power rule for exponents.
3 Use the power of a product rule for exponents.
4 Use the quotient rule for exponents.
5 Use the zero exponent rule.
6 Use the power of a quotient rule for exponents.
7 Evaluate functions by using the rules for exponents.

We have used exponents to represent repeated multiplication of a particular factor. For example, the expression 3^7 is used to represent a product in which 3 is a factor 7 times.

$$3^7 = 3 \cdot 3 \cdot 3 \cdot 3 \cdot 3 \cdot 3 \cdot 3$$

Recall that in the expression 3^7, the number 3 is called the base and the number 7 is called the exponent.

In this section, we will introduce several properties of exponents and learn how to apply these properties to simplify expressions involving exponents.

Product Rule

Objective **1**   Use the product rule for exponents.

Product Rule

For any base x, $x^m \cdot x^n = x^{m+n}$.

When multiplying two expressions with the same base, we keep the base and add the exponents. Consider the example $x^3 \cdot x^7$. We know that $x^3 = x \cdot x \cdot x$ and $x^7 = x \cdot x \cdot x \cdot x \cdot x \cdot x \cdot x$. So $x^3 \cdot x^7$ can be rewritten as $(x \cdot x \cdot x) \cdot (x \cdot x \cdot x \cdot x \cdot x \cdot x \cdot x)$.

Since x is being repeated as a factor 10 times, this expression can be rewritten as x^{10}.

$$x^3 \cdot x^7 = x^{3+7} = x^{10}$$

> *A Word of Caution*   When multiplying two expressions with the same base, we keep the base and add the exponents. *Do not multiply the exponents.*
>
> $$x^3 \cdot x^7 = x^{10}, \text{ not } x^{21}$$

EXAMPLE **1**   Simplify $x^2 \cdot x^9$.

Solution

Since we are multiplying and the two bases are the same, we keep the base and add the exponents.

$$x^2 \cdot x^9 = x^{2+9} \qquad \text{Keep the base and add the exponents.}$$
$$= x^{11} \qquad \text{Add.}$$

EXAMPLE 2 Simplify $2 \cdot 2^5 \cdot 2^4$.

Solution

Quick Check 1
Simplify.
a) $x^5 \cdot x^4$ b) $3^3 \cdot 3^2 \cdot 3$

In this example, we are multiplying three expressions that have the same base. To do this, we keep the base and add the three exponents together. Keep in mind that if an exponent is not written for a factor, then the exponent is a 1.

$$2 \cdot 2^5 \cdot 2^4 = 2^{1+5+4} \quad \text{Keep the base and add the exponents.}$$
$$= 2^{10} \quad \text{Add.}$$
$$= 1024 \quad \text{Raise 2 to the 10}^{\text{th}} \text{ power.}$$

EXAMPLE 3 Simplify $(x + y)^4 \cdot (x + y)^9$.

Solution

Quick Check 2
Simplify
$(a - 6)^{10} \cdot (a - 6)^{17}$.

The base for these expressions is the sum $x + y$.

$$(x + y)^4 \cdot (x + y)^9 = (x + y)^{4+9} \quad \text{Keep the base and add the exponents.}$$
$$= (x + y)^{13} \quad \text{Add.}$$

EXAMPLE 4 Simplify $(x^2y^7)(x^3y)$.

Solution

When we have a product involving more than one base, the associative and commutative properties of multiplication can be used to simplify the product.

$$(x^2y^7)(x^3y) = x^2y^7x^3y \quad \text{Write without parentheses.}$$
$$= x^2x^3y^7y \quad \text{Group like bases together.}$$
$$= x^5y^8 \quad \text{For each base, add the exponents and keep the base.}$$

In future examples, we will not show the application of the associative and commutative properties of multiplication. Look for bases that are the same, add their exponents, and write each base to the calculated power.

Quick Check 3
Simplify $(a^6b^5)(a^9b^5)$.

Power Rule

Objective 2 Use the power rule for exponents. The second property of exponents involves raising an expression with an exponent to a power.

Power Rule

$$\text{For any base } x, (x^m)^n = x^{m \cdot n}.$$

In essence, this property tells us that when we raise a power to a power, we keep the base and multiply the exponents. To show why this is true, consider the expression $(x^3)^7$.

Whenever we raise an expression to the seventh power, this means that the expression is repeated as a factor seven times. So $(x^3)^7 = x^3 \cdot x^3 \cdot x^3 \cdot x^3 \cdot x^3 \cdot x^3 \cdot x^3$. Applying the first property from this section, $x^3 \cdot x^3 \cdot x^3 \cdot x^3 \cdot x^3 \cdot x^3 \cdot x^3 = x^{3+3+3+3+3+3+3}$ or x^{21}. This exponent could be found by multiplying 3 by 7.

$$(x^3)^7 = x^{3 \cdot 7} = x^{21}$$

> **A Word of Caution**   When raising an expression with an exponent to another power, we keep the base and multiply the exponents. *Do not add the exponents.*
>
> $$(x^3)^7 = x^{21}, \text{ not } x^{10}$$

EXAMPLE 5   Simplify $(x^5)^4$.

Solution

In this example, we are raising a power to another power, so we keep the base and multiply the exponents.

$$(x^5)^4 = x^{5 \cdot 4} \qquad \text{Keep the base and multiply the exponents.}$$
$$= x^{20} \qquad \text{Multiply.}$$

Quick Check 4
Simplify $(x^9)^7$.

EXAMPLE 6   Simplify $(x^3)^9(x^7)^8$.

Solution

To simplify this expression, we must use both properties that have been introduced in this section.

$$(x^3)^9(x^7)^8 = x^{3 \cdot 9} x^{7 \cdot 8} \qquad \text{Apply the power rule.}$$
$$= x^{27} x^{56} \qquad \text{Multiply.}$$
$$= x^{83} \qquad \text{Keep the base and add the exponents.}$$

Quick Check 5
Simplify $(x^4)^6(x^5)^2$.

Power of a Product Rule

Objective 3   Use the power of a product rule for exponents.   The third property introduced in this section involves raising a product to a power.

Power of a Product Rule

For any bases x and y, $(xy)^n = x^n y^n$.

This property tells us that when we raise a product to a power, we may raise each factor to that power. Consider the expression $(xy)^5$. We know that this can be rewritten as $xy \cdot xy \cdot xy \cdot xy \cdot xy$. Using the associative and commutative properties of multiplication, we can rewrite this expression as $x \cdot x \cdot x \cdot x \cdot x \cdot y \cdot y \cdot y \cdot y \cdot y$, which simplifies to $x^5 y^5$.

EXAMPLE 7   Simplify $(x^2 y^3)^6$.

Solution

This example combines two properties. When we raise each factor to the sixth power, we are raising a power to a power and therefore multiply the exponents.

Quick Check 6
Simplify $(a^5 b^3)^8$.

$$(x^2 y^3)^6 = (x^2)^6(y^3)^6 \qquad \text{Raise each factor to the 6th power.}$$
$$= x^{12} y^{18} \qquad \text{Keep the base and multiply the exponents.}$$

EXAMPLE 8   Simplify $(x^2y^3)^7(3x^5)^2$.

Solution

In this example, we begin by raising each factor to the appropriate power. Then we multiply expressions with the same base.

$$(x^2y^3)^7(3x^5)^2 = x^{14}y^{21} \cdot 3^2x^{10} \qquad \text{Raise each base to the appropriate power.}$$
$$= 9x^{24}y^{21} \qquad \text{Simplify.}$$

It is a good idea to write the variables in an expression in alphabetical order, as shown in this example, although it is not necessary.

Quick Check **7**

Simplify
$(4a^2b^3)^3(a^4b^2c^7)^5$.

A Word of Caution This property applies only when we raise a *product* to a power, not a sum or a difference. Although $(xy)^2 = x^2y^2$, a similar approach does not work for $(x + y)^2$ or $(x - y)^2$.

$$(x + y)^2 \neq x^2 + y^2$$
$$(x - y)^2 \neq x^2 - y^2$$

For example, let $x = 12$ and $y = 5$. In this case, $(x + y)^2$ is equal to 17^2 or 289, but $x^2 + y^2$ is equal to $12^2 + 5^2$ or 169. This shows that $(x + y)^2$ is not equal to $x^2 + y^2$ in general.

Quotient Rule

Objective 4   Use the quotient rule for exponents. The next property involves fractions containing the same base in their numerators and denominators. It also applies to dividing exponential expressions with the same base.

Quotient Rule

$$\text{For any base } x, \frac{x^m}{x^n} = x^{m-n} \ (x \neq 0).$$

Note that this property has a restriction; it does not apply when the base is 0. This is because division by 0 is undefined. This property tells us that when we are dividing two expressions with the same base, we subtract the exponent of the denominator from the exponent of the numerator. Consider the expression $\frac{x^7}{x^3}$. This can be rewritten as $\frac{x \cdot x \cdot x \cdot x \cdot x \cdot x \cdot x}{x \cdot x \cdot x}$. From our previous work, we know that this fraction can be simplified by dividing out common factors in the numerator and denominator.

$$\frac{\cancel{x} \cdot \cancel{x} \cdot \cancel{x} \cdot x \cdot x \cdot x \cdot x}{\cancel{x} \cdot \cancel{x} \cdot \cancel{x}} = x^4$$

One way to think of this is that the number of factors in the numerator has been reduced by three, because there were three factors in the denominator. The property tells us that we can simplify this expression by subtracting the exponents, which produces the same result.

$$\frac{x^7}{x^3} = x^{7-3} = x^4$$

EXAMPLE ►9   Simplify $\dfrac{y^9}{y^2}$. (Assume $y \neq 0$.)

Solution

Since the bases are the same, we keep the base and subtract the exponents.

$$\dfrac{y^9}{y^2} = y^{9-2} \qquad \text{Keep the base and subtract the exponents.}$$
$$= y^7 \qquad \text{Subtract.}$$

EXAMPLE ►10   Simplify $a^{21} \div a^3$. (Assume $a \neq 0$.)

Solution

Although in a different form than the previous example, $a^{21} \div a^3$ is equivalent to $\dfrac{a^{21}}{a^3}$. We keep the base and subtract the exponents.

$$a^{21} \div a^3 = a^{21-3} \qquad \text{Keep the base and subtract the exponents.}$$
$$= a^{18} \qquad \text{Subtract.}$$

Quick Check 8
Simplify. (Assume $x \neq 0$.)

a) $\dfrac{x^{20}}{x^4}$ b) $x^{24} \div x^8$

> **A Word of Caution**   When dividing two expressions with the same base, keep the base and subtract the exponents. *Do not divide the exponents.*
>
> $$a^{21} \div a^3 = a^{18}, \text{ not } a^7$$

EXAMPLE ►11   Simplify $\dfrac{16x^{10}y^9}{2x^7y^4}$. (Assume $x, y \neq 0$.)

Solution

Although we will subtract the exponents for the bases x and y, we will not subtract the numerical factors 16 and 2. We simplify numerical factors in the same way that we originally did in Chapter 1, so $\frac{16}{2} = 8$.

$$\dfrac{16x^{10}y^9}{2x^7y^4} = 8x^3y^5 \qquad \text{Simplify } \tfrac{16}{2}. \text{ Keep the variable bases and subtract the exponents.}$$

Quick Check 9
Simplify $\dfrac{20a^{10}b^{12}c^{15}}{5a^2b^6c^9}$. (Assume $a, b, c \neq 0$.)

Zero Exponent Rule

Objective 5   **Use the zero exponent rule.**   Consider the expression $\dfrac{2^7}{2^7}$. We know that any number (except 0) divided by itself is equal to 1. The previous property also tells us that we can subtract the exponents in this expression, so $\dfrac{2^7}{2^7} = 2^{7-7} = 2^0$.

Since we have shown that $\dfrac{2^7}{2^7}$ is equal to both 2^0 and 1, then the expression 2^0 must be equal to 1. The following property involves raising a base to the power of zero.

Zero Exponent Rule

For any base x, $x^0 = 1$ $(x \neq 0)$.

EXAMPLE ▸ 12   Simplify 7^0.

Solution

By applying the zero exponent rule we see that $7^0 = 1$.

> ***A Word of Caution***   When we raise a nonzero base to the power of zero, the result is 1, not 0.
>
> $$7^0 = 1, \text{ not } 0$$

The expressions $(4x)^0$ and $4x^0$ are different $(x \neq 0)$. In the first expression, the base that is being raised to the power of zero is $4x$. So $(4x)^0 = 1$, since any nonzero base raised to the power of zero is equal to 1. In the second expression, the base that is being raised to the power of zero is x, not $4x$. The expression $4x^0$ simplifies to be $4 \cdot 1$ or 4.

***Quick Check* 10**
Simplify. (Assume $x \neq 0$.)
a) 124^0 b) $(3x^2)^0$ c) $9x^0$

Power of a Quotient Rule

Objective 6   Use the power of a quotient rule for exponents. The power of a quotient rule is similar to the property involving raising a product to a power.

Power of a Quotient Rule

For any bases x and y, $\left(\dfrac{x}{y}\right)^n = \dfrac{x^n}{y^n}$ $(y \neq 0)$.

To raise a quotient to a power, we raise both the numerator and the denominator to that power. The restriction $y \neq 0$ is again due to the fact that division by 0 is undefined. Consider the expression $\left(\frac{r}{s}\right)^3$, where $s \neq 0$. The property tells us that this is equivalent to $\dfrac{r^3}{s^3}$, and here's why:

$$\left(\frac{r}{s}\right)^3 = \frac{r}{s} \cdot \frac{r}{s} \cdot \frac{r}{s} \qquad \text{Repeat } \frac{r}{s} \text{ as a factor three times.}$$

$$= \frac{r \cdot r \cdot r}{s \cdot s \cdot s} \qquad \text{Use the definition of multiplication for fractions.}$$

$$= \frac{r^3}{s^3} \qquad \text{Rewrite the numerator and denominator using exponents.}$$

A Word of Caution   When raising a quotient to a power, raise both the numerator and the denominator to that power. Do not just raise the numerator to that power.

$$\left(\frac{r}{s}\right)^3 = \frac{r^3}{s^3}, \text{ not } \frac{r^3}{s} \qquad (\text{Assume } s \neq 0.)$$

EXAMPLE 13   Simplify $\left(\dfrac{x^2y^3}{z^4}\right)^8$. (Assume $z \neq 0$.)

Solution

We begin by raising each factor in the numerator and denominator to the eighth power, and then we simplify the resulting expression.

$$\left(\frac{x^2y^3}{z^4}\right)^8 = \frac{(x^2)^8(y^3)^8}{(z^4)^8} \qquad \text{Raise each factor in the numerator and denominator to the } 8^{th} \text{ power.}$$
$$= \frac{x^{16}y^{24}}{z^{32}} \qquad \text{Keep the bases and multiply the exponents.}$$

Quick Check 11 Simplify $\left(\dfrac{ab^7}{c^3d^4}\right)^5$. (Assume $c, d \neq 0$.)

Here is a brief summary of the properties introduced in this section:

Properties of Exponents

1. **Product Rule** For any base x, $x^m \cdot x^n = x^{m+n}$.
2. **Power Rule** For any base x, $(x^m)^n = x^{m \cdot n}$.
3. **Power of a Product Rule** For any bases x and y, $(xy)^n = x^n y^n$.
4. **Quotient Rule** For any base x, $\dfrac{x^m}{x^n} = x^{m-n}$ $(x \neq 0)$.
5. **Zero Exponent Rule** For any base x, $x^0 = 1$ $(x \neq 0)$.
6. **Power of a Quotient Rule** For any bases x and y, $\left(\dfrac{x}{y}\right)^n = \dfrac{x^n}{y^n}$ $(y \neq 0)$.

Objective 7   **Evaluate functions by using the rules for exponents.** We conclude this section by investigating functions containing exponents.

EXAMPLE 14   For the function $f(x) = x^4$, evaluate $f(3)$.

Solution

We substitute 3 for x and simplify the resulting expression.

$$f(3) = (3)^4 \qquad \text{Substitute 3 for } x.$$
$$= 81 \qquad \text{Simplify.}$$

Quick Check 12 For the function $f(x) = x^6$ evaluate $f(-4)$.

EXAMPLE 15   For the function $f(x) = x^3$, evaluate $f(a^5)$.

Solution

We substitute a^5 for x and simplify.

$$\begin{aligned} f(a^5) &= (a^5)^3 && \text{Substitute } a^5 \text{ for } x.\\ &= a^{5\cdot 3} && \text{Simplify using the power rule.}\\ &= a^{15} && \text{Multiply.} \end{aligned}$$

Quick Check 13 For the function $f(x) = x^7$, evaluate $f(a^8)$.

Building Your Study Strategy Test Taking, 1 **Preparing Completely**   The first test-taking skill we will discuss is preparing yourself completely. As legendary basketball coach John Wooden once said, "Failing to prepare is preparing to fail." Start preparing for the exam well in advance; do not plan to study (or cram) on the night before the exam. To prepare adequately, you should know what the format of the exam will be. What topics will be covered on the test? Approximately how many questions will there be?

Review your old homework assignments, notes, and note cards, spending more time on problems or concepts that you struggled with before. Work through the chapter review and chapter test in the text, to identify any areas of weakness for you. If you are having trouble with a particular type of problem, go back to that section in the text and review.

Make a practice test for yourself, and take it under test conditions without using your text or notes. Allow yourself the same amount of time that you will be allowed for the actual test, so you know if you are working fast enough.

Get a good night's sleep on the night before the exam. Tired students do not think as well as students who are rested. Also, be sure to eat properly before an exam. Hungry students can be distracted during an exam.

Vocabulary

1. The product rule states that for any base x, _____ .

2. The power rule states that for any base x, _____ .

3. The power of a product rule states that for any bases x and y, _____ .

4. The quotient rule states that for any nonzero base x, _____ .

5. The zero exponent rule states that for any nonzero base x, _____ .

6. The power of a quotient rule states that for any bases x and y, where $y \neq 0$, _____ .

Simplify.

7. $2^2 \cdot 2^3$ 8. $5^5 \cdot 5$ 9. $x^9 \cdot x^7$

10. $a^3 \cdot a^6$ 11. $m^{13} \cdot m^{22}$ 12. $n^{11} \cdot n^{40}$

13. $b^5 \cdot b^6 \cdot b^3$ 14. $x^8 \cdot x^6 \cdot x$

15. $(2x - 3)^4 \cdot (2x - 3)^{10}$

16. $(4x + 1)^{12} \cdot (4x + 1)^9$

17. $(x^6 y^5)(x^2 y^8)$ 18. $(5x^8 y^8)(2x^5 y^{11})$

Find the missing factor.

19. $x^4 \cdot ? = x^9$ 20. $x^7 \cdot ? = x^{10}$

21. $? \cdot b^5 = b^6$ 22. $? \cdot a^{11} = a^{20}$

Simplify.

23. $(x^2)^5$ 24. $(x^3)^3$ 25. $(a^7)^8$

26. $(b^{12})^5$ 27. $(2^3)^2$ 28. $(3^2)^5$

29. $(x^3)^7(x^4)^2$ 30. $(x^2)^8(x^5)^3$

Find the missing exponent.

31. $(x^6)^? = x^{42}$ 32. $(p^?)^8 = p^{112}$

33. $(x^3)^8(x^?)^4 = x^{44}$ 34. $(x^9)^?(x^{11})^{12} = x^{195}$

Simplify.

35. $(5x)^3$ 36. $(3a)^6$

37. $(-4y^3)^3$

38. $(10x^5)^5$

39. $(x^3 y^4)^7$ 40. $(x^8 y^3)^6$

41. $(3x^4 y)^4$

42. $(2x^5 y^6)^8$

43. $(x^2 y^5 z^3)^7 (y^4 z^5)^8$

44. $(a^{10} b^{15} c^{20})^7 (a^{11} b^7 c^3)^8$

Find the missing exponent(s).

45. $(x^5 y^4)^? = x^{35} y^{28}$

46. $(c^7 d^{11})^? = c^{21} d^{33}$

47. $(2s^? t^?)^4 = 16 s^{24} t^{52}$

48. $(m^? n^?)^9 = m^{54} n^{216}$

Simplify. (Assume all variables are nonzero.)

49. $\dfrac{x^7}{x^5}$ 50. $\dfrac{x^{11}}{x^3}$ 51. $d^{35} \div d^5$

52. $z^{87} \div z^{86}$ 53. $\dfrac{32x^8}{8x^3}$ 54. $\dfrac{91y^{15}}{7y^3}$

55. $\dfrac{(a + 5b)^{13}}{(a + 5b)^{11}}$

56. $\dfrac{(2x - 17)^{11}}{(2x - 17)^5}$

57. $\dfrac{a^8 b^6}{a^3 b}$ 58. $\dfrac{28 r^6 s^7 t^9}{7 r^3 s^7 t^6}$

Find the missing exponent(s). (Assume all variables are nonzero.)

59. $\dfrac{x^{14}}{x^?} = x^2$ 60. $\dfrac{y^?}{y^8} = y^9$

61. $m^? \div m^{11} = m^{19}$ 62. $\dfrac{x^{15} y^?}{x^? y^{25}} = x^7 y^6$

Simplify. (Assume all variables are nonzero.)

63. 9^0 64. -8^0 65. $(-16)^0$

66. 5^0 67. $\dfrac{3}{7^0}$ 68. $\left(\dfrac{4}{13}\right)^0$

69. $(13x)^0$ 70. $(22x^5 y^{12} z^9)^0$

Simplify. (Assume all variables are nonzero.)

71. $\left(\dfrac{2}{5}\right)^4$ **72.** $\left(\dfrac{x}{3}\right)^5$ **73.** $\left(\dfrac{a}{b}\right)^7$

74. $\left(\dfrac{m}{n}\right)^9$ **75.** $\left(\dfrac{x^2}{y^3}\right)^7$ **76.** $\left(\dfrac{y^8}{x^4}\right)^8$

77. $\left(\dfrac{2a^6}{b^7}\right)^3$ **78.** $\left(\dfrac{x^{13}}{5y^9}\right)^5$

79. $\left(\dfrac{a^5b^4}{b^2c^7}\right)^9$ **80.** $\left(\dfrac{17x^5y^{19}z^{24}}{32a^{11}b^{70}c^{33}}\right)^0$

Find the missing exponent(s). (Assume all variables are nonzero.)

81. $\left(\dfrac{3}{4}\right)^? = \dfrac{81}{256}$ **82.** $\left(\dfrac{x^4}{y^7}\right)^? = \dfrac{x^{12}}{y^{21}}$

83. $\left(\dfrac{a^3b^6}{c^8}\right)^? = \dfrac{a^{36}b^{72}}{c^{96}}$ **84.** $\left(\dfrac{ab^8}{c^5d^9}\right)^? = \dfrac{a^{15}b^{120}}{c^{75}d^{135}}$

Evaluate the given function.

85. $f(x) = x^2, f(4)$ **86.** $g(x) = x^3, g(9)$

87. $g(x) = x^4, g(-2)$ **88.** $f(x) = x^{10}, f(-1)$

89. $f(x) = x^6, f(a^4)$ **90.** $f(x) = x^5, f(b^7)$

Mixed Practice, 91–104

Simplify.

91. $(a^7)^6$ **92.** $(13a)^0$ **93.** $\left(\dfrac{5x^4}{2y^7}\right)^3$

94. $(x^6y^9z^2)^7$ **95.** $\dfrac{x^{13}}{x}$ **96.** $\dfrac{x^{100}}{x^{100}}$

97. $\left(\dfrac{a^8b^9}{2c^2}\right)^5$ **98.** $c^8 \cdot c^{12}$

99. $(b^{10})^2$ **100.** $\dfrac{x^8y^{13}}{xy^6}$ **101.** $(x^{15})^6$

102. $(2x^9yz^3)^4$ **103.** $\left(\dfrac{3a^2b^9}{7cd^4}\right)^4$

104. $(6x^7y^9z)^3$

The number of feet traveled by a free-falling object in t seconds is given by the function $f(t) = 16t^2$.

105. If a ball is dropped from the top of a building, how far would it fall in 3 seconds?

106. How far would a skydiver fall in 7 seconds?

The area of a square with side x is given by the function $A(x) = x^2$.

107. Use the function to find the area of a square if each side is 90 feet long.

108. Use the function to find the area of a square if each side is 17 meters long.

The area of a circle with radius r is given by the function $A(r) = \pi r^2$.

109. Use the function to find the area of the base of a circular storage tank if the radius is 14 feet long. Use $\pi \approx 3.14$.

110. Columbus Circle in New York City is a traffic circle. The radius of the inner circle is 107 feet. Use the function to find the area of the inner circle. Use $\pi \approx 3.14$.

Writing in Mathematics

Answer in complete sentences.

111. Which expression is equal to x^9: $(x^5)^4$ or $x^5 \cdot x^4$? Explain your answer.

112. Explain why the power of a product rule, $(xy)^n = x^n y^n$, does not apply to the expression $(x + y)^2$. In other words, why is $(x + y)^2$ not equal to $x^2 + y^2$? Explain your answer.

113. Explain, in your own words, why $8^0 = 1$.

114. Explain the difference between the expressions $\dfrac{a^3}{b}$ and $\left(\dfrac{a}{b}\right)^3$.

115. *Solutions Manual* * Write a solutions manual page for the following problem:
Simplify $(3x^2y^3z^5)^3(4x^5yz^3)^2$.

116. *Newsletter* * Write a newsletter explaining how to use the six properties of exponents: product rule, power rule, power of a product rule, quotient rule, zero exponent rule, and power of a quotient rule.

*See Appendix B for details and sample answers.

5.2

Negative Exponents; Scientific Notation

1. **Understand negative exponents.**
2. **Use the rules of exponents to simplify expressions containing negative exponents.**
3. **Convert numbers from standard notation to scientific notation.**
4. **Convert numbers from scientific notation to standard notation.**
5. **Perform arithmetic operations using numbers in scientific notation.**
6. **Use scientific notation to solve applied problems.**

Negative Exponents

Objective 1   Understand negative exponents. In this section, we introduce the concept of a **negative exponent**.

Negative Exponents

$$\text{For any nonzero base } x, \ x^{-n} = \frac{1}{x^n}.$$

For example, $2^{-3} = \dfrac{1}{2^3}$ or $\dfrac{1}{8}$.

Let's examine this definition. Consider the expression $\dfrac{2^4}{2^7}$. If we apply the property $\dfrac{x^m}{x^n} = x^{m-n}$ from the previous section, we see that $\dfrac{2^4}{2^7}$ simplifies to 2^{-3}.

$$\frac{2^4}{2^7} = 2^{4-7}$$
$$= 2^{-3}$$

If we simplify $\dfrac{2^4}{2^7}$ by dividing out common factors, the result is $\dfrac{1}{2^3}$.

$$\frac{2^4}{2^7} = \frac{\overset{1}{\cancel{2}} \cdot \overset{1}{\cancel{2}} \cdot \overset{1}{\cancel{2}} \cdot \overset{1}{\cancel{2}}}{\underset{1}{\cancel{2}} \cdot \underset{1}{\cancel{2}} \cdot \underset{1}{\cancel{2}} \cdot \underset{1}{\cancel{2}} \cdot 2 \cdot 2 \cdot 2}$$
$$= \frac{1}{2^3}$$

Since $\dfrac{2^4}{2^7}$ is equal to both 2^{-3} and $\dfrac{1}{2^3}$, $2^{-3} = \dfrac{1}{2^3}$.

A Word of Caution   Raising a positive base to a negative exponent is not the same as raising the opposite of that base to a positive power.

$$2^{-3} = \frac{1}{2^3}, \text{ not } -2^3$$

EXAMPLE 1   Rewrite the expression 7^{-2} without using negative exponents and simplify.

Solution

When raising a number to a negative exponent, we begin by rewriting the expression without negative exponents. We finish by actually raising the base to the appropriate positive power.

Quick Check 1

Rewrite 4^{-3} without using negative exponents and simplify.

$$7^{-2} = \frac{1}{7^2} \qquad \text{Rewrite without negative exponents.}$$

$$= \frac{1}{49} \qquad \text{Simplify.}$$

EXAMPLE 2   Rewrite the expression $9x^{-6}$ without using negative exponents. (Assume $x \neq 0$.)

Solution

The base in this example is x. The exponent does not apply to the 9, since there are no parentheses. When we rewrite the expression without a negative exponent, the number 9 is unaffected. Rather than writing x^{-6} as $\frac{1}{x^6}$ first, we can move the factor directly to the denominator of a fraction whose numerator is 9 by changing the sign of its exponent.

$$9x^{-6} = \frac{9}{x^6}$$

Quick Check 2

Rewrite the expression $a^5 b^{-4}$ without using negative exponents. (Assume $b \neq 0$.)

Note the difference between the expressions $9x^{-6}$ and $(9x)^{-6}$. In the first expression, the base x is being raised to the power of -6, leaving the 9 in the numerator when the expression is rewritten without negative exponents. The base in the second expression is $9x$, so we rewrite $(9x)^{-6}$ as $\frac{1}{(9x)^6}$.

EXAMPLE 3   Rewrite the expression $\frac{x^8}{z^{-5}}$ without using negative exponents. (Assume $z \neq 0$.)

Solution

In this expression, we have a factor in the denominator with a negative exponent. To change the sign of its exponent to a positive number, we move it to the numerator to join the other factor.

$$\frac{x^8}{z^{-5}} = x^8 z^5$$

EXAMPLE 4   Rewrite the expression $\frac{a^7 b^{-3}}{c^{-5} d^{12}}$ without using negative exponents. (Assume all variables are nonzero.)

Solution

Of the four factors in this example, two of them (b^{-3} and c^{-5}) need to be rewritten without using negative exponents. This can be done by changing the sign of their exponents and rewriting the factors on the other side of the fraction bar.

$$\frac{a^7 b^{-3}}{c^{-5} d^{12}} = \frac{a^7 c^5}{b^3 d^{12}}$$

Quick Check   3 Rewrite without using negative exponents. (Assume all variables are nonzero.)

a) $\dfrac{x^{12}}{y^{-7}}$ b) $\dfrac{x^{-2} y^4}{z^{-1} w^{-9}}$

Using the Rules of Exponents with Negative Exponents

Objective   2   Use the rules of exponents to simplify expressions containing negative exponents. All the properties of exponents described in the last section hold true for negative exponents. When we are simplifying expressions involving negative exponents, there are two general routes we can take. We can choose to apply the appropriate property first, and then we can rewrite the expression without negative exponents. On the other hand, in some circumstances it will be more convenient to rewrite the expression without negative exponents before attempting to apply the appropriate property.

EXAMPLE   5   Simplify the expression $x^{11} \cdot x^{-6}$. Write the result without using negative exponents. (Assume $x \neq 0$.)

Solution

This example uses the product rule $x^m \cdot x^n = x^{m+n}$.

$$\begin{aligned} x^{11} \cdot x^{-6} &= x^{11+(-6)} \qquad \text{Keep the base and add the exponents.} \\ &= x^5 \qquad\qquad \text{Simplify.} \end{aligned}$$

Quick Check   4

Simplify the expression $x^{-13} \cdot x^6$. Write the result without using negative exponents. (Assume $x \neq 0$.)

An alternative approach would be to rewrite the expression without negative exponents first.

$$\begin{aligned} x^{11} \cdot x^{-6} &= \frac{x^{11}}{x^6} \qquad \text{Rewrite without negative exponents.} \\ &= x^5 \qquad \text{Keep the base and subtract the exponents.} \end{aligned}$$

Use the approach that seems clearer to you.

EXAMPLE   6   Simplify the expression $(y^3)^{-2}$. Write the result without using negative exponents. (Assume $y \neq 0$.)

Solution

This example uses the power rule $(x^m)^n = x^{m \cdot n}$.

$$\begin{aligned} (y^3)^{-2} &= y^{-6} \qquad \text{Keep the base and multiply the exponents.} \\ &= \frac{1}{y^6} \qquad \text{Rewrite without using negative exponents.} \end{aligned}$$

Quick Check   5 Simplify the expression $(x^{-6})^{-7}$. Write the result without using negative exponents. (Assume $x \neq 0$.)

EXAMPLE 7   Simplify the expression $(2a^{-7}b^9)^{-4}$. Write the result without using negative exponents. (Assume $a, b \neq 0$.)

Solution

This example uses the power of a product rule $(xy)^n = x^n y^n$.

$$
\begin{aligned}
(2a^{-7}b^9)^{-4} &= 2^{-4}a^{-7(-4)}b^{9(-4)} && \text{Raise each factor to the power of } -4. \\
&= 2^{-4}a^{28}b^{-36} && \text{Simplify each exponent.} \\
&= \frac{a^{28}}{2^4 b^{36}} && \text{Rewrite without using negative exponents.} \\
&= \frac{a^{28}}{16 b^{36}} && \text{Simplify } 2^4.
\end{aligned}
$$

> **Quick Check 6**
>
> Simplify the expression $(9ab^{-3}c^2)^{-2}$. Write the result without using negative exponents. (Assume $a, b, c \neq 0$.)

EXAMPLE 8   Simplify the expression $\dfrac{x^3}{x^{15}}$. Write the result without using negative exponents. (Assume $x \neq 0$.)

Solution

This example uses the quotient rule $\dfrac{x^m}{x^n} = x^{m-n}$.

$$
\begin{aligned}
\frac{x^3}{x^{15}} &= x^{-12} && \text{Keep the base and subtract the exponents.} \\
&= \frac{1}{x^{12}} && \text{Rewrite without using negative exponents.}
\end{aligned}
$$

> **Quick Check 7**
>
> Simplify the expression $\dfrac{x^8}{x^{14}}$. Write the result without using negative exponents. (Assume $x \neq 0$.)

EXAMPLE 9   Simplify the expression $\left(\dfrac{r^4}{s^2}\right)^{-3}$. Write the result without using negative exponents. (Assume $r, s \neq 0$.)

Solution

This example uses the power of a quotient rule $\left(\dfrac{x}{y}\right)^n = \dfrac{x^n}{y^n}$.

$$
\begin{aligned}
\left(\frac{r^4}{s^2}\right)^{-3} &= \frac{r^{4(-3)}}{s^{2(-3)}} && \text{Raise the numerator and the denominator to the power of } -3. \\
&= \frac{r^{-12}}{s^{-6}} && \text{Multiply.} \\
&= \frac{s^6}{r^{12}} && \text{Rewrite without using negative exponents.}
\end{aligned}
$$

> **Quick Check 8**
>
> Simplify the expression $\left(\dfrac{x^5}{y^{-4}}\right)^{-4}$. Write the result without using negative exponents. (Assume $x, y \neq 0$.)

Scientific Notation

Objective 3   Convert numbers from standard notation to scientific notation.
Scientific notation is used to represent numbers that are very large, such as 93,000,000, or very small, such as 0.0000324.

Rewriting a Number in Scientific Notation

> To convert a number to scientific notation, we rewrite it in the form $a \times 10^b$, where $1 \le a < 10$ and b is an integer.

First, consider the following table listing several powers of 10:

10^5	10^4	10^3	10^2	10^1	10^0	10^{-1}	10^{-2}	10^{-3}	10^{-4}	10^{-5}
100,000	10,000	1000	100	10	1	0.1	0.01	0.001	0.0001	0.00001

Notice that all of the positive powers of 10 are numbers that are 10 or higher. All of the negative powers of 10 are numbers between 0 and 1.

To convert a number to scientific notation, we first move the decimal point so that it immediately follows the first nonzero digit in the number. Count the number of decimal places that the decimal point moves. This will give us the power of 10 when the number is written in scientific notation. If the original number was 10 or larger, then the exponent is positive, but if the original number was between 0 and 1, then the exponent is negative.

EXAMPLE 10   Convert 0.000321 to scientific notation.

Solution

Move the decimal point so that it follows the first nonzero digit, which is 3.

$$0.000321$$

To do this, we must move the decimal point four places to the right. Since 0.000321 is less than 1, this exponent must be negative.

$$0.000321 = 3.21 \times 10^{-4}$$

EXAMPLE 11   Convert 24,000,000,000 to scientific notation.

Solution

Move the decimal point so that it follows the digit 2. To do this, we must move the decimal point 10 places to the left. Since 24,000,000,000 is 10 or larger, the exponent will be positive.

$$24,000,000,000 = 2.4 \times 10^{10}$$

Quick Check   **9** Convert to scientific notation.

a) 0.0046

b) 3,570,000

Objective 4 **Convert numbers from scientific notation to standard notation.**

EXAMPLE ▶12   Convert 5.28×10^4 from scientific notation to standard notation.

Solution

$$5.28 \times 10^4 = 5.28 \times 10,000 \qquad \text{Rewrite } 10^4 \text{ as } 10,000.$$
$$= 52,800 \qquad\qquad \text{Multiply.}$$

To multiply a decimal number by 10^4 or 10,000, we move the decimal point 4 places to the right. Note that the power of 10 is positive in this example.

EXAMPLE ▶13   Convert 2.0039×10^{-5} from scientific notation to standard notation.

Solution

$$2.0039 \times 10^{-5} = 2.0039 \times 0.00001 \qquad \text{Rewrite } 10^{-5} \text{ as } 0.00001.$$
$$= 0.000020039 \qquad\qquad \text{Multiply.}$$

To multiply a decimal number by 10^{-5} or 0.00001, we move the decimal point five places to the left. Note that the power of 10 is negative in this example.

If you are unsure about which direction to move the decimal point, try thinking about whether we are making the number larger or smaller. Multiplying by positive powers of 10 makes the number larger, so move the decimal point to the right. Multiplying by negative powers of 10 makes the number smaller, so move the decimal point to the left.

> **Quick Check 10**
> Convert from scientific notation to standard notation.
>
> a) 3.2×10^6
> b) 7.21×10^{-4}

Objective 5 **Perform arithmetic operations using numbers in scientific notation.** When we are performing calculations involving very large or very small numbers, using scientific notation can be very convenient.

EXAMPLE ▶14   Multiply $(2.2 \times 10^7)(2.8 \times 10^{13})$. Express your answer using scientific notation.

Solution

When multiplying two numbers that are in scientific notation, we may first multiply the two decimal numbers. We then multiply the powers of 10 using the product rule, $x^m \cdot x^n = x^{m+n}$.

$$(2.2 \times 10^7)(2.8 \times 10^{13}) = (2.2)(2.8)(10^7)(10^{13}) \qquad \text{Reorder the factors.}$$
$$= 6.16 \times 10^{20} \qquad\qquad\qquad \text{Multiply decimal numbers.}$$
$$\qquad\qquad\qquad\qquad\qquad\qquad\qquad\quad \text{Add the exponents for the base 10.}$$

> **Quick Check 11** Multiply $(5.8 \times 10^4)(1.2 \times 10^9)$. Express your answer using scientific notation.

Using Your Calculator   The TI-84 can help us perform calculations with numbers in scientific notation. To enter a number in scientific notation, we use the second function **EE** above the key labeled [.]. For example, to enter the number 2.2×10^7, type 2.2 [2nd] [.] 7. Here is the screen shot showing how to multiply $(2.2 \times 10^7)(2.8 \times 10^{13})$.

```
2.2E7*2.8E13
         6.16E20
```

EXAMPLE 15   Divide $(1.2 \times 10^{-6}) \div (2.4 \times 10^9)$. Express your answer using scientific notation.

Solution

To divide numbers that are in scientific notation, we begin by dividing the decimal numbers. We then divide the powers of 10 separately, using the property $\dfrac{x^m}{x^n} = x^{m-n}$.

$$(1.2 \times 10^{-6}) \div (2.4 \times 10^9) = \frac{1.2 \times 10^{-6}}{2.4 \times 10^9} \qquad \text{Rewrite as a fraction.}$$

$$= \frac{1.2}{2.4} \times 10^{-6-9} \qquad \begin{array}{l}\text{Divide the decimal numbers.} \\ \text{Subtract the exponents for the} \\ \text{powers of 10.}\end{array}$$

$$= 0.5 \times 10^{-15} \qquad \begin{array}{l}\text{Divide 1.2 by 2.4. Simplify the} \\ \text{exponent.}\end{array}$$

Although this is the correct quotient, our answer is not in scientific notation, as the number 0.5 does not have a nonzero digit to the left of the decimal point. In order to rewrite 0.5 as 5.0, we move the decimal point one place to the right. This will decrease the exponent by one from -15 to -16. $(1.2 \times 10^{-6}) \div (2.4 \times 10^9) = 5.0 \times 10^{-16}$.

Quick Check   12 Divide $(9.9 \times 10^5) \div (3.3 \times 10^{-7})$. Express your answer using scientific notation.

Applications

Objective   6   Use scientific notation to solve applied problems. We finish with an applied problem using scientific notation.

EXAMPLE 16   At its farthest point, the planet Mars is 155 million miles from the Sun. How many seconds does it take light from the Sun to reach Mars if the speed of light is 1.86×10^5 miles per second?

Solution

To determine the length of time that a trip takes, we divide the distance traveled by the rate of speed. To solve this problem, we need to divide the distance of 155 million

(155,000,000) miles by the speed of light. We will convert the distance to scientific notation, which is 1.55×10^8.

$$\frac{1.55 \times 10^8}{1.86 \times 10^5} \approx 0.833 \times 10^3$$

Divide the decimal numbers; approximate with a calculator. Subtract the exponents for the powers of 10.

To write this result in scientific notation, we move the decimal point one place to the right. This decreases the exponent by one. The light will reach Mars in approximately 8.33×10^2 or 833 seconds, or 13 minutes and 53 seconds.

Quick Check 13
How many seconds does it take light to travel 1,488,000,000 miles? (The speed of light is 1.86×10^5 miles per second.)

Building Your Study Strategy Test Taking, 2 **Write Down Information**   As soon as you receive your test, write down any formulas, rules, or procedures that will help you during the exam. Once you have written these down, you can refer to them as you work through the test. For example, write down a table or formula that you use to solve a particular type of word problem while it is still fresh in your memory. That way, when you reach the appropriate problem on the test, you will not need to worry about not being able to remember the table or formula.

By writing all of this information on your test, you will reduce memorization difficulties that arise when taking a test. However, if you do not understand the material or how to use what you have written down, then you will struggle on the test. There is no substitute for understanding what you are doing.

EXERCISES 5.2

Vocabulary

1. For any *nonzero* base x, $x^{-n} = $ _____.

2. A number is in _____ if it is in the form $a \times 10^b$ where $1 \le a < 10$ and b is an integer.

Rewrite the expression without using negative exponents. (Assume all variables represent nonzero real numbers.)

3. 4^{-2}

4. 3^{-2}

5. 6^{-3}

6. 5^{-3}

7. -7^{-3}

8. -2^{-9}

9. a^{-10}

10. a^{-8}

11. $4x^{-3}$

12. $2x^{-9}$

13. $-5m^{-19}$

14. $-12m^{-6}$

15. $\dfrac{1}{x^{-5}}$

16. $\dfrac{1}{x^{-7}}$

17. $\dfrac{3}{y^{-4}}$

18. $\dfrac{x^7}{y^{-5}}$

19. $\dfrac{x^{-11}}{y^{-2}}$

20. $\dfrac{x^{-10}}{y^4}$

21. $\dfrac{-8a^{-6}b^5}{c^7}$

22. $\dfrac{-10x^3y^{-5}}{z^{-4}}$

23. $\dfrac{5a^{-6}b^{-9}}{c^{-4}d^5}$

24. $\dfrac{a^3b^{-3}}{c^{-4}d^{-9}}$

Simplify the expression. Write the result without using negative exponents. (Assume all variables represent nonzero real numbers.)

25. $x^{-7} \cdot x^{10}$ **26.** $x^8 \cdot x^{-3}$ **27.** $a^{-5} \cdot a^{-11}$

28. $b^{-9} \cdot b^{-7}$ **29.** $m^4 \cdot m^{-8}$

30. $n^{-11} \cdot n^9$ **31.** $(x^5)^{-4}$ **32.** $(x^{10})^{-5}$

33. $(x^{-6})^{-7}$ **34.** $(x^{-9})^{-2}$

35. $(x^5 y^4 z^{-6})^{-2}$ **36.** $(x^{-3} y^3 z^{-2})^{-7}$

37. $(4a^{-5} b^{-8} z^2)^{-3}$

38. $(-2x^4 b^{-6} c^{-7})^5$

39. $\dfrac{x^{-4}}{x^9}$ **40.** $\dfrac{x^5}{x^{-5}}$

41. $\dfrac{a^9}{a^{-3}}$ **42.** $\dfrac{a^{-8}}{a}$

43. $\dfrac{y^{-5}}{y^{-3}}$ **44.** $\dfrac{y^{-4}}{y^{-11}}$

45. $\dfrac{x^7}{x^{18}}$ **46.** $\dfrac{x^3}{x^{12}}$

47. $\dfrac{x^5 \cdot x^{-11}}{x^{-6}}$ **48.** $\dfrac{x^{-13}}{x^{-17} \cdot x^4}$

49. $\left(\dfrac{x^3}{y^4}\right)^{-5}$ **50.** $\left(\dfrac{2x^8}{y^5}\right)^{-3}$

51. $\left(\dfrac{a^4 b^7}{3c^5 d^{-8}}\right)^{-2}$

52. $\left(\dfrac{x^{-8} y^5}{wz^{-6}}\right)^{-8}$

Convert the given number to standard notation.

53. 3.07×10^{-7}

54. 2.3×10^5

55. 8.935×10^9

56. 6.001×10^{-8}

57. 9.021×10^4

58. 7.0×10^{10}

Convert the given number to scientific notation.

59. 0.0000045   **60.** 125,000

61. 47,000,000   **62.** 0.0031

63. 0.000000006

64. 5,000,000,000

Perform the following calculations. Express your answer using scientific notation.

65. $(3.2 \times 10^7)(2.9 \times 10^{11})$

66. $(1.5 \times 10^{-8})(6.1 \times 10^{-9})$

67. $(2.0088 \times 10^9) \div (5.4 \times 10^{-6})$

68. $(2.6 \times 10^{-13}) \div (6.5 \times 10^{10})$

69. $(5.32 \times 10^{-15})(7.8 \times 10^3)$

70. $(6.0 \times 10^{13}) \div (7.5 \times 10^{-12})$

71. $(87,000,000,000)(0.000002)$

72. $0.0000000064 \div 160,000,000,000$

73. If a computer can perform a calculation in 0.0000000008 second, how long would it take to perform 42,000,000,000,000 calculations?

74. If a computer can perform a calculation in 0.0000000008 second, how many calculations can it perform in 2 minutes (120 seconds)?

75. The speed of light is 1.86×10^5 miles per second. How far can light from the Sun travel in 10 minutes?

76. The speed of light is 1.86×10^5 miles per second. At its farthest point, Pluto is 4,555,000,000 miles from the Sun. How many seconds does it take light from the Sun to reach Pluto?

77. If the mass of a typical star is 2.2×10^{33} grams and there are approximately 500,000,000,000 stars in the Milky Way galaxy, what is the total mass of these stars?

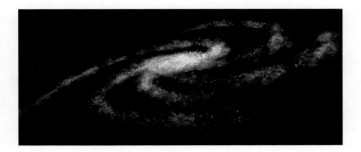

78. The speed of light is 1.86×10^5 miles per second. How far can light from the Sun travel in 1 year? (This distance is often referred to as a "light-year.")

79. For the fiscal year ending in 2006, Microsoft showed total revenues of $\$4.428 \times 10^{10}$ and Apple showed total revenues of $\$1.931 \times 10^{10}$. What was the combined revenue of the two companies for the fiscal year that ended in 2006? (*Source:* Stargeek.com)

80. Burger King restaurants serve 7.7×10^6 customers daily. How many customers do Burger King restaurants serve in one year? (*Source:* Burger King Corporation)

81. The Intel Pentium 4 chip performs 4.2×10^7 calculations per second. How many calculations can it perform in one hour (3600 seconds)? (*Source:* Innovation Watch)

82. The Intel Pentium 4 chip performs 4.2×10^7 calculations per second. How many calculations can it perform in one year (31,536,000 seconds)? (*Source:* Innovation Watch)

83. The pie chart shows the percent of Burger King's business that comes from take-out sales and drive-thru sales, as well as from customers who eat in the restaurant. If Burger King's sales totaled $\$1.11 \times 10^{10}$ in 2003, what were the total sales to drive-thru customers? (*Source:* Burger King Corporation)

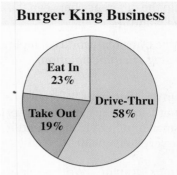

Burger King Business

Eat In 23%
Drive-Thru 58%
Take Out 19%

84. The pie chart shows the percent of U.S. college students who received some type of financial aid in 2003–2004. If there were approximately 1.73×10^7 U.S. college students in 2003–2004, how many of them received some type of financial aid? (*Source:* U.S. Department of Education, National Center for Educational Statistics, 2003–2004)

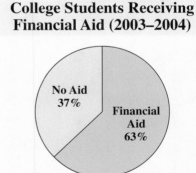

College Students Receiving Financial Aid (2003–2004)

No Aid 37%
Financial Aid 63%

Writing in Mathematics

Answer in complete sentences.

85. Explain the difference between the expressions 4^{-2} and -4^2.

86. Describe three real-world examples of numbers using scientific notation.

87. Explain the statement "When we are performing calculations involving very large or very small numbers, using scientific notation can be very convenient."

88. *Solutions Manual** Write a solutions manual page for the following problem:

    Simplify $(6a^{-3}b^5c^{-4})^{-2}$.

89. *Newsletter** Write a newsletter explaining how to convert to and from scientific notation.

*See Appendix B for details and sample answers.

QUICK REVIEW EXERCISES

Section 5.2

Simplify.

1. $8x + 3x$

2. $9a - 16a$

3. $-9x^2 + 4x + 3x^2 + 7x$

4. $10y^3 - y^2 + 5y + 13 + 2y^3 + 6y^2 - 18y - 50$

Objectives

1 **Identify polynomials and understand the vocabulary used to describe them.**
2 **Evaluate polynomials.**
3 **Add and subtract polynomials.**
4 **Understand polynomials in several variables.**

Polynomials in a Single Variable

Objective 1 **Identify polynomials and understand the vocabulary used to describe them.** Many real-world phenomena cannot be described by linear expressions and linear functions. For example, if a car is traveling at x miles per hour on dry pavement, the distance in feet required for the car to come to a complete stop can be approximated by the expression $0.06x^2 + 1.1x + 0.02$. This nonlinear expression is an example of a **polynomial.**

Polynomials

A **polynomial in a single variable x** is a sum of **terms** of the form ax^n, where a is a real number and n is a whole number.

Here are some examples of a polynomial in a single variable x:

$$x^2 - 9x + 14 \qquad 4x^5 - 13 \qquad x^7 - 4x^5 + 8x^2 + 13x$$

An expression is *not* a polynomial if it contains a term with a variable that is raised to a power other than a whole number (such as $x^{3/4}$ or x^{-2}), or if it has a term with a variable in a denominator $\left(\text{such as } \dfrac{5}{x^3}\right)$.

Degree of a Term

The **degree of each term** in a polynomial in a single variable is equal to the variable's exponent.

For example, the degree of the term $8x^9$ is 9. A **constant term** does not contain a variable. The degree of a constant term is 0, because a constant term such as -15 can be rewritten as $-15x^0$. The polynomial $4x^5 - 9x^4 + 2x^3 - 7x - 11$ has five terms, and their degrees are 5, 4, 3, 1, and 0 respectively.

Term	$4x^5$	$-9x^4$	$2x^3$	$-7x$	-11
Degree	5	4	3	1	0

A polynomial with only one term is called a **monomial.** A **binomial** is a polynomial that has two terms, while a **trinomial** is a polynomial that has three terms. (We do not have special names to describe polynomials with four or more terms.) Here are some examples:

Monomial	Binomial	Trinomial
$2x$	$x^2 - 25$	$x^2 - 15x - 76$
$9x^2$	$7x + 3$	$4x^2 + 12x + 9$
$-4x^3$	$5x^3 - 135$	$x^6 + 11x^5 - 26x^4$

EXAMPLE 1 Classify the polynomial as a monomial, binomial, or trinomial. List the degree of each term.

a) $16x^4 - 81$ **b)** $-13x^7$ **c)** $x^6 - 7x^3 - 18x$

Solution

a) $16x^4 - 81$ has two terms, so it is a binomial. The degrees of its two terms are 4 and 0.
b) $-13x^7$ has only one term, so it is a monomial. The degree of this term is 7.
c) $x^6 - 7x^3 - 18x$ is a trinomial as it has three terms. The degrees of those terms are 6, 3, and 1.

Quick Check **1**
Classify as a monomial, binomial, or trinomial. List the degree of each term.

a) $x^2 - 9x + 20$
b) $x^2 - 81$
c) $-92x^7$

The **coefficient** of a term is the numerical part of a term. When determining the coefficient of a term, be sure to include its sign. For the polynomial $8x^3 - 6x^2 - 5x + 13$, the four terms have coefficients 8, -6, -5, and 13 respectively.

Term	$8x^3$	$-6x^2$	$-5x$	13
Coefficient	8	-6	-5	13

EXAMPLE 2 For the trinomial $x^2 - 9x + 18$, determine the coefficient of each term.

Solution

The first term of this trinomial, x^2, has a coefficient of 1. Even though we do not see a coefficient in front of the term, the coefficient is 1 because $x^2 = 1 \cdot x^2$. The second term, $-9x$, has a coefficient of -9. Finally, the coefficient of the constant term is 18.

Quick Check **2**
For the polynomial $x^3 - 6x^2 - 11x + 32$, determine the coefficient of each term.

We write polynomials in **descending order,** writing the term of highest degree first, followed by the term of next highest degree, and so on. For example, the polynomial $3x^2 - 9 - 5x^5 + 7x$ is written in descending order as $-5x^5 + 3x^2 + 7x - 9$. The term of highest degree is called the **leading term,** and its coefficient is called the **leading coefficient.**

Degree of a Polynomial

The degree of the leading term is also called the **degree of the polynomial.**

EXAMPLE 3   Rewrite the polynomial $7x^2 - 9 + 3x^5 - x$ in descending order and identify the leading term, the leading coefficient, and the degree of the polynomial.

Quick Check 3
Rewrite the polynomial $x + 9 + 4x^2 - x^3$ in descending order and identify the leading term, the leading coefficient, and the degree of the polynomial.

Solution

To write a polynomial in descending order, we write the terms according to their degree from highest to lowest. In descending order, this polynomial is $3x^5 + 7x^2 - x - 9$. The leading term is $3x^5$, and the leading coefficient is 3. The leading term has degree 5, so this polynomial has degree 5.

Evaluating Polynomials

Objective 2   Evaluate polynomials. To **evaluate a polynomial** for a particular value of a variable, we substitute the value for the variable in the polynomial and simplify the resulting expression.

EXAMPLE 4   Evaluate $4x^5 - 7x^3 + 12x^2 - 13x$ for $x = -2$.

Solution

$$4(-2)^5 - 7(-2)^3 + 12(-2)^2 - 13(-2) \qquad \text{Substitute } -2 \text{ for } x.$$
$$= 4(-32) - 7(-8) + 12(4) - 13(-2) \qquad \text{Perform operations involving exponents.}$$
$$= -128 + 56 + 48 + 26 \qquad \text{Multiply.}$$
$$= 2 \qquad \text{Simplify.}$$

Quick Check 4
Evaluate $x^4 - 5x^3 - 6x^2 + 10x - 21$ for $x = -3$.

A **polynomial function** is a function that is described by a polynomial, such as $f(x) = x^3 - 5x^2 - 5x + 7$. Linear functions of the form $f(x) = mx + b$ are first degree polynomial functions.

EXAMPLE 5   For the polynomial function $f(x) = x^2 - 9x + 16$, find $f(3)$.

Solution

Recall that the notation $f(3)$ tells us to substitute 3 for x in the function. After substituting we simplify the resulting expression.

$$f(3) = (3)^2 - 9(3) + 16 \qquad \text{Substitute 3 for } x.$$
$$= 9 - 9(3) + 16 \qquad \text{Perform operations involving exponents.}$$
$$= 9 - 27 + 16 \qquad \text{Multiply.}$$
$$= -2 \qquad \text{Simplify.}$$

Quick Check 5
For the polynomial function $f(x) = x^3 + 12x^2 - 21x$, find $f(-5)$.

Adding and Subtracting Polynomials

Objective 3   Add and subtract polynomials. Just as we can perform arithmetic operations combining two numbers, we can perform operations combining two polynomials, such as adding and subtracting.

Adding Polynomials

To add two polynomials, we combine their like terms.

Recall that two terms are like terms if they have the same variables with the same exponents.

EXAMPLE ▸ 6   Add $(7x^2 - 5x - 8) + (4x^2 - 9x - 17)$.

Solution

To add these polynomials, we drop the parentheses and combine like terms.

$$(7x^2 - 5x - 8) + (4x^2 - 9x - 17)$$
$$= 7x^2 - 5x - 8 + 4x^2 - 9x - 17 \qquad \text{Rewrite without using parentheses.}$$
$$= 11x^2 - 14x - 25 \qquad \text{Combine like terms. } 7x^2 + 4x^2 = 11x^2,$$
$$-5x - 9x = -14x, -8 - 17 = -25$$

Quick Check **6** $(x^3 - x^2 - 9x + 25) + (x^2 + 12x + 144)$.

A Word of Caution   The terms $11x^2$ and $-14x$ are not like terms, as the variables do not have the same exponents.

> To subtract one polynomial from another, such as $(x^2 + 5x - 9) - (3x^2 - 2x + 15)$, we change the sign of each term that is being subtracted and then combine like terms. Changing the sign of each term in the parentheses being subtracted is equivalent to applying the distributive property with -1.

EXAMPLE ▸ 7   Subtract $(x^2 + 5x - 9) - (3x^2 - 2x + 15)$.

Solution

We remove the parentheses by changing the sign of each term in the polynomial that is being subtracted. We then can combine like terms as follows:

$$(x^2 + 5x - 9) - (3x^2 - 2x + 15)$$
$$= x^2 + 5x - 9 - 3x^2 + 2x - 15 \qquad \text{Distribute to remove parentheses.}$$
$$= -2x^2 + 7x - 24 \qquad \text{Combine like terms.}$$

Quick Check **7** $(5x^2 + 6x + 27) - (3x^3 + 6x^2 - 9x + 22)$.

A Word of Caution   When subtracting one polynomial from another polynomial, be sure to change the sign of each term in the polynomial that is being subtracted.

EXAMPLE ▸ 8   Given the polynomial functions $f(x) = x^2 + 5x + 7$ and $g(x) = 2x^2 - 7x + 11$, find $f(x) - g(x)$.

Solution

We will substitute the appropriate expressions for $f(x)$ and $g(x)$, and simplify the resulting expression.

$$f(x) - g(x) = (x^2 + 5x + 7) - (2x^2 - 7x + 11) \quad \text{Substitute.}$$
$$= x^2 + 5x + 7 - 2x^2 + 7x - 11 \quad \text{Distribute to remove parentheses.}$$
$$= -x^2 + 12x - 4 \quad \text{Combine like terms.}$$

Quick Check 8 Given the polynomial functions
$$f(x) = 4x^2 + 30x - 45 \text{ and } g(x) = -2x^2 + 17x + 52, \text{ find } f(x) - g(x).$$

Polynomials in Several Variables

Objective 4 **Understand polynomials in several variables.** While the polynomials that we have examined to this point contained a single variable, some polynomials contain two or more variables. Here are some examples:

$$x^2y^2 + 3xy - 10 \qquad 7a^3 - 8a^2b + 3ab^2 - 15b^3 \qquad x^2 + 2xh + h^2 + 5x + 5h$$

We evaluate polynomials in several variables by substituting values for each variable and simplifying the resulting expression.

EXAMPLE 9 Evaluate the polynomial $x^2 - 8xy + 2y^2$ for $x = 5$ and $y = -4$.

Solution

We will substitute 5 for x and -4 for y. Be careful to substitute the right value for each variable.

$$x^2 - 8xy + 2y^2$$
$$(5)^2 - 8(5)(-4) + 2(-4)^2 \quad \text{Substitute 5 for } x \text{ and } -4 \text{ for } y.$$
$$= 25 - 8(5)(-4) + 2(16) \quad \text{Perform operations involving exponents.}$$
$$= 25 + 160 + 32 \quad \text{Multiply.}$$
$$= 217 \quad \text{Add.}$$

Quick Check 9
Evaluate the polynomial $b^2 - 4ac$ for $a = 1$, $b = -9$, and $c = -52$.

For a polynomial in several variables, the degree of a term is equal to the sum of the exponents of its variable factors. For example, the degree of the term $3x^2y^5$ is 2 + 5 or 7.

EXAMPLE 10 Find the degree of each term in the polynomial $a^4b^3 - 3a^7b^5 + 9ab^6$. In addition, find the degree of the polynomial.

Quick Check 10
Find the degree of each term in the polynomial $x^5 - 3x^2y^6 + 8x^9y$. In addition, find the degree of the polynomial.

Solution

The first term, a^4b^3, has degree 7. The second term, $-3a^7b^5$, has degree 12. The third term, $9ab^6$, has degree 7. The degree of a polynomial is equal to the highest degree of any of its terms, so the degree of this polynomial is 12.

To add or subtract polynomials in several variables, we need to combine like terms. Two terms are like terms if they have the same variables with the same exponents.

EXAMPLE ▸11   Add $4xy^2 + 7x^3y^5$ and $9x^2y - 3x^3y^5$.

Solution

Quick Check 11
Add $5x^4y^2 - 6x^3y^3$ and $8x^4y^2 + 2x^3y^3$.

$(4xy^2 + 7x^3y^5) + (9x^2y - 3x^3y^5)$ Write as a sum.
$= 4xy^2 + 7x^3y^5 + 9x^2y - 3x^3y^5$ Rewrite without parentheses.
$= 4xy^2 + 4x^3y^5 + 9x^2y$ Combine like terms ($7x^3y^5$ and $-3x^3y^5$).

Building Your Study Strategy **Test Taking, 3 Read the Test**   In the same way that you would begin to solve a word problem, you should begin to take a test by briefly reading through it. This will give you an idea of how many problems you have to solve, how many word problems there are, and roughly how much time you can devote to each problem. It is a good idea to establish a schedule, such as "I need to be done with 12 problems by the time half of the class period is over." This way you will know whether you need to speed up during the second half of the exam or if you have plenty of time.

Not all problems are assigned the same point value. Identify which problems are worth more points than others. It is important not to have to rush through the problems that have higher point values because you didn't notice that they were worth more until you got to them at the end of the test.

EXERCISES *5.3* ▸

Vocabulary

1. A(n) _____ in a single variable x is a sum of terms of the form ax^n, where a is a real number and n is a whole number.

2. For a polynomial in a single variable x, the _____ of a term is equal to its exponent.

3. A polynomial with one term is called a(n) _____.

4. A polynomial with two terms is called a(n) _____.

5. A polynomial with three terms is called a(n) _____.

6. The _____ of a term is the numerical part of a term.

7. When the terms of a polynomial are written from highest degree to lowest degree, the polynomial is said to be in _____.

8. The _____ of a polynomial is the term that has the greatest degree.

List the degree of each term in the given polynomial.

9. $9x^4 - 7x^2 + 3x - 8$

10. $-4x^3 + 10x^2 + 5x + 4$

11. $13x - 8x^7 - 11x^4$

12. $5x^5 - 3x^4 + 2x^3 + x^2 - 15$

List the coefficient of each term in the given polynomial.

13. $7x^2 + x - 15$

14. $3x^3 + x^2 - 8x - 13$

15. $10x^5 - 17x^4 + 6x^3 - x^2 + 2$

16. $-5x^3 - 12x^2 - 9x - 4$

Identify the given polynomial as a monomial, binomial, or trinomial.

17. $x^2 - 9x + 20$ **18.** $15x^9$

19. $4x - 7$ **20.** $2x^5 + 13x^4$

21. $-8x^3$ **22.** $5x^5 - 8x^3 + 17x$

Rewrite the polynomial in descending order. Identify the leading term, the leading coefficient, and the degree of the polynomial.

23. $8x - 7 + 3x^2$

24. $13 - 9x - x^2$

25. $6x^2 - 11x + 2x^4 + 10$

26. $2x^5 - x^6 + 24 - 3x^3 + x - 5x^2$

Evaluate the polynomial for the given value of the variable.

27. $x^2 + 5x + 8$ for $x = 2$

28. $2x^2 + 3x + 7$ for $x = 5$

29. $x^3 - 9x - 10$ for $x = -4$

30. $x^5 - 8x^2 + 15x$ for $x = -3$

31. $-3x^4 - 8x^2 + 21x$ for $x = 3$

32. $-x^4 + 3x^3 + 12$ for $x = 6$

Evaluate the given polynomial function.

33. $f(x) = x^2 - 7x - 10, f(8)$

34. $f(x) = x^2 + 12x - 13, f(-9)$

35. $g(x) = 3x^3 - 8x^2 + 5x + 9, g(-5)$

36. $g(x) = -2x^3 - 5x^2 + 10x + 35, g(6)$

Add or subtract.

37. $(5x^2 + 8x - 11) + (3x^2 - 14x + 14)$

38. $(2x^2 - 9x - 35) + (8x^2 - 6x + 17)$

39. $(4x^2 - 7x + 30) - (2x^2 + 10x - 50)$

40. $(x^2 + 12x - 42) - (6x^2 - 19x - 23)$

41. $(2x^4 + 7x^2 - 13x - 20) + (x^3 - 5x^2 - 14x)$

42. $(3x^5 - 5x^3 + 2x^2) - (8x^5 - 5x^3 - 2x^2 + 7)$

43. $(x^3 - 5x^2 + 17x) - (4x^2 - 3x + 16)$

44. $(3x^2 + 10x - 11) + (-4x^3 - 9x^2 + 32x)$

45. $(2x^9 - 5x^4 + 7x^2) - (-7x^6 + 4x^5 - 12)$

46. $(9x^3 + x^2 - x - 13) - (9x^3 + x^2 - x - 13)$

Find the missing polynomial.

47. $(3x^2 + 8x + 11) + ? = 7x^2 + 5x - 2$

48. $(2x^3 - 4x^2 - 7x + 19) + ? = x^2 + 3x + 24$

49. $(3x^4 - 5x^3 + 6x + 12) - ? =$
$x^4 - 2x^3 - x^2 - 5x + 1$

50. $? - (6x^3 - 11x^2 - 16x + 22) =$
$-x^3 + 19x^2 - 12x + 9$

For the given functions $f(x)$ and $g(x)$, find $f(x) + g(x)$ and $f(x) - g(x)$.

51. $f(x) = 7x^2 + 10x + 3, g(x) = 5x^2 - 9x + 6$

52. $f(x) = 2x^2 + 11x - 5, g(x) = -x^2 - 5x + 20$

53. $f(x) = x^3 - 8x - 31, g(x) = 4x^3 + x^2 + 3x + 25$

54. $f(x) = 2x^3 - 4x^2 + 8x - 16, g(x) = x^3 + 12x^2 - 15$

Evaluate the polynomial.

55. $x^2 - 7xy + 10y^2$ for $x = 2$ and $y = 5$

56. $3x^2 + 11xy + y^2$ for $x = 4$ and $y = 3$

57. $b^2 - 4ac$ for $a = -9, b = -2,$ and $c = 4$

58. $b^2 - 4ac$ for $a = 5, b = -1,$ and $c = 6$

59. $3x^2yz^3 - 4xy^2z - 6x^4y^2z$ for $x = 4$, $y = 5$, and $z = -2$

60. $7x^2yz + 9xy^2z - 10xyz^2$ for $x = 9$, $y = 3$, and $z = -2$

For each polynomial, list the degree of each term and the degree of the polynomial.

61. $6x^5y^5 - 7x^3y^3 + 9x^2y$

62. $-4x^8y^6 + 3x^3y^7 + x^5y^{11}$

63. $x^5y^2z - 5x^3y^3z^3 + 6x^2y^5z^4$

64. $a^{12}b^9c^{10} + 2a^{11}b^6c^{16} + 4a^{10}b^{10}c^3$

Add or subtract.

65. $(15x^2y + 8xy^2 - 7x^2y^3) + (-8x^2y + 3xy^2 + 4x^2y^3)$

66. $(x^4y - 5x^2y^3 - 14y^5) - (6x^4y + 11x^2y^3 - 7y^5)$

67. $(a^3b^2 + 11a^5b + 24a^2b^3) - (19a^2b^3 - 2a^5b + 15a^3b^2)$

68. $(9m^4n^2 - 13m^3n^3 + 11m^2n^4) + (-4m^2n^4 - 3m^4n^2 - m^3n^3)$

69. $(2x^3yz^2 - xy^4z^3 - 10x^2y^2z^5) - (4x^2yz^3 + 14xy^4z^3 - 3x^2y^2z^5)$

70. $(22x^3y^9 - 21x^6y^4 - 7x^8y^2) - (22x^3y^9 - 21x^6y^4 - 7x^8y^2)$

71. Ellen is the yearbook advisor at a high school, and she determines that the revenue generated by all yearbook sales can be approximated by the function $R(x) = -6x^2 + 240x$, where x represents the price of a yearbook in dollars. How much money will be generated if the yearbook is sold for $30?

72. The average cost per shirt, in dollars, to produce x T-shirts is given by the function $f(x) = 0.00015x^2 - 0.06x + 10.125$. What is the average cost per shirt to produce 250 T-shirts?

73. The number of students earning master's degrees in mathematics in the United States in a particular year can be approximated by the function $g(x) = -11x^2 + 477x - 1104$, where x represents the number of years after 1970. Use the function to estimate the number

of students who earned a master's degree in mathematics in the United States in the year 2007. (*Source:* U.S. Department of Education, National Center for Education Statistics)

74. The number of births, in thousands, in the United States in a particular year can be approximated by the function $f(x) = 6x^2 - 72x + 4137$, where x represents the number of years after 1990. Use the function to estimate the number of births in the United States in the year 2008. (*Source:* National Center for Health Statistics, U.S. Department of Health and Human Services)

Writing in Mathematics

Answer in complete sentences.

75. Explain why it is a good idea to use parentheses when evaluating a polynomial for negative values of a variable. Do you feel that using parentheses is a good idea for evaluating any polynomial?

76. A classmate made the following error when subtracting two polynomials.

$(3x^2 + 2x - 7) - (x^2 - x + 9)$
$= 3x^2 + 2x - 7 - x^2 - x + 9$

Explain the error to your classmate, and give advice on how to avoid making this error.

77. *Solutions Manual* [*] Write a solutions manual page for the following problem:

Simplify $(3x^2 + 4x - 7) - (2x^2 - 6x + 19)$.

78. *Newsletter* [*] Write a newsletter that explains the terms and definitions related to polynomials.

*See Appendix B for details and sample answers.

5.4
Multiplying Polynomials

1 Multiply monomials.
2 Multiply a monomial by a polynomial.
3 Multiply polynomials.
4 Find special products.

Multiplying Monomials

Objective 1 **Multiply monomials.** Now that we have learned how to add and subtract polynomials, we will explore multiplication of polynomials. We begin by learning how to multiply monomials.

Multiplying Monomials

When multiplying monomials, we begin by multiplying their coefficients. Then we multiply the variable factors, using the property of exponents which states that for any real number x, $x^m \cdot x^n = x^{m+n}$.

EXAMPLE 1   Multiply $3x \cdot 5x$.

Solution

We multiply the coefficients first, and then we multiply the variables.

$$
\begin{aligned}
3x \cdot 5x &= 3 \cdot 5 \cdot x \cdot x && \text{Multiply coefficients, then multiply variables.} \\
&= 15x^{1+1} && \text{Multiply variables using the property } x^m \cdot x^n = x^{m+n}. \\
&= 15x^2 && \text{Simplify the exponent.}
\end{aligned}
$$

Quick Check 1
Multiply $7x \cdot 8x$.

EXAMPLE 2   Multiply $2x^4y^3 \cdot 3x^2y^7z^4$.

Solution

The procedure for multiplying monomials containing more than one variable is the same as multiplying monomials containing a single variable. After multiplying the coefficients, we multiply the variables one variable at a time.

$$2x^4y^3 \cdot 3x^2y^7z^4 = 6x^6y^{10}z^4 \qquad \text{Multiply coefficients, then multiply variables.}$$

Notice that the variable z was only a factor of one of the monomials. The exponent of z was not changed.

Quick Check 2
Multiply $5x^6yz^7 \cdot 12x^8z^6$.

Multiplying a Monomial by a Polynomial

Objective 2 **Multiply a monomial by a polynomial.** We now advance to multiplying a monomial by a polynomial containing two or more terms, such as $3x(2x^2 - 5x + 2)$. To do this, we use the distributive property $a(b + c) = ab + ac$. To find the product

$3x(2x^2 - 5x + 2)$, we multiply the monomial $3x$ by each term of the polynomial $2x^2 - 5x + 2$.

Multiplying a Monomial by a Polynomial

> To multiply a monomial by a polynomial containing two or more terms, multiply the monomial by each term of the polynomial.

EXAMPLE ▶3 Multiply $3x(2x^2 - 5x + 2)$.

Solution

We begin by distributing the monomial $3x$ to each term of the polynomial.

$$3x(2x^2 - 5x + 2) = 3x \cdot 2x^2 - 3x \cdot 5x + 3x \cdot 2 \qquad \text{Distribute } 3x.$$
$$= 6x^3 - 15x^2 + 6x \qquad \text{Multiply.}$$

Quick Check **3**

Multiply
$4x(7x^3 - 6x^2 + 5x - 8).$

Although we will continue to show the distribution of the monomial, your goal should be to perform this task mentally.

EXAMPLE ▶4 Multiply $-8x^3(x^5 - 4x^4 - 9x^2)$.

Solution

Notice that the coefficient of the monomial being distributed is negative. We must distribute $-8x^3$, and multiplying by this negative term changes the sign of each term in the polynomial.

$$-8x^3(x^5 - 4x^4 - 9x^2) = (-8x^3)(x^5) - (-8x^3)(4x^4) - (-8x^3)(9x^2)$$
$$\text{Distribute } -8x^3.$$
$$= -8x^8 - (-32x^7) - (-72x^5) \qquad \text{Multiply.}$$
$$= -8x^8 + 32x^7 + 72x^5 \qquad \text{Simplify.}$$

Quick Check **4** Multiply $-3x^5(-2x^4 + x^3 - x^2 - 15x + 21).$

> **A Word of Caution** When multiplying a polynomial by a term with a negative coefficient, be sure to change the sign of each term in the polynomial.

Multiplying Polynomials

Objective 3 Multiply polynomials.

Multiplying Two Polynomials

> To multiply two polynomials when each contains two or more terms, we multiply each term in the first polynomial by each term in the second polynomial.

Suppose that we wanted to multiply $(x + 9)(x + 7)$. We could distribute the factor $(x + 9)$ to each term in the second polynomial as follows:

$$(x + 9)(x + 7) = (x + 9) \cdot x + (x + 9) \cdot 7$$

We could then perform the two multiplications by first distributing x in the first product and then distributing 7 in the second product.

$$(x + 9)(x + 7) = (x + 9) \cdot x + (x + 9) \cdot 7$$
$$= x \cdot x + 9 \cdot x + x \cdot 7 + 9 \cdot 7$$
$$= x^2 + 9x + 7x + 63$$
$$= x^2 + 16x + 63$$

We end up with each term in the first polynomial being multiplied by each term in the second polynomial.

EXAMPLE 5   Multiply $(x + 6)(x + 4)$.

Solution

We begin by taking the first term, x, in the first polynomial and multiplying it by each term in the second polynomial. We then repeat this for the second term, 6, in the first polynomial.

$$(x + 6)(x + 4) = x \cdot x + x \cdot 4 + 6 \cdot x + 6 \cdot 4$$

Distribute the term x from the first polynomial; then distribute the 6 from the first polynomial.

$$= x^2 + 4x + 6x + 24$$ Multiply.
$$= x^2 + 10x + 24$$ Combine like terms.

Quick Check 5
Multiply
$(x + 11)(x + 9)$.

We often refer to the process of multiplying a binomial by another binomial as "FOIL." FOIL is an acronym for **F**irst, **O**uter, **I**nner, **L**ast, which describes the four multiplications that occur when we multiply two binomials. Here are the four multiplications that were performed in the previous example:

First	**O**uter	**I**nner	**L**ast
$x \cdot x$	$x \cdot 4$		
$(x + 6)(x + 4)$	$(x + 6)(x + 4)$	$(x + 6)(x + 4)$	$(x + 6)(x + 4)$
		$6 \cdot x$	$6 \cdot 4$

The term FOIL applies only when multiplying a binomial by another binomial. If we are multiplying a binomial by a trinomial, there are six multiplications that must be performed and the term FOIL cannot be used. Keep in mind that each term in the first polynomial must be multiplied by each term in the second polynomial.

EXAMPLE 6   Multiply $(x - 6)(4x - 5)$.

Solution

When multiplying polynomials we must be careful with our signs. When we distribute the second term of $x - 6$ to the second polynomial, we must distribute a negative 6.

$$(x - 6)(4x - 5) = x \cdot 4x - x \cdot 5 - 6 \cdot 4x + 6 \cdot 5$$

Distribute (FOIL). The product of two negative numbers is positive: $(-6)(-5) = 6 \cdot 5$.

Quick Check 6
Multiply
$(5x - 2)(2x - 9)$.

$$= 4x^2 - 5x - 24x + 30$$ Multiply.
$$= 4x^2 - 29x + 30$$ Combine like terms.

EXAMPLE 7   Given the functions $f(x) = 3x - 2$ and $g(x) = x^2 + 4x - 7$, find $f(x) \cdot g(x)$.

Solution

We will substitute the appropriate expressions for $f(x)$ and $g(x)$ in parentheses, and simplify the resulting expression. We need to multiply each term in the first polynomial by each term in the second polynomial.

$$f(x) \cdot g(x) = (3x - 2)(x^2 + 4x - 7) \qquad \text{Substitute for } f(x)$$
$$\text{and } g(x).$$
$$= 3x \cdot x^2 + 3x \cdot 4x - 3x \cdot 7 - 2 \cdot x^2 - 2 \cdot 4x + 2 \cdot 7 \qquad \text{Distribute.}$$
$$= 3x^3 + 12x^2 - 21x - 2x^2 - 8x + 14 \qquad \text{Multiply.}$$
$$= 3x^3 + 10x^2 - 29x + 14 \qquad \text{Combine like terms.}$$

Quick Check 7
Given the functions
$f(x) = x + 8$ and
$g(x) = 3x^2 + 6x - 2$,
find $f(x) \cdot g(x)$.

Special Products

Objective 4   Find special products.   We finish this section by examining some special products. The first special product is of the form $(a + b)(a - b)$. In words, this is the product of the sum and the difference of two terms. Here is the multiplication:

$$(a + b)(a - b) = a^2 - ab + ab - b^2 \qquad \text{Distribute.}$$
$$= a^2 - b^2 \qquad \text{Combine like terms.}$$

Notice that when we were combining like terms, two of the terms were opposites. This left us with only two terms. We can use this result to assist us any time we multiply two binomials of the form $(a + b)(a - b)$.

$$(a + b)(a - b) = a^2 - b^2$$

EXAMPLE 8   Multiply $(x + 7)(x - 7)$.

Solution

$$(x + 7)(x - 7) = x^2 - 7^2 \qquad \text{Multiply using the pattern}$$
$$(a + b)(a - b) = a^2 - b^2.$$
$$= x^2 - 49 \qquad \text{Simplify.}$$

Quick Check 8
Multiply
$(x + 10)(x - 10)$.

EXAMPLE 9   Multiply $(3x - 5)(3x + 5)$.

Solution

Although the difference is listed first, we can still multiply using the same pattern.

$$(3x - 5)(3x + 5) = (3x)^2 - 5^2 \qquad \text{Multiply using the pattern}$$
$$(a + b)(a - b) = a^2 - b^2.$$
$$= 9x^2 - 25 \qquad \text{Simplify.}$$

Quick Check 9
Multiply
$(2x + 7)(2x - 7)$.

The other special product we will examine is the square of a binomial, such as $(x + 3)^2$ or $(9x + 4)^2$. Here are the patterns for squaring binomials of the form $a + b$ and $a - b$:

$$(a + b)^2 = a^2 + 2ab + b^2$$
$$(a - b)^2 = a^2 - 2ab + b^2$$

Let's derive the first of these patterns.

$$(a + b)^2 = (a + b)(a + b)$$ To square a binomial, we multiply it by itself.
$$= a^2 + ab + ab + b^2$$ Distribute (FOIL).
$$= a^2 + 2ab + b^2$$ Combine like terms.

The second pattern can be derived in the same fashion.

EXAMPLE 10 Multiply $(x - 3)^2$.

Solution

We can use the pattern $(a - b)^2 = a^2 - 2ab + b^2$, substituting x for a and 3 for b.

$$a^2 - 2ab + b^2$$
$$x^2 - 2 \cdot x \cdot 3 + 3^2$$ Substitute x for a and 3 for b.
$$= x^2 - 6x + 9$$ Multiply.

Quick Check **10**
Multiply $(5x - 8)^2$.

EXAMPLE 11 Multiply $(2x + 5)^2$.

Solution

We can use the pattern $(a + b)^2 = a^2 + 2ab + b^2$, substituting $2x$ for a and 5 for b.

$$a^2 + 2ab + b^2$$
$$(2x)^2 + 2 \cdot (2x) \cdot 5 + 5^2$$ Substitute $2x$ for a and 5 for b.
$$= 4x^2 + 20x + 25$$ Multiply.

Quick Check **11**
Multiply $(x + 6)^2$.

Although the patterns developed for these three special products may help to save time when multiplying, keep in mind that we can find products of these types by multiplying as we did earlier in this section. When we square a binomial, we can start by rewriting the expression as the product of the binomial and itself. For example, we could rewrite $(8x - 7)^2$ as $(8x - 7)(8x - 7)$ and then multiply.

Building Your Study Strategy **Test Taking, 4 Easier Problems First** When you take a test, work on the easiest problems first, saving the more difficult problems for later. One benefit to this approach is that you will gain confidence as you progress through the test, so you are confident when you attempt to solve a difficult problem. Students who struggle on a problem on the test may lose confidence and miss later problems that they know how to do. By completing the easier problems quickly first, you will save time to work on the few difficult problems.

One exception to this practice is to begin the test by first attempting the most difficult type of problem for you. For example, if a particular type of word problem has been difficult for you to solve throughout your preparation for the exam, you may want to focus on how to solve that type of problem just before arriving for the test. With the solution of this type of problem committed to your short-term memory, your best opportunity for success comes at the beginning of the test prior to solving any other problems.

Vocabulary

1. To multiply a monomial by another monomial, multiply the coefficients and _____ the exponents of the variable factors.

2. To multiply a monomial by a polynomial, _____ the monomial to each term in the polynomial.

3. To multiply a polynomial by another polynomial, _____ each term in the first polynomial by each term in the second polynomial.

4. The acronym _____ can be used when multiplying a binomial by another binomial.

Multiply.

5. $5x \cdot 7x^3$

6. $8x^2 \cdot 3x^2$

7. $-6x^5 \cdot 3x^4$

8. $-2x^6(-11x^9)$

9. $12x^3y^4 \cdot 8x^5y^2$

10. $7x^{10}y^7 \cdot 6x^{13}y^9$

11. $-5x^3y^2z^4 \cdot 14xz^5w^4$

12. $3a^5b^8c \cdot 9bc^7d^4$

13. $2x^5 \cdot 6x^3 \cdot 7x^2$

14. $-12x^8 \cdot 5x^{10} \cdot 6x^9$

Find the missing monomial.

15. $3x^4 \cdot ? = 15x^7$

16. $8x^6 \cdot ? = 56x^{11}$

17. $2x^9 \cdot ? = -24x^{18}$

18. $-6x^{10} \cdot ? = 84x^{23}$

Multiply.

19. $2(5x - 8)$

20. $7(3x + 2)$

21. $-4(2x - 9)$

22. $-3(4x + 7)$

23. $6x(3x + 5)$

24. $9x(4x - 13)$

25. $x^3(3x^2 - 4x + 7)$

26. $-x^3(2x^5 + 9x^4 - 6x)$

27. $2xy^2(3x^2 - 6xy + 7y^2)$

28. $11x^4y^3(2xy - 5x^4y^2 - 7x)$

Multiply.

29. $(x + 2)(x + 3)$

30. $(x + 6)(x - 2)$

31. $(x - 9)(x - 10)$

32. $(x - 3)(x - 8)$

33. $(2x + 1)(x - 5)$

34. $(5x - 4)(2x + 5)$

35. $(3x + 2)(x^2 - 5x - 9)$

36. $(7x - 4)(49x^2 + 28x + 16)$

37. $(x^2 - 7x + 10)(x^2 + 3x - 40)$

38. $(2x^2 - 5x - 8)(x^2 + 3x + 9)$

39. $(x + 2y)(x - 4y)$

40. $(x - 6y)(x - 7y)$

Find the missing factor.

41. $?(2x^2 - 7x - 10) = 6x^5 - 21x^4 - 30x^3$

42. $?(3x^4 + 9x^2 + 16) = 18x^9 + 54x^7 + 96x^5$

43. $4x^4(?) = 12x^7 - 20x^6 - 48x^4$

44. $8x^9(?) = 56x^{17} + 104x^{14} - 96x^{11}$

45. $(x + ?)(x + 5) = x^2 + 8x + 15$

46. $(x - 7)(x + ?) = x^2 + 3x - 70$

47. $(x + ?)(x + ?) = x^2 + 9x + 18$

48. $(? - 5)(? + 3) = 2x^2 + x - 15$

For the given functions $f(x)$ and $g(x)$, find $f(x) \cdot g(x)$.

49. $f(x) = x - 9, g(x) = x + 2$

50. $f(x) = 4x^3, g(x) = x^2 - 8x - 14$

51. $f(x) = 6x^5$, $g(x) = -3x^4$

52. $f(x) = 2x^7$, $g(x) = 11x^6$

53. $f(x) = x^3 - 5x^2 + 8x + 3$, $g(x) = -4x^5$

54. $f(x) = 7x + 4$, $g(x) = 6x - 13$

Find the special product using the appropriate formula.

55. $(x + 8)(x - 8)$

56. $(x - 13)(x + 13)$

57. $(2x + 5)(2x - 5)$

58. $(4x - 9)(4x + 9)$

59. $(x + 6)^2$

60. $(x - 1)^2$

61. $(2x - 3)^2$

62. $(4x + 7)^2$

Find the missing factor or term.

63. $(x + 9) \cdot ? = x^2 - 81$

64. $(5x + 7) \cdot ? = 25x^2 - 49$

65. $(?)^2 = x^2 - 12x + 36$

66. $(?)^2 = 4x^2 + 20x + 25$

Mixed Practice, 67–84

Multiply.

67. $(x + 7)(x - 1)$

68. $2x^5 \cdot 3x^2$

69. $5x(x^2 - 8x - 9)$

70. $(x + 2)(x - 2)$

71. $-7x^2y^3 \cdot 4x^4y^6$

72. $(x - 7)^2$

73. $-5x^6(-4x^8)$

74. $(2x + 15)(3x - 7)$

75. $(5x + 3)(5x - 3)$

76. $-3x^2(x^2 - 10x - 13)$

77. $(x + 13)^2$

78. $(x + 8)(3x^2 + 7x - 6)$

79. $2x^2y(3x^2 - x^4y^3 - 7y)$

80. $9x^3(-6x^3)$

81. $(4x - 9)^2$

82. $(6 + 5x)^2$

83. $(x^2 + 3x + 4)(x^2 - 5x + 4)$

84. $-6x^5(-x^5 + 3x^3 - 11)$

Writing in Mathematics

Answer in complete sentences.

85. Explain the difference between simplifying the expression $8x^2 + 3x^2$ and simplifying the expression $(8x^2)(3x^2)$. Discuss how the coefficients and exponents are handled differently.

86. Explain why the term "FOIL" applies only when we multiply a binomial by another binomial and not when we multiply polynomials that have more than two terms. Explain how to multiply a binomial by a trinomial.

87. To simplify the expression $(x + 8)^2$, we can either multiply $(x + 8)(x + 8)$ or use the special product formula $(a + b)^2 = a^2 + 2ab + b^2$. Which method do you prefer? Explain your answer.

88. *Solutions Manual** Write a solutions manual page for the following problem:

Simplify $(2x - 3)(5x^2 + 4x - 9)$.

89. *Newsletter** Write a newsletter explaining how to multiply two binomials.

See Appendix B for details and sample answers.

5.5
Dividing Polynomials

1. **Divide a monomial by a monomial.**
2. **Divide a polynomial by a monomial.**
3. **Divide a polynomial by a polynomial by using long division.**
4. **Use placeholders when dividing a polynomial by a polynomial.**

Dividing Monomials by Monomials

Objective 1 Divide a monomial by a monomial. In this section, we will learn to divide a polynomial by another polynomial. We will begin by reviewing how to divide a monomial by another monomial, such as $\dfrac{4x^5}{2x^2}$ or $\dfrac{30a^5b^4}{6ab^2}$.

Dividing a Monomial by a Monomial

> To divide a monomial by another monomial, we divide the coefficients first. Then we divide the variables, using the quotient rule $\dfrac{x^m}{x^n} = x^{m-n}$, assuming that no variable in the denominator is equal to 0.

EXAMPLE 1 Divide $\dfrac{18x^5}{3x^3}$. (Assume $x \neq 0$.)

Solution

$$\frac{18x^5}{3x^3} = 6x^{5-3} \qquad \text{Divide coefficients. Subtract exponents of } x.$$
$$= 6x^2 \qquad \text{Simplify the exponent.}$$

We can check our quotients by using multiplication. If $\dfrac{18x^5}{3x^3} = 6x^2$, then we know that $3x^3 \cdot 6x^2$ should equal $18x^5$.

Quick Check 1
Divide $\dfrac{32x^9}{4x^4}$.
(Assume $x \neq 0$.)

Check:

$$3x^3 \cdot 6x^2 = 18x^{3+2} \qquad \text{Multiply coefficients. Keep the base and add the exponents.}$$
$$= 18x^5 \qquad \text{Simplify the exponent.}$$

Our quotient of $6x^2$ checks.

EXAMPLE 2 Divide $\dfrac{24x^3y^2}{3xy^2}$. (Assume $x, y \neq 0$.)

Solution

Quick Check 2
Divide $\dfrac{40x^3y^{11}}{8xy^2}$. (Assume $x, y \neq 0$.)

When there is more than one variable, as in this example, we divide the coefficients and then divide the variables one at a time.

$$\frac{24x^3y^2}{3xy^2} = 8x^2y^0 \qquad \text{Divide coefficients and subtract exponents.}$$
$$= 8x^2 \qquad \text{Rewrite without } y \text{ as a factor. } (y^0 = 1)$$

Dividing Polynomials by Monomials

Objective 2 Divide a polynomial by a monomial. Now we move on to dividing a polynomial by a monomial, such as $\dfrac{3x^5 - 9x^3 - 18x^2}{3x}$.

Dividing a Polynomial by a Monomial

To divide a polynomial by a monomial, divide each term of the polynomial by the monomial.

EXAMPLE 3   Divide $\dfrac{15x^2 + 10x - 5}{5}$.

Solution

We will divide each term in the numerator by 5.

$$\frac{15x^2 + 10x - 5}{5} = \frac{15x^2}{5} + \frac{10x}{5} - \frac{5}{5} \qquad \text{Divide each term in the numerator by 5.}$$
$$= 3x^2 + 2x - 1 \qquad \text{Divide.}$$

If $\dfrac{15x^2 + 10x - 5}{5} = 3x^2 + 2x - 1$, then $5(3x^2 + 2x - 1)$ should equal $15x^2 + 10x - 5$. We can use this to check our work.

Quick Check 3

Divide $\dfrac{7x^2 - 21x - 49}{7}$.

Check:
$$5(3x^2 + 2x - 1) = 5 \cdot 3x^2 + 5 \cdot 2x - 5 \cdot 1 \qquad \text{Distribute.}$$
$$= 15x^2 + 10x - 5 \qquad \text{Multiply.}$$

Our quotient of $3x^2 + 2x - 1$ checks.

EXAMPLE 4   Divide $\dfrac{3x^5 - 9x^3 - 18x^2}{3x}$. (Assume $x \neq 0$.)

Solution

Quick Check 4

Divide
$\dfrac{48x^{10} + 12x^7 + 30x^5}{6x^2}$.
(Assume $x \neq 0$.)

In this example, we will divide each term in the numerator by $3x$.

$$\frac{3x^5 - 9x^3 - 18x^2}{3x}$$
$$= \frac{3x^5}{3x} - \frac{9x^3}{3x} - \frac{18x^2}{3x} \qquad \text{Divide each term in the numerator by } 3x.$$
$$= x^4 - 3x^2 - 6x \qquad \text{Divide.}$$

Dividing a Polynomial by a Polynomial (Long Division)

Objective 3 **Divide a polynomial by a polynomial using long division.** To divide a polynomial by another polynomial containing at least two terms, we use a procedure similar to long division. Before outlining this procedure, let's review some of the terms associated with long division.

$$3 \longleftarrow \text{Quotient}$$
$$\text{Divisor} \longrightarrow 2\overline{)6} \longleftarrow \text{Dividend}$$

Suppose that we were asked to divide $\dfrac{x^2 - 10x + 16}{x - 2}$. The polynomial in the numerator is the dividend and the polynomial in the denominator is the divisor. We may rewrite this division as $x - 2\overline{)x^2 - 10x + 16}$. We must be sure to write both the divisor and the dividend in descending order. We perform the division using the following steps:

Division by a Polynomial

1. Divide the term in the dividend with the highest degree by the term in the divisor with the highest degree. Add this result to the quotient.
2. Multiply the monomial obtained in step 1 by the divisor, writing the result underneath the dividend. (Align like terms vertically.)
3. Subtract the product obtained in step 2 from the dividend. (Recall that to subtract a polynomial from another polynomial, we change the signs of its terms and then combine like terms.)
4. Repeat steps 1–3 with the result of step 3 as the new dividend. Keep repeating this procedure until the degree of the new dividend is less than the degree of the divisor.

EXAMPLE 5   Divide $\dfrac{x^2 - 10x + 16}{x - 2}$.

Solution

We begin by writing $x - 2\overline{)x^2 - 10x + 16}$. We divide the term in the dividend with the highest degree (x^2) by the term in the divisor with the highest degree (x). Since $\dfrac{x^2}{x} = x$, we will write x in the quotient and multiply x by the divisor $x - 2$, writing this product underneath the dividend.

$$
\begin{array}{r}
x \\
x - 2\overline{)x^2 - 10x + 16} \\
x^2 - 2x
\end{array}
\qquad \text{Multiply } x(x - 2).
$$

To subtract, we change the signs of the second polynomial and then combine like terms.

$$
\begin{array}{r}
x \\
x - 2\overline{)x^2 - 10x + 16} \\
\underset{-}{x^2} \underset{+}{\not/} 2x \quad \downarrow \\
\hline
-8x + 16
\end{array}
\qquad \text{Change the signs and combine like terms.}
$$

We now begin the process again by dividing $-8x$ by x, which equals -8. Multiply -8 by $x - 2$ and subtract.

$$
\begin{array}{r}
x - 8 \\
x - 2 \overline{)\, x^2 - 10x + 16} \\
\underline{x^2 + 2x} \quad\;\; \downarrow \\
-8x + 16 \\
\underline{+\;8x - 16} \\
0
\end{array}
$$

Multiply -8 by $x - 2$.
Subtract the product.

The remainder of 0 tells us that we are finished, since its degree is less than the degree of the divisor $x - 2$. The expression written above the division box $(x - 8)$ is the quotient.

$$
\frac{x^2 - 10x + 16}{x - 2} = x - 8
$$

We can check our work by multiplying the quotient $(x - 8)$ by the divisor $(x - 2)$, which should equal the dividend $(x^2 - 10x + 16)$.

Check:

$$
\begin{aligned}
(x - 8)(x - 2) &= x^2 - 2x - 8x + 16 \\
&= x^2 - 10x + 16
\end{aligned}
$$

Our division checks; the quotient is $x - 8$.

Quick Check **5** Divide $\dfrac{x^2 + 13x + 36}{x + 4}$.

In the previous example, the remainder of 0 also tells us that $x - 2$ divides into $x^2 - 10x + 16$ evenly, so $x - 2$ is a **factor** of $x^2 - 10x + 16$. The quotient, $x - 8$, is also a factor of $x^2 - 10x + 16$.

In the next example, we will learn how to write a quotient when there is a nonzero remainder.

EXAMPLE ▸ 6   Divide $x^2 - 10x + 9$ by $x - 3$.

Solution

Since the divisor and dividend are already written in descending order, we may begin to divide.

$$
\begin{array}{r}
x - 7 \\
x - 3 \overline{)\, x^2 - 10x + 9} \\
\underline{x^2 + 3x} \quad\;\; \downarrow \\
-7x + \;\; 9 \\
\underline{+\;7x - 21} \\
-12
\end{array}
$$

Multiply x by $x - 3$ and subtract.

Multiply -7 by $x - 3$ and subtract.

The remainder is -12. After the quotient, we write a fraction with the remainder in the numerator and the divisor in the denominator. Since the remainder is negative, we subtract this fraction from the quotient. If the remainder had been positive, we would add this fraction to the quotient.

$$\frac{x^2 - 10x + 9}{x - 3} = x - 7 - \frac{12}{x - 3}$$

Quick Check 6 Divide $\dfrac{x^2 - 3x - 8}{x + 7}$.

EXAMPLE 7 Divide $6x^2 + 17x + 17$ by $2x + 3$.

Solution

Since the divisor and dividend are already written in descending order, we may begin to divide.

$$
\begin{array}{r}
3x + 4 \\
2x + 3\overline{\smash{)}6x^2 + 17x + 17} \\
\underline{6x^2 + 9x}\phantom{{}+17} \downarrow \\
8x + 17 \\
\underline{8x + 12} \\
5
\end{array}
$$

Multiply $3x$ by $2x + 3$ and subtract.

Multiply 4 by $2x + 3$ and subtract.

The remainder is 5. We write the quotient, and add a fraction with the remainder in the numerator and the divisor in the denominator.

$$\frac{6x^2 + 17x + 17}{2x + 3} = 3x + 4 + \frac{5}{2x + 3}$$

Quick Check 7 Divide $\dfrac{12x^2 - 7x + 4}{3x - 4}$.

Using Placeholders when Dividing a Polynomial by a Polynomial

Objective 4 Use placeholders when dividing a polynomial by a polynomial.
Suppose that we wanted to divide $x^3 - 12x - 11$ by $x - 3$. Notice that the dividend is missing an x^2 term. When this is the case, we add the term $0x^2$ as a **placeholder.** We add placeholders to dividends that are missing terms of a particular degree.

EXAMPLE 8   Divide $(x^3 - 12x - 11) \div (x - 3)$.

Solution

The degree of the dividend is 3, and each degree lower than 3 must be represented in the dividend. We will add the term $0x^2$ as a placeholder. The divisor does not have any missing terms.

$$
\begin{array}{r}
x^2 + 3x - 3 \\
x - 3\overline{\smash{)}x^3 + 0x^2 - 12x - 11} \\
\end{array}
$$

Multiply x^2 by $x - 3$ and subtract.

Multiply $3x$ by $x - 3$ and subtract.

Multiply -3 by $x - 3$ and subtract.

$$
(x^3 - 12x - 11) \div (x - 3) = x^2 + 3x - 3 - \frac{20}{x - 3}.
$$

Quick Check 8

Divide $\dfrac{x^3 - 3x^2 - 9}{x - 2}$.

Building Your Study Strategy Test Taking, 5 **Review Your Test**   Try to leave yourself enough time to review the test at the end of the test period. Check for careless errors, which can cost you a fair number of points. Also be sure that your answers make sense within the context of the problem. For example, if the question asks how tall a person is and your answer is 68 feet tall, then chances are that something has gone astray.

Check for problems, or parts of problems, that you may have skipped and left blank. There is no worse feeling in a math class than getting a test back and realizing that you simply forgot to do one or more of the problems.

Take all of the allotted time to review your test. There is no reward for turning in a test early, and the more that you work on the test, the more likely it is that you will find a mistake or a problem where you can gain points. Keep in mind that a majority of the students who turn in tests early do so because they are not prepared and cannot do several of the problems.

EXERCISES 5.5

Vocabulary

1. To divide a polynomial by a monomial, divide each _____ by the monomial.

2. To divide a polynomial by another polynomial containing at least two terms, we use _____.

3. If one polynomial divides evenly into a second polynomial, then the first polynomial is a(n) _____ of the second polynomial.

4. If the polynomial in the dividend is missing a term of a particular degree, then we use a(n) _____ in the dividend.

Divide. Assume all variables are nonzero.

5. $\dfrac{30x^5}{6x^3}$

6. $\dfrac{14x^8}{7x^2}$

7. $\dfrac{-55x^{11}}{5x}$

8. $\dfrac{24x^9}{8x^8}$

9. $\dfrac{34x^5y^4}{2xy^3}$

10. $\dfrac{20x^9y^9}{4x^9y^3}$

11. $(15x^6) \div (3x^2)$

12. $(70x^{13}) \div (10x^5)$

13. $\dfrac{12x^6}{8x^4}$

14. $\dfrac{-35x^9}{25x^5}$

15. $\dfrac{7x^{12}}{21x^4}$

16. $\dfrac{3x^{15}}{18x^7}$

Find the missing monomial. Assume $x \neq 0$.

17. $\dfrac{?}{5x^4} = 2x^7$

18. $\dfrac{?}{3x^6} = -15x^8$

19. $\dfrac{24x^9}{?} = -6x^4$

20. $\dfrac{20x^9}{?} = \dfrac{5x^4}{3}$

Divide. Assume all variables are nonzero.

21. $\dfrac{24x^2 + 28x - 16}{4}$

22. $\dfrac{3x^2 - 33x}{3}$

23. $\dfrac{5x^4 + 35x^3 - 45x^2 + 15x}{5x}$

24. $\dfrac{22x^5 + 26x^4 - 14x^2}{2x}$

25. $(6x^7 + 9x^6 + 15x^5) \div (3x)$

26. $(-14x^8 + 21x^6 + 7x^4) \div (7x)$

27. $\dfrac{20x^6 - 30x^4}{-10x^2}$

28. $\dfrac{-8x^5 + 2x^4 - 6x^3 - 2x^2}{-2x^2}$

29. $\dfrac{x^6y^6 - x^4y^5 + x^2y^4}{xy^2}$

30. $\dfrac{2a^{10}b^7 - 8a^3b^6 - 10a^5b}{2a^3b}$

Find the missing dividend or divisor. Assume $x \neq 0$.

31. $\dfrac{24x^3 - 48x^2 + 36x}{?} = 4x^2 - 8x + 6$

32. $\dfrac{10x^7 - 50x^5 + 35x^3}{?} = 2x^5 - 10x^3 + 7x$

33. $\dfrac{?}{2x^3} = 3x^4 - 4x^2 - 9$

34. $\dfrac{?}{7x^5} = -x^2 - 9x + 1$

Divide using long division.

35. $\dfrac{x^2 + 10x + 21}{x + 7}$

36. $\dfrac{x^2 + 11x + 18}{x + 2}$

37. $\dfrac{x^2 - 7x - 30}{x - 10}$

38. $\dfrac{x^2 + x - 30}{x - 5}$

39. $(x^2 - 15x + 56) \div (x - 7)$

40. $(x^2 - 22x + 119) \div (x - 13)$

41. $\dfrac{x^2 - 4x - 29}{x - 8}$

42. $\dfrac{x^2 + 7x - 7}{x + 5}$

43. $\dfrac{x^3 - 11x^2 - 37x + 14}{x + 1}$

44. $\dfrac{x^3 - x^2 - 24x - 19}{x + 3}$

45. $\dfrac{2x^2 + 3x - 32}{x + 5}$

46. $\dfrac{3x^2 + 26x + 24}{x + 9}$

47. $\dfrac{12x^2 - 11x + 15}{3x - 5}$

48. $\dfrac{14x^2 - 3x - 39}{7x + 2}$

49. $\dfrac{x^2 - 100}{x - 10}$

50. $(4x^2 - 121) \div (2x + 11)$

51. $\dfrac{x^4 + 2x^2 - 15x + 32}{x - 3}$

52. $\dfrac{x^5 - 9x^3 - 17x^2 + x - 15}{x + 8}$

53. $\dfrac{x^3 - 125}{x - 5}$

54. $\dfrac{x^3 + 343}{x + 7}$

Find the missing dividend or divisor.

55. $\dfrac{?}{x + 8} = x - 3$

56. $\dfrac{?}{x - 5} = x - 11$

57. $\dfrac{x^2 + 10x - 39}{?} = x - 3$

58. $\dfrac{6x^2 - 35x + 49}{?} = 3x - 7$

59. Is $x + 9$ a factor of $x^2 + 28x + 171$?

60. Is $x - 6$ a factor of $x^2 + 4x - 54$?

61. Is $2x - 5$ a factor of $4x^2 - 8x - 15$?

62. Is $3x + 4$ a factor of $9x^2 + 24x + 16$?

Mixed Practice, 63–80

Divide. Assume all denominators are nonzero.

63. $\dfrac{x^2 - 8x + 9}{x - 3}$

64. $\dfrac{x^2 + 3x - 15}{x + 7}$

65. $\dfrac{x^3 + 8x - 19}{x + 4}$

66. $\dfrac{8x^9}{2x^3}$

67. $\dfrac{12x^6 - 8x^4 + 20x^3 - 4x^2}{4x^2}$

68. $\dfrac{x^2 - 4x - 192}{x + 12}$

69. $\dfrac{8x^2 + 42x - 25}{x + 6}$

70. $\dfrac{4x^2 - 25x - 100}{x - 8}$

71. $\dfrac{20x^2 - 51x - 6}{4x - 3}$

72. $\dfrac{6x^2 - 19x + 30}{2x - 5}$

73. $\dfrac{27x^9 - 18x^7 - 6x^6 - 3x^3}{-3x^2}$

74. $\dfrac{6x^2 + x - 176}{3x - 16}$

75. $\dfrac{8x^3 - 86x - 63}{4x + 10}$

76. $\dfrac{16a^4b^8c^7}{4ab^7c^7}$

77. $\dfrac{6x^3 - 37x^2 + 11x + 153}{2x - 9}$

78. $\dfrac{6x^5 - 12x^4 + 18x^3}{4x^3}$

79. $\dfrac{21x^9y^6z^{17}}{-7x^7y^5z^{12}}$

80. $\dfrac{x^7 + 127}{x + 2}$

Writing in Mathematics

Answer in complete sentences.

81. *Solutions Manual*[*] Write a solutions manual page for the following problem:

$$\text{Divide } \dfrac{6x^2 - 17x - 19}{2x - 7}.$$

82. *Newsletter*[*] Write a newsletter explaining long division of polynomials using placeholders.

*See Appendix B for details and sample answers.

Chapter 5 Summary

Section 5.1—Topic	Chapter Review Exercises
Simplifying Expressions by Using the Rules of Exponents	1–10

Section 5.2—Topic	Chapter Review Exercises
Simplifying Expressions with Negative Exponents	11–24
Rewriting Numbers by Using Scientific Notation	25–27
Rewriting Numbers That Are in Scientific Notation	28–30
Performing Calculations Involving Scientific Notation	31–32
Solving Applied Problems by Using Scientific Notation	33–34

Section 5.3—Topic	Chapter Review Exercises
Evaluating Polynomials	35–40
Adding and Subtracting Polynomials	41–46
Evaluating Polynomial Functions	47–50
Adding and Subtracting Polynomial Functions	51–52

Section 5.4—Topic	Chapter Review Exercises
Multiplying Polynomials	53–68
Multiplying Polynomial Functions	69–70

Section 5.5—Topic	Chapter Review Exercises
Dividing Polynomials	71–80

Summary of Chapter 5 Study Strategies

Attending class each day and doing all of your homework does not guarantee that you will earn a good grade when you take the exam. Through careful preparation and the adoption of the test-taking strategies introduced in this chapter, you can maximize your grade on a math exam. Here is a summary of the points that have been introduced:

- Make the most of your time before the exam.
- Write down important facts as soon as you get your test.
- Briefly read through the test before you work any problems.
- Begin by solving the easier problems first.
- Review your test as thoroughly as possible before turning it in.

Simplify the expression. Write the result without using negative exponents. (Assume all variables represent nonzero real numbers.) [5.1–5.2]

1. $\dfrac{x^9}{x^3}$

2. $(4x^7)^3$

3. $7x^0$

4. $\left(\dfrac{5x^9}{2y^2}\right)^4$

5. $7x^7 \cdot 4x^{12}$

6. $3x^8y^5 \cdot 8x^8y^3$

7. $(-6a^5b^3c^6)^2$

8. $15^0 - 3x^0$

9. $(x^{10}y^{13}z^4)^4$

10. $\dfrac{x^{16}}{x^6}$

11. 5^{-2}

12. 10^{-3}

13. $7x^{-4}$

14. $-6x^{-6}$

15. $\dfrac{-8}{y^{-5}}$

16. $(3x^{-4})^3$

17. $(2x^{-5})^{-2}$

18. $\dfrac{x^{10}}{x^{-5}}$

19. $\left(\dfrac{a^{-4}}{b^{-7}}\right)^3$

20. $m^{13} \cdot m^{-19}$

21. $(4x^{-5}y^4z^{-7})^{-3}$

22. $\left(\dfrac{x^{-11}}{y^{-6}}\right)^6$

23. $a^{-15} \cdot a^{-5}$

24. $\dfrac{x^{-20}}{x^{-9}}$

Rewrite in scientific notation. [5.2]

25. $1{,}400{,}000{,}000$

26. 0.0000000000021

27. 0.000005002

Convert from scientific to standard notation. [5.2]

28. 1.23×10^{-4}

29. 4.075×10^7

30. 6.1275×10^{12}

Perform the following calculations. Express your answer using scientific notation. [5.2]

31. $(2.5 \times 10^7)(6.0 \times 10^6)$

32. $(3.0 \times 10^{-13}) \div (1.2 \times 10^5)$

33. The mass of a hydrogen atom is 1.66×10^{-24} grams. What is the mass of 1,500,000,000,000 hydrogen atoms? [5.2]

34. If a computer can perform a calculation in 0.000000000005 second, how many seconds would it take to perform 3,000,000,000,000,000 calculations? [5.2]

Evaluate the polynomial for the given value of the variable. [5.3]

35. $x^2 - 8x + 15$ for $x = -5$

36. $x^2 - 11x - 29$ for $x = 2$

37. $-x^2 + 16x - 30$ for $x = 6$

38. $-3x^2 - 4x + 7$ for $x = -7$

39. $5x^2 + 10x + 12$ for $x = -3$

40. $7x^2 - 14x + 35$ for $x = 8$

Worked-out solutions to Review Exercises marked with can be found on page AN–19.

Add or subtract. [5.3]

41. $(x^2 + 3x - 15) + (x^2 - 9x + 6)$

42. $(3x^2 - 31) + (5x^2 - 19x - 19)$

43. $(x^2 - 6x - 13) - (x^2 - 14x + 8)$

44. $(2x^2 + 8x - 7) - (x^2 - x + 15)$

45. $(3x^3 + x^2 - 10) - (6x^2 - 13x + 20)$

46. $(10x^2 - 21x - 35) - (4x^3 + 6x^2 - 15x + 80)$

Evaluate the given polynomial function. [5.3]

47. $f(x) = x^2 - 25, f(-4)$

48. $f(x) = 2x^2 - 6x + 17, f(5)$

49. $f(x) = x^2 - 7x + 10, f(a^4)$

50. $f(x) = x^2 + 3x - 30, f(2a^3)$

Given the functions $f(x)$ and $g(x)$, find $f(x) + g(x)$ and $f(x) - g(x)$. [5.3]

51. $f(x) = 3x^2 - 8x - 9, g(x) = 7x^2 + 6x + 40$

52. $f(x) = x^2 + 5x - 30, g(x) = -8x^2 - 15x + 6$

Multiply. [5.4]

53. $(x - 5)^2$

54. $(x - 4)(x + 13)$

55. $4x(3x^2 - 7x - 16)$

56. $(2x - 7)^2$

57. $5x^7 \cdot 3x^6$

58. $(x - 8)(x + 8)$

59. $(3x + 10)(2x - 13)$

60. $-9x^5 \cdot 2x$

61. $(6x + 5)(6x - 5)$

62. $(x + 10)^2$

63. $(4x + 7)^2$

64. $-8x^2(3x^5 - 7x^4 - 6x^3)$

65. $(x + 9)(x^2 - 6x + 12)$

66. $(x - 2)(3x^2 - 5x - 9)$

67. $(10x^5)(-3x^7)(4x^6)$

68. $-2x^4(-5x^2 - 11x + 12)$

Given the functions $f(x)$ and $g(x)$, find $f(x) \cdot g(x)$. [5.4]

69. $f(x) = x + 10, g(x) = x - 7$

70. $f(x) = -6x^6, g(x) = -4x^2 - 9x + 20$

Divide. Assume all denominators are nonzero. [5.5]

71. $\dfrac{6x^4 - 8x^3 + 20x^2}{2x^2}$

72. $\dfrac{3x^2 - 8x - 25}{x - 4}$

73. $\dfrac{4x^5y^7}{-2x^5y}$

74. $\dfrac{x^3 + 3x^2 - 15}{x + 5}$

75. $\dfrac{x^4 + 81}{x + 3}$

76. $\dfrac{18x^3y^7z^6}{3x^3y^6z}$

77. $\dfrac{8x^2 + 87x + 171}{x + 9}$

78. $\dfrac{6x^2 - 30x - 84}{x - 7}$

79. $\dfrac{15x^2 - 43x - 238}{5x + 14}$

80. $\dfrac{16x^2 + 2x + 20}{8x - 3}$

Simplify the expression. Write the result without using negative exponents. (Assume all variables represent nonzero real numbers.)

1. $\dfrac{x^{10}}{x^2}$

2. $\left(\dfrac{x^5}{3y^4}\right)^6$

3. $(x^9 y^5 z^8)^8$

4. 4^{-4}

5. $(2x^{-7})^5$

6. $\dfrac{x^4}{x^{-13}}$

7. $x^{22} \cdot x^{-15}$

8. $(3xy^{-6}z^5)^{-2}$

Rewrite in scientific notation.

9. 23,500,000

Convert to standard notation.

10. 4.7×10^{-8}

Perform the following calculations. Express your answer using scientific notation.

11. $(4.3 \times 10^{13})(1.8 \times 10^{-9})$

Evaluate the polynomial for the given value of the variable.

12. $x^2 + 12x - 38$ for $x = -8$

Add or subtract.

13. $(x^2 - 6x - 32) - (5x^2 + 8x - 33)$

14. $(4x^2 - 3x - 15) + (-2x^2 + 17x - 49)$

Evaluate the given polynomial function.

15. $f(x) = x^2 + 3x - 14$, $f(-7)$

Multiply.

16. $5x^3(6x^2 + 8x - 17)$

17. $(x + 6)(x - 6)$

18. $(5x - 9)(4x + 7)$

Divide. Assume all denominators are nonzero.

19. $\dfrac{21x^7 + 33x^6 - 15x^5}{3x^2}$

20. $\dfrac{6x^2 - 11x + 38}{x + 3}$

Mathematicians in History
Carl Friedrich Gauss

$\mathcal{C}$arl Friedrich Gauss was a German mathematician who lived in the 18th and 19th centuries and was one of the most prominent and prolific mathematicians of his day. Gauss made significant contributions to the fields of analysis, probability, statistics, number theory, geometry, and astronomy, among others. The Prussian mathematician Leopold Kronecker once said of him, "Almost everything, which the mathematics of our century has brought forth in the way of original scientific ideas, attaches to the name of Gauss." Gauss once said, "It is not knowledge, but the act of learning, not possession but the act of getting there, which grants the greatest enjoyment."

Write a one-page summary (*or* make a poster) of the life of Carl Friedrich Gauss and his accomplishments. Also, look up Gauss's most famous quotes and list your favorite quote.

Interesting issues

- Where and when was Carl Friedrich Gauss born?
- Gauss displayed incredible genius at an early age. Relay the story of how Gauss found his father's accounting error at only three years of age.
- At what age did Gauss find a shortcut for summing the integers from 1 to 100?
- Gauss's motto was "Few, but ripe." In your own words, explain the meaning of this motto.
- In 1798, Gauss showed how to construct a regular 17-gon by ruler and compass. This was the first major advance in the area of regular polygons in approximately 2000 years. What is a regular 17-gon?
- Gauss earned his doctorate in 1799 with a proof of what major mathematical theorem?
- Gauss gained notoriety in the field of astronomy by accurately predicting the orbit of the newly discovered asteroid Ceres by inventing the method of least squares. Describe the method of least squares.
- In 1821, Gauss built the first heliotrope. What is a heliotrope?
- Describe the circumstances that led to Gauss's death.

At a recent school carnival, George agreed to be launched out of a cannon to become the world's first flying mathematician. He landed after 3.4625 seconds. George's height, in feet, after t seconds is given by the function $h(t) = -16t^2 + 55.4t$.

a) Complete the following chart listing George's height at 0.5-second intervals:

Time (in seconds)	Height (in feet)
0	
0.5	
1	
1.5	
2	
2.5	
3	
3.5	

b) Create a coordinate system in which the horizontal axis represents time (in seconds) and the vertical axis represents height (in feet). Plot these eight points and connect them to see George's course of flight through the air. (Your graph should be a smooth curve.)

Use the table and graph to answer the following questions.

c) What was George's height when he was launched out of the cannon?

d) What was George's height after 0.5 seconds?

e) How long did it take George to reach a height of 39.4 feet?

f) What was George's height when he landed on the ground? How long did it take (from the time of launch) for George to land on the ground?

g) When did George reach his maximum height? What was his maximum height?

Factoring and Quadratic Equations

In this chapter, we will begin to learn how to solve quadratic equations. A quadratic equation is an equation of the form $ax^2 + bx + c = 0$, where a, b, and c are real numbers and $a \neq 0$. To solve a quadratic equation, we will attempt to rewrite the quadratic expression $ax^2 + bx + c$ as a product of two expressions. In other words, we will factor the quadratic expression. The first five sections of this chapter focus on factoring techniques.

Once we have learned how to factor polynomials, we will learn how to solve quadratic equations and to apply quadratic equations to real-world problems. We will also examine quadratic functions, which are functions of the form

$$f(x) = ax^2 + bx + c,$$

where $a \neq 0$.

Study Strategy **Overcoming Math Anxiety** *In this chapter, we will focus on how to overcome math anxiety, a condition shared by many students. Math anxiety can prevent a student from learning mathematics, and it can seriously impact a student's performance on quizzes and exams.*

6.1

An Introduction to Factoring; the Greatest Common Factor; Factoring by Grouping

Objectives

1. Find the greatest common factor (GCF) of two or more integers.
2. Find the GCF of two or more variable terms.
3. Factor the GCF out of each term of a polynomial.
4. Factor a common binomial factor out of a polynomial.
5. Factor a polynomial by grouping.

In the previous chapter, we learned about polynomials; in this section we will begin to learn how to **factor** a polynomial. A polynomial has been **factored** when it is represented as the product of two or more polynomials.

Greatest Common Factor

Objective **1**   Find the greatest common factor (GCF) of two or more integers.
Before we begin to learn how to factor polynomials, we will go over the procedure for finding the **greatest common factor (GCF)** of two or more integers. The GCF of two or more integers is the largest whole number that is a factor of each integer.

Finding the GCF of Two or More Integers

> Find the prime factorization of each integer.
> Write the prime factors that are common to each integer; the GCF is the product of these prime factors.

EXAMPLE 1   Find the GCF of 30 and 42.

Solution

The prime factorization of 30 is $2 \cdot 3 \cdot 5$ and the prime factorization of 42 is $2 \cdot 3 \cdot 7$. (For a refresher on finding the prime factorization of a number, refer back to Section 1.3.) The prime factors that they share in common are 2 and 3, so the GCF is $2 \cdot 3$ or 6. (This means that 6 is the greatest number that divides evenly into both 30 and 42.)

Quick Check **1**
Find the GCF of 16 and 20.

EXAMPLE 2   Find the GCF of 108, 504, and 720.

Solution

We begin with the prime factorizations.

$$108 = 2^2 \cdot 3^3 \qquad 504 = 2^3 \cdot 3^2 \cdot 7 \qquad 720 = 2^4 \cdot 3^2 \cdot 5$$

The only primes that are factors of all three numbers are 2 and 3. The smallest power of 2 that is a factor of any of the three numbers is 2^2, and the smallest power of 3 that is a factor is 3^2. The GCF is $2^2 \cdot 3^2$ or 36.

Quick Check **2**
Find the GCF of 48, 120, and 156.

Note that if two or more integers do not have any common prime factors, then their GCF is 1. For example, the GCF of 8, 12, and 15 is 1, since there are no prime numbers that are factors of all three numbers.

Objective 2 **Find the GCF of two or more variable terms.** We can also find the GCF of two or more variable terms. For a variable factor to be included in the GCF, it must be a factor of each term. As with prime factors, the exponent for a variable factor used in the GCF is the smallest exponent that can be found for that variable in any one term.

EXAMPLE 3 Find the GCF of $6x^2$ and $15x^4$.

Solution

The GCF of the two coefficients is 3. The variable x is a factor of each term, and its smallest exponent is 2. The GCF of these two terms is $3x^2$.

EXAMPLE 4 Find the GCF of $a^5b^4c^2$, a^7b^3, $a^6b^9c^5$, and $a^3b^8c^3$.

Solution

Only the variables a and b are factors of all the terms. (The variable c is not a factor of the second term.) The smallest power of a in any one term is 3, and this is the case for the variable b as well. The GCF is a^3b^3.

Quick Check 3

Find the GCF.

a) $12x^5$ and $28x^3$
b) $x^3y^4z^9$, $x^6y^2z^{10}$, and x^7z^4

Factoring Out the Greatest Common Factor

Objective 3 **Factor the GCF out of each term of a polynomial.** The first step for factoring any polynomial is to factor out the GCF of all of the terms. This process uses the distributive property, and you may think of it as "undistributing" the GCF from each term. Consider the polynomial $4x^2 + 8x + 20$. The GCF of these three terms is 4, and the polynomial can be rewritten as the product $4(x^2 + 2x + 5)$. Notice that the GCF has been factored out of each term, and the polynomial inside the parentheses is the polynomial that we would multiply by 4 in order to equal $4x^2 + 8x + 20$.

$$4(x^2 + 2x + 5) = 4 \cdot x^2 + 4 \cdot 2x + 4 \cdot 5 \qquad \text{Distribute.}$$
$$= 4x^2 + 8x + 20 \qquad \text{Multiply.}$$

EXAMPLE 5 Factor $5x^3 - 30x^2 + 10x$ by factoring out the GCF.

Solution

We begin by finding the GCF, which is $5x$. The next task is to fill in the missing terms of the polynomial inside the parentheses so that the product of $5x$ and that polynomial is $5x^3 - 30x^2 + 10x$.

$$5x(? - ? + ?)$$

Quick Check 4

Factor $6x^4 - 42x^3 - 90x^2$ by factoring out the GCF.

To find the missing terms, we can divide each term of the original polynomial by $5x$.

$$5x^3 - 30x^2 + 10x = 5x(x^2 - 6x + 2)$$

We can check our answer by multiplying $5x(x^2 - 6x + 2)$, which should equal $5x^3 - 30x^2 + 10x$. The check is left to the reader.

EXAMPLE 6 Factor $6x^6 + 8x^4 + 2x^3$ by factoring out the GCF.

Solution

The GCF for these three terms is $2x^3$. Notice that this is also the third term.

$$6x^6 + 8x^4 + 2x^3 = 2x^3(3x^3 + 4x + 1) \qquad \text{Factor out the GCF.}$$

When a term in a polynomial is the GCF of the polynomial, factoring out the GCF leaves a 1 in that term's place. Why? Because we need to determine what we multiply $2x^3$ by in order to equal $2x^3$, and $2x^3 \cdot 1 = 2x^3$.

Quick Check 5

Factor
$15x^7 - 30x^5 + 3x^4$ by factoring out the GCF.

A Word of Caution Be sure to write a 1 in the place of a term that was the GCF when factoring out the GCF from a polynomial.

Objective 4 Factor a common binomial factor out of a polynomial.
Occasionally, the GCF of two terms will be a binomial or some other polynomial with more than one term. Consider the expression $8x(x - 5) + 3(x - 5)$, which contains the two terms $8x(x - 5)$ and $3(x - 5)$. Each term has the binomial $x - 5$ as a factor. This common factor can be factored out of this expression.

EXAMPLE 7 Factor $x(x + 3) + 7(x + 3)$ by factoring out the GCF.

Solution

The binomial $x + 3$ is a common factor for these two terms. We begin by factoring out this common factor.

$$x\underbrace{(x + 3)} + 7\underbrace{(x + 3)}$$

$$= (x + 3)(? + ?)$$

After factoring out $x + 3$, what factors remain in the first term? The only factor that remains is x. Using the same method we see that the only factor remaining in the second term is 7.

$$x(x + 3) + 7(x + 3) = (x + 3)(x + 7)$$

EXAMPLE 8 Factor $9x(2x + 3) - 8(2x + 3)$ by factoring out the GCF.

Quick Check 6

Factor by factoring out the GCF.

a) $5x(x - 9) + 14(x - 9)$
b) $7x(x - 8) - 6(x - 8)$

Solution

The GCF of these two terms is $2x + 3$, which can be factored out as follows:

$$9x(2x + 3) - 8(2x + 3) = (2x + 3)(9x - 8) \qquad \text{Factor out the common factor } 2x + 3.$$

Notice that the second term was negative, which led to $9x - 8$, rather than $9x + 8$, as the other factor.

Again, factoring out the GCF will be our first step in factoring any polynomial. Often this will make our factoring easier, and sometimes the expression cannot be factored without factoring out the GCF.

Factoring by Grouping

Objective 5 Factor a polynomial by grouping. We now turn our attention to a factoring technique known as **factoring by grouping.** Consider the polynomial $3x^3 + 33x^2 + 7x + 77$. The GCF for the four terms is 1, so we cannot factor the polynomial by factoring out the GCF. However, the first pair of terms has a common factor of $3x^2$ and the second pair of terms has a common factor of 7. If we factor $3x^2$ out of the first two terms and 7 out of the last two terms, we produce the following:

$$3x^2(x + 11) + 7(x + 11)$$

Notice that the two resulting terms have a common factor of $x + 11$. This common binomial factor can then be factored out.

$$
\begin{aligned}
&3x^3 + 33x^2 + 7x + 77 \\
&= 3x^2(x + 11) + 7(x + 11) \qquad \text{Factor } 3x^2 \text{ from the first two terms and 7 from} \\
&\qquad\qquad\qquad\qquad\qquad\qquad\; \text{the last two terms.} \\
&= (x + 11)(3x^2 + 7) \qquad\quad \text{Factor out the common binomial factor } x + 11.
\end{aligned}
$$

Factoring a Polynomial with Four Terms by Grouping

> Factor a common factor out of the first two terms and another common factor out of the last two terms. If the two "groups" share a common binomial factor, this binomial can then be factored out to complete the factoring of the polynomial.

If the two "groups" do not share a common binomial factor, we can try rearranging the terms of the polynomial in a different order. If we cannot find two "groups" that share a common factor, then the polynomial cannot be factored by this method.

EXAMPLE 9   Factor $5x^3 - 30x^2 + 4x - 24$ by grouping.

Solution

We first check for a factor that is common to each of the four terms. Since there is no common factor other than 1, we proceed to factoring by grouping:

$$
\begin{aligned}
5x^3 - 30x^2 + 4x - 24 &= 5x^2(x - 6) + 4x - 24 \qquad \text{Factor } 5x^2 \text{ out of first} \\
&\qquad\qquad\qquad\qquad\qquad\quad\;\; \text{two terms.}\\
&= 5x^2(x - 6) + 4(x - 6) \qquad \text{Factor 4 out of last}\\
&\qquad\qquad\qquad\qquad\qquad\quad\;\; \text{two terms.}\\
&= (x - 6)(5x^2 + 4) \qquad\qquad\;\; \text{Factor out the common}\\
&\qquad\qquad\qquad\qquad\qquad\quad\;\; \text{factor } x - 6.
\end{aligned}
$$

Quick Check 7

Factor $x^3 + 4x^2 + 7x + 28$ by grouping.

As was the case earlier, we can check our factoring by multiplying the two factors together. The check is left to the reader.

EXAMPLE 10   Factor $8x^2 + 32x - 7x - 28$ by grouping.

Solution

Since the four terms have no common factors other than 1, we factor by grouping. After factoring $8x$ out of the first two terms, we must factor a *negative* 7 out of the last two terms. If we factored a positive 7 out of the last two terms, rather than a negative 7, the two binomial factors would be different.

$$8x^2 + 32x - 7x - 28 = 8x(x + 4) - 7(x + 4)$$

Factor the common factor $8x$ out of the first two terms and the common factor -7 out of the last two terms.

$$= (x + 4)(8x - 7)$$

Factor out the common factor $x + 4$.

When the third of the four terms is negative, this often indicates that a negative common factor will need to be factored from the last two terms.

Quick Check 8

Factor
$2x^2 - 10x - 9x + 45$
by grouping.

A Word of Caution   Factoring a negative common factor from two terms when factoring by grouping is often necessary.

EXAMPLE 11   Factor $10x^3 + 50x^2 + x + 5$ by grouping.

Solution

Again, the four terms have no common factor other than 1, so we may proceed to factor this polynomial by grouping. Notice that the last two terms have no common factor other than 1. When this is the case, we will actually factor out the common factor of 1 from the two terms.

$$10x^3 + 50x^2 + x + 5 = 10x^2(x + 5) + 1(x + 5)$$

Factor the common factors of $10x^2$ and 1 from the first two terms and the last two terms, respectively.

$$= (x + 5)(10x^2 + 1)$$

Factor out the common factor $x + 5$.

Quick Check 9

Factor
$4x^3 - 36x^2 + x - 9$
by grouping.

In the next example, the four terms have a common factor other than 1. We will factor out the GCF from the polynomial before attempting to use factoring by grouping.

EXAMPLE 12   Factor $2x^3 - 6x^2 - 20x + 60$ completely.

Solution

The four terms share a common factor of 2, and we will factor this out before proceeding with factoring by grouping.

$$2x^3 - 6x^2 - 20x + 60 = 2(x^3 - 3x^2 - 10x + 30)$$

Factor out the common factor 2.

$$= 2[x^2(x - 3) - 10(x - 3)]$$

Factor out the common factor x^2 from the first two terms. Factor out the common factor -10 from the last two terms.

$$= 2(x - 3)(x^2 - 10)$$

Factor out the common factor $x - 3$.

If we had not factored out the common factor 2 before factoring this polynomial by grouping, we would have factored $2x^3 - 6x^2 - 20x + 60$ to be $(x - 3)(2x^2 - 20)$. If we stop here, we have not factored this polynomial completely, since $2x^2 - 20$ has a common factor of 2.

Quick Check **10**

Factor
$3x^2 + 24x - 12x - 96$
by grouping.

Occasionally, despite our best efforts, a polynomial cannot be factored. Here is an example of just such a polynomial. Consider the polynomial $x^2 - 5x + 3x + 15$. The first two terms have a common factor of x and the last two terms have a common factor of 3.

$$x^2 - 5x + 3x + 15 = x(x - 5) + 3(x + 5)$$

Since the two binomials are not the same, we cannot factor a common factor out of the two terms. Reordering the terms $-5x$ and $3x$ leads to the same problem. The polynomial cannot be factored.

There are instances when factoring by grouping fails, yet the polynomial can be factored by other techniques. For example, the polynomial $x^3 + 2x^2 - 7x - 24$ cannot be factored by grouping. However, it can be shown that $x^3 + 2x^2 - 7x - 24$ is equal to $(x - 3)(x^2 + 5x + 8)$.

Building Your Study Strategy Overcoming Math Anxiety, 1 **Understanding Math Anxiety** If you have been avoiding taking a math class, if you panic when asked a mathematical question, or if you feel that you cannot learn mathematics, then you may have a condition known as math anxiety. Like many other anxiety-related conditions, math anxiety may be traced back to an event that first triggered the negative feelings. If you are going to overcome your anxiety, the first step is to understand the cause of your anxiety.

Many students can recall one embarrassing moment in their life that may be the initial cause of their anxiety. Were you ridiculed as a child by a teacher or another student when you couldn't solve a math problem? Were you expected to be a mathematical genius like a parent or older sibling but were not able to measure up to their reputation? One negative event can start math anxiety, and the journey to overcome math anxiety can begin with a single success. If you are able to complete a homework assignment, or show improvement on a quiz or exam, then celebrate your success. Let this be your vindication, and consider your slate to have been wiped clean.

Vocabulary

1. A polynomial has been _____ when it is represented as the product of two or more polynomials.

2. The _____ of two or more integers is the largest whole number that is a factor of each integer.

3. The exponent for a variable factor used in the GCF is the _____ exponent that can be found for that variable in any one term.

4. The process of factoring a polynomial by first breaking the polynomial into two sets of terms is called factoring by _____.

Find the GCF.

5. $6, 8$

6. $15, 21$

7. $30, 42$

8. $5, 50$

9. $16, 40, 60$

10. $8, 24, 27$

11. x^3, x^7

12. y^8, y^4

13. a^2b^3, a^5b^2

14. $a^8b^3c^4, a^5bc^2, a^3b^4c^5$

15. $4x^4, 6x^3$

16. $27x^5, 24x^{11}$

17. $15a^5b^2c, 25a^2c^3, 5a^3bc^2$

18. $30x^3y^4z^5, 12x^6y^{11}z^3, 24y^9z$

Factor the GCF out of the given expression.

19. $7x - 14$

20. $3a + 12$

21. $5x^2 + 4x$

22. $9x^2 - 20x$

23. $8x^3 + 20x$

24. $11x^5 - 33x^3$

25. $7x^2 + 56x + 70$

26. $9a^2 - 6a + 36$

27. $18x^6 - 14x^4 - 8x^3$

28. $18a^3 - 27a^5 + 9a^4$

29. $m^2n^3 - m^5n^2 + m^3n$

30. $x^5y^5 + x^3y^2 - x^2y^4$

31. $16a^9b^7 - 72a^5b^5 - 80a^3b^6$

32. $25r^3s^4 + 10r^2s^5 - 15r^4s$

33. $-8x^3 + 12x^2 - 16x$

34. $-18a^5 - 30a^3 + 24a^2$

35. $5x(2x - 7) + 8(2x - 7)$

36. $x(4x + 3) + 4(4x + 3)$

37. $x(3x + 11) - 7(3x + 11)$

38. $4x(x - 9) - 9(x - 9)$

39. $2x(x + 10) + (x + 10)$

40. $3x(x - 2) - (x - 2)$

Factor by grouping.

41. $x^2 + 8x + 7x + 56$

42. $x^2 + 3x + 9x + 27$

43. $x^2 - 5x + 4x - 20$

44. $x^2 - 6x + 9x - 54$

45. $x^2 + 2x - 9x - 18$

46. $x^2 - 8x - 5x + 40$

47. $3x^2 + 15x + 4x + 20$

48. $4x^2 - 24x + 7x - 42$

49. $7x^2 + 28x - 6x - 24$

50. $5x^2 - 25x - 12x + 60$

51. $3x^2 + 21x + x + 7$

52. $4x^2 - 48x + x - 12$

53. $2x^2 - 10x - x + 5$

54. $9x^2 - 81x - x + 9$

55. $x^3 + 8x^2 + 6x + 48$

56. $2x^3 - 12x^2 + 5x - 30$

57. $4x^3 + 12x + 3x^2 + 9$

58. $7x^3 + 49x + 4x^2 + 28$

59. $2x^2 + 10x + 6x + 30$

60. $3x^2 + 27x + 21x + 189$

Find a polynomial that has the given factor.

61. $x - 5$

62. $x + 3$

63. $2x + 9$

64. $3x - 4$

Writing in Mathematics

Answer in complete sentences.

65. When you are factoring a polynomial, how do you check your work?

66. Give an example of a four-term polynomial that can be factored by grouping. Use this example to explain the process of factoring by grouping.

67. Explain why it is necessary to factor a negative common factor from the last two terms of the polynomial $x^3 - 5x^2 - 7x + 35$ in order to factor it by grouping.

68. *Solutions Manual* * Write a solutions manual page for the following problem:

Factor $x^3 - 6x^2 - 7x + 42$.

69. *Newsletter* * Write a newsletter explaining how to factor out a GCF from a polynomial.

*See Appendix B for details and sample answers.

QUICK REVIEW EXERCISES

Section 6.1

Multiply.

1. $(x + 7)(x + 4)$

2. $(x - 9)(x - 11)$

3. $(x + 5)(x - 8)$

4. $(x - 10)(x + 6)$

6.2

Factoring Trinomials of the Form $x^2 + bx + c$

Objectives

1. Factor a trinomial of the form $x^2 + bx + c$ when c is positive.
2. Factor a trinomial of the form $x^2 + bx + c$ when c is negative.
3. Factor a perfect square trinomial.
4. Determine that a trinomial is prime.
5. Factor a trinomial by first factoring out a common factor.
6. Factor a trinomial in several variables.

In this section, we will learn how to factor trinomials of degree 2 with a leading coefficient of 1. Some examples of this type of polynomial are $x^2 + 12x + 32$, $x^2 - 9x + 14$, $x^2 + 8x - 20$, and $x^2 - x - 12$.

Factoring Trinomials of the Form $x^2 + bx + c$

If $x^2 + bx + c$, where b and c are integers, is factorable, it can be factored as the product of two binomials of the form $(x + m)(x + n)$, where m and n are integers.

We will begin by multiplying two binomials of the form $(x + m)(x + n)$ and using the result to help us learn to factor trinomials of the form $x^2 + bx + c$.

Multiply the two binomials $x + 4$ and $x + 8$.

$$(x + 4)(x + 8) = x^2 + 8x + 4x + 32 \qquad \text{Distribute.}$$
$$= x^2 + 12x + 32 \qquad \text{Combine like terms.}$$

The two binomials have a product that is a trinomial of the form $x^2 + bx + c$ with $b = 12$ and $c = 32$. Notice that the two numbers 4 and 8 have a product of 32 and a sum of 12; in other words, their product is equal to c and their sum is equal to b.

$$4 \cdot 8 = 32$$
$$4 + 8 = 12$$

We will use this pattern to help us factor trinomials of the form $x^2 + bx + c$. We will look for two integers m and n with a product equal to c and a sum equal to b. If we can find two such integers, then the trinomial factors as $(x + m)(x + n)$.

Factoring $x^2 + bx + c$ when c Is Positive

Objective 1 Factor a trinomial of the form $x^2 + bx + c$ when c is positive.

EXAMPLE 1 Factor $x^2 + 9x + 18$.

Solution

We begin, as always, by looking for common factors. The terms in this trinomial have no common factors other than 1. We are looking for two integers m and n whose product is 18 and whose sum is 9. Here are the factors of 18:

Factors	$1 \cdot 18$	$2 \cdot 9$	$3 \cdot 6$
Sum	19	11	9

The pair of factors that have a sum of 9 are 3 and 6. The trinomial $x^2 + 9x + 18$ factors to be $(x + 3)(x + 6)$. Be aware that this could also be written as $(x + 6)(x + 3)$.

We can check our work by multiplying $x + 3$ by $x + 6$. If the product equals $x^2 + 9x + 18$, then we have factored correctly. The check is left to the reader.

EXAMPLE 2   Factor $x^2 - 11x + 24$.

Solution

The major difference between this trinomial and the previous one is that the x term has a negative coefficient. We are looking for two integers m and n whose product is 24 and whose sum is -11. Since the product of these two integers is positive, they must have the same sign. The sum of these two integers is negative, so each integer must be negative. Here are the negative factors of 24:

Factors	$(-1)(-24)$	$(-2)(-12)$	$(-3)(-8)$	$(-4)(-6)$
Sum	-25	-14	-11	-10

Quick Check **1**

Factor.

a) $x^2 + 7x + 10$
b) $x^2 - 11x + 30$

The pair that has a sum of -11 is -3 and -8. $x^2 - 11x + 24 = (x - 3)(x - 8)$.

From the previous examples, we see that when the constant term c is positive, such as in $x^2 + 9x + 18$ and $x^2 - 11x + 24$, then the two numbers we are looking for (m and n) will both have the same sign. Both numbers will be positive if b is positive, and both numbers will be negative if b is negative.

Factoring $x^2 + bx + c$ when c Is Negative

Objective 2   Factor a trinomial of the form $x^2 + bx + c$ when c is negative. If the product of two numbers is negative, one of the numbers must be negative and the other number must be positive. So if the constant term c is negative, then m and n must have opposite signs.

EXAMPLE 3   Factor $x^2 + 8x - 33$.

Solution

These three terms do not have any common factors, so we look for two integers m and n with a product of -33 and a sum of 8. Since the product is negative, we are looking for one negative number and one positive number. Here are the factors of -33, along with their sums:

Factors	$-1 \cdot 33$	$-3 \cdot 11$	$-11 \cdot 3$	$-33 \cdot 1$
Sum	32	8	-8	-32

The pair of integers we are looking for are -3 and 11. $x^2 + 8x - 33 = (x - 3)(x + 11)$.

EXAMPLE 4   Factor $x^2 - 3x - 40$.

Solution

Since there are no common factors to factor out, we are looking for two integers m and n that have a product of -40 and a sum of -3. Here are the factors of -40:

Factors	$-1 \cdot 40$	$-2 \cdot 20$	$-4 \cdot 10$	$-5 \cdot 8$	$-8 \cdot 5$	$-10 \cdot 4$	$-20 \cdot 2$	$-40 \cdot 1$
Sum	39	18	6	3	-3	-6	-18	-39

Our integers m and n are -8 and 5. $x^2 - 3x - 40 = (x - 8)(x + 5)$.

Quick Check **2** Factor.

a) $x^2 + 9x - 36$ b) $x^2 - x - 42$

It is not necessary to list each set of factors as we did in the previous examples. We can often find the two integers m and n quickly through trial and error.

EXAMPLE 5 Factor.

a) $x^2 - 10x - 24$

Solution

The two integers whose product is -24 and sum is -10 are -12 and 2.

$$x^2 - 10x - 24 = (x - 12)(x + 2)$$

b) $x^2 + 10x + 24$

Solution

This example is similar to the last example, but we are looking for two integers that have a product of *positive* 24 instead of -24. Also, the sum of these two integers is 10 instead of -10. The two integers whose product is 24 and whose sum is 10 are 4 and 6.

$$x^2 + 10x + 24 = (x + 4)(x + 6)$$

c) $x^2 + 10x - 24$

Quick Check **3**

Factor.

a) $x^2 + 5x - 6$
b) $x^2 + 5x + 6$
c) $x^2 - 5x + 6$
d) $x^2 - 5x - 6$

Solution

The two integers that have a product of -24 and a sum of 10 are -2 and 12.

$$x^2 + 10x - 24 = (x - 2)(x + 12)$$

d) $x^2 - 10x + 24$

Solution

The integers -4 and -6 have a product of 24 and a sum of -10.

$$x^2 - 10x + 24 = (x - 4)(x - 6)$$

Factoring a Perfect Square Trinomial

Objective **3** Factor a perfect square trinomial.

EXAMPLE 6 Factor $x^2 - 6x + 9$.

Solution

Quick Check **4**

Factor $x^2 + 10x + 25$.

We begin by looking for two integers that have a product of 9 and a sum of -6. Since the product of -3 and -3 is 9 and their sum is -6, $x^2 - 6x + 9 = (x - 3)(x - 3)$. Notice that the same factor is listed twice. We can rewrite this as $(x - 3)^2$.

When a trinomial factors to equal the square of a binomial, we call it a **perfect square trinomial.**

Trinomials That Are Prime

Objective 4   Determine that a trinomial is prime. Not every trinomial of the form $x^2 + bx + c$ can be factored. For example, there may not be a pair of integers that have a product of c and a sum of b. In this case, we say that the trinomial cannot be factored; it is **prime.** For example, $x^2 + 9x + 12$ is prime because we cannot find two numbers with a product of 12 and a sum of 9.

EXAMPLE 7   Factor $x^2 + 11x - 18$.

Solution

As there are no common factors to factor out, we are looking for two integers m and n that have a product of -18 and a sum of 11. Here are the factors of -18.

Factors	$-1 \cdot 18$	$-2 \cdot 9$	$-3 \cdot 6$	$-6 \cdot 3$	$-9 \cdot 2$	$-18 \cdot 1$
Sum	17	7	3	-3	-7	-17

Quick Check 5
Factor $x^2 + 9x - 20$.

There are not two integers that have a product of -18 and a sum of 11, so the trinomial $x^2 + 11x - 18$ is prime.

Factoring a Trinomial Whose Terms Have a Common Factor

Objective 5   Factor a trinomial by first factoring out a common factor. We now turn our attention to factoring trinomials whose terms contain common factors other than 1. After we factor out a common factor, we will attempt to factor the remaining polynomial factor.

EXAMPLE 8   Factor $4x^2 + 48x + 80$.

Solution

The GCF of these three terms is 4.

> *A Word of Caution*   When factoring a polynomial, begin by factoring out the GCF of *all* of the terms.

After factoring out the GCF, we have $4(x^2 + 12x + 20)$. We now try to factor the trinomial $x^2 + 12x + 20$ by finding two integers that have a product of 20 and a sum of 12. The integers that satisfy these conditions are 2 and 10.

Quick Check 6
Factor $5x^2 + 40x + 60$.

$$4x^2 + 48x + 80 = 4(x^2 + 12x + 20) \qquad \text{Factor out the GCF 4.}$$
$$= 4(x + 2)(x + 10) \qquad \text{Factor } x^2 + 12x + 20.$$

A Word of Caution When we factor out a common factor from a polynomial, that common factor *must* be written in all following stages of factoring. *Don't "lose" the common factor!*

EXAMPLE 9 Factor $-x^2 - 7x + 60$.

Solution

The leading coefficient for this trinomial is -1, but to factor the trinomial using our technique, the leading coefficient must be a *positive* 1. We begin by factoring out a -1 from each term, which will change the sign of each term.

$$-x^2 - 7x + 60 = -(x^2 + 7x - 60)$$ Factor out -1, so that the leading coefficient is 1.

$$= -(x - 5)(x + 12)$$ Factor $x^2 + 7x - 60$ by finding two integers whose product is -60 and whose sum is 7.

Quick Check 7
Factor $-x^2 + 14x - 48$.

If a trinomial's leading coefficient is not 1, or if the degree of the trinomial is not 2, then we should factor out a common factor so that the trinomial factor is of the form $x^2 + bx + c$. At this point in the text, this is the only type of trinomial we know how to factor. We will learn techniques to factor other trinomials in the next section.

Factoring Trinomials in Several Variables

Objective 6 Factor a trinomial in several variables. Trinomials in several variables can also be factored. In the next example, we will explore the similarities and the differences between factoring trinomials in two variables and trinomials in one variable.

EXAMPLE 10 Factor $x^2 + 13xy + 36y^2$.

Solution

These three terms have no common factors, and the leading coefficient is 1. Suppose that the variable *y* were not included; in other words, suppose that we were asked to factor $x^2 + 13x + 36$. After looking for two integers whose product is 36 and whose sum is 13, we would see that $x^2 + 13x + 36 = (x + 4)(x + 9)$. Now we can examine the changes that are necessary because of the second variable *y*. Note that $4 \cdot 9 = 36$, not $36y^2$. However, $4y \cdot 9y$ does equal $36y^2$, and $x \cdot 9y + x \cdot 4y$ does equal $13xy$. This trinomial factors to be $(x + 4y)(x + 9y)$.

Quick Check 8
Factor $x^2 - 4xy - 32y^2$.

We can factor a trinomial in two or more variables by first ignoring all of the variables except for the first variable. After determining how the trinomial factors for the first variable, we can then determine where the rest of the variables appear in the factored form of the trinomial. We can check that our factoring is correct by multiplying, making sure that the product is equal to the trinomial.

EXAMPLE 11   Factor $p^2r^2 - 9pr - 10$.

Solution

Quick Check 9

Factor
$x^2y^2 + 10xy - 24$.

If the variable r did not appear in the trinomial, and we were factoring $p^2 - 9p - 10$, we would look for two integers that have a product of -10 and a sum of -9. The two integers are -10 and 1, so the polynomial $p^2 - 9p - 10 = (p - 10)(p + 1)$. Now we work on the variable r. Since the first term in the trinomial is p^2r^2, the first term in each binomial must be pr. $p^2r^2 - 9pr - 10 = (pr - 10)(pr + 1)$.

> *Building Your Study Strategy* Overcoming Math Anxiety, 2 **Mathematical Autobiography** One effective tool for understanding your past difficulties in mathematics and how those past difficulties are hindering your ability to learn mathematics today is to write a mathematical autobiography. Write down your successes and your failures, as well as the reasons why you think you succeeded or failed. Go back into your mathematical past as far as you can remember. Write down how friends, relatives, and teachers affected you.
>
> Let your autobiography sit for a few days and then read it over. Read it on an analytical level as if another person had written it and detach yourself from it personally. Look for patterns, and think about strategies to reverse the bad patterns and continue any good patterns that you can find.

EXERCISES 6.2

Vocabulary

1. A polynomial of the form $x^2 + bx + c$ is a second-degree trinomial with a _____ of 1.

2. A _____ trinomial is a trinomial whose two binomial factors are identical.

3. A polynomial is _____ if it cannot be factored.

4. Before attempting to factor a second-degree trinomial, it is wise to factor out any _____.

Factor completely. If the polynomial cannot be factored, write "prime."

5. $x^2 - 8x - 20$

6. $x^2 - 2x - 15$

7. $x^2 + 5x - 36$

8. $x^2 - 10x - 39$

9. $x^2 + 12x + 24$

10. $x^2 - 12x + 36$

11. $x^2 + 14x + 48$

12. $x^2 + 14x - 33$

13. $x^2 + 13x - 30$

14. $x^2 - 13x + 30$

15. $x^2 - 5x + 36$

16. $x^2 - 15x + 54$

17. $x^2 - 17x - 60$

18. $x^2 - 17x + 60$

19. $x^2 + 11x - 12$

20. $x^2 + 15x + 56$

21. $x^2 + 14x + 49$

22. $x^2 - 4x - 12$

23. $x^2 + 20x + 91$

24. $x^2 + 14x + 13$

25. $x^2 - 24x + 144$

26. $x^2 + 5x - 14$

27. $x^2 - 13x + 40$

28. $x^2 - 5x - 24$

29. $x^2 - 13x - 30$

30. $x^2 + 18x + 81$

31. $x^2 - 9x + 20$

32. $x^2 - 11x - 28$

33. $x^2 - 16x + 60$

34. $x^2 + 16x + 48$

35. $5x^2 + 30x - 55$

36. $-x^2 - 7x + 18$

37. $21x^2 + 147x + 210$

38. $9x^2 - 45x - 450$

39. $-3x^2 + 48x - 192$

40. $8x^2 - 128x + 504$

41. $x^3 + 8x^2 + 16x$

42. $-x^4 + x^3 + 30x^2$

43. $-3x^5 - 6x^4 + 240x^3$

44. $2x^3 - 42x^2 + 196x$

45. $10x^8 + 20x^7 - 990x^6$

46. $35x^4 - 70x^3 + 35x^2$

47. $x^2 + xy - 42y^2$

48. $3x^2 - 36xy - 84y^2$

49. $x^2y^2 - 7xy + 12$

50. $x^2 - 20xy + 100y^2$

51. $x^2 - 10xy - 39y^2$

52. $x^2y^2 - 6xy - 7$

53. $5x^5 + 50x^4y + 105x^3y^2$

54. $x^2y^2 + 11xy - 60$

Mixed Practice, 55–72

Factor completely, using the appropriate technique. (If the polynomial cannot be factored, write "prime.")

55. $x^4 + 9x^3 + 5x + 45$

56. $x^2 - 26x + 169$

57. $x^2 + x + 30$

58. $x^2 - 17x + 42$

59. $x^2 + 22x + 40$

60. $x^2 + 10x$

61. $x^2 + 40x + 400$

62. $x^2 + 4x - 60$

63. $4x^2 + 8x - 96$

64. $x^2 + 16x + 15$

65. $x^4 - 15x^3$

66. $x^2 + 2x - 120$

67. $x^2 - 16x + 63$

68. $x^2 - 7x - 78$

69. $x^2 + 10x - 25$

70. $x^3 - 12x^2 - 6x + 72$

71. $x^2 - 21x - 100$

72. $-x^2 + 17x - 60$

Find the missing value such that the given binomial is a factor of the given polynomial.

73. $x^2 + 10x + ?, x + 2$

74. $x^2 - 16x + ?, x - 7$

75. $x^2 + 11x - ?, x - 4$

76. $x^2 - 7x - ?, x + 17$

Writing in Mathematics

Answer in complete sentences.

77. Explain why it is a good idea to factor a common factor of -1 from the expression $-x^2 + 7x + 30$ before attempting to factor the trinomial.

78. ***Solutions Manual*** * Write a solutions manual page for the following problem:

Factor $3x^2 - 9x - 84$.

79. ***Newsletter*** * Write a newsletter explaining how to factor a trinomial of the form $x^2 + bx + c$.

*See Appendix B for details and sample answers.

QUICK REVIEW EXERCISES

Section 6.2

Multiply.

1. $(2x + 5)(x + 4)$

2. $(3x - 8)(2x - 7)$

3. $(x + 6)(3x - 4)$

4. $(4x - 3)(2x + 9)$

6.3

Factoring Trinomials of the Form $ax^2 + bx + c$, Where $a \neq 1$

Objectives

1 Factor a trinomial of the form $ax^2 + bx + c$, where $a \neq 1$, by grouping.

2 Factor a trinomial of the form $ax^2 + bx + c$, where $a \neq 1$, by trial and error.

In this section, we continue to learn how to factor second-degree trinomials, focusing on trinomials with a leading coefficient that is not 1. Some examples of this type of trinomial are $2x^2 + 13x + 20$, $8x^2 - 22x - 21$, and $3x^2 - 11x + 8$.

There are two methods for factoring this type of trinomial that we will examine: factoring by grouping and factoring by trial and error. Your instructor may prefer one of these methods to the other, or an altogether different method, and may ask you to use one method only. However, if you are allowed to use either method, use the method that you feel more comfortable using.

Factoring by Grouping

Objective 1 Factor a trinomial of the form $ax^2 + bx + c$, where $a \neq 1$, by grouping. We begin with factoring by grouping. Consider the product $(3x + 2)(2x + 5)$.

$$(3x + 2)(2x + 5) = 6x^2 + 15x + 4x + 10 \qquad \text{Distribute.}$$
$$= 6x^2 + 19x + 10 \qquad \text{Combine like terms.}$$

Since $(3x + 2)(2x + 5) = 6x^2 + 19x + 10$, we know that $6x^2 + 19x + 10$ can be factored as $(3x + 2)(2x + 5)$. By rewriting the middle term $19x$ as $15x + 4x$, we can factor by grouping. The important skill is being able to determine that $15x + 4x$ is the correct way to rewrite $19x$, instead of $16x + 3x$, $9x + 10x$, $22x - 3x$, or any other two terms whose sum is $19x$.

Factoring $ax^2 + bx + c$, where $a \neq 1$, by Grouping

1. Multiply $a \cdot c$.
2. Find two integers with a product of $a \cdot c$ and a sum of b.
3. Rewrite the term bx as two terms, using the two integers found in step 2.
4. Factor the resulting polynomial by grouping.

To factor the polynomial $6x^2 + 19x + 10$ by grouping, we begin by multiplying $6 \cdot 10$, which equals 60. We now look for two integers with a product of 60 and a sum of 19, and those two integers are 15 and 4. We can then rewrite $19x$ as $15x + 4x$. (These terms could be written in the opposite order and would still lead to the correct factoring.) We then factor the polynomial $6x^2 + 15x + 4x + 10$ by grouping.

$$6x^2 + 19x + 10 = 6x^2 + 15x + 4x + 10 \qquad \text{Rewrite } 19x \text{ as } 15x + 4x.$$
$$= 3x(2x + 5) + 2(2x + 5) \qquad \text{Factor the common factor } 3x \text{ from the first two terms and the common factor 2 from the last two terms.}$$
$$= (2x + 5)(3x + 2) \qquad \text{Factor out the common factor } 2x + 5.$$

Now we will examine several examples using this technique.

EXAMPLE 1 Factor $2x^2 + 7x + 6$ by grouping.

Solution

We first check to see whether there are any common factors; and in this case, there are not. We proceed with factoring by grouping. Since $2 \cdot 6 = 12$, we look for two integers with a product of 12 and a sum of 7, which is the coefficient of the middle term. The integers 3 and 4 meet these criteria, so we rewrite $7x$ as $3x + 4x$.

$$2x^2 + 7x + 6 = 2x^2 + 3x + 4x + 6$$ Find two integers with a product of 12 and a sum of 7. Rewrite $7x$ as $3x + 4x$.

$$= x(2x + 3) + 2(2x + 3)$$ Factor the common factor x from the first two terms and the common factor 2 from the last two terms.

Quick Check 1

Factor $3x^2 + 13x + 12$ by grouping.

$$= (2x + 3)(x + 2)$$ Factor out the common factor $2x + 3$.

We check that this factoring is correct by multiplying $2x + 3$ by $x + 2$, which should equal $2x^2 + 7x + 6$. The check is left to the reader.

EXAMPLE 2 Factor $5x^2 - 7x + 2$ by grouping.

Solution

We begin by multiplying $5 \cdot 2$ as there are no common factors other than 1. We look for two integers with a product of 10 and a sum of -7. The integers are -5 and -2, so we rewrite $-7x$ as $-5x - 2x$ and then factor by grouping.

$$5x^2 - 7x + 2 = 5x^2 - 5x - 2x + 2$$ Find two integers with a product of 10 and a sum of -7. Rewrite $-7x$ as $-5x - 2x$.

Quick Check 2

Factor $4x^2 - 25x + 36$ by grouping.

$$= 5x(x - 1) - 2(x - 1)$$ Factor the common factor $5x$ from the first two terms and the common factor -2 from the last two terms.

$$= (x - 1)(5x - 2)$$ Factor out the common factor $x - 1$.

EXAMPLE 3 Factor $12x^2 + 11x - 15$ by grouping.

Solution

The terms have no common factor other than 1, so we begin by multiplying $12(-15)$, which equals -180. We need to find two integers with a product of -180 and a sum of 11. We can start by listing the different factors of 180. Since we know that one of the integers will be positive and one will be negative, we look for two factors in this list that have a difference of 11.

$$1 \cdot 180 \quad 2 \cdot 90 \quad 3 \cdot 60 \quad 4 \cdot 45 \quad 5 \cdot 36 \quad 6 \cdot 30 \quad 9 \cdot 20 \quad 10 \cdot 18 \quad 12 \cdot 15$$

Notice that the pair 9 and 20 has a difference of 11. The two integers are -9 and 20; their product is -180 and their sum is 11. We rewrite $11x$ as $-9x + 20x$ and then factor by grouping.

$$12x^2 + 11x - 15 = 12x^2 - 9x + 20x - 15$$

Find two integers with a product of -180 and a sum of 11. Rewrite $11x$ as $-9x + 20x$.

$$= 3x(4x - 3) + 5(4x - 3)$$

Factor the common factor $3x$ out of the first two terms and the common factor 5 out of the last two terms.

Quick Check 3

Factor $16x^2 - 38x + 21$ by grouping.

$$= (4x - 3)(3x + 5)$$

Factor out the common factor $4x - 3$.

Factoring by Trial and Error

Objective 2 **Factor a trinomial of the form $ax^2 + bx + c$, where $a \neq 1$, by trial and error.** We now turn our attention to the method of trial and error. We need to factor a trinomial $ax^2 + bx + c$ into the following form:

$$(\text{variable term} + \text{constant})(\text{variable term} + \text{constant})$$

The product of the variable terms will be ax^2, and the product of the constants will equal c. We will use these facts to get started by listing all the factors of ax^2 and c.

Consider the trinomial $2x^2 + 7x + 6$, which has no common factors other than 1. We begin by listing all the factors of $2x^2$ and 6.

Factors of $2x^2$: $x \cdot 2x$.   Factors of 6: $1 \cdot 6$, $2 \cdot 3$.

Since there is only one pair of factors with a product of $2x^2$, we know that if this trinomial is factorable, it will be of the form $(2x + ?)(x + ?)$. We will substitute the different factors of 6 in all possible orders in place of the question marks, satisfying the following products.

$$\overset{2x^2}{(2x + ?)(x + ?)}_{6}$$

We will keep substituting factors of 6 until the middle term of the product of the two binomials is $7x$. Rather than fully distributing for each trial, we need to look only at the two products shown in the following graphic.

$$(2x + ?)(x + ?)$$

When the sum of these two products is $7x$, we have found the correct factors.

We begin by using the factors 1 and 6, which leads to the binomial factors $(2x + 1)(x + 6)$. Since the middle term for these factors is $13x$, we have not found the correct factors. We then switch the positions of 1 and 6 and try again. However, when we multiply out $(2x + 6)(x + 1)$, the middle term equals $8x$, not $7x$. We need to try another pairing, so we try the other factors of 6, which are 2 and 3. When we multiply out $(2x + 3)(x + 2)$, the middle term is $7x$, so these are the correct factors:

$$\overset{4x}{(2x + 3)(x + 2)}_{3x}$$

$$2x^2 + 7x + 6 = (2x + 3)(x + 2).$$

EXAMPLE ▶ 4   Factor $10x^2 + 13x - 3$ by trial and error.

Solution

There are no common factors other than 1, so we begin by listing the factors of $10x^2$ and -3.

Factors of $10x^2$: $x \cdot 10x$, $2x \cdot 5x$
Factors of -3: $-1 \cdot 3$, $-3 \cdot 1$

We are looking for two factors that produce the middle term $13x$ when multiplied. We will begin by using x and $10x$ for the variable terms and -1 and 3 for the constants.

Factors	Middle Term
$(x - 1)(10x + 3)$	$-7x$
$(x + 3)(10x - 1)$	$29x$

Neither of these middle terms is equal to $13x$, so we have not found the correct factoring. Also, since neither middle term is the opposite of $13x$, switching the signs of the constants 1 and 3 will not lead to the correct factors either. So $(x + 1)(10x - 3)$ and $(x - 3)(10x + 1)$ are not correct. We switch to the pair $2x$ and $5x$.

Factors	Middle Term
$(2x - 1)(5x + 3)$	x
$(2x + 3)(5x - 1)$	$13x$

The second of these middle terms is the one that we are looking for, so $10x^2 + 13x - 3 = (2x + 3)(5x - 1)$.

Quick Check ◀ **4**
Factor $2x^2 + x - 28$ by trial and error.

Now we will try to factor another trinomial by trial and error, using a trinomial with terms that have more factors.

EXAMPLE ▶ 5   Factor $12x^2 + 11x - 15$ by trial and error.

Solution

Since there are no common factors to factor out, we begin by listing the factors of $12x^2$ and -15.

Factors of $12x^2$: $x \cdot 12x$, $2x \cdot 6x$, $3x \cdot 4x$
Factors of -15: $-1 \cdot 15$, $-15 \cdot 1$, $-3 \cdot 5$, $-5 \cdot 3$

At first glance, this may seem like it will take a while, but there are some ways that we can shorten the process. For example, we may skip over any pairings resulting in a binomial factor in which both terms have a common factor other than 1, such as $3x - 3$. If the terms of a binomial factor had a common factor other than 1, then the original polynomial would also have a common factor, but we know that $12x^2 + 11x - 15$ does not have a common factor that can be factored out.

If the pair of constants -1 and 15 does not produce the desired middle term and does not produce the opposite of the desired middle term, then we do not need to use the pair of constants -15 and 1. Instead, we could move on to the next pair of constants. Eventually, through trial and error, we find that $12x^2 + 11x - 15 = (3x + 5)(4x - 3)$.

Quick Check 5
Factor $20x^2 + 17x - 24$ by trial and error.

> **A Word of Caution**   If a trinomial has a leading coefficient other than 1, be sure to check if the three terms have a common factor. If there is a common factor, we may be able to factor a trinomial using the techniques from Section 6.2.

EXAMPLE 6   Factor $8x^2 + 40x + 48$.

Solution

The three terms have a common factor of 8. After we factor out this common factor, the trinomial factor will be $x^2 + 5x + 6$, which is a trinomial with a leading coefficient of 1. We look for two integers with a product of 6 and a sum of 5. Those two integers are 2 and 3.

Quick Check 6
Factor $9x^2 + 27x - 162$.

$$8x^2 + 40x + 48 = 8(x^2 + 5x + 6)$$
$$= 8(x + 2)(x + 3)$$

Factor out the common factor 8.
To factor $x^2 + 5x + 6$ we need to find two integers with a product of 6 and a sum of 5. The two integers are 2 and 3.

Factoring out the common factor in the previous example makes factoring the trinomial much more manageable. If we were to try factoring by grouping without factoring out the common factor, we would begin by multiplying 8 by 48, which equals 384. We would look for two integers with a product of 384 and a sum of 40, which would be challenging. If we tried to factor by trial and error, $8x^2$ has two pairs of factors $(x \cdot 8x, 2x \cdot 4x)$, but 48 has several pairs of factors $(1 \cdot 48, 2 \cdot 24, 3 \cdot 16, 4 \cdot 12, 6 \cdot 8)$. Factoring by trial and error would be tedious at best.

> **Building Your Study Strategy** Overcoming Math Anxiety, 3 **Test Anxiety**
> Many students confuse test anxiety with math anxiety. If you feel that you can learn mathematics and you understand the material, but you "freeze up" on tests, then you may have test anxiety, especially if this is true in your classes besides your math class. Speak to your academic counselor about this. Many colleges offer seminars or short-term classes about how to reduce test anxiety; ask your counselor whether such a course would be appropriate for you.

Vocabulary

1. A second-degree trinomial with a leading coefficient not equal to 1 can be written in the form _____.

2. The two techniques for factoring trinomials of the form $ax^2 + bx + c$, where $a \neq 1$, are called _____ and _____.

Factor completely.

3. $5x^2 - 18x - 8$

4. $3x^2 + 11x + 6$

5. $2x^2 + 9x + 4$

6. $7x^2 - 25x + 12$

7. $3x^2 - x - 4$

8. $4x^2 - 13x + 10$

9. $4x^2 + 4x - 3$

10. $21x^2 - 18x - 3$

11. $6x^2 + 7x + 2$

12. $8x^2 + 2x - 21$

13. $9x^2 - 3x - 20$

14. $10x^2 - 29x + 10$

15. $20x^2 + 161x + 8$

16. $18x^2 + 93x - 16$

17. $3x^2 + 13x - 10$

18. $20x^2 + 47x + 24$

19. $12x^2 + 11x - 56$

20. $24x^2 + 55x - 24$

21. $16x^2 - 24x + 9$

22. $16x^2 + 66x - 27$

23. $15x^2 - 24x - 12$

24. $15x^2 + 32x - 60$

25. $16x^2 + 24x - 40$

26. $15x^2 + 45x - 150$

27. $16x^2 - 64x + 64$

28. $40x^2 - 100x - 60$

29. $15x^2 - 25x - 560$

30. $-12x^2 - 36x + 120$

Mixed Practice, 31–66

Factor completely, using the appropriate technique. If the polynomial cannot be factored, write "prime."

31. $x^2 + 14x + 13$

32. $x^2 - 10x - 11$

33. $18x - 9$

34. $3x^2 + 33x + 48$

35. $2x^2 - 7x + 6$

36. $x^2 + 30x + 225$

37. $x^2 + 19x - 20$

38. $32x^2 - 20x - 25$

39. $x^5 - 6x^4$

40. $x^2 - 8x - 48$

41. $4x^2 - 20x$

42. $x^2 + 17x - 60$

43. $-12x^2 - 23x - 10$

44. $4x^2 + 20x + 21$

45. $9x^2 - 30x + 13$

46. $3x^2 - 7x - 6$

47. $x^6 + 3x^4 + 8x^2 + 24$

48. $24x^3 + 168x^2 + 54x + 378$

49. $x^3 + 2x^2 - 28x - 56$

50. $x^2 + 25x + 150$

51. $x^2 - 24x + 144$

52. $x^2 - 3x + 2$

53. $9x^2 - 45x - 54$

54. $3x^3 - 30x^2 - 7x + 70$

55. $-5x^2 + 28x + 12$

56. $x^2 - 6x + 5$

57. $2x^2 - 15x - 27$

58. $x^2 + 8x + 16$

59. $x^2 + 12x - 45$

60. $3x^5 + 6x^4$

61. $-3x^2 - 12x + 96$

62. $x^2 - 17x + 60$

63. $15x^2 + 16x - 15$

64. $18x^2 + 51x + 8$

65. $2x^2 + 9x + 4$

66. $x^2 - 7x - 44$

(**Writing in Mathematics**)

Answer in complete sentences.

67. Compare the two different factoring methods for trinomials of the form $ax^2 + bx + c$, where $a \neq 1$, factoring by grouping and factoring by trial and error. Which method do you prefer? Explain your reasoning.

68. Explain why it is a good idea to factor out a common factor from a trinomial of the form $ax^2 + bx + c$ before trying other factoring techniques.

69. *Solutions Manual** Write a solutions manual page for the following problem:

Factor $6x^2 - 35x + 36$.

70. *Newsletter** Write a newsletter explaining how to factor a trinomial of the form $ax^2 + bx + c$, where $a \neq 1$, either by grouping or by trial and error.

*See Appendix B for details and sample answers.

QUICK REVIEW EXERCISES ▶

Section 6.3

Multiply.

1. $(x + 8)(x - 8)$

2. $(4x + 3)(4x - 3)$

3. $(x + 8)(x^2 - 8x + 64)$

4. $(2x - 5)(4x^2 + 10x + 25)$

6.4

Factoring Special Binomials

1 Factor a difference of squares.
2 Factor the factors of a difference of squares.
3 Factor a difference of cubes.
4 Factor a sum of cubes.

Difference of Squares

Objective 1 Factor a difference of squares. Recall the special product $(a + b)(a - b) = a^2 - b^2$ from Section 5.4. The binomial $a^2 - b^2$ is a **difference of squares.** Based on this special product, a difference of squares factors in the following manner:

Difference of Squares

$$a^2 - b^2 = (a + b)(a - b)$$

To identify a binomial as a difference of squares, we must verify that each term is a perfect square. Variable factors must have exponents that are multiples of 2 such as x^2, y^4, and a^6. Any constant must also be a perfect square. Here are the first 10 perfect squares:

$$1^2 = 1 \quad 2^2 = 4 \quad 3^2 = 9 \quad 4^2 = 16 \quad 5^2 = 25$$
$$6^2 = 36 \quad 7^2 = 49 \quad 8^2 = 64 \quad 9^2 = 81 \quad 10^2 = 100$$

EXAMPLE 1 Factor $x^2 - 49$.

Solution

This binomial is a difference of squares, as it can be rewritten as $(x)^2 - (7)^2$. Rewriting the binomial in this form helps us to use the formula $a^2 - b^2 = (a + b)(a - b)$.

$$\begin{aligned} x^2 - 49 &= (x)^2 - (7)^2 & \text{Rewrite each term as a square.} \\ &= (x + 7)(x - 7) & \text{Factor using the formula for a difference of} \\ & & \text{squares, } a^2 - b^2 = (a + b)(a - b). \end{aligned}$$

We can check our work by multiplying $x + 7$ by $x - 7$, which should equal $x^2 - 49$. The check is left to the reader.

Quick Check 1
Factor $x^2 - 25$.

Although we will continue to write each term as a perfect square, if you can factor a difference of squares without writing each term as a perfect square, then do not feel that you *must* rewrite each term as a perfect square before proceeding.

EXAMPLE 2 Factor $64a^{10} - 25b^4$.

Solution

The first term of this binomial is a square, as 64 is a square and the exponent for the variable factor a is even. The first term can be rewritten as $(8a^5)^2$. In a similar fashion, $25b^4$ can be rewritten as $(5b^2)^2$. This binomial is a difference of squares.

Quick Check 2
Factor $81x^{12} - 100y^{16}$.

$$\begin{aligned} 64a^{10} - 25b^4 &= (8a^5)^2 - (5b^2)^2 & \text{Rewrite each term as a square.} \\ &= (8a^5 + 5b^2)(8a^5 - 5b^2) & \text{Factor as a difference of squares.} \end{aligned}$$

This binomial $8x^2 - 81$ is not a difference of squares because the coefficient of the first term (8) is not a square. Since there are no common factors other than 1, this binomial cannot be factored.

EXAMPLE 3 Factor $3x^2 - 108$.

Solution

Although this binomial is not a difference of squares (3 and 108 are not squares), the common factor of 3 can be factored out. After we factor out the common factor 3, we have $3x^2 - 108 = 3(x^2 - 36)$. The binomial in parentheses is a difference of squares and can be factored accordingly.

$$\begin{aligned} 3x^2 - 108 &= 3(x^2 - 36) && \text{Factor out the common factor 3.} \\ &= 3[(x)^2 - (6)^2] && \text{Rewrite each term in the binomial factor} \\ & && \text{as a square.} \\ &= 3(x + 6)(x - 6) && \text{Factor as a difference of squares.} \end{aligned}$$

Quick Check 3
Factor $7x^2 - 175$.

If a binomial is a **sum of squares,** then it cannot be factored unless the two terms have a common factor. Some examples of a sum of squares are $x^2 + 49$, $4a^2 + 9b^2$, and $64a^{10} + 25b^4$.

Sum of Squares

A sum of squares, $a^2 + b^2$, cannot be factored.

Objective 2 Factor the factors of a difference of squares. Occasionally, after you have factored a difference of squares, one or more of the factors may still be factorable. For example, consider the binomial $x^4 - 81$, which factors to be $(x^2 + 9)(x^2 - 9)$. The first factor, $x^2 + 9$, is a sum of squares and cannot be factored. However, the second factor, $x^2 - 9$, is a difference of squares and can be factored further.

EXAMPLE 4 Factor $x^4 - 81$ completely.

Solution

This is a difference of squares and we factor it accordingly. After we factor the binomial as a difference of squares, one of the binomial factors $(x^2 - 9)$ is a difference of squares and must be factored as well.

$$\begin{aligned} x^4 - 81 &= (x^2)^2 - (9)^2 && \text{Rewrite each term as a perfect square.} \\ &= (x^2 + 9)(x^2 - 9) && \text{Factor as a difference of squares.} \\ &= (x^2 + 9)[(x)^2 - (3)^2] && \text{Rewrite each term in the binomial} \\ & && x^2 - 9 \text{ as a square.} \\ &= (x^2 + 9)(x + 3)(x - 3) && \text{Factor } x^2 - 9 \text{ as a difference of squares.} \end{aligned}$$

Quick Check 4
Factor $x^4 - y^4$.

A Word of Caution When factoring a difference of squares, check the binomial factor containing a difference to see if it can be factored.

Difference of Cubes

Objective **3** **Factor a difference of cubes.** Another special binomial that can be factored is a **difference of cubes**. In this case, both terms are perfect cubes rather than squares. Here is the formula for factoring a difference of cubes.

Difference of Cubes

$$a^3 - b^3 = (a - b)(a^2 + ab + b^2)$$

Before proceeding, let's multiply out $(a - b)(a^2 + ab + b^2)$ to show that it actually equals $a^3 - b^3$.

$$
\begin{aligned}
(a - b)(a^2 + ab + b^2) &= a \cdot a^2 + a \cdot ab + a \cdot b^2 - b \cdot a^2 - b \cdot ab - b \cdot b^2 && \text{Distribute.} \\
&= a^3 + a^2b + ab^2 - a^2b - ab^2 - b^3 && \text{Multiply.} \\
&= a^3 - b^3 && \text{Combine like terms.}
\end{aligned}
$$

We see that the formula is correct.

For a term to be a cube, its variable factors must have exponents that are multiples of 3, such as x^3, y^6, and z^9. Also, each constant factor must be a perfect cube. Here are the first 10 perfect cubes:

$$1^3 = 1 \qquad 2^3 = 8 \qquad 3^3 = 27 \qquad 4^3 = 64 \qquad 5^3 = 125$$
$$6^3 = 216 \qquad 7^3 = 343 \qquad 8^3 = 512 \qquad 9^3 = 729 \qquad 10^3 = 1000$$

We need to memorize the formula for factoring a difference of cubes. There are some patterns that can help us to remember the formula. A difference of cubes, $a^3 - b^3$, has two factors: a binomial $(a - b)$ and a trinomial $(a^2 + ab + b^2)$. The binomial looks just like the difference of cubes without the cubes, including the sign between the terms. The signs between the three terms in the trinomial are both addition signs. To find the actual terms in the trinomial factor, the following diagram may be helpful.

$$a^3 - b^3 = (a - b)(a^2 + ab + b^2)$$

1st   2nd

1st · 1st   2nd · 2nd

1st · 2nd

EXAMPLE 5   Factor $x^3 - 27$.

Solution

We begin by looking for common factors other than 1 that the two terms share, but there are none. This binomial is not a difference of squares, as the exponent of the variable term is not a multiple of 2 and the constant 27 is not a square. The binomial is, however, a difference of cubes. We can rewrite the term x^3 as $(x)^3$ and we can rewrite 27 as $(3)^3$. Rewriting $x^3 - 27$ as $(x)^3 - (3)^3$ will help us identify the terms that are in the binomial and trinomial factors.

$$x^3 - 27 = (x)^3 - (3)^3$$

Rewrite each term as a perfect cube.

$$= (x - 3)(x \cdot x + x \cdot 3 + 3 \cdot 3)$$

Factor as a difference of cubes. The binomial factor is the same as $(x)^3 - (3)^3$ without the cubes. Treating x as the 1st term in the binomial factor and 3 as the 2nd term, the terms in the trinomial are
1st · 1st + 1st · 2nd + 2nd · 2nd

$$= (x - 3)(x^2 + 3x + 9)$$

Simplify each term in the trinomial factor.

Quick Check 5

Factor $x^3 - 125$.

A Word of Caution   A difference of cubes $x^3 - y^3$ cannot be factored as $(x - y)^3$.

EXAMPLE 6   Factor $y^{15} - 64$.

Solution

Although the number 64 is a square, this binomial is not a difference of squares because y^{15} is not a square. For a term to be a square, its variable factors must have exponents that are multiples of 2. This binomial is a difference of cubes. The exponent in the first term is a multiple of 3, and the number 64 is equal to 4^3.

$$y^{15} - 64 = (y^5)^3 - (4)^3$$

Rewrite each term as a perfect cube.

$$= (y^5 - 4)(y^5 \cdot y^5 + y^5 \cdot 4 + 4 \cdot 4)$$

Factor as a difference of cubes.

$$= (y^5 - 4)(y^{10} + 4y^5 + 16)$$

Simplify each term in the trinomial factor.

Quick Check 6

Factor $x^{12} - 8$.

A Word of Caution   When factoring a difference of cubes $a^3 - b^3$, do not attempt to factor the trinomial factor $a^2 + ab + b^2$.

Sum of Cubes

Objective 4 Factor a sum of cubes. Unlike a sum of squares, a **sum of cubes** can be factored. Here is the formula:

Sum of Cubes

$$a^3 + b^3 = (a + b)(a^2 - ab + b^2)$$

Notice that the terms in the factors are the same as the factors in a *difference* of cubes, with the exception of some of their signs. When factoring a sum of cubes, the binomial factor is a sum rather than a difference. The sign of the middle term in the trinomial is negative rather than positive. The last term in the trinomial factor is positive, just as it was in the formula for a difference of cubes. The diagram shows the differences between the formula for a difference of cubes and a sum of cubes.

$$a^3 - b^3 = (a - b)(a^2 + ab + b^2)$$

$$a^3 + b^3 = (a + b)(a^2 - ab + b^2)$$

EXAMPLE 7   Factor $z^3 + 125$.

Solution

This binomial is a sum of cubes and can be rewritten as $(z)^3 + (5)^3$. We then can factor the sum of cubes using the formula.

$$
\begin{aligned}
z^3 + 125 &= (z)^3 + (5)^3 \\
&= (z + 5)(z \cdot z - z \cdot 5 + 5 \cdot 5) \\
&= (z + 5)(z^2 - 5z + 25)
\end{aligned}
$$

Rewrite each term as a perfect cube.
Factor as a sum of cubes.
Simplify each term in the trinomial factor.

Quick Check **7**
Factor $x^3 + 216$.

A Word of Caution   A sum of cubes $x^3 + y^3$ cannot be factored as $(x + y)^3$.

We finish this section by summarizing the strategies for factoring binomials.

Factoring Binomials

- Factor out the GCF if there is a common factor other than 1.
- Determine whether both terms are perfect squares.
  If the binomial is a sum of squares, it cannot be factored.
  If the binomial is a difference of squares, we can factor it by using the formula $a^2 - b^2 = (a + b)(a - b)$. Keep in mind that some of the resulting factors can be differences of squares as well, which will need to be further factored.
- If both terms are not perfect squares, determine whether they both are perfect cubes.
  If the binomial is a difference of cubes, we factor using the formula $a^3 - b^3 = (a - b)(a^2 + ab + b^2)$.
  If the binomial is a sum of cubes, we factor using the formula $a^3 + b^3 = (a + b)(a^2 - ab + b^2)$.

Building Your Study Strategy Overcoming Math Anxiety, 4 **Taking the Right Course**   Many students think they have math anxiety because they constantly feel that they are in over their heads, when in fact they may not be prepared for their particular class. Taking a course without having the prerequisite skills is not a good idea, and success will be difficult to achieve. For example, if you are taking an elementary algebra course, but are struggling with performing arithmetic operations on fractions, order of operations problems, and operations with positive and negative numbers, then perhaps you would be better off taking a prealgebra class.

   If this situation applies to you, talk to your instructor. Your instructor will be able to advise you whether you should be taking another course or whether you have the skills to be successful in this course.

Vocabulary

1. A binomial of the form $a^2 - b^2$ is a _____.

2. A numerical factor is a _____ if it can be written as n^2 for some integer n.

3. A variable factor is a perfect square if its exponents are _____.

4. A binomial of the form $a^2 + b^2$ is a _____.

5. A binomial of the form $a^3 - b^3$ is a _____.

6. A binomial of the form $a^3 + b^3$ is a _____.

Factor completely. If the polynomial cannot be factored, write "prime."

7. $x^2 - 4$

8. $x^2 - 9$

9. $x^2 - 36$

10. $a^2 - 64$

11. $100x^2 - 81$

12. $49x^2 - 64$

13. $x^2 - 2$

14. $x^2 - 18$

15. $a^2 - 16b^2$

16. $m^2 - 9n^2$

17. $25x^2 - 64y^2$

18. $36x^2 - 121y^2$

19. $16 - x^2$

20. $144 - x^2$

21. $3x^2 - 75$

22. $8x^2 - 392$

23. $x^2 + 25$

24. $x^2 + 1$

25. $5x^2 + 20$

26. $4x^2 + 100y^2$

27. $x^4 - 16$

28. $x^4 - 1$

29. $x^4 - 81y^2$

30. $x^8 - 64y^4$

31. $x^3 - 1$

32. $x^3 - y^3$

33. $x^3 - 8$

34. $a^3 - 216$

35. $1000 - y^3$

36. $343 - x^3$

37. $x^3 - 8y^3$

38. $y^3 - 27x^3$

39. $8a^3 - 729b^3$

40. $64r^3 - 125s^3$

41. $x^3 + y^3$

42. $x^3 + 27$

43. $x^3 + 8$

44. $b^3 + 64$

45. $x^6 + 27$

46. $y^{12} + 125$

47. $64m^3 + n^3$

48. $x^3 + 343y^3$

49. $6x^3 - 162$

50. $72a^3 + 243b^3$

51. $16x^4 + 250xy^3$

52. $64a^7b^{11} - a^4b^5$

Mixed Practice, 53–88

Factor completely, using the appropriate technique. If the polynomial cannot be factored, write "prime."

53. $x^2 - 16x + 64$

54. $4x^3 + 16x^2 + x + 4$

55. $x^2 + 16x + 28$

56. $2x^2 + 16x - 18$

57. $14x - 6$

58. $x^2 - 8x + 33$

59. $6x^2 - 5x - 25$

60. $-8x^2 + 32x + 18$

61. $x^4 - 2x^3 - 63x^2$

62. $x^2 + 18x + 77$

63. $-8x^3 - 72$

64. $25 - x^2$

65. $-6x^2 + 60x - 144$

66. $x^2 - 36x$

67. $x^2 - 10x - 39$

68. $x^2 + 4x - 140$

69. $25x^2 - 36y^8$

70. $x^2 + 3x - 40$

71. $2x^2 + x - 13$

72. $x^3 - 216y^3$

73. $x^2 - 64y^6$

74. $x^3 - 64y^6$

75. $9x^5 - 24x^4 - 45x^2$

76. $20x^2 - 61x + 36$

77. $4x^2 + 196$

78. $36x^2 + 6x - 210$

79. $x^2 + 12x + 35$

80. $343x^3 - 8y^6z^9$

81. $3x^2 - 13x - 38$

82. $2x^2 + 17x + 26$

83. $x^3 - 10x^2 - 3x + 30$

84. $x^2 + 16y^2$

85. $x^3 + 125y^3$

86. $125x^6 - 64y^9$

87. $9x^7 + 30x^6 + 25x^5$

88. $x^3 - 5x^2 + 2x - 10$

Determine which factoring technique is appropriate for the given polynomial.

89. ___ $8x^2 - 26x + 15$ **a.** factor out GCF

90. ___ $9a^2b^9 + 6a^4b^5 - 15a^6b^3$ **b.** factoring by grouping

91. ___ $x^2 + 121$ **c.** trinomial, leading coefficient of 1

92. ___ $x^2 - 3x - 54$ **d.** trinomial, leading coefficient other than 1

93. ___ $125x^3 - 216$ **e.** difference of squares

94. ___ $x^3 + 1331y^3$ **f.** sum of squares

95. ___ $x^3 - 3x^2 - 18x + 54$ **g.** difference of cubes

96. ___ $x^2 - 64$ **h.** sum of cubes

Writing in Mathematics

Answer in complete sentences.

97. Explain how to identify a binomial as a difference of squares.

98. A student was asked to factor $4 - x^2$. The student's answer was $(x + 2)(x - 2)$. Was the student correct? If not, what was the student's error? Explain fully.

99. Explain how to identify a binomial as a difference of cubes or as a sum of cubes.

100. Explain, in your own words, how to remember the formula for factoring a difference of cubes.

101. *Solutions Manual* * Write a solutions manual page for the following problem:

  Factor $8x^3 - 27y^3$.

102. *Newsletter* * Write a newsletter explaining how to factor a difference of squares.

*See Appendix B for details and sample answers.

6.5

Factoring Polynomials: A General Strategy

Objective **1** **Understand the strategy for factoring a general polynomial.** In this section, we will review the different factoring techniques introduced in this chapter and develop a general strategy for factoring any polynomial. Keep in mind that a polynomial must be factored completely, meaning each of its polynomial factors cannot be factored further. For example, if we factored the trinomial $4x^2 + 24x + 32$ to be $(2x + 4)(2x + 8)$, this would not be factored completely because the two terms in each binomial factor have a common factor of 2 that must be factored out as follows:

$$4x^2 + 24x + 32 = (2x + 4)(2x + 8)$$
$$= 2(x + 2) \cdot 2(x + 4)$$
$$= 4(x + 2)(x + 4)$$

If we factor out all common factors as our first step when factoring a polynomial, we can avoid having to factor out common factors at the end of the process. We could have factored out the common factor of 4 from the polynomial $4x^2 + 24x + 32$ at the very beginning. In addition to ensuring that the polynomial has been factored completely, factoring out common factors will often make our work easier, and will occasionally allow us to factor polynomials that we would not be able to factor otherwise.

Here is a general strategy for factoring any polynomial:

Factoring Polynomials

> **1.** Factor out any common factors.
>
> **2.** Determine the number of terms in the polynomial.
>   **(a)** If there are only **two terms,** check to see if the binomial is one of the special binomials discussed in Section 6.4.
>   - Difference of Squares: $a^2 - b^2 = (a + b)(a - b)$
>   - Sum of Squares: $a^2 + b^2$ is not factorable
>   - Difference of Cubes: $a^3 - b^3 = (a - b)(a^2 + ab + b^2)$
>   - Sum of Cubes: $a^3 + b^3 = (a + b)(a^2 - ab + b^2)$
>   **(b)** If there are **three terms,** try to factor the trinomial using the techniques of Sections 6.2 and 6.3.
>   - $x^2 + bx + c = (x + m)(x + n)$: find two integers m and n with a product of c and a sum of b.
>   - $ax^2 + bx + c$, where $a \neq 1$, factor either by grouping or by trial and error. To use factoring by grouping, refer to objective 1 in Section 6.3. To factor by trial and error, refer to objective 2 in Section 6.3.
>   **(c)** If there are **four terms,** try factoring by grouping, discussed in Section 6.1.
>
> **3.** After the polynomial has been factored, be sure that any factor with two or more terms does not have any common factors other than 1. If there are common factors, factor them out.
>
> **4.** Check your factoring through multiplication.

We will now factor several polynomials of various forms. Some of the examples will have twists that we did not see in the previous sections. The focus will be on identifying the best technique.

EXAMPLE ▶ **1**   Factor $-8x^2 - 80x + 192$ completely.

Solution

These three terms have a common factor of -8, and after we factor it out, we have $-8(x^2 + 10x - 24)$. This is a quadratic trinomial with a leading coefficient of 1, so we factor it using the method introduced in Section 6.2.

Quick Check ◀ **1**

Factor $-6x^2 + 54x + 60$ completely.

$$\begin{aligned} -8x^2 - 80x + 192 &= -8(x^2 + 10x - 24) \\ &= -8(x + 12)(x - 2) \end{aligned}$$

Factor out the GCF -8.
Find two integers with a product of -24 and a sum of 10. The integers are 12 and -2.

EXAMPLE ▶ **2**   Factor $343r^9 - 64s^6t^{12}$ completely.

Solution

There are no common factors other than 1, so we begin by noticing that this is a binomial. This is not a difference of squares, as $343r^9$ cannot be rewritten as a perfect square. It is, however, a difference of cubes. The variable factors all have exponents that are multiples of 3, and the coefficients are perfect cubes. (Recall the list of the first 10 perfect cubes given in Section 6.4: $343 = 7^3$ and $64 = 4^3$.) We begin by rewriting each term as a perfect cube.

$$\begin{aligned} 343r^9 - 64s^6t^{12} &= (7r^3)^3 - (4s^2t^4)^3 \\ &= (7r^3 - 4s^2t^4)(7r^3 \cdot 7r^3 + 7r^3 \cdot 4s^2t^4 + 4s^2t^4 \cdot 4s^2t^4) \\ \\ \\ &= (7r^3 - 4s^2t^4)(49r^6 + 28r^3s^2t^4 + 16s^4t^8) \end{aligned}$$

Rewrite each term as a perfect cube.
Factor as a difference of cubes. (Recall the pattern for the trinomial factor: 1st · 1st + 1st · 2nd + 2nd · 2nd.)
Simplify each term in the trinomial factor.

For a difference or sum of cubes, check to be sure that the binomial factor cannot be factored further as one of our special binomials.

Quick Check ▶ **2** Factor $216x^{12}y^{21} + 125$ completely.

EXAMPLE ▶ **3**   Factor $2x^3 - 22x^2 - 8x + 88$ completely.

Solution

The four terms have a common factor of 2, so we will first factor out the GCF. Then we will try to factor by grouping. If we have trouble factoring the polynomial as written, we can always rearrange the order of its terms.

$$2x^3 - 22x^2 - 8x + 88 = 2(x^3 - 11x^2 - 4x + 44)$$
$$= 2[x^2(x - 11) - 4(x - 11)]$$

Factor out the GCF.

Factor the common factor x^2 from the first two terms and factor the common factor -4 from the last two terms.

$$= 2(x - 11)(x^2 - 4)$$

Factor out the common factor $x - 11$.

Quick Check 3

Factor
$x^4 + 5x^3 - 8x - 40$
completely.

Notice that the factor $x^2 - 4$ is a difference of squares and must be factored further.

$$2(x - 11)(x^2 - 4) = 2(x - 11)(x + 2)(x - 2)$$

Factor the difference of squares.
$x^2 - 4 = (x + 2)(x - 2)$

If a binomial is both a difference of squares and a difference of cubes, such as $x^6 - 64$, to factor it completely we must start by factoring it as a difference of squares. The next example illustrates this.

EXAMPLE 4   Factor $x^6 - 64$ completely.

Solution

There are no common factors to factor out, so we begin by factoring this binomial as a difference of squares.

$$x^6 - 64 = (x^3)^2 - (8)^2$$ Rewrite each term as a perfect square.
$$= (x^3 + 8)(x^3 - 8)$$ Factor as a difference of squares.

Notice that each factor can be factored further. The binomial $x^3 + 8$ is a sum of cubes and the binomial $x^3 - 8$ is a difference of cubes. We use the fact that

Quick Check 4

Factor $x^{12} - y^6$
completely.

$x^3 + 8 = (x + 2)(x^2 - 2x + 4)$ and $x^3 - 8 = (x - 2)(x^2 + 2x + 4)$ to complete the factoring.

$$x^6 - 64 = (x + 2)(x^2 - 2x + 4)(x - 2)(x^2 + 2x + 4).$$

Note what would have occurred if we had first factored $x^6 - 64$ as a difference of cubes.

$$x^6 - 64 = (x^2)^3 - (4)^3$$ Rewrite each term as a perfect cube.
$$= (x^2 - 4)(x^4 + 4x^2 + 16)$$ Factor the difference of cubes.
$$= (x + 2)(x - 2)(x^4 + 4x^2 + 16)$$ Factor the difference of squares.

At this point, we do not know how to factor $x^4 + 4x^2 + 16$, so the technique of Example 4 has factored the binomial $x^6 - 64$ more completely.

EXAMPLE 5   Factor $x^{12} + 1$ completely.

Solution

Quick Check 5

Factor $x^6 + 64$
completely.

The two terms do not have a common factor other than 1, so we need to determine whether this binomial matches one of our special forms. The binomial can be rewritten as a sum of cubes, $(x^4)^3 + (1)^3$.

$$x^{12} + 1 = (x^4)^3 + (1)^3$$ Rewrite each term as a perfect cube.
$$= (x^4 + 1)(x^8 - x^4 + 1)$$ Factor as a sum of cubes.

EXAMPLE ▸ 6   Factor $15x^5 - 55x^4 + 40x^3$ completely.

Solution

The three terms have a common factor of $5x^3$, so we begin by factoring this out of the trinomial. After we have done this, we can factor the trinomial factor $3x^2 - 11x + 8$ using either of the techniques in Section 6.3. We will factor it by grouping, although you may prefer to use trial and error.

$$15x^5 - 55x^4 + 40x^3 = 5x^3(3x^2 - 11x + 8)$$

Factor out the common factor $5x^3$.

$$= 5x^3(3x^2 - 3x - 8x + 8)$$

To factor by grouping we need to find two integers whose product is equal to $3 \cdot 8 = 24$ and whose sum is -11. The two integers are -3 and -8, so we rewrite the term $-11x$ as $-3x - 8x$.

$$= 5x^3[3x(x - 1) - 8(x - 1)]$$

Factor the common factor $3x$ from the first two terms and factor the common factor -8 from the last two terms.

$$= 5x^3(x - 1)(3x - 8)$$

Factor out the common factor $x - 1$.

Quick Check ◂ 6

Factor $8x^2 + 42x - 36$ completely.

Building Your Study Strategy Overcoming Math Anxiety, **5 Study Skills**
Many students who perform poorly in their math class attribute their performance to math anxiety, when, in reality, poor study skills are to blame. Be sure that you are giving your fullest possible effort, including

- working with a study group;
- reading the text before the material is covered in class;
- rereading the text after the material is covered in class;
- completing each homework assignment;
- making note cards for particularly difficult problems or procedures;
- getting help if there is a problem that you do not understand;
- asking your instructor questions when you do not understand; and
- seeing a tutor.

EXERCISES 6.5

Vocabulary

1. The first step for factoring a trinomial is to factor out _____, if there are any.

2. A binomial of the form $a^2 - b^2$ is a _____.

3. A binomial of the form $a^2 + b^2$ is a _____.

4. A binomial of the form $a^3 - b^3$ is a _____.

5. A binomial of the form $a^3 + b^3$ is a _____.

6. Factoring can be checked by _____ the factors.

Factor completely. If the polynomial cannot be factored, write "prime."

7. $27x^3 - 8y^3$

8. $x^2 - 8x - 65$

9. $4x^3 - 28x^2 - x + 7$

10. $x^2 - 10x + 20$

11. $x^6 - 1$

12. $-10x^2 + 70x + 180$

13. $x^2 - 18x + 45$

14. $x^2 - 64$

15. $x^6 + 64y^{12}$

16. $x^2 + 17x - 30$

17. $4x^2 + 56x + 192$

18. $18x^2 - 60x + 42$

19. $3x^2 + 10x - 48$

20. $x^3 - 1000y^{18}$

21. $x^2 + 13x + 42$

22. $x^4 + 81y^2$

23. $x^4 + 16x^3 + x + 16$

24. $9x^2 + 54x + 72$

25. $9x^2 - 30x + 25$

26. $a^5b^7 - 3a^4b^6 + 8a^2b^9$

27. $x^2 - 14xy + 49y^2$

28. $x^2 + 7xy - 60y^2$

29. $x^2 + 4x - 165$

30. $x^7 + 18x^5 - 36x$

31. $12x + 20$

32. $x^3 - 9x^2 - 4x + 36$

33. $x^2 + 13x - 14$

34. $x^8 - 16y^6$

35. $6x^2 - 29x - 5$

36. $16x^{10} + 49y^6$

37. $84x^2 + 4x - 20$

38. $x^2y^2 + 17xy + 16$

39. $7x^2 + 29x + 4$

40. $1 + 25a^2b^8$

41. $9x^3 + 18x^2 - 4x - 8$

42. $7x^3 + 189$

43. $x^2 - 6x - 280$

44. $x^{16} - 1$

45. $4x^2 + 4x - 224$

46. $12x^2 - 41x + 22$

47. $x^6 - y^6$

48. $8 + x^{15}$

49. $-x^2 + 7x - 12$

50. $2x^2 - 3x - 12$

51. $6x^2 + 11x - 10$

52. $x^2 - 23x + 42$

53. $x^6 + 729$

54. $9x^2 - 29x + 6$

55. $27x^9 + 125y^{15}$

56. $x^4 - 64$

57. $x^2 - 3x - 154$

58. $8x^2 + 19x + 6$

59. $x^2 + 20x + 100$

60. $x^{21} - 1$

61. $6x^2 + 75x + 225$

62. $x^4 - 25x^2$

Find the missing term(s).

63. $x^2 + 26x - 120 = (x - 4)(x + ?)$

64. $27x^3 + 512y^9z^{12} = (3x + 8y^3z^4)(9x^2 - 24xy^3z^4 + ?)$

65. $x^2 - 1444 = (x + ?)(x - ?)$

66. $28x^2 - 15x - 247 = (4x - 13)(?)$

67. $2058x^3 - 750z^6 = 6(?)(49x^2 + 35xz^2 + 25z^4)$

68. $x^6 - 46{,}656 = (x + 6)(x^2 - 6x + 36)(?)$
$(x^2 + 6x + 36)$

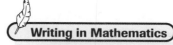

Writing in Mathematics

Answer in complete sentences.

69. *Solutions Manual** Write a solutions manual page for the following problem:

Factor.

- $x^3 + 3x^2 - 10x - 30$
- $x^2 - 5x - 36$
- $3x^2 + 7x - 20$
- $9x^2 - 25$
- $x^3 - 343$
- $x^3 + 125$

70. *Newsletter** Write a newsletter explaining the following strategies for factoring polynomials:

- Factoring out the GCF
- Factoring by grouping
- Factoring trinomials of the form $x^2 + bx + c$
- Factoring polynomials of the form $ax^2 + bx + c$, where $a \neq 1$, by grouping
- Factoring polynomials of the form $ax^2 + bx + c$, where $a \neq 1$, by trial and error
- Factoring a difference of squares
- Factoring a difference of cubes
- Factoring a sum of cubes

*See Appendix B for details and sample answers.

QUICK REVIEW EXERCISES

Section 6.5

Solve.

1. $x - 7 = 0$

2. $x + 3 = 0$

3. $5x - 12 = 0$

4. $2x + 29 = 0$

6.6

Solving Quadratic Equations by Factoring

Objectives

1. Solve an equation by using the zero-factor property of real numbers.
2. Solve a quadratic equation by factoring.
3. Solve a quadratic equation that is not in standard form.
4. Solve a quadratic equation with coefficients that are fractions.
5. Find a quadratic equation, given its solutions.

Quadratic Equations

A **quadratic equation** is an equation that can be written as $ax^2 + bx + c = 0$, where a, b, and c are real numbers and $a \neq 0$. This form is the **standard form of a quadratic equation.**

We have already learned to solve linear equations ($ax + b = 0$). The difference between these two types of equations is that a quadratic equation has a second-degree term, ax^2. Because of the second-degree term, the techniques used to solve linear equations will not work for quadratic equations.

The Zero-Factor Property of Real Numbers

Objective 1 Solve an equation by using the zero-factor property of real numbers. To solve a quadratic equation, we will use the **zero-factor property of real numbers.**

Zero-Factor Property of Real Numbers

If $a \cdot b = 0$, then $a = 0$ or $b = 0$.

The principle behind this property is that if two or more unknown numbers have a product of zero, then at least one of the numbers must be zero. This property only holds true when the product is equal to 0, not for any other numbers.

EXAMPLE 1 Use the zero-factor property to solve the equation $(x + 3)(x - 7) = 0$.

Solution

In this example, the product of two unknown numbers, $x + 3$ and $x - 7$, is equal to 0. The zero-factor property tells us that either $x + 3 = 0$ or $x - 7 = 0$. Essentially, we have taken an equation that we did not know how to solve yet, $(x + 3)(x - 7) = 0$, and rewritten it as two linear equations that we do know how to solve.

$$(x + 3)(x - 7) = 0$$

$x + 3 = 0$	or	$x - 7 = 0$	Set each factor equal to 0.
$x = -3$	or	$x = 7$	Solve each linear equation.

There are two solutions to this equation, -3 and 7. We write these solutions in a solution set as $\{-3, 7\}$.

EXAMPLE 2   Use the zero-factor property to solve the equation $x(3x - 5) = 0$.

Solution

Applying the zero-factor property gives us the equations $x = 0$ and $3x - 5 = 0$. The first equation, $x = 0$, is already solved, so we only need to solve the equation $3x - 5 = 0$ to find the second solution.

$$x(3x - 5) = 0$$

$x = 0$	or	$3x - 5 = 0$	Set each factor equal to 0.
$x = 0$	or	$3x = 5$	Solve each linear equation.
$x = 0$	or	$x = \dfrac{5}{3}$	

Quick Check   1
Use the zero-factor property to solve the equation.

a) $(x - 2)(x + 8) = 0$
b) $x(4x + 9) = 0$

The solution set for this equation is $\{0, \frac{5}{3}\}$.

If an equation has the product of more than two factors equal to 0, such as $(x + 1)(x - 2)(x + 4) = 0$, we set each factor equal to 0 and solve. The solution set to this equation is $\{-1, 2, -4\}$.

Solving Quadratic Equations by Factoring

Objective 2   Solve a quadratic equation by factoring. To solve a quadratic equation, we will use the following procedure.

Solving Quadratic Equations by Factoring

1. Write the equation in standard form: $ax^2 + bx + c = 0$. *We need to collect all the terms on one side of the equation. It helps to collect all the terms so that the coefficient of the squared term is positive.*
2. Factor the expression $ax^2 + bx + c$ completely. *If you are struggling with factoring, refer back to Sections 6.1 through 6.5.*
3. Set each factor equal to 0 and solve the resulting equations. *Each of these equations should be a linear equation.*
4. Finish by checking the solutions. *This is an excellent opportunity to catch mistakes in factoring.*

EXAMPLE 3   Solve $x^2 + 7x - 30 = 0$.

Solution

This equation is already in standard form, so we begin by factoring the expression $x^2 + 7x - 30$.

$x^2 + 7x - 30 = 0$			
$(x + 10)(x - 3) = 0$			Factor.
$x + 10 = 0$	or	$x - 3 = 0$	Set each factor equal to 0.
$x = -10$	or	$x = 3$	Solve each equation.

The solution set is $\{-10, 3\}$. Now we will check our solutions.

Check ($x = -10$)	Check ($x = 3$)
$x^2 + 7x - 30 = 0$	$x^2 + 7x - 30 = 0$
$(-10)^2 + 7(-10) - 30 = 0$	$(3)^2 + 7(3) - 30 = 0$
$100 + 7(-10) - 30 = 0$	$9 + 7(3) - 30 = 0$
$100 - 70 - 30 = 0$	$9 + 21 - 30 = 0$
$0 = 0$	$0 = 0$

Quick Check 2
Solve $x^2 - x - 20 = 0$.

The check shows that our solutions are correct.

> **A Word of Caution** Pay close attention to the directions of a problem. If you are asked to solve a quadratic equation, do not just factor the quadratic expression and stop. If you are asked to factor a quadratic expression, do not set each factor equal to 0 and solve the resulting equations.

EXAMPLE 4   Solve $3x^2 - 18x - 48 = 0$.

Solution

The equation is in standard form. To factor $3x^2 - 18x - 48$, we begin by factoring out the common factor 3.

$$3x^2 - 18x - 48 = 0$$
$$3(x^2 - 6x - 16) = 0 \qquad \text{Factor out the GCF.}$$
$$3(x - 8)(x + 2) = 0 \qquad \text{Factor } x^2 - 6x - 16.$$
$$x - 8 = 0 \quad \text{or} \quad x + 2 = 0 \qquad \text{Set each factor containing a variable equal to 0. We can ignore the numerical factor 3.}$$
$$x = 8 \quad \text{or} \quad x = -2 \qquad \text{Solve each equation.}$$

The solution set is $\{8, -2\}$. The check is left to the reader.

We do not need to set a numerical factor equal to 0, because such an equation will not have a solution. For example, the equation $3 = 0$ has no solution. The zero-factor property tells us that at least one of the three factors must equal 0; we just know that it cannot be the factor 3 that is equal to 0.

Quick Check 3
Solve
$4x^2 + 60x + 224 = 0$.

> **A Word of Caution** A common numerical factor does not affect the solutions of a quadratic equation.

EXAMPLE 5   Solve $2x^2 - 7x - 15 = 0$.

Solution

When factoring the expression $2x^2 - 7x - 15$, there is not a common factor other than 1 that can be factored out. So the leading coefficient of this trinomial is not 1. We can factor by trial and error or by grouping. We will use factoring by grouping. We look for two integers whose product is equal to $2 \cdot 15$ or 30 and whose sum is -7. The two integers are

−10 and 3, so we can rewrite the term $-7x$ as $-10x + 3x$ and then factor by grouping. (Refer to Section 6.3 to review this technique.)

$$2x^2 - 7x - 15 = 0$$
$$2x^2 - 10x + 3x - 15 = 0 \qquad \text{Rewrite } -7x \text{ as } -10x + 3x.$$
$$2x(x - 5) + 3(x - 5) = 0 \qquad \text{Factor the common factor } 2x \text{ from the}$$

first two terms and factor the common
factor 3 from the last two terms.

$$(x - 5)(2x + 3) = 0 \qquad \text{Factor out the common factor } x - 5.$$
$$x - 5 = 0 \qquad \text{or} \qquad 2x + 3 = 0 \qquad \text{Set each factor equal to 0.}$$
$$x = 5 \qquad \text{or} \qquad 2x = -3 \qquad \text{Solve each equation.}$$
$$x = 5 \qquad \text{or} \qquad x = -\frac{3}{2}$$

Quick Check 4   The solution set is $\{5, -\frac{3}{2}\}$. The check is left to the reader.

Solve
$6x^2 - 23x + 7 = 0.$

Objective 3   **Solve a quadratic equation that is not in standard form.** We now turn our attention to equations that are not already in standard form. In each of the examples that follows, the check of the solutions is left to the reader.

EXAMPLE 6   Solve $x^2 = 49$.

Solution

We begin by rewriting this equation in standard form. This can be done by subtracting 49 from each side of the equation. Once this has been done, the expression to be factored in this example is a difference of squares.

$$x^2 = 49$$
$$x^2 - 49 = 0 \qquad \text{Subtract 49.}$$
$$(x + 7)(x - 7) = 0 \qquad \text{Factor.}$$
$$x + 7 = 0 \qquad \text{or} \qquad x - 7 = 0 \qquad \text{Set each factor equal to 0.}$$
$$x = -7 \qquad \text{or} \qquad x = 7 \qquad \text{Solve each equation.}$$

The solution set is $\{-7, 7\}$.

EXAMPLE 7   Solve $x^2 + 25 = 10x$.

Solution

To rewrite this equation in standard form, we need to subtract $10x$ so that all terms will be on the left side of the equation. When we subtract $10x$, we must be sure to write the terms in descending order.

$$x^2 + 25 = 10x$$
$$x^2 - 10x + 25 = 0 \qquad \text{Subtract } 10x.$$
$$(x - 5)(x - 5) = 0 \qquad \text{Factor.}$$
$$x - 5 = 0 \qquad \text{or} \qquad x - 5 = 0 \qquad \text{Set each factor equal to 0.}$$
$$x = 5 \qquad \text{or} \qquad x = 5 \qquad \text{Solve each equation.}$$

Notice that both solutions are identical. In this case, we only need to write the repeated

Quick Check 5   solution once. The solution set is $\{5\}$.

Solve $x^2 + 4x = 45$.

Occasionally, we will need to simplify one or both sides of an equation in order to rewrite the equation in standard form. For instance, to solve the equation $x(x + 9) = 10$, we must first multiply x by $x + 9$. You may be wondering why we would want to multiply out the left side, since it is already factored. Although it is factored, the product is equal to 10, not 0.

EXAMPLE ▸ 8   Solve $x(x + 9) = 10$.

Solution

As mentioned, we must first multiply x by $x + 9$. Then we can rewrite the equation in standard form.

$$
\begin{array}{ll}
x(x + 9) = 10 & \\
x^2 + 9x = 10 & \text{Multiply.} \\
x^2 + 9x - 10 = 0 & \text{Subtract 10.} \\
(x + 10)(x - 1) = 0 & \text{Factor.} \\
x + 10 = 0 \quad \text{or} \quad x - 1 = 0 & \text{Set each factor equal to 0.} \\
x = -10 \quad \text{or} \quad x = 1 & \text{Solve each equation.}
\end{array}
$$

Quick Check 6   The solution set is $\{-10, 1\}$.

Solve $x(x - 3) = 70$.

> **A Word of Caution**   Be sure that the equation you are solving is written as a product equal to 0 before setting each factor equal to 0 and solving.

Objective 4 **Solve a quadratic equation with coefficients that are fractions.**
If an equation contains fractions, clearing those fractions makes it easier to factor the quadratic expression. We can clear the fractions by multiplying both sides of the equation by the LCM of the denominators.

EXAMPLE ▸ 9   Solve $\frac{1}{6}x^2 + x + \frac{4}{3} = 0$.

Solution

The LCM for these two denominators is 6, so we can clear the fractions by multiplying both sides of the equation by 6.

$$\frac{1}{6}x^2 + x + \frac{4}{3} = 0$$

$$6 \cdot \left(\frac{1}{6}x^2 + x + \frac{4}{3}\right) = 6 \cdot 0 \qquad \text{Multiply both sides by 6, which is the LCM of the denominators.}$$

$$\overset{1}{\cancel{6}} \cdot \frac{1}{\cancel{6}}x^2 + 6 \cdot x + \overset{2}{\cancel{6}} \cdot \frac{4}{\underset{1}{\cancel{3}}} = 6 \cdot 0 \qquad \text{Distribute and divide out common factors.}$$

$$
\begin{array}{ll}
x^2 + 6x + 8 = 0 & \text{Multiply.} \\
(x + 2)(x + 4) = 0 & \text{Factor.} \\
x + 2 = 0 \quad \text{or} \quad x + 4 = 0 & \text{Set each factor equal to 0.} \\
x = -2 \quad \text{or} \quad x = -4 & \text{Solve each equation.}
\end{array}
$$

Quick Check 7   The solution set is $\{-2, -4\}$.

Solve $\frac{2}{45}x^2 + \frac{4}{15}x - \frac{6}{5} = 0$.

Finding a Quadratic Equation, Given Its Solutions

Objective 5 **Find a quadratic equation, given its solutions.** If we know the two solutions to a quadratic equation, then we can determine an equation with these solutions. The next example illustrates this process.

EXAMPLE 10   Find a quadratic equation in standard form, with integer coefficients, that has the solution set $\{4, -9\}$.

Solution

We know that $x = 4$ is a solution to the equation. This tells us that $x - 4$ is a factor of the quadratic expression. Similarly, knowing that $x = -9$ is a solution tells us that $x + 9$ is a factor of the quadratic expression. Multiplying these two factors will give us a quadratic equation with these two solutions.

$x = 4$	or	$x = -9$	Begin with the solutions.
$x - 4 = 0$	or	$x + 9 = 0$	Rewrite each equation so the right side is equal to 0.
$(x - 4)(x + 9) = 0$			Write an equation that has these two expressions as factors.
$x^2 + 5x - 36 = 0$			Multiply.

A quadratic equation that has the solution set $\{4, -9\}$ is $x^2 + 5x - 36 = 0$.

Quick Check   8

Find a quadratic equation in standard form, with integer coefficients, that has the solution set $\{-6, 8\}$.

Notice that we say "*a*" quadratic equation rather than "*the*" quadratic equation. There are infinitely many quadratic equations with integer coefficients that have this solution set. For example, multiplying both sides of our equation by 2 gives us the equation $2x^2 + 10x - 72 = 0$, which has the same solution set.

Building Your Study Strategy Overcoming Math Anxiety, 6 **Relaxation** Learning to relax in stressful situations will help you keep your anxiety under control. By relaxing, you will be able to avoid the physical symptoms associated with math anxiety such as feeling tense and uncomfortable, sweating, shortness of breath, and an accelerated heartbeat.

Wanting to relax and being able to relax are often two different things. Many students use deep-breathing exercises. Close your eyes and take a deep breath. Try to clear your mind and relax your entire body. Slowly exhale, and then repeat the process until you feel relaxed.

There are other relaxation methods that may work for you, but you must make sure that they are appropriate in class. For instance, primal scream therapy would be inappropriate during an exam.

EXERCISES 6.6

Vocabulary

1. A _____ is an equation that can be written as $ax^2 + bx + c = 0$, where a, b, and c are real numbers and $a \neq 0$.

2. A quadratic equation is in standard form if it is in the form _____, where $a \neq 0$.

3. The _____ property of real numbers states that if $a \cdot b = 0$, then $a = 0$ or $b = 0$.

4. Once a quadratic expression has been set equal to 0 and factored, we _____ and solve.

Solve.

5. $(x + 7)(x - 100) = 0$

6. $(x - 2)(x - 10) = 0$

7. $x(x - 12) = 0$

8. $x(x + 5) = 0$

9. $9(x - 4)(x - 8) = 0$

10. $-2(x + 2)(x - 16) = 0$

11. $(2x + 7)(x - 5) = 0$

12. $(x + 1)(5x - 18) = 0$

13. $x^2 - 5x + 4 = 0$

14. $x^2 - 11x + 24 = 0$

15. $x^2 + 11x + 30 = 0$

16. $x^2 + 7x + 12 = 0$

17. $x^2 + 6x - 40 = 0$

18. $x^2 - 3x - 88 = 0$

19. $x^2 - 12x - 45 = 0$

20. $x^2 + 7x - 44 = 0$

21. $x^2 - 16x + 64 = 0$

22. $x^2 + 12x + 36 = 0$

23. $x^2 - 81 = 0$

24. $x^2 - 4 = 0$

25. $x^2 + 7x = 0$

26. $x^2 - 10x = 0$

27. $5x^2 - 22x = 0$

28. $14x^2 + 13x = 0$

29. $2x^2 + 8x - 154 = 0$

30. $7x^2 - 7x - 630 = 0$

31. $-x^2 + 14x - 13 = 0$

32. $-x^2 + 5x + 66 = 0$

33. $-4x^2 - 44x + 168 = 0$

34. $-9x^2 - 81x - 126 = 0$

35. $27x^2 - 3x - 2 = 0$

36. $6x^2 + 7x - 68 = 0$

37. $x^2 - 2x = 35$

38. $x^2 + 8x = -15$

39. $x^2 = 8x - 16$

40. $x^2 - 13x = 48$

41. $x^2 + 13x = 6x - 10$

42. $x^2 - x = 3x + 12$

43. $2x^2 - x - 13 = x^2 + 11x + 15$

44. $x^2 + 5x = 2x^2 + 7x - 8$

45. $x^2 = 9$

46. $x^2 = 25$

47. $x^2 + 11x = 11x + 36$

48. $x^2 - 3x = -3x + 100$

49. $x(x + 4) = 4(x + 16)$

50. $x(x + 7) = (4x + 3) + (3x + 13)$

51. $x(x + 13) = -40$

52. $x(x - 2) = 24$

53. $(x - 2)(x - 5) = 40$

54. $(x + 9)(x - 7) = -55$

55. $(x + 2)(x + 3) = (x + 7)(x - 4)$

56. $(x - 6)(x - 4) = (x + 12)(x - 10)$

57. $\dfrac{1}{6}x^2 - \dfrac{5}{4}x - \dfrac{9}{4} = 0$

58. $\dfrac{1}{2}x^2 + \dfrac{11}{12}x + \dfrac{1}{3} = 0$

59. $\dfrac{2}{15}x^2 - \dfrac{7}{10}x + \dfrac{2}{3} = 0$

60. $\dfrac{3}{4}x^2 - \dfrac{3}{4}x - \dfrac{5}{6} = 0$

Find a quadratic equation with integer coefficients that has the given solution set.

61. $\{4, 5\}$

62. $\{7, -10\}$

63. $\{0, 6\}$

64. $\{4\}$

65. $\{8, -8\}$

66. $\left\{-4, \dfrac{5}{4}\right\}$

67. $\left\{-\dfrac{3}{7}, \dfrac{7}{3}\right\}$

68. $\left\{-\dfrac{2}{5}, -\dfrac{9}{2}\right\}$

69. Use the fact that $x = \frac{7}{2}$ is a solution to the equation $6x^2 + 7x - 98 = 0$ to find the other solution to the equation.

70. Use the fact that $x = \frac{13}{3}$ is a solution to the equation $3x^2 - 28x + 65 = 0$ to find the other solution to the equation.

For the given quadratic equation, one of its solutions has been provided. Use it to find the other solution.

71. $x^2 + 6x - 667 = 0$, $x = 23$

72. $x^2 - 78x + 1517 = 0$, $x = 41$

73. $15x^2 + 11x - 532 = 0$, $x = -\dfrac{19}{3}$

74. $48x^2 - 154x - 735 = 0$, $x = \dfrac{35}{6}$

Writing in Mathematics

Answer in complete sentences.

75. Explain the procedure for solving a quadratic equation by factoring. Give an example to illustrate.

76. Explain the zero-factor property of real numbers. Describe how this property is used when solving quadratic equations by factoring.

77. When solving a quadratic equation by factoring, explain why we should not factor until one side of the quadratic equation is set equal to 0.

78. Explain why the common factor of 2 has no effect on the solutions of the equation $2(x - 7)(x + 5) = 0$.

79. *Solutions Manual** Write a solutions manual page for the following problem:

Solve $\dfrac{1}{6}x^2 - \dfrac{1}{3}x - 4 = 0$.

80. *Newsletter** Write a newsletter explaining how to solve an equation of the form $ax^2 + bx + c = 0$, where $a \neq 0$.

See Appendix B for details and sample answers.

QUICK REVIEW EXERCISES

Section 6.6

Evaluate the given function.

1. $f(x) = 3x + 8$, $f(5)$

2. $f(x) = -10x + 21$, $f(-8)$

3. $f(x) = -7x - 19$, $f(2b + 3)$

4. $f(x) = x^2 + 7x - 20$, $f(6)$

6.7
Quadratic Functions

Objectives

1. Evaluate quadratic functions.
2. Solve equations involving quadratic functions.
3. Solve applied problems involving quadratic functions.

We first investigated linear functions in Chapter 3. A linear function is a function of the form $f(x) = mx + b$. In this section we turn our attention to **quadratic functions.**

Quadratic Functions

A quadratic function is a function of the form $f(x) = ax^2 + bx + c$, where $a \neq 0$.

Evaluating Quadratic Functions

Objective 1 Evaluate quadratic functions. We begin our investigation of quadratic functions by learning to evaluate these functions for particular values of the variable. For example, if we are asked to find $f(3)$, we are being asked to evaluate the function $f(x)$ at $x = 3$. To do this, we substitute 3 for the variable x and simplify the resulting expression.

EXAMPLE 1   For the function $f(x) = x^2 - 5x - 8$, find $f(-4)$.

Solution

We will substitute -4 for the variable x and then simplify.

$$\begin{aligned} f(-4) &= (-4)^2 - 5(-4) - 8 && \text{Substitute } -4 \text{ for } x. \\ &= 16 - 5(-4) - 8 && \text{Square } -4.\ (-4)(-4) = 16 \\ &= 16 + 20 - 8 && \text{Multiply.} \\ &= 28 && \text{Simplify.} \end{aligned}$$

Quick Check 1
For the function
$f(x) = x^2 - 9x + 405$,
find $f(-6)$.

Solving Equations Involving Quadratic Functions

Objective 2 Solve equations involving quadratic functions. Now that we have learned to solve quadratic equations, we can solve equations involving quadratic functions. Suppose that we were trying to find all values x for which some function $f(x)$ was equal to 0. We replace $f(x)$ by its formula and solve the resulting quadratic equation.

EXAMPLE 2   Let $f(x) = x^2 + 11x - 26$. Find all values x for which $f(x) = 0$.

Solution

We begin by replacing the function by its formula, and then we solve the resulting equation.

$$\begin{aligned} f(x) &= 0 \\ x^2 + 11x - 26 &= 0 && \text{Replace } f(x) \text{ by its formula} \\ && & x^2 + 11x - 26. \\ (x + 13)(x - 2) &= 0 && \text{Factor.} \\ x + 13 = 0 \quad &\text{or} \quad x - 2 = 0 && \text{Set each factor equal to 0.} \\ x = -13 \quad &\text{or} \quad x = 2 && \text{Solve each equation.} \end{aligned}$$

The two values x for which $f(x) = 0$ are -13 and 2.

Quick Check **2** Let $f(x) = x^2 - 17x + 72$. Find all values x for which $f(x) = 0$.

EXAMPLE **3**   Let $f(x) = x^2 - 24$. Find all values x for which $f(x) = 25$.

Solution

After setting the function equal to 25, we need to rewrite the equation in standard form in order to solve it.

$$f(x) = 25$$
$$x^2 - 24 = 25 \qquad\qquad \text{Replace } f(x) \text{ by its formula.}$$
$$x^2 - 49 = 0 \qquad\qquad \text{Subtract 25.}$$
$$(x + 7)(x - 7) = 0 \qquad\qquad \text{Factor.}$$
$$x + 7 = 0 \qquad \text{or} \qquad x - 7 = 0 \qquad \text{Set each factor equal to 0.}$$
$$x = -7 \qquad \text{or} \qquad x = 7 \qquad \text{Solve each equation.}$$

The two values x for which $f(x) = 25$ are -7 and 7.

Quick Check **3** Let $f(x) = x^2 - 6x + 20$. Find all values x for which $f(x) = 92$.

Applications

Objective **3** **Solve applied problems involving quadratic functions.** We conclude the section with an applied problem that requires interpreting the graph of a quadratic function, a U-shaped curve called a **parabola.** Here are some examples. Note that these graphs are not linear.

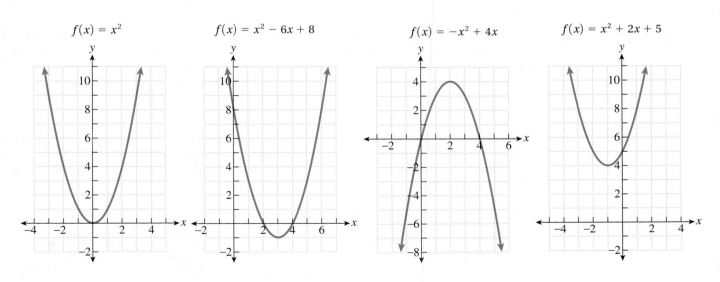

$f(x) = x^2$ $\qquad\qquad$ $f(x) = x^2 - 6x + 8$ $\qquad\qquad$ $f(x) = -x^2 + 4x$ $\qquad\qquad$ $f(x) = x^2 + 2x + 5$

EXAMPLE 4   A rocket is fired into the air from the ground at a speed of 96 feet per second. Its height, in feet, after t seconds is given by the function $h(t) = -16t^2 + 96t$, which is graphed as follows:

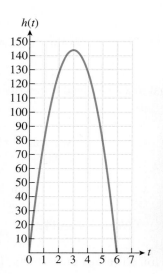

Use the graph to answer the following questions:

a) How high above the ground is the rocket after 5 seconds?

Solution   To answer this question, we need to find the function's value when $t = 5$. Looking at the graph, we see that it passes through the point $(5, 80)$. The height of the rocket is 80 feet after 5 seconds have passed.

b) How long will it take for the rocket to land on the ground?

Solution   When the rocket lands on the ground, its height is 0 feet. We need to find the values of t for which $h(t) = 0$. There are two such values, $t = 0$ and $t = 6$. The time $t = 0$ represents the instant the rocket was fired into the air, so the time that it takes to land on the ground is represented by $t = 6$. It takes the rocket 6 seconds to land on the ground.

c) After how many seconds does the rocket reach its greatest height?

Solution   On the graph, we are looking for the point at which the function reaches its highest point. (This point where the parabola changes from rising to falling is called the **vertex** of the parabola.) We see that this point corresponds to a t value of 3, so it reaches its greatest height after 3 seconds.

d) What is the greatest height that the rocket reaches?

Solution   From the graph, the best that we can say is that the maximum height is between 140 feet and 150 feet. However, since we know that the object reaches its greatest height after 3 seconds, we can evaluate the function $h(t) = -16t^2 + 96t$ at $t = 3$.

$$h(t) = -16t^2 + 96t$$

$$\begin{aligned}
h(3) &= -16(3)^2 + 96(3) & \text{Replace } t \text{ by 3.}\\
&= -16(9) + 96(3) & \text{Square 3.}\\
&= -144 + 288 & \text{Multiply.}\\
&= 144 & \text{Simplify.}
\end{aligned}$$

The rocket reaches a height of 144 feet.

Quick Check **4** A ball is thrown into the air from the top of a building 48 feet above the ground at a speed of 32 feet per second. Its height, in feet, after t seconds is given by the function $h(t) = -16t^2 + 32t + 48$, which is graphed as follows.

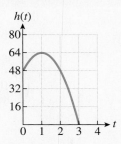

Use the graph to answer the following questions.

a) How high above the ground is the ball after 2 seconds?
b) How long will it take for the ball to land on the ground?
c) After how many seconds does the ball reach its greatest height?
d) What is the greatest height that the ball reaches?

Building Your Study Strategy Overcoming Math Anxiety, 7 **Positive Attitude**
Math anxiety causes many students to think negatively about math. A positive attitude will lead to more success than a negative attitude. Jot down some of the negative thoughts that you have about math and your ability to learn and understand mathematics. Write each of these thoughts on a note card, then on the other side of the note card, write the exact opposite statement. For example, if you write "I am not smart enough to learn mathematics," on the other side of the card, you should write "I am smart. I do well in all of my classes, and I can learn mathematics also!" Once you have finished your note cards, cycle through the positive thoughts on a regular basis.

A little confidence will go a long way towards improving your performance in class. Have the confidence to sit in the front row and ask questions during class, knowing that having your questions answered will increase your understanding and your chances for a better grade on the next exam. Finally, each time you have a success, no matter how small it may be, reward yourself.

Vocabulary

1. A _____ is a function that can be expressed as $f(x) = ax^2 + bx + c$, where $a \neq 1$.

2. The graph of a quadratic function is a U-shaped curve called a _____.

Evaluate the given function.

3. $f(x) = x^2 + 6x + 10, f(3)$

4. $f(x) = x^2 + 2x + 15, f(6)$

5. $f(x) = x^2 + 8x + 5, f(-2)$

6. $f(x) = x^2 + 4x + 20, f(-4)$

7. $f(x) = x^2 - 7x + 10, f(5)$

8. $f(x) = x^2 - 10x - 39, f(8)$

9. $g(x) = x^2 + 12x - 45, g(-15)$

10. $g(x) = x^2 - 11x + 22, g(-10)$

11. $g(x) = 3x^2 + 2x - 14, g(7)$

12. $g(x) = 5x^2 - 4x + 8, g(-9)$

13. $f(x) = -2x^2 - 9x + 8, f(-3)$

14. $f(x) = -4x^2 + 16x + 13, f(4)$

15. $h(t) = -16t^2 + 96t + 32, h(4)$

16. $h(t) = -4.9t^2 + 26t + 10, h(2)$

17. $f(x) = x^2 + 13x + 30, f(3a)$

18. $f(x) = x^2 - 5x + 17, f(4a)$

19. $f(x) = x^2 - 9x - 21, f(a + 2)$

20. $f(x) = x^2 + 14x - 36, f(a - 6)$

21. Let $f(x) = x^2 + 4x - 60$. Find all values x for which $f(x) = 0$.

22. Let $f(x) = x^2 - 11x + 30$. Find all values x for which $f(x) = 0$.

23. Let $f(x) = x^2 + 13x + 42$. Find all values x for which $f(x) = 0$.

24. Let $f(x) = x^2 - 7x - 78$. Find all values x for which $f(x) = 0$.

25. Let $f(x) = x^2 - 23x + 144$. Find all values x for which $f(x) = 18$.

26. Let $f(x) = x^2 + 14x + 14$. Find all values x for which $f(x) = -31$.

27. Let $f(x) = x^2 + 11$. Find all values x for which $f(x) = 60$.

28. Let $f(x) = x^2 + 27$. Find all values x for which $f(x) = 108$.

A **fixed point** for a function $f(x)$ is a value a for which $f(a) = a$. For example, if $f(5) = 5$, then $x = 5$ is a fixed point for the function $f(x)$.

Find all fixed points for the following functions by setting the function equal to x and solving the resulting equation for x.

29. $f(x) = x^2 + 7x - 16$

30. $f(x) = x^2 - 5x + 8$

31. $f(x) = x^2 + 14x + 40$

32. $f(x) = x^2 - 9x + 25$

33. A ball is dropped from an airplane flying at an altitude of 400 feet. The ball's height above the ground, in feet, after t seconds is given by the function $h(t) = 400 - 16t^2$, whose graph is shown.

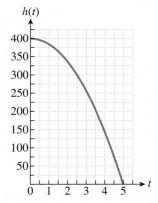

Use the function and its graph to answer the following questions.

a) How long will it take for the ball to land on the ground?

b) How high above the ground is the ball after 2 seconds?

34. A football is kicked up with an initial velocity of 64 feet/second. The football's height above the ground, in feet, after t seconds is given by the function $h(t) = -16t^2 + 64t$, whose graph is shown.

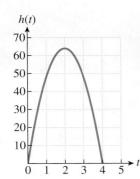

Use the function and its graph to answer the following questions:

a) Use the function to determine how high above the ground the football is after 1 second.

b) How long will it take for the football to land on the ground?

c) After how many seconds does the football reach its greatest height?

d) What is the greatest height that the football reaches?

35. A man is standing on a cliff 240 feet above a beach. He throws a rock up off the cliff with an initial velocity of 32 feet/second. The rock's height above the beach, in feet, after t seconds is given by the function $h(t) = -16t^2 + 32t + 240$, whose graph is shown.

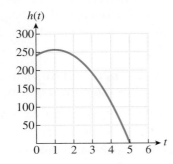

Use the function and its graph to answer the following questions:

a) Use the function to determine how high above the beach the rock is after 3 seconds.

b) How long will it take for the rock to land on the beach?

c) After how many seconds does the rock reach its greatest height above the beach?

d) What is the greatest height above the beach that the rock reaches?

36. The average price in dollars to produce x lawn chairs is given by the function
$$f(x) = 0.00000016x^2 - 0.0024x + 14.55,$$
whose graph is shown.

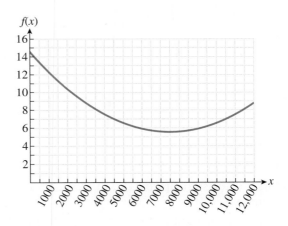

Use the function and its graph to answer the following questions.

a) Use the function to determine the average cost to produce 3000 lawn chairs.

b) What number of lawn chairs corresponds to the lowest average cost of production?

c) What is the lowest average cost that is possible?

37. A teenager starts a company selling personalized coffee mugs. The profit function, in dollars, for producing and selling x mugs is $f(x) = -0.4x^2 + 16x - 70$, whose graph is shown.

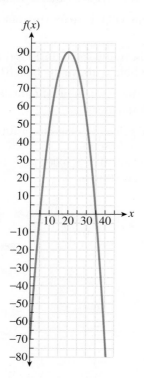

a) What are the start-up costs for the teenager's company?

b) How many mugs must the teenager sell before she breaks even?

c) How many mugs will give the maximum profit?

d) What will the profit be if she sells 25 mugs?

38. A peach farmer must determine how many peaches to thin from his trees. After he thins the peaches, the trees will produce fewer peaches, but they will be larger and of better quality. The expected profit, in dollars, if x percent of the peaches are removed during thinning is given by the function
$f(x) = -38x^2 + 2280x + 244{,}625$,
whose graph is shown.

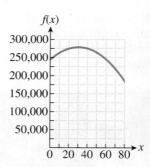

Use the function and its graph to answer the following questions:

a) What will the profit be if the farmer does not thin any peaches?

b) Use the function to determine the profit if 10% of the peaches are thinned.

c) What percent of the peaches must be thinned to produce the maximum profit?

d) What is the maximum potential profit?

Writing in Mathematics

Answer in complete sentences.

39. Suppose that the height, in feet, of a projectile after t seconds is given by a function $h(t)$. If we are asked to find the instant that the projectile lands on the ground, explain why we set $h(t)$ equal to 0 and solve for t.

40. Suppose that the height, in feet, of a projectile after t seconds is given by a function $h(t)$. If we are asked to find the height of the projectile after 8 seconds, explain why we evaluate $h(8)$.

41. ***Solutions Manual*** * Write a solutions manual page for the following problem:

For the function $f(x) = x^2 + 6x + 15$, solve the equation $f(x) = 31$.

42. ***Newsletter*** * Write a newsletter explaining how to evaluate a quadratic function.

*See Appendix B for details and sample answers.

6.8

Applications of Quadratic Equations and Quadratic Functions

1 Solve applied problems involving consecutive integers.
2 Solve applied problems involving the product of two unknown numbers.
3 Solve applied geometry problems.
4 Solve applied problems involving projectile motion.
5 Solve applied problems involving the sum of the first *n* natural numbers.

In this section, we will learn to solve applied problems that involve quadratic equations. When problems become more complex, solving them by trial and error becomes difficult at best, so it is important to have a procedure to use. Even though some of the examples may be simple enough to solve by trial and error, they help us to learn the new procedure.

Consecutive Integer Problems

Objective 1   Solve applied problems involving consecutive integers. We will begin with problems involving consecutive integers. Recall that consecutive integers follow the pattern $x, x + 1, x + 2$, and so on. Consecutive even integers, as well as consecutive odd integers, follow the pattern $x, x + 2, x + 4$, and so on.

EXAMPLE 1   The product of two consecutive positive integers is 56. Find the two integers.

Solution

We begin by creating a table of unknowns. It is important to represent each unknown in terms of the same variable. In this problem, we are looking for two consecutive positive integers. We can let x represent the first integer. Since the integers are consecutive, we can represent the second integer by $x + 1$.

> ***Unknowns***
>
> First: x
> Second: $x + 1$

We know the product is 56, which leads to the equation $x(x + 1) = 56$.

$$x(x + 1) = 56$$ Multiply x by $x + 1$.
$$x^2 + x = 56$$ Multiply x by $x + 1$.
$$x^2 + x - 56 = 0$$ Subtract 56.
$$(x + 8)(x - 7) = 0$$ Factor.
$$x + 8 = 0 \quad \text{or} \quad x - 7 = 0$$ Set each factor equal to 0.
$$x = -8 \quad \text{or} \quad x = 7$$ Solve each equation.

Since we know the integers are positive, we omit the solution $x = -8$. We use the table of our unknowns with the solution $x = 7$ to find the solution to our problem.

First: $x = 7$
Second: $x + 1 = 7 + 1 = 8$

Quick Check **1**

The product of two consecutive positive integers is 132. Find the two integers.

The two integers are 7 and 8. The product of these two integers is 56.

A Word of Caution   Be sure to check the practicality of your answers to an applied problem. In the previous example, the integers that we were looking for were positive, so we omitted a solution that led to negative integers.

Problems Involving the Product of Two Unknown Numbers

Objective **2**   Solve applied problems involving the product of two unknown numbers.

EXAMPLE ▶**2**   The sum of two numbers is 16 and their product is 55. Find the two numbers.

Solution

There are two unknowns in this problem: the two numbers. We let x represent the first number. Since the sum of the two numbers is 16, we represent the second number by $16 - x$.

Unknowns

First Number: x
Second Number: $16 - x$

Knowing that the product of these two numbers is 55 leads to the equation $x(16 - x) = 55$, which we now solve for x.

$$x(16 - x) = 55$$
$$16x - x^2 = 55 \qquad \text{Multiply.}$$
$$0 = x^2 - 16x + 55 \qquad \text{Collect all terms on the right side of the}$$
$$\qquad\qquad\qquad\qquad\qquad \text{equation. This makes the leading}$$
$$\qquad\qquad\qquad\qquad\qquad \text{coefficient positive.}$$
$$0 = (x - 5)(x - 11) \qquad \text{Factor.}$$
$$x - 5 = 0 \quad \text{or} \quad x - 11 = 0 \qquad \text{Set each factor equal to 0.}$$
$$x = 5 \quad \text{or} \quad x = 11 \qquad \text{Solve each equation.}$$

We return to the table of unknowns with the first solution, $x = 5$.

First Number: $x = 5$
Second Number: $16 - x = 16 - 5 = 11$

For this solution, the two numbers are 5 and 11. If we use the second solution, $x = 11$, we find the same two numbers.

First Number: $x = 11$
Second Number: $16 - x = 16 - 11 = 5$

Quick Check 2
The sum of two numbers is 21 and their product is 108. Find the two numbers.

So the two numbers are 5 and 11. Their product is indeed 55.

Geometry Problems

Objective 3 **Solve applied geometry problems.** The next example involves the area of a rectangle. Recall that the area of a rectangle is equal to the product of its length and width.

$$\text{Area} = \text{Length} \cdot \text{Width}$$

EXAMPLE 3 The area of a rectangle is 108 square meters. If the length of the rectangle is 3 meters more than its width, find the dimensions of the rectangle.

Solution

For this problem, the unknowns are the length and the width of the rectangle. Since the length is defined in terms of the width, we let w represent the width. The length can be represented by the expression $w + 3$ as it is 3 meters longer than the width.

Unknowns
Length: $w + 3$
Width: w

Since the product of the length and the width is equal to the area of the rectangle, the equation we need to solve is $(w + 3) \cdot w = 108$.

$$
\begin{aligned}
(w + 3) \cdot w &= 108 \\
w^2 + 3w &= 108 &&\text{Multiply.} \\
w^2 + 3w - 108 &= 0 &&\text{Subtract 108.} \\
(w + 12)(w - 9) &= 0 &&\text{Factor.} \\
w + 12 = 0 \quad &\text{or} \quad w - 9 = 0 &&\text{Set each factor equal to 0.} \\
w = -12 \quad &\text{or} \quad w = 9 &&\text{Solve each equation.}
\end{aligned}
$$

Quick Check 3
The area of a rectangle is 105 square feet. If the width of the rectangle is 8 feet less than its length, find the dimensions of the rectangle.

We omit the solution $w = -12$, as a rectangle cannot have a negative width or length. We can use the solution $w = 9$ and the table of unknowns to find the length and the width of the rectangle.

Length: $w + 3 = 9 + 3 = 12$
Width: $w = 9$

The length of the rectangle is 12 meters and the width is 9 meters. We can verify that the area of a rectangle with these dimensions is 108 square meters.

EXAMPLE 4   A homeowner has installed an in-ground spa in the backyard. The spa is rectangular in shape, with a length that is 4 feet more than its width. The homeowner put a one-foot-wide concrete border around the spa. If the area covered by the spa and the border is 96 square feet, find the dimensions of the spa.

Solution

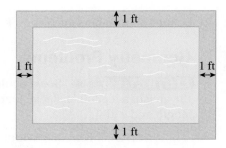

The unknowns are the length and width of the rectangular spa. We will let x represent the width of the spa.

> ***Unknowns***
> Length: $x + 4$
> Width: x

Since we know the area covered by the spa and the border, our equation must involve the outer rectangle in the picture. The width of the border can be represented by $x + 2$ because we need to add 2 feet to the width of the spa (x). The width of the spa is increased by 1 foot on both sides. In a similar fashion, the length of the border is $x + 6$ because we add a foot to the length of the spa on both sides.

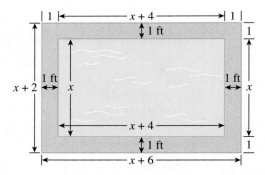

The equation we will solve is $(x + 6)(x + 2) = 96$, since the area covered is 96 square feet.

$$(x + 6)(x + 2) = 96$$
$$x^2 + 2x + 6x + 12 = 96 \qquad \text{Multiply.}$$
$$x^2 + 8x + 12 = 96 \qquad \text{Combine like terms.}$$
$$x^2 + 8x - 84 = 0 \qquad \text{Subtract 96.}$$
$$(x + 14)(x - 6) = 0 \qquad \text{Factor.}$$
$$x + 14 = 0 \quad \text{or} \quad x - 6 = 0 \qquad \text{Set each factor equal to 0.}$$
$$x = -14 \quad \text{or} \qquad x = 6 \qquad \text{Solve each equation.}$$

Quick Check **4**

A homeowner has poured a rectangular concrete slab in her backyard to use as a barbecue area. The length is 3 feet more than its width. There is a 2-foot-wide flower bed around the barbecue area. If the area covered by the barbecue area and the flower bed is 270 square feet, find the dimensions of the barbecue area.

We will omit the solution $x = -14$, as the dimensions of the spa cannot be negative. We use the solution $x = 6$ and the table of unknowns to find the length and the width of the rectangle.

> Length: $x + 4 = 6 + 4 = 10$
> Width: $x = 6$

The length of the spa is 10 feet and the width of the spa is 6 feet. We can verify that the area covered by the spa and border is indeed 96 square feet.

Projectile Problems

Objective **4** **Solve applied problems involving projectile motion.** The height, in feet, of a projectile after t seconds can be found using the function $h(t) = -16t^2 + v_0 t + s$, where v_0 is the initial velocity of the projectile and s is the initial height.

> $$h(t) = -16t^2 + v_0 t + s$$
> t: Time (seconds)
> v_0: Initial Velocity (feet/second)
> s: Initial Height (feet)

This formula does not cover projectiles that continue to propel themselves, such as a rocket with an engine.

EXAMPLE 5   A cannonball is fired from a platform that is 256 feet above the ground. The initial velocity of the cannonball is 96 feet per second.

a) Find the function $h(t)$ that gives the height of the cannonball in feet after t seconds.

Solution

Since the initial velocity is 96 feet per second and the initial height is 256 feet, the function is $h(t) = -16t^2 + 96t + 256$.

b) How long will it take for the cannonball to land on the ground?

Solution

The cannonball's height when it lands on the ground is 0 feet, so we set the function equal to 0 and solve for the time t in seconds.

$-16t^2 + 96t + 256 = 0$	Set $h(t)$ equal to 0.
$0 = 16t^2 - 96t - 256$	Collect all terms on the right side of the equation so the leading coefficient is positive.
$0 = 16(t^2 - 6t - 16)$	Factor out the GCF (16).
$0 = 16(t - 8)(t + 2)$	Factor $t^2 - 6t - 16$.
$t - 8 = 0$ or $t + 2 = 0$	Set each variable factor equal to 0.
$t = 8$ or $t = -2$	Solve each equation.

We will omit the negative solution $t = -2$, as the time must be positive.
It takes the cannonball 8 seconds to land on the ground.

c) What is the domain of the function $h(t)$?

Solution

Since the cannonball lands on the ground after 8 seconds, any length of time greater than 8 seconds does not apply. The domain of the function is $0 \leq t \leq 8$, which can be written in interval notation as $[0, 8]$.

Quick Check **5** A model rocket is launched from the ground with an initial velocity of 144 feet per second.

a) Find the function $h(t)$ that gives the height of the rocket in feet after t seconds.
b) How long will it take for the rocket to land on the ground?
c) What is the domain of the function $h(t)$?

Problems Involving the Sum of the First n Natural Numbers

Objective 5 **Solve applied problems involving the sum of the first n natural numbers.** The famous German mathematician Carl Gauss found a function for totaling the natural numbers from 1 to n, which is $f(n) = \frac{1}{2}n^2 + \frac{1}{2}n$. We use this function in the last example of the section.

EXAMPLE 6 **a)** Use Gauss's function to find the sum of the natural numbers from 1 to 46.

Solution

We evaluate Gauss's function at $n = 46$.

$$f(46) = \frac{1}{2}(46)^2 + \frac{1}{2}(46) \qquad \text{Substitute 46 for } n.$$

$$= \frac{1}{2}(2116) + \frac{1}{2}(46) \qquad \text{Square 46.}$$

$$= 1058 + 23 \qquad \text{Simplify each term.}$$
$$= 1081 \qquad \text{Add.}$$

The sum of the first 46 natural numbers is 1081.

b) Find a natural number n such that the sum of the first n natural numbers is 78.

Solution

We begin by setting the function $f(n)$ equal to 78 and solving for n.

$$\frac{1}{2}n^2 + \frac{1}{2}n = 78 \qquad \text{Set the function equal to 78.}$$

$$2\left(\frac{1}{2}n^2 + \frac{1}{2}n\right) = 2 \cdot 78 \qquad \text{Multiply both sides of the equation by 2 to clear the fractions.}$$

$$n^2 + n = 156 \qquad \text{Distribute and simplify each term.}$$
$$n^2 + n - 156 = 0 \qquad \text{Collect all terms on the left side of the equation.}$$
$$(n + 13)(n - 12) = 0 \qquad \text{Factor.}$$
$$n + 13 = 0 \quad \text{or} \quad n - 12 = 0 \qquad \text{Set each factor equal to 0.}$$
$$n = -13 \quad \text{or} \quad n = 12 \qquad \text{Solve each equation.}$$

We omit the solution $n = -13$ because -13 is not a natural number. The sum of the first 12 natural numbers is 78.

Quick Check **6** a) Use Gauss's function to find the sum of the natural numbers from 1 to 100.
             b) Find a natural number n such that the sum of the first n natural numbers is 45.

Building Your Study Strategy Overcoming Math Anxiety, 8 **Procrastination**
Some students put off doing their math homework or studying due to negative feelings for mathematics. Procrastination is your enemy. Convince yourself that you can do it and get started. If you wait until you are tired, you will not be able to give your best effort. If you do not do the homework assignment, you will have trouble following the next day's material. Try scheduling a time to devote to mathematics each day, and stick to your schedule.

EXERCISES 6.8

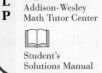

F
O
R

E
X
T
R
A

H
E
L
P

MyMathLab
MathXL
Interactmath.com

MathXL
Tutorials on CD

Video Lectures
on CD

Addison-Wesley
Math Tutor Center

Student's
Solutions Manual

(Vocabulary)

1. Three consecutive integers can be represented by _____, _____, and _____.

2. Three consecutive odd integers can be represented by _____, _____, and _____.

3. State the formula for the area of a rectangle.

4. The function _____ gives the height of a projectile with initial velocity v_0 and initial height s in feet after t seconds.

5. Two consecutive positive integers have a product of 182. Find the integers.

6. Two consecutive positive integers have a product of 380. Find the integers.

7. Two consecutive even positive integers have a product of 288. Find the integers.

8. Two consecutive even negative integers have a product of 80. Find the integers.

9. Two consecutive odd negative integers have a product of 99. Find the integers.

10. Two consecutive odd positive integers have a product of 399. Find the integers.

11. The product of two consecutive positive integers is 27 more than 5 times the larger integer. Find the integers.

12. The product of two consecutive odd positive integers is 14 more than 3 times the larger integer. Find the integers.

13. The product of two consecutive even positive integers is 398 more than the sum of the two integers. Find the integers.

14. The product of two consecutive positive integers is 131 more than the sum of the two integers. Find the integers.

15. One positive number is 5 more than a second number, and their product is 66. Find the two numbers.

16. One positive number is 9 more than a second number, and their product is 112. Find the two numbers.

17. One positive number is 4 less than 3 times a second number, and their product is 84. Find the two numbers.

18. One positive number is 3 more than twice a second number, and their product is 189. Find the two numbers.

19. The sum of two numbers is 32 and their product is 240. Find the two numbers.

20. The sum of two numbers is 24 and their product is 95. Find the two numbers.

21. The sum of two numbers is 17 and their product is 42. Find the two numbers.

22. The sum of two numbers is 20 and their product is 91. Find the two numbers.

23. The difference of two positive numbers is 3 and their product is 88. Find the two numbers.

24. The difference of two positive numbers is 9 and their product is 220. Find the two numbers.

25. Gabriela is 5 years older than her sister Zorayda. If the product of their ages is 126, how old is Gabriela?

26. Marco is 8 years younger than his brother Paolo. If the product of their ages is 105, how old is Marco?

27. Clyde's age is 5 years more than twice Bonnie's age. If the product of their ages is 168, how old is Clyde?

28. Edith's age is 1 year less than 3 times Archie's age. If the product of their ages is 70, how old is Edith?

29. The length of a rectangular rug is 3 feet more than its width. If the area of the rug is 40 square feet, find the length and width of the rug.

30. The length of a rectangular swimming pool is twice its width, and the area covered by the pool is 800 square feet. Find the length and width of the swimming pool.

31. The length of a rectangular classroom is 3 times its width. If the area covered by the classroom is 432 square feet, find the length and width of the classroom.

32. The width of a rectangular photo frame is 8 centimeters less than its length. If the area of this rectangle is 240 square centimeters, find the length and width of the frame.

33. Jose has a rectangular garden in his backyard that covers 105 square meters. The length of the garden is 1 meter more than twice its width. Find the dimensions of the garden.

34. The height of a doorway is 7 feet less than 5 times its width. If the area of the doorway is 24 square feet, find the dimensions of the doorway.

35. The area of a rectangular lawn is 960 square feet. If the length of the lawn is 8 feet less than twice the width of the lawn, find the dimensions of the lawn.

36. A rectangular soccer field has an area of 8800 square meters. If the length of the field is 30 meters more than its width, find the dimensions of the field.

37. A photo's width is 3 inches less than its length. A border of 1 inch is placed around the photo, and the area covered by the photo and its border is 70 square inches. Find the dimensions of the photo itself.

38. The length of a rectangular quilt is 4 inches more than its width. After a 2-inch border is placed around the quilt, its area is 320 square inches. Find the original dimensions of the quilt.

39. A rectangular flower garden has a length that is 1 foot less than twice its width. A 2-foot brick border is added around the garden, and the area of the garden and brick border is a total of 117 square feet. Find the dimensions of the garden without the brick border.

40. Larry has a rectangular pen for his pet goats that is adjacent to his barn, as shown in the following diagram:

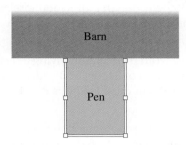

The side adjacent to the barn is 5 feet less than the side that projects out from the barn. He decides to expand the pen by 5 feet in all three directions, as shown in the following diagram:

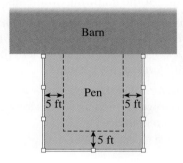

This expansion makes the area of the pen 625 square feet. Find the dimensions of the original pen.

For the following exercises, recall that the area of a square with side s is s^2 and that the area of a triangle with base b and height h is $\frac{1}{2}bh$.

41. The area of the floor in a square room is 64 square meters. Find the length of a wall of the room.

42. A dartboard is enclosed in a square that has an area of 1600 square centimeters. Find the length of a side of the square.

43. The height of a triangular sail is 5 feet more than twice its base. If the sail is made of 84 square feet of fabric, find the base and height of the triangle.

44. The three towns of Visalia, Hanford, and Fresno form a triangle. The base of the triangle extends from Visalia to Hanford, and the height of the triangle extends from Visalia to Fresno, as shown in the following diagram.

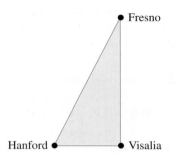

The distance from Visalia to Fresno is twice the distance from Visalia to Hanford. If the area of the triangle formed by these three cities is 400 square miles, find the distance from Visalia to Fresno.

For problems 45–54, use the following function:
$$h(t) = -16t^2 + v_0t + s.$$

45. A projectile is launched from a building 320 feet tall. If the initial velocity of the projectile was 128 feet per second, how long will it take the projectile to land on the ground?

46. Standing on a platform, Eesaeng throws a softball upward from a height 48 feet above the ground at a speed of 32 feet per second. How long will it take for the softball to land on the ground?

47. Standing on the edge of a cliff, Fernando throws a rock upward from a height 80 feet above the water at a speed of 64 feet per second. How long will it take for the rock to land in the water?

48. A projectile is launched from the top of a building 384 feet tall with an initial velocity of 160 feet per second. How long will it take for the projectile to land on the ground?

49. A projectile is launched from a platform 40 feet above the ground with an initial velocity of 64 feet per second. After how many seconds is the projectile at a height of 88 feet?

50. Standing atop a building, Angela throws a rock upward from a height 55 feet above the ground with an initial velocity of 32 feet per second. How long will it take before the rock is 7 feet above the ground?

51. A projectile is launched from the ground with an initial velocity of 160 feet per second. How long will it take for the projectile to land on the ground?

52. A projectile is launched from the ground with an initial velocity of 112 feet per second. How long will it take for the projectile to land on the ground?

53. A projectile is launched upward from the ground, and it lands on the ground after 8 seconds. What was the initial velocity of the projectile?

54. A projectile is launched upward from an initial height of 432 feet, and it lands on the ground after 9 seconds. What was the initial velocity of the projectile?

The height, in feet, of a free-falling object t seconds after being dropped from an initial height s can be found using the function $h(t) = -16t^2 + s$.

55. An object is dropped from a helicopter 144 feet above the ground. How long will it take for the object to land on the ground?

56. An object is dropped from an airplane that is 400 feet above the ground. How long will it take for the object to land on the ground?

Use Gauss's function for totaling the natural numbers from 1 to n, $f(n) = \frac{1}{2}n^2 + \frac{1}{2}n$, in the following exercises.

57. Find a natural number n such that the sum of the first n natural numbers is 36.

58. Find a natural number n such that the sum of the first n natural numbers is 55.

59. Find a natural number n such that the sum of the first n natural numbers is 210.

60. Find a natural number n such that the sum of the first n natural numbers is 325.

Writing in Mathematics

Answer in complete sentences.

61. Explain why the first three odd integers cannot be represented as x, $x + 1$, and $x + 3$.

62. Write a word problem involving a rectangle with a length of 24 feet and width of 16 feet. Your problem must lead to a quadratic equation.

63. Write a word problem involving a rectangle that leads to the equation $(x + 9)(x + 5) = 192$. Explain how you created your problem.

64. *Solutions Manual** Write a solutions manual page for the following problem:

A cannonball is fired upward with an initial velocity of 96 feet per second from a building with a height of 256 feet. When will the cannonball land on the ground?

65. *Newsletter** Write a newsletter explaining how to solve applied problems involving the area of a rectangle.

*See Appendix B for details and sample answers.

Chapter 6 Summary

Section 6.1—Topic	Chapter Review Exercises
Factoring Out the GCF	1–4
Factoring by Grouping	5–8

Section 6.2—Topic	Chapter Review Exercises
Factoring Trinomials of the Form $x^2 + bx + c$	9–16

Section 6.3—Topic	Chapter Review Exercises
Factoring Trinomials of the Form $ax^2 + bx + c$, where $a \neq 1$	17–20

Section 6.4—Topic	Chapter Review Exercises
Factoring Binomials: Difference of Squares, Difference of Cubes, Sum of Cubes	21–26

Section 6.5—Topic	Chapter Review Exercises
Factoring Polynomials	1–26

Section 6.6—Topic	Chapter Review Exercises
Solving Quadratic Equations by Factoring	27–42
Finding a Quadratic Equation Given Its Solution Set	43–46

Section 6.7—Topic	Chapter Review Exercises
Evaluating Quadratic Functions	47–50
Solving Equations Involving Quadratic Functions	51–54
Solving Applications by Using the Graph of a Quadratic Function	55–56

Section 6.8—Topic	Chapter Review Exercises
Solving Applications of Quadratic Equations	57–60

Summary of Chapter 6 Study Strategies

Math anxiety can hinder your success in a mathematics class, but it can be overcome.
- The first step is to understand what has caused your anxiety to begin with.
- Relaxation techniques can help you to overcome the physical symptoms of math anxiety, allowing you to give your best effort.
- Developing a positive attitude and confidence in your abilities will help as well.
- Avoid procrastinating.
- Many students believe that they have math anxiety, but poor performance can be caused by other factors as well. Do you completely understand the material, only to "freeze up" on the tests? Does this happen in other classes as well? If so, you may have test anxiety, not math anxiety.
- Some students do poorly because they are taking classes for which they are not prepared. Discuss your correct placement with your instructor and your academic counselor if you feel that this may be the case for you.
- Finally, be sure that poor study skills are not the cause of your difficulties. If you are not giving your fullest effort, you cannot expect to learn and understand mathematics.

Factor completely. If the polynomial cannot be factored, write "prime." [6.1–6.5]

1. $3x - 21$

2. $x^2 - 20x$

3. $5x^7 + 15x^4 - 20x^3$

4. $6a^5b^2 - 10a^3b^3 + 15a^4b$

5. $x^3 + 6x^2 + 4x + 24$

6. $x^3 + 9x^2 - 7x - 63$

7. $x^3 - 3x^2 - 11x + 33$

8. $x^3 - 3x^2 - 9x + 27$

9. $x^2 - 8x + 16$

10. $x^2 + 3x - 28$

11. $x^2 - 11x + 24$

12. $-5x^2 - 20x + 160$

13. $x^2 + 14xy + 45y^2$

14. $x^2 + 9x - 23$

15. $x^2 + 17x + 30$

16. $x^2 - x - 42$

17. $10x^2 - 83x + 24$

18. $4x^2 - 3x - 10$

19. $18x^2 + 12x - 6$

20. $4x^2 - 12x + 9$

21. $3x^2 - 75$

22. $4x^2 - 25y^2$

23. $x^3 + 8y^3$

24. $x^2 + 121$

25. $x^3 - 216$

26. $7x^3 - 189$

Solve. [6.6]

27. $(x - 5)(x + 7) = 0$

28. $(3x - 8)(2x + 1) = 0$

29. $x^2 - 25 = 0$

30. $x^2 - 14x + 40 = 0$

Worked-out solutions to Review Exercises marked with can be found on page AN–22.

31. $x^2 + 16x + 64 = 0$

32. $x^2 + 3x - 108 = 0$

33. $x^2 - 9x - 70 = 0$

34. $x^2 - 16x = 0$

35. $x^2 + 19x + 60 = 0$

36. $x^2 - 18x + 80 = 0$

37. $x^2 - 15x - 16 = 0$

38. $2x^2 - 7x - 60 = 0$

39. $7x^2 - 252 = 0$

40. $4x^2 - 60x + 144 = 0$

41. $x(x + 10) = -21$

42. $(x + 4)(x + 5) = 72$

Find a quadratic equation with integer coefficients that has the given solution set. [6.6]

43. $\{6, 7\}$

44. $\{-4, 4\}$

45. $\left\{2, -\dfrac{3}{5}\right\}$

46. $\left\{-\dfrac{5}{4}, \dfrac{2}{7}\right\}$

Evaluate the given quadratic function. [6.7]

47. $f(x) = x^2, f(-3)$

48. $f(x) = x^2 - 8x, f(4)$

49. $f(x) = x^2 + 10x + 24, f(-8)$

50. $f(x) = 3x^2 + 8x - 19, f(5)$

For the given function $f(x)$, find all values x for which $f(x) = 0$. [6.7]

51. $f(x) = x^2 - 36$

52. $f(x) = x^2 + 3x - 54$

53. $f(x) = x^2 + 12x + 20$

54. $f(x) = x^2 - 7x - 44$

55. A man is standing on a cliff above a beach. He throws a rock upward from a height 128 feet above the beach with an initial velocity of 32 feet/second. The rock's height above the beach, in feet, after t seconds is given by the function $h(t) = -16t^2 + 32t + 128$, whose graph is shown. [6.7]

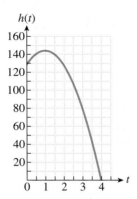

h(t)

Use the function and its graph to answer the following questions:

a) Use the graph to determine how high above the beach the rock is after three seconds.

b) How long will it take for the rock to land on the beach?

c) After how many seconds does the rock reach its greatest height above the beach?

d) Use the function to determine the greatest height above the beach that the rock reaches.

56. The average price in dollars for a manufacturer to produce x toys is given by the function $f(x) = 0.0001x^2 - 0.08x + 20.85$, whose graph is shown. [6.7]

f(x)

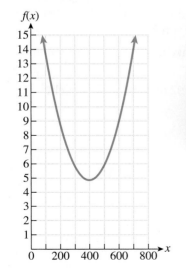

Use the function and its graph to answer the following questions:

a) Use the function to determine the average cost to produce 800 toys.

b) What number of toys corresponds to the lowest average cost of production?

c) Use the function to determine the lowest average cost that is possible.

57. Two consecutive even positive integers have a product of 288. Find the two integers. [6.8]

58. Rosa is 8 years younger than Dale. If the product of their ages is 105, how old is each person? [6.8]

59. The length of a rectangle is 3 meters less than twice the width. The area of the rectangle is 104 square meters. Find the length and the width of the rectangle. [6.8]

60. A ball is thrown upward with an initial velocity of 80 feet per second from the edge of a cliff that is 384 feet above a river. Use the function $h(t) = -16t^2 + v_0t + s$. [6.8]

a) How long will it take for the ball to land in the river?

b) When is the ball 480 feet above the river?

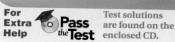
For Extra Help | Pass *the* Test | Test solutions are found on the enclosed CD.

Factor completely.

1. $x^2 - 25x$

2. $x^3 - 3x^2 - 5x + 15$

3. $x^2 - 9x + 20$

4. $x^2 - 14x - 72$

5. $3x^2 + 60x + 57$

6. $6x^2 - x - 5$

7. $x^2 - 25y^2$

8. $x^3 - 216y^3$

Solve.

9. $x^2 - 100 = 0$

10. $x^2 - 13x + 36 = 0$

11. $x^2 + 7x - 60 = 0$

12. $(x + 2)(x - 9) = 60$

Find a quadratic equation with integer coefficients that has the given solution set.

13. $\{-2, 2\}$

14. $\left\{-6, \dfrac{9}{5}\right\}$

Evaluate the given quadratic function.

15. $f(x) = x^2 - 12x + 35$, $f(-5)$

16. $f(x) = -x^2 + 2x + 17$, $f(7)$

For the given function $f(x)$, find all values x for which $f(x) = 0$.

17. $f(x) = x^2 + 14x + 48$

18. The average price in dollars for a manufacturer to produce x baseball hats is given by the function

$f(x) = 0.00000006x^2 - 0.00096x + 5.95$, whose graph is shown below.

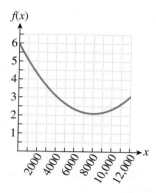

Use the function and its graph to answer the following questions:

a) Use the function to determine the average cost to produce 6500 baseball hats.

b) What number of baseball hats corresponds to the lowest average cost?

c) Use the function to determine the lowest average cost that is possible.

19. The length of a rectangle is 5 feet more than three times the width. The area of the rectangle is 100 square feet. Find the length and the width of the rectangle.

20. A ball is thrown upward with an initial velocity of 32 feet per second from the edge of a cliff 240 feet above a beach. Use the function $h(t) = -16t^2 + v_0 t + s$.

a) How long will it take for the ball to land on the beach?

b) When is the ball 192 feet above the beach?

Mathematicians in History
Sir Isaac Newton

Sir Isaac Newton was an English mathematician and scientist who lived in the 17th and 18th centuries. His work with motion and gravity helped us to better understand our world, as well as our solar system. Alexander Pope once said "Nature and Nature's laws lay hid in the night; God said, Let Newton be! And all was light."

Write a one-page summary (*or* make a poster) of the life of Sir Isaac Newton and his accomplishments.

Interesting issues:

* Where and when was Sir Isaac Newton born?
* Describe Newton's upbringing, and his relationship with his mother and stepfather.
* It has been said that an apple was the inspiration for Newton's ideas about the force of gravity. Explain how the apple is believed to have inspired Newton's ideas.
* In a letter to Robert Hooke, Newton wrote "If I have been able to see further, it was only because I stood on the shoulders of giants." Explain what this statement means.
* Newton is often referred to as the "Father of Calculus." What is calculus?
* What are Newton's three laws of motion? Explain what they mean in your own words.
* What did Newton invent for his pets?
* Where was Newton buried?

Stretch Your Thinking ❭ Chapter 6

In the year 1997, 23,100 Americans had weight-loss surgery. In the year 2002 there were 63,100 such surgeries, and in the year 2003 there were an estimated 103,200 surgeries.

(*Source:* American Society for Bariatric Surgery)

1. Do the data, as plotted here, suggest that the number of surgeries is increasing in a linear fashion? Explain why or why not in your own words.

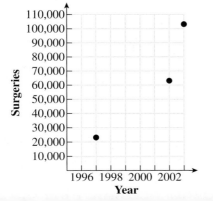

2. The number of weight-loss surgeries in a particular year can be modeled by the quadratic function $f(x) = 5350x^2 - 18,750x + 23,100$, where x represents the number of years after 1997. Use the function to predict the number of weight-loss surgeries in the year 2015. Does this number of surgeries seem reasonable? Why or why not?

Solve the system by the method of your choice. If the system is inconsistent and has no solution, state this. If the system is dependent, write the form of the solution for any real number x. [4.2, 4.3]

1. $4x + y = 22$
   $-3x + 8y = 1$

2. $y = 3x - 2$
   $2x - 3y = -15$

3. $9x + 8y = 25$
   $5x - 6y = -7$

4. $4x + y = 6$
   $\frac{3}{2}x + \frac{3}{5}y = \frac{9}{2}$

5. A minor league baseball team charges adults $5 and children $4 for admission. If 1050 people paid a total of $5000 to see last night's game, how many children were at the game? [4.4]

6. Bill's backyard is rectangular. The fence that runs around the backyard is 130 feet long. If the length of the backyard is 10 feet less than twice the width, find the dimensions of the backyard. [4.4]

7. Mya deposits a total of $15,000 in two different certificates of deposit (CDs). One CD pays 3% annual interest, while the other pays 2.5% annual interest. If Mya earned $430 in interest during the first year, how much did she put in each CD? [4.4]

Solve the system of inequalities. [4.5]

8. $y < x + 4$
   $y \geq 4x - 2$

Simplify the expression. Write the result without using negative exponents. (Assume all variables represent nonzero real numbers.) [5.1/5.2]

9. $\dfrac{x^{11}}{x^4}$

10. $12m^{10}n^7 \cdot 9m^6n^{14}$

11. $\left(\dfrac{5a^6b^{11}}{c^7}\right)^4$

12. $\dfrac{t^{12}}{t^{-9}}$

13. $(a^{-6}b^7)^{-8}$

14. $x^{-16} \cdot x^{-19}$

Rewrite in scientific notation. [5.2]

15. 330,000,000,000

16. 0.000000000000714

17. If a computer can perform a calculation in 0.000000000005 second, how long would it take to perform 200,000,000 calculations? [5.2]

Evaluate the polynomial for the given value of the variable. [5.3]

18. $x^2 + 11x - 16$ for $x = -7$

Add or subtract. [5.3]

19. $(x^2 + 6x - 20) + (x^2 - 13x - 39)$

20. $(x^2 - 15x - 9) - (2x^2 - 3x - 31)$

Evaluate the given function. [5.3]

21. $f(x) = x^2 + 5x - 37, f(8)$

Multiply. [5.4]

22. $3x(5x^2 - 8x + 12)$

23. $(4x - 7)(3x - 8)$

Divide. [5.5]

24. $\dfrac{10x^{12} + 24x^9 - 18x^6}{2x^5}$

25. $\dfrac{x^2 + 13x + 39}{x + 6}$

Factor completely. [6.1–6.5]

26. $x^3 - 4x^2 + 6x - 24$

27. $2x^2 - 26x + 80$

28. $x^2 + 5x - 36$

29. $2x^2 - x - 10$

30. $x^3 - 125$

31. $x^2 - 81$

Solve. [6.6]

32. $x^2 - 49 = 0$

33. $x^2 - 11x + 24 = 0$

34. $x^2 + 14x + 49 = 0$

35. $x(x + 8) = 20$

Find a quadratic equation with integer coefficients that has the given solution set. [6.6]

36. $\{-2, 7\}$

For the given function $f(x)$, find all values x for which $f(x) = 0$. [6.7]

37. $f(x) = x^2 - 100$

38. $f(x) = x^2 + 2x - 35$

39. A man is standing on a cliff above a beach. He throws a rock upward from a height of 192 feet above the beach with an initial velocity of 64 feet/second. The rock's height above the beach, in feet, after t seconds is given by the function $h(t) = -16t^2 + 64t + 192$, which is graphed as follows:

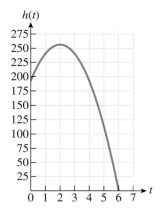

Use the function and its graph to answer the following questions: [6.7]

a) Use the function to determine how high above the beach the rock is after 3 seconds.

b) How long will it take for the rock to land on the beach?

c) After how many seconds does the rock reach its greatest height above the beach?

d) Use the function to determine the greatest height above the beach that the rock reaches.

40. Two consecutive even positive integers have a product of 168. Find the two integers. [6.8]

41. The length of a rectangle is 5 meters more than twice the width. The area of the rectangle is 133 square meters. Find the length and the width of the rectangle. [6.8]

Rational Expressions and Equations

In this chapter we will examine rational expressions, which are fractions whose numerator and denominator are polynomials. We will learn to simplify, add, subtract, multiply, and divide rational expressions. Rational expressions are involved in many applied problems, such as determining the maximum load that a wooden beam can support, finding the illumination from a light source, and solving work-rate problems.

Study Strategy   **Preparing for a Cumulative Exam**   *In this chapter we will focus on how to prepare for a cumulative exam, such as a final exam. Although some of the strategies are similar to those used to prepare for a chapter test or quiz, there are some differences as well.*

7.1
Rational Expressions and Functions

A **rational expression** is a quotient of two polynomials, such as $\dfrac{x^2 + 15x + 44}{x^2 - 16}$. The denominator of a rational expression must not be zero, as division by zero is undefined. A major difference between rational expressions and linear or quadratic expressions is that a rational expression has one or more variables in the denominator.

Evaluating Rational Expressions

Objective 1 Evaluate rational expressions. We can evaluate a rational expression for a particular value of the variable just as we evaluated polynomials. We substitute the value for the variable in the expression and then simplify. When simplifying, we evaluate the numerator and denominator separately and then simplify the resulting fraction.

EXAMPLE 1 Evaluate the rational expression $\dfrac{x^2 + 2x - 24}{x^2 - 7x + 12}$ for $x = -5$.

Solution

We begin by substituting -5 for x.

$$\frac{(-5)^2 + 2(-5) - 24}{(-5)^2 - 7(-5) + 12}$$ Substitute -5 for x.

$$= \frac{25 - 10 - 24}{25 + 35 + 12}$$ Simplify the numerator and denominator separately.

$$= \frac{-9}{72}$$ Simplify the numerator and denominator.

$$= -\frac{1}{8}$$ Simplify.

Quick Check 1
Evaluate the rational expression $\dfrac{x^2 - 3x - 18}{x^2 - 5x - 6}$ for $x = -4$.

Finding Values for Which a Rational Expression Is Undefined

Objective 2 Find the values for which a rational expression is undefined. Rational expressions are undefined for values of the variable that cause the denominator to equal 0, as division by 0 is undefined. In general, to find the values for which a rational expression is undefined, we set the denominator equal to 0, ignoring the numerator, and solve the resulting equation.

EXAMPLE 2   Find the values for which the rational expression $\dfrac{3}{5x+4}$ is undefined.

Solution

We begin by setting the denominator, $5x + 4$, equal to 0. Then we solve for x.

$$5x + 4 = 0 \qquad \text{Set the denominator equal to 0.}$$
$$5x = -4 \qquad \text{Subtract 4.}$$
$$x = -\dfrac{4}{5} \qquad \text{Divide by 5.}$$

The expression $\dfrac{3}{5x+4}$ is undefined for $x = -\dfrac{4}{5}$.

Quick Check 2
Find the values for which $\dfrac{3}{2x-7}$ is undefined.

EXAMPLE 3   Find the values for which $\dfrac{x^2 - 7x + 12}{x^2 + 5x - 36}$ is undefined.

Solution

We begin by setting the denominator $x^2 + 5x - 36$ equal to 0, ignoring the numerator. Notice that the resulting equation is quadratic and can be solved by factoring.

$$x^2 + 5x - 36 = 0 \qquad \text{Set the denominator equal to 0.}$$
$$(x + 9)(x - 4) = 0 \qquad \text{Factor } x^2 + 5x - 36.$$
$$x + 9 = 0 \quad \text{or} \quad x - 4 = 0 \qquad \text{Set each factor equal to 0.}$$
$$x = -9 \quad \text{or} \quad x = 4 \qquad \text{Solve.}$$

Quick Check 3
Find the values for which $\dfrac{x-4}{x^2 - 8x - 9}$ is undefined.

The expression $\dfrac{x^2 - 7x + 12}{x^2 + 5x - 36}$ is undefined when $x = -9$ or $x = 4$.

Simplifying Rational Expressions to Lowest Terms

Objective 3   **Simplify rational expressions to lowest terms.** Rational expressions are often referred to as *algebraic fractions*. As with numerical fractions, we will learn to simplify rational expressions to lowest terms. In later sections, we will learn to add, subtract, multiply, and divide rational expressions.

We simplified a numerical fraction to lowest terms by dividing out factors that were common to the numerator and denominator. For example, consider the fraction $\dfrac{30}{84}$. To simplify this fraction, we could begin by factoring the numerator and denominator.

$$\frac{30}{84} = \frac{2 \cdot 3 \cdot 5}{2 \cdot 2 \cdot 3 \cdot 7}$$

The numerator and denominator have common factors of 2 and 3, which are divided out to simplify the fraction to lowest terms.

$$\frac{\overset{1}{\cancel{2}} \cdot \overset{1}{\cancel{3}} \cdot 5}{\underset{1}{\cancel{2}} \cdot 2 \cdot \underset{1}{\cancel{3}} \cdot 7} = \frac{5}{2 \cdot 7} \text{ or } \frac{5}{14}$$

Simplifying Rational Expressions

To simplify a rational expression to lowest terms, we first factor the numerator and denominator completely. Then we divide out any common factors in the numerator and denominator.

If P, Q, and R are polynomials, $Q \neq 0$, and $R \neq 0$, then $\dfrac{PR}{QR} = \dfrac{P}{Q}$.

EXAMPLE 4 Simplify the rational expression $\dfrac{10x^3}{8x^5}$. (Assume $x \neq 0$.)

Solution

This rational expression has a numerator and denominator that are monomials. In this case, we can simplify the expression using the properties of exponents developed in Chapter 5.

Quick Check 4
Simplify the rational expression $\dfrac{15x^4}{21x^3}$.
(Assume $x \neq 0$.)

$$\frac{10x^3}{8x^5} = \frac{\overset{5}{\cancel{10}}x^3}{\underset{4}{\cancel{8}}x^5} \qquad \text{Divide out the common factor 2.}$$

$$= \frac{5x^3}{4x^5} \qquad \text{Simplify.}$$

$$= \frac{5}{4x^2} \qquad \text{Divide numerator and denominator by } x^3.$$

EXAMPLE 5 Simplify $\dfrac{(x-3)(x+7)}{(x+7)(x+2)(x-3)}$. (Assume the denominator is nonzero.)

Solution

In this example, the numerator and denominator have already been factored. Notice that they share the common factors $x-3$ and $x+7$.

Quick Check 5
Simplify
$\dfrac{(x+9)(x-2)(x-8)}{(x-2)(x-8)(x+3)}$.
(Assume the denominator is nonzero.)

$$\frac{(x-3)(x+7)}{(x+7)(x+2)(x-3)} = \frac{\overset{1}{\cancel{(x-3)}}\,\overset{1}{\cancel{(x+7)}}}{\underset{1}{\cancel{(x+7)}}(x+2)\underset{1}{\cancel{(x-3)}}} \qquad \text{Divide out common factors.}$$

$$= \frac{1}{x+2} \qquad \text{Simplify.}$$

Notice that both factors were divided out of the numerator. Be careful to note that the numerator is equal to 1 in such a situation.

A Word of Caution If we divide out each factor in the numerator of a rational expression when simplifying the expression, we must be sure to write a 1 in the numerator.

EXAMPLE 6 Simplify $\dfrac{x^2 + 8x - 20}{x^2 - 7x + 10}$. (Assume the denominator is nonzero.)

Solution

The trinomials in the numerator and denominator must be factored before we can simplify this expression. (For a review of factoring techniques, you may refer to Sections 6.1 through 6.5.)

$$\frac{x^2 + 8x - 20}{x^2 - 7x + 10} = \frac{(x - 2)(x + 10)}{(x - 2)(x - 5)}$$ Factor numerator and denominator.

$$= \frac{\overset{1}{(x - 2)}(x + 10)}{\underset{1}{(x - 2)}(x - 5)}$$ Divide out common factors.

$$= \frac{x + 10}{x - 5}$$ Simplify.

> *Quick Check* **6**
> Simplify $\dfrac{x^2 + 10x + 24}{x^2 - 2x - 48}$.
> (Assume the denominator is nonzero.)

A Word of Caution When we are simplifying a rational expression, we must be very careful that we divide out only expressions that are common factors of the numerator and denominator. We cannot *reduce* individual terms in the numerator and denominator as in the following examples.

$$\frac{x + 8}{x - 6} \ne \frac{x + \overset{4}{\cancel{8}}}{x - \underset{3}{\cancel{6}}}$$

$$\frac{x^2 - 25}{x^2 - 36} \ne \frac{\overset{1}{\cancel{x^2}} - 25}{\underset{1}{\cancel{x^2}} - 36}$$

To avoid making this mistake, keep in mind that we must factor the numerator and denominator completely before attempting to divide out common factors.

EXAMPLE 7 Simplify $\dfrac{x^2 - 8x}{2x^2 - 17x + 8}$. (Assume the denominator is nonzero.)

Solution

The polynomials in the numerator and denominator must be factored before we can simplify this expression. The numerator $x^2 - 8x$ has a common factor of x that must be factored out first.

$$x^2 - 8x = x(x - 8)$$

The denominator $2x^2 - 17x + 8$ is a trinomial with a leading coefficient that is not equal to 1, and can be factored by grouping or by trial-and-error. (For a review of these factoring techniques, you may refer to Section 6.3.)

$$2x^2 - 17x + 8 = (2x - 1)(x - 8)$$

Now we can simplify the rational expression.

$$\frac{x^2 - 8x}{2x^2 - 17x + 8} = \frac{x(x - 8)}{(2x - 1)(x - 8)}$$ Factor the numerator and denominator.

$$= \frac{x(x \overset{1}{\cancel{- 8}})}{(2x - 1)(\underset{1}{\cancel{x - 8}})}$$ Divide out common factors.

$$= \frac{x}{2x - 1}$$ Simplify.

> **Quick Check 7**
>
> Simplify $\dfrac{3x^2 + 16x + 5}{x^2 + 8x + 15}$.
> (Assume the denominator is nonzero.)

Identifying Factors in the Numerator and Denominator That Are Opposites

Objective 4 Identify factors that are opposites of each other. Two expressions of the form $a - b$ and $b - a$ are **opposites**. Subtraction in the opposite order produces the opposite result. Consider the expressions $a - b$ and $b - a$ when $a = 10$ and $b = 4$. In this case, $a - b = 10 - 4$ or 6 and $b - a = 4 - 10$ or -6. We can also see that $a - b$ and $b - a$ are opposites by noting that their sum, $(a - b) + (b - a)$, is equal to 0.

This is useful to know when simplifying rational expressions. The rational expression $\dfrac{a - b}{b - a}$ simplifies to -1, as any fraction whose numerator is the opposite of its denominator is equal to -1. If a rational expression has a factor in the numerator that is the opposite of a factor in the denominator, these two factors can be divided out to equal -1, as in the next example. We write the -1 in the numerator.

EXAMPLE 8 Simplify $\dfrac{49 - x^2}{x^2 - 12x + 35}$. (Assume the denominator is nonzero.)

Solution

We begin by factoring the numerator and denominator completely.

$$\frac{49 - x^2}{x^2 - 12x + 35} = \frac{(7 + x)(7 - x)}{(x - 5)(x - 7)}$$ Factor the numerator and denominator.

$$= \frac{(7 + x)(\overset{-1}{\cancel{7 - x}})}{(x - 5)(\underset{1}{\cancel{x - 7}})}$$ Divide out the opposite factors.

$$= -\frac{7 + x}{x - 5}$$ Simplify, writing the negative sign in front of the fraction.

> **Quick Check 8**
>
> Simplify $\dfrac{x^2 + 8x - 9}{1 - x^2}$.
> (Assume the denominator is nonzero.)

A Word of Caution Two expressions of the form $a + b$ and $b + a$ are not opposites but are equal to each other. Addition in the opposite order produces the same result. When we divide two expressions of the form $a + b$ and $b + a$, the result is 1, not -1. For example, $\dfrac{x + 2}{2 + x} = 1$.

Rational Functions

> A **rational function** $r(x)$ is a function of the form $r(x) = \dfrac{f(x)}{g(x)}$, where $f(x)$ and $g(x)$ are polynomials and $g(x) \neq 0$.

Objective 5 **Evaluate rational functions.** We begin our investigation of rational functions by learning to evaluate them.

EXAMPLE 9 For $r(x) = \dfrac{x^2 + 6x - 17}{x^2 - 5x - 22}$, find $r(8)$.

Solution

We begin by substituting 8 for x in the function.

$$
\begin{aligned}
r(8) &= \frac{(8)^2 + 6(8) - 17}{(8)^2 - 5(8) - 22} && \text{Substitute 8 for } x. \\
&= \frac{64 + 48 - 17}{64 - 40 - 22} && \text{Simplify each term in the numerator} \\
&&& \text{and denominator.} \\
&= \frac{95}{2} && \text{Simplify the numerator and denominator.}
\end{aligned}
$$

Quick Check 9

For
$r(x) = \dfrac{x^2 - 3x - 12}{x^2 + 7x - 15}$,
find $r(-6)$.

Finding the Domain of a Rational Function

Objective 6 **Find the domain of a rational function.** Rational functions differ from linear functions and quadratic functions in that the domain of a rational function is not always the set of real numbers. We have to exclude any value that causes the function to be undefined, namely any value for which the denominator is equal to 0. Suppose that the function $r(x)$ was undefined for $x = 6$. Then the domain of $r(x)$ is the set of all real numbers except 6. This can be expressed in interval notation as $(-\infty, 6) \cup (6, \infty)$, which is the union of the set of all real numbers that are less than 6 with the set of all real numbers that are greater than 6.

EXAMPLE 10 Find the domain of $r(x) = \dfrac{x^2 - 32x + 60}{x^2 - 9x}$.

Solution

We begin by setting the denominator equal to 0 and solving for x.

$$
\begin{aligned}
x^2 - 9x &= 0 && \text{Set the denominator equal to 0.} \\
x(x - 9) &= 0 && \text{Factor } x^2 - 9x. \\
x = 0 \quad \text{or} \quad x - 9 &= 0 && \text{Set each factor equal to 0.} \\
x = 0 \quad \text{or} \quad x &= 9 && \text{Solve each equation.}
\end{aligned}
$$

Quick Check 10

Find the domain of
$r(x) = \dfrac{x^2 + 11x + 24}{x^2 - 4x - 45}$.

The domain of the function is the set of all real numbers except 0 and 9. In interval notation, this can be written as $(-\infty, 0) \cup (0, 9) \cup (9, \infty)$.

EXERCISES 7.1 ❯

Vocabulary

1. A(n) _____ is a quotient of two polynomials.

2. Rational expressions are undefined for values of the variable that cause the _____ to equal 0.

3. A rational expression is said to be in _____ if its numerator and denominator do not have any common factors.

4. Two expressions of the form $a - b$ and $b - a$ are _____.

5. A _____ $r(x)$ is a function of the form $r(x) = \dfrac{f(x)}{g(x)}$, where $f(x)$ and $g(x)$ are polynomials and $g(x) \neq 0$.

6. The _____ of a rational function excludes all values for which the function is undefined.

Evaluate the rational expression for the given value of the variable.

7. $\dfrac{6}{x + 4}$ for $x = 4$

8. $\dfrac{9}{x - 20}$ for $x = 5$

9. $\dfrac{x + 3}{x - 8}$ for $x = -25$

10. $\dfrac{x + 1}{x + 13}$ for $x = -7$

11. $\dfrac{x^2 - 13x - 48}{x^2 - 6x - 12}$ for $x = 3$

12. $\dfrac{x^2 + 5x - 23}{x^2 - 7x + 20}$ for $x = 2$

13. $\dfrac{x^2 - 6x + 15}{x^2 + 14x - 7}$ for $x = -9$

14. $\dfrac{x^2 + 10x - 56}{x - 4}$ for $x = -7$

Find all values of the variable for which the rational expression is undefined.

15. $\dfrac{9}{x - 3}$

16. $\dfrac{x - 5}{10x - 13}$

17. $\dfrac{x + 6}{x(x - 8)}$

18. $\dfrac{x + 2}{(x + 2)(x + 3)}$

19. $\dfrac{x - 10}{x^2 - 6x - 16}$

20. $\dfrac{x + 1}{x^2 + 11x + 24}$

21. $\dfrac{7x}{x^2 - 49}$

22. $\dfrac{x - 3}{x^2 - 9x}$

Simplify the given rational expression. (Assume all denominators are nonzero.)

23. $\dfrac{3x^5}{9x^8}$

24. $\dfrac{10x^7}{12x^{10}}$

25. $\dfrac{x + 5}{(x + 6)(x + 5)}$

26. $\dfrac{(x-7)(x+7)}{x-7}$

27. $\dfrac{(x+2)(x-5)}{(x-2)(x-5)}$

28. $\dfrac{(x-9)(x-3)}{(x-3)(x+3)}$

29. $\dfrac{x^2-4x-32}{x^2-15x+56}$

30. $\dfrac{x^2-3x}{x^2-8x+15}$

31. $\dfrac{x^2-4x-45}{x^2+11x+30}$

32. $\dfrac{x^2+6x-40}{x^2-10x+24}$

33. $\dfrac{x^2+9x+8}{x^2-3x-4}$

34. $\dfrac{x^2-36}{x^2+9x+18}$

35. $\dfrac{3x^2-25x-50}{x^2-3x-70}$

36. $\dfrac{x^2+7x+12}{4x^2+9x-9}$

37. $\dfrac{3x^2-6x-105}{x^2-11x+28}$

38. $\dfrac{2x^2+14x+20}{5x^2-25x-70}$

Determine whether the two given binomials are or are not opposites.

39. $x+5$ and $5+x$

40. $x-11$ and $11-x$

41. $14-x$ and $x-14$

42. $3x-2$ and $2x-3$

43. $2x+13$ and $-2x-13$

44. $6x-7$ and $6x+7$

Simplify the given rational expression. (Assume all denominators are nonzero.)

45. $\dfrac{3x-21}{(5+x)(7-x)}$

46. $\dfrac{(8+x)(8-x)}{x^2-2x-48}$

47. $\dfrac{x^2-3x-54}{81-x^2}$

48. $\dfrac{5x-x^2}{x^2+2x-35}$

49. $\dfrac{9-x^2}{x^2+12x+27}$

50. $\dfrac{x^2-x-72}{36-4x}$

Evaluate the given rational function.

51. $r(x)=\dfrac{20}{x^2-5x+10}$, $r(-5)$

52. $r(x)=\dfrac{x}{x^2-13x-20}$, $r(8)$

53. $r(x)=\dfrac{x^2+10x+24}{x^2-5x-66}$, $r(4)$

54. $r(x)=\dfrac{x^2-100}{x^2+13x+30}$, $r(-3)$

55. $r(x)=\dfrac{x^3-7x^2-11x+20}{x^2+8x-20}$, $r(10)$

56. $r(x)=\dfrac{x^3+8x^2+17x+10}{x^3+3x^2-18x-40}$, $r(2)$

Find the domain of the given rational function.

57. $r(x)=\dfrac{x^2+18x+77}{x^2+10x}$

58. $r(x)=\dfrac{x^2-16x+60}{x^2-9x+8}$

59. $r(x)=\dfrac{x^2+2x-3}{x^2-2x-15}$

60. $r(x)=\dfrac{x^2+7x-8}{x^2-64}$

61. $r(x) = \dfrac{x^2 + 4x - 60}{x^2 + 3x - 18}$

62. $r(x) = \dfrac{x^2 - 13x + 36}{x^2 + 14x + 45}$

Identify the given function as a linear function, a quadratic function, or a rational function.

63. $f(x) = x^2 - 11x + 30$

64. $f(x) = 3x - 8$

65. $f(x) = \dfrac{x^2}{x^3 - 5x^2 + 11x - 35}$

66. $f(x) = \dfrac{x^2 + 3x + 8}{x^2 - 5x - 5}$

67. $f(x) = \dfrac{1}{5}x + \dfrac{4}{9}$

68. $f(x) = -\dfrac{1}{2}x^2 + 3x - \dfrac{1}{7}$

Use the given graph of a rational function $r(x)$ to solve the problems that follow.

69.

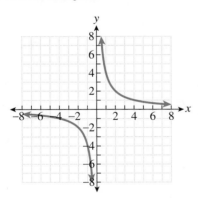

a) Find $r(1)$.

b) Find all values x such that $r(x) = -4$.

70.

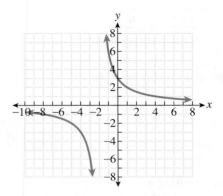

a) Find $r(0)$.

b) Find all values x such that $r(x) = -1$.

71.

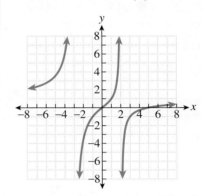

a) Find $r(-4)$.

b) Find all values x such that $r(x) = 0$.

72.

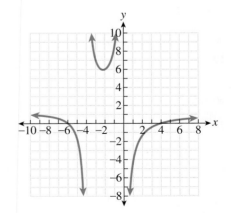

a) Find $r(1)$.

b) Find all values x such that $r(x) = 0$.

Writing in Mathematics

Answer in complete sentences.

73. Explain how to find the values for which a rational expression is undefined.

74. Is the rational expression $\dfrac{(x + 2)(x - 7)}{(x - 5)(x + 2)}$ undefined for the value $x = -2$? Explain your answer.

75. Explain how to determine whether two factors are opposites. Use examples.

76. *Solutions Manual*[*] Write a solutions manual page for the following problem:

Find the values for which the rational expression $\dfrac{x^2 + 3x - 28}{x^2 - 16}$ *is undefined.*

77. *Newsletter*[*] Write a newsletter that explains how to simplify a rational expression.

*See Appendix B for details and sample answers.

QUICK REVIEW EXERCISES

Section 7.1

Simplify.

1. $\dfrac{5}{12} + \dfrac{4}{15}$

2. $\dfrac{17}{24} - \dfrac{3}{10}$

3. $\dfrac{14}{45} \cdot \dfrac{165}{308}$

4. $\dfrac{20}{99} \div \dfrac{35}{51}$

7.2
Multiplication and Division of Rational Expressions

1 Multiply two rational expressions.
2 Multiply two rational functions.
3 Divide a rational expression by another rational expression.
4 Divide a rational function by another rational function.

Multiplying Rational Expressions

Objective **1** **Multiply two rational expressions.** In this section, we will learn how to multiply and divide rational expressions. Multiplying rational expressions is similar to multiplying numerical fractions. Suppose that we needed to multiply $\frac{4}{9} \cdot \frac{21}{10}$. Before multiplying, we can divide out factors common to one of the numerators and one of the denominators. For example, the first numerator (4) and the second denominator (10) have a common factor of 2 that can be divided out of each. The second numerator (21) and the first denominator (9) have a common factor of 3 that can be divided out as well.

$$\frac{4}{9} \cdot \frac{21}{10} = \frac{2 \cdot 2}{3 \cdot 3} \cdot \frac{3 \cdot 7}{2 \cdot 5}$$ Factor each numerator and denominator.

$$= \frac{\overset{1}{\cancel{2}} \cdot 2}{\underset{1}{\cancel{3}} \cdot 3} \cdot \frac{\overset{1}{\cancel{3}} \cdot 7}{\underset{1}{\cancel{2}} \cdot 5}$$ Divide out common factors.

$$= \frac{2 \cdot 7}{3 \cdot 5}$$ Multiply remaining factors.

$$= \frac{14}{15}$$

Multiplying Rational Expressions

$$\frac{A}{B} \cdot \frac{C}{D} = \frac{AC}{BD} \quad (B \neq 0 \text{ and } D \neq 0)$$

To multiply two rational expressions, we will begin by factoring each numerator and denominator completely. After dividing out factors common to a numerator and a denominator, we will express the product of the two rational expressions as a single rational expression, leaving the numerator and denominator in factored form.

EXAMPLE 1   Multiply: $\dfrac{x^2 - 11x + 30}{x^2 + 5x - 24} \cdot \dfrac{x^2 - 9}{x^2 - 2x - 15}$.

Solution

We begin by factoring each numerator and denominator. Then we divide out common factors.

$$\frac{x^2 - 11x + 30}{x^2 + 5x - 24} \cdot \frac{x^2 - 9}{x^2 - 2x - 15}$$

$$= \frac{(x-5)(x-6)}{(x+8)(x-3)} \cdot \frac{(x+3)(x-3)}{(x-5)(x+3)}$$ Factor numerators and denominators completely.

$$= \frac{\overset{1}{\cancel{(x-5)}}(x-6)}{(x+8)\underset{1}{\cancel{(x-3)}}} \cdot \frac{\overset{1}{\cancel{(x+3)}}\overset{1}{\cancel{(x-3)}}}{\underset{1}{\cancel{(x-5)}}\underset{1}{\cancel{(x+3)}}}$$ Divide out common factors.

$$= \frac{x-6}{x+8}$$ Simplify.

Quick Check   **1** Multiply: $\dfrac{x^2 - 3x - 18}{x^2 - 9x + 20} \cdot \dfrac{x^2 - 3x - 10}{x^2 + 13x + 30}$.

EXAMPLE 2   Multiply: $\dfrac{x^2 - 10x}{x^2 - 2x - 8} \cdot \dfrac{16 - x^2}{x^2 - 9x - 10}$.

Solution

Again, we begin by completely factoring both numerators and denominators.

$$\frac{x^2 - 10x}{x^2 - 2x - 8} \cdot \frac{16 - x^2}{x^2 - 9x - 10}$$

$$= \frac{x(x-10)}{(x+2)(x-4)} \cdot \frac{(4+x)(4-x)}{(x+1)(x-10)}$$ Factor completely.

$$= \frac{x\overset{1}{\cancel{(x-10)}}}{(x+2)\underset{1}{\cancel{(x-4)}}} \cdot \frac{(4+x)\overset{-1}{\cancel{(4-x)}}}{(x+1)\underset{1}{\cancel{(x-10)}}}$$ Divide out common factors. Notice that the factors $4 - x$ and $x - 4$ are opposites, and divide out to equal -1.

$$= \frac{-x(4+x)}{(x+2)(x+1)}$$ Multiply remaining factors.

$$= -\frac{x(4+x)}{(x+2)(x+1)}$$ Write the negative sign in the numerator in front of the fraction.

Quick Check   **2** Multiply: $\dfrac{36 - x^2}{x^2 + 9x + 14} \cdot \dfrac{x^2 - 7x - 18}{x^2 - 15x + 54}$.

Here is a summary of the procedure for multiplying rational expressions:

Multiplying Rational Expressions

- Completely factor each numerator and each denominator.
- Divide out factors that are common to a numerator and a denominator, as well as factors in a numerator and denominator that are opposites.
- Multiply the remaining factors, leaving the numerator and denominator in factored form.

Objective **2**   **Multiply two rational functions.**

EXAMPLE **3**   For $f(x) = \dfrac{x^2 - 3x - 18}{x^2 + 8x + 16}$ and $g(x) = \dfrac{x^2 + 3x - 4}{9 - x^2}$, find $f(x) \cdot g(x)$.

Solution

We replace $f(x)$ and $g(x)$ by their formulas and proceed to multiply.

$$f(x) \cdot g(x) = \frac{x^2 - 3x - 18}{x^2 + 8x + 16} \cdot \frac{x^2 + 3x - 4}{9 - x^2} \qquad \text{Replace } f(x) \text{ and } g(x) \text{ by}$$
their formulas.

$$= \frac{(x - 6)(x + 3)}{(x + 4)(x + 4)} \cdot \frac{(x + 4)(x - 1)}{(3 + x)(3 - x)} \qquad \text{Factor completely.}$$

$$= \frac{(x - 6)\overset{1}{\cancel{(x + 3)}}}{\cancel{(x + 4)}(x + 4)} \cdot \frac{\overset{1}{\cancel{(x + 4)}}(x - 1)}{\underset{1}{\cancel{(3 + x)}}(3 - x)} \qquad \text{Divide out common factors.}$$

$$= \frac{(x - 6)(x - 1)}{(x + 4)(3 - x)} \qquad \text{Multiply remaining factors.}$$

Quick Check   **3** For $f(x) = \dfrac{x + 4}{49 - x^2}$ and $g(x) = \dfrac{x^2 - 10x + 21}{x^2 + x - 12}$, find $f(x) \cdot g(x)$.

Dividing a Rational Expression by Another Rational Expression

Objective **3**   **Divide a rational expression by another rational expression.**
Dividing a rational expression by another rational expression is similar to dividing a numerical fraction by another numerical fraction. We replace the divisor, which is the rational expression by which we are dividing, by its reciprocal and then multiply.

Dividing Rational Expressions

$$\frac{A}{B} \div \frac{C}{D} = \frac{A}{B} \cdot \frac{D}{C} \qquad (B \neq 0, C \neq 0, \text{ and } D \neq 0)$$

EXAMPLE **4**   Divide: $\dfrac{x^2 + 13x + 42}{x^2 + x - 20} \div \dfrac{x^2 + 8x + 12}{x^2 - 4x}$.

Solution

We begin by inverting the divisor and multiplying. We then factor each numerator and denominator completely.

$$\frac{x^2 + 13x + 42}{x^2 + x - 20} \div \frac{x^2 + 8x + 12}{x^2 - 4x}$$

$$= \frac{x^2 + 13x + 42}{x^2 + x - 20} \cdot \frac{x^2 - 4x}{x^2 + 8x + 12} \qquad \text{Invert the divisor and multiply.}$$

$$= \frac{(x+6)(x+7)}{(x+5)(x-4)} \cdot \frac{x(x-4)}{(x+2)(x+6)}$$ Factor completely.

$$= \frac{\overset{1}{(x+6)}(x+7)}{(x+5)\underset{1}{(x-4)}} \cdot \frac{x\overset{1}{(x-4)}}{(x+2)\underset{1}{(x+6)}}$$ Divide out common factors.

$$= \frac{x(x+7)}{(x+5)(x+2)}$$ Multiply remaining factors.

Quick Check 4 Divide: $\dfrac{x^2-4}{x^2-5x+4} \div \dfrac{x^2-6x-16}{x^2+3x-28}$.

EXAMPLE 5 Divide: $\dfrac{x^2-14x-15}{x^2+2x-35} \div \dfrac{x^2-1}{5-x}$.

Solution

We rewrite the problem as a multiplication problem by inverting the divisor. We then factor each numerator and denominator completely before dividing out common factors. (You may want to factor at the same time you invert the divisor.)

$$\frac{x^2-14x-15}{x^2+2x-35} \div \frac{x^2-1}{5-x}$$

$$= \frac{x^2-14x-15}{x^2+2x-35} \cdot \frac{5-x}{x^2-1}$$ Invert the divisor and multiply.

$$= \frac{(x-15)(x+1)}{(x+7)(x-5)} \cdot \frac{5-x}{(x+1)(x-1)}$$ Factor completely.

$$= \frac{(x-15)\overset{1}{(x+1)}}{(x+7)\underset{1}{(x-5)}} \cdot \frac{\overset{-1}{5-x}}{\underset{1}{(x+1)}(x-1)}$$ Divide out common factors. Note that $5-x$ and $x-5$ are opposites.

$$= -\frac{x-15}{(x+7)(x-1)}$$ Multiply remaining factors.

Quick Check 5 Divide: $\dfrac{x^2-13x+36}{x^2-x-12} \div \dfrac{9-x}{x^2+8x+15}$.

Here is a summary of the procedure for dividing rational expressions.

Dividing Rational Expressions

- Invert the divisor and change the operation from division to multiplication.
- Completely factor each numerator and each denominator.
- Divide out factors that are common to a numerator and a denominator, and factors in a numerator and denominator that are opposites.
- Multiply the remaining factors, leaving the numerator and denominator in factored form.

Objective 4 Divide a rational function by another rational function.

EXAMPLE 6 For $f(x) = \dfrac{x^2 - x - 12}{x + 7}$ and $g(x) = x^2 + 3x - 28$, find $f(x) \div g(x)$.

Solution

Replace $f(x)$ and $g(x)$ by their formulas and divide. Treat $g(x)$ as a rational function with a denominator of 1. We can see that its reciprocal is $\dfrac{1}{x^2 + 3x - 28}$.

$f(x) \div g(x)$

$= \dfrac{x^2 - x - 12}{x + 7} \div x^2 + 3x - 28$ Replace $f(x)$ and $g(x)$ by their formulas.

$= \dfrac{x^2 - x - 12}{x + 7} \cdot \dfrac{1}{x^2 + 3x - 28}$ Invert the divisor and multiply.

$= \dfrac{(x - 4)(x + 3)}{x + 7} \cdot \dfrac{1}{(x + 7)(x - 4)}$ Factor completely.

$= \dfrac{\cancel{(x - 4)}^{1}(x + 3)}{x + 7} \cdot \dfrac{1}{(x + 7)\cancel{(x - 4)}_{1}}$ Divide out common factors.

$= \dfrac{x + 3}{(x + 7)^2}$ Multiply remaining factors.

Quick Check 6 For $f(x) = \dfrac{x^2 - 12x + 36}{x^2 - 11x + 10}$ and $g(x) = x^2 - 8x + 12$, find $f(x) \div g(x)$.

Building Your Study Strategy **Preparing for a Cumulative Exam, 2 Old Quizzes and Tests** To prepare for a cumulative exam, begin by reviewing old exams and quizzes. Review and understand any mistakes that you made on the exam. By understanding what you did wrong, you are minimizing your chances of making the same mistakes on the cumulative exam.

Once you understand the errors that you made, try reworking all of the problems on the exam or quiz. You should do this without referring to your notes or textbook, to help you learn which topics you need to review and which topics you have under control. You need to focus your time on the topics you are struggling with and spend less time on the topics that you understand; retaking your tests will help you determine which topics you need to spend more time on.

Vocabulary

1. To multiply two rational expressions, begin by
_____ each numerator and denominator
completely.

2. When you are multiplying two rational expressions,
_____ out factors common to a numerator and
a denominator.

3. When you are multiplying two rational expressions,
once the numerator and the denominator do not
share any common factors, express the product as a
single rational expression, leaving the numerator and
denominator in _____ form.

4. When you are dividing by a rational expression,
replace the divisor by its _____ and then
multiply.

Multiply.

5. $\dfrac{x+2}{(x-2)(x-4)} \cdot \dfrac{(x-4)(x+5)}{(x-3)(x+2)}$

6. $\dfrac{x+1}{x+10} \cdot \dfrac{(x+10)(x-6)}{(x-6)(x-1)}$

7. $\dfrac{x^2+18x+72}{x^2+2x-63} \cdot \dfrac{x^2-6x-7}{x^2+10x-24}$

8. $\dfrac{x^2+8x+16}{x^2+14x+48} \cdot \dfrac{x^2+4x-60}{x^2+x-12}$

9. $\dfrac{x^2-x}{x^2+2x-99} \cdot \dfrac{x^2+13x+22}{x^2+9x-10}$

10. $\dfrac{x^2-144}{x^2-16x+63} \cdot \dfrac{x^2-7x-18}{x^2+8x-48}$

11. $\dfrac{x^2-8x-33}{x^2-8x+16} \cdot \dfrac{2x^2-7x-4}{x^2+5x+6}$

12. $\dfrac{x^2+12x+20}{x^2-13x+40} \cdot \dfrac{x^2-x-56}{x^2+18x+80}$

13. $\dfrac{x^2-11x}{x^2+5x+4} \cdot \dfrac{x^2+10x+9}{x^2+6x}$

14. $\dfrac{x^2-6x-55}{36-x^2} \cdot \dfrac{x^2-4x-12}{2x^2+15x+25}$

15. $\dfrac{4x^2+48x+128}{x^2-6x+5} \cdot \dfrac{x^2+4x-45}{6x^2+36x+48}$

16. $\dfrac{x^2+22x+120}{x^2+11x+28} \cdot \dfrac{x^2+4x-21}{x^2+14x+24}$

17. $\dfrac{4-x^2}{x^2-x-30} \cdot \dfrac{x^2+9x+20}{x^2+7x-18}$

18. $\dfrac{x^2+3x-70}{5x^2+6x-8} \cdot \dfrac{x^2+9x+14}{7x-x^2}$

19. $\dfrac{x^2-11x+24}{x^2-7x+12} \cdot \dfrac{x^2+4x-32}{x^2-64}$

20. $\dfrac{2x^2+24x+22}{8x^2+104x+320} \cdot \dfrac{x^2+3x-10}{x^2+7x+6}$

21. $\dfrac{2x^2 - x - 36}{x^2 + x - 90} \cdot \dfrac{9x - x^2}{x^2 - 2x - 24}$

22. $\dfrac{x^2 - 3x - 10}{x^2 + 2x} \cdot \dfrac{x^2 + 8x}{x^2 + 3x - 40}$

For the given functions $f(x)$ and $g(x)$, find $f(x) \cdot g(x)$.

23. $f(x) = \dfrac{2}{x - 3}, g(x) = \dfrac{x^2 - 12x + 27}{x^2 - 5x - 36}$

24. $f(x) = \dfrac{x^2 - 9x + 20}{x^2 + x - 2}, g(x) = \dfrac{x^2 + 6x - 16}{x^2 - 25}$

25. $f(x) = \dfrac{x^2 - 13x + 22}{x^2 + 11x - 12}, g(x) = \dfrac{x^2 + 4x - 5}{x^2 - 6x + 8}$

26. $f(x) = \dfrac{x^2 + 5x - 84}{x^2 - x - 12}, g(x) = \dfrac{x^2 + 11x + 24}{x^2 - 9x + 14}$

27. $f(x) = \dfrac{x^2 + 6x - 16}{x^2 - x - 20}, g(x) = \dfrac{x^2 + 6x - 55}{x^2 + 17x + 72}$

28. $f(x) = x + 7, g(x) = \dfrac{x^2 + 3x - 28}{x^2 + 14x + 49}$

Find the missing numerator and denominator.

29. $\dfrac{(x + 5)(x - 2)}{(2x + 1)(x - 7)} \cdot \dfrac{?}{?} = \dfrac{(x + 5)}{(2x + 1)}$

30. $\dfrac{x^2 - 5x - 6}{x^2 + 15x + 54} \cdot \dfrac{?}{?} = \dfrac{(x + 1)(x + 5)}{(x + 6)^2}$

31. $\dfrac{x^2 - 10x + 16}{x^2 + 4x - 77} \cdot \dfrac{?}{?} = -\dfrac{x^2 - 11x + 24}{x^2 - 5x - 14}$

32. $\dfrac{x^2 + 4x}{x^2 - 81} \cdot \dfrac{?}{?} = \dfrac{x^2 - 6x}{x^2 - 7x - 18}$

Divide.

33. $\dfrac{(2x + 3)(x - 6)}{(3x + 1)(x - 8)} \div \dfrac{2x + 3}{x - 8}$

34. $\dfrac{(x + 3)(x - 4)}{x - 9} \div \dfrac{(x - 4)(x + 1)}{(x - 9)(x - 2)}$

35. $\dfrac{x^2 + 2x - 15}{x^2 - 49} \div \dfrac{x^2 + 7x - 30}{x^2 + 4x - 21}$

36. $\dfrac{x^2 + 19x + 88}{x^2 - 4x + 4} \div \dfrac{x^2 + 5x - 24}{x^2 - x - 6}$

37. $\dfrac{x^2 - 13x + 42}{x^2 + 8x - 20} \div \dfrac{x^2 - 3x - 28}{x^2 + 2x - 8}$

38. $\dfrac{x^2 - 2x - 80}{x^2 - 3x - 18} \div \dfrac{x^2 - 17x + 70}{x^2 + 4x + 3}$

39. $\dfrac{4x^2 + 4x - 80}{x^2 + 12x + 35} \div \dfrac{x^2 + 5x - 36}{3x^2 + 15x - 42}$

40. $\dfrac{x^2 - 13x + 30}{x^2 - 9x + 18} \div \dfrac{x^2 - 15x + 50}{x^2 - 11x + 30}$

41. $\dfrac{x^2 + 6x - 7}{x^2 + 16x + 55} \div \dfrac{x^2 + 13x + 42}{x^2 + 7x - 44}$

42. $\dfrac{x^2 + 13x + 36}{x^2 - 12x + 32} \div \dfrac{x^2 + 6x - 27}{16 - x^2}$

43. $\dfrac{x^2 - 4x - 5}{x^2 + x - 30} \div \dfrac{x^2 - x - 2}{x^2 + 4x - 12}$

44. $\dfrac{x^2 - 12x + 20}{x^2 + 6x + 5} \div \dfrac{x^2 - 11x + 18}{x^2 - 3x - 40}$

45. $\dfrac{x^2 - 2x - 63}{x^2 - 4x - 21} \div \dfrac{81 - x^2}{x^2 - 7x - 30}$

46. $\dfrac{x^2 - 6x - 27}{x^2 + 3x - 4} \div \dfrac{x^2 - 18x + 81}{2x^2 - 9x + 7}$

47. $\dfrac{x^2 + 10x + 24}{4x^2 - 19x - 5} \div \dfrac{x^2 + 16x + 60}{x^2 + 5x - 50}$

48. $\dfrac{x^2 - 15x + 54}{x^2 - 16x + 48} \div \dfrac{x^2 - 14x + 48}{x^2 - 2x - 8}$

49. $\dfrac{x^2 + 16x + 63}{2x - x^2} \div \dfrac{x^2 + x - 72}{x^2 - 5x + 6}$

50. $\dfrac{x^2 - 12x + 36}{x^2 + 8x + 15} \div \dfrac{6x^2 - x^3}{x^2 + 5x}$

For the given functions $f(x)$ and $g(x)$, find $f(x) \div g(x)$.

51. $f(x) = \dfrac{x^2 + 8x - 9}{x^2 - 10x + 25}, g(x) = x + 9$

52. $f(x) = 2x + 14, g(x) = \dfrac{x^2 + 15x + 56}{x^2 - 10x + 21}$

53. $f(x) = \dfrac{x^2 + 3x - 54}{x^2 - 2x - 3}, g(x) = \dfrac{x^2 + 9x}{x^2 - 7x - 8}$

54. $f(x) = \dfrac{x^2 - 14x + 24}{x^2 + x - 6}, g(x) = \dfrac{x^2 + 2x - 80}{x^2 + 13x + 30}$

55. $f(x) = \dfrac{2x^2 - 5x}{x^2 - 1}, g(x) = \dfrac{2x^2 + x - 15}{x^2 - 8x - 9}$

56. $f(x) = \dfrac{x^2 + 4x + 4}{x^2 + 14x + 40}, g(x) = \dfrac{3x^2 + 7x + 2}{5x^2 + 19x - 4}$

Find the missing numerator and denominator.

57. $\dfrac{(x + 9)(x - 6)}{(x - 3)(x - 7)} \div \dfrac{?}{?} = \dfrac{x + 9}{x - 7}$

58. $\dfrac{x^2 - 5x + 4}{x^2 - 14x + 45} \div \dfrac{?}{?} = \dfrac{(x - 1)(x - 6)}{(x - 9)^2}$

59. $\dfrac{x^2 - 10x + 9}{x^2 + 7x} \div \dfrac{?}{?} = \dfrac{x^2 - 13x + 12}{x^2 + 11x + 28}$

60. $\dfrac{x^2 + 12x + 36}{x^2 - 100} \div \dfrac{?}{?} = \dfrac{x^2 - 36}{x^2 + 10x}$

Writing in Mathematics

Answer in complete sentences.

61. Explain the similarities between dividing numerical fractions and dividing rational expressions. Are there any differences?

62. Explain why the restrictions $B \neq 0$, $C \neq 0$, and $D \neq 0$ are necessary when dividing $\dfrac{A}{B} \div \dfrac{C}{D}$.

63. *Solutions Manual* * Write a solutions manual page for the following problem:

$$Divide\, \dfrac{x^2 - 6x - 27}{x^2 + 4x - 5} \div \dfrac{9x - x^2}{x^2 - 25}.$$

64. *Newsletter* * Write a newsletter that explains how to multiply two rational expressions.

*See Appendix B for details and sample answers.

QUICK REVIEW EXERCISES

Section 7.2

Simplify.

1. $(7x + 8) + (2x - 5)$

2. $(x^2 + 3x - 40) + (x^2 - 9x + 11)$

3. $(8x - 15) - (5x + 27)$

4. $(x^2 - 4x + 6) - (3x^2 - 2x + 35)$

Objectives

1 Add rational expressions that have the same denominator.
2 Subtract rational expressions that have the same denominator.
3 Add or subtract rational expressions that have opposite denominators.

Now that we have learned how to multiply and divide rational expressions, we move on to addition and subtraction. We know from our work with numerical fractions that two fractions must have the same denominator before we can add or subtract them. The same holds true for rational expressions. In this section, we will begin with rational expressions that already have the same denominator, and in the next section we will learn how to add and subtract rational expressions that have different denominators.

Adding Rational Expressions That Have the Same Denominator

Objective 1 **Add rational expressions that have the same denominator.** To add fractions that have the same denominator, we add the numerators and place the result over the common denominator. We will follow the same procedure when adding two rational expressions. Of course, we should check that our result is in simplest terms.

Adding Rational Expressions That Have the Same Denominator

$$\frac{A}{C} + \frac{B}{C} = \frac{A + B}{C} \qquad C \neq 0$$

EXAMPLE 1 Add: $\dfrac{7}{x + 3} + \dfrac{5}{x + 3}$.

Solution

These two fractions have the same denominator, so we add the two numerators and place the result over the common denominator $x + 3$.

$$\frac{7}{x + 3} + \frac{5}{x + 3} = \frac{7 + 5}{x + 3} \qquad \text{Add numerators, placing the sum over the common denominator.}$$

$$= \frac{12}{x + 3} \qquad \text{Simplify the numerator.}$$

Quick Check 1
Add: $\dfrac{9}{2x + 7} + \dfrac{12}{2x + 7}$.

The numerator and denominator do not have any common factors, so this is our final result. A common error is to attempt to divide a common factor out of 12 in the numerator and 3 in the denominator, but the number 3 is a term of the denominator and not a factor.

EXAMPLE 2 Add: $\dfrac{6x - 7}{x^2 + 5x + 6} + \dfrac{2x + 23}{x^2 + 5x + 6}$.

Solution

The two denominators are the same, so we may add.

$$\frac{6x - 7}{x^2 + 5x + 6} + \frac{2x + 23}{x^2 + 5x + 6}$$

$$= \frac{(6x - 7) + (2x + 23)}{x^2 + 5x + 6} \qquad \text{Add numerators.}$$

$$= \frac{8x + 16}{x^2 + 5x + 6} \qquad \text{Combine like terms.}$$

$$= \frac{8\overset{1}{\cancel{(x + 2)}}}{\cancel{(x + 2)}(x + 3)} \qquad \begin{array}{l}\text{Factor the numerator and denominator} \\ \text{and divide out the common factor.}\end{array}$$

$$= \frac{8}{x + 3} \qquad \text{Simplify.}$$

Quick Check   2 Add: $\dfrac{2x + 54}{x^2 + 4x - 32} + \dfrac{3x - 14}{x^2 + 4x - 32}$.

EXAMPLE 3   Add: $\dfrac{x^2 - 10x}{x^2 - 5x - 36} + \dfrac{7x - 54}{x^2 - 5x - 36}$.

Solution

The denominators are the same, so we add the numerators and then simplify.

$$\frac{x^2 - 10x}{x^2 - 5x - 36} + \frac{7x - 54}{x^2 - 5x - 36}$$

$$= \frac{(x^2 - 10x) + (7x - 54)}{x^2 - 5x - 36} \qquad \text{Add numerators.}$$

$$= \frac{x^2 - 3x - 54}{x^2 - 5x - 36} \qquad \text{Combine like terms.}$$

$$= \frac{(x + 6)\overset{1}{\cancel{(x - 9)}}}{(x + 4)\underset{1}{\cancel{(x - 9)}}} \qquad \begin{array}{l}\text{Factor the numerator and denominator} \\ \text{and divide out the common factor.}\end{array}$$

$$= \frac{x + 6}{x + 4} \qquad \text{Simplify.}$$

Quick Check   3 Add: $\dfrac{x^2 + 6x + 4}{x^2 + 4x - 21} + \dfrac{5x + 24}{x^2 + 4x - 21}$.

Subtracting Rational Expressions
That Have the Same Denominator

Objective 2 Subtract rational expressions that have the same denominator.
Subtracting two rational expressions that have the same denominator is just like adding them, except that we subtract the two numerators rather than adding them.

Subtracting Rational Expressions That Have the Same Denominator

$$\frac{A}{C} - \frac{B}{C} = \frac{A - B}{C} \qquad C \neq 0$$

EXAMPLE ▶ **4**   Subtract: $\dfrac{9}{2x + 8} - \dfrac{5}{2x + 8}$.

Solution

Since the two fractions have the same denominator, we subtract the numerators and place the result over the common denominator.

$$\frac{9}{2x + 8} - \frac{5}{2x + 8} = \frac{4}{2x + 8}$$
Subtract numerators, place difference over the denominator.

$$= \frac{\overset{2}{\cancel{4}}}{\underset{1}{\cancel{2}}(x + 4)}$$
Factor the denominator and divide out common factors.

$$= \frac{2}{x + 4}$$
Simplify.

Quick Check **4** Subtract: $\dfrac{25}{6x + 36} - \dfrac{17}{6x + 36}$.

EXAMPLE ▶ **5**   Subtract: $\dfrac{5x}{x - 2} - \dfrac{10}{x - 2}$.

Solution

The two denominators are the same, so we may subtract these two rational expressions.

$$\frac{5x}{x - 2} - \frac{10}{x - 2} = \frac{5x - 10}{x - 2}$$
Subtract numerators.

$$= \frac{5\overset{1}{\cancel{(x - 2)}}}{\underset{1}{\cancel{x - 2}}}$$
Factor the numerator and divide out the common factor.

Quick Check **5**

Subtract:
$\dfrac{3x - 7}{x - 9} - \dfrac{20}{x - 9}$.

$$= 5$$
Simplify.

Notice that the denominator $x - 2$ was also a factor of the numerator, leaving a denominator of 1.

EXAMPLE ▶ **6**   Subtract: $\dfrac{2x^2 - 3x - 9}{7x - x^2} - \dfrac{x^2 + 9x - 44}{7x - x^2}$.

Solution

The denominators are the same, so we can subtract these two rational expressions. When the numerator of the second fraction has more than one term, we must remember that we are subtracting the whole numerator and not just the first term. When we subtract the numerators and place the difference over the common denominator, it is a good idea to write each numerator within parentheses. This will remind us to subtract each term in the second numerator.

$$\frac{2x^2 - 3x - 9}{7x - x^2} - \frac{x^2 + 9x - 44}{7x - x^2}$$

$$= \frac{(2x^2 - 3x - 9) - (x^2 + 9x - 44)}{7x - x^2}$$

Subtract numerators.
Change the sign of each term in the second set of parentheses by distributing -1.

$$= \frac{2x^2 - 3x - 9 - x^2 - 9x + 44}{7x - x^2}$$

$$= \frac{x^2 - 12x + 35}{7x - x^2}$$

Combine like terms.
Factor the numerator and denominator and divide out the common factor. The -1 results from the fact that $x - 7$ and $7 - x$ are opposites.

$$= \frac{\overset{-1}{\cancel{(x - 7)}}(x - 5)}{x\underset{1}{\cancel{(7 - x)}}}$$

$$= -\frac{x - 5}{x}$$

Simplify.

Quick Check **6** Subtract: $\dfrac{3x^2 + 6x - 37}{64 - x^2} - \dfrac{2x^2 + 17x - 61}{64 - x^2}$.

A Word of Caution When subtracting a rational expression that has a numerator containing more than one term, be sure to subtract the *entire* numerator and not just the first term. One way to remember this is by placing the numerators inside parentheses.

Adding or Subtracting Rational Expressions That Have Opposite Denominators

Objective 3 **Add or subtract rational expressions that have opposite denominators.** Consider the expression $\dfrac{10}{x - 2} + \dfrac{3}{2 - x}$. Are the two denominators the same?

No, but they are opposites. We can rewrite the denominator $2 - x$ as its opposite $x - 2$ if we also rewrite the operation (addition) as its opposite (subtraction). In other words, we can rewrite the expression $\dfrac{10}{x - 2} + \dfrac{3}{2 - x}$ as $\dfrac{10}{x - 2} - \dfrac{3}{x - 2}$. Once the denominators are the same, we can subtract the numerators.

EXAMPLE 7 Add: $\dfrac{7x}{2x - 16} + \dfrac{56}{16 - 2x}$.

Solution

The two denominators are opposites, so we may change the second denominator to $2x - 16$ by changing the operation from addition to subtraction.

$$\frac{7x}{2x - 16} + \frac{56}{16 - 2x}$$

$$= \frac{7x}{2x - 16} - \frac{56}{2x - 16}$$

Rewrite the second denominator as $2x - 16$ by changing the operation from addition to subtraction.

$$= \frac{7x - 56}{2x - 16}$$

Subtract the numerators.

Quick Check 7

Add: $\dfrac{x}{5x - 20} + \dfrac{4}{20 - 5x}$.

$$= \frac{7\cancel{(x - 8)}}{2\underset{1}{\cancel{(x - 8)}}}$$

Factor the numerator and denominator and divide out the common factor.

$$= \frac{7}{2}$$

Simplify.

EXAMPLE 8   Subtract: $\dfrac{x^2 - 4x}{x - 3} - \dfrac{11x - 30}{3 - x}$.

Solution

These two denominators are opposites, so we begin by rewriting the second rational expression in such a way that the two rational expressions have the same denominator.

$$\frac{x^2 - 4x}{x - 3} - \frac{11x - 30}{3 - x}$$

Rewrite the second denominator as $x - 3$ by changing the operation from subtraction to addition.

$$= \frac{x^2 - 4x}{x - 3} + \frac{11x - 30}{x - 3}$$

$$= \frac{(x^2 - 4x) + (11x - 30)}{x - 3}$$

Add the numerators.

Quick Check 8

Subtract:

$\dfrac{x^2 - 2x - 11}{x^2 - 16} - \dfrac{7x - 25}{16 - x^2}$.

$$= \frac{x^2 + 7x - 30}{x - 3}$$

Combine like terms.

$$= \frac{(x + 10)\overset{1}{\cancel{(x - 3)}}}{\underset{1}{\cancel{x - 3}}}$$

Factor the numerator and divide out the common factor.

$$= x + 10$$

Simplify.

Building Your Study Strategy **Preparing for a Cumulative Exam, 3 Homework and Notes**   When you are studying a particular topic, look back at your old homework for that topic. If you struggled with a particular type of problem, your homework will reflect that. Your homework should contain some notes to yourself about how to do certain problems or mistakes to avoid.

   Look over your class notes from that topic. Your notes should include examples of the problems in that section of the textbook, as well as pointers from your instructor. You can also review that section, paying close attention to the examples and any procedures developed in that section.

   If you have been creating note cards during the semester, they will be most advantageous to you at this time. These note cards should focus on problems that you considered difficult at the time and contain strategies for solving these types of problems. A quick glance at these note cards will really speed up your review of a particular topic.

EXERCISES 7.3

MyMathLab
MathXL
Interactmath.com
MathXL
Tutorials on CD
Video Lectures on CD
Tutor Center
Addison-Wesley
Math Tutor Center
Student's
Solutions Manual

Vocabulary

1. To add fractions that have the same denominator, we add the _____ and place the result over the common denominator.

2. To subtract fractions that have the same denominator, we _____ the numerators and place the result over the common denominator.

3. When subtracting a rational expression that has a numerator containing more than one term, subtract the entire numerator and not just the _____.

4. When adding two rational expressions that have opposite denominators, we can replace the second denominator with its opposite by changing the addition to _____.

Add.

5. $\dfrac{5}{x + 3} + \dfrac{8}{x + 3}$

6. $\dfrac{9}{x - 6} + \dfrac{7}{x - 6}$

7. $\dfrac{x}{x - 5} + \dfrac{10}{x - 5}$

8. $\dfrac{3x}{x + 14} + \dfrac{26}{x + 14}$

9. $\dfrac{2x}{x + 10} + \dfrac{20}{x + 10}$

10. $\dfrac{3x}{4x + 28} + \dfrac{21}{4x + 28}$

11. $\dfrac{x}{x^2 + 7x - 18} + \dfrac{9}{x^2 + 7x - 18}$

12. $\dfrac{x}{x^2 + 12x + 35} + \dfrac{7}{x^2 + 12x + 35}$

13. $\dfrac{2x + 4}{x^2 + 14x + 33} + \dfrac{x + 5}{x^2 + 14x + 33}$

14. $\dfrac{x + 14}{x^2 + 7x + 6} + \dfrac{3x + 10}{x^2 + 7x + 6}$

15. $\dfrac{x^2 + 3x + 11}{x^2 - 2x - 48} + \dfrac{12x + 43}{x^2 - 2x - 48}$

16. $\dfrac{x^2 + 5x + 20}{x^2 - 5x - 14} + \dfrac{5x - 4}{x^2 - 5x - 14}$

17. $\dfrac{x^2 - 16x - 45}{x^2 - 3x - 18} + \dfrac{x^2 - 2x - 27}{x^2 - 3x - 18}$

18. $\dfrac{x^2 - 11x - 34}{x^2 + 16x + 64} + \dfrac{x^2 + 17x - 46}{x^2 + 16x + 64}$

For the given rational functions $f(x)$ and $g(x)$, find $f(x) + g(x)$.

19. $f(x) = \dfrac{3}{x}, g(x) = \dfrac{5}{x}$

20. $f(x) = \dfrac{x}{5x + 30}, g(x) = \dfrac{6}{5x + 30}$

21. $f(x) = \dfrac{x^2 + 3x - 15}{x^2 - 4x - 45}, g(x) = \dfrac{4x + 25}{x^2 - 4x - 45}$

22. $f(x) = \dfrac{2x - 21}{x^2 + 20x + 99}, g(x) = \dfrac{x^2 + 2x - 56}{x^2 + 20x + 99}$

Subtract.

23. $\dfrac{13}{x + 1} - \dfrac{9}{x + 1}$

24. $\dfrac{6}{x - 4} - \dfrac{11}{x - 4}$

25. $\dfrac{x}{x + 7} - \dfrac{7}{x + 7}$

26. $\dfrac{3x}{2x + 5} - \dfrac{9}{2x + 5}$

27. $\dfrac{4x}{x - 3} - \dfrac{12}{x - 3}$

28. $\dfrac{2x}{3x - 15} - \dfrac{10}{3x - 15}$

29. $\dfrac{3x - 7}{x - 8} - \dfrac{x + 9}{x - 8}$

30. $\dfrac{5x - 23}{x - 9} - \dfrac{3x - 5}{x - 9}$

31. $\dfrac{x}{x^2 - 13x + 30} - \dfrac{3}{x^2 - 13x + 30}$

32. $\dfrac{x}{x^2 + 2x - 48} - \dfrac{6}{x^2 + 2x - 48}$

33. $\dfrac{4x - 7}{x^2 - 16} - \dfrac{2x - 15}{x^2 - 16}$

34. $\dfrac{6x + 1}{x^2 - 16x + 63} - \dfrac{3x + 22}{x^2 - 16x + 63}$

35. $\dfrac{x^2 - 7x + 10}{x^2 + 4x - 12} - \dfrac{5x - 10}{x^2 + 4x - 12}$

36. $\dfrac{x^2 + 3x - 18}{x^2 - 8x + 7} - \dfrac{6x + 10}{x^2 - 8x + 7}$

37. $\dfrac{2x^2 + x - 4}{x^2 + 8x} - \dfrac{x^2 + 9x - 4}{x^2 + 8x}$

38. $\dfrac{3x^2 - 7x - 19}{x^2 - 6x - 27} - \dfrac{2x^2 + 7x - 64}{x^2 - 6x - 27}$

39. $\dfrac{(x - 6)(x + 3)}{x^2 - 2x - 15} - \dfrac{9(x - 5)}{x^2 - 2x - 15}$

40. $\dfrac{(2x + 9)(x + 4)}{x^2 - 2x - 24} - \dfrac{(x + 6)(x - 6)}{x^2 - 2x - 24}$

For the given rational functions $f(x)$ and $g(x)$, find $f(x) - g(x)$.

41. $f(x) = \dfrac{12}{x + 8}, g(x) = \dfrac{23}{x + 8}$

42. $f(x) = \dfrac{21}{x - 7}, g(x) = \dfrac{3x}{x - 7}$

43. $f(x) = \dfrac{x^2 + 3x - 5}{x^2 - 11x + 18}, g(x) = \dfrac{8x + 9}{x^2 - 11x + 18}$

44. $f(x) = \dfrac{2x^2 + 10x - 13}{x^2 + 15x + 50}, g(x) = \dfrac{x^2 + 11x + 17}{x^2 + 15x + 50}$

Add or subtract.

45. $\dfrac{2x}{x - 5} + \dfrac{10}{5 - x}$ **46.** $\dfrac{5x}{2x - 14} + \dfrac{35}{14 - 2x}$

47. $\dfrac{x^2 + 8x}{x - 1} - \dfrac{2x - 11}{1 - x}$

48. $\dfrac{x^2 - 8x - 10}{x - 3} + \dfrac{x - 28}{3 - x}$

49. $\dfrac{x^2 - 3x - 9}{x - 9} + \dfrac{x^2 - 7x + 27}{9 - x}$

50. $\dfrac{3x^2 - 14x - 45}{x - 6} - \dfrac{-2x^2 + 14x + 9}{6 - x}$

$\boxed{\textbf{Mixed Practice, 51–68}}$

Add or subtract.

51. $\dfrac{x^2 - 3x + 5}{2x - 20} + \dfrac{4x + 35}{20 - 2x}$

52. $\dfrac{x^2 - 7x + 7}{x^2 + x - 2} + \dfrac{2x - 3}{x^2 + x - 2}$

53. $\dfrac{2x^2 + x - 9}{x^2 - 8x + 15} - \dfrac{x^2 + 10x - 29}{x^2 - 8x + 15}$

54. $\dfrac{x^2 - 7x - 50}{x - 8} - \dfrac{x^2 + 3x - 46}{8 - x}$

55. $\dfrac{4x - 15}{x^2 - 3x} + \dfrac{x^2 + x - 9}{x^2 - 3x}$

56. $\dfrac{4x^2 - 6x + 3}{x^2 + 17x + 70} - \dfrac{3x^2 - 6x + 52}{x^2 + 17x + 70}$

57. $\dfrac{x^2 - 3x + 45}{x^2 - 81} + \dfrac{x^2 + 29x + 27}{x^2 - 81}$

58. $\dfrac{x^2 - 8x + 6}{x^2 + 10x + 21} + \dfrac{5x - 24}{x^2 + 10x + 21}$

59. $\dfrac{x^2 - 3x + 6}{x^2 - 12x + 32} - \dfrac{3x + 22}{x^2 - 12x + 32}$

60. $\dfrac{10}{5x - 25} + \dfrac{2x}{25 - 5x}$

61. $\dfrac{2x^2 - 15x + 39}{x^2 + 3x - 4} + \dfrac{x^2 - 18x - 9}{x^2 + 3x - 4}$

62. $\dfrac{2x^2 - 3x + 20}{x^2 + 6x - 55} - \dfrac{x^2 + 9x - 15}{x^2 + 6x - 55}$

63. $\dfrac{6x}{3x - 2} - \dfrac{4}{2 - 3x}$

64. $\dfrac{x^2 + 20x + 45}{x^2 + 15x + 56} - \dfrac{3x - 27}{x^2 + 15x + 56}$

65. $\dfrac{5x^2 + 3x - 13}{x^2 - 8x - 9} - \dfrac{4x^2 - 7x - 22}{x^2 - 8x - 9}$

66. $\dfrac{5x - 12}{x^2 - 8x + 12} + \dfrac{x^2 + 8x - 18}{x^2 - 8x + 12}$

67. $\dfrac{2x^2 + 5x + 20}{x^2 + 5x + 6} + \dfrac{3x^2 - 7x - 4}{x^2 + 5x + 6} - \dfrac{4x^2 - 2x + 25}{x^2 + 5x + 6}$

68. $\dfrac{x^2 + 3x - 7}{x^2 - 2x - 8} - \dfrac{x^2 + 5x - 9}{x^2 - 2x - 8} + \dfrac{x^2 - 6x + 14}{x^2 - 2x - 8}$

Find the missing numerator.

69. $\dfrac{?}{x + 6} + \dfrac{18}{x + 6} = 3$

70. $\dfrac{x^2 + 5x}{x - 2} - \dfrac{?}{x - 2} = x + 7$

71. $\dfrac{x^2 + 7x + 17}{(x + 8)(x + 5)} + \dfrac{?}{(x + 8)(x + 5)} = \dfrac{x + 3}{x + 5}$

72. $\dfrac{x^2 - 5x - 24}{(x + 2)(x - 4)} - \dfrac{?}{(x + 2)(x - 4)} = \dfrac{x + 2}{x - 4}$

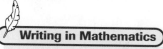

Writing in Mathematics

Answer in complete sentences.

73. Explain how to determine whether two rational expressions have opposite denominators.

74. Explain why it is a good idea to use parentheses when subtracting a rational expression that has more than one term in the numerator.

75. *Solutions Manual** Write a solutions manual page for the following problem:

Find $\dfrac{x^2 + 3x + 3}{x^2 + 6x - 16} - \dfrac{12x - 11}{x^2 + 6x - 16}$.

76. *Newsletter** Write a newsletter which explains how to add two rational expressions that have the same denominator.

*See Appendix B for details and sample answers.

Addition and Subtraction of Rational Expressions That Have Different Denominators

Objectives

1 Find the least common denominator (LCD) of two or more rational expressions.
2 Add or subtract rational expressions that have different denominators.

The Least Common Denominator of Two or More Rational Expressions

Objective 1 Find the least common denominator (LCD) of two or more rational expressions. If two numerical fractions do not have the same denominator, we cannot add or subtract the fractions until we rewrite them as equivalent fractions with a common denominator. The same holds true for rational expressions that have different denominators. We will begin this section by learning how to find the **least common denominator (LCD)** for two or more rational expressions. Then we will learn how to add or subtract rational expressions that have different denominators.

Finding the LCD of Two Rational Expressions

Begin by completely factoring each denominator, and then identify each expression that is a factor of one or both denominators. The LCD is the product of these factors.

If an expression is a repeated factor of one or more of the denominators, then we repeat it as a factor in the LCD as well. The exponent used for this factor is the greatest power to which the factor is raised in any one denominator.

EXAMPLE 1 Find the LCD of $\dfrac{7}{24a^3b}$ and $\dfrac{9}{16a}$.

Solution

> **Quick Check 1**
> Find the LCD of $\dfrac{10}{9r^2s^3}$ and $\dfrac{1}{12rs^6}$.

We begin with the coefficients 24 and 16. The smallest number into which both divide evenly is 48. Moving on to variable factors in the denominator, we see that the variables a and b are factors of one or both denominators. Note that the variable a is raised to the third power in the first denominator, so the LCD must contain a factor of a^3. The LCD is $48a^3b$.

EXAMPLE 2 Find the LCD of $\dfrac{8}{x^2 - 4x + 3}$ and $\dfrac{9}{x^2 - 9}$.

Solution

> **Quick Check 2**
> Find the LCD of $\dfrac{10}{x^2 - 13x + 40}$ and $\dfrac{3}{x^2 - 4x - 32}$.

We begin by factoring each denominator.

$$\frac{8}{x^2 - 4x + 3} = \frac{8}{(x - 1)(x - 3)} \qquad \frac{9}{x^2 - 9} = \frac{9}{(x + 3)(x - 3)}$$

The factors in the denominators are $x - 1$, $x - 3$, and $x + 3$. Since no expression is repeated as a factor in any one denominator, the LCD is $(x - 1)(x - 3)(x + 3)$.

EXAMPLE 3 Find the LCD of $\dfrac{2}{x^2 - 2x - 35}$ and $\dfrac{x}{x^2 + 10x + 25}$.

Solution

Again, we begin by factoring each denominator.

$$\frac{2}{x^2 - 2x - 35} = \frac{2}{(x - 7)(x + 5)} \qquad \frac{x}{x^2 + 10x + 25} = \frac{x}{(x + 5)(x + 5)}$$

The two expressions that are factors are $x - 7$ and $x + 5$; the factor $x + 5$ is repeated twice in the second denominator. So the LCD must have $x + 5$ as a factor twice as well. The LCD is $(x - 7)(x + 5)(x + 5)$, or $(x - 7)(x + 5)^2$.

Quick Check 3
Find the LCD of $\dfrac{x + 8}{x^2 - 81}$ and $\dfrac{x - 7}{x^2 + 18x + 81}$.

Adding or Subtracting Rational Expressions That Have Different Denominators

Objective 2 Add or subtract rational expressions that have different denominators. To add or subtract two rational expressions that do not have the same denominator, we begin by finding the LCD. We then convert each rational expression to an equivalent rational expression that has the LCD as its denominator. We can then add or subtract as we did in the previous section. As always, we should attempt to simplify the resulting rational expression.

EXAMPLE 4 Add: $\dfrac{2a}{3} + \dfrac{3a}{7}$.

Solution

The LCD of these two fractions is 21. We will rewrite each fraction as an equivalent fraction whose denominator is 21. Since $\frac{7}{7} = 1$, multiplying the first fraction by $\frac{7}{7}$ will produce an equivalent fraction whose denominator is 21. In a similar fashion, we multiply the second fraction by $\frac{3}{3}$. Once both fractions have the same denominator, we add the numerators and write the sum above the common denominator.

$$\frac{2a}{3} + \frac{3a}{7} = \frac{2a}{3} \cdot \frac{7}{7} + \frac{3a}{7} \cdot \frac{3}{3}$$

Multiply the first fraction by $\frac{7}{7}$ and the second fraction by $\frac{3}{3}$ to rewrite each fraction with a denominator of 21.

$$= \frac{14a}{21} + \frac{9a}{21}$$

Multiply numerators and denominators.

$$= \frac{23a}{21}$$

Add the numerators and place the sum over the common denominator.

EXAMPLE 5 Add: $\dfrac{8}{x + 6} + \dfrac{2}{x - 7}$.

Solution

The two denominators are not the same, so we begin by finding the LCD for these two rational expressions. Each denominator has a single factor, and the LCD is the product of these two denominators. The LCD is $(x + 6)(x - 7)$. We will multiply $\dfrac{8}{x + 6}$ by $\dfrac{x - 7}{x - 7}$ to write it as an equivalent fraction whose denominator is the LCD. We need to multiply $\dfrac{2}{x - 7}$ by $\dfrac{x + 6}{x + 6}$ to write it as an equivalent fraction whose denominator is the LCD.

$$\dfrac{8}{x + 6} + \dfrac{2}{x - 7}$$

$$= \dfrac{8}{x + 6} \cdot \dfrac{x - 7}{x - 7} + \dfrac{2}{x - 7} \cdot \dfrac{x + 6}{x + 6}$$

Multiply to rewrite each expression as an equivalent rational expression that has the LCD as its denominator.

$$= \dfrac{8x - 56}{(x + 6)(x - 7)} + \dfrac{2x + 12}{(x - 7)(x + 6)}$$

Distribute in each numerator, but do not distribute in the denominators.

$$= \dfrac{(8x - 56) + (2x + 12)}{(x + 6)(x - 7)}$$

Add the numerators, writing the sum over the common denominator.

$$= \dfrac{10x - 44}{(x + 6)(x - 7)}$$

Combine like terms.

$$= \dfrac{2(5x - 22)}{(x + 6)(x - 7)}$$

Factor the numerator.

Since the numerator and denominator do not have any common factors, this rational expression cannot be simplified any further.

Quick Check   4

Add: $\dfrac{5}{x - 2} + \dfrac{6}{x + 6}$.

A Word of Caution   When we add two rational expressions, we cannot simply add the two numerators together and place their sum over the sum of the two denominators.

$$\dfrac{8}{x + 6} + \dfrac{2}{x - 7} \neq \dfrac{8 + 2}{(x + 6) + (x - 7)}$$

We must first find a common denominator and rewrite each rational expression as an equivalent expression whose denominator is equal to the common denominator.

When adding or subtracting rational expressions, we simplify the numerator, but leave the denominator in factored form. After we simplify the numerator, we factor if possible and check the denominator for common factors that can be divided out.

EXAMPLE ▶ 6   Subtract: $\dfrac{x}{x^2 + 8x + 15} - \dfrac{5}{x^2 + 12x + 35}$.

Solution

In this example, we must factor each denominator to find the LCD.

$$\dfrac{x}{x^2 + 8x + 15} - \dfrac{5}{x^2 + 12x + 35}$$

$$= \dfrac{x}{(x + 3)(x + 5)} - \dfrac{5}{(x + 5)(x + 7)}$$

Factor each denominator. The LCD is $(x + 3)(x + 5)(x + 7)$.

$$= \frac{x}{(x+3)(x+5)} \cdot \frac{x+7}{x+7} - \frac{5}{(x+5)(x+7)} \cdot \frac{x+3}{x+3}$$

Multiply to rewrite each expression as an equivalent rational expression that has the LCD as its denominator.

$$= \frac{x^2 + 7x}{(x+3)(x+5)(x+7)} - \frac{5x + 15}{(x+5)(x+7)(x+3)}$$

Distribute in each numerator.

$$= \frac{(x^2 + 7x) - (5x + 15)}{(x+3)(x+5)(x+7)}$$

Subtract the numerators, writing the difference over the LCD.

$$= \frac{x^2 + 7x - 5x - 15}{(x+3)(x+5)(x+7)}$$

Distribute.

$$= \frac{x^2 + 2x - 15}{(x+3)(x+5)(x+7)}$$

Combine like terms.

$$= \frac{\overset{1}{\cancel{(x+5)}}(x-3)}{(x+3)\underset{1}{\cancel{(x+5)}}(x+7)}$$

Factor the numerator and divide out the common factor.

$$= \frac{x-3}{(x+3)(x+7)}$$

Simplify.

We must be careful to subtract the entire second numerator, not just the first term. In other words, the subtraction must change the sign of each term in the second numerator before we combine like terms. Using parentheses when subtracting the two numerators will help us remember to do this.

Quick Check **5** Add: $\dfrac{x}{x^2 - 6x + 5} + \dfrac{1}{x^2 + 2x - 3}$.

EXAMPLE 7 Subtract: $\dfrac{x-6}{x^2-1} - \dfrac{5}{x^2 - 4x + 3}$.

Solution

Notice that the first numerator contains a binomial. We must be careful when multiplying to create equivalent rational expressions with the LCD as their denominators. We begin by factoring each denominator to find the LCD.

$$\frac{x-6}{x^2-1} - \frac{5}{x^2 - 4x + 3}$$

$$= \frac{x-6}{(x+1)(x-1)} - \frac{5}{(x-3)(x-1)}$$

Factor each denominator. The LCD is $(x+1)(x-1)(x-3)$.

$$= \frac{x-6}{(x+1)(x-1)} \cdot \frac{x-3}{x-3} - \frac{5}{(x-3)(x-1)} \cdot \frac{x+1}{x+1}$$

Multiply to rewrite each expression as an equivalent rational expression with the LCD as its denominator.

$$= \frac{x^2 - 9x + 18}{(x+1)(x-1)(x-3)} - \frac{5x+5}{(x-3)(x-1)(x+1)}$$

Distribute in each numerator.

$$= \frac{(x^2 - 9x + 18) - (5x + 5)}{(x+1)(x-1)(x-3)}$$

Subtract the numerators, writing the difference over the LCD.

$$= \frac{x^2 - 9x + 18 - 5x - 5}{(x+1)(x-1)(x-3)}$$

Distribute.

$$= \frac{x^2 - 14x + 13}{(x+1)(x-1)(x-3)}$$

Combine like terms.

$$= \frac{\overset{1}{\cancel{(x-1)}}(x - 13)}{(x+1)\underset{1}{\cancel{(x-1)}}(x-3)}$$

Factor the numerator and divide out the common factor.

$$= \frac{x - 13}{(x+1)(x-3)}$$

Simplify.

Quick Check   **6**   Add: $\dfrac{x+5}{x^2 + 8x + 12} + \dfrac{x+9}{x^2 - 36}$.

Building Your Study Strategy **Preparing for a Cumulative Exam, 4 Problems from Your Instructor**   Some of the most valuable materials for preparing for a cumulative exam are the materials provided by your instructor, such as a review sheet for the final, a practice final, or a list of problems from the text to review. If your instructor feels that these problems are important enough for you to review, then they are definitely important enough to appear on the cumulative exam.

As you work through your study schedule, try to solve your instructor's problems when you review that section or chapter. Your performance on these problems will show you which topics you understand and which topics require further study.

One mistake that students make is trying to memorize the solution of each and every problem; this will not help you solve a similar problem on the exam. If you are having trouble with a certain problem, write down the steps to solve that problem on a note card. Review your note cards for a short period each day before the exam. This will help you understand how to solve a problem, and it increases your chance of solving a similar problem on the exam.

MyMathLab

MathXL

Interactmath.com

MathXL
Tutorials on CD

Video Lectures
on CD

Tutor
Center

Addison-Wesley
Math Tutor Center

Student's
Solutions Manual

F
O
R

E
X
T
R
A

H
E
L
P

Vocabulary

1. The _____ of two rational expressions is an expression that is a product of each factor of the two denominators.

2. To add two rational expressions that have different denominators, we begin by converting each rational expression to a(n) _____ rational expression that has the LCD as its denominator.

Find the LCD of the given rational expressions.

3. $\dfrac{5}{2a}, \dfrac{1}{3a}$

4. $\dfrac{9}{8x^2}, \dfrac{7}{6x^3}$

5. $\dfrac{6}{x-9}, \dfrac{8}{x+5}$

6. $\dfrac{x}{x-7}, \dfrac{x+2}{2x+3}$

7. $\dfrac{2}{x^2-9}, \dfrac{x+7}{x^2-9x+18}$

8. $\dfrac{x-6}{x^2-3x-28}, \dfrac{x+1}{x^2-x-20}$

9. $\dfrac{3}{x^2+7x+10}, \dfrac{x-1}{x^2+4x+4}$

10. $\dfrac{x-10}{x^2-12x+36}, \dfrac{8}{x^2-36}$

Add or subtract.

11. $\dfrac{3x}{4} + \dfrac{x}{6}$

12. $\dfrac{7x}{12} - \dfrac{3x}{20}$

13. $\dfrac{8}{5a} - \dfrac{2}{3a}$

14. $\dfrac{1}{8b} + \dfrac{6}{7b}$

15. $\dfrac{4}{m^2 n^3} + \dfrac{5}{m^4 n}$

16. $\dfrac{6}{st^5} + \dfrac{7}{t^6}$

17. $\dfrac{5}{x+2} + \dfrac{3}{x+4}$

18. $\dfrac{7}{x+5} + \dfrac{2}{x-8}$

19. $\dfrac{10}{x+3} - \dfrac{4}{x-3}$

20. $\dfrac{6}{x} - \dfrac{9}{x-4}$

21. $\dfrac{6}{x+4} + \dfrac{3}{x-2}$

22. $\dfrac{5}{x+5} - \dfrac{5}{x-6}$

23. $\dfrac{3}{(x+5)(x+2)} + \dfrac{4}{(x+2)(x-2)}$

24. $\dfrac{8}{(x-6)(x-10)} - \dfrac{6}{(x-8)(x-10)}$

25. $\dfrac{7}{(x+1)(x-6)} - \dfrac{3}{(x-3)(x-6)}$

26. $\dfrac{5}{(x-8)(x+2)} + \dfrac{1}{(x+2)(x+4)}$

27. $\dfrac{4}{x^2-6x-27} - \dfrac{2}{x^2-12x+27}$

28. $\dfrac{9}{x^2-11x+28} + \dfrac{3}{x^2-7x+12}$

29. $\dfrac{5}{x^2-15x+50} + \dfrac{9}{x^2-x-20}$

30. $\dfrac{4}{x^2+12x} - \dfrac{2}{x^2+18x+72}$

31. $\dfrac{3}{x^2-13x+40} - \dfrac{4}{x^2-12x+32}$

32. $\dfrac{2}{x^2+x-6} - \dfrac{4}{x^2-4x-21}$

33. $\dfrac{5}{2x^2+3x-2} + \dfrac{7}{2x^2-9x+4}$

34. $\dfrac{7}{2x^2-15x-27} - \dfrac{3}{2x^2-3x-9}$

35. $\dfrac{x+6}{x^2-9} - \dfrac{1}{x^2+7x+12}$

36. $\dfrac{x-2}{x^2+x-56} + \dfrac{4}{x^2+10x+16}$

37. $\dfrac{x-3}{x^2-x-20} + \dfrac{6}{x^2-2x-24}$

38. $\dfrac{x-10}{x^2-2x-3} - \dfrac{3}{x^2-9x+18}$

39. $\dfrac{x-7}{x^2+5x-24} + \dfrac{3}{x^2-9}$

40. $\dfrac{x+5}{x^2+2x-24} - \dfrac{4}{x^2+16x+60}$

41. $\dfrac{x+14}{x^2+5x-50} - \dfrac{8}{x^2+15x+50}$

42. $\dfrac{x-6}{x^2-16} - \dfrac{2}{x^2+8x+16}$

43. $\dfrac{x-3}{x^2+17x+70} + \dfrac{10}{x^2+10x}$

44. $\dfrac{x+9}{x^2-8x} + \dfrac{8}{x^2-16x+64}$

45. $\dfrac{x+10}{x^2+5x-66} - \dfrac{7}{x^2+20x+99}$

46. $\dfrac{x-4}{x^2+20x+96} - \dfrac{14}{x^2+6x-16}$

Find the missing numerator.

47. $\dfrac{?}{x^2+x-2} + \dfrac{2}{x^2-4x+3} = \dfrac{5}{x^2-x-6}$

48. $\dfrac{?}{x^2+7x+10} + \dfrac{4}{x^2-x-6} = \dfrac{13x-7}{(x+5)(x+2)(x-3)}$

49. $\dfrac{?}{x^2+2x-24} + \dfrac{x}{x^2-3x-54} = \dfrac{x-6}{x^2-13x+36}$

50. $\dfrac{?}{x^2-x-12} + \dfrac{x}{x^2-12x+32} = \dfrac{x+14}{x^2-5x-24}$

Writing in Mathematics

Answer in complete sentences.

51. Explain how to find the LCD of two rational expressions. Use an example to illustrate the process.

52. Here is a student's solution to a problem on an exam. Describe the student's error, and provide the correct solution. Assuming that the problem was worth 10 points, how many points would you give to the student for this solution? Explain your reasoning.

$$\dfrac{5}{(x+3)(x+2)} + \dfrac{7}{(x+3)(x+6)}$$
$$= \dfrac{5}{(x+3)(x+2)} \cdot \dfrac{x+6}{x+6} + \dfrac{7}{(x+3)(x+6)} \cdot \dfrac{x+2}{x+2}$$
$$= \dfrac{5(x+6)+7(x+2)}{(x+3)(x+2)(x+6)}$$
$$= \dfrac{5(x+6)+7(x+2)}{(x+3)(x+2)(x+6)}$$
$$= \dfrac{12}{x+3}$$

53. *Solutions Manual** * Write a solutions manual page for the following problem:

Add: $\dfrac{x+6}{x^2+9x+20} + \dfrac{4}{x^2+6x+8}$.

54. *Newsletter** * Write a newsletter that explains how to subtract two rational expressions that have different denominators.

*See Appendix B for details and sample answers.

7.5
Complex Fractions

1 Simplify complex numerical fractions.
2 Simplify complex fractions containing variables.

Complex Fractions

A **complex fraction** is a fraction or rational expression containing one or more fractions in its numerator or denominator. Here are some examples:

$$\frac{\dfrac{1}{2}+\dfrac{5}{3}}{\dfrac{10}{3}-\dfrac{7}{4}} \qquad \frac{x+5}{x-9} \over 1-\dfrac{5}{x} \qquad \frac{\dfrac{1}{2}+\dfrac{1}{x}}{\dfrac{1}{4}-\dfrac{1}{x^2}} \qquad \frac{1+\dfrac{3}{x}-\dfrac{28}{x^2}}{1+\dfrac{7}{x}}$$

Objective 1 Simplify complex numerical fractions. To simplify a complex fraction, we must rewrite it so that its numerator and denominator do not contain fractions. We can do this by finding the LCD of all fractions within the complex fraction and then multiplying the numerator and denominator by this LCD. This will clear the fractions within the complex fraction. We finish by simplifying the resulting rational expression, if possible.

We begin with a complex fraction made up of numerical fractions.

EXAMPLE 1 Simplify the complex fraction $\dfrac{1+\dfrac{4}{3}}{\dfrac{2}{3}+\dfrac{11}{4}}$.

Solution

The LCD of the three denominators (3, 3, and 4) is 12, so we begin by multiplying the complex fraction by $\frac{12}{12}$. Notice that when we multiply by $\frac{12}{12}$, we are really multiplying by 1, which does not change the value of the original expression.

$$\frac{1+\dfrac{4}{3}}{\dfrac{2}{3}+\dfrac{11}{4}} = \frac{12}{12} \cdot \frac{1+\dfrac{4}{3}}{\dfrac{2}{3}+\dfrac{11}{4}} \qquad \text{Multiply the numerator and denominator by the LCD.}$$

$$= \frac{12 \cdot 1 + \overset{4}{\cancel{12}} \cdot \dfrac{4}{\underset{1}{\cancel{3}}}}{\overset{4}{\cancel{12}} \cdot \dfrac{2}{\underset{1}{\cancel{3}}} + \overset{3}{\cancel{12}} \cdot \dfrac{11}{\underset{1}{\cancel{4}}}} \qquad \text{Distribute and divide out common factors.}$$

$$= \frac{12+16}{8+33} \qquad \text{Multiply.}$$

$$= \frac{28}{41} \qquad \text{Simplify the numerator and denominator.}$$

There is another method for simplifying complex fractions. We can rewrite the numerator as a single fraction by adding $1 + \frac{4}{3}$, which would equal $\frac{7}{3}$.

$$1 + \frac{4}{3} = \frac{3}{3} + \frac{4}{3} = \frac{7}{3}$$

We can also rewrite the denominator as a single fraction by adding $\frac{2}{3} + \frac{11}{4}$, which would equal $\frac{41}{12}$.

$$\frac{2}{3} + \frac{11}{4} = \frac{8}{12} + \frac{33}{12} = \frac{41}{12}$$

Once the numerator and denominator are single fractions, we can rewrite $\frac{\frac{7}{3}}{\frac{41}{12}}$ as a division problem $\frac{7}{3} \div \frac{41}{12}$ and simplify from there. This method would produce the same result of $\frac{28}{41}$.

$$\frac{7}{3} \div \frac{41}{12} = \frac{7}{\overset{}{\underset{1}{3}}} \cdot \frac{\overset{4}{12}}{41} = \frac{28}{41}$$

In most of the remaining examples, we will use the LCD method, as the technique is somewhat similar to the technique used to solve rational equations in the next section.

Quick Check 1
Simplify the complex fraction $\dfrac{\frac{2}{5} + \frac{3}{8}}{\frac{1}{4} + \frac{7}{10}}$.

Simplifying Complex Fractions Containing Variables

Objective 2   Simplify complex fractions containing variables.

EXAMPLE 2   Simplify $\dfrac{1 - \frac{9}{x^2}}{1 + \frac{3}{x}}$.

Solution

The LCD for the two simple fractions with denominators x and x^2 is x^2, so we will begin by multiplying the complex fraction by $\frac{x^2}{x^2}$. Then, once we have cleared the fractions, resulting in a rational expression, we simplify the rational expression by factoring and dividing out common factors.

$$\frac{1 - \frac{9}{x^2}}{1 + \frac{3}{x}} = \frac{x^2}{x^2} \cdot \frac{1 - \frac{9}{x^2}}{1 + \frac{3}{x}} \qquad \text{Multiply the numerator and denominator by the LCD.}$$

$$= \frac{x^2 \cdot 1 - \overset{1}{x^2} \cdot \frac{9}{x^2}}{x^2 \cdot 1 + \overset{x}{x^2} \cdot \frac{3}{x}} \qquad \text{Distribute and divide out common factors, clearing the fractions.}$$

$$= \frac{x^2 - 9}{x^2 + 3x} \qquad \text{Multiply.}$$

$$= \frac{(x + 3)(x - 3)}{x(x + 3)}$$ Factor the numerator and denominator and divide out the common factor.

$$= \frac{x - 3}{x}$$ Simplify.

Quick Check **2** Simplify $\dfrac{1 + \dfrac{3}{x} - \dfrac{10}{x^2}}{1 - \dfrac{2}{x}}$.

EXAMPLE **3** Simplify $\dfrac{\dfrac{1}{6} + \dfrac{1}{x}}{\dfrac{1}{36} - \dfrac{1}{x^2}}$.

Solution

We begin by multiplying the numerator and denominator by the LCD of all denominators. In this case, the LCD is $36x^2$.

$$\frac{\dfrac{1}{6} + \dfrac{1}{x}}{\dfrac{1}{36} - \dfrac{1}{x^2}} = \frac{36x^2}{36x^2} \cdot \frac{\dfrac{1}{6} + \dfrac{1}{x}}{\dfrac{1}{36} - \dfrac{1}{x^2}}$$ Multiply the numerator and denominator by the LCD.

$$= \frac{36x^2 \cdot \dfrac{1}{6} + 36x^2 \cdot \dfrac{1}{x}}{36x^2 \cdot \dfrac{1}{36} - 36x^2 \cdot \dfrac{1}{x^2}}$$ Distribute and divide out common factors.

$$= \frac{6x^2 + 36x}{x^2 - 36}$$ Multiply.

Quick Check **3**

Simplify $\dfrac{\dfrac{1}{64} - \dfrac{1}{x^2}}{\dfrac{1}{8} - \dfrac{1}{x}}$.

$$= \frac{6x(x + 6)}{(x + 6)(x - 6)}$$ Factor the numerator and denominator and divide out the common factor.

$$= \frac{6x}{x - 6}$$ Simplify.

EXAMPLE **4** Simplify $\dfrac{\dfrac{3}{x + 1} - \dfrac{2}{x + 2}}{\dfrac{x}{x + 2} + \dfrac{6}{x + 1}}$.

Solution

The LCD for the four simple fractions is $(x + 1)(x + 2)$.

$$\frac{\dfrac{3}{x + 1} - \dfrac{2}{x + 2}}{\dfrac{x}{x + 2} + \dfrac{6}{x + 1}} = \frac{(x + 1)(x + 2)}{(x + 1)(x + 2)} \cdot \frac{\dfrac{3}{x + 1} - \dfrac{2}{x + 2}}{\dfrac{x}{x + 2} + \dfrac{6}{x + 1}}$$ Multiply the numerator and denominator by the LCD.

$$= \frac{(x+1)(x+2) \cdot \dfrac{3}{x+1} - (x+1)(x+2) \cdot \dfrac{2}{x+2}}{(x+1)(x+2) \cdot \dfrac{x}{x+2} + (x+1)(x+2) \cdot \dfrac{6}{x+1}}$$ Distribute and divide out common factors, clearing the fractions.

$$= \frac{3(x+2) - 2(x+1)}{x(x+1) + 6(x+2)}$$ Multiply.

$$= \frac{3x + 6 - 2x - 2}{x^2 + x + 6x + 12}$$ Distribute.

$$= \frac{x+4}{x^2 + 7x + 12}$$ Combine like terms.

$$= \frac{x+4}{(x+3)(x+4)}$$ Factor the denominator and divide out the common factor.

$$= \frac{1}{x+3}$$ Simplify.

Quick Check 4

Simplify $\dfrac{\dfrac{6}{x-4} + \dfrac{5}{x+7}}{\dfrac{x+2}{x-4}}$.

EXAMPLE 5

Simplify $\dfrac{\dfrac{x^2 + 5x - 14}{x^2 - 9}}{\dfrac{x^2 - 6x + 8}{x^2 + 8x + 15}}$.

Solution

In this case, it will be easier to rewrite the complex fraction as a division problem rather than multiplying the numerator and denominator by the LCD. This is the best technique when we have a complex fraction with a single rational expression in its numerator and a single rational expression in its denominator. We have used this method when dividing rational expressions.

$$\frac{\dfrac{x^2 + 5x - 14}{x^2 - 9}}{\dfrac{x^2 - 6x + 8}{x^2 + 8x + 15}}$$

$$= \frac{x^2 + 5x - 14}{x^2 - 9} \div \frac{x^2 - 6x + 8}{x^2 + 8x + 15}$$ Rewrite as a division problem.

$$= \frac{x^2 + 5x - 14}{x^2 - 9} \cdot \frac{x^2 + 8x + 15}{x^2 - 6x + 8}$$ Invert the divisor and multiply.

$$= \frac{(x+7)(x-2)}{(x+3)(x-3)} \cdot \frac{(x+3)(x+5)}{(x-2)(x-4)}$$ Factor each numerator and denominator.

$$= \frac{(x+7)(x-2)}{(x+3)(x-3)} \cdot \frac{(x+3)(x+5)}{(x-2)(x-4)}$$ Divide out common factors.

$$= \frac{(x+7)(x+5)}{(x-3)(x-4)}$$ Simplify.

Quick Check 5

Simplify $\dfrac{\dfrac{x^2 - 2x - 48}{x^2 + 7x - 30}}{\dfrac{x^2 + 7x + 6}{x^2 - 5x + 6}}$.

Building Your Study Strategy **Preparing for a Cumulative Exam, 5**
Cumulative Review Exercises Use your cumulative review exercises to prepare for a cumulative exam. These exercise sets contain problems representative of the material covered in the previous several chapters.

After you have been preparing for a while, try these exercises without referring to your notes, note cards, or the text. In this way, you can determine which topics require more study. If you made a mistake while solving a problem, make note of the mistake and how to avoid it in the future. If there are problems that you do not recognize or know how to begin, then these topics will require additional study. Ask your instructor or a tutor for help.

EXERCISES 7.5 ❯

Vocabulary

1. A(n) _____ is a fraction or rational expression containing one or more fractions in its numerator or denominator.

2. To simplify a complex fraction, multiply the numerator and denominator by the _____ of all fractions within the complex fraction.

Simplify the complex fraction.

3. $\dfrac{\dfrac{2}{5} - \dfrac{1}{4}}{\dfrac{9}{10} + \dfrac{5}{2}}$

4. $\dfrac{\dfrac{3}{7} + \dfrac{2}{3}}{\dfrac{16}{21} - \dfrac{2}{7}}$

5. $\dfrac{2 - \dfrac{3}{8}}{\dfrac{5}{4} + \dfrac{1}{3}}$

6. $\dfrac{\dfrac{5}{6} - 3}{\dfrac{3}{4} + \dfrac{11}{12}}$

7. $\dfrac{x + \dfrac{3}{5}}{x + \dfrac{4}{7}}$

8. $\dfrac{x - \dfrac{2}{9}}{x + \dfrac{5}{4}}$

9. $\dfrac{6 + \dfrac{15}{x}}{x + \dfrac{5}{2}}$

10. $\dfrac{x + \dfrac{3}{7}}{14 + \dfrac{6}{x}}$

11. $\dfrac{14 - \dfrac{4}{x}}{21 - \dfrac{6}{x}}$

12. $\dfrac{10 + \dfrac{8}{x}}{25 + \dfrac{20}{x}}$

13. $\dfrac{2 + \dfrac{8}{x}}{1 - \dfrac{16}{x^2}}$

14. $\dfrac{5 - \dfrac{20}{x^2}}{1 - \dfrac{2}{x}}$

15. $\dfrac{\dfrac{6}{x + 3} - \dfrac{2}{x + 5}}{\dfrac{x + 6}{x + 3}}$

16. $\dfrac{\dfrac{7}{x + 4} - \dfrac{1}{x - 2}}{\dfrac{5x - 15}{x - 2}}$

17. $\dfrac{\dfrac{4}{x} + \dfrac{8}{x - 6}}{1 + \dfrac{4}{x - 6}}$

18. $\dfrac{\dfrac{3}{x - 1} - \dfrac{2}{x + 2}}{1 + \dfrac{6}{x + 2}}$

19. $\dfrac{1 + \dfrac{4}{x} - \dfrac{32}{x^2}}{1 + \dfrac{13}{x} + \dfrac{40}{x^2}}$

20. $\dfrac{1 - \dfrac{9}{x} + \dfrac{18}{x^2}}{1 + \dfrac{2}{x} - \dfrac{15}{x^2}}$

21. $\dfrac{\dfrac{8}{x^2} - \dfrac{8}{x} + 2}{1 - \dfrac{4}{x^2}}$

22. $\dfrac{3 - \dfrac{18}{x}}{\dfrac{1}{x} - \dfrac{42}{x^2} + 1}$

23. $\dfrac{\dfrac{x^2 + 9x + 14}{x^2 - 2x - 63}}{\dfrac{x^2 - 3x - 10}{x^2 - 15x + 54}}$

24. $\dfrac{\dfrac{x^2 - x - 20}{x^2 - 9}}{\dfrac{x^2 + 12x + 32}{x^2 - 4x - 21}}$

25. $\dfrac{\dfrac{x^2 + 2x}{x^2 - 11x + 24}}{\dfrac{x^2 + 11x + 18}{x^2 - 7x + 12}}$

26. $\dfrac{\dfrac{x^2 + 19x + 88}{x^2 - 4x - 5}}{\dfrac{x^2 + 5x - 66}{x^2 + x - 30}}$

27. $\dfrac{\dfrac{1}{x^2 + 11x + 10}}{\dfrac{1}{x^2 - 6x - 7}}$

28. $\dfrac{\dfrac{1}{x^2 - 16x + 60}}{\dfrac{1}{x^2 - 5x - 50}}$

Mixed Practice, 29–52

Simplify the given rational expression, using the techniques developed in Sections 7.1 through 7.5.

29. $\dfrac{x^2 + 4x}{x^2 - x - 42} + \dfrac{x - 6}{x^2 - x - 42}$

30. $\dfrac{x^2 - 14x + 40}{x^2 + 4x + 3} \div \dfrac{x^2 - 6x + 8}{x^2 + 10x + 9}$

31. $\dfrac{x + 3}{x^2 - 2x - 24} + \dfrac{5}{x^2 - 8x + 12}$

32. $\dfrac{6}{x^2 - 2x - 35} - \dfrac{4}{x^2 + 2x - 15}$

33. $\dfrac{x^2 + 7x}{3x^2 - 19x + 20} \div \dfrac{x^2 + 12x + 35}{x^2 - 3x - 10}$

34. $\dfrac{x^2 - 36}{x^2 + 12x + 36}$

35. $\dfrac{x + 5}{x^2 - 49} - \dfrac{6}{x^2 - 7x}$

36. $\dfrac{x^2 + 3x - 40}{x^2 + x - 6} \cdot \dfrac{x^2 - 7x - 30}{2x^2 + 17x + 8}$

37. $\dfrac{5x}{x - 2} + \dfrac{10}{2 - x}$

38. $\dfrac{1 - \dfrac{11}{x} + \dfrac{18}{x^2}}{1 - \dfrac{2}{x} - \dfrac{63}{x^2}}$

39. $\dfrac{\dfrac{7}{x + 2} - \dfrac{2}{x - 3}}{\dfrac{x - 5}{x + 2}}$

40. $\dfrac{3}{x^2 - x - 2} + \dfrac{5}{x^2 + 7x + 6}$

41. $\dfrac{3}{x^2 + 2x - 80} - \dfrac{1}{x^2 - 11x + 24}$

42. $\dfrac{x^2 + 6x - 16}{x^2 - 4x - 45} \cdot \dfrac{2x^2 + 13x + 15}{x^2 - 4x + 4}$

43. $\dfrac{\dfrac{1}{8} + \dfrac{1}{x}}{\dfrac{1}{64} - \dfrac{1}{x^2}}$

44. $\dfrac{x^2 + 9x}{x - 5} - \dfrac{6x + 50}{5 - x}$

45. $\dfrac{8}{x^2 + 7x + 12} + \dfrac{4}{x^2 + 10x + 24}$

46. $\dfrac{8x}{2x - 5} + \dfrac{20}{5 - 2x}$

47. $\dfrac{x^2 + 18x + 77}{x^2 - 4x - 32} \cdot \dfrac{x^2 + 6x + 8}{x^2 + 9x + 14}$

48. $\dfrac{x + 3}{x^2 - 2x - 3} + \dfrac{3}{x^2 - 8x + 15}$

49. $\dfrac{x^2 + 7x}{x^2 - 2x - 3} - \dfrac{4x + 18}{x^2 - 2x - 3}$

50. $\dfrac{x + \dfrac{5}{8}}{x - \dfrac{7}{2}}$

51. $\dfrac{x^2 - 10x}{x^2 + 10x - 11} \div \dfrac{3x^2 - 37x + 70}{x^2 + 4x - 5}$

52. $\dfrac{x + 5}{x^2 - 3x + 2} - \dfrac{8}{x^2 - 8x + 12}$

Writing in Mathematics

Answer in complete sentences.

53. Explain what a complex fraction is. Compare and contrast complex fractions and the rational expressions found in Section 7.1.

54. One method for simplifying complex fractions is to rewrite the complex fraction as one rational expression divided by another rational expression. When is this the most efficient way to simplify a complex fraction? Give an example.

55. ***Solutions Manual*** * Write a solutions manual page for the following problem:

Simplify $\dfrac{\dfrac{3}{x - 9} + \dfrac{6}{x - 3}}{3 - \dfrac{x + 5}{x - 3}}$.

56. ***Newsletter*** * Write a newsletter that explains how to simplify complex fractions.

See Appendix B for details and sample answers.

Objectives

1 **Solve rational equations.**
2 **Solve literal equations containing rational expressions.**

Solving Rational Equations

Objective 1　Solve rational equations. In this section, we will learn how to solve **rational equations,** which are equations containing at least one rational expression. The main goal is to rewrite the equation as an equivalent equation that does not contain a rational expression. We then solve the equation using methods developed in earlier chapters.

In Chapter 2, we learned how to solve an equation containing fractions such as the equation $\frac{1}{4}x - \frac{3}{5} = \frac{9}{10}$. We began by finding the LCD of all fractions and then multiplied both sides of the equation by that LCD to clear the equation of fractions. We will employ the same technique in this section. There is a major difference, though, when solving equations containing a variable in a denominator. Occasionally, we will find a solution that causes one of the rational expressions in the equation to be undefined. If a denominator of a rational expression is equal to 0 when the value of a solution is substituted for the variable, then the solution must be omitted and is called an **extraneous solution.** We must check each solution that we find to make sure that it is not an extraneous solution.

Solving Rational Equations

1. Find the LCD of all denominators in the equation.
2. Multiply both sides of the equation by the LCD to clear the equation of fractions.
3. Solve the resulting equation.
4. Check for extraneous solutions.

EXAMPLE 1 Solve $\dfrac{4}{x} + \dfrac{1}{3} = \dfrac{5}{6}$.

Solution

We begin by finding the LCD of these three fractions, which is $6x$. Now we multiply both sides of the equation by the LCD to clear the equation of fractions. Once we have done this, we can solve the resulting equation.

$$\frac{4}{x} + \frac{1}{3} = \frac{5}{6}$$

$$6x \cdot \left(\frac{4}{x} + \frac{1}{3} \right) = 6x \cdot \frac{5}{6} \qquad \text{Multiply both sides of the equation by the LCD.}$$

$$6\overset{1}{\cancel{x}} \cdot \frac{4}{\cancel{x}} + \overset{2}{\cancel{6}}x \cdot \frac{1}{\cancel{3}} = \overset{1}{\cancel{6}}x \cdot \frac{5}{\cancel{6}} \qquad \text{Distribute and divide out common factors.}$$

$$24 + 2x = 5x \qquad \text{Multiply. The resulting equation is linear.}$$

$$24 = 3x \qquad \text{Subtract } 2x \text{ to collect all variable terms on one side of the equation.}$$

$$8 = x \qquad \text{Divide both sides by 3.}$$

Check:

$$\frac{4}{(8)} + \frac{1}{3} = \frac{5}{6} \qquad \text{Substitute 8 for } x.$$

$$\frac{1}{2} + \frac{1}{3} = \frac{5}{6} \qquad \text{Simplify the fraction } \frac{4}{8}. \text{ The LCD of these fractions is 6.}$$

$$\frac{3}{6} + \frac{2}{6} = \frac{5}{6} \qquad \text{Write each fraction with a common denominator of 6.}$$

$$\frac{5}{6} = \frac{5}{6} \qquad \text{Add.}$$

Since the solution $x = 8$ does not make any rational expression in the original equation undefined, this value is a valid solution. The solution set is $\{8\}$.

Quick Check 1

Solve $\dfrac{6}{x} - \dfrac{1}{8} = \dfrac{7}{40}$.

When checking whether a solution is an extraneous solution, we need only determine whether the solution causes the LCD to equal 0. If the LCD is equal to 0 for this solution, then one or more rational expressions are undefined and the solution is an extraneous solution. Also, if the LCD is equal to 0, then we have multiplied both sides of the equation by 0. The multiplication property of equality says that we can multiply both sides of an equation by any *nonzero* number without affecting the equality of both sides. In the previous example, the only solution that could possibly be an extraneous solution is $x = 0$, because that is the only value of x for which the LCD is equal to 0.

EXAMPLE 2 Solve $x - 5 - \dfrac{36}{x} = 0$.

Solution

The LCD in this example is x. The LCD is equal to 0 only if $x = 0$. If we find that $x = 0$ is a solution, then we must omit that solution as an extraneous solution.

$$x - 5 - \frac{36}{x} = 0$$

$$x \cdot \left(x - 5 - \frac{36}{x}\right) = x \cdot 0 \qquad \text{Multiply each side of the equation by the LCD, } x.$$

$$x \cdot x - x \cdot 5 - \overset{1}{\cancel{x}} \cdot \frac{36}{\underset{1}{\cancel{x}}} = 0 \qquad \text{Distribute and divide out common factors.}$$

$$x^2 - 5x - 36 = 0 \qquad \text{Multiply. The resulting equation is quadratic.}$$

$$(x - 9)(x + 4) = 0 \qquad \text{Factor.}$$

$$x = 9 \quad \text{or} \quad x = -4 \qquad \text{Set each factor equal to 0 and solve.}$$

You may verify that neither solution causes the LCD to equal 0. The solution set is $\{-4, 9\}$.

Quick Check 2

Solve $1 = \dfrac{5}{x} + \dfrac{24}{x^2}$.

A Word of Caution When solving a rational equation, the use of the LCD is completely different than when we are adding or subtracting rational expressions. We use the LCD to clear the denominators of the rational expressions when solving a rational equation. When we are adding or subtracting rational expressions, we rewrite each expression as an equivalent expression whose denominator is the LCD.

EXAMPLE 3 Solve $\dfrac{x}{x+2} - 5 = \dfrac{3x+4}{x+2}$.

Solution

The LCD is $x + 2$, so we will begin to solve this equation by multiplying both sides of the equation by $x + 2$.

$$\frac{x}{x+2} - 5 = \frac{3x+4}{x+2}$$ The LCD is $x + 2$.

$$(x+2)\cdot\left(\frac{x}{x+2} - 5\right) = (x+2)\cdot\frac{3x+4}{x+2}$$ Multiply both sides by the LCD.

$$(x+2)\cdot\frac{x}{x+2} - (x+2)\cdot 5 = (x+2)\cdot\frac{3x+4}{x+2}$$ Distribute and divide out common factors.

$$x - 5x - 10 = 3x + 4$$ Multiply. The resulting equation is linear.

$$-4x - 10 = 3x + 4$$ Combine like terms.

$$-10 = 7x + 4$$ Add $4x$.

$$-14 = 7x$$ Subtract 4.

$$-2 = x$$ Divide both sides by 7.

Quick Check 3

Solve

$\dfrac{7}{x-4} + 3 = \dfrac{2x-1}{x-4}$.

The LCD is equal to 0 when $x = -2$, and two rational expressions in the original equation are undefined when $x = -2$. This solution is an extraneous solution, and since there are no other solutions, this equation has no solution. Recall that we write the solution set as $\varnothing$ when there is no solution.

EXAMPLE 4 Solve $\dfrac{x+9}{x^2+9x+8} = \dfrac{2}{x^2+2x-48}$.

Solution

We begin by factoring the denominators to find the LCD.

$$\frac{x+9}{x^2+9x+8} = \frac{2}{x^2+2x-48}$$

$$\frac{x+9}{(x+1)(x+8)} = \frac{2}{(x+8)(x-6)}$$ The LCD is $(x+1)(x+8)(x-6)$.

$$(x+1)(x+8)(x-6)\cdot\frac{x+9}{(x+1)(x+8)} = (x+1)(x+8)(x-6)\cdot\frac{2}{(x+8)(x-6)}$$

Multiply by the LCD. Divide out common factors.

$$(x-6)(x+9) = 2(x+1)$$ Multiply remaining factors.

$$x^2 + 3x - 54 = 2x + 2$$ Multiply.

$$x^2 + x - 56 = 0$$ Collect all terms on the left side. The resulting equation is quadratic.

$$(x+8)(x-7) = 0$$ Factor.

$$x = -8 \quad\text{or}\quad x = 7$$ Set each factor equal to 0 and solve.

The solution $x = -8$ is an extraneous solution, because it makes the LCD equal to 0. The reader may verify that the solution $x = 7$ checks. The solution set for this equation is $\{7\}$.

Quick Check   **4**   Solve $\dfrac{x+2}{x^2-3x-54} = \dfrac{2}{x^2-12x+27}$.

EXAMPLE ▶ **5**   Solve $\dfrac{x+10}{x^2+4x-5} - \dfrac{1}{x-3} = \dfrac{x-6}{x^2-4x+3}$.

Solution

$$\frac{x+10}{x^2+4x-5} - \frac{1}{x-3} = \frac{x-6}{x^2-4x+3}$$

$$\frac{x+10}{(x+5)(x-1)} - \frac{1}{x-3} = \frac{x-6}{(x-1)(x-3)} \qquad \text{The LCD is } (x+5)(x-1)(x-3).$$

$$(x+5)(x-1)(x-3) \cdot \left(\frac{x+10}{(x+5)(x-1)} - \frac{1}{x-3} \right) = (x+5)(x-1)(x-3) \cdot \frac{x-6}{(x-1)(x-3)}$$

Multiply by the LCD.

$$\cancel{(x+5)}\cancel{(x-1)}(x-3) \cdot \frac{x+10}{\cancel{(x+5)}\cancel{(x-1)}} - (x+5)(x-1)\cancel{(x-3)} \cdot \frac{1}{\cancel{(x-3)}}$$

$$= (x+5)\cancel{(x-1)}\cancel{(x-3)} \cdot \frac{x-6}{\cancel{(x-1)}\cancel{(x-3)}} \qquad \begin{array}{l}\text{Distribute and divide out common}\\ \text{factors.}\end{array}$$

$$(x-3)(x+10) - (x+5)(x-1) = (x+5)(x-6) \qquad \text{Multiply remaining factors.}$$

$$(x^2+7x-30) - (x^2+4x-5) = x^2-x-30 \qquad \text{Multiply.}$$

$$x^2+7x-30-x^2-4x+5 = x^2-x-30 \qquad \text{Distribute.}$$

$$3x-25 = x^2-x-30 \qquad \text{Combine like terms.}$$

$$0 = x^2-4x-5 \qquad \begin{array}{l}\text{Collect all terms on the right side.}\\ \text{The resulting equation is quadratic.}\end{array}$$

$$0 = (x+1)(x-5) \qquad \text{Factor.}$$

$$x = -1 \quad \text{or} \quad x = 5 \qquad \text{Set each factor equal to 0 and solve.}$$

Check to verify that neither solution is extraneous. The solution set is $\{-1, 5\}$.

Quick Check   **5**   Solve $\dfrac{x+3}{x^2+x-12} - \dfrac{3}{x^2-2x-3} = \dfrac{x+7}{x^2+5x+4}$.

Literal Equations

Objective 2   **Solve literal equations containing rational expressions.** Recall that a literal equation is an equation containing two or more variables, and we solve the equation for one of the variables by isolating that variable on one side of the equation. In this section, we will learn how to solve literal equations containing one or more rational expressions.

EXAMPLE ▶ 6 Solve the literal equation $\dfrac{1}{x} + \dfrac{1}{y} = \dfrac{2}{5}$ for x.

Solution

We begin by multiplying by the LCD ($5xy$) to clear the equation of fractions.

$$\frac{1}{x} + \frac{1}{y} = \frac{2}{5}$$

$$5xy \cdot \left(\frac{1}{x} + \frac{1}{y}\right) = 5xy \cdot \frac{2}{5} \qquad \text{Multiply by the LCD, } 5xy.$$

$$5\overset{1}{\cancel{x}}y \cdot \frac{1}{\cancel{x}} + 5x\overset{1}{\cancel{y}} \cdot \frac{1}{\cancel{y}} = \overset{1}{\cancel{5}}xy \cdot \frac{2}{\cancel{5}} \qquad \text{Distribute and divide out common factors.}$$

$$5y + 5x = 2xy \qquad \text{Multiply remaining factors.}$$

Notice that there are two terms that contain the variable for which we are solving. We need to collect both of these terms on the same side of the equation and then factor x out of those terms. This will allow us to divide and isolate x.

$$5y + 5x = 2xy$$

$$5y = 2xy - 5x \qquad \begin{array}{l}\text{Subtract } 5x \text{ to collect all terms with } x \\ \text{on the right side of the equation.}\end{array}$$

$$5y = x(2y - 5) \qquad \text{Factor out the common factor } x.$$

$$\frac{5y}{2y - 5} = \frac{x(\cancel{2y - 5})}{(\cancel{2y - 5})} \qquad \text{Divide both sides by } 2y - 5 \text{ to isolate } x.$$

$$\frac{5y}{2y - 5} = x \qquad \text{Simplify.}$$

$$x = \frac{5y}{2y - 5} \qquad \begin{array}{l}\text{Rewrite with the variable we are solving for} \\ \text{on the left side of the equation.}\end{array}$$

> **Quick Check 6**
> Solve the literal equation
> $\dfrac{2}{x} + \dfrac{3}{y} = \dfrac{4}{z}$ for x.

As in Chapter 2, we will rewrite our solution so that the variable for which we are solving is on the left side.

EXAMPLE ▶ 7 Solve the literal equation $\dfrac{1}{a + h} = \dfrac{1}{b} - 5$ for a.

Solution

We begin by multiplying both sides of the equation by the LCD, which is $(a + h)b$.

$$\frac{1}{a + h} = \frac{1}{b} - 5$$

$$(a + h)b \cdot \frac{1}{a + h} = (a + h)b \cdot \left(\frac{1}{b} - 5\right) \qquad \begin{array}{l}\text{Multiply both sides by the} \\ \text{LCD.}\end{array}$$

$$\overset{1}{\cancel{(a + h)}}b \cdot \frac{1}{\cancel{(a + h)}} = (a + h)\overset{1}{\cancel{b}} \cdot \frac{1}{\cancel{b}} - (a + h)b \cdot 5 \qquad \begin{array}{l}\text{Distribute and divide out} \\ \text{common factors.}\end{array}$$

$$b = a + h - 5b(a + h) \qquad \begin{array}{l}\text{Multiply remaining factors.} \\ \text{Distribute.}\end{array}$$

$$b = a + h - 5ab - 5bh$$

$$b - h + 5bh = a - 5ab \qquad \begin{array}{l}\text{Subtract } h \text{ and add } 5bh \text{ so} \\ \text{only the terms containing} \\ a \text{ are on the right side.}\end{array}$$

Quick Check 7

Solve the literal equation
$\frac{1}{x-4} + \frac{1}{y} = \frac{3}{4}$ for x.

$b - h + 5bh = a(1 - 5b)$ Factor out the common factor a.

$\frac{b - h + 5bh}{1 - 5b} = \frac{a(1 - 5b)}{(1 - 5b)}$ Divide both sides by $1 - 5b$ to isolate a.

$a = \frac{b - h + 5bh}{1 - 5b}$ Rewrite with a on the left side.

Building Your Study Strategy **Preparing for a Cumulative Exam, 6 Applied Problems**   Many students have a difficult time with applied problems on a cumulative exam. Some students are unable to recognize what type of problem an applied problem is, and others cannot recall how to start to solve that type of problem. You will find the following strategy helpful:

- Make a list of the different applied problems that you have covered this semester. Create a study sheet for each type of problem.
- Write down an example or two of each type of problem.
- List the steps necessary to solve each type of problem.

Review these study sheets frequently as the exam approaches. This should help you identify the applied problems on the exam, as well as to remember how to solve the problems.

EXERCISES 7.6

Vocabulary

1. A(n) _____ is an equation containing at least one rational expression.

2. To solve a rational equation, begin by multiplying both sides of the equation by the _____ of the denominators in the equation.

3. A(n) _____ of a rational equation is a solution that causes one or more of the rational expressions in the equation to be undefined.

4. A(n) _____ equation is an equation containing two or more variables.

Solve.

5. $\frac{x}{3} - \frac{7}{4} = \frac{9}{2}$

6. $\frac{x}{5} + \frac{1}{6} = \frac{8}{3}$

7. $\frac{3x}{4} + \frac{5}{6} = \frac{x}{12} + \frac{7}{8}$

8. $\frac{6x}{7} - \frac{9}{2} = \frac{x}{4} + \frac{3}{14}$

9. $\frac{9}{x} - \frac{3}{4} = \frac{7}{8}$

10. $\frac{9}{10} + \frac{6}{x} = \frac{17}{12}$

11. $\frac{1}{x} = \frac{21}{80}$

12. $\dfrac{1}{2x} + 7 = \dfrac{1}{3x} + \dfrac{1}{6x}$

13. $x - 5 + \dfrac{12}{x} = 2$

14. $x - \dfrac{25}{x} = 2 + \dfrac{23}{x}$

15. $\dfrac{x}{2} + \dfrac{6}{x} = 4$

16. $\dfrac{x}{3} + 2 - \dfrac{9}{x} = 0$

17. $\dfrac{8x - 3}{x + 7} = \dfrac{2x + 15}{x + 7}$

18. $\dfrac{6x + 13}{x - 1} = \dfrac{4x + 15}{x - 1}$

19. $\dfrac{7x - 11}{5x - 2} = \dfrac{2x - 9}{5x - 2}$

20. $\dfrac{x^2 + 3x}{x + 8} = \dfrac{3x + 64}{x + 8}$

21. $8 + \dfrac{6}{x - 4} = \dfrac{x - 12}{x - 4}$

22. $1 + \dfrac{2x - 7}{x + 3} = \dfrac{4x - 1}{x + 3}$

23. $\dfrac{3x + 4}{x + 2} - 7 = \dfrac{x}{x + 2}$

24. $\dfrac{6x + 7}{5x - 8} - \dfrac{2x - 13}{5x - 8} = 3$

25. $x + \dfrac{x + 11}{x - 6} = \dfrac{8x - 31}{x - 6}$

26. $x + \dfrac{3x - 10}{x + 1} = \dfrac{9x + 14}{x + 1}$

27. $\dfrac{3}{x - 7} = \dfrac{7}{x + 5}$

28. $\dfrac{6}{x + 4} = \dfrac{9}{2x - 3}$

29. $\dfrac{6}{x + 2} = \dfrac{25}{3x + 13}$

30. $\dfrac{2}{x} = \dfrac{11}{4x - 9}$

31. $\dfrac{x + 1}{x + 13} = \dfrac{3}{x + 4}$

32. $\dfrac{x - 3}{x + 9} = \dfrac{6}{x + 2}$

33. $\dfrac{x - 1}{3x - 7} = \dfrac{x + 1}{2x + 7}$

34. $\dfrac{2x}{x + 9} = \dfrac{x - 2}{x - 1}$

35. $\dfrac{3}{x + 2} - \dfrac{1}{x + 1} = \dfrac{x + 3}{x^2 + 3x + 2}$

36. $\dfrac{2}{x} + \dfrac{3}{x + 2} = \dfrac{7x - 8}{x^2 + 2x}$

37. $\dfrac{4}{x + 4} + \dfrac{3}{x - 4} = \dfrac{24}{x^2 - 16}$

38. $\dfrac{x}{x + 5} + \dfrac{3}{x - 7} = \dfrac{36}{x^2 - 2x - 35}$

39. $\dfrac{2}{x^2 - x - 2} + \dfrac{10}{x^2 - 2x - 3} = \dfrac{x + 12}{x^2 - x - 2}$

40. $\dfrac{20}{x^2 + 12x + 27} - \dfrac{4}{x^2 + 14x + 45} = \dfrac{x + 10}{x^2 + 8x + 15}$

41. $\dfrac{x - 8}{x - 5} + \dfrac{x - 9}{x - 4} = \dfrac{x + 7}{x^2 - 9x + 20}$

42. $\dfrac{x}{x + 3} + \dfrac{x - 4}{x - 3} = \dfrac{9x - 5}{x^2 - 9}$

43. $\dfrac{4}{x^2 + 4x - 5} + \dfrac{x + 9}{x^2 - 1} = \dfrac{41}{x^2 + 6x + 5}$

44. $\dfrac{x + 4}{x^2 + 5x - 14} + \dfrac{2}{x^2 + 3x - 10} = \dfrac{38}{x^2 + 12x + 35}$

45. $\dfrac{x + 3}{x^2 - 4x - 12} + \dfrac{x - 11}{x^2 - 2x - 24} = \dfrac{x + 1}{x^2 + 6x + 8}$

46. $\dfrac{x - 2}{x^2 + 13x + 40} + \dfrac{x + 5}{x^2 + 7x - 8} = \dfrac{x + 3}{x^2 + 4x - 5}$

47. $\dfrac{3x - 4}{x^2 - 10x + 21} - \dfrac{x - 8}{x^2 - 18x + 77} = \dfrac{x - 5}{x^2 - 14x + 33}$

48. $\dfrac{5x + 4}{x^2 + x - 90} - \dfrac{1}{x - 9} = \dfrac{3x - 2}{x^2 - 100}$

Solve for the specified variable.

49. $L = \dfrac{A}{W}$ for W

50. $b = \dfrac{2A}{h}$ for A

51. $y = \dfrac{x}{2x + 5}$ for x

52. $y = \dfrac{3x - 7}{2x}$ for x

53. $\dfrac{x}{r} + \dfrac{y}{2r} = 1$ for r

54. $r = \dfrac{d}{t}$ for t

55. $\dfrac{1}{a} + \dfrac{1}{b} = \dfrac{1}{c}$ for b

56. $\dfrac{a}{b} + \dfrac{c}{d} = \dfrac{e}{f}$ for b

57. $m = \dfrac{y - y_1}{x - x_1}$ for x

58. $\dfrac{1}{R} = \dfrac{1}{R_1} + \dfrac{1}{R_2}$ for R

59. $n = \dfrac{n_1 + n_2}{n_1 \cdot n_2}$ for n_1

60. $p = \dfrac{x_1 + x_2}{n_1 + n_2}$ for n_1

61. If $x = 9$ is a solution to the equation
$$\dfrac{x - 4}{x - 1} + \dfrac{7}{x + 3} = \dfrac{?}{x^2 + 2x - 3},$$

a) Find the constant in the missing numerator.
b) Find the other solution to the equation.

62. If $x = 12$ is a solution to the equation
$$\dfrac{x - 8}{x - 5} + \dfrac{?}{x - 2} = \dfrac{7x + 19}{x^2 - 7x + 10},$$

a) Find the constant in the missing numerator.
b) Find the other solution to the equation.

Writing in Mathematics

Answer in complete sentences.

63. Explain how to determine whether a solution is an extraneous solution.

64. We use the LCD when we add two rational expressions, as well as when we solve a rational equation. Explain how the LCD is used differently for these two types of problems.

65. A student was asked to solve the rational equation $\dfrac{1}{x} + \dfrac{1}{y} = \dfrac{3}{8}$ for x, and his answer was $x = \dfrac{3xy - 8y}{8}$. Explain why this is incorrect, then explain how to solve the equation for x.

66. *Solutions Manual*[*] Write a solutions manual page for the following problem:

Solve
$$\dfrac{x - 6}{x^2 - 9} + \dfrac{5}{x^2 + 4x - 21} = \dfrac{2}{x^2 + 10x + 21}.$$

67. *Newsletter*[*] Write a newsletter that explains how to solve rational equations.

*See Appendix B for details and sample answers.

Applications of Rational Equations

Objectives

1. Solve applied problems involving the reciprocal of a number.
2. Solve applied work–rate problems.
3. Solve applied uniform motion problems.
4. Solve variation problems.

In this section, we will look at applied problems requiring the use of rational equations to solve them. We begin with problems involving reciprocals.

Reciprocals

Objective 1 Solve applied problems involving the reciprocal of a number.

EXAMPLE 1 The sum of the reciprocal of a number and $\frac{1}{3}$ is $\frac{1}{2}$. Find the number.

Solution

There is only one unknown in this problem, and we will let x represent the unknown number.

Unknown
Number: x

The reciprocal of this number can be written as $\frac{1}{x}$. We are told that the sum of this reciprocal and $\frac{1}{3}$ is $\frac{1}{2}$, which leads to the equation $\frac{1}{x} + \frac{1}{3} = \frac{1}{2}$.

$$\frac{1}{x} + \frac{1}{3} = \frac{1}{2} \qquad \text{The LCD is } 6x.$$

$$6x \cdot \left(\frac{1}{x} + \frac{1}{3} \right) = 6x \cdot \frac{1}{2} \qquad \text{Multiply both sides by the LCD.}$$

$$6\overset{1}{x} \cdot \frac{1}{\overset{}{\underset{1}{x}}} + \overset{2}{6}x \cdot \frac{1}{\overset{}{\underset{1}{3}}} = \overset{3}{6}x \cdot \frac{1}{\overset{}{\underset{1}{2}}} \qquad \text{Distribute and divide out common factors.}$$

$$6 + 2x = 3x \qquad \text{Multiply remaining factors. The resulting equation is linear.}$$

$$6 = x \qquad \text{Subtract } 2x.$$

The unknown number is 6. The reader should verify that $\frac{1}{6} + \frac{1}{3} = \frac{1}{2}$.

Quick Check 1
The sum of the reciprocal of a number and $\frac{3}{8}$ is $\frac{19}{40}$. Find the number.

EXAMPLE 2 One positive number is 4 larger than another positive number. If the reciprocal of the smaller number is added to six times the reciprocal of the larger number, the sum is 1. Find the two numbers.

Solution

In this problem, there are two unknown numbers. If we let x represent the smaller number, then we can write the larger number as $x + 4$.

Unknowns

Smaller Number: x

Larger Number: $x + 4$

The reciprocal of the smaller number is $\dfrac{1}{x}$ and six times the reciprocal of the larger number is $6 \cdot \dfrac{1}{x+4}$ or $\dfrac{6}{x+4}$. This leads to the equation $\dfrac{1}{x} + \dfrac{6}{x+4} = 1$.

$$\frac{1}{x} + \frac{6}{x+4} = 1 \qquad \text{The LCD is } x(x+4).$$

$$x(x+4) \cdot \left(\frac{1}{x} + \frac{6}{x+4}\right) = x(x+4) \cdot 1 \qquad \text{Multiply both sides by the LCD.}$$

$$\overset{1}{\cancel{x}}(x+4) \cdot \frac{1}{\cancel{x}} + x\cancel{(x+4)} \cdot \frac{6}{\underset{1}{\cancel{(x+4)}}} = x(x+4) \cdot 1 \qquad \text{Distribute and divide out common factors.}$$

$$x + 4 + 6x = x(x+4) \qquad \text{Multiply remaining factors.}$$

$$x + 4 + 6x = x^2 + 4x \qquad \text{Distribute. The resulting equation is quadratic.}$$

$$7x + 4 = x^2 + 4x \qquad \text{Combine like terms.}$$

$$0 = x^2 - 3x - 4 \qquad \text{Collect all terms on the right side of the equation.}$$

$$0 = (x+1)(x-4) \qquad \text{Factor.}$$

$$x = -1 \quad \text{or} \quad x = 4 \qquad \text{Set each factor equal to 0 and solve.}$$

Since we were told that the numbers must be positive, we can omit the solution $x = -1$. We now return to the table of unknowns to find the two numbers.

Smaller Number: $x = 4$

Larger Number: $x + 4 = 4 + 4 = 8$

The two numbers are 4 and 8. The reader should verify that $\frac{1}{4} + 6 \cdot \frac{1}{8} = 1$.

Quick Check 2
One positive number is 9 less than another positive number. If two times the reciprocal of the smaller number is added to four times the reciprocal of the larger number, the sum is 1. Find the two numbers.

Work-Rate Problems

Objective 2 **Solve applied work-rate problems.** Now we turn our attention to work-rate problems. These problems usually involve two or more people or objects working together to perform a job, such as two people painting a room together or two copy machines processing an exam. In general, the equation we will be solving corresponds to the following:

| Portion of the job completed by person 1 | + | Portion of the job completed by person 2 | = | 1 (Completed job) |

To determine the portion of the job completed by each person, we must know the **rate** at which the person works. If it takes Kim 4 hours to paint a room, how much of the room could she paint in one hour? She could paint $\frac{1}{4}$ of the room in 1 hour, and this is her working rate. In general, the work rate for a person is equal to the reciprocal of the time it takes for that person to complete the whole job. If we multiply a person's work rate by the time that person has been working, then this tells us the portion of the job that the person has completed.

EXAMPLE 3   Working alone, Kim can paint a room in 4 hours. Greg can paint the same room in only 3 hours. How long would it take the two of them to paint the room if they work together?

Solution

A good approach to any work-rate problem is to start with the following table and fill in the information:

Person	Time to Complete the Job Alone	Work Rate	Time Working	Portion of the Job Completed
Person 1				
Person 2				

- Since we know it would take Kim 4 hours to paint the room, her work rate is $\frac{1}{4}$ room per hour. Similarly, Greg's work rate is $\frac{1}{3}$ room per hour.
- The unknown in this problem is the amount of time they will be working together, which we will represent by the variable t.
- Finally, to determine the portion of the job completed by each person, we multiply the person's work rate by the time they have been working.

Person	Time to Complete the Job Alone	Work Rate	Time Working	Portion of the Job Completed
Kim	4 hours	$\frac{1}{4}$	t	$\frac{t}{4}$
Greg	3 hours	$\frac{1}{3}$	t	$\frac{t}{3}$

To find the equation we need to solve, we add the portion of the room painted by Kim to the portion of the room painted by Greg and set this sum equal to 1. The equation for this problem is $\frac{t}{4} + \frac{t}{3} = 1$.

$$\frac{t}{4} + \frac{t}{3} = 1 \qquad \text{The LCD is 12.}$$

$$12 \cdot \left(\frac{t}{4} + \frac{t}{3} \right) = 12 \cdot 1 \qquad \text{Multiply both sides by the LCD.}$$

Quick Check 3
Javier's new printer can print a complete set of brochures in 20 minutes. His old printer can print the set of brochures in 35 minutes. How long would it take the two printers to print the set of brochures if they work together?

$$\cancel{12}^{3} \cdot \frac{t}{\cancel{4}_{1}} + \cancel{12}^{4} \cdot \frac{t}{\cancel{3}_{1}} = 12 \cdot 1 \qquad \text{Distribute and divide out common factors.}$$

$$3t + 4t = 12 \qquad \text{Multiply remaining factors.}$$

$$7t = 12 \qquad \text{Combine like terms.}$$

$$t = \frac{12}{7} \quad \text{or} \quad 1\frac{5}{7} \qquad \text{Divide both sides by 7.}$$

It would take them $1\frac{5}{7}$ hours to paint the room if they worked together. To convert the answer to hours and minutes we multiply $\frac{5}{7}$ of an hour by 60 minutes per hour, which is approximately 43 minutes. It would take them approximately one hour and 43 minutes. The check of this solution is left to the reader.

EXAMPLE 4 Two drainpipes working together can drain a tank in 6 hours. Working alone, the smaller pipe would take 9 hours longer than the larger pipe to drain the tank. How long would it take the smaller pipe alone to drain the tank?

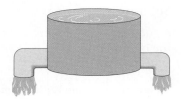

Solution

In this example, we know the amount of time it would take the pipes to drain the tank if they were working together, but we do not know how long it would take each pipe working alone. If we let t represent the time, in hours, that it takes the larger pipe to drain the tank, then the smaller pipe takes $t + 9$ hours to drain the tank. We will begin with the following table:

Pipe	Time to Complete the Job Alone	Work Rate	Time Working	Portion of the Job Completed
Smaller	$t + 9$ hours	$\frac{1}{t + 9}$	6 hours	$\frac{6}{t + 9}$
Larger	t hours	$\frac{1}{t}$	6 hours	$\frac{6}{t}$

Adding the portion of the tank drained by the smaller pipe in 6 hours to the portion of the tank drained by the larger pipe, we find that the sum equals 1. The equation is $\frac{6}{t + 9} + \frac{6}{t} = 1$.

$$\frac{6}{t + 9} + \frac{6}{t} = 1 \qquad \text{The LCD is } t(t + 9).$$

$$t(t + 9) \cdot \left(\frac{6}{t + 9} + \frac{6}{t} \right) = t(t + 9) \cdot 1 \qquad \text{Multiply both sides by the LCD.}$$

$$t\cancel{(t+9)}^{\,1} \cdot \frac{6}{\cancel{(t+9)}_{1}} + \cancel{t}^{\,1}(t+9) \cdot \frac{6}{\cancel{t}_{1}} = t(t+9) \cdot 1$$

Distribute and divide out common factors.

$$6t + 6(t+9) = t(t+9)$$ Multiply remaining factors.

$$6t + 6t + 54 = t^2 + 9t$$ Multiply. The resulting equation is quadratic.

$$12t + 54 = t^2 + 9t$$ Combine like terms.

$$0 = t^2 - 3t - 54$$ Collect all terms on the right side of the equation.

$$0 = (t-9)(t+6)$$ Factor.

$$t = 9 \quad \text{or} \quad t = -6$$ Set each factor equal to 0 and solve.

Since the time spent by each pipe must be positive, we may immediately omit the solution $t = -6$. The reader may check the solution $t = 9$ by verifying that $\frac{6}{9+9} + \frac{6}{9} = 1$. We now use $t = 9$ to find the amount of time it would take the smaller pipe to drain the tank. Since $t + 9$ represents the amount of time it would take for the smaller pipe to drain the tank, it would take $9 + 9$, or 18, hours to drain the tank.

Quick Check 4

Two drainpipes working together can drain a tank in 6 hours. Working alone, the smaller pipe would take 16 hours longer than the larger pipe to drain the tank. How long would it take the smaller pipe alone to drain the tank?

Uniform Motion Problems

Objective 3   Solve applied uniform motion problems. We now turn our attention to uniform motion problems. Recall that if an object is moving at a constant rate of speed for a certain amount of time, then the distance traveled by the object is equal to the product of its rate of speed and the length of time it traveled. The formula we used earlier in the text is rate · time = distance, or $r \cdot t = d$. If we solve this formula for the time traveled, we have

$$\text{time} = \frac{\text{distance}}{\text{rate}} \quad \text{or} \quad t = \frac{d}{r}$$

EXAMPLE 5   Nick drove 24 miles, one way, to deliver a package. On the way home he drove 20 miles per hour faster than he did on his way to deliver the package. If the total driving time was 1 hour for the entire trip, find Nick's driving speed on the way home.

Solution

- We know that the distance traveled in each direction is 24 miles.
- Nick's speed on the way home was 20 miles per hour faster than it was on the way to deliver the package. We will let r represent his rate of speed on the way to deliver the package, and we can represent his rate of speed on the way home as $r + 20$.
- To find an expression for the time spent on each part of the trip, we divide the distance by the rate.

We can summarize this information in a table:

	Distance (*d*)	Rate (*r*)	Time (*t*)
To Deliver Package	24 miles	r	$\dfrac{24}{r}$
Return Trip	24 miles	$r + 20$	$\dfrac{24}{r + 20}$

The equation we need to solve comes from the fact that the driving time on the way to deliver the package plus the driving time on the way home is equal to 1 hour. If we add the time spent on the way to deliver the package $\left(\dfrac{24}{r}\right)$ to the time spent on the way home $\left(\dfrac{24}{r + 20}\right)$, this will be equal to 1. The equation we need to solve is $\dfrac{24}{r} + \dfrac{24}{r + 20} = 1$.

$$\frac{24}{r} + \frac{24}{r + 20} = 1 \qquad \text{The LCD is } r(r + 20).$$

$$r(r + 20) \cdot \left(\frac{24}{r} + \frac{24}{r + 20}\right) = r(r + 20) \cdot 1 \qquad \text{Multiply both sides by the LCD.}$$

$$\overset{1}{\cancel{r}}(r + 20) \cdot \frac{24}{\underset{1}{\cancel{r}}} + r\overset{1}{\cancel{(r + 20)}} \cdot \frac{24}{\underset{1}{\cancel{(r + 20)}}} = r(r + 20) \cdot 1 \qquad \begin{array}{l}\text{Distribute and divide out} \\ \text{common factors.}\end{array}$$

$$24(r + 20) + 24r = r(r + 20) \cdot 1 \qquad \text{Multiply remaining factors.}$$

$$24r + 480 + 24r = r^2 + 20r \qquad \text{Multiply.}$$

$$48r + 480 = r^2 + 20r \qquad \text{Combine like terms.}$$

$$0 = r^2 - 28r - 480 \qquad \begin{array}{l}\text{Collect all terms on the right side} \\ \text{of the equation by subtracting} \\ 48r \text{ and } 480.\end{array}$$

$$0 = (r - 40)(r + 12) \qquad \text{Factor.}$$

$$r = 40 \quad \text{or} \quad r = -12 \qquad \begin{array}{l}\text{Set each factor equal to 0} \\ \text{and solve.}\end{array}$$

Quick Check **5**

Rehema drove 300 miles to pick up a friend and then returned home. On the way home, she drove 15 miles per hour faster than she did on her way to pick up her friend. If the total driving time for the trip was 9 hours for the entire trip, find Rehema's driving speed on the way home.

We omit the solution $r = -12$, as Nick's speed cannot be negative. The expression for Nick's driving speed on the way home is $r + 20$, so his speed on the way home was $40 + 20$, or 60, miles per hour. The reader can check the solution by verifying that $\dfrac{24}{40} + \dfrac{24}{60} = 1$.

EXAMPLE 6   Gretchen has taken her kayak to the Pawuxet River, which flows downstream at a rate of 3 kilometers per hour. If the time it takes Gretchen to paddle 12 kilometers upstream is 1 hour longer than the time it takes for her to paddle 12 kilometers downstream, find the speed that Gretchen can paddle in still water.

Solution

We will let r represent the speed that Gretchen can paddle in still water. Since the current of the river is 3 kilometers per hour, Gretchen's kayak travels at a speed of $r - 3$ kilometers per hour when she is paddling upstream. This is because the current is

pushing against the kayak. Gretchen travels at a speed of $r + 3$ kilometers per hour when she is paddling downstream, as the current is flowing in the same direction as the kayak. The equation we will solve involves the time spent paddling upstream and downstream. To find expressions in terms of r for the time spent in each direction, we divide the distance (12 km) by the rate of speed. Here is a table containing the relevant information:

	Distance	Rate	Time
Upstream	12 km	$r - 3$	$\frac{12}{r-3}$
Downstream	12 km	$r + 3$	$\frac{12}{r+3}$

We are told that the time needed to paddle 12 kilometers upstream is 1 hour more than the time needed to paddle 12 kilometers downstream. In other words, the time spent paddling upstream is equal to the time spent paddling downstream plus 1 hour, or $\frac{12}{r-3} = \frac{12}{r+3} + 1$.

$$\frac{12}{r-3} = \frac{12}{r+3} + 1 \qquad \text{The LCD is } (r-3)(r+3).$$

$$(r-3)(r+3) \cdot \frac{12}{r-3} = (r-3)(r+3) \cdot \left(\frac{12}{r+3} + 1\right) \qquad \begin{array}{l}\text{Multiply both sides by}\\ \text{the LCD.}\end{array}$$

$$(\overset{1}{\cancel{r-3}})(r+3) \cdot \frac{12}{\cancel{(r-3)}} = (r-3)\overset{1}{\cancel{(r+3)}} \cdot \frac{12}{\cancel{(r+3)}} + (r-3)(r+3) \cdot 1 \qquad \begin{array}{l}\text{Distribute and divide out}\\ \text{common factors.}\end{array}$$

$$12(r+3) = 12(r-3) + (r-3)(r+3) \qquad \text{Multiply remaining factors.}$$

$$12r + 36 = 12r - 36 + r^2 - 9 \qquad \text{Multiply.}$$

$$12r + 36 = r^2 + 12r - 45 \qquad \begin{array}{l}\text{Combine like terms, writing}\\ \text{the right side in descending}\\ \text{order.}\end{array}$$

$$0 = r^2 - 81 \qquad \begin{array}{l}\text{Collect all terms on the right}\\ \text{side of the equation by}\\ \text{subtracting } 12r \text{ and } 36.\end{array}$$

$$0 = (r+9)(r-9) \qquad \text{Factor.}$$

$$r = -9 \quad \text{or} \quad r = 9 \qquad \begin{array}{l}\text{Set each factor equal to } 0\\ \text{and solve.}\end{array}$$

Quick Check 6

Lucy took her canoe to a river that flows downstream at a rate of 2 miles per hour. She paddled 8 miles downstream, and then returned back to the camp she started from. If the round trip took her 3 hours, find the speed that Lucy can paddle in still water.

We omit the negative solution, as the speed of the kayak in still water must be positive. Gretchen's kayak travels at a speed of 9 kilometers per hour in still water. The reader can check this solution by verifying that $\frac{12}{9-3} = \frac{12}{9+3} + 1$.

Variation

Objective 4   **Solve variation problems.** In the remaining examples, we will investigate the concept of **variation** between two or more quantities. Two quantities are said to **vary directly** if an increase in one quantity produces a proportional increase in the other quantity and a decrease in one quantity produces a proportional decrease in the other quantity. For example, suppose that you have a part-time job that pays by the hour. The hours you work in a week and the amount of money you earn (before taxes) vary directly. As the hours you work increase, the amount of money you earn increases by the same factor. If there is a decrease in the number of hours you work, the amount of money that you earn decreases by the same factor.

Direct Variation

If a quantity y varies directly as a quantity x, then the two quantities are related by the equation
$$y = kx$$
where k is called the **constant of variation.**

For example, if you are paid \$12 per hour at your part-time job, then the amount of money you earn (y) and the number of hours you work (x) are related by the equation $y = 12x$. The value of k in this situation is 12, and it tells us that each time x increases by 1 hour, y increases by \$12.

EXAMPLE 7 y varies directly as x. If $y = 42$ when $x = 7$, find y when $x = 13$.

Solution

Since the variation is direct, we will use the equation $y = kx$. We begin by finding k. Using $y = 42$ when $x = 7$, we can use the equation $42 = k \cdot 7$ to find k.

$$y = kx$$
$$42 = k \cdot 7 \qquad \text{Substitute 42 for } y \text{ and 7 for } x \text{ into } y = kx.$$
$$6 = k \qquad \text{Divide both sides by 7.}$$

Now we use this value of k to find y when $x = 13$.

$$y = 6 \cdot 13 \qquad \text{Substitute 6 for } k \text{ and 13 for } x \text{ into } y = kx.$$
$$y = 78 \qquad \text{Multiply.}$$

EXAMPLE 8   The distance a train travels varies directly as the time it is traveling. If a train can travel 424 miles in 8 hours, how far can it travel in 15 hours?

Solution

The first step in a variation problem is to find k. We will let y represent the distance traveled and x represent the time. To find the variation constant, we will substitute the related information (424 miles in 8 hours) into $y = kx$, the equation for direct variation.

$$y = kx$$
$$424 = k \cdot 8 \qquad \text{Substitute 424 for } y \text{ and 8 for } x \text{ into } y = kx.$$
$$53 = k \qquad \text{Divide both sides by 8.}$$

Now we will use this variation constant to find the distance traveled in 15 hours.

$$y = 53 \cdot 15 \qquad \text{Substitute 53 for } k \text{ and 15 for } x \text{ into } y = kx.$$
$$y = 795 \qquad \text{Multiply.}$$

The train travels 795 miles in 15 hours.

Quick Check **7**
The tuition a college student pays varies directly as the number of units the student is taking. If a student pays \$288 to take 12 units in a semester, how much would a student pay to take 15 units?

In some cases, an increase in one quantity produces a *decrease* in another quantity. In this case, we say that the quantities **vary inversely.** For example, suppose that you and some of your friends are going to buy a birthday gift for someone, splitting the cost equally. As the number of people who are contributing increases, the cost for each person decreases. The cost for each person varies inversely as the number of people contributing.

Inverse Variation

> If a quantity y varies inversely as a quantity x, then the two quantities are related by the equation
>
> $$y = \frac{k}{x}$$
>
> where k is the constant of variation.

EXAMPLE 9 y varies inversely as x. If $y = 10$ when $x = 8$, find y when $x = 5$.

Solution

Since the quantities vary inversely, we will use the equation $y = \dfrac{k}{x}$. We begin by finding k.

$$y = \frac{k}{x}$$

$$10 = \frac{k}{8} \qquad \text{Substitute 10 for } y \text{ and 8 for } x \text{ into } y = \tfrac{k}{x}.$$

$$80 = k \qquad \text{Multiply both sides by 8.}$$

Now we use this value of k to find y when $x = 5$.

$$y = \frac{80}{5} \qquad \text{Substitute 80 for } k \text{ and 5 for } x \text{ into } y = \tfrac{k}{x}.$$

$$y = 16 \qquad \text{Divide.}$$

EXAMPLE 10   The time required to drive from Visalia to San Francisco varies inversely as the speed of the car. If it takes 3 hours to make the drive at 70 miles per hour, how long would it take to make the drive at 60 miles per hour?

Solution

Again, we begin by finding k. We will let y represent the time required to drive from Visalia to San Francisco and x represent the average speed of the car. To find k, we will substitute the related information (3 hours to make the trip at 70 miles per hour) into the equation for inverse variation.

$$y = \frac{k}{x}$$

$$3 = \frac{k}{70} \qquad \text{Substitute 3 for } y \text{ and 70 for } x \text{ into } y = \tfrac{k}{x}.$$

$$210 = k \qquad \text{Multiply both sides by 70.}$$

Now we will use this variation constant to find the time required to drive from Visalia to San Francisco at 60 miles per hour.

$$y = \frac{210}{60} \qquad \text{Substitute 210 for } k \text{ and 60 for } x \text{ into } y = \tfrac{k}{x}.$$

$$y = 3.5 \qquad \text{Divide.}$$

It would take 3.5 hours to drive from Visalia to San Francisco at 60 miles per hour.

Quick Check **8** The time required to complete the Boston Marathon varies inversely as the speed of the runner. If it takes 131 minutes to run the marathon at an average speed of 5 miles per hour, how long would it take to finish the marathon at an average speed of 4 miles per hour?

Often, a quantity varies depending on two or more variables. A quantity y is said to **vary jointly** as two quantities x and z if it varies directly as the product of these two quantities. The equation in such a case is $y = kxz$.

EXAMPLE 11 y varies jointly as x and the square of z. If $y = 2400$ when $x = 8$ and $z = 10$, find y when $x = 12$ and $z = 20$.

Solution

We begin by finding k, using the equation $y = kxz^2$.

$$y = kxz^2$$

Quick Check **9**

y varies jointly as x and the square root of z. If $y = 1000$ when $x = 50$ and $z = 25$, find y when $x = 72$ and $z = 9$.

$$2400 = k \cdot 8 \cdot 10^2 \qquad \text{Substitute 2400 for } y, 8 \text{ for } x, \text{ and } 10 \text{ for } z \text{ into } y = kxz^2.$$
$$2400 = k \cdot 800 \qquad \text{Simplify.}$$
$$3 = k \qquad \text{Divide both sides by 800.}$$

Now we use this value of k to find y when $x = 12$ and $z = 20$.

$$y = 3 \cdot 12 \cdot 20^2 \qquad \text{Substitute 3 for } k, 12 \text{ for } x, \text{ and } 20 \text{ for } z \text{ into } y = kxz^2.$$
$$y = 14{,}400 \qquad \text{Simplify.}$$

Building Your Study Strategy **Preparing for a Cumulative Exam, 7 Study Groups** If you have participated in a study group throughout the semester, now is not the time to start studying exclusively on your own. Have each student in the group bring problems that he or she is struggling with or feels are important, and work through them as a group.

At the end of your group study session, try to write a practice exam as a group. In trying to determine which problems to include on the practice exam, you are really focusing on which types of problems are most likely to appear on your cumulative exam.

In some groups, members bring problems and challenge others in the group to solve them. Although this can be an effective way to study, it is not a good idea to do this the night before the exam. You may find that if you are unable to solve two or three problems in a row, your confidence may be shattered, and you need plenty of confidence when you sit down to take an exam. Spend the night before the exam alone, reviewing what you have done.

Vocabulary

1. The _____ of a number x is $\frac{1}{x}$.

2. The _____ of a person is the amount of work he or she does in one unit of time.

3. If an object is moving at a constant rate of speed for a certain amount of time, then the length of time it traveled is equal to the distance traveled by the object _____ by its rate of speed.

4. If a canoe can travel at a rate of r miles per hour in still water and a river's current is c miles per hour, then the rate of speed that the canoe travels at while moving upstream is given by the expression _____.

5. Two quantities are said to _____ if an increase in one quantity produces a proportional increase in the other quantity.

6. Two quantities are said to _____ if an increase in one quantity produces a proportional decrease in the other quantity.

7. The sum of the reciprocal of a number and $\frac{2}{3}$ is $\frac{13}{15}$. Find the number.

8. The sum of the reciprocal of a number and $\frac{4}{7}$ is $\frac{43}{63}$. Find the number.

9. The sum of four times the reciprocal of a number and $\frac{3}{4}$ is $\frac{49}{44}$. Find the number.

10. The sum of seven times the reciprocal of a number and $\frac{5}{8}$ is $\frac{3}{2}$. Find the number.

11. The difference of the reciprocal of a number and $\frac{5}{9}$ is $-\frac{41}{90}$. Find the number.

12. The difference of the reciprocal of a number and $\frac{3}{10}$ is $\frac{1}{5}$. Find the number.

13. One positive number is three larger than another positive number. If four times the reciprocal of the smaller number is added to three times the reciprocal of the larger number, the sum is 1. Find the two numbers.

14. One positive number is six less than another positive number. If two times the reciprocal of the smaller number is added to five times the reciprocal of the larger number, the sum is 1. Find the two numbers.

15. One positive number is 10 less than another positive number. If the reciprocal of the smaller number is added to three times the reciprocal of the larger number, the sum is $\frac{1}{4}$. Find the two numbers.

16. One positive number is nine larger than another positive number. If the reciprocal of the smaller number is added to four times the reciprocal of the larger number, the sum is $\frac{2}{3}$. Find the two numbers.

17. One copy machine can run off copies in 15 minutes. A newer machine can do the same job in 10 minutes. How long would it take the two machines, working together, to make all the necessary copies?

18. Dylan can mow a lawn in 40 minutes, while Alycia takes 50 minutes to mow the same lawn. If Dylan and Alycia work together, using two lawn mowers, how long would it take them to mow the lawn?

19. Tina can completely weed her vegetable garden in 1 hour and 30 minutes. Her friend Marisa can do the same task in 2 hours. If they worked together, how long would it take Tina and Marisa to weed the vegetable garden?

20. One hose can fill a 40,000-gallon swimming pool in 18 hours. A hose from the neighbor's house can fill a swimming pool of that size in 24 hours. If the two hoses run at the same time, how long would they take to fill the swimming pool?

21. After a room has been prepared for painting, Andria can paint the entire room in 2 hours. Tonya can do the same job in 3 hours, while it would take Linda 5 hours to paint the entire room. If the three friends work together, how long would it take them to paint an entire room?

22. A company has printed out 500 surveys to be put into preaddressed envelopes and mailed out. Dusty can fill all of the envelopes in 5 hours, Felipe can do the job in 6 hours, and Grady can do it in 9 hours. If all three work together, how long will it take them to stuff the 500 envelopes?

23. Two drainpipes working together can drain a pool in 12 hours. Working alone, the smaller pipe would take 7 hours longer than the larger pipe to drain the pool. How long would it take the smaller pipe alone to drain the pool?

24. Working together, Rosa and Dianne can plant 20 flats of pansies in 3 hours. If Rosa were working alone, it would take her 8 hours longer than it would take Dianne to plant all 20 flats. How long would it take Rosa to plant the pansies by herself?

25. Steve and his assistant Ross can wire a house in 6 hours. If they worked alone, Ross would take 9 hours longer than Steve to wire the house. How long would it take Ross to wire the house?

26. Earl and John can clean an entire building in 4 hours. Earl can clean the entire building by himself in 6 fewer hours than John can. How long would it take Earl to clean the building by himself?

27. Working together, Sarah and Jeff can feed all of the animals at the zoo in 2 hours. Jeff uses a motorized cart, whereas Sarah uses a pushcart to carry the food. If each person were to feed all of the animals without the help of the other, it would take Sarah twice as long as it would take Jeff. How long would it take Sarah to feed all of the animals?

28. A small pipe takes three times longer to fill a tank than a larger pipe takes. If both pipes are working at the same time, it takes 12 minutes to fill the tank. How long would it take the smaller pipe, working alone, to fill the tank?

29. When he works alone, Brent takes 2 hours longer to buff the gymnasium floor than Tracy takes when she works alone. After working for 3 hours, Brent quits and Tracy takes over. If it takes Tracy 2 hours to finish the rest of the floor, how long would it have taken Brent to buff the entire floor?

30. Frida can wallpaper a room in 3 fewer hours than Gerardo can. After they worked together for 2 hours, Frida had to leave. It took Gerardo an additional 4 hours to finish wallpapering the room. How long would it take Frida to wallpaper a room by herself?

31. Vang is training for a triathlon. He rides his bicycle at a speed that is 15 miles per hour faster than his running speed. If Vang can cycle 40 miles in the same amount of time that it takes him to run 16 miles, what is Vang's running speed?

32. Ariel drives 15 miles per hour faster than Sharon does. Ariel can drive 100 miles in the same amount of time that it takes for Sharon to drive 80 miles. Find Ariel's driving speed.

33. Jared is training to run a marathon. Today he ran 14 miles in 2 hours. After running the first 9 miles at a certain speed, he increased his speed by 4 miles per hour for the remaining 5 miles. Find the speed at which Jared was running for the first 9 miles.

34. Kyung ran 7 miles this morning. After running the first 3 miles, she increased her speed by 2 miles per hour. If it took her exactly 1 hour to finish her run, find the speed at which she was running for the last 4 miles.

35. A salmon is swimming in a river that is flowing downstream at a speed of 10 kilometers per hour. The salmon can swim 3 kilometers upstream in the

same time that it would take to swim 5 kilometers downstream. What is the speed of the salmon in still water?

36. Denae is swimming in a river that flows downstream at a speed of 0.5 meter per second. It takes her the same amount of time to swim 500 meters upstream as it does to swim 1000 meters downstream. Find Denae's swimming speed in still water.

37. An airplane flies 72 miles with a 30-mph tailwind and then flies back into the 30-mph wind. If the time for the round trip was 1 hour, find the speed of the airplane in calm air.

38. Selma is kayaking in a river that flows downstream at a rate of 1 mile per hour. Selma paddles 5 miles downstream and then turns around and paddles 6 miles upstream, and the trip takes 3 hours.

   a) How fast can Selma paddle in still water?

   b) Selma is now 1 mile upstream of her starting point. How many minutes will it take her to paddle back to her starting point?

39. y varies directly as x. If $y = 30$ when $x = 3$, find y when $x = 8$.

40. y varies directly as x. If $y = 78$ when $x = 6$, find y when $x = 15$.

41. y varies directly as x. If $y = 60$ when $x = 16$, find y when $x = 24$.

42. y varies directly as x. If $y = 80$ when $x = 35$, find y when $x = 21$.

43. y varies inversely as x. If $y = 8$ when $x = 7$, find y when $x = 4$.

44. y varies inversely as x. If $y = 6$ when $x = 12$, find y when $x = 9$.

45. y varies inversely as x. If $y = 20$ when $x = 4$, find y when $x = 15$.

46. y varies inversely as x. If $y = 15$ when $x = 18$, find y when $x = 100$.

47. y varies directly as the square of x. If $y = 150$ when $x = 5$, find y when $x = 7$.

48. y varies inversely as the square of x. If $y = 4$ when $x = 6$, find y when $x = 3$.

49. y varies jointly as x and z. If $y = 270$ when $x = 10$ and $z = 6$, find y when $x = 3$ and $z = 8$.

50. y varies directly as x and inversely as z. If $y = 78$ when $x = 26$ and $z = 7$, find y when $x = 30$ and $z = 18$.

51. Sam's gross pay varies directly as the number of hours he works. In a week that he worked 28 hours, his gross pay was $238. What would Sam's gross pay be if he worked 32 hours?

52. The height that a ball bounces varies directly as the height from which it is dropped. If a ball dropped from a height of 54 inches bounces 24 inches, how high would the ball bounce if it were dropped from a height of 72 inches?

53. *Ohm's law.* In a circuit, the electric current (in amperes) varies directly as the voltage. If the current is 8 amperes when the voltage is 24 volts, find the current when the voltage is 9 volts.

54. *Hooke's law.* The distance that a hanging object stretches a spring varies directly as the mass of the object. If a 5-kilogram weight stretches a spring by 32 centimeters, how far would a 2-kilogram weight stretch the spring?

55. The amount of money that each person must contribute to buy a retirement gift for a coworker varies inversely as the number of people contributing. If 10 people would each have to contribute $24, how much would each person have to contribute if there were 16 people?

56. The maximum load that a wooden beam can support varies inversely as its length. If a beam that is 8 feet long can support 700 pounds, what is the maximum load that can be supported by a beam that is 5 feet long?

57. In a circuit, the electric current (in amperes) varies inversely as the resistance (in ohms). If the current is 30 amperes when the resistance is 3 ohms, find the current when the resistance is 5 ohms.

58. *Boyle's law.* The volume of a gas varies inversely as the pressure upon it. The volume of a gas is 128 cubic centimeters when it is under a pressure of 50 kilograms per square centimeter. Find the volume when the pressure is reduced to 40 kilograms per square centimeter.

59. The illumination of an object varies inversely as the square of its distance from the source of light. If a light source provides an illumination of 30 foot-candles at a distance of 10 feet, find the illumination at a distance of 20 feet. (One foot-candle is the amount of illumination produced by a standard candle at a distance of 1 foot.)

60. The illumination of an object varies inversely as the square of its distance from the source of light. If a light source provides an illumination of 80 foot-candles at a distance of 5 feet, find the illumination at a distance of 40 feet.

Writing in Mathematics

Answer in complete sentences.

61. Write a work-rate word problem that can be solved by the equation $\dfrac{t}{8} + \dfrac{t}{10} = 1$. Explain how you created your problem.

62. Write a work-rate word problem that can be solved by the equation $\dfrac{12}{t+8} + \dfrac{2}{t} = 1$. Explain how you created your problem.

63. Write a uniform-motion word problem that can be solved by the equation $\dfrac{10}{r-4} + \dfrac{10}{r+4} = 6$. Explain how you created your problem.

64. ***Solutions Manual*** * Write a solutions manual page for the following problem:

Mauricio drives 150 miles to his job each day. Due to a rainstorm on his trip home, he had to drive 25 miles per hour slower than he drove on his way to work, and it took him 1 hour longer to get home than it did to get to work. How fast was Mauricio driving on his way home?

65. ***Newsletter*** * Write a newsletter that explains how to solve work-rate problems.

*See Appendix B for details and sample answers.

Chapter 7 Summary

Summary of Chapter 7 Study Strategies

- When preparing for a cumulative exam, start approximately two weeks before the exam by determining what material will be on the exam.
- Begin your studying by going over your old exams and quizzes. Although you should rework each problem, pay particular attention to the problems you got wrong on the exam or quiz. Also, look over your old homework assignments for problems you struggled with and review these topics.
- Use review materials from your instructor and the cumulative review exercises in the text to check your progress. Try to work the exercises without referring to your notes, note cards, or text. This will give you a clear assessment as to which problems require further study.
- Create study sheets for each type of applied problem that may appear on the exam. These sheets should contain examples of these types of problems, the procedure for solving them, and the solutions of your examples.
- Continue to meet regularly with your study group. If your group has helped you to succeed on the previous exams, then they will be able to help you succeed on this exam as well.

Evaluate the rational expression for the given value of the variable. [7.1]

1. $\dfrac{3}{x-9}$ for $x = -6$

2. $\dfrac{x-7}{x+13}$ for $x = 19$

3. $\dfrac{x^2 + 9x - 11}{x^2 + 2x - 5}$ for $x = 3$

4. $\dfrac{x^2 - 5x - 15}{x^2 + 17x + 66}$ for $x = -4$

Find all values for which the rational expression is undefined. [7.1]

5. $\dfrac{-6}{x+9}$

6. $\dfrac{x^2 + 9x + 18}{x^2 - 3x - 54}$

Simplify the given rational expression. (Assume all denominators are nonzero.) [7.1]

7. $\dfrac{x+7}{x^2 + 4x - 21}$

8. $\dfrac{x^2 + 6x - 40}{x^2 - 9x + 20}$

9. $\dfrac{36 - x^2}{x^2 - 7x + 6}$

10. $\dfrac{x^2 + 6x + 9}{4x^2 + 15x + 9}$

Evaluate the given rational function. [7.1]

11. $r(x) = \dfrac{x+7}{x^2 + 8x - 15}$, $r(3)$

12. $r(x) = \dfrac{x^2 - 15x}{x^2 - 11x + 18}$, $r(-2)$

Find the domain of the given rational function. [7.1]

13. $r(x) = \dfrac{x^2 + 3x - 40}{x^2 + 6x}$

14. $r(x) = \dfrac{x^2 + 7x - 8}{x^2 - 5x + 4}$

Multiply. [7.2]

15. $\dfrac{x+9}{x+5} \cdot \dfrac{x^2 - 7x}{x^2 + 2x - 63}$

16. $\dfrac{x^2 + 14x + 45}{x^2 - 6x - 7} \cdot \dfrac{x^2 + 9x + 8}{x^2 + 3x - 10}$

17. $\dfrac{x^2 - 4x}{x^2 - 15x + 54} \cdot \dfrac{x^2 - x - 72}{x^2 + 8x}$

18. $\dfrac{16 - x^2}{x+2} \cdot \dfrac{7x^2 + 13x - 2}{x^2 + 2x - 24}$

For the given functions $f(x)$ and $g(x)$, find $f(x) \cdot g(x)$. [7.2]

19. $f(x) = \dfrac{x-6}{x-1}$, $g(x) = \dfrac{x^2 + 6x + 5}{x^2 - 3x - 18}$

20. $f(x) = \dfrac{x^2 + 16x + 64}{x^2 + x - 42}$, $g(x) = \dfrac{49 - x^2}{x^2 + 11x + 24}$

Divide. [7.2]

21. $\dfrac{x^2 + 9x + 18}{x^2 - 7x + 12} \div \dfrac{x+3}{x-3}$

22. $\dfrac{x^2 - x - 6}{x^2 - 19x + 88} \div \dfrac{x^2 + 10x + 16}{x^2 + 4x - 96}$

23. $\dfrac{x^2 - 3x - 4}{x^2 + x} \div \dfrac{3x^2 - 10x - 8}{x^2 - 1}$

24. $\dfrac{9 - x^2}{x - 7} \div \dfrac{x^2 - 8x + 15}{x^2 - 5x - 14}$

For the given functions $f(x)$ and $g(x)$, find $f(x) \div g(x)$. [7.2]

25. $f(x) = \dfrac{x^2 + 3x}{x^2 + 10x + 25}$, $g(x) = \dfrac{x^2 - 6x - 27}{x^2 - x - 30}$

26. $f(x) = \dfrac{x^2 + 8x + 12}{x^2 + 5x + 4}$, $g(x) = \dfrac{x^2 - 4x - 60}{x^2 + 7x + 12}$

Worked-out solutions to Review Exercises marked with ⬤ can be found on page AN–25.

Add or subtract. [7.3/7.4]

27. $\dfrac{5}{x+6} + \dfrac{7}{x+6}$

28. $\dfrac{x^2 + 10x}{x^2 + 15x + 56} - \dfrac{5x + 24}{x^2 + 15x + 56}$

29. $\dfrac{x^2 - 6x - 13}{x^2 - x - 20} + \dfrac{10x - 32}{x^2 - x - 20}$

30. $\dfrac{5x - 34}{x - 7} - \dfrac{x - 8}{7 - x}$

31. $\dfrac{2x^2 + 7x - 3}{x - 2} + \dfrac{x^2 + 3x + 9}{2 - x}$

32. $\dfrac{x^2 - 3x - 12}{x^2 - 25} - \dfrac{2x - 18}{25 - x^2}$

33. $\dfrac{5}{x^2 + 3x - 4} + \dfrac{2}{x^2 - 4x + 3}$

34. $\dfrac{6}{x^2 - 10x + 16} - \dfrac{1}{x^2 - 15x + 56}$

35. $\dfrac{x}{x^2 - 5x - 50} - \dfrac{4}{x^2 - 14x + 40}$

36. $\dfrac{x - 5}{x^2 + 14x + 33} + \dfrac{4}{x^2 + 20x + 99}$

37. $\dfrac{x + 1}{x^2 + 2x - 24} + \dfrac{x - 5}{x^2 - 6x + 8}$

For the given rational functions $f(x)$ and $g(x)$, find $f(x) + g(x)$. [7.4]

38. $f(x) = \dfrac{x + 3}{x^2 - x - 2}, g(x) = \dfrac{5}{x^2 - 7x + 10}$

For the given rational functions $f(x)$ and $g(x)$, find $f(x) - g(x)$. [7.4]

39. $f(x) = \dfrac{x + 2}{x^2 + 4x - 5}, g(x) = \dfrac{1}{x^2 + 12x + 35}$

Simplify the complex fraction. [7.5]

40. $\dfrac{1 - \dfrac{9}{x}}{1 - \dfrac{81}{x^2}}$

41. $\dfrac{\dfrac{x - 5}{x - 3} - \dfrac{3}{x - 7}}{\dfrac{x^2 + 7x - 44}{x^2 - 10x + 21}}$

42. $\dfrac{1 + \dfrac{2}{x} - \dfrac{35}{x^2}}{1 - \dfrac{7}{x} + \dfrac{10}{x^2}}$

43. $\dfrac{\dfrac{x^2 - 8x + 15}{x^2 - 11x + 18}}{\dfrac{x^2 + x - 30}{x^2 - 6x - 27}}$

Solve. [7.6]

44. $\dfrac{9}{x} - \dfrac{3}{4} = \dfrac{7}{8}$

45. $x - 5 + \dfrac{12}{x} = 2$

46. $\dfrac{3}{x - 7} = \dfrac{7}{x + 5}$

47. $\dfrac{2x - 11}{x - 1} = \dfrac{x - 7}{x + 3}$

48. $\dfrac{3}{x + 2} - \dfrac{1}{x + 1} = \dfrac{x + 3}{x^2 + 3x + 2}$

49. $\dfrac{x}{x + 5} + \dfrac{3}{x - 7} = \dfrac{36}{x^2 - 2x - 35}$

50. $\dfrac{2}{x^2 - x - 2} + \dfrac{10}{x^2 - 2x - 3} = \dfrac{x + 12}{x^2 - x - 2}$

51. $\dfrac{20}{x^2 + 12x + 27} - \dfrac{4}{x^2 + 14x + 45} = \dfrac{x + 10}{x^2 + 8x + 15}$

Solve for the specified variable. [7.6]

52. $\dfrac{1}{2} = \dfrac{A}{bh}$ for h

53. $y = \dfrac{3x}{4x - 7}$ for x

54. $\dfrac{x}{2r} - \dfrac{y}{3r} = \dfrac{1}{5}$ for r

55. The sum of the reciprocal of a number and $\frac{5}{6}$ is $\frac{23}{24}$. Find the number. [7.7]

56. One positive number is five more than another positive number. If three times the reciprocal of the smaller number is added to four times the reciprocal of the larger number, the sum is 1. Find the two numbers. [7.7]

57. Rob can mow a lawn in 30 minutes, while Sean takes 60 minutes to mow the same lawn. If Rob and Sean work together, using two lawn mowers, how long would it take them to mow the lawn? [7.7]

58. Two pipes, working together, can fill a tank in 28 minutes. Working alone, the smaller pipe takes 42 minutes longer than the larger pipe to fill the tank. How long would it take the smaller pipe alone to fill the tank? [7.7]

59. Linda had to make a 275-mile drive to Omaha. After driving the first 100 miles, she increased her speed by 15 miles per hour. If the drive took her exactly 4 hours, find the speed at which she was driving for the first 100 miles. [7.7]

60. Dominique is kayaking in a river that is flowing downstream at a speed of 2 miles per hour. Dominique can paddle 10 miles downstream in the same amount of time that she can paddle 5 miles upstream. What is the speed that Dominique can paddle in still water? [7.7]

61. The number of calories in a glass of milk varies directly as the amount of milk. If a 12-ounce serving of milk has 195 calories, how many calories are there in an 8-ounce glass of milk? [7.7]

62. The distance required for a car to stop after applying the brakes varies directly as the square of the speed of the car. If it takes 100 feet for a car traveling at 40 miles per hour to stop, how far would it take for a car traveling 60 miles per hour to come to a stop? [7.7]

63. The maximum load that a wooden beam can support varies inversely as its length. If a beam that is 6 feet long can support 900 pounds, what is the maximum load that can be supported by a beam that is 8 feet long? [7.7]

64. The illumination of an object varies inversely as the square of its distance from the source of light. If a light source provides an illumination of 20 foot-candles at a distance of 12 feet, find the illumination at a distance of 6 feet. [7.7]

For
Extra
Help

Pass
*the*Test

Test solutions
are found on the
enclosed CD.

Evaluate the rational expression for the given value of the variable.

1. $\dfrac{10}{x-5}$ for $x = -7$

Find all values for which the rational expression is undefined.

2. $\dfrac{x^2 + 5x - 14}{x^2 + 10x + 21}$

Simplify the given rational expression. (Assume all denominators are nonzero.)

3. $\dfrac{x^2 - 8x + 15}{x^2 + 7x - 30}$

Evaluate the rational function.

4. $r(x) = \dfrac{x^2 + 3x - 13}{x^2 - 4x - 20}, \; r(-3)$

Multiply.

5. $\dfrac{x^2 + x - 42}{x^2 + 6x - 7} \cdot \dfrac{x^2 - 2x + 1}{x^2 - 9x + 18}$

Divide.

6. $\dfrac{x^2 - 10x + 25}{x^2 + 11x + 24} \div \dfrac{25 - x^2}{x^2 + 3x - 40}$

Add or subtract.

7. $\dfrac{5x + 3}{x - 6} - \dfrac{2x + 21}{x - 6}$

8. $\dfrac{2x^2 + 6x - 11}{x - 4} + \dfrac{x^2 + 2x + 21}{4 - x}$

9. $\dfrac{6}{x^2 + 8x + 7} + \dfrac{5}{x^2 - 3x - 4}$

10. $\dfrac{x - 4}{x^2 - 13x + 40} - \dfrac{8}{x^2 - 10x + 16}$

Simplify the complex fraction.

11. $\dfrac{1 + \dfrac{4}{x} - \dfrac{45}{x^2}}{1 + \dfrac{2}{x} - \dfrac{35}{x^2}}$

Simplify.

12. $\dfrac{x + 5}{x^2 + 3x - 54} \cdot \dfrac{x^2 - 11x + 30}{x^2 + 16x + 55}$

13. $\dfrac{x^2 - 3x - 40}{x^2 + 7x + 6} \div \dfrac{x^2 - 8x}{x^2 + 10x + 9}$

14. $\dfrac{x^2 + 7x - 14}{x^2 + 6x + 8} + \dfrac{3x + 38}{x^2 + 6x + 8}$

15. $\dfrac{x - 7}{x^2 + 7x - 8} - \dfrac{2}{x^2 + x - 2}$

Solve.

16. $\dfrac{15}{x} + \dfrac{3}{8} = \dfrac{19}{24}$

17. $\dfrac{x + 2}{x^2 + 5x + 4} - \dfrac{4}{x^2 + x - 12} = \dfrac{2}{x^2 - 2x - 3}$

Solve for the specified variable.

18. $x = \dfrac{2y}{7y - 5}$ for y

19. Two pipes working together can fill a tank in 40 minutes. Working alone, the smaller pipe would take 18 minutes longer than the larger pipe to fill the tank. How long would it take the smaller pipe alone to fill the tank?

20. Preparing for a race, Greg went for a 25-mile bicycle ride. After the first 10 miles, he increased his speed by 10 miles per hour. If the ride took him exactly 1 hour, find the speed at which he was riding for the first 10 miles.

Mathematicians in History
John Nash

$\mathcal{J}$ohn Nash is an American mathematician whose research has greatly affected mathematics and economics, as well as many other fields. When he was applying to go to graduate school at Princeton, one of his math professors said quite simply, "This man is a genius."

Write a one-page summary (*or* make a poster) of the life of John Nash and his accomplishments.

Interesting issues:

- Where and when was John Nash born?
- Describe Nash's childhood, as well as his life as a college student.
- What mental illness struck Nash in the late 1950s?
- Nash was influential in the field of game theory. What is game theory?
- In 1994, Nash won the Nobel Prize for Economics. Exactly what did Nash win the prize for?
- One of Nash's nicknames is "The Phantom of Fine Hall." Why was this nickname chosen for him?
- What color sneakers did Nash wear?
- Sylvia Nasar wrote a biography of Nash's life, which was made into an Academy Award–winning movie. What was the title of the book and movie?
- What actor played John Nash in the movie?

The weight W of an object varies inversely as the square of the distance d from the center of the Earth. At sea level (3978 miles from the center of the Earth), a person weighs 150 pounds. The formula used to compute the weight of this person at different distances from the center of the Earth is $W = \dfrac{2{,}373{,}672{,}600}{d^2}$.

a) Use the given formula to calculate the weight of this person at different distances from the center of the Earth. Round to the nearest tenth of a pound.

b) What do you notice about the weight of the person as the distance from the center of the Earth increases?

c) Why do you think the weight changes with increasing distance?

d) As accurately as possible, plot these points on an axis system in which the horizontal axis represents the distance and the vertical axis represents the weight. Do the points support your observations from part (b)?

e) How far from the center of the Earth would this 150-pound person have to travel to weigh 100 pounds?

f) How far from the center of the Earth would this 150-pound person have to travel to weigh ·75 pounds?

g) How far from the center of the Earth would this 150-pound person have to travel to weigh 0 pounds?

	Distance from Center of Earth	Weight
On top of the tallest building, Taipei 101 (Taipei, Taiwan)	3978.316 miles	
On top of the tallest structure, KVLY-TV mast (Mayville, ND)	3978.391 miles	
In an airplane	3984.629 miles	
In the space station	4201.402 miles	
Halfway to the moon	123,406 miles	
Halfway to Mars	4,925,728 miles	

CHAPTER 8

Radical Expressions and Equations

In this chapter we will investigate radical expressions and equations, and their applications. Among the applications are the method for finding the speed of a tsunami and the way to determine the speed a car was traveling by measuring the skid marks left by its tires.

Study Strategy ⟨ Note Taking ⟩ *If you have poor note-taking skills, then you will not learn as much during the class period as you might be able to if your skills were better. In this chapter we will focus on how to take notes in a math class. We will discuss how to be more efficient when taking notes, what material to include in your notes, and how to rework your notes.*

8.1
Square Roots and Radical Notation

1. **Find the square root of a number.**
2. **Simplify the square root of a variable expression.**
3. **Approximate the square root of a number using a calculator.**
4. **Find nth roots.**
5. **Multiply radical expressions.**
6. **Divide radical expressions.**
7. **Evaluate radical functions.**
8. **Find the domain of a radical function.**

Square Roots

Consider the equation $x^2 = 36$. There are two solutions to this equation, $x = 6$ and $x = -6$.

Square Root

> A number a is a **square root** of a number b if $a^2 = b$.

The numbers 6 and -6 are square roots of 36, since $6^2 = 36$ and $(-6)^2 = 36$. The number 6 is the positive square root of 36, while the number -6 is the negative square root of 36.

Objective 1　Find the square root of a number.

Principal Square Root

> The **principal square root** of b, denoted $\sqrt{b}$, for $b > 0$, is the positive number a such that $a^2 = b$.

The expression $\sqrt{b}$ is called a **radical expression.** The sign $\sqrt{}$ is called a **radical sign,** while the expression contained inside the radical sign is called the **radicand.**

Note: Unless otherwise requested, simplifying the square root of a number means finding the principal square root of that number.

EXAMPLE ▶ 1　Simplify $\sqrt{49}$.

Solution

▶ We are looking for a positive number a such that $a^2 = 49$. The number is 7, so $\sqrt{49} = 7$.

EXAMPLE 2   Simplify $\sqrt{\dfrac{4}{25}}$.

Solution

Since $\left(\dfrac{2}{5}\right)^2 = \dfrac{4}{25}$, $\sqrt{\dfrac{4}{25}} = \sqrt{\left(\dfrac{2}{5}\right)^2} = \dfrac{2}{5}$.

EXAMPLE 3   Simplify $-\sqrt{121}$.

Solution

Quick Check 1

Simplify.

a) $\sqrt{25}$ b) $\sqrt{\dfrac{1}{81}}$
c) $-\sqrt{49}$

In this example we are looking for the negative square root of 121, which is -11. So $-\sqrt{121} = -11$. We can first find the principal square root of 121 and then make it negative.

The principal square root of a negative number, such as $\sqrt{-121}$, is not a real number because there is no real number a such that $a^2 = -121$. We will learn in section 8.6 that $\sqrt{-121}$ is called an **imaginary number.**

Square Roots of Variable Expressions

Objective 2 **Simplify the square root of a variable expression.** Now we turn our attention toward simplifying radical expressions containing variables, such as $\sqrt{x^8}$.

Simplifying $\sqrt{a^2}$

For any real number a, $\sqrt{a^2} = |a|$.

You may be wondering why the absolute value bars are necessary. For any nonnegative number a, $\sqrt{a^2} = a$. For example, $\sqrt{8^2} = 8$. The absolute value bars are necessary to include negative values of a. Suppose that $a = -10$. Then $\sqrt{a^2} = \sqrt{(-10)^2}$, which is equal to $\sqrt{100}$ or 10. So $\sqrt{a^2}$ equals the opposite of a. In either case, the principal square root of a^2 will be a nonnegative number.

$$\sqrt{a^2} = \begin{cases} a & \text{if} \quad a \geq 0 \\ -a & \text{if} \quad a < 0 \end{cases}$$

The absolute value bars address both cases. We must use absolute values when dealing with variables, because we do not know if the variable is negative.

EXAMPLE 4   Simplify $\sqrt{x^{10}}$.

Solution

We begin by rewriting the radicand as a square. Note that $x^{10} = (x^5)^2$.

$$\sqrt{x^{10}} = \sqrt{(x^5)^2} \qquad \text{Rewrite radicand as a square.}$$
$$= |x^5| \qquad \text{Simplify.}$$

Since we do not know if x is negative, the absolute value bars are necessary.

EXAMPLE 5   Simplify $\sqrt{36z^{14}}$.

Solution

Again, we rewrite the radicand as a square.

Quick Check 2
Simplify.
a) $\sqrt{b^{26}}$ b) $\sqrt{49a^6}$

$$\sqrt{36z^{14}} = \sqrt{(6z^7)^2} \qquad \text{Rewrite radicand as a square.}$$
$$= |6z^7| \qquad \text{Simplify.}$$

Since we know that 6 is a positive number we can remove it from the absolute value bars. This allows us to write the expression as $6|z^7|$.

EXAMPLE 6   Simplify $\sqrt{25x^8}$.

Solution

We rewrite the radicand as a square.

$$\sqrt{25x^8} = \sqrt{(5x^4)^2} \qquad \text{Rewrite radicand as a square.}$$
$$= |5x^4| \qquad \text{Simplify.}$$

Quick Check 3
Simplify $\sqrt{16x^{20}}$.

Since we know that 5 is a positive number, we can remove it from the absolute value bars. x^4 cannot be negative either, since any number raised to an even power cannot be negative, so it may be removed from the absolute value bars as well.

$$\sqrt{25x^8} = 5x^4$$

From this point on, we will assume that all variable factors in a radicand represent non-negative numbers. This eliminates the need to use absolute value bars when simplifying radical expressions whose radicand contains variables.

Approximating Square Roots Using a Calculator

Objective 3 **Approximate the square root of a number using a calculator.**
Consider the expression $\sqrt{12}$. There is no positive integer that is the square root of 12. In such a case we can use a calculator to approximate the radical expression. All calculators have a function for calculating square roots. We see that $\sqrt{12} \approx 3.464$, rounded to the nearest thousandth. The symbol $\approx$ is read as *is approximately equal to,* so the principal square root of 12 is approximately equal to 3.464. If we square 3.464 it is equal to 11.999296, which is very close to 12.

EXAMPLE 7   Approximate $\sqrt{20}$ to the nearest thousandth, using a calculator.

Solution

Quick Check 4
Approximate $\sqrt{175}$ to the nearest thousandth, using a calculator.

Since $4^2 = 16$ and $5^2 = 25$, we know that $\sqrt{20}$ must be a number between 4 and 5.

$$\sqrt{20} \approx 4.472$$

Using Your Calculator   We can approximate square roots using the TI-84.

*n*th Roots

Objective **4**   **Find *n*th roots.** Now we move on to discuss roots other than square roots.

Principal *n*th Root

> For any positive integer $n > 1$ and any number b, if $a^n = b$ and a and b both have the same sign, then a is the **principal *n*th root** of b, denoted $a = \sqrt[n]{b}$. The number n is called the **index** of the radical.

A square root has an index of 2, and the radical is written without the index. If the index is 3, this is called a **cube root**. If n is even, then the principal *n*th root is a nonnegative number, but if n is odd, the principal *n*th root is of the same sign as the radicand. As with square roots, an even root of a negative number is not a real number. For example, $\sqrt[6]{-64}$ is not a real number. However, an odd root of a negative number, such as $\sqrt[3]{-64}$, is a negative real number.

EXAMPLE **8**   Simplify $\sqrt[3]{27}$.

Solution

We are looking for a number that, when cubed, is equal to 27. Since $27 = 3^3$, $\sqrt[3]{27} = 3$.

EXAMPLE **9**   Simplify $\sqrt[5]{1024}$.

Solution

In this example we are looking for a number that, when raised to the fifth power, is equal to 1024. Since $1024 = 4^5$, $\sqrt[5]{1024} = 4$.

Using Your Calculator   The function for calculating *n*th roots on the TI-84 can be found by pressing the MATH key and selecting option 5 under the MATH menu.

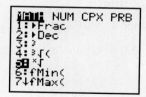

continued

Press the index for the radical, then the nth root function, and <u>then</u> the radicand inside a set of parentheses. Here is the screen that would calculate $\sqrt[5]{1024}$.

```
5 × √(1024)
                    4
```

EXAMPLE ▶10   Simplify $\sqrt[3]{-125}$.

Solution

Notice that the radicand is negative. So we are looking for a negative number that, when raised to the third power, is equal to -125. (If the <u>index</u> was even, then this expression would not be a real number.) Since $(-5)^3 = -125$, $\sqrt[3]{-125} = -5$.

For any nonnegative number x, $\sqrt[n]{x^n} = x$.

> **Quick Check 5**
> Simplify.
> a) $\sqrt[3]{1000}$ b) $\sqrt[4]{1296}$
> c) $\sqrt[3]{-512}$

EXAMPLE ▶11   Simplify $\sqrt[5]{x^{35}}$. (Assume that x is nonnegative.)

Solution

We begin by rewriting the radicand as an expression raised to the fifth power. We can rewrite x^{35} as $(x^7)^5$.

$$\sqrt[5]{x^{35}} = \sqrt[5]{(x^7)^5} \qquad \text{Rewrite radicand as an expression raised to the fifth power.}$$
$$= x^7 \qquad \text{Simplify.}$$

> **Quick Check 6**
> Simplify $\sqrt[8]{x^{56}}$. (Assume that x is nonnegative.)

EXAMPLE ▶12   Simplify $\sqrt[3]{8a^{12}b^{42}}$.

Solution

We begin by rewriting the radicand as a cube. We can rewrite $8a^{12}b^{42}$ as $(2a^4b^{14})^3$.

$$\sqrt[3]{8a^{12}b^{42}} = \sqrt[3]{(2a^4b^{14})^3} \qquad \text{Rewrite radicand as a cube.}$$
$$= 2a^4b^{14} \qquad \text{Simplify.}$$

> **Quick Check 7**
> Simplify $\sqrt[3]{-64x^{12}y^{30}}$. (Assume that x and y are nonnegative.)

Multiplying Radical Expressions

Objective 5   Multiply radical expressions. We know that $\sqrt{9} \cdot \sqrt{100} = 3 \cdot 10$ or 30. We also know that $\sqrt{9 \cdot 100} = \sqrt{900}$ or 30. Therefore $\sqrt{9} \cdot \sqrt{100} = \sqrt{9 \cdot 100}$. If two radical expressions with nonnegative radicands have the same index, then we can multiply the two expressions by multiplying the two radicands and writing the product inside the same radical.

Product Rule for Radicals

For any root n, if $\sqrt[n]{a}$ and $\sqrt[n]{b}$ are both real numbers, then $\sqrt[n]{a} \cdot \sqrt[n]{b} = \sqrt[n]{ab}$.

For example, the product of $\sqrt{2}$ and $\sqrt{8}$ is $\sqrt{16}$, which simplifies to equal 4.

EXAMPLE ▶13   Multiply: $\sqrt{12m} \cdot \sqrt{3m}$. (Assume that m is nonnegative.)

Solution

Quick Check 8
Multiply: $\sqrt{7b^7} \cdot \sqrt{28b^3}$. (Assume that b is nonnegative.)

Since both radicals are square roots, we can multiply the radicands.

$$\sqrt{12m} \cdot \sqrt{3m} = \sqrt{36m^2} \qquad \text{Multiply the radicands.}$$
$$\qquad\qquad\qquad = 6m \qquad\quad \text{Simplify the square root.}$$

EXAMPLE ▶14   Multiply: $8\sqrt{5} \cdot 4\sqrt{5}$.

Solution

Quick Check 9
Multiply: $9\sqrt{2} \cdot 11\sqrt{2}$.

We multiply the factors in front of the radicals together, and multiply the radicands together.

$$8\sqrt{5} \cdot 4\sqrt{5} = 32\sqrt{25} \qquad \text{Multiply the factors in front of the radicals } (8 \cdot 4).$$
$$\qquad\qquad\qquad\qquad \text{Multiply the radicands.}$$
$$\qquad\qquad\quad = 32 \cdot 5 \qquad \text{Simplify the square root.}$$
$$\qquad\qquad\quad = 160 \qquad\;\, \text{Multiply.}$$

Dividing Rational Expressions

Objective 6 **Divide radical expressions.** We can rewrite the quotient of two radical expressions that have the same index as the quotient of the two radicands inside the same radical.

Quotient Rule for Radicals

For any root n, if $\sqrt[n]{a}$ and $\sqrt[n]{b}$ are both real numbers, and $b \neq 0$, then $\dfrac{\sqrt[n]{a}}{\sqrt[n]{b}} = \sqrt[n]{\dfrac{a}{b}}$.

EXAMPLE ▶15   Simplify $\dfrac{\sqrt{175}}{\sqrt{7}}$.

Solution

Since both radicals are square roots, we begin by dividing the radicands. We write the quotient of the radicands under a single square root.

Quick Check 10
Simplify $\dfrac{\sqrt{468}}{\sqrt{13}}$.

$$\frac{\sqrt{175}}{\sqrt{7}} = \sqrt{\frac{175}{7}} \qquad \text{Rewrite as the square root of the quotient of the radicands.}$$
$$\qquad\quad = \sqrt{25} \qquad \text{Divide.}$$
$$\qquad\quad = 5 \qquad\;\; \text{Simplify the square root.}$$

Radical Functions

Objective **7** **Evaluate radical functions.** A **radical function** is a function that involves radicals, such as $f(x) = \sqrt{x - 4} + 3$.

EXAMPLE **16**   For the radical function $f(x) = \sqrt{x - 7} + 6$, find $f(32)$.

Solution

To evaluate this function, we substitute 32 for x and simplify.

$$
\begin{aligned}
f(32) &= \sqrt{32 - 7} + 6 && \text{Substitute 32 for } x. \\
&= \sqrt{25} + 6 && \text{Simplify radicand.} \\
&= 5 + 6 && \text{Take the square root of 25.} \\
&= 11 && \text{Add.}
\end{aligned}
$$

> **Quick Check** **11**
> For the radical function $f(x) = \sqrt{2x - 5} + 9$, find $f(27)$.

Finding the Domain of Radical Functions

Objective **8** **Find the domain of a radical function.** Radical functions involving even roots are different from the functions we have seen to this point, in that their domain is restricted. To find the domain of a radical function involving an even root, we need to find the values of the variable that make the radicand nonnegative. In other words, set the radicand greater than or equal to 0 and solve. Remember that square roots are considered to be even roots with an index of 2.

EXAMPLE **17**   Find the domain of the radical function $f(x) = \sqrt{x - 13} - 8$. Express your answer in interval notation.

Solution

As the radical has an even index, we begin by setting the radicand $(x - 13)$ greater than or equal to 0. We solve this inequality to find the domain.

$$
\begin{aligned}
x - 13 &\geq 0 && \text{Set the radicand greater than or equal to 0.} \\
x &\geq 13 && \text{Add 13 to both sides.}
\end{aligned}
$$

> **Quick Check** **12**
> Find the domain of the radical function $f(x) = \sqrt[4]{x + 3} - 9$. Express your answer in interval notation.

The domain of the function is $[13, \infty)$.

The domain of a radical function involving odd roots is the set of all real numbers. For example, the domain of $f(x) = \sqrt[5]{2x + 3} + 7$ is $(-\infty, \infty)$.

Building Your Study Strategy Note Taking, 1 **Choosing a Seat** One important yet frequently overlooked aspect of effective note taking is choosing the appropriate seat in the classroom. You must sit in a location where you can clearly see the entire board, preferably in the center of the classroom. Students who sit in the front rows on the left or right side of the classroom may struggle to see material on the opposite side of the room. This problem is compounded in classrooms equipped with a whiteboard, as these types of boards can create a glare. Finally, try not to sit behind anyone who will obstruct your vision, such as an extremely tall person.

Location also impacts your ability to hear your instructor clearly. You will find that your instructor speaks towards the middle of the classroom, so finding a seat toward the center of the classroom should help you to hear everything your instructor says. Try not to sit close to students who talk to each other during class; their discussions may distract you from what your instructor is saying.

In summary, try to choose your classroom seat in the same fashion that you choose a seat at a movie theater. Be sure that you can see and hear everything.

EXERCISES 8.1

Vocabulary

1. A number a is a(n) _____ of a number b if $a^2 = b$.

2. The _____ square root of b, denoted $\sqrt{b}$, for $b > 0$, is the positive number a such that $a^2 = b$.

3. For any positive integer $n > 1$ and any number b, if $a^n = b$ and a and b both have the same sign, then a is the principal _____ of b, denoted $a = \sqrt[n]{b}$.

4. For the expression $a = \sqrt[n]{b}$, n is called the _____ of the radical.

5. The expression contained inside a radical is called the _____.

6. A radical with an index of 3 is also known as a(n) _____ root.

7. A(n) _____ is a function that involves radicals.

8. The domain of a radical function involving an even root consists of values of the variable for which the radicand is _____.

Simplify the radical expression. Indicate if the expression is not a real number.

9. $\sqrt{16}$

10. $\sqrt{36}$

11. $\sqrt{9}$

12. $\sqrt{144}$

13. $\sqrt{\dfrac{1}{100}}$

14. $\sqrt{\dfrac{1}{81}}$

15. $\sqrt{\dfrac{49}{64}}$

16. $\sqrt{\dfrac{36}{121}}$

17. $-\sqrt{25}$

18. $-\sqrt{9}$

19. $\sqrt{-169}$

20. $\sqrt{-81}$

Simplify the radical expression. Where appropriate, include absolute values.

21. $\sqrt{a^{20}}$

22. $\sqrt{m^{18}}$

23. $\sqrt{x^{22}}$

24. $\sqrt{x^{24}}$

25. $\sqrt{9x^{10}}$

26. $\sqrt{49b^2}$

27. $\sqrt{\dfrac{9}{64}x^8}$

28. $\sqrt{\dfrac{1}{25}n^{16}}$

F O R

E X T R A

H E L P

MyMathLab

Math XP

Interactmath.com

MathXL
Tutorials on CD

Video Lectures
on CD

Tutor
Center

Addison-Wesley
Math Tutor Center

Student's
Solutions Manual

29. $\sqrt{x^{26}y^{14}}$

30. $\sqrt{a^{30}b^{18}}$

31. $\sqrt{a^{16}b^8c^{12}}$

32. $\sqrt{x^{24}y^2z^{22}}$

Find the missing number or expression. Assume that all variables represent nonnegative real numbers.

33. $\sqrt{?} = 7$

34. $\sqrt{?} = 25$

35. $\sqrt{?} = 5a$

36. $\sqrt{?} = 4x^6$

Approximate to the nearest thousandth, using a calculator.

37. $\sqrt{70}$

38. $\sqrt{300}$

39. $\sqrt{247}$

40. $\sqrt{1255}$

41. $\sqrt{0.18}$

42. $\sqrt{0.02}$

Simplify the radical expression. Assume that all variables represent nonnegative real numbers.

43. $\sqrt[3]{64}$

44. $\sqrt[3]{-8}$

45. $\sqrt[4]{81}$

46. $\sqrt[4]{625}$

47. $\sqrt[3]{x^{15}}$

48. $\sqrt[4]{n^{60}}$

49. $\sqrt[3]{-125x^{33}}$

50. $\sqrt[3]{343a^{24}}$

51. $\sqrt[4]{256m^{20}n^{24}}$

52. $\sqrt[3]{729s^{66}t^{42}}$

Simplify.

53. $\sqrt{8}\cdot\sqrt{2}$

54. $\sqrt{10}\cdot\sqrt{10}$

55. $\sqrt{8}\cdot\sqrt{50}$

56. $\sqrt{3}\cdot\sqrt{147}$

57. $\dfrac{\sqrt{252}}{\sqrt{7}}$

58. $\dfrac{\sqrt{200}}{\sqrt{8}}$

59. $\dfrac{\sqrt{2187}}{\sqrt{27}}$

60. $\dfrac{\sqrt{1815}}{\sqrt{15}}$

61. $\sqrt[3]{4}\cdot\sqrt[3]{16}$

62. $\sqrt[3]{4}\cdot\sqrt[3]{250}$

63. $\sqrt[4]{125}\cdot\sqrt[4]{405}$

64. $\sqrt[5]{81}\cdot\sqrt[5]{729}$

65. $\dfrac{\sqrt[3]{104}}{\sqrt[3]{13}}$

66. $\dfrac{\sqrt[3]{4374}}{\sqrt[3]{6}}$

67. $3\sqrt{7}\cdot2\sqrt{7}$

68. $9\sqrt{5}\cdot8\sqrt{5}$

69. $4\sqrt{63}\cdot\sqrt{7}$

70. $5\sqrt{169}\cdot4\sqrt{81}$

Find the missing number.

71. $\sqrt{8}\cdot\sqrt{?} = 24$

72. $\sqrt{15}\cdot\sqrt{?} = 30$

73. $\dfrac{\sqrt{?}}{\sqrt{6}} = 15$

74. $\dfrac{\sqrt{?}}{\sqrt{8}} = 7$

Evaluate the radical function. (Round to the nearest thousandth if necessary.)

75. $f(x) = \sqrt{x+8} - 4$, find $f(41)$

76. $f(x) = \sqrt{2x-5} + 14$, find $f(63)$

77. $f(x) = \sqrt{4x-3} - 6$, find $f(10)$

78. $f(x) = \sqrt{x+37} - 11$, find $f(-8)$

79. $f(x) = \sqrt{x^2-5x+57}$, find $f(8)$

80. $f(x) = \sqrt{x^2+10x+280}$, find $f(-6)$

Find the domain of the radical function. Express your answer in interval notation.

81. $f(x) = \sqrt{x-5}$

82. $f(x) = \sqrt{3x+12} - 6$

83. $f(x) = \sqrt{4x - 13} + 4$

84. $f(x) = \sqrt{8x - 20} - 7$

85. $f(x) = \sqrt[4]{3x} + 10$

86. $f(x) = \sqrt[6]{6x + 16}$

87. $f(x) = \sqrt[3]{x - 9} + 5$

88. $f(x) = \sqrt[5]{3x + 4} - 300$

Writing in Mathematics

Answer in complete sentences.

89. Which of the following are real numbers, and which are not real numbers: $-\sqrt{64}$, $\sqrt{-64}$, $-\sqrt[3]{64}$, $\sqrt[3]{-64}$? Explain your reasoning.

90. If we do not know whether x is a nonnegative number, explain why $\sqrt{x^2} = |x|$ rather than x.

91. Explain how to find the domain of a radical function. Use examples.

92. True or False: For any real number x, $\sqrt{x} \geq \sqrt[3]{x}$. Explain your reasoning.

93. *Solutions Manual** Write a solutions manual page for the following problem:
Simplify $\sqrt[3]{125x^9y^6}$.

94. *Newsletter** Write a newsletter that explains how to simplify the principal square root of a variable expression.

*See Appendix B for details and sample answers.

1   Simplify radical expressions using the product property.
2   Add or subtract radical expressions containing like radicals.
3   Simplify radical expressions before adding or subtracting.

Simplifying Radical Expressions Using the Product Property

Objective   1   Simplify radical expressions using the product property. A radical expression is considered simplified if the radicand contains no factors with exponents greater than or equal to the index of the radical. For example, $\sqrt[3]{x^5}$ is not simplified because there is an exponent inside the radical that is greater than the index of the radical. The goal for simplifying radical expressions is to remove as many factors as possible from the radicand. We will use the product property for radical expressions to help us with this.

We could rewrite $\sqrt[3]{x^5}$ as $\sqrt[3]{x^3 \cdot x^2}$, and then rewrite this radical expression as the product of two radicals. Using the product property for radicals, $\sqrt[n]{a} \cdot \sqrt[n]{b} = \sqrt[n]{ab}$, we know that $\sqrt[3]{x^3 \cdot x^2} = \sqrt[3]{x^3} \cdot \sqrt[3]{x^2}$. The reason for rewriting $\sqrt[3]{x^5}$ as $\sqrt[3]{x^3} \cdot \sqrt[3]{x^2}$ is that $\sqrt[3]{x^3}$ equals x.

$$\sqrt[3]{x^5} = \sqrt[3]{x^3 \cdot x^2}$$
$$= \sqrt[3]{x^3} \cdot \sqrt[3]{x^2}$$
$$= x \cdot \sqrt[3]{x^2}$$

Now the radicand contains no factors with an exponent that is greater than or equal to the index 3 and is simplified.

Simplifying Radical Expressions

- Completely factor any numerical factors in the radicand.
- Rewrite each factor as a product of two factors. The exponent for the first factor should be the largest multiple of the radical's index that is less than or equal to the factor's original exponent.
- Use the product property to remove factors from the radicand.

EXAMPLE   1   Simplify $\sqrt[4]{x^{39}}$. (Assume that x is nonnegative.)

Solution

We begin by rewriting x^{39} as a product of two factors. The largest multiple of the index (4) that is less than or equal to the exponent for this factor (39) is 36, so we will rewrite x^{39} as $x^{36} \cdot x^3$.

Quick Check   1

Simplify $\sqrt[5]{x^{23}}$. (Assume that x is nonnegative.)

$$\sqrt[4]{x^{39}} = \sqrt[4]{x^{36} \cdot x^3}$$ Rewrite x^{39} as the product of two factors.

$$= \sqrt[4]{x^{36}} \cdot \sqrt[4]{x^3}$$ Use the product property of radicals to rewrite the radical as the product of two radicals.

$$= x^9 \sqrt[4]{x^3}$$ Simplify the radical.

EXAMPLE 2 Simplify $\sqrt{a^6b^9c^{15}}$. (Assume that all variables represent nonnegative real numbers.)

Solution

Again, we begin by rewriting factors as a product of two factors. In this example, the exponent of the factor a is a multiple of the index 2. We do not need to rewrite this factor as the product of two factors.

$$\sqrt{a^6b^9c^{15}} = \sqrt{a^6(b^8 \cdot b)(c^{14} \cdot c)} \qquad \text{Rewrite factors.}$$
$$= \sqrt{a^6b^8c^{14}} \cdot \sqrt{bc} \qquad \text{Rewrite as the product of two radicals.}$$
$$= a^3b^4c^7\sqrt{bc} \qquad \text{Simplify the radical.}$$

Quick Check 2

Simplify $\sqrt{x^5yz^{10}w^{13}}$. (Assume that all variables represent nonnegative real numbers.)

EXAMPLE 3 Simplify $\sqrt{54}$.

Solution

We begin by rewriting 54 using its prime factorization $(2 \cdot 3^3)$.

$$\sqrt{54} = \sqrt{2 \cdot 3^3} \qquad \text{Factor 54.}$$
$$= \sqrt{2 \cdot (3^2 \cdot 3)} \qquad \text{Rewrite } 3^3 \text{ as } 3^2 \cdot 3.$$
$$= \sqrt{3^2} \cdot \sqrt{2 \cdot 3} \qquad \text{Rewrite as the product of two radicals.}$$
$$= 3\sqrt{6} \qquad \text{Simplify.}$$

We could have used a different tactic to simplify this square root. The largest factor of 54 that is a perfect square is 9, so we could begin by rewriting 54 as $9 \cdot 6$. Since we know that the square root of 9 is 3, we can factor 9 out of the radicand and write it as 3 in front of the radical.

Quick Check 3

Simplify $\sqrt{75}$.

$$\sqrt{54} = \sqrt{9 \cdot 6}$$
$$= 3\sqrt{6}$$

EXAMPLE 4 Simplify $\sqrt[4]{288}$.

Solution

We begin by factoring 288 to be $2^5 \cdot 3^2$.

$$\sqrt[4]{288} = \sqrt[4]{2^5 \cdot 3^2} \qquad \text{Factor 288.}$$
$$= \sqrt[4]{(2^4 \cdot 2) \cdot 3^2} \qquad \text{Rewrite } 2^5 \text{ as } 2^4 \cdot 2.$$
$$= \sqrt[4]{2^4} \cdot \sqrt[4]{2 \cdot 3^2} \qquad \text{Rewrite as the product of two radicals.}$$
$$= 2\sqrt[4]{18} \qquad \text{Simplify.}$$

Quick Check 4

Simplify $\sqrt[3]{594}$.

There is an alternate approach for simplifying radical expressions. Suppose we were trying to simplify $\sqrt[5]{a^{48}}$. Using the previous method, we would arrive at the answer $a^9\sqrt[5]{a^3}$.

$$\sqrt[5]{a^{48}} = \sqrt[5]{a^{45} \cdot a^3} \qquad \text{Rewrite } a^{48} \text{ as } a^{45} \cdot a^3, \text{ since 45 is the highest multiple of 5 that is less than or equal to 48.}$$
$$= \sqrt[5]{a^{45}} \cdot \sqrt[5]{a^3} \qquad \text{Rewrite as the product of two radicals.}$$
$$= a^9\sqrt[5]{a^3} \qquad \text{Simplify.}$$

We know that for every 5 times that a is repeated as a factor in the radicand, we can take a^5 out of the radicand and write it as a in front of the radical. We need to determine how many groups of 5 can be removed from the radicand, using division. Notice that if we divide the exponent 48 by the index 5, the quotient is 9 with a remainder of 3. When we divide the exponent of a factor in the radicand by the index of the radical, the quotient tells us the exponent of the factor removed from the radicand and the remainder tells us the exponent of the factor remaining in the radicand.

An Alternate Approach for Simplifying $\sqrt[n]{x^p}$

- Divide p by n: $\frac{p}{n} = q + \frac{r}{n}$
- The quotient q tells us how many times x will be a factor in front of the radical.
- The remainder r tells us how many times x will remain as a factor in the radicand.

$$\sqrt[n]{x^p} = x^q \sqrt[n]{x^r}$$

EXAMPLE 5 Simplify $\sqrt[5]{a^{22}b^{31}c^{15}d^3}$. Assume that all variables represent nonnegative real numbers.

Solution

We will work with one factor at a time, beginning with a. The index, 5, divides into the exponent, 22, four times with a remainder of two. This tells us we can write a^4 as a factor in front of the radical and can write a^2 in the radicand.

$$\sqrt[5]{a^{22}b^{31}c^{15}d^3} = a^4 \sqrt[5]{a^2 b^{31}c^{15}d^3}$$

For the factor b, $31 \div 5 = 6$ with a remainder of 1. We will write b^6 as a factor in front of the radical, and b^1 or b as a factor in the radicand.

For the factor c, $15 \div 5 = 3$ with a remainder of 0. We will write c^3 as a factor in front of the radical, and since the remainder is 0, we will not write c as a factor in the radicand. Finally, for the factor d, the index does not divide into 3, so d^3 remains as a factor in the radicand.

$$\sqrt[5]{a^{22}b^{31}c^{15}d^3} = a^4 b^6 c^3 \sqrt[5]{a^2 b d^3}$$

Quick Check 5
Simplify $\sqrt[6]{x^{43}y^{17}z^{12}}$. (Assume that all variables are nonnegative real numbers.)

Adding and Subtracting Radical Expressions Containing Like Radicals

Objective 2 Add or subtract radical expressions containing like radicals.

Like Radicals

Two radical expressions are called **like radicals** if they have the same index and the same radicand.

The radical expressions $5\sqrt[3]{4x}$ and $9\sqrt[3]{4x}$ are like radicals because they have the same index (3) and the same radicand $(4x)$. Here are some examples of radical expressions that are not like radicals:

$$\sqrt{5} \text{ and } \sqrt[3]{5} \qquad \text{The two radicals have different indices.}$$

$$\sqrt[4]{7x^2y^3} \text{ and } \sqrt[4]{7x^3y^2} \qquad \text{The two radicands are different.}$$

We can add and subtract radical expressions by combining like radicals similar to the way we combine like terms. We add or subtract the coefficients in front of the like radicals.

$$6\sqrt{2} + 3\sqrt{2} = 9\sqrt{2}$$

EXAMPLE ▶ 6   Simplify $7\sqrt{x} - 10\sqrt{x}$.

Solution

Notice that the radicals are like radicals. We can combine these two expressions by subtracting the coefficients.

$$7\sqrt{x} - 10\sqrt{x} = -3\sqrt{x} \qquad \text{Combine the like radicals by subtracting the}$$
$$\text{coefficients and keeping the radical.}$$

> **Quick Check ◀ 6**
> Simplify
> $16\sqrt{2xy} + 11\sqrt{2xy}$.

EXAMPLE ▶ 7   Simplify $5\sqrt{7} + 4\sqrt{10} + \sqrt{10} - 8\sqrt{7}$.

Solution

There are two pairs of like radicals in this example. There are two radical expressions containing $\sqrt{7}$ and two radical expressions containing $\sqrt{10}$.

$$5\sqrt{7} + 4\sqrt{10} + \sqrt{10} - 8\sqrt{7} = -3\sqrt{7} + 4\sqrt{10} + \sqrt{10} \qquad \text{Subtract } 5\sqrt{7} - 8\sqrt{7}.$$
$$= -3\sqrt{7} + 5\sqrt{10} \qquad \text{Add } 4\sqrt{10} + \sqrt{10}.$$

> **Quick Check ◀ 7**
> Simplify $6\sqrt{3} - 11\sqrt{5}$
> $- 8\sqrt{5} - 16\sqrt{3}$.

Objective 3   Simplify radical expressions before adding or subtracting. Are the expressions $\sqrt{24}$ and $\sqrt{54}$ like radicals? We must simplify each radical completely before we can determine whether the two expressions are like radicals. In this case, $\sqrt{24} = 2\sqrt{6}$ and $\sqrt{54} = 3\sqrt{6}$, so $\sqrt{24}$ and $\sqrt{54}$ are like radicals.

EXAMPLE ▶ 8   Simplify $\sqrt{45} + 7\sqrt{5}$.

Solution

We begin by simplifying each radical completely. Since 45 can be written as $9 \cdot 5$, $\sqrt{45}$ can be simplified to be $3\sqrt{5}$. We could also use the prime factorization of 45 ($3^2 \cdot 5$) to simplify $\sqrt{45}$.

> **Quick Check ◀ 8**
> Simplify $\sqrt{28} + \sqrt{7}$.

$$\sqrt{45} + 7\sqrt{5} = \sqrt{9 \cdot 5} + 7\sqrt{5} \qquad \text{Factor 45.}$$
$$= 3\sqrt{5} + 7\sqrt{5} \qquad \text{Simplify } \sqrt{9 \cdot 5}.$$
$$= 10\sqrt{5} \qquad \text{Add.}$$

EXAMPLE 9   Simplify $\sqrt{32} - \sqrt{50} + \sqrt{162}$.

Solution

In this example we must simplify all three radicals before proceeding.

Quick Check 9
Simplify
$\sqrt{12} + \sqrt{27} - \sqrt{147}$.

$$\sqrt{32} - \sqrt{50} + \sqrt{162} = \sqrt{16 \cdot 2} - \sqrt{25 \cdot 2} + \sqrt{81 \cdot 2}$$ Factor each radicand.
$$= 4\sqrt{2} - 5\sqrt{2} + 9\sqrt{2}$$ Simplify each radical.
$$= 8\sqrt{2}$$ Combine like radicals.

Building Your Study Strategy **Note Taking, 2** **Note-Taking Speed**   "I'm a slow writer. There's no way I can copy down all of the notes in time." This is a common concern for students. Try to focus on writing down enough information so that you will understand the material later, rather than feeling you must copy every single word your instructor has written or spoken. Then rewrite your notes, filling in any blanks or adding explanations, as soon as possible after class.

Your handwriting should be legible, but don't feel that it must be as neat as possible. Try to use abbreviations whenever possible. Write your notes using phrases rather than complete sentences. These ideas should help you improve your speed.

With your instructor's approval, you may want to try using a tape recorder to record a lecture. After class you can use the tape to help fill in holes in your notes.

If you are falling behind in your note taking, leave space in your notes. You can always borrow a classmate's notes to get the information you missed.

EXERCISES *8.2*

Vocabulary

1. Two radical expressions are called _____ if they have the same index and the same radicand.

2. To add radical expressions with like radicals, add the _____ of the radicals and place the sum in front of the like radical.

Simplify.

3. $\sqrt{75}$

4. $\sqrt{48}$

5. $\sqrt{98}$

6. $\sqrt{275}$

7. $\sqrt{720}$

8. $\sqrt{648}$

9. $\sqrt[3]{54}$

10. $\sqrt[3]{56}$

11. $\sqrt[3]{384}$

12. $\sqrt[3]{1080}$

13. $\sqrt[4]{1701}$

14. $\sqrt[5]{800}$

Simplify the radical expression. Assume that all variables represent nonnegative real numbers.

15. $\sqrt{x^7}$

16. $\sqrt{b^{11}}$

17. $\sqrt[3]{a^{23}}$

18. $\sqrt[3]{x^{70}}$

19. $\sqrt[4]{m^{34}}$

20. $\sqrt[5]{n^{73}}$

21. $\sqrt{x^{13}y^{20}}$

22. $\sqrt{x^{19}y^{11}}$

23. $\sqrt{ab^{10}}$

24. $\sqrt{a^{16}b^{13}}$

25. $\sqrt{x^{17}y^6z^{31}}$

26. $\sqrt{x^{16}y^9z^{12}}$

27. $\sqrt[3]{a^{15}b^{10}c^5}$

28. $\sqrt[3]{x^{37}y^{24}z^{11}}$

29. $\sqrt[5]{x^{23}y^{32}z^{55}}$

30. $\sqrt[6]{a^{54}b^{36}c^{13}}$

31. $\sqrt{18x^7y^6}$

32. $\sqrt{75x^{10}y^{13}}$

33. $\sqrt[3]{48a^{17}b^{10}}$

34. $\sqrt[4]{405m^{24}n^{19}}$

48. $14\sqrt[3]{x} - 2\sqrt[3]{x} + 17\sqrt[3]{x}$

49. $5\sqrt{a} + 9\sqrt{b} - 2\sqrt{b} - 11\sqrt{a}$

50. $18\sqrt[3]{x} + 7\sqrt[3]{y} - 12\sqrt[3]{x} - 23\sqrt[3]{y}$

51. $\sqrt{12} + 15\sqrt{3}$

52. $9\sqrt{2} + \sqrt{32}$

53. $5\sqrt{8} - 13\sqrt{18} + \sqrt{98}$

54. $7\sqrt{7} + \sqrt{63} - 10\sqrt{175}$

55. $3\sqrt[3]{16} + 4\sqrt[3]{54} + 11\sqrt[3]{2}$

56. $8\sqrt[3]{24} - 19\sqrt[3]{3} + \sqrt[3]{192}$

57. $3\sqrt{50} - 9\sqrt{24} - 11\sqrt{18} + 6\sqrt{150}$

58. $4\sqrt{108} + 13\sqrt{5} + 17\sqrt{45} - 3\sqrt{12}$

Add or subtract. Assume that all variables represent nonnegative real numbers.

35. $9\sqrt{2} + 8\sqrt{2}$

36. $6\sqrt{5} - 13\sqrt{5}$

37. $-3\sqrt[3]{6} + 17\sqrt[3]{6}$

38. $10\sqrt[4]{15} + 11\sqrt[4]{15}$

39. $8\sqrt{10} - 12\sqrt{10} + 9\sqrt{10}$

40. $-19\sqrt{7} + 4\sqrt{7} - \sqrt{7}$

41. $7\sqrt{14} - 9\sqrt{21} - 6\sqrt{21} - \sqrt{14}$

42. $3\sqrt{13} + 10\sqrt{11} - 8\sqrt{13} + 12\sqrt{11}$

43. $\left(7\sqrt{5} + 3\sqrt{2}\right) - \left(2\sqrt{5} - 8\sqrt{2}\right)$

44. $\left(10\sqrt{2} - 7\sqrt{6}\right) - \left(16\sqrt{2} + 4\sqrt{6}\right)$

45. $10\sqrt{a} + 8\sqrt{a}$

46. $15\sqrt{y} - 33\sqrt{y}$

47. $\sqrt[5]{x} - 12\sqrt[5]{x}$

Find the missing radical expression.

59. $\left(2\sqrt{5} + 9\sqrt{2}\right) + (?) = 10\sqrt{5} + 13\sqrt{2}$

60. $\left(7\sqrt{3} - 2\sqrt{11}\right) - (?) = 4\sqrt{3} - 13\sqrt{11}$

61. $\left(\sqrt{75} - 2\sqrt{18}\right) - (?) = 9\sqrt{3} + 7\sqrt{2}$

62. $\left(3\sqrt{40} + 5\sqrt{180}\right) + (?) = 2\sqrt{10} + 38\sqrt{5}$

63. List four expressions involving a sum or difference of radicals and that are equivalent to $6\sqrt{3}$.

64. List four expressions involving a sum or difference of radicals and that are equivalent to $-9\sqrt{2}$.

65. List four expressions involving a sum or difference of radicals and that are equivalent to $11\sqrt{5} - 2\sqrt{3}$. At least one of your terms must contain $\sqrt{45}$.

66. List four expressions involving a sum or difference of radicals and that are equivalent to $-13\sqrt{2} - 7\sqrt{3}$. At least one of your terms must contain $\sqrt{18}$ and another term must contain $\sqrt{75}$.

Writing in Mathematics

Answer in complete sentences.

67. Explain how to simplify $\sqrt{360}$.

68. Explain how to determine whether radical expressions are like radicals.

69. *Solutions Manual** Write a solutions manual page for the following problem:

Find $9\sqrt{12} + 6\sqrt{75}$.

70. *Newsletter** Write a newsletter that explains how to simplify a radical of the form $\sqrt[n]{x^p}$.

*See Appendix B for details and sample answers.

8.3

Multiplying and Dividing Radical Expressions

Objectives

1   Multiply radical expressions.
2   Use the distributive property to multiply radical expressions.
3   Multiply radical expressions that have two or more terms.
4   Multiply radical expressions that are conjugates.
5   Rationalize a denominator that has one term.
6   Rationalize a denominator that has two terms.

Multiplying Two Radical Expressions

Objective 1   Multiply radical expressions.   In Section 8.1 we learned how to multiply one radical by another, as well as how to divide one radical by another.

Multiplying and Dividing Radicals with the Same Index

For any nonnegative expressions x and y and any positive integer $n > 1$,

$$\sqrt[n]{x} \cdot \sqrt[n]{y} = \sqrt[n]{xy} \text{ and } \frac{\sqrt[n]{x}}{\sqrt[n]{y}} = \sqrt[n]{\frac{x}{y}}.$$

In this section we will build on that knowledge and learn how to multiply and divide expressions containing two or more radicals. We begin with a review of multiplying radicals.

EXAMPLE 1   Multiply: $\sqrt{6} \cdot \sqrt{54}$.

Solution

Since the index is the same for each radical, and neither radicand is a perfect square, we begin by multiplying the two radicands.

$$\begin{aligned} \sqrt{6} \cdot \sqrt{54} &= \sqrt{324} && \text{Multiply the radicands.} \\ &= 18 && \text{Simplify the radical.} \end{aligned}$$

EXAMPLE 2   Multiply: $\sqrt[4]{x^{19}y^9} \cdot \sqrt[4]{x^{16}y^{11}}$. (Assume all variables are nonnegative.)

Solution

Since there are powers in each radical greater than the index of that radical, we could simplify each radical first. However, we would then have to multiply and simplify the radical again. A more efficient approach is to multiply first and then simplify only once.

$$\begin{aligned} \sqrt[4]{x^{19}y^9} \cdot \sqrt[4]{x^{16}y^{11}} &= \sqrt[4]{x^{35}y^{20}} && \text{Multiply the radicands by adding the} \\ & && \text{exponents for each factor.} \\ &= x^8 y^5 \sqrt[4]{x^3} && \text{Simplify the radical.} \end{aligned}$$

EXAMPLE 3   Multiply: $7\sqrt{15} \cdot 8\sqrt{6}$.

Solution

Since the index is the same for each radical, we can multiply the radicands together. We will multiply the factors in front of each radical, 7 and 8, by each other as well. After multiplying, we finish by simplifying the radical completely.

$$7\sqrt{15} \cdot 8\sqrt{6} = 56\sqrt{90}$$ Multiply factors in front of the radicals and multiply the radicands.

$$= 56\sqrt{2 \cdot 3^2 \cdot 5}$$ Factor the radicand.

$$= 56 \cdot 3\sqrt{2 \cdot 5}$$ Simplify the radical.

$$= 168\sqrt{10}$$ Multiply.

Quick Check 1 **Multiply. (Assume all variables are nonnegative.)**

a) $\sqrt{20} \cdot \sqrt{5}$ b) $\sqrt[3]{a^7 b^8 c^{11}} \cdot \sqrt[3]{a^2 b^8 c^5}$ c) $4\sqrt{3} \cdot 6\sqrt{42}$

It is important to note that whenever we multiply the square root of an expression by the square root of the same expression, the product is equal to the expression itself as long as the expression is nonnegative.

Multiplying a Square Root by Itself

For any nonnegative x, $\sqrt{x} \cdot \sqrt{x} = x$.

Using the Distributive Property with Radical Expressions

Objective 2 Use the distributive property to multiply radical expressions.

Now we will use the distributive property to multiply radical expressions.

EXAMPLE 4   Multiply: $\sqrt{2}(\sqrt{22} - \sqrt{2})$.

Solution

We begin by distributing $\sqrt{2}$ to each term in the parentheses, and then we multiply the radicals as in the previous examples.

$$\sqrt{2}(\sqrt{22} - \sqrt{2}) = \sqrt{2} \cdot \sqrt{22} - \sqrt{2} \cdot \sqrt{2}$$ Distribute $\sqrt{2}$.

$$= \sqrt{44} - 2$$ Multiply. Recall that $\sqrt{2} \cdot \sqrt{2} = 2$.

$$= 2\sqrt{11} - 2$$ Simplify the radical.

Quick Check 2
Multiply:
$\sqrt{18}(\sqrt{40} + \sqrt{6})$.

EXAMPLE ▶5 Multiply: $\sqrt[3]{x^5y^4}\left(\sqrt[3]{x^8y^5} + \sqrt[3]{x^{14}y^7}\right)$. Assume that x and y are nonnegative.

Solution

We begin by using the distributive property. Since each radical has an index of 3, we can then multiply the radicals.

$$\sqrt[3]{x^5y^4}\left(\sqrt[3]{x^8y^5} + \sqrt[3]{x^{14}y^7}\right) = \sqrt[3]{x^5y^4} \cdot \sqrt[3]{x^8y^5} + \sqrt[3]{x^5y^4} \cdot \sqrt[3]{x^{14}y^7} \qquad \text{Distribute } \sqrt[3]{x^5y^4}.$$

$$= \sqrt[3]{x^{13}y^9} + \sqrt[3]{x^{19}y^{11}} \qquad \text{Multiply by adding exponents for each factor.}$$

$$= x^4y^3 \sqrt[3]{x} + x^6y^3 \sqrt[3]{xy^2} \qquad \text{Simplify each radical.}$$

Since the radicals are not like radicals, we cannot simplify this expression any further.

Quick Check 3 Multiply: $\sqrt[3]{x^7y^{11}}\left(\sqrt[3]{x^8y^8} - \sqrt[3]{x^{16}y^{13}}\right)$. Assume that x and y are nonnegative.

Multiplying Radical Expressions That Have at Least Two Terms

Objective 3 Multiply radical expressions that have two or more terms.

EXAMPLE ▶6 Multiply: $\left(\sqrt{2} + 5\sqrt{3}\right)\left(6\sqrt{8} - 7\sqrt{12}\right)$.

Solution

We begin by multiplying each term in the first set of parentheses by each term in the second set of parentheses, using the distributive property. Since there are two terms in each set of parentheses, we can use the FOIL technique. Multiply factors outside a radical by factors outside a radical, and multiply radicands by radicands.

$$\left(\sqrt{2} + 5\sqrt{3}\right)\left(6\sqrt{8} - 7\sqrt{12}\right) = 6\sqrt{2 \cdot 8} - 7\sqrt{2 \cdot 12} + 5 \cdot 6\sqrt{3 \cdot 8} - 5 \cdot 7\sqrt{3 \cdot 12}$$

Distribute.

$$= 6\sqrt{16} - 7\sqrt{24} + 30\sqrt{24} - 35\sqrt{36} \qquad \text{Multiply.}$$

$$= 6 \cdot 4 - 7 \cdot 2\sqrt{6} + 30 \cdot 2\sqrt{6} - 35 \cdot 6 \qquad \text{Simplify each radical.}$$

$$= 24 - 14\sqrt{6} + 60\sqrt{6} - 210 \qquad \text{Multiply.}$$

$$= -186 + 46\sqrt{6} \qquad \text{Combine like terms and like radicals.}$$

EXAMPLE 7   Multiply: $\left(\sqrt{13} - \sqrt{10}\right)^2$.

Solution

To square any binomial, we multiply it by itself.

$$\left(\sqrt{13} - \sqrt{10}\right)^2 = \left(\sqrt{13} - \sqrt{10}\right)\left(\sqrt{13} - \sqrt{10}\right)$$ Square the binomial $\sqrt{13} - \sqrt{10}$ by multiplying it by itself.

$$= \sqrt{13} \cdot \sqrt{13} - \sqrt{13} \cdot \sqrt{10} - \sqrt{10} \cdot \sqrt{13} + \sqrt{10} \cdot \sqrt{10}$$ Distribute.

$$= 13 - \sqrt{130} - \sqrt{130} + 10$$ Multiply.

$$= 23 - 2\sqrt{130}$$ Combine like terms.

Quick Check **4** **Multiply.**

   a) $\left(8\sqrt{5} + \sqrt{2}\right)\left(3\sqrt{5} + 4\sqrt{2}\right)$ b) $\left(\sqrt{6} + \sqrt{7}\right)^2$

A Word of Caution   Whenever we square a binomial, such as $\left(\sqrt{13} - \sqrt{10}\right)^2$, we must multiply the binomial by itself. We cannot simply square each term.

$$(a + b)^2 \neq a^2 + b^2$$

Multiplying Conjugates

Objective 4 **Multiply radical expressions that are conjugates.** The expressions $\sqrt{13} + \sqrt{5}$ and $\sqrt{13} - \sqrt{5}$ are called **conjugates.** Two expressions are conjugates if they are of the form $x + y$ and $x - y$. Notice that the two terms are the same, with the exception of the sign of the second term.

The multiplication of two conjugates follows a pattern. Let's look at the product $\left(\sqrt{x} + \sqrt{y}\right)\left(\sqrt{x} - \sqrt{y}\right)$.

$$\left(\sqrt{x} + \sqrt{y}\right)\left(\sqrt{x} - \sqrt{y}\right) = x - \sqrt{xy} + \sqrt{xy} - y$$ Distribute. Note that $\sqrt{x} \cdot \sqrt{x} = x$ and $\sqrt{y} \cdot \sqrt{y} = y$.

$$= x - y$$ Combine the two opposite terms $-\sqrt{xy}$ and $\sqrt{xy}$.

Whenever we multiply conjugates, the two middle terms will be opposites of each other and therefore their sum is 0. We can multiply the first term in the first set of parentheses by the first term in the second set of parentheses, then multiply the second term in the first set of parentheses by the second term in the second set of parentheses, and place a minus sign between them.

Multiplication of Two Conjugates

$$(a + b)(a - b) = a^2 - b^2$$

EXAMPLE ▸8   Multiply: $(3\sqrt{2} + 8\sqrt{3})(3\sqrt{2} - 8\sqrt{3})$.

Solution

These two expressions are conjugates so we multiply them accordingly.

$(3\sqrt{2} + 8\sqrt{3})(3\sqrt{2} - 8\sqrt{3}) = 3\sqrt{2} \cdot 3\sqrt{2} - 8\sqrt{3} \cdot 8\sqrt{3}$ Multiply first term by first term and second term by second term.

$$= 9 \cdot 2 - 64 \cdot 3$$ Multiply.
$$= 18 - 192$$ Multiply.
$$= -174$$ Subtract.

Quick Check **5** Multiply: $(8\sqrt{11} - 5\sqrt{7})(8\sqrt{11} + 5\sqrt{7})$.

Rationalizing the Denominator

Objective 5 **Rationalize a denominator that has one term.** Earlier in this chapter we introduced the criterion for determining whether a radical was simplified. We stated that for a radical to be simplified, its index must be greater than any exponent within the radical. There are two other rules that we now add.

- There can be no fractions in a radicand.
- There can be no radicals in the denominator of a fraction.

For example, we would not consider the following expressions simplified: $\sqrt{\frac{3}{10}}$, $\frac{9}{\sqrt{2}}$, $\frac{6}{\sqrt{4} - \sqrt{3}}$, and $\frac{\sqrt{6} + \sqrt{12}}{\sqrt{3} - 8}$. The process of rewriting an expression without a radical in its denominator is called **rationalizing the denominator**.

The rational expression $\frac{\sqrt{16}}{\sqrt{49}}$ is not simplified, as there is a radical in the denominator. However, we know that $\sqrt{49} = 7$, so we can simplify the denominator in such a way that it no longer contains a radical.

$$\frac{\sqrt{16}}{\sqrt{49}} = \frac{4}{7}$$ Simplify the numerator and denominator.

The radical expression $\sqrt{\dfrac{75}{3}}$ is not simplified, as there is a fraction inside the radical.

We can simplify $\dfrac{75}{3}$ to be 25, rewriting the radical without a fraction inside.

$$\sqrt{\dfrac{75}{3}} = \sqrt{25} = 5 \qquad \text{Simplify the fraction first, then } \sqrt{25}.$$

Suppose that we needed to simplify $\dfrac{\sqrt{13}}{\sqrt{3}}$. We cannot simplify $\sqrt{3}$, and the fraction itself cannot be simplified either. In such a case, we will multiply both the numerator and denominator by an expression that will allow us to rewrite the denominator without a radical. Then we simplify.

EXAMPLE 9   Rationalize the denominator: $\dfrac{\sqrt{13}}{\sqrt{3}}$.

Solution

If we multiply the denominator by $\sqrt{3}$, the denominator would equal 3 and the denominator will be rationalized.

$$\dfrac{\sqrt{13}}{\sqrt{3}} = \dfrac{\sqrt{13}}{\sqrt{3}} \cdot \dfrac{\sqrt{3}}{\sqrt{3}} \qquad \text{Multiply by } \dfrac{\sqrt{3}}{\sqrt{3}}, \text{ which makes the denominator equal}$$

Quick Check 6

Rationalize the denominator: $\dfrac{\sqrt{21}}{\sqrt{2}}$.

to 3. Multiplying by $\dfrac{\sqrt{3}}{\sqrt{3}}$ is equivalent to multiplying by 1.

$$= \dfrac{\sqrt{39}}{3} \qquad \text{Multiply.}$$

Since $\sqrt{39}$ cannot be simplified, this expression cannot be simplified further.

EXAMPLE 10   Rationalize the denominator: $\dfrac{8}{\sqrt{18}}$.

Solution

At first glance we might think that multiplying the numerator and denominator by $\sqrt{18}$ is the correct way to proceed. However, if we multiply the numerator and denominator by $\sqrt{2}$, the radicand in the denominator will be 36, which is a perfect square.

$$\dfrac{8}{\sqrt{18}} = \dfrac{8}{\sqrt{18}} \cdot \dfrac{\sqrt{2}}{\sqrt{2}} \qquad \text{Multiply by } \dfrac{\sqrt{2}}{\sqrt{2}} \text{ to make the radicand in the}$$

denominator a perfect square.

$$= \dfrac{8\sqrt{2}}{\sqrt{36}} \qquad \text{Multiply.}$$

$$= \dfrac{8\sqrt{2}}{6} \qquad \text{Simplify the radical in the denominator.}$$

$$= \frac{\overset{4}{\cancel{8}}\sqrt{2}}{\underset{3}{\cancel{6}}} \qquad \text{Divide out the common factor 2.}$$

$$= \frac{4\sqrt{2}}{3} \qquad \text{Simplify.}$$

Multiplying by $\dfrac{\sqrt{18}}{\sqrt{18}}$ would also be valid, but the subsequent process of simplifying would be difficult. One way to determine the best expression by which to multiply is to completely factor the radicand in the denominator. In this example, $18 = 2 \cdot 3^2$. The factor 3 is already a perfect square, but the factor 2 is not. Multiplying by $\sqrt{2}$ makes the factor 2 a perfect square as well.

Quick Check 7

Rationalize the denominator: $\dfrac{10}{\sqrt{45}}$.

EXAMPLE 11   Rationalize the denominator: $\sqrt{\dfrac{x^9 y^5 z^6}{x^{15} y^8 z^3}}$. (Assume that all variables represent nonnegative real numbers.)

Solution

Notice that the numerator and denominator have common factors. We will begin by simplifying the fraction to lowest terms.

$$\sqrt{\frac{x^9 y^5 z^6}{x^{15} y^8 z^3}} = \sqrt{\frac{z^3}{x^6 y^3}} \qquad \text{Divide out common factors and simplify.}$$

$$= \frac{\sqrt{z^3}}{\sqrt{x^6 y^3}} \qquad \text{Rewrite as the quotient of two square roots.}$$

$$\text{Notice that the factor } x^6 \text{ is already a perfect square, but } y^3 \text{ is not.}$$

$$= \frac{\sqrt{z^3}}{\sqrt{x^6 y^3}} \cdot \frac{\sqrt{y}}{\sqrt{y}} \qquad \text{Multiply by } \frac{\sqrt{y}}{\sqrt{y}}.$$

$$= \frac{\sqrt{yz^3}}{\sqrt{x^6 y^4}} \qquad \text{Multiply.}$$

$$= \frac{z\sqrt{yz}}{x^3 y^2} \qquad \text{Simplify both radicals.}$$

Quick Check 8

Rationalize the denominator:

$\sqrt{\dfrac{18a^{13}b^8}{27a^{18}b^{11}}}$. (Assume that all variables represent nonnegative real numbers.)

Rationalizing a Denominator That Has Two Terms

Objective 6   **Rationalize a denominator that has two terms.** In the previous examples, each denominator had only one term. If a denominator is a binomial that contains one or two square roots, then we rationalize the denominator by multiplying the numerator and denominator by the conjugate of the denominator. For example, consider the expression $\dfrac{9}{\sqrt{19} + \sqrt{13}}$. We know from earlier in the section that multiplying $\sqrt{19} + \sqrt{13}$ by its conjugate $\sqrt{19} - \sqrt{13}$ will produce a product that does not contain a radical.

EXAMPLE 12   Rationalize the denominator: $\dfrac{9}{\sqrt{19} + \sqrt{13}}$.

Solution

Since this denominator is a binomial, we will multiply the numerator and denominator by the conjugate of the denominator.

$$\dfrac{9}{\sqrt{19} + \sqrt{13}} = \dfrac{9}{\sqrt{19} + \sqrt{13}} \cdot \dfrac{\sqrt{19} - \sqrt{13}}{\sqrt{19} - \sqrt{13}}$$

Multiply the numerator and denominator by the conjugate of the denominator $(\sqrt{19} - \sqrt{13})$.

$$= \dfrac{9\sqrt{19} - 9\sqrt{13}}{\sqrt{19} \cdot \sqrt{19} - \sqrt{13} \cdot \sqrt{13}}$$

Multiply. Since $\sqrt{19} + \sqrt{13}$ and $\sqrt{19} - \sqrt{13}$ are conjugates, we only need to multiply $\sqrt{19}$ by $\sqrt{19}$ and $\sqrt{13}$ by $\sqrt{13}$.

$$= \dfrac{9\sqrt{19} - 9\sqrt{13}}{19 - 13}$$

Simplify the products in the denominator.

$$= \dfrac{9\sqrt{19} - 9\sqrt{13}}{6}$$

Subtract.

$$= \dfrac{9(\sqrt{19} - \sqrt{13})}{6}$$

Factor the numerator.

$$= \dfrac{\overset{3}{\cancel{9}}(\sqrt{19} - \sqrt{13})}{\underset{2}{\cancel{6}}}$$

Divide out the common factor 3.

$$= \dfrac{3(\sqrt{19} - \sqrt{13})}{2}$$

Simplify.

Quick Check **9**
Rationalize the denominator:
$\dfrac{\sqrt{42}}{\sqrt{7} - \sqrt{3}}$.

EXAMPLE 13   Rationalize the denominator: $\dfrac{6\sqrt{5} - 5\sqrt{2}}{3\sqrt{5} + \sqrt{2}}$.

Solution

Since the denominator is a binomial, we begin by multiplying the numerator and denominator by the conjugate of the denominator, which is $3\sqrt{5} - \sqrt{2}$.

$$\dfrac{6\sqrt{5} - 5\sqrt{2}}{3\sqrt{5} + \sqrt{2}} = \dfrac{6\sqrt{5} - 5\sqrt{2}}{3\sqrt{5} + \sqrt{2}} \cdot \dfrac{3\sqrt{5} - \sqrt{2}}{3\sqrt{5} - \sqrt{2}}$$

Multiply the numerator and denominator by the conjugate of the denominator.

$$= \dfrac{6\sqrt{5} \cdot 3\sqrt{5} - 6\sqrt{5} \cdot \sqrt{2} - 5\sqrt{2} \cdot 3\sqrt{5} + 5\sqrt{2} \cdot \sqrt{2}}{3\sqrt{5} \cdot 3\sqrt{5} - \sqrt{2} \cdot \sqrt{2}}$$

Multiply. We must multiply each term in the first numerator by each term in the second numerator. To multiply the denominators we can take advantage of the fact that the two denominators are conjugates.

$$= \frac{18 \cdot 5 - 6\sqrt{10} - 15\sqrt{10} + 5 \cdot 2}{9 \cdot 5 - 2}$$ Simplify each product.

$$= \frac{90 - 6\sqrt{10} - 15\sqrt{10} + 10}{45 - 2}$$ Multiply.

$$= \frac{100 - 21\sqrt{10}}{43}$$ Combine like terms and like radicals.

Quick Check 10
Rationalize the denominator:
$$\frac{3\sqrt{6} + 2\sqrt{3}}{2\sqrt{6} + \sqrt{3}}.$$

Building Your Study Strategy Note Taking, 3 **Things to Include**   What belongs in your notes? Begin by including anything that your instructor writes on the board. If your instructor solves several problems of the same type on the board, make a note of this fact as well; this should alert you that this type of problem is important. Instructors also give verbal cues when they feel that something should be taken down in your notes. Your instructor may pause to give you enough time to finish writing in your notebook or may repeat the same phrase to make sure that you accurately write down the statement in your notes.

Some instructors will warn the class that a particular topic, problem, or step is difficult. When you hear this, your instructor is telling you to take the best notes that you can because you will need them later.

Finally, keep an open ear for what may be the most important six-word phrase in a math class: "This will be on the test." When you hear this, your instructor is trying to make sure that you are prepared for the exam. Try writing the words "ON TEST," circled, next to the notes you take.

EXERCISES 8.3

Vocabulary

1. For any nonnegative x, $\sqrt{x} \cdot \sqrt{x} =$ _____.

2. Two expressions of the form $x + y$ and _____ are called conjugates.

3. The process of rewriting an expression without a radical in its denominator is called _____ the denominator.

4. To rationalize a denominator containing two terms and at least one square root, multiply the numerator and denominator by the _____ of the denominator.

Multiply. Assume that all variables represent nonnegative real numbers.

5. $\sqrt{36} \cdot \sqrt{25}$

6. $\sqrt{4} \cdot \sqrt{64}$

7. $\sqrt{5} \cdot \sqrt{20}$

8. $\sqrt{6} \cdot \sqrt{54}$

9. $\sqrt[3]{6} \cdot \sqrt[3]{36}$

10. $\sqrt[3]{20} \cdot \sqrt[3]{14}$

11. $4\sqrt{32} \cdot 5\sqrt{8}$

12. $5\sqrt{18} \cdot 6\sqrt{96}$

13. $\sqrt{x^{17}} \cdot \sqrt{x^5}$

14. $\sqrt{x^{15}} \cdot \sqrt{x}$

15. $\sqrt{m^{13}n^{12}} \cdot \sqrt{m^5 n^{11}}$

16. $\sqrt{s^7 t} \cdot \sqrt{s^3 t^{13}}$

17. $\sqrt[4]{x^7 y^{12}} \cdot \sqrt[4]{x^{15} y^{19}}$

18. $\sqrt[5]{x^{16} y^{22}} \cdot \sqrt[5]{x^{24} y^{34}}$

Multiply. Assume that all variables represent nonnegative real numbers.

19. $\sqrt{5}\left(\sqrt{15} + \sqrt{10}\right)$

20. $\sqrt{21}\left(\sqrt{7} - \sqrt{3}\right)$

21. $\sqrt{18}\left(\sqrt{75} + \sqrt{6}\right)$

22. $\sqrt{10}\left(\sqrt{15} - \sqrt{30}\right)$

23. $\sqrt{22}\left(7\sqrt{14} - 3\sqrt{6}\right)$

24. $9\sqrt{6}\left(4\sqrt{3} - 17\sqrt{2}\right)$

25. $\sqrt{m^3 n^5}\left(\sqrt{m^7 n^6} + \sqrt{m^8 n^{15}}\right)$

26. $\sqrt{xy^7}\left(\sqrt{x^9 y^3} - \sqrt{x^{25} y^{16}}\right)$

27. $\sqrt[3]{a^8 b^5 c^{17}}\left(\sqrt[3]{a^5 b^7 c^{10}} + \sqrt[3]{a^{16} b^{19} c^{11}}\right)$

28. $\sqrt[3]{xy^8 z^{13}}\left(\sqrt[3]{x^8 y^{11} z^{16}} - \sqrt[3]{x^{13} yz^{10}}\right)$

Multiply.

29. $\left(\sqrt{3} + \sqrt{7}\right)\left(\sqrt{3} - \sqrt{2}\right)$

30. $\left(\sqrt{6} + \sqrt{12}\right)\left(\sqrt{6} - \sqrt{3}\right)$

31. $\left(\sqrt{7} - \sqrt{3}\right)\left(\sqrt{7} + \sqrt{3}\right)$

32. $\left(\sqrt{21} + \sqrt{10}\right)\left(\sqrt{7} - \sqrt{40}\right)$

33. $\left(9\sqrt{3} + 4\sqrt{2}\right)\left(3\sqrt{2} - \sqrt{3}\right)$

34. $\left(2\sqrt{3} + 7\sqrt{2}\right)\left(5\sqrt{3} + 3\sqrt{2}\right)$

35. $\left(5 + 9\sqrt{3}\right)\left(8 - \sqrt{3}\right)$

36. $\left(10 + 9\sqrt{2}\right)\left(13 - 6\sqrt{2}\right)$

37. $\left(7\sqrt{2} - 4\sqrt{7}\right)^2$

38. $\left(5 - 2\sqrt{3}\right)^2$

Multiply the conjugates.

39. $\left(\sqrt{15} - \sqrt{6}\right)\left(\sqrt{15} + \sqrt{6}\right)$

40. $\left(\sqrt{33} + \sqrt{17}\right)\left(\sqrt{33} - \sqrt{17}\right)$

41. $\left(4\sqrt{10} + 3\sqrt{7}\right)\left(4\sqrt{10} - 3\sqrt{7}\right)$

42. $\left(7\sqrt{11} - 2\sqrt{30}\right)\left(7\sqrt{11} + 2\sqrt{30}\right)$

43. $\left(6 - 13\sqrt{3}\right)\left(6 + 13\sqrt{3}\right)$

44. $\left(9\sqrt{2} + 14\right)\left(9\sqrt{2} - 14\right)$

Simplify. Assume that all variables represent nonnegative real numbers.

45. $\dfrac{\sqrt{81}}{\sqrt{25}}$

46. $\dfrac{\sqrt{36}}{\sqrt{121}}$

47. $\sqrt{\dfrac{18}{x^{10} y^8}}$

48. $\sqrt{\dfrac{b^{13}}{a^{18} c^2}}$

49. $\sqrt{\dfrac{132}{11}}$

50. $\sqrt{\dfrac{168}{6}}$

51. $\sqrt{\dfrac{40}{45}}$

52. $\sqrt{\dfrac{264}{150}}$

Rationalize the denominator and simplify. Assume that all variables represent nonnegative real numbers.

53. $\dfrac{\sqrt{45}}{\sqrt{2}}$

54. $\dfrac{\sqrt{28}}{\sqrt{3}}$

55. $\dfrac{1}{\sqrt{6}}$

56. $\dfrac{11}{\sqrt{11}}$

57. $\sqrt{\dfrac{8}{5a^5}}$

58. $\sqrt{\dfrac{63}{2n^{14}}}$

59. $\sqrt[3]{\dfrac{s^{13}}{3r^4 t^5}}$

60. $\sqrt[3]{\dfrac{3x^{17}}{5y^{14} z^{28}}}$

61. $\dfrac{6}{\sqrt{20}}$

62. $\dfrac{5\sqrt{2}}{\sqrt{27}}$

63. $\dfrac{a^3 b^4}{\sqrt{a^{11} b^4 c^9}}$

64. $\dfrac{7x^5}{\sqrt{14x^3 y}}$

Rationalize the denominator.

65. $\dfrac{6}{\sqrt{15} - \sqrt{11}}$

66. $\dfrac{15}{\sqrt{18} + \sqrt{6}}$

67. $\dfrac{7\sqrt{2}}{\sqrt{2} + \sqrt{6}}$

68. $\dfrac{6\sqrt{3}}{\sqrt{15} - \sqrt{6}}$

69. $\dfrac{15\sqrt{7}}{\sqrt{11} - 2}$

70. $\dfrac{4\sqrt{2}}{8 - \sqrt{6}}$

71. $\dfrac{4\sqrt{3} + 3\sqrt{2}}{5\sqrt{3} - \sqrt{2}}$

72. $\dfrac{\sqrt{5} + \sqrt{2}}{4\sqrt{5} - 3\sqrt{2}}$

73. $\dfrac{\sqrt{3} + \sqrt{27}}{\sqrt{3} - \sqrt{27}}$

74. $\dfrac{\sqrt{7} - \sqrt{10}}{\sqrt{10} - \sqrt{7}}$

75. $\dfrac{6\sqrt{3} - \sqrt{5}}{4\sqrt{3} - 2\sqrt{5}}$

76. $\dfrac{\sqrt{18} - 3\sqrt{8}}{5\sqrt{18} - \sqrt{8}}$

Mixed Practice, 77–94

Simplify. Assume that all variables represent nonnegative real numbers.

77. $\sqrt{\dfrac{24}{147}}$

78. $\sqrt{24x^7} \cdot \sqrt{21x^3}$

79. $\left(5\sqrt{3} + \sqrt{10}\right)^2$

80. $\dfrac{15}{\sqrt{27}}$

81. $\sqrt[3]{9a^7b^4c^7} \cdot \sqrt[3]{9a^{10}b^5c^{17}}$

82. $\sqrt[3]{\dfrac{3x^5y^{16}}{81x^{14}y^2}}$

83. $\dfrac{\sqrt{7} - 3\sqrt{5}}{6\sqrt{7} + \sqrt{5}}$

84. $\sqrt{24}\left(6\sqrt{3} - 11\sqrt{15}\right)$

85. $\left(4\sqrt{14} - 3\sqrt{2}\right)\left(5\sqrt{14} + 6\sqrt{7}\right)$

86. $\dfrac{15\sqrt{3}}{\sqrt{33}}$

87. $\sqrt{30x^{13}}\left(3\sqrt{10x^2} - \sqrt{5x^8}\right)$

88. $\left(7\sqrt{7} - 2\sqrt{8}\right)\left(7\sqrt{7} + 2\sqrt{8}\right)$

89. $\sqrt{\dfrac{a^{12}b^{11}}{63c^7}}$

90. $\dfrac{8\sqrt{5} + 9}{4 - 3\sqrt{5}}$

91. $\dfrac{6x^5}{\sqrt[3]{4x^4}}$

92. $\left(13\sqrt{3} - 7\sqrt{2}\right)\left(8\sqrt{3} + 9\sqrt{2}\right)$

93. $\left(12\sqrt{5} - \sqrt{8}\right)\left(12\sqrt{5} + \sqrt{8}\right)$

94. $\left(9\sqrt{6} - 5\sqrt{3}\right)^2$

95. Develop a general formula for the product $\left(\sqrt{a} + \sqrt{b}\right)^2$.

96. Develop a general formula for the product $\left(a\sqrt{b} + c\sqrt{d}\right)^2$.

97. List four different pairs of conjugates whose product is 15.

98. List four different pairs of conjugates whose product is 53.

99. State a sum of two radical expressions that, when squared, equals $66 + 24\sqrt{6}$.

100. State a difference of two radical expressions that, when squared, equals $95 - 20\sqrt{15}$.

Writing in Mathematics

Answer in complete sentences.

101. Explain how to determine whether two radical expressions are conjugates.

102. Explain how to multiply conjugates. Use an example to illustrate the process.

103. *Solutions Manual** Write a solutions manual page for the following problem:

Simplify $\left(5\sqrt{2} + 4\sqrt{3}\right)\left(8\sqrt{2} - 3\sqrt{3}\right).$

104. *Newsletter** Write a newsletter that explains how to rationalize a denominator containing two terms and at least one square root.

*See Appendix B for details and sample answers.

QUICK REVIEW EXERCISES

Section 8.3

Find the prime factorization of the given number.

1. 144

2. 1400

3. 1024

4. 29,106

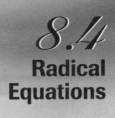

8.4
Radical Equations

Objectives

1 Solve radical equations.
2 Solve equations containing radical functions.
3 Solve equations in which a radical is equal to a variable expression.
4 Solve equations containing two radicals.

Solving Radical Equations

Objective 1 Solve radical equations. A **radical equation** is an equation containing one or more radicals. Here are some examples of radical equations.

$$\sqrt{x} = 9 \qquad\qquad \sqrt[3]{2x-5} = 3 \qquad\qquad x + \sqrt{x} = 20$$

$$\sqrt[4]{3x-8} = \sqrt[4]{2x+11} \qquad \sqrt{x-4} + \sqrt{x+8} = 6$$

In this section we will learn how to solve radical equations. We will find a way to convert a radical equation to an equivalent equation that we already know how to solve.

Raising Equal Numbers to the Same Power

> If two numbers a and b are equal, then for any n, $a^n = b^n$.

If we raise two equal numbers to the same power, they remain equal to each other. We will use this fact to solve radical equations.

Solving Radical Equations

> - Isolate one radical containing the variable on one side of the equation.
> - Raise both sides of the equation to the *nth* power, where n is the index of the radical. *For any nonnegative number x and any integer $n > 1$,* $\left(\sqrt[n]{x}\right)^n = x$.
> - If the resulting equation does not contain a radical, solve the equation. If the resulting equation does contain a radical, begin the process again by isolating the radical on one side of the equation.
> - Check the solution(s).

It is crucial that we check all solutions when solving a radical equation, as raising both sides of an equation to a power can introduce **extraneous solutions.** An extraneous solution is a solution to the equation that results when we raise both sides to a certain power, but it is not a solution to the original equation. Raising both sides of an equation to an odd power will not introduce extraneous solutions.

EXAMPLE 1   Solve $\sqrt{x+7} = 4$.

Solution

Since the radical $\sqrt{x+7}$ is already isolated on the left side of the equation, we may begin by squaring both sides of the equation.

$$\sqrt{x + 7} = 4$$

$$\left(\sqrt{x + 7}\right)^2 = 4^2 \qquad \text{Square both sides.}$$

$$x + 7 = 16 \qquad \text{Simplify.}$$

$$x = 9 \qquad \text{Subtract 7.}$$

We need to check this solution against the original equation.

$$\sqrt{(9) + 7} = 4 \qquad \text{Substitute 9 for } x.$$

$$\sqrt{16} = 4 \qquad \text{Add.}$$

$$4 = 4 \qquad \text{Simplify the square root.}$$

▸ The solution $x = 9$ checks, so the solution set is $\{9\}$.

EXAMPLE 2   Solve $\sqrt[3]{x - 8} - 3 = 4$.

Solution

In this example we begin by isolating the radical.

$$\sqrt[3]{x - 8} - 3 = 4$$

$$\sqrt[3]{x - 8} = 7 \qquad \text{Add 3 to isolate the radical.}$$

$$\left(\sqrt[3]{x - 8}\right)^3 = 7^3 \qquad \text{Raise both sides of the equation to the third power.}$$

$$x - 8 = 343 \qquad \text{Simplify.}$$

$$x = 351 \qquad \text{Add 8.}$$

Now we check the solution, using the original equation.

$$\sqrt[3]{(351) - 8} - 3 = 4 \qquad \text{Substitute 351 for } x.$$

$$\sqrt[3]{343} - 3 = 4 \qquad \text{Simplify the radicand.}$$

$$7 - 3 = 4 \qquad \text{Simplify the cube root.}$$

$$4 = 4 \qquad \text{Subtract. The solution checks.}$$

Quick Check 1

Solve.

a) $\sqrt{x - 5} = 6$

b) $\sqrt[3]{x + 2} + 8 = 3$

The solution set is $\{351\}$. Because we raised both sides of the equation to an odd power, we did not introduce any extraneous solutions.

EXAMPLE 3   Solve $\sqrt{x + 2} + 6 = 3$.

Solution

We begin by isolating $\sqrt{x + 2}$ on the left side of the equation.

$$\sqrt{x + 2} + 6 = 3$$

$$\sqrt{x + 2} = -3 \qquad \text{Subtract 6 to isolate the square root.}$$

$$\left(\sqrt{x + 2}\right)^2 = (-3)^2 \qquad \text{Square both sides of the equation.}$$

$$x + 2 = 9 \qquad \text{Simplify.}$$

$$x = 7 \qquad \text{Subtract 2.}$$

Now we check this solution, using the original equation.

$$\sqrt{7 + 2} + 6 = 3 \qquad \text{Substitute 7 for } x.$$
$$\sqrt{9} + 6 = 3 \qquad \text{Simplify the radicand.}$$
$$3 + 6 = 3 \qquad \text{Simplify the square root. The principal square root of 9 is}$$
$$\qquad\qquad\qquad 3, \text{ not } -3.$$
$$9 = 3 \qquad \text{Add.}$$

Quick Check 2

Solve
$$\sqrt{2x + 7} - 6 = -13.$$

This solution does not check, so it is an extraneous solution. The equation has no solution; the solution set is $\varnothing$.

If we obtain an equation in which an even root is equal to a negative number, such as $\sqrt{x + 2} = -3$, this equation will not have any solutions. This is because the principal even root of a number, if it exists, cannot be negative.

Solving Equations Involving Radical Functions

Objective 2   Solve equations containing radical functions.

EXAMPLE 4   For $f(x) = \sqrt{3x - 5} + 9$, find all values x for which $f(x) = 13$.

Solution

We begin by setting the function equal to 13.

$$f(x) = 13$$
$$\sqrt{3x - 5} + 9 = 13 \qquad \text{Replace } f(x) \text{ by its formula.}$$
$$\sqrt{3x - 5} = 4 \qquad \text{Subtract 9 to isolate the radical.}$$
$$\left(\sqrt{3x - 5}\right)^2 = 4^2 \qquad \text{Square both sides.}$$
$$3x - 5 = 16 \qquad \text{Simplify.}$$
$$3x = 21 \qquad \text{Add 5.}$$
$$x = 7 \qquad \text{Divide both sides by 3.}$$

Quick Check 3

For
$$f(x) = \sqrt{2x + 7} - 5,$$
find all values x for which $f(x) = 4$.

It is left to the reader to verify that the solution is not an extraneous solution. $f(7) = 13$.

Solving Equations in Which a Radical Is Equal to a Variable Expression

Objective 3   Solve equations in which a radical is equal to a variable expression. After we isolated the radical in all of the previous examples, the resulting equation had a radical equal to a constant number. In the next example, we will learn how to solve equations that result in a radical equal to a variable expression.

EXAMPLE 5   Solve $\sqrt{2x + 15} = x$.

Solution

Since the radical is already isolated, we begin by squaring both sides. This will result in a quadratic equation, which we solve by collecting all terms on one side of the equation and factoring.

$$\sqrt{2x + 15} = x$$
$$(\sqrt{2x + 15})^2 = x^2 \qquad \text{Square both sides.}$$
$$2x + 15 = x^2 \qquad \text{Simplify.}$$
$$0 = x^2 - 2x - 15 \qquad \text{Collect all terms on the right side of the}$$
$$\text{equation by subtracting } 2x \text{ and } 15.$$
$$0 = (x - 5)(x + 3) \qquad \text{Factor.}$$
$$x = 5 \quad \text{or} \quad x = -3 \qquad \text{Set each factor equal to 0 and solve.}$$

Now we check both solutions.

$x = 5$	$x = -3$
$\sqrt{2(5) + 15} = 5$	$\sqrt{2(-3) + 15} = -3$
$\sqrt{10 + 15} = 5$	$\sqrt{-6 + 15} = -3$
$\sqrt{25} = 5$	$\sqrt{9} = -3$
$5 = 5$	$3 = -3$
True	False

Quick Check 4
Solve $\sqrt{7x - 12} = x$.

The solution $x = -3$ is an extraneous solution. The solution set is $\{5\}$.

EXAMPLE 6   Solve $\sqrt{2x - 5} + 4 = x$.

Solution

We begin by isolating the radical.

$$\sqrt{2x - 5} + 4 = x$$
$$\sqrt{2x - 5} = x - 4 \qquad \text{Subtract 4 to isolate the radical.}$$
$$(\sqrt{2x - 5})^2 = (x - 4)^2 \qquad \text{Square both sides.}$$
$$2x - 5 = (x - 4)(x - 4) \qquad \text{Square the binomial by multiplying it by itself.}$$
$$2x - 5 = x^2 - 8x + 16 \qquad \text{Multiply. The resulting equation is quadratic.}$$
$$0 = x^2 - 10x + 21 \qquad \text{Collect all terms on the right side of the equation.}$$
$$0 = (x - 3)(x - 7) \qquad \text{Factor.}$$
$$x = 3 \text{ or } x = 7 \qquad \text{Set each factor equal to 0 and solve.}$$

Now we check both solutions.

$x = 3$	$x = 7$
$\sqrt{2(3) - 5} + 4 = 3$	$\sqrt{2(7) - 5} + 4 = 7$
$\sqrt{1} + 4 = 3$	$\sqrt{9} + 4 = 7$
$1 + 4 = 3$	$3 + 4 = 7$
$5 = 3$	$7 = 7$
False	True

Quick Check 5
Solve $\sqrt{3x + 6} = x$.

The solution $x = 3$ is an extraneous solution. The solution set is $\{7\}$.

Solving Radical Equations Containing Two Radicals

Objective 4 Solve equations containing two radicals.

EXAMPLE ▶ 7   Solve $\sqrt[3]{2x + 9} = \sqrt[3]{4x - 5}$.

Solution

We will raise both sides to the third power. Since both radicals have the same index, this will result in an equation that does not contain a radical.

$$\sqrt[3]{2x + 9} = \sqrt[3]{4x - 5}$$

$\left(\sqrt[3]{2x + 9}\right)^3 = \left(\sqrt[3]{4x - 5}\right)^3$	Raise both sides to the third power.
$2x + 9 = 4x - 5$	Simplify.
$9 = 2x - 5$	Subtract $2x$ from both sides.
$14 = 2x$	Add 5.
$7 = x$	Divide both sides by 2.

Quick Check 6

Solve
$\sqrt[4]{5x + 1} = \sqrt[4]{4x + 17}$.

It is left to the reader to verify that the solution is not an extraneous solution. This solution checks, and the solution set is $\{7\}$.

Occasionally equations containing two square roots will still contain a square root after we have squared both sides. This will require us to square both sides a second time.

EXAMPLE ▶ 8   Solve $\sqrt{x + 8} - \sqrt{x + 1} = 1$.

Solution

We must begin by isolating one of the two radicals on the left side of the equation. We will isolate $\sqrt{x + 8}$, as it is positive.

$$\sqrt{x + 8} - \sqrt{x + 1} = 1$$

$\sqrt{x + 8} = 1 + \sqrt{x + 1}$	Add $\sqrt{x + 1}$ to isolate the radical $\sqrt{x + 8}$ on the left side.
$\left(\sqrt{x + 8}\right)^2 = \left(1 + \sqrt{x + 1}\right)^2$	Square both sides.
$x + 8 = \left(1 + \sqrt{x + 1}\right)\left(1 + \sqrt{x + 1}\right)$	Square the binomial on the right side by multiplying it by itself.
$x + 8 = 1 \cdot 1 + 1 \cdot \sqrt{x + 1} + 1 \cdot \sqrt{x + 1} + \sqrt{x + 1} \cdot \sqrt{x + 1}$	Distribute.
$x + 8 = 1 + 2\sqrt{x + 1} + x + 1$	Simplify.
$x + 8 = 2 + 2\sqrt{x + 1} + x$	Combine like terms.
$6 = 2\sqrt{x + 1}$	Subtract 2 and x to isolate the radical.
$3 = \sqrt{x + 1}$	Divide both sides by 2.
$3^2 = \left(\sqrt{x + 1}\right)^2$	Square both sides.

$$9 = x + 1$$ Simplify.
$$8 = x$$ Subtract 1.

Quick Check 7

Solve.

$$\sqrt{x - 5} - \sqrt{x - 8} = 1.$$

It is left to the reader to verify that the solution is not an extraneous solution. The solution set is $\{8\}$.

Building Your Study Strategy Note Taking, 4 **Being an Active Learner**   Some students fall into the trap of mechanically taking notes, without thinking about the material at all. In a math class, there can be no learning without thinking. Try to be an active learner while taking notes. Do your best to understand each statement that you write down in your notes.

   When your instructor writes a problem on the board, try to solve it yourself in your notes. When you have finished solving the problem, compare your solution to your instructor's solution. If you make a mistake, or if your solution varies from your instructor's solution, you can make editor's notes on your solution.

EXERCISES 8.4

Vocabulary

1. A(n) _____ equation is an equation containing one or more radicals.

2. If two numbers a and b are equal, then for any n, $a^n =$ _____.

3. To solve a radical equation, first _____ one radical containing the variable on one side of the equation.

4. A(n) _____ solution is a solution to the equation that results when we raise both sides to a certain power, but it is not a solution to the original equation.

Solve. (Check for extraneous solutions.)

5. $\sqrt{x + 3} = 7$

6. $\sqrt{x - 8} = 2$

7. $\sqrt{2x - 3} = 5$

8. $\sqrt{4x + 9} = 11$

9. $\sqrt[3]{x - 6} = 5$

10. $\sqrt[4]{x + 3} = 4$

11. $\sqrt{3x - 8} = -2$

12. $\sqrt[3]{2x + 11} = -5$

13. $\sqrt{x + 4} - 7 = 3$

14. $\sqrt{x - 9} - 8 = 6$

15. $\sqrt{4x - 7} - 5 = 4$

16. $\sqrt{5x + 8} + 13 = 6$

17. $\sqrt[4]{x + 1} + 7 = 10$

18. $\sqrt[6]{x - 9} - 9 = -7$

19. $\sqrt[6]{2x + 7} + 14 = 8$

20. $\sqrt[3]{2x + 9} + 8 = 5$

21. $\sqrt{x^2 + 9x + 3} = 5$

22. $\sqrt{x^2 + 3x - 24} = 4$

23. For the function $f(x) = \sqrt{3x + 49}$, find all values x for which $f(x) = 10$.

24. For the function $f(x) = \sqrt{x + 3} + 5$, find all values x for which $f(x) = 12$.

25. For the function $f(x) = \sqrt[4]{2x - 3} - 2$, find all values x for which $f(x) = 1$.

26. For the function $f(x) = \sqrt[3]{5x + 21} + 8$, find all values x for which $f(x) = 4$.

27. For the function $f(x) = \sqrt{x^2 - 17x + 39}$, find all values x for which $f(x) = 3$.

28. For the function $f(x) = \sqrt{x^2 + 6x - 15}$, find all values x for which $f(x) = 5$.

Solve. (Check for extraneous solutions.)

29. $x = \sqrt{7x - 10}$

30. $x = \sqrt{3x + 54}$

31. $\sqrt{2x + 43} = x + 4$

32. $\sqrt{x + 9} = x + 7$

33. $\sqrt{5x + 40} + 2 = x$

34. $\sqrt{15x - 29} - 1 = x$

35. $\sqrt{46 - 6x} + 9 = x$

36. $\sqrt{2x + 45} - 5 = x$

37. $2x + 3 = \sqrt{16x + 44}$

38. $3x - 4 = \sqrt{10x + 24}$

39. $3x = \sqrt{7x^2 + 11x + 40} - 2$

40. $2x = 7 + \sqrt{2x^2 - 12x + 25}$

41. $\sqrt{3x + 8} = \sqrt{x + 30}$

42. $\sqrt[3]{4x - 9} = \sqrt[3]{2x - 19}$

43. $\sqrt[4]{x^2 - 8x + 10} = \sqrt[4]{2x - 11}$

44. $\sqrt{3x^2 + 11x - 9} = \sqrt{2x^2 - 2x + 21}$

45. $\sqrt{5x^2 - 3x + 8} = \sqrt{3x^2 - 14x + 14}$

46. $\sqrt{2x^2 - 5x - 6} = \sqrt{x^2 + 7x - 33}$

47. $\sqrt{x + 9} = \sqrt{x - 3} + 2$

48. $\sqrt{x + 5} = \sqrt{x - 11} + 2$

49. $\sqrt{x - 9} + \sqrt{5x - 14} = 7$

50. $\sqrt{4x - 15} + 3 = \sqrt{6x}$

51. $\sqrt{3x - 5} = 5 + \sqrt{2x + 3}$

52. $\sqrt{5x + 1} = 2 - \sqrt{5x - 1}$

53. List four different radical equations that have $x = 6$ as a solution.

54. List four different radical equations that have $x = -8$ as a solution.

Writing in Mathematics

Answer in complete sentences.

55. What is an extraneous solution to an equation? Explain how to determine whether a solution to a radical equation is actually an extraneous solution.

56. *Solutions Manual** Write a solutions manual page for the following problem:

Solve $\sqrt{3x - 2} = x - 2$.

57. *Newsletter** Write a newsletter that explains how to solve a radical equation.

*See Appendix B for details and sample answers.

1 Solve applied problems involving a pendulum and its period.

2 Solve other applied problems involving radicals.

A Pendulum and Its Period

Objective 1 Solve applied problems involving a pendulum and its period. The **period** of a pendulum is the amount of time it takes to swing from one extreme to the other and then back again. The period T of a pendulum in seconds can be found using the formula $T = 2\pi\sqrt{\dfrac{L}{32}}$, where L is the length of the pendulum in feet.

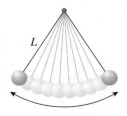

EXAMPLE ▸ 1 A pendulum has a length of 2 feet. Find its period, rounded to the nearest hundredth of a second.

Solution

We substitute 2 for L in the formula and simplify to find the period T.

$$T = 2\pi\sqrt{\frac{L}{32}}$$

$$T = 2\pi\sqrt{\frac{2}{32}} \qquad \text{Substitute 2 for } L.$$

$$T \approx 1.57 \qquad \text{Approximate using a calculator.}$$

The period of a pendulum that is 2 feet long is approximately 1.57 seconds.

> **Quick Check 1**
> A pendulum has a length of 5 feet. Find its period, rounded to the nearest hundredth of a second.

EXAMPLE ▸ 2 If a pendulum has a period of 3 seconds, find its length in feet. Round to the nearest hundredth of a foot.

Solution

In this example we substitute 3 for T and solve for L. To solve this equation for L, we isolate the radical and then square both sides.

$$T = 2\pi\sqrt{\frac{L}{32}}$$

$$3 = 2\pi\sqrt{\frac{L}{32}} \qquad \text{Substitute 3 for } T.$$

$$\frac{3}{2\pi} = \frac{2\pi\sqrt{\dfrac{L}{32}}}{2\pi} \qquad \text{Divide by } 2\pi \text{ to isolate the radical.}$$

$$\frac{3}{2\pi} = \sqrt{\frac{L}{32}} \qquad \text{Simplify.}$$

$$\left(\frac{3}{2\pi}\right)^2 = \left(\sqrt{\frac{L}{32}}\right)^2 \qquad \text{Square both sides.}$$

$$\frac{9}{4\pi^2} = \frac{L}{32} \qquad \text{Simplify.}$$

$$32 \cdot \frac{9}{4\pi^2} = 32 \cdot \frac{L}{32} \qquad \text{Multiply by 32 to isolate } L.$$

$$\overset{8}{\cancel{32}} \cdot \frac{9}{\underset{1}{\cancel{4}\pi^2}} = \overset{1}{\cancel{32}} \cdot \frac{L}{\underset{1}{\cancel{32}}} \qquad \text{Divide out common factors.}$$

$$\frac{72}{\pi^2} = L \qquad \text{Simplify.}$$

$$L \approx 7.30 \qquad \text{Approximate using a calculator.}$$

Quick Check 2
If a pendulum has a period of 4 seconds, find its length in feet. Round to the nearest hundredth of a foot.

The length of the pendulum is approximately 7.30 feet.

Other Applications Involving Radicals

Objective 2 Solve other applied problems involving radicals.

EXAMPLE 3 A vehicle made 225 feet of skid marks on the asphalt before crashing. The speed, s, in miles per hour, that the vehicle was traveling when it started skidding can be approximated by the formula $s = \sqrt{30df}$, where d represents the length of the skid marks in feet and f represents the drag factor of the road. If the drag factor for asphalt is 0.75, find the speed the car was traveling. Round to the nearest mile per hour.

Solution

We begin by substituting 225 for d and 0.75 for f.

$$s = \sqrt{30df}$$
$$s = \sqrt{30(225)(0.75)} \qquad \text{Substitute 225 for } d \text{ and 0.75 for } f.$$
$$s = \sqrt{5062.5} \qquad \text{Simplify the radicand.}$$
$$s \approx 71 \qquad \text{Approximate, using a calculator.}$$

The car was traveling at approximately 71 miles per hour when it started skidding.

Quick Check 3 A vehicle made 300 feet of skid marks on the asphalt before crashing. The speed, s, that the vehicle was traveling in miles per hour when it started skidding can be approximated by the formula $s = \sqrt{30df}$, where d represents the length of the skid marks in feet and f represents the drag factor of the road. If the drag factor for asphalt is 0.75, find the speed that the car was traveling when it started skidding. Round to the nearest mile per hour.

> *Building Your Study Strategy* Note Taking, 5 **Formatting Your Notes**   How should you format your notes? This is really a matter of personal preference, but it is important to come up with a system that works for you that you consistently use.
>
> A good note-taking system takes advantage of the margins for specific tasks. You can use the left margin for writing down key words or as a place to denote important material. You can then use the right margin to clarify steps in problems or to take down advice from your instructor.

EXERCISES 8.5

(Vocabulary)

1. An object suspended from a support so it swings freely back and forth under the influence of gravity is called a(n) _____.

2. The _____ of a pendulum is the amount of time it takes to swing from one extreme to the other and then back again.

For Exercises 3–8, use the formula $T = 2\pi\sqrt{\dfrac{L}{32}}$.

3. A pendulum has a length of 3 feet. Find its period, rounded to the nearest hundredth of a second.

4. A pendulum has a length of 1 foot. Find its period, rounded to the nearest hundredth of a second.

5. A pendulum has a length of 1.5 feet. Find its period, rounded to the nearest hundredth of a second.

6. A pendulum has a length of 2.7 feet. Find its period, rounded to the nearest hundredth of a second.

7. If a pendulum has a period of 1 second, find its length in feet. Round to the nearest tenth of a foot.

8. If a pendulum has a period of 2.5 seconds, find its length in feet. Round to the nearest tenth of a foot.

If a pendulum has a length of L inches, then its period in seconds, T, can be found using the formula $T = 2\pi\sqrt{\dfrac{L}{384}}$.

9. A pendulum has a length of 25 inches. Find its period, rounded to the nearest hundredth of a second.

10. If a pendulum has a period of 0.3 seconds, find its length in inches. Round to the nearest tenth of an inch.

If a pendulum has a length of L meters, then its period in seconds, T, can be found using the formula

$$T = 2\pi\sqrt{\dfrac{L}{9.8}}.$$

11. A pendulum has a length of 4 meters. Find its period, rounded to the nearest hundredth of a second.

12. If a pendulum has a period of 1.8 seconds, find its length in meters. Round to the nearest tenth of a meter.

Skid mark analysis is one way to estimate the speed a car was traveling prior to an accident. The speed, s, that the vehicle was traveling in miles per hour can be approximated by the formula $s = \sqrt{30df}$, where d represents the length of the skid marks in feet and f represents the drag factor of the road.

13. A vehicle that was involved in an accident made 140 feet of skid marks on the asphalt before crashing. If the drag factor for asphalt is 0.75, find the speed the car was traveling when it started skidding. Round to the nearest mile per hour.

14. A vehicle that was involved in an accident made 320 feet of skid marks on the asphalt before crashing. If the drag factor for asphalt is 0.75, find the speed the car was traveling when it started skidding. Round to the nearest mile per hour.

15. A vehicle that was involved in an accident made 205 feet of skid marks on a concrete road before crashing. If the drag factor for concrete is 0.95, find the speed the car was traveling when it started skidding. Round to the nearest mile per hour.

16. A vehicle that was involved in an accident made 90 feet of skid marks on a concrete road before crashing. If the drag factor for concrete is 0.95, find the speed the car was traveling when it started skidding. Round to the nearest mile per hour.

17. A vehicle that was involved in an accident on an asphalt road was traveling at a speed of 75 mph when it started skidding. If the drag factor for asphalt is 0.75, find the length of the skid marks made by the car. Round to the nearest foot.

18. A vehicle that was involved in an accident on a concrete road was traveling at a speed of 60 mph when it started skidding. If the drag factor for concrete is 0.95, find the length of the skid marks made by the car. Round to the nearest foot.

A water tank has a hole at the bottom. The rate r at which water flows out of the hole in gallons per minute can be found using the formula $r = 19.8\sqrt{d}$, where d represents the depth of the water in the tank in feet.

19. Find the rate of water flow if the depth of water in the tank is 64 feet.

20. Find the rate of water flow if the depth of water in the tank is 25 feet.

21. Find the rate of water flow if the depth of water in the tank is 40 feet. Round to the nearest tenth of a gallon per minute.

22. Find the rate of water flow if the depth of water in the tank is 18 feet. Round to the nearest tenth of a gallon per minute.

23. If water is flowing out of the tank at a rate of 50 gallons per minute, find the depth of water in the tank. Round to the nearest tenth of a foot.

24. If water is flowing out of the tank at a rate of 75 gallons per minute, find the depth of water in the tank. Round to the nearest tenth of a foot.

The hull speed of a sailboat is the maximum speed that the hull can attain from wind power. The hull speed, h, in knots can be calculated using the formula $h = 1.34\sqrt{L}$, where L is the length of the waterline in feet.

25. Find the hull speed of a sailboat with a waterline of 30 feet. Round to the nearest tenth of a knot.

26. Find the hull speed of a sailboat with a waterline of 24 feet. Round to the nearest tenth of a knot.

27. If the hull speed of a sailboat is 8 knots, find the length of its waterline. Round to the nearest tenth of a foot.

28. If the hull speed of a sailboat is 9.2 knots, find the length of its waterline. Round to the nearest tenth of a foot.

A tsunami is a great sea wave caused by natural phenomena such as earthquakes or volcanic activity. The speed of a tsunami, s, in miles per hour at any particular point, can be found from the formula $s = 308.29\sqrt{d}$, where d is the depth of the ocean in miles at that point.

29. Find the speed of a tsunami if the depth of the water is 5 miles. Round to the nearest mile per hour.

30. Find the speed of a tsunami if the depth of the water is 0.2 miles. Round to the nearest mile per hour.

31. If a tsunami is traveling at 500 miles per hour, what is the depth of the ocean at that location? Round to the nearest tenth of a mile.

32. If a tsunami is traveling at 375 miles per hour, what is the depth of the ocean at that location? Round to the nearest tenth of a mile.

33. What is the length of a pendulum which has a period that is three times as long as the period of a pendulum whose length is 2.2 feet? Round to the nearest tenth of a foot.

34. A vehicle made skid marks on the asphalt that were twice as long as the skid marks that would be made by a car traveling at 60 mph when it started skidding. If the drag factor for asphalt is 0.75, find the speed the vehicle was traveling when the skid started. Round to the nearest mile per hour.

Writing in Mathematics

Answer in complete sentences.

35. Write a real-world word problem that involves finding the speed of a car in an accident using skid mark analysis. Solve your problem, explaining each step of the process.

36. Does the length of a car's skid marks vary directly or vary inversely as the speed the car was traveling? Explain your reasoning.

37. *Solutions Manual*[*] Write a solutions manual page for the following problem:

If a pendulum has a period of 1.45 seconds, find its length in feet. Round to the nearest tenth of a foot.

38. *Newsletter*[*] Write a newsletter that explains how to find the period of a pendulum, given its length in feet.

*See Appendix B for details and sample answers.

8.6
Complex Numbers

Objectives

1 **Rewrite square roots of negative numbers as imaginary numbers.**
2 **Add and subtract complex numbers.**
3 **Multiply imaginary numbers.**
4 **Multiply complex numbers.**
5 **Divide by a complex number.**
6 **Divide by an imaginary number.**

Imaginary Numbers

Objective 1 Rewrite square roots of negative numbers as imaginary numbers. Section 8.1 stated that the square root of a negative number, such as $\sqrt{-1}$ or $\sqrt{-36}$, was not a real number. This is because there is no real number that equals a negative number when it is squared. The square root of a negative number is an **imaginary number.**

We define the **imaginary unit** i to be a number that is equal to $\sqrt{-1}$. The number i has the property that $i^2 = -1$.

Imaginary Unit i

$$i = \sqrt{-1}$$
$$i^2 = -1$$

All imaginary numbers can be expressed in terms of i because $\sqrt{-1}$ is a factor of every imaginary number.

EXAMPLE 1   Express $\sqrt{-36}$ in terms of i.

Solution

Whenever we have a square root with a negative radicand, we begin by factoring out i. Then we simplify the resulting square root.

$$\sqrt{-36} = \sqrt{36(-1)} \qquad \text{Rewrite } -36 \text{ as } 36(-1).$$
$$= \sqrt{36} \cdot \sqrt{-1} \qquad \text{Rewrite as the product of two square roots.}$$
$$= \sqrt{36}\, i \qquad \text{Rewrite } \sqrt{-1} \text{ as } i.$$
$$= 6i \qquad \text{Simplify the square root.}$$

As you become more experienced at working with imaginary numbers, you may wish to combine a few of the previous steps into one step.

EXAMPLE 2   Express $\sqrt{-90}$ in terms of i.

Solution

$$
\begin{aligned}
\sqrt{-90} &= \sqrt{90} \cdot \sqrt{-1} && \text{Rewrite as the product of two square roots.} \\
&= 3\sqrt{10}\ i && \text{Simplify the square root. Rewrite } \sqrt{-1} \text{ as } i. \\
&= 3i\sqrt{10} && \text{Rewrite with } i \text{ in front of the radical.}
\end{aligned}
$$

Quick Check 1

Express in terms of i.

a) $\sqrt{-81}$

b) $\sqrt{-44}$

A Word of Caution   After rewriting the square root of a negative number, such as $\sqrt{-90}$, as an imaginary number, be sure that i does not appear in the radicand. Instead, i should be written in front of the radical.

Complex Numbers

The set of imaginary numbers, together with the set of real numbers, are subsets of the set of **complex numbers.**

Complex Numbers

> A **complex number** is a number of the form $a + bi$, where a and b are real numbers.

Real numbers are complex numbers for which $b = 0$, while imaginary numbers are complex numbers for which $a = 0$ but $b \neq 0$. We often associate the word complex with something that is difficult, but here complex refers to the fact that these numbers are made up of two parts.

Real and Imaginary Parts of a Complex Number

> For the complex number $a + bi$, the number a is the **real part** and the number b is the **imaginary part.**

Addition and Subtraction of Complex Numbers

Objective 2 Add and subtract complex numbers. We now focus on operations involving complex numbers. We add two complex numbers by adding the two real parts and adding the two imaginary parts. We can follow a similar technique for subtraction.

EXAMPLE 3   Simplify $(8 + 3i) + (9 - 6i)$.

Solution

We begin by removing the parentheses, and then we combine the two real parts and the two imaginary parts of these complex numbers.

$$
\begin{aligned}
(8 + 3i) + (9 - 6i) &= 8 + 3i + 9 - 6i && \text{Remove parentheses.} \\
&= 17 - 3i && \text{Combine the two real parts. Combine} \\
& && \text{the two imaginary parts.}
\end{aligned}
$$

Notice that the process of adding these two complex numbers is similar to simplifying the expression $(8 + 3x) + (9 - 6x)$.

EXAMPLE 4 Simplify $(-7 + 10i) - (4 - 3i)$.

Solution

Quick Check 2
Simplify
a) $(7 + 6i) + (-4 + 8i)$
b) $(5 - 4i) - (9 + 7i)$

We must distribute the negative sign to both parts of the second complex number, just as we distributed negative signs when we subtracted variable expressions.

$$(-7 + 10i) - (4 - 3i) = -7 + 10i - 4 + 3i \qquad \text{Distribute.}$$
$$= -11 + 13i \qquad \text{Combine the two real parts and combine the two imaginary parts.}$$

Multiplying Imaginary Numbers

Objective 3 Multiply imaginary numbers. Before learning to multiply two complex numbers, we will discuss the multiplication of two imaginary numbers. Suppose that we wanted to multiply $6i$ by $9i$. Just as $6x \cdot 9x = 54x^2$, the product $6i \cdot 9i$ is equal to $54i^2$. However, recall that $i^2 = -1$. So this product is $54(-1)$ or -54. When multiplying two imaginary numbers, we substitute -1 for i^2.

EXAMPLE 5 Multiply: $5i \cdot 12i$.

Solution

$$5i \cdot 12i = 60i^2 \qquad \text{Multiply.}$$
$$= 60(-1) \qquad \text{Rewrite } i^2 \text{ as } -1.$$
$$= -60 \qquad \text{Multiply.}$$

Quick Check 3
Multiply: $7i \cdot 8i$.

EXAMPLE 6 Multiply: $\sqrt{-40} \cdot \sqrt{-18}$.

Solution

Although it may be tempting to multiply -40 by -18 and combine the two square roots into one, we cannot do this. The property $\sqrt{x} \cdot \sqrt{y} = \sqrt{xy}$ only holds true if either x or y is nonnegative. We must rewrite each square root as an imaginary number before multiplying.

$$\sqrt{-40} \cdot \sqrt{-18} = i\sqrt{40} \cdot i\sqrt{18} \qquad \text{Rewrite each radical as an imaginary number.}$$
$$= i\sqrt{2^3 \cdot 5} \cdot i\sqrt{2 \cdot 3^2} \qquad \text{Factor each radicand.}$$
$$= i^2\sqrt{2^4 \cdot 3^2 \cdot 5} \qquad \text{Multiply the two radicands.}$$
$$= 2^2 \cdot 3\, i^2\sqrt{5} \qquad \text{Simplify the square root.}$$
$$= -12\sqrt{5} \qquad \text{Rewrite } i^2 \text{ as } -1 \text{ and simplify.}$$

Quick Check 4
Multiply: $\sqrt{-14} \cdot \sqrt{-50}$.

When we multiply $\sqrt{40}$ by $\sqrt{18}$, our work will be easier if we factor 40 and 18 before multiplying, rather than trying to simplify $\sqrt{720}$.

Multiplying Complex Numbers

Objective 4 Multiply complex numbers. We multiply two complex numbers by using the distributive property. Often a product of two complex numbers will contain a term with i^2, and we will rewrite i^2 as -1.

EXAMPLE 7   Multiply: $6i(5 + 4i)$.

Solution

We begin by multiplying $6i$ by both terms in the parentheses.

$$
\begin{aligned}
6i(5 + 4i) &= 6i \cdot 5 + 6i \cdot 4i && \text{Distribute.} \\
&= 30i + 24i^2 && \text{Multiply.} \\
&= 30i - 24 && \text{Rewrite } i^2 \text{ as } -1 \text{ and simplify.} \\
&= -24 + 30i && \text{Rewrite in the form } a + bi.
\end{aligned}
$$

Quick Check 5
Multiply: $-2i(6 - 7i)$.

EXAMPLE 8   Multiply: $(3 + 4i)(2 - 5i)$.

Solution

In this example we must multiply each term in the first set of parentheses by each term in the second set.

$$
\begin{aligned}
(3 + 4i)(2 - 5i) &= 3 \cdot 2 - 3 \cdot 5i + 4i \cdot 2 - 4i \cdot 5i && \text{Distribute (FOIL).} \\
&= 6 - 15i + 8i - 20i^2 && \text{Multiply.} \\
&= 6 - 7i + 20 && \text{Simplify } -15i + 8i. \\
&&& \text{Rewrite } i^2 \text{ as } -1 \text{ and} \\
&&& \text{simplify.} \\
&= 26 - 7i && \text{Combine like terms.}
\end{aligned}
$$

Quick Check 6
Multiply: $(5 - 4i)(6 + i)$.

EXAMPLE 9   Multiply: $(4 + 7i)^2$.

Solution

Recall that we square a binomial by multiplying it by itself.

$$
\begin{aligned}
(4 + 7i)^2 &= (4 + 7i)(4 + 7i) && \text{Multiply } 4 + 7i \text{ by itself.} \\
&= 4 \cdot 4 + 4 \cdot 7i + 7i \cdot 4 + 7i \cdot 7i && \text{Distribute.} \\
&= 16 + 28i + 28i + 49i^2 && \text{Multiply.} \\
&= 16 + 56i - 49 && \text{Add } 28i + 28i. \text{ Rewrite } i^2 \text{ as } -1 \\
&&& \text{and simplify.} \\
&= -33 + 56i && \text{Combine like terms.}
\end{aligned}
$$

Quick Check 7
Multiply: $(10 - 3i)^2$.

Two complex numbers of the form $a + bi$ and $a - bi$ are conjugates, and their product will always be equal to $a^2 + b^2$.

$$
\begin{aligned}
(a + bi)(a - bi) &= a^2 - abi + abi - b^2i^2 && \text{Distribute.} \\
&= a^2 - b^2i^2 && \text{Combine like terms.} \\
&= a^2 + b^2 && \text{Rewrite } i^2 \text{ as } -1 \text{ and simplify.}
\end{aligned}
$$

EXAMPLE 10   Multiply: $(9 + 4i)(9 - 4i)$.

Solution

We will use the fact that $(a + bi)(a - bi) = a^2 + b^2$ for two complex numbers that are conjugates.

$$(9 + 4i)(9 - 4i) = 9^2 + 4^2 \qquad \text{The product equals } a^2 + b^2.$$
$$= 81 + 16 \qquad \text{Square 9 and 4.}$$
$$= 97 \qquad \text{Add.}$$

Quick Check 8

Multiply:
$(7 + 6i)(7 - 6i)$.

Dividing by a Complex Number

Objective 5   **Divide by a complex number.** Since the imaginary number i is a square root ($\sqrt{-1}$), a simplified expression cannot contain i in its denominator. We will use conjugates to rewrite the expression without i in the denominator, in a procedure similar to rationalizing a denominator (Section 8.3).

EXAMPLE 11   Simplify $\dfrac{4}{5 + i}$.

Solution

We begin by multiplying the numerator and denominator by the conjugate of the denominator, which is $5 - i$.

$$\frac{4}{5 + i} = \frac{4}{5 + i} \cdot \frac{5 - i}{5 - i} \qquad \text{Multiply the numerator and denominator by the conjugate of the denominator.}$$

$$= \frac{20 - 4i}{5^2 + 1^2} \qquad \text{Multiply the numerators by distributing 4 to both terms in the second numerator. Multiply the denominators by using the fact that } (a + bi)(a - bi) = a^2 + b^2.$$

$$= \frac{20 - 4i}{26} \qquad \text{Simplify the denominator.}$$

$$= \frac{\overset{1}{2}(10 - 2i)}{\underset{13}{26}} \qquad \text{Factor the numerator. Divide out factors common to the numerator and denominator.}$$

$$= \frac{10 - 2i}{13} \qquad \text{Simplify.}$$

$$= \frac{10}{13} - \frac{2}{13}i \qquad \text{Rewrite in the form } a + bi.$$

Quick Check 9

Simplify $\dfrac{6}{5 + 3i}$.

Answers in the back of the text will be written in both forms: as a single fraction and in $a + bi$ form. Your instructor will let you know which form is preferred for your class.

EXAMPLE 12   Simplify $\dfrac{7 + 4i}{2 + 3i}$.

Solution

In this example the numerator has two terms, and we must multiply by using the distributive property.

$$\frac{7 + 4i}{2 + 3i} = \frac{7 + 4i}{2 + 3i} \cdot \frac{2 - 3i}{2 - 3i}$$

Multiply the numerator and denominator by the conjugate of the denominator.

$$= \frac{14 - 21i + 8i - 12i^2}{2^2 + 3^2}$$

Multiply.

$$= \frac{14 - 13i - 12i^2}{4 + 9}$$

Combine like terms in the numerator. Square 2 and 3 in the denominator.

$$= \frac{14 - 13i + 12}{13}$$

Rewrite i^2 as -1 and simplify. Simplify the denominator.

$$= \frac{26 - 13i}{13}$$

Combine like terms.

$$= \frac{\overset{1}{\cancel{13}}(2 - i)}{\underset{1}{\cancel{13}}}$$

Factor the numerator. Divide out factors common to the numerator and denominator.

$$= 2 - i$$

Simplify.

Quick Check 10
Simplify $\dfrac{5 - 2i}{1 - 3i}$.

Note that the expression $\dfrac{1 + 6i}{2 + 5i}$ is equivalent to $(1 + 6i) \div (2 + 5i)$. If we are asked to divide a complex number by another complex number, we begin by rewriting the expression as a fraction and then proceed as in the previous example.

Dividing by an Imaginary Number

Objective 6   Divide by an imaginary number.   When a denominator is an imaginary number, we multiply the numerator and denominator by i to rewrite the fraction without i in the denominator.

EXAMPLE 13   Simplify $\dfrac{7 - 5i}{2i}$.

Solution

Since the denominator is an imaginary number, we will begin by multiplying the fraction by $\dfrac{i}{i}$.

$$\frac{7 - 5i}{2i} = \frac{7 - 5i}{2i} \cdot \frac{i}{i}$$

Multiply the numerator and denominator by i.

$$= \frac{7i - 5i^2}{2i^2}$$

Multiply.

$$= \frac{7i + 5}{-2} \qquad \text{Rewrite } i^2 \text{ as } -1 \text{ and simplify.}$$

$$= -\frac{(7i + 5)}{2} \qquad \text{Factor } -1 \text{ out of the denominator.}$$

$$= \frac{-7i - 5}{2} \qquad \text{Distribute.}$$

$$= -\frac{5}{2} - \frac{7}{2}i \qquad \text{Rewrite in } a + bi \text{ form.}$$

Quick Check 11

Simplify $\dfrac{6 + 7i}{5i}$.

If the coefficient of the imaginary number is positive, it is a good idea to multiply by $\dfrac{-i}{-i}$ so the denominator will be positive, rather than negative.

Building Your Study Strategy **Note Taking, 6 Rewriting Your Notes**   A good strategy is to rewrite your notes as soon as possible after class. The simple task of rewriting your notes serves as a review of the material that was covered in class. If you rewrite your notes while the material is still fresh, then you have a much better chance of being able to actually read what you have written.

As you rework your notes, take the time to supplement them. Replace a brief definition with the full definition from the text. Add notes to clarify what you wrote down in class, including notes to yourself explaining each step of the solution to a problem. This includes any calculator steps that are necessary. Each time a new term or procedure is introduced in your notes, add it to your set of note cards.

Finally, consider creating a page in your notes that contains the problems (without solutions) your instructor solved during class. When you begin to review for the exam, this list of problems will be a good review sheet.

EXERCISES 8.6

Vocabulary

1. The square root of a negative number is a(n) _____ number.

2. The imaginary unit i is a number that is equal to _____.

3. The number i has the property that $i^2 =$ _____.

4. A(n) _____ number is a number of the form $a + bi$, where a and b are real numbers.

5. For the complex number $a + bi$, the number a is the _____ part.

6. For the complex number $a + bi$, the number b is the _____ part.

Express in terms of i.

7. $\sqrt{-9}$

8. $\sqrt{-25}$

9. $\sqrt{-49}$

10. $\sqrt{-64}$

11. $-\sqrt{-169}$

12. $-\sqrt{-225}$

13. $\sqrt{-48}$

14. $\sqrt{-180}$

15. $\sqrt{-243}$

16. $-\sqrt{-700}$

Add or subtract the complex numbers.

17. $(5 + 4i) + (9 + 3i)$

18. $(9 - 2i) + (4 + 7i)$

19. $(-6 + 7i) + (-11 - 5i)$

20. $(3 - 10i) + (-8 - 13i)$

21. $(11 + 8i) - (6 + 7i)$

22. $(15 + 4i) - (8 + i)$

23. $(8 - 5i) - (8 + 3i)$

24. $(-7 + 6i) - (9 + 6i)$

25. $(9 - 11i) - (2 + 3i) + (10 - 6i)$

26. $(2 + 7i) - (7 - 3i) - (14 + 9i)$

Find the missing complex number.

27. $(6 + 5i) + ? = 13 + 2i$

28. $(13 - 4i) + ? = 5 - 9i$

29. $(7 + 2i) - ? = 16 - 7i$

30. $? - (9 - 6i) = 14 + i$

Multiply.

31. $6i \cdot 3i$

32. $2i \cdot 17i$

33. $-8i \cdot 9i$

34. $-10i \cdot i$

35. $\sqrt{-9} \cdot \sqrt{-25}$

36. $\sqrt{-49} \cdot \sqrt{-49}$

37. $\sqrt{-6} \cdot \sqrt{-75}$

38. $\sqrt{-27} \cdot \sqrt{-12}$

39. $\sqrt{-35} \cdot \sqrt{-140}$

40. $\sqrt{-8} \cdot \sqrt{-126}$

Find two imaginary numbers with the given product. (Answers may vary.)

41. -54

42. -20

43. 32

44. 15

Multiply.

45. $4i(3 - 5i)$

46. $i(6 + 7i)$

47. $-2i(-8 + 3i)$

48. $-5i(2 + 7i)$

49. $(3 + i)(4 - 5i)$

50. $(6 - 2i)(7 - 3i)$

51. $(5 - 2i)^2$

52. $(9 + 4i)(4 - 9i)$

53. $(5 + 7i)(5 - 7i)$

54. $(8 + i)^2$

55. $(10 + 9i)^2$

56. $(3 - 2i)(3 + 2i)$

Multiply the conjugates, using the fact that $(a + bi)(a - bi) = a^2 + b^2$.

57. $(3 + 4i)(3 - 4i)$

58. $(13 - 2i)(13 + 2i)$

59. $(11 - 10i)(11 + 10i)$

60. $(7 + i)(7 - i)$

Rationalize the denominator.

61. $\dfrac{12}{3 + i}$

62. $\dfrac{10}{4 - 3i}$

63. $\dfrac{3i}{2 - 7i}$

64. $\dfrac{8i}{7 + 3i}$

65. $\dfrac{3 + 2i}{5 + 4i}$

66. $\dfrac{5 - i}{3 + 8i}$

67. $\dfrac{6 - 4i}{7 - i}$

68. $\dfrac{10 + 3i}{4 - 3i}$

Rationalize the denominator.

69. $\dfrac{5}{i}$

70. $\dfrac{4}{7i}$

71. $\dfrac{4 - 9i}{6i}$

72. $\dfrac{13 + 6i}{3i}$

73. $\dfrac{5 + i}{-4i}$

74. $\dfrac{7 - 5i}{-8i}$

Divide.

75. $2i \div (7 - 5i)$

76. $(12 - 9i) \div (3i)$

77. $(3 - 5i) \div (2 + i)$

78. $(6 - 7i) \div (9 - 2i)$

Find the missing complex number.

79. $(3 + 2i)(?) = 16 - 11i$

80. $(5 - i)(?) = 27 + 31i$

81. $\dfrac{?}{4 - 6i} = 2 + i$

82. $\dfrac{?}{8 + 3i} = 5 - 6i$

Mixed Practice, 83–104

Simplify.

83. $(8 + 3i)(7 - 2i)$

84. $(7 + 8i)(7 - 8i)$

85. $\dfrac{11 + 2i}{3i}$

86. $(6 + 11i) - (9 + 4i)$

87. $3i \div (4 + 11i)$

88. $\dfrac{7 + 4i}{6 - i}$

89. $(5 - 9i)(5 + 9i)$

90. $(23 - 19i) + (-6 + 27i)$

91. $\sqrt{-8} \cdot \sqrt{-150}$

92. $(10 - 6i) \div (4i)$

93. $\dfrac{12i}{9 - 5i}$

94. $(13 - 8i)^2$

95. $\sqrt{-150}$

96. $(10 + 3i)(1 - 4i)$

97. $(8 + 5i)^2$

98. $5i \cdot 11i$

99. $(4 - 12i) + (7 + 6i)$

100. $\sqrt{-27} \cdot \sqrt{-6}$

101. $-7i \cdot 8i$

102. $\dfrac{4 - 9i}{2i}$

103. $(21 - 30i) - (60 - 43i)$

104. $\sqrt{-252}$

Writing in Mathematics

Answer in complete sentences.

105. Explain how adding two complex numbers is similar to adding two variable expressions.

106. *Solutions Manual*[*] Write a solutions manual page for the following problem:

Simplify $\sqrt{-84}$.

107. *Newsletter*[*] Write a newsletter that explains how to multiply complex numbers.

*See Appendix B for details and sample answers.

8.7 Rational Exponents

Objectives

1 Simplify expressions containing exponents of the form $1/n$.
2 Simplify expressions containing exponents of the form m/n.
3 Simplify expressions containing rational exponents.
4 Simplify expressions containing negative rational exponents.

Rational Exponents of the Form $1/n$

Objective 1   Simplify expressions containing exponents of the form $1/n$.
We have used exponents to represent repeated multiplication. For example, x^n tells us that the base x is a factor n times.

$$x^3 = x \cdot x \cdot x$$
$$x^6 = x \cdot x \cdot x \cdot x \cdot x \cdot x$$

We run into a problem with this definition when we encounter an expression with a fractional exponent such as $x^{1/2}$. Saying that the base x is a factor $\frac{1}{2}$ times does not make any sense. Using the properties of exponents introduced in Chapter 5, we know that $x^{1/2} \cdot x^{1/2} = x^{1/2 + 1/2}$ or x. If we multiply $x^{1/2}$ by itself, the result is x. The same is true when we multiply $\sqrt{x}$ by itself, suggesting that $x^{1/2} = \sqrt{x}$.

> For any integer $n > 1$, we define $a^{1/n}$ to be the *nth* root of a, or $\sqrt[n]{a}$.

EXAMPLE 1   Rewrite $27^{1/3}$ as a radical expression and simplify if possible.

Solution

$$27^{1/3} = \sqrt[3]{27}$$ Rewrite as a radical expression. An exponent of $1/3$ is equivalent to a third root.
$$= 3$$ Simplify.

EXAMPLE 2   Rewrite $(625x^8)^{1/4}$ as a radical expression and simplify if possible.

Solution

$$(625x^8)^{1/4} = \sqrt[4]{625x^8}$$ Rewrite as a radical expression.
$$= 5x^2$$ Simplify.

Quick Check   1 Rewrite as a radical expression and simplify if possible.

a) $81^{1/2}$
b) $(-27x^{15})^{1/3}$

EXAMPLE 3   Rewrite $(x^{40}y^{15}z^5)^{1/5}$ as a radical expression and simplify if possible.

Solution

$$(x^{40}y^{15}z^5)^{1/5} = \sqrt[5]{x^{40}y^{15}z^5} \qquad \text{Rewrite as a radical expression}$$
$$= x^8y^3z \qquad \text{Simplify.}$$

Quick Check **2**
Rewrite $(a^{40}b^{64}c^{24})^{1/4}$
as a radical expression
and simplify if possible.
Assume that all variables
represent nonnegative
real numbers.

For any negative number x and even integer n, $x^{1/n}$ is not a real number. For example, $(-64)^{1/6}$ is not a real number, because $(-64)^{1/6} = \sqrt[6]{-64}$ and an even root of a negative number is not a real number.

EXAMPLE 4   Rewrite $\sqrt[5]{x}$ using rational exponents.

Solution

Quick Check **3**
Rewrite $\sqrt[8]{a}$ using
rational exponents.

$$\sqrt[5]{x} = x^{1/5} \qquad \text{Rewrite using the definition } \sqrt[n]{x} = x^{1/n}.$$

Rational Exponents of the Form *m/n*

Objective 2   Simplify expressions containing exponents of the form *m/n*.
We now turn our attention to rational exponents of the form m/n, for any integers m and $n > 1$. The expression $a^{m/n}$ can be rewritten as $(a^{1/n})^m$, which is equivalent to $(\sqrt[n]{a})^m$.

> For any integers m and n, $n > 1$, we define $a^{m/n}$ to be $(\sqrt[n]{a})^m$. This is also equivalent to $\sqrt[n]{a^m}$. The denominator in the exponent, n, is the root. The numerator in the exponent, m, is the power to which we raise this radical.
>
> $$\overset{\text{power}}{\overset{\displaystyle\searrow}{\underset{\displaystyle\nearrow}{\underset{\text{root}}{}}}}$$
> $$a^{m/n}$$

EXAMPLE 5   Rewrite $(81x^{16})^{3/4}$ as a radical expression and simplify if possible.

Quick Check **4**
Rewrite $(64x^{30})^{5/6}$ as
a radical expression
and simplify if possible.
Assume that x is a
nonnegative real
number.

Solution

$$(81x^{16})^{3/4} = (\sqrt[4]{81x^{16}})^3 \qquad \text{Rewrite as a radical expression.}$$
$$= (3x^4)^3 \qquad \text{Simplify the radical.}$$
$$= 27x^{12} \qquad \text{Raise } 3x^4 \text{ to the third power.}$$

EXAMPLE 6   Rewrite $\sqrt[4]{x^3}$ using rational exponents.

Solution

Quick Check **5**
Rewrite $\sqrt[14]{x^5}$ using
rational exponents.

$$\sqrt[4]{x^3} = x^{3/4} \qquad \text{Rewrite using the definition } \sqrt[n]{x^m} = x^{m/n}.$$

Simplifying Expressions Containing Rational Exponents

Objective 3 **Simplify expressions containing rational exponents.** The properties of exponents developed in Chapter 5 for integer exponents are true for fractional exponents as well. Here is a summary of the properties.

Properties of Exponents

1. For any base x, $x^m \cdot x^n = x^{m+n}$.
2. For any base x, $(x^m)^n = x^{m \cdot n}$.
3. For any bases x and y, $(xy)^n = x^n y^n$.
4. For any base x, $\dfrac{x^m}{x^n} = x^{m-n}$ $(x \neq 0)$.
5. For any base x, $x^0 = 1$ $(x \neq 0)$.
6. For any bases x and y, $\left(\dfrac{x}{y}\right)^n = \dfrac{x^n}{y^n}$ $(y \neq 0)$.
7. For any base x, $x^{-n} = \dfrac{1}{x^n}$. $(x \neq 0)$.

EXAMPLE 7   Simplify the expression $x^{5/4} \cdot x^{7/6}$. Assume that x is a nonnegative real number.

Solution

When multiplying two expressions with the same base, we add the exponents and keep the base.

$$
\begin{aligned}
x^{5/4} \cdot x^{7/6} &= x^{5/4 + 7/6} && \text{Add the exponents, keep the base.}\\
&= x^{15/12 + 14/12} && \text{Rewrite the fractions with a common denominator.}\\
&= x^{29/12} && \text{Add.}
\end{aligned}
$$

Quick Check 6

Simplify the expression $x^{2/3} \cdot x^{5/9}$. Assume that x is a nonnegative real number.

EXAMPLE 8   Simplify the expression $(x^{3/8})^{10/9}$. Assume that x is a nonnegative real number.

Solution

When raising an exponential expression to another power, we multiply the exponents and keep the base.

$$
\begin{aligned}
(x^{3/8})^{10/9} &= x^{3/8 \cdot 10/9} && \text{Multiply the exponents, keep the base.}\\
&= x^{5/12} && \text{Multiply.}
\end{aligned}
$$

Quick Check 7

Simplify the expression $(x^{4/5})^{3/2}$. Assume that x is a nonnegative real number.

EXAMPLE 9 Simplify the expression $\left(\dfrac{a^{10}b^6}{c^8 d^2}\right)^{5/2}$, $(c \neq 0, d \neq 0)$. Assume that all variables represent nonnegative real numbers.

Solution

We begin by raising each factor to the $\frac{5}{2}$ power.

$$\left(\frac{a^{10}b^6}{c^8 d^2}\right)^{5/2} = \frac{a^{10 \cdot 5/2} b^{6 \cdot 5/2}}{c^{8 \cdot 5/2} d^{2 \cdot 5/2}}$$
Raise each factor to the $\frac{5}{2}$ power by multiplying each exponent by $\frac{5}{2}$.

$$= \frac{a^{25} b^{15}}{c^{20} d^5}$$
Multiply exponents.

Quick Check 8
Simplify the expression
$\left(\dfrac{x^{15}}{y^5 z^{10}}\right)^{2/5}$, $(y \neq 0, z \neq 0)$.

Simplifying Expressions Containing Negative Rational Exponents

Objective 4 Simplify expressions containing negative rational exponents.

EXAMPLE 10 Simplify the expression $243^{-4/5}$.

Solution

We begin by rewriting the expression with a positive exponent.

$$243^{-4/5} = \frac{1}{243^{4/5}}$$
Rewrite the expression with a positive exponent.

$$= \frac{1}{\left(\sqrt[5]{243}\right)^4}$$
Rewrite in radical notation.

$$= \frac{1}{3^4}$$
Simplify the radical.

$$= \frac{1}{81}$$
Raise 3 to the fourth power.

Quick Check 9
Simplify the expression
$25^{-5/2}$.

> **Building Your Study Strategy** Note Taking, 7 **Write Down Each Step** When your instructor solves a problem during class, copy down every step your instructor writes. Many students feel that they completely understand each step of a solution when it is written on the board, only to realize that they have forgotten when they start to do their homework. Writing down each step prevents this.
>
> If you become confused about a problem, writing down each step will help you when you try to make sense of the material after class. Leave yourself some room for adding explanations later. Place a large question mark in the margin of your notes next to steps you do not understand, and ask your instructor, study group, or tutor for clarification after class.
>
> If you ask your instructor for clarification about a particular step, write down the explanation in your notes right next to the step in question.

Vocabulary

1. For any integer $n > 1$, $a^{1/n} =$ _____.

2. For any integers m and n, $n > 1$, $a^{m/n} =$ _____.

Rewrite each radical expression using rational exponents.

3. $\sqrt[6]{x}$

4. $\sqrt[7]{a}$

5. $\sqrt{10}$

6. $\sqrt[4]{29}$

7. $4\sqrt[5]{x}$

8. $\sqrt[5]{4x}$

Rewrite as a radical expression and simplify if possible. Assume that all variables represent nonnegative real numbers.

9. $36^{1/2}$

10. $49^{1/2}$

11. $32^{1/5}$

12. $256^{1/4}$

13. $(-216)^{1/3}$

14. $(-3125)^{1/5}$

15. $(x^{28})^{1/4}$

16. $(x^{33})^{1/3}$

17. $(a^{42}b^{35})^{1/7}$

18. $(a^{38}b^{24})^{1/2}$

19. $(81x^{16}y^{24})^{1/4}$

20. $(125x^{21}y^{30}z^3)^{1/3}$

21. $(-27x^9y^{12}z^{15})^{1/3}$

22. $(-243x^5y^{20}z^{55})^{1/5}$

Rewrite each radical expression using rational exponents.

23. $(\sqrt[7]{x})^4$

24. $(\sqrt[6]{x})^5$

25. $(\sqrt[10]{a})^9$

26. $(\sqrt[5]{b})^{13}$

27. $\sqrt[8]{x^3}$

28. $\sqrt[4]{x^{19}}$

29. $(\sqrt[3]{5x^2})^8$

30. $(\sqrt[7]{2x^5})^6$

Rewrite as a radical expression and simplify if possible. Assume that all variables represent nonnegative real numbers.

31. $9^{3/2}$

32. $16^{5/4}$

33. $1000^{5/3}$

34. $32^{3/5}$

35. $(x^{18})^{2/3}$

36. $(x^{20})^{7/4}$

37. $(625a^{12}b^{16})^{3/4}$

38. $(49x^{20}y^{32}z^{44})^{3/2}$

39. $(-27x^{15}y^{21}z^3)^{4/3}$

40. $(16a^{12}b^8c^{12})^{9/4}$

Simplify the expression. Assume that all variables represent nonnegative real numbers.

41. $x^{3/8} \cdot x^{1/8}$

42. $x^{5/4} \cdot x^{9/4}$

43. $x^{2/3} \cdot x^{1/4}$

44. $x^{3/8} \cdot x^{4/7}$

45. $x^{7/4} \cdot x^{2/5}$

46. $x^{5/6} \cdot x^{7/30}$

47. $a^{2/3} \cdot a^{1/4} \cdot a^{1/6}$

48. $a^{3/5} \cdot a^{3/2} \cdot a^{7/10}$

49. $x^{3/7}y^{2/5} \cdot x^{1/2}y^{3/4}$

50. $x^{3/4}y^{3/8} \cdot x^{5/2}y^{2/3}$

51. $(x^{1/4})^{2/3}$

52. $(x^{5/6})^{3/8}$

53. $(m^{3/5})^{3/5}$

54. $(n^{3/4})^{2/9}$

55. $(x^6)^{7/8}$

56. $(x^{11/6})^8$

57. $\dfrac{x^{3/5}}{x^{1/10}} \ (x \neq 0)$

58. $\dfrac{x^{5/6}}{x^{7/12}} \ (x \neq 0)$

59. $\dfrac{x^{9/8}}{x^{2/3}} \ (x \neq 0)$

60. $\dfrac{x^{4/5}}{x^{11/30}} \ (x \neq 0)$

61. $\dfrac{x^{9/13}}{x^{9/13}} \ (x \neq 0)$

62. $\dfrac{x^{2/7}}{x^{2/7}} \ (x \neq 0)$

63. $(x^{5/8})^0 \ (x \neq 0)$

64. $(x^{4/9})^0 \ (x \neq 0)$

65. $8^{-1/3}$

66. $25^{-1/2}$

67. $32^{-2/5}$

68. $343^{-2/3}$

69. $27^{-2/3} \cdot 27^{-5/3}$

70. $16^{-3/4} \cdot 16^{-5/4}$

71. $12^{-2/5} \cdot 12^{-3/5}$

72. $10^{-4/3} \cdot 10^{-8/3}$

73. $\dfrac{125^{1/3}}{125^{5/3}}$

74. $\dfrac{32^{3/5}}{32^{9/5}}$

75. $\dfrac{64^{10/3}}{64^{14/3}}$

76. $\dfrac{9^{5/2}}{9^7}$

77. List 4 expressions containing rational exponents of the form $\dfrac{1}{n}$ that are equivalent to 3.

78. List 4 expressions containing rational exponents of the form $\dfrac{m}{n}$ that are equivalent to 8.

79. List 4 expressions containing rational exponents that are equivalent to $7x^4$.

80. List 4 expressions containing rational exponents that are equivalent to $a^9 b^{12} c^{15}$.

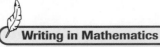

Writing in Mathematics

Answer in complete sentences.

81. Is $-16^{1/2}$ a real number? Explain your answer.

82. Is $16^{-1/2}$ a real number? Explain your answer.

83. *Solutions Manual* * Write a solutions manual page for the following problem:

Simplify $(16x^8)^{3/4}$.

84. *Newsletter* * Write a newsletter that explains how to rewrite an expression with rational exponents in radical form.

*See Appendix B for details and sample answers.

Summary of Chapter 8 Study Strategies

- A good set of class notes is one of the most valuable study resources you can create. To ensure that you create the best set of notes that you can, begin by choosing your seat wisely. Be sure that you can see the entire board and that you can clearly hear the instructor from your location.
- Write down each step of the solution to a problem, whether you feel the step is necessary or not. Often solutions seem clear while sitting in class, but will not be so clear when you sit down to work the homework exercises. If you find that you are completely lost in class, do your best to write down every step so that you can try to make sense of the steps after class when reworking your notes.
- If you find that you are having trouble keeping up with your instructor, it is important that you write down enough information so that you will be able to understand the material after class has ended. Try using abbreviations when appropriate, and write down phrases rather than complete sentences. The use of a tape recorder, with your instructor's permission, can help you fill in missing sections in your notes. A classmate can help you fill in missing notes as well. If your instructor gives you a cue that material is important, be sure that you write that material in your notes regardless of where you are in your notes. You can catch up on missing material after class.
- Reworking your notes after class serves as an excellent review of what was covered during the class period, as well as giving you an opportunity to make your notes more legible and user friendly. During this time you can fill in any holes in your notes and supplement your notes with extra material.

Simplify the radical expression. Assume that all variables represent nonnegative real numbers. [8.1]

1. $\sqrt{81}$

2. $\sqrt{196}$

3. $\sqrt[3]{-125}$

4. $\sqrt[4]{x^{20}}$

5. $\sqrt{25x^{18}}$

6. $\sqrt[3]{64x^{21}y^{18}z^{30}}$

Approximate to the nearest thousandth, using a calculator. [8.1]

7. $\sqrt{60}$ 8. $\sqrt{239}$

Evaluate the radical function. (Round to the nearest thousandth if necessary.) [8.1]

9. $f(x) = \sqrt{3x + 4}, f(20)$

10. $f(x) = \sqrt{x^2 - 9x + 30}, f(-10)$

Find the domain of the radical function. Express your answer in interval notation. [8.1]

11. $f(x) = \sqrt{5x - 15}$

12. $f(x) = \sqrt[4]{3x + 7}$

Simplify the radical expression. Assume that all variables represent nonnegative real numbers. [8.2]

13. $\sqrt{54}$

14. $\sqrt[3]{x^{17}}$

15. $\sqrt[4]{a^{23}b^{18}c^{12}}$

16. $\sqrt[3]{440x^{25}y^{35}}$

17. $\sqrt{150x^{33}y^{14}z}$

18. $\sqrt{63r^{15}s^{20}t^{39}}$

Add or subtract. Assume that all variables represent nonnegative real numbers. [8.2]

19. $7\sqrt{5} + 9\sqrt{5}$

20. $12\sqrt[3]{x} - 4\sqrt[3]{x}$

21. $9\sqrt{12} + 7\sqrt{147}$

22. $8\sqrt{50} - 10\sqrt{48} + 6\sqrt{18}$

Multiply. Assume that all variables represent nonnegative real numbers. [8.3]

23. $\sqrt[3]{6x^2} \cdot \sqrt[3]{45x^8}$

24. $7\sqrt{2}(9\sqrt{6} - 4\sqrt{10})$

25. $(2\sqrt{7} + \sqrt{3})(3\sqrt{7} + 2\sqrt{3})$

26. $(5\sqrt{5} - 9\sqrt{2})(3\sqrt{5} + 6\sqrt{2})$

27. $(4\sqrt{3} + 11\sqrt{2})^2$

28. $(3\sqrt{5} + 4\sqrt{3})(3\sqrt{5} - 4\sqrt{3})$

Simplify. Assume that all variables represent nonnegative real numbers. [8.3]

29. $\dfrac{\sqrt{168x^9}}{\sqrt{2x}}$

30. $\sqrt{\dfrac{99a^{13}b^{12}c^{20}}{44a^7b^5c^{34}}}$

Rationalize the denominator and simplify. Assume that all variables represent nonnegative real numbers. [8.3]

31. $\sqrt{\dfrac{12}{5}}$ 32. $\dfrac{10}{\sqrt{6}}$

33. $\sqrt{\dfrac{7}{20}}$ 34. $\dfrac{x^7y^3z^8}{\sqrt{x^2yz^5}}$

35. $\dfrac{6}{\sqrt{13} + \sqrt{5}}$

36. $\dfrac{9\sqrt{2}}{2\sqrt{6} - \sqrt{14}}$

37. $\dfrac{3\sqrt{5} + 4\sqrt{2}}{5\sqrt{5} - \sqrt{2}}$

38. $\dfrac{4 + 2\sqrt{3}}{8 - 3\sqrt{3}}$

Solve. [8.4]

39. $\sqrt{2x + 11} = 5$

40. $\sqrt[3]{8x - 9} = 4$

41. $\sqrt{4x - 15} + 3 = x$

42. $\sqrt{2x + 31} = x - 2$

43. $\sqrt{5x - 8} = \sqrt{2x + 25}$

44. $\sqrt{5x - 4} = 2 + \sqrt{x + 8}$

Worked-out solutions to Review Exercises marked with can be found on page AN–27.

45. For the function $f(x) = \sqrt{4x - 23}$, find all values x for which $f(x) = 5$. [8.4]

46. For the function $f(x) = \sqrt[3]{x^2 - 7x + 34}$, find all values x for which $f(x) = 4$. [8.4]

47. A pendulum has a length of 4.5 feet. Find its period, rounded to the nearest hundredth of a second. Use the formula $T = 2\pi\sqrt{\dfrac{L}{32}}$. [8.5]

48. A vehicle involved in an accident made 320 feet of skid marks on the asphalt. The speed, s, that the vehicle was traveling in miles per hour when it started skidding can be approximated by the formula $s = \sqrt{30df}$, where d represents the length of the skid marks in feet and f represents the drag factor of the road. If the drag factor for asphalt is 0.75, find the speed the car was traveling. Round to the nearest mile per hour. [8.5]

Express in terms of i. [8.6]

49. $\sqrt{-16}$

50. $\sqrt{-56}$

51. $\sqrt{-882}$

52. $-\sqrt{-325}$

Add or subtract the complex numbers. [8.6]

53. $(9 + 4i) + (3 + 10i)$

54. $(10 - 7i) + (-6 + 4i)$

55. $(5 - 4i) - (9 + 7i)$

56. $(8 + 5i) - (7 - 5i)$

Multiply. [8.6]

57. $4i \cdot 9i$

58. $\sqrt{-14} \cdot \sqrt{-18}$

59. $5i(8 + 3i)$

60. $(9 - 5i)(2 + 7i)$

61. $(5 + 4i)^2$

62. $(8 + 3i)(8 - 3i)$

Rationalize the denominator. [8.6]

63. $\dfrac{4}{5 + 3i}$

64. $\dfrac{10i}{3 - 4i}$

65. $\dfrac{5 - i}{7 + 4i}$

66. $\dfrac{8}{i}$

Rewrite each radical expression using rational exponents. [8.7]

67. $\sqrt[3]{x}$

68. $(\sqrt[8]{a})^5$

69. $(\sqrt{m})^{11}$

70. $\sqrt[4]{x^9}$

Rewrite as a radical expression and simplify if possible. Assume that all variables represent nonnegative real numbers. [8.7]

71. $81^{1/4}$

72. $(32x^{20})^{3/5}$

Simplify the expression. Assume that all variables represent nonnegative real numbers. [8.7]

73. $x^{7/3} \cdot x^{1/6}$

74. $(x^{2/3})^{9/10}$

75. $\dfrac{x^{7/10}}{x^{1/2}}$

76. $36^{-1/2}$

77. $27^{-4/3}$

78. $\dfrac{16^{3/4}}{16^{9/4}}$

For Extra Help Pass the Test

Test solutions are found on the enclosed CD.

Simplify the radical expression. Assume that all variables represent nonnegative real numbers.

1. $\sqrt{36a^{14}b^{12}}$

2. For $f(x) = \sqrt{x^2 - 4x + 32}$, evaluate $f(-5)$. Round to the nearest thousandth.

Simplify the radical expression. Assume that all variables represent nonnegative real numbers.

3. $\sqrt{88}$

4. $\sqrt[5]{x^{29}y^{17}z^{40}}$

Add or subtract. Assume that all variables represent nonnegative real numbers.

5. $5\sqrt{8} + 9\sqrt{18} - 2\sqrt{32}$

Multiply. Assume that all variables represent nonnegative real numbers.

6. $7\sqrt{3}(4\sqrt{6} - 3\sqrt{15})$

7. $(8\sqrt{2} + \sqrt{3})(6\sqrt{2} - 5\sqrt{3})$

Rationalize the denominator and simplify. Assume that all variables represent nonnegative real numbers.

8. $\dfrac{a^5b^7}{\sqrt{a^3b^{10}}}$

9. $\dfrac{9\sqrt{6} + 2\sqrt{5}}{\sqrt{6} - 3\sqrt{5}}$

Solve.

10. $\sqrt[3]{2x + 1} = 7$

11. $\sqrt{3x + 4} + 2 = x$

12. If a pendulum has a period of 3.2 seconds, find its length, rounded to the nearest tenth of a foot. Use the formula $T = 2\pi\sqrt{\dfrac{L}{32}}$.

13. Express $\sqrt{-68}$ in terms of i.

Add or subtract the complex numbers.

14. $(12 - 7i) - (31 + 6i)$

Multiply.

15. $(10 + 3i)(10 - 3i)$

Rationalize the denominator.

16. $\dfrac{7 + 6i}{3 - i}$

Simplify the expression. Assume that all variables represent nonnegative real numbers.

17. $x^{3/10} \cdot x^{1/2}$

18. $\dfrac{n^{5/6}}{n^{3/8}}$

19. $216^{-2/3}$

Mathematicians in History
Pythagoras of Samos

Pythagoras of Samos, often referred to simply as Pythagoras, was the leader of a society known as the Pythagoreans. This society was partially religious and partially scientific in nature. One of their mathematical achievements is one of the first proofs of a theorem that related the lengths of the sides of a right triangle. This theorem is known as the Pythagorean theorem.

Write a one-page summary (*or* make a poster) of the life of Pythagoras, the mathematical achievements of Pythagoras and his society, and the beliefs of the Pythagoreans.

Interesting issues:

- What was the "semicircle?"
- Who were the mathematikoi, and by what rules did they lead their lives?
- What number did the Pythagoreans consider to be the "best" number, and why?
- Which society was the first to know of the Pythagorean theorem?

The Pythagoreans believed that all relationships could be expressed numerically as a ratio of two integers. They discovered, however, that the diagonal of a square whose side has length 1 was an irrational number. This number is $\sqrt{2}$. This discovery rocked the foundation of their system of beliefs, and they swore each other to secrecy regarding this discovery. A Pythagorean named Hippasus told others outside of the society of this irrational number. What was the fate of Hippasus?

Have you ever watched a child swing back and forth on a playground swing? Do you remember seeing in the movies a person swinging a pocket-watch back and forth in order to put another person into a trance? This motion of the swing and pocket-watch is called a pendulum. Today, you will be creating your own pendulum and using the data you collect in order to discover a mathematical model.

Chart 1

Length of Pendulum (Inches)	Number of Seconds for 10 Swings					Average Time
	Trial 1	Trial 2	Trial 3	Trial 4	Trial 5	

Step 1: Form groups of no more than 4 or 5 students. Each group should have a yard stick or ruler, a piece of string (approximately 2 yards long), a weight tied to the end of the string (hardware washers work well), and a stop watch (or some method for counting seconds).

Step 2: Measure a length for your pendulum (in inches) and fasten your string to the edge of a desk or table at this length. For example you might want to start with a pendulum that is 20 inches. Write the length of your pendulum in Chart 1.

Step 3: Pull the pendulum back and release. Calculate the number of seconds it takes for your pendulum to make 10 swings (note a swing is one complete arc back and forth). Record the time in Chart 1. Repeat this process

three to five times for the same length. Calculate the average time needed for this given length.

Step 4: Change the length of your pendulum and record the length in Chart 1. Record the time needed for 10 swings. Repeat this process three to five times for the same length. Calculate the average time needed for this given length.

Step 5: Repeat Step 4 a few more times for different lengths of the pendulum.

Step 6: What did you notice about the length of the pendulum in relation to the length of time needed for 10 swings? Did the time increase as the length increased? Or did the time decrease as the length increased?

Chart 2

Length of Pendulum	Square Root Length of Pendulum	Average Time for 10 Swings	Average Time for 1 Swing	Average Time for 1 Swing Divided by Square Root Length of Pendulum

Step 7: Fill in Chart 2.

Step 8: What do you notice about the last column? As it turns out, these values should be relatively the same and should be around 0.32. There is actually a formula that uses square roots and that relates the time it takes for a pendulum to make one swing and the length of the

pendulum. This formula is $T = 2\pi\sqrt{\dfrac{L}{384}}$ where T is the time for one swing in seconds and L is the length of the pendulum in inches.

Step 9: See how close your times were to the times the formula would predict. Were your actual times close to the predicted times?

Quadratic Equations

In this chapter, we will learn how to apply techniques other than factoring to solve quadratic equations. One important goal of the chapter is determining which technique is the most efficient way to solve a particular equation. We will also apply these techniques to new application problems that previously could not be solved because they produce equations which are not factorable. Later in the chapter, we will examine the graph of a quadratic equation. The graph of a quadratic equation is called a parabola and is U-shaped. The techniques used to graph linear equations, along with some new techniques, will be used in graphing quadratic equations.

Study Strategy   **Time Management**   *When asked why they are having difficulties in a particular class, many students claim that they simply do not have enough time. A great number of these students do have enough time but lack the* time-management skills *to make the most of their time. It seems* as if they fritter and waste the hours in an offhand way. *In this chapter we will discuss effective time-management strategies, focusing on how to make more efficient use of available time.*

9.1

Solving Quadratic Equations by Extracting Square Roots; Completing the Square

Objectives

1 Review solving quadratic equations by factoring.

2 Solve quadratic equations by extracting square roots.

3 Solve quadratic equations by extracting square roots from a linear expression that is squared.

4 Solve quadratic equations by completing the square.

5 Find a quadratic equation given its solutions.

Solving Quadratic Equations by Factoring

Objective 1 **Review solving quadratic equations by factoring.** In Chapter 6 we learned how to solve quadratic equations by factoring. We begin this section with a review of this technique. We start by simplifying both sides of the equation completely. This includes distributing and combining like terms when possible. After that, we need to set the equation equal to 0 by collecting all of the terms on one side of the equation. Recall that it is a good idea to move the terms to the side of the equation that will produce a positive second-degree term. After the equation contains an expression equal to 0, we need to factor the expression. Finally, we set each factor equal to 0 and solve the resulting equation.

EXAMPLE 1   Solve $x^2 - 8x = 48$.

Solution

$$x^2 - 8x = 48$$
$$x^2 - 8x - 48 = 0 \qquad \text{Subtract 48 to collect all terms on the left side.}$$
$$(x - 12)(x + 4) = 0 \qquad \text{Factor.}$$
$$x - 12 = 0 \quad \text{or} \quad x + 4 = 0 \qquad \text{Set each factor equal to 0.}$$
$$x = 12 \quad \text{or} \quad x = -4 \qquad \text{Solve the resulting equations.}$$

▶ The solution set is $\{-4, 12\}$.

EXAMPLE 2   Solve $x^2 = 81$.

Solution

$$x^2 = 81$$
$$x^2 - 81 = 0 \qquad \text{Set the equation equal to 0.}$$
$$(x + 9)(x - 9) = 0 \qquad \text{Factor. (Difference of Squares)}$$
$$x + 9 = 0 \quad \text{or} \quad x - 9 = 0 \qquad \text{Set each factor equal to 0.}$$
$$x = -9 \quad \text{or} \quad x = 9 \qquad \text{Solve the resulting equations.}$$

The solution set is $\{-9, 9\}$.

Quick Check 1

Solve.

a) $x^2 = 9x - 18$

b) $x^2 = 100$

Solving Quadratic Equations by Extracting Square Roots

Objective 2 **Solve quadratic equations by extracting square roots.** In the preceding example, we were trying to find a number x that, when squared, equals 81. The principal square root of 81 is 9, which gives us one of the two solutions. In an equation that has a squared term equal to a constant, we can take the square root of both sides of the equation to find the solution, as long as we take both the positive and negative square root of the constant. The symbol " $\pm$ " is used to represent both the positive and negative square root, and is read as "plus or minus." For example, $x = \pm 9$ means $x = 9$ or $x = -9$. This technique for solving quadratic equations is called **extracting square roots.**

Extracting Square Roots

1. Isolate the squared term.
2. Take the square root of each side. (Remember to take both the positive and negative $(\pm)$ square root of the constant.)
3. Simplify the square root.
4. Solve by isolating the variable.

Now we will use extracting square roots to solve an equation that we would not have been able to solve by factoring.

EXAMPLE 3   Solve $x^2 - 40 = 0$.

Solution

The expression $x^2 - 40$ cannot be factored because 40 is not a perfect square. We proceed to solve this equation by extracting square roots.

$$x^2 - 40 = 0$$
$$x^2 = 40 \qquad \text{Add 40 to isolate the squared term.}$$
$$\sqrt{x^2} = \pm\sqrt{40} \qquad \text{Take the square root of each side.}$$
$$x = \pm 2\sqrt{10} \qquad \text{Simplify the square root. } \sqrt{40} = \sqrt{4 \cdot 10} = 2\sqrt{10}$$

The solution set is $\{-2\sqrt{10}, 2\sqrt{10}\}$. Using a calculator, we find that the solutions are approximately ± 6.32.

A Word of Caution   When you take the square root of both sides of an equation, do not forget to use the symbol $\pm$.

This technique can be used to find complex solutions to equations, as well as real solutions.

EXAMPLE ▶4   Solve $x^2 = -25$.

Solution

If we added the 25 to the left side of the equation, then the resulting equation would be $x^2 + 25 = 0$. The expression $x^2 + 25$ is not factorable (recall that the sum of two squares is not factorable). The only way for us to find a solution would be to take the square roots of both sides of the original equation. We proceed to solve this equation by extracting square roots.

$$x^2 = -25$$
$$\sqrt{x^2} = \pm\sqrt{-25} \qquad \text{Take the square root of each side.}$$
$$x = \pm 5i \qquad \text{Simplify the square root.}$$

The solution set is $\{-5i, 5i\}$.

Quick Check 2

Solve.
a) $x^2 = 18$
b) $x^2 + 24 = 0$

Objective 3 Solve quadratic equations by extracting square roots from a linear expression that is squared. Extracting square roots is an excellent technique to solve quadratic equations whenever the equation is made up of a squared term and a constant term. Consider the equation $(2x + 9)^2 = 25$. In order to use factoring to solve this equation, we would first have to square $2x + 9$, then collect all terms on the left side of the equation by subtracting 25, and hope that the resulting expression can be factored. Extracting square roots is a more efficient way to solve this equation.

EXAMPLE ▶5   Solve $(2x + 9)^2 = 25$.

Solution

$$(2x + 9)^2 = 25$$
$$\sqrt{(2x + 9)^2} = \pm\sqrt{25} \qquad \text{Take the square root of each side.}$$
$$2x + 9 = \pm 5 \qquad \text{Simplify the square root. Note that the square root of } (2x + 9)^2 \text{ is } 2x + 9.$$
$$2x = -9 \pm 5 \qquad \text{Subtract 9.}$$
$$x = \frac{-9 \pm 5}{2} \qquad \text{Divide by 2.}$$

$\dfrac{-9 - 5}{2} = -7$ and $\dfrac{-9 + 5}{2} = -2$, so the solution set is $\{-7, -2\}$.

EXAMPLE ▶6   Solve $(4x + 3)^2 + 18 = 10$.

Solution

$$(4x + 3)^2 + 18 = 10$$
$$(4x + 3)^2 = -8 \qquad \text{Subtract 18 to isolate the squared term.}$$
$$\sqrt{(4x + 3)^2} = \pm\sqrt{-8} \qquad \text{Take the square root of each side.}$$

$$4x + 3 = \pm 2i\sqrt{2}$$ Simplify the square root.

$$4x = -3 \pm 2i\sqrt{2}$$ Subtract 3.

$$x = \frac{-3 \pm 2i\sqrt{2}}{4}$$ Divide both sides by 4.

Quick Check   3

Solve.
a) $(3x - 7)^2 = 64$
b) $4(2x - 1)^2 - 23 = 25$

Since we cannot simplify the numerator in this case, the solution set is

$$\left\{ \frac{-3 - 2i\sqrt{2}}{4}, \frac{-3 + 2i\sqrt{2}}{4} \right\}.$$

Solving Quadratic Equations by Completing the Square

Objective 4   Solve quadratic equations by completing the square. We can solve any quadratic equation by converting it to an equation that sets a squared term equal to a constant. Rewriting the equation in this form allows us to solve it by extracting square roots. We will now examine a procedure for doing this called **completing the square**. This procedure is used for an equation such as $x^2 - 6x - 27 = 0$, which has both a second-degree term (x^2) and a first-degree term ($-6x$). In order for us to use this technique, the leading coefficient must be positive 1.

The first step is to isolate the variable terms on one side of the equation with the constant term on the other side. This can be done by adding 27 to both sides of the equation. The resulting equation is $x^2 - 6x = 27$.

The next step is to add a number to both sides of the equation that makes the side of the equation containing the variable terms into a perfect square trinomial. To do this, we take half of the coefficient of the first-degree term, which in this case is -6, and square it.

$$\left(\frac{-6}{2} \right)^2 = (-3)^2 = 9$$

After we add 9 to both sides of the equation, the resulting equation is $x^2 - 6x + 9 = 36$. The expression on the left side of the equation, $x^2 - 6x + 9$, is a perfect square trinomial and can be factored as $(x - 3)^2$. The resulting equation, $(x - 3)^2 = 36$, is in the correct form for extracting square roots. (This equation will be solved completely in the following example.)

Here is the procedure for solving a quadratic equation by completing the square, provided that the coefficient of the squared term is 1.

Completing the Square

1. Isolate all variable terms on one side of the equation, with the constant term on the other side of the equation.
2. Identify the coefficient of the first-degree term. Divide that number by two, square it, and add that to both sides of the equation.
3. Factor the resulting perfect square trinomial.
4. Take the square root of each side of the equation. Be sure to include $\pm$ in front of the constant.
5. Solve the resulting equation.

Here is the full solution for the example $x^2 - 6x - 27 = 0$.

EXAMPLE ▸ 7   Solve $x^2 - 6x - 27 = 0$ by completing the square.

Solution

$$x^2 - 6x - 27 = 0$$

$$x^2 - 6x = 27 \qquad \text{Add 27.}$$

$$x^2 - 6x + 9 = 27 + 9 \qquad \left(\tfrac{-6}{2}\right) = (-3)^2 = 9. \text{ Add 9 to complete the square.}$$

$$(x - 3)^2 = 36 \qquad \text{Simplify. Factor the left side.}$$

$$\sqrt{(x - 3)^2} = \pm\sqrt{36} \qquad \text{Take the square root of each side.}$$

$$x - 3 = \pm 6 \qquad \text{Simplify the square root.}$$

$$x = 3 \pm 6 \qquad \text{Add 3.}$$

▸ $3 + 6 = 9$ and $3 - 6 = -3$, so the solution set is $\{-3, 9\}$.

Note that the equation in the previous example could have been solved with less work by factoring $x^2 - 6x - 27$. Always use factoring when possible, because it often leads to the quickest and most direct solutions.

EXAMPLE ▸ 8   Solve $x^2 + 10x + 30 = 0$ by completing the square.

Solution

The expression $x^2 + 10x + 30$ does not factor, so we proceed with completing the square.

$$x^2 + 10x + 30 = 0$$

$$x^2 + 10x = -30 \qquad \text{Subtract 30.}$$

$$x^2 + 10x + 25 = -30 + 25 \qquad \text{Half of 10 is 5, which equals 25 when squared:}$$
$$\left(\tfrac{10}{2}\right)^2 = (5)^2 = 25. \text{ Add 25.}$$

$$x^2 + 10x + 25 = -5 \qquad \text{Simplify.}$$

$$(x + 5)^2 = -5 \qquad \text{Factor the left side.}$$

$$\sqrt{(x + 5)^2} = \pm\sqrt{-5} \qquad \text{Take the square root of each side.}$$

$$x + 5 = \pm i\sqrt{5} \qquad \text{Simplify the square root.}$$

$$x = -5 \pm i\sqrt{5} \qquad \text{Subtract 5.}$$

Quick Check ◂ **4**

Solve by completing the square.

a) $x^2 + 8x + 15 = 0$
b) $x^2 - 6x + 20 = 0$

The solution set is $\{-5 - i\sqrt{5}, -5 + i\sqrt{5}\}$.

In the first two examples of completing the square, the coefficient of the first-degree term was an even integer. If this is not the case, we must use fractions to complete the square.

EXAMPLE ▸ 9   Solve $x^2 - 7x - 12 = 0$ by completing the square.

Solution

The expression $x^2 - 7x - 12$ does not factor, so we proceed with completing the square.

$$x^2 - 7x - 12 = 0$$

$$x^2 - 7x = 12 \qquad \text{Add 12.}$$

$$x^2 - 7x + \frac{49}{4} = 12 + \frac{49}{4}$$

Half of -7 is $-\frac{7}{2}$, which equals $\frac{49}{4}$ when squared: $\left(-\frac{7}{2}\right)^2 = \frac{49}{4}$. Add $\frac{49}{4}$.

$$x^2 - 7x + \frac{49}{4} = \frac{97}{4}$$

Add 12 and $\frac{49}{4}$ by rewriting 12 as a fraction whose denominator is 4: $12 + \frac{49}{4} = \frac{48}{4} + \frac{49}{4} = \frac{97}{4}$.

$$\left(x - \frac{7}{2}\right)^2 = \frac{97}{4}$$

Factor.

$$\sqrt{\left(x - \frac{7}{2}\right)^2} = \pm\sqrt{\frac{97}{4}}$$

Take the square root of each side.

$$x - \frac{7}{2} = \pm\frac{\sqrt{97}}{2}$$

Simplify the square root.

$$x = \frac{7}{2} \pm \frac{\sqrt{97}}{2}$$

Add $\frac{7}{2}$

Quick Check 5

Solve $x^2 + 5x + 13 = 0$ by completing the square.

The solution set is $\left\{\dfrac{7 - \sqrt{97}}{2}, \dfrac{7 + \sqrt{97}}{2}\right\}$.

A Word of Caution   In order to solve a quadratic equation by completing the square, the leading coefficient must be positive 1. If the leading coefficient is not equal to 1, then divide both sides of the equation by the leading coefficient.

EXAMPLE 10   Solve $2x^2 + 24x + 66 = 0$ by completing the square.

Solution

The leading coefficient is not equal to 1, so we begin by dividing each term on both sides of the equation by 2. We are then able to solve this equation by completing the square.

$$2x^2 + 24x + 66 = 0$$

$$\frac{2x^2 + 24x + 66}{2} = \frac{0}{2}$$

Divide both sides of the equation by 2, to make the leading coefficient equal to 1.

$$x^2 + 12x + 33 = 0$$

Simplify.

$$x^2 + 12x = -33$$

Subtract 33.

$$x^2 + 12x + 36 = -33 + 36$$

Half of 12 is 6, which equals 36 when squared: $\left(\frac{12}{2}\right)^2 = (6)^2 = 36$.

$$x^2 + 12x + 36 = 3$$

Simplify.

$$(x + 6)^2 = 3$$

Factor.

$$\sqrt{(x+6)^2} = \pm\sqrt{3}$$ Take the square root of each side.

$$x + 6 = \pm\sqrt{3}$$ Simplify.

$$x = -6 \pm \sqrt{3}$$ Subtract 6.

Quick Check **6**
Solve $3x^2 - 24x + 24 = 0$
by completing the square.

The solution set is $\{-6 - \sqrt{3}, -6 + \sqrt{3}\}$.

Finding a Quadratic Equation Given Its Solutions

Objective 5 Find a quadratic equation given its solutions.

EXAMPLE 11 Find a quadratic equation in standard form, with integer coefficients, that has the solution set $\{-\frac{2}{5}, 3\}$.

Solution

Knowing that $x = -\frac{2}{5}$ is a solution tells us that $5x + 2$ is a factor of the quadratic expression:

$$x = -\frac{2}{5}$$

$$5 \cdot x = -\frac{2}{\cancel{5}} \cdot \cancel{5}$$ Multiply both sides by 5 to clear fractions.

$$5x = -2$$ Simplify.

$$5x + 2 = 0$$ Add 2.

Similarly, knowing that $x = 3$ is a solution to the equation tells us that $x - 3$ is a factor of the quadratic expression.

Multiplying these two factors together will give us a quadratic equation with these two solutions.

$$x = -\frac{2}{5} \quad \text{or} \quad x = 3$$ Begin with the solutions.

$$5x + 2 = 0 \quad \text{or} \quad x - 3 = 0$$ Rewrite each equation so the right side is equal to 0.

$$(5x + 2)(x - 3) = 0$$ Write an equation that has these two expressions as factors.

$$5x^2 - 15x + 2x - 6 = 0$$ Multiply.

$$5x^2 - 13x - 6 = 0$$ Simplify.

Quick Check **7**
Find a quadratic
equation in standard
form, with integer
coefficients, that has the
solution set $\{-7, \frac{3}{4}\}$.

A quadratic equation that has the solution set $\{-\frac{2}{5}, 3\}$ is $5x^2 - 13x - 6 = 0$.

Notice that we say "a" quadratic equation rather than "the" quadratic equation. There are infinitely many quadratic equations with integer coefficients that have this solution set. For example, multiplying both sides of our equation by 2 gives us the equation $10x^2 - 26x - 12 = 0$, which has the same solution set.

EXAMPLE 12   Find a quadratic equation in standard form, with integer coefficients, that has the solution set $\{-3i, 3i\}$.

Solution

We know that $x = -3i$ is a solution to the equation. This tells us that $x + 3i$ is a factor of the quadratic expression. Similarly, knowing that $x = 3i$ is a solution tells us that $x - 3i$ is a factor of the quadratic expression. Multiplying these two factors together will give us a quadratic equation with these two solutions.

$x = -3i$ or $x = 3i$	Begin with the solutions.
$x + 3i = 0$ or $x - 3i = 0$	Rewrite each equation so the right side is equal to 0.
$(x + 3i)(x - 3i) = 0$	Write an equation that has these two expressions as factors.
$x^2 - 3ix + 3ix - 9i^2 = 0$	Multiply.
$x^2 + 9 = 0$	Simplify. Note that

Quick Check   8

Find a quadratic equation in standard form, with integer coefficients, that has the solution set $\{-2i, 2i\}$.

A quadratic equation that has the solution set $\{-3i, 3i\}$ is $x^2 + 9 = 0$.

> *Building Your Study Strategy* **Time Management, 1 Keeping Track of Your Time**   The first step in achieving effective time management is keeping track of your time over a one-week period. Mark down the times that you spend in class, working, cooking, watching TV, eating, sleeping, spending time with family, socializing, getting tutorial help, getting ready for school, and, most importantly, studying. This will give you a good idea of how much extra time you have to devote to studying, as well as how much time you devote to other activities. Most students do not realize how much time they waste in a day.
>
> Once you have examined your current schedule, try to schedule more time for studying math. You should be spending between 2 and 4 hours studying per week for each hour that you spend in the classroom. Try to schedule some time each day, rather than cramming it all into the weekend. Leave some blank time periods for flexibility; these periods can be used for emergency study sessions for any of your classes.

Vocabulary

1. The method of solving an equation by taking the square root of both sides is called solving by _____.

2. In order to solve an equation by extracting square roots, first isolate the expression that is _____.

3. The method of solving an equation by rewriting the variable expression as a perfect square binomial is called solving by _____.

4. To solve the equation $x^2 + bx = c$ by completing the square, first add _____ to both sides of the equation.

22. $x^2 = -50$

23. $x^2 + 15 = 47$

24. $x^2 + 12 = -12$

25. $(x - 4)^2 = 81$

26. $(x + 7)^2 = 24$

27. $(x + 5)^2 = -18$

28. $(x - 6)^2 = -49$

29. $(x + 2)^2 + 11 = -37$

30. $(x - 8)^2 - 18 = 46$

31. $2(x - 10)^2 - 21 = 29$

32. $3(x + 4)^2 + 16 = 7$

33. $\left(x - \dfrac{1}{4}\right)^2 = \dfrac{25}{16}$

34. $\left(x + \dfrac{3}{7}\right)^2 = \dfrac{4}{49}$

Solve by factoring.

5. $x^2 - 7x - 18 = 0$

6. $x^2 + 12x + 20 = 0$

7. $x^2 - 9x + 14 = 0$

8. $x^2 + 5x - 24 = 0$

9. $x^2 + 8x + 7 = 0$

10. $x^2 + 15x + 54 = 0$

11. $x^2 - 5x = 50$

12. $x^2 - 28 = 3x$

13. $x^2 - 49 = 0$

14. $x^2 - 4 = 0$

15. $x^2 - 7x = 9x - 55$

16. $x^2 + 12x - 9 = 5x + 21$

Solve by extracting square roots.

17. $x^2 = 25$

18. $x^2 = 9$

19. $x^2 - 80 = 0$

20. $x^2 - 96 = 0$

21. $x^2 = -36$

Fill in the missing term that makes the expression a perfect square trinomial, and factor the resulting expression.

35. $x^2 + 10x + $ _____

36. $x^2 - 4x + $ _____

37. $x^2 - 9x + $ _____

38. $x^2 + 15x + $ _____

39. $x^2 - 12x + $ _____

40. $x^2 - 20x + $ _____

41. $x^2 + \dfrac{1}{6}x + $ _____

42. $x^2 - x + $ _____

Solve by completing the square.

43. $x^2 - 6x - 40 = 0$

44. $x^2 + 8x - 48 = 0$

45. $x^2 + 2x + 13 = 0$

46. $x^2 - 10x + 20 = 0$

47. $x^2 + 6x = 16$

48. $x^2 + 4x = 45$

49. $x^2 + 48 = 14x$

50. $x^2 - 27 = 6x$

51. $x^2 + 9x + 18 = 0$

52. $x^2 + 11x - 42 = 0$

53. $x^2 + 5x = 12$

54. $x^2 - x = 16$

55. $x^2 + \frac{1}{2}x - 3 = 0$

56. $x^2 - \frac{23}{6}x + \frac{10}{3} = 0$

57. $2x^2 - 24x + 64 = 0$

58. $3x^2 - 6x - 12 = 0$

59. $5x^2 - 30x + 65 = 0$

60. $4x^2 - 64x + 268 = 0$

Find a quadratic equation with integer coefficients that has the following solution set.

61. $\{-2, 4\}$

62. $\{-7, -5\}$

63. $\{3, 10\}$

64. $\{-4, 0\}$

65. $\left\{\frac{1}{4}, 3\right\}$

66. $\left\{\frac{2}{3}, \frac{3}{4}\right\}$

67. $\{-5, 5\}$

68. $\{-9, 9\}$

69. $\{-7i, 7i\}$

70. $\{-2i, 2i\}$

Mixed Practice, 71–94

Solve by any method (factoring, extracting square roots, or completing the square).

71. $x^2 - 13x - 30 = 0$

72. $x^2 + 20 = 0$

73. $x^2 + 2x - 39 = 0$

74. $(2x + 7)^2 = 1$

75. $x^2 + 11x - 12 = 0$

76. $x^2 - 8x = -20$

77. $(x - 2)^2 + 19 = 27$

78. $x^2 - 10x - 24 = 0$

79. $x^2 - 4x = 28$

80. $x^2 + 60 = 14x$

81. $x^2 + 2x - 35 = 0$

82. $(x - 6)^2 - 9 = 16$

83. $x^2 + 10x + 40 = 0$

84. $x^2 - 17x + 72 = 0$

85. $2(x + 5)^2 + 30 = -24$

86. $x^2 - 56 = 0$

87. $x^2 + 5x - 13 = 0$

88. $x^2 + 8x + 25 = 0$

89. $x(2x - 3) + 7(2x - 3) = 0$

90. $(x - 9)(x - 9) = 36$

91. $-16x^2 + 80x + 96 = 0$

92. $\left(x + \frac{b}{2a}\right)^2 = \frac{b^2 - 4ac}{4a^2}$ where a, b, and c are constants and $a \neq 0$.

93. $\frac{2}{5}x^2 + \frac{4}{5}x - \frac{48}{5} = 0$

94. $x^2 - 6x - 8 = 0$

For each of the following rational expressions, list the values that are excluded from the domain. (Recall that a real number is excluded from the domain of a rational expression if it causes the denominator to equal 0 when it is substituted into the expression.)

95. $\dfrac{9}{x^2 - 5x - 36}$

96. $\dfrac{12}{x^2 + 5x}$

97. $\dfrac{x - 6}{x^2 - 24}$

98. $\dfrac{x + 4}{x^2 + 36}$

99. $\dfrac{x^2 + x - 20}{(x - 3)^2 - 25}$

100. $\dfrac{x^2 + 3x - 17}{2(2x - 3)^2 - 72}$

101. Fill in the blank in the equation
$x^2 - 2x + 9 = $ _____ so that the equation has

   a) $x = 10$ as a solution
   b) $x = -4$ as a solution

102. Fill in the blank in the equation
$x^2 + 7x - 30 = $ _____ so that the equation has

   a) $x = 9$ as a solution
   b) $x = -6$ as a solution

103. Fill in the blank in the equation
$x^2 + 10x - 22 = $ _____ so that the equation has

   a) two real solutions

   b) two nonreal solutions

104. Fill in the blank in the equation
$x^2 - 4x + 40 = $ _____ so that the equation has

   a) two real solutions

   b) two nonreal solutions

Writing in Mathematics

Answer in complete sentences.

105. Explain why we use the symbol $\pm$ when we take the square root of each side of an equation.

106. If the coefficient of the second-degree term is not 1, we divide both sides of the equation by that coefficient before attempting to complete the square. Explain why dividing both sides of the equation by this coefficient does not affect the solutions of the equation.

107. **Solutions Manual**[*] Write a solutions manual page for the following problem:

Solve $(x - 6)^2 + 20 = -28$ *by extracting square roots.*

108. **Newsletter**[*] Write a newsletter explaining how to solve a quadratic equation by completing the square.

*See Appendix B for details and sample answers.

Objectives

1 Derive the quadratic formula.

2 Identify the coefficients a, b, and c of a quadratic equation.

3 Solve quadratic equations using the quadratic formula.

4 Use the discriminant to determine the number and types of solutions to a quadratic equation.

5 Use the discriminant to determine whether a quadratic expression is factorable.

The Quadratic Formula

Objective 1 **Derive the quadratic formula.** Completing the square to solve a quadratic equation can be tedious. In this section we will develop and use an alternative, the **quadratic formula,** which is a numerical formula that gives the solutions to *any* quadratic equation.

We derive the quadratic formula by solving the general equation $ax^2 + bx + c = 0$ (where a, b, and c are real numbers and $a \neq 0$) by completing the square. Recall that the coefficient of the second-degree term must be equal to 1, so the first step is to divide both sides of the equation $ax^2 + bx + c = 0$ by a.

$$ax^2 + bx + c = 0$$

$$x^2 + \frac{b}{a}x + \frac{c}{a} = 0 \qquad \text{Divide both sides by } a.$$

$$x^2 + \frac{b}{a}x = -\frac{c}{a} \qquad \text{Subtract the constant term,}$$

$$x^2 + \frac{b}{a}x + \frac{b^2}{4a^2} = -\frac{c}{a} + \frac{b^2}{4a^2} \qquad \text{Half of } \frac{b}{a} \text{ is } \frac{b}{2a} \left(\frac{b}{2a}\right)^2 = \frac{b^2}{4a^2}. \text{ Add } \frac{b^2}{4a^2} \text{ to both sides of the equation.}$$

$$x^2 + \frac{b}{a}x + \frac{b^2}{4a^2} = \frac{b^2 - 4ac}{4a^2} \qquad \text{Simplify the right side of the equation as a single fraction by rewriting } -\frac{c}{a} \text{ as } -\frac{4ac}{4a^2}.$$

$$\left(x + \frac{b}{2a}\right)^2 = \frac{b^2 - 4ac}{4a^2} \qquad \text{Factor the left side of the equation.}$$

$$\sqrt{\left(x + \frac{b}{2a}\right)^2} = \pm\sqrt{\frac{b^2 - 4ac}{4a^2}} \qquad \text{Take the square root of both sides.}$$

$$x + \frac{b}{2a} = \pm\frac{\sqrt{b^2 - 4ac}}{\sqrt{4a^2}} \qquad \text{Simplify the square root on the left side. Rewrite the right side as the quotient of two square roots.}$$

$$x + \frac{b}{2a} = \pm\frac{\sqrt{b^2 - 4ac}}{2a} \qquad \text{Simplify the square root in the denominator.}$$

$$x = -\frac{b}{2a} \pm \frac{\sqrt{b^2 - 4ac}}{2a}$$ Subtract $\frac{b}{2a}$.

$$x = \frac{-b \pm \sqrt{b^2 - 4ac}}{2a}$$ Rewrite as a single fraction.

The Quadratic Formula

Given a general quadratic equation $ax^2 + bx + c = 0$ (where a, b, and c are real numbers and $a \neq 0$), the solutions to this equation are given by the quadratic formula:

$$x = \frac{-b \pm \sqrt{b^2 - 4ac}}{2a}.$$

If we can identify the coefficients a, b, and c in a quadratic equation, then we can find the solutions to the equation by substituting these values for a, b, and c in the quadratic formula.

Identifying the Coefficients to Be Used in the Quadratic Formula

Objective 2 Identify the coefficients a, b, and c of a quadratic equation. The first step in using the quadratic formula is to identify the coefficients a, b, and c. We can do this only after our equation is in standard form: $ax^2 + bx + c = 0$.

EXAMPLE ▸1   For the quadratic equation, identify a, b, and c.

a) $x^2 - 4x - 9 = 0$

Solution

$a = 1$, $b = -4$, and $c = -9$. Be sure to include the negative sign when identifying coefficients that are negative.

b) $2x^2 + 6x = 13$

Solution

Before identifying our coefficients, we must convert this equation to standard form: $2x^2 + 6x - 13 = 0$. So $a = 2$, $b = 6$, and $c = -13$.

c) $x^2 - 9 = 0$

Solution

In this example, there is no first-degree term. In a case like this, $b = 0$. The other coefficients are $a = 1$ and $c = -9$.

Quick Check   **1 ▸** For the given quadratic equation, identify a, b, and c.

a) $x^2 + 10x - 24 = 0$ **b)** $4x^2 - 11 = 5x$ **c)** $x^2 = -25$

Solving Quadratic Equations by Using the Quadratic Formula

Objective 3 Solve quadratic equations using the quadratic formula. Now we will solve several equations using the quadratic formula.

EXAMPLE ▶ 2   Solve $x^2 - 4x - 60 = 0$.

Solution

We can use the quadratic formula with $a = 1$, $b = -4$, and $c = -60$.

$$x = \frac{-b \pm \sqrt{b^2 - 4ac}}{2a}$$

$$x = \frac{4 \pm \sqrt{(-4)^2 - 4(1)(-60)}}{2(1)}$$
Substitute 1 for a, -4 for b, and -60 for c. In using the formula, it is a good idea to think of the $-b$ in the numerator as the "opposite of b." The opposite of -4 is 4.

$$x = \frac{4 \pm \sqrt{16 + 240}}{2}$$
Simplify each term in the radicand.

$$x = \frac{4 \pm \sqrt{256}}{2}$$
Add.

$$x = \frac{4 \pm 16}{2}$$
Simplify the square root.

$\dfrac{4 + 16}{2} = 10$ and $\dfrac{4 - 16}{2} = -6$, so the solution set is $\{-6, 10\}$.

Note that the equation in the previous example could have been solved much more quickly by factoring $x^2 - 4x - 60 = 0$ to be $(x - 10)(x + 6)$ and then solving. Use factoring whenever you can, treating the quadratic formula as an alternative.

EXAMPLE ▶ 3   Solve $5x^2 + 12x = -6$.

Solution

To solve this equation, we must rewrite the equation in standard form by collecting all terms on the left side of the equation: $5x^2 + 12x + 6 = 0$.

If an equation does not appear to factor easily, it is wise to proceed directly to the quadratic formula rather than trying to factor an expression that may not be factorable. Here $a = 5$, $b = 12$, and $c = 6$.

$$x = \frac{-12 \pm \sqrt{(12)^2 - 4(5)(6)}}{2(5)}$$
Substitute 5 for a, 12 for b, and 6 for c in the quadratic formula.

$$x = \frac{-12 \pm \sqrt{24}}{10}$$
Simplify the radicand.

$$x = \frac{-12 \pm 2\sqrt{6}}{10}$$
Simplify the square root.

$$x = \frac{\overset{1}{2}(-6 \pm \sqrt{6})}{\underset{5}{10}}$$ Factor the numerator and divide out the common factor.

$$x = \frac{-6 \pm \sqrt{6}}{5}$$ Simplify.

The solution set is $\left\{ \dfrac{-6 - \sqrt{6}}{5}, \dfrac{-6 + \sqrt{6}}{5} \right\}$. These solutions are approximately -1.69 and -0.71, respectively.

Quick Check 2 Solve by using the quadratic formula.

a) $x^2 - 8x + 15 = 0$ b) $2x^2 + 13x = -16$

Using Your Calculator   Here is the screen that shows how to approximate $\dfrac{-6 - \sqrt{6}}{5}$ and $\dfrac{-6 + \sqrt{6}}{5}$.

```
(-6-√(6))/5
      -1.689897949
(-6+√(6))/5
      -.7101020514
```

EXAMPLE 4   Solve $x^2 - 9x + 25 = 0$.

Solution

The quadratic expression in this equation does not factor, so we use the quadratic formula with $a = 1$, $b = -9$, and $c = 25$.

$$x = \frac{9 \pm \sqrt{(-9)^2 - 4(1)(25)}}{2(1)}$$ Substitute 1 for a, -9 for b, and 25 for c in the quadratic formula.

$$x = \frac{9 \pm \sqrt{-19}}{2}$$ Simplify the radicand.

$$x = \frac{9 \pm i\sqrt{-19}}{2}$$ Simplify the square root.

Quick Check 3 The solution set is $\left\{ \dfrac{9 - i\sqrt{19}}{2}, \dfrac{9 + i\sqrt{19}}{2} \right\}$.

Solve $x^2 + 5x + 13 = 0$.

If an equation has coefficients that are fractions, multiply each side of the equation by the LCD to clear the equation of fractions. If we can solve the equation by factoring, it will be easier to factor without the fractions involved. If we cannot solve by factoring, the quadratic formula will be easier to simplify using integers rather than fractions.

EXAMPLE 5 Solve $\frac{1}{3}x^2 - 2x + \frac{1}{2} = 0$.

Solution

The first step is to clear the fractions by multiplying each side of the equation by the LCD, which in this example is 6.

$$6 \cdot \left(\frac{1}{3}x^2 - 2x + \frac{1}{2} \right) = 6 \cdot 0 \qquad \text{Multiply both sides of the equation by the LCD.}$$

$$2x^2 - 12x + 3 = 0 \qquad \text{Distribute and simplify.}$$

Now we can use the quadratic formula with $a = 2$, $b = -12$, and $c = 3$.

$$x = \frac{12 \pm \sqrt{(-12)^2 - 4(2)(3)}}{2(2)} \qquad \begin{array}{l} \text{Substitute 2 for } a, -12 \text{ for } b, \text{ and 3 for } c \text{ in} \\ \text{the quadratic formula.} \end{array}$$

$$x = \frac{12 \pm \sqrt{120}}{4} \qquad \text{Simplify the radicand.}$$

$$x = \frac{12 \pm 2\sqrt{30}}{4} \qquad \text{Simplify the square root.}$$

$$x = \frac{\overset{1}{\cancel{2}}(6 \pm \sqrt{30})}{\underset{2}{\cancel{4}}} \qquad \begin{array}{l} \text{Factor the numerator and divide out} \\ \text{common factors.} \end{array}$$

$$x = \frac{6 \pm \sqrt{30}}{2} \qquad \text{Simplify.}$$

The solution set is $\left\{ \dfrac{6 - \sqrt{30}}{2}, \dfrac{6 + \sqrt{30}}{2} \right\}$. These solutions are approximately 0.26 and 5.74, respectively.

Quick Check 4

Solve
$\dfrac{1}{12}x^2 + \dfrac{2}{3}x + \dfrac{3}{4} = 0.$

In addition to clearing any fractions, make sure that the coefficient (a) of the second-degree term is positive. If this term is negative, you can collect all terms on the other side of the equation or multiply each side of the equation by negative 1.

EXAMPLE 6 Solve $-3x^2 + 17x - 10 = 0$.

Solution

We begin by rewriting the equation so that a is positive, because factoring a quadratic expression is more convenient when the leading coefficient is positive. Also, you may find that simplifying the quadratic formula is easier when the leading coefficient is positive.

$$0 = 3x^2 - 17x + 10 \qquad \text{Collect all terms on the right side of the equation.}$$

Often, when the leading coefficient is not 1, factoring can be difficult or time consuming if the expression is factorable at all. In such a situation, go right to the quadratic formula. In this example, we can use the quadratic formula with $a = 3$, $b = -17$, and $c = 10$.

$$x = \frac{17 \pm \sqrt{(-17)^2 - 4(3)(10)}}{2(3)} \qquad \begin{array}{l}\text{Substitute 3 for } a, -17 \text{ for } b, \text{ and 10 for } c \text{ in} \\ \text{the quadratic formula.}\end{array}$$

$$x = \frac{17 \pm \sqrt{169}}{6} \qquad \text{Simplify the radicand.}$$

$$x = \frac{17 \pm 13}{6} \qquad \text{Simplify the square root.}$$

$$\frac{17 + 13}{6} = 5 \text{ and } \frac{17 - 13}{6} = \frac{2}{3}, \text{ so the solution set is } \left\{\frac{2}{3}, 5\right\}.$$

Quick Check 5

Solve
$-15x^2 + 17x + 4 = 0$.

Although the quadratic formula can be used to solve *any* quadratic equation, it is not always the most efficient way to solve a particular equation. We should always check to see if factoring or extracting square roots is feasible before using the quadratic formula.

Using the Discriminant to Determine the Number and Type of Solutions to a Quadratic Equation

Objective 4 **Use the discriminant to determine the number and types of solutions to a quadratic equation.** In the quadratic formula, the expression $b^2 - 4ac$ is called the **discriminant.** The discriminant can give us some information about our solutions. If the discriminant is negative ($b^2 - 4ac < 0$), then the equation has two complex solutions. This is because we take the square root of a negative number in the quadratic formula. It is important to know whether an equation has no solutions that are real numbers when working on an applied problem. (It will also be important when graphing quadratic equations, which is covered later in this chapter.) The following chart summarizes what the discriminant tells us about the number and type of solutions in an equation.

Solutions to a Quadratic Equation Based on the Discriminant

$b^2 - 4ac$	Number and Type of Solutions
Negative	Two Complex Solutions
Zero	One Real Solution
Positive	Two Real Solutions

EXAMPLE ▶ 7 For each equation, use the discriminant of the quadratic formula to determine the number and type of solutions.

a) $x^2 - 5x + 5 = 0$

Solution

$$(-5)^2 - 4(1)(5) = 5 \qquad \text{Substitute 1 for } a, -5 \text{ for } b, \text{ and 5 for } c \text{ into the formula } b^2 - 4ac.$$

Since the discriminant is positive, this equation has two real solutions.

b) $x^2 + 8x + 24 = 0$

Solution

$$(8)^2 - 4(1)(24) = -32 \qquad \text{Substitute 1 for } a, 8 \text{ for } b, \text{ and 24 for } c.$$

The discriminant is negative. This equation has two complex solutions.

c) $4x^2 - 28x + 49 = 0$

Solution

$$(-28)^2 - 4(4)(49) = 0 \qquad \text{Substitute 4 for } a, -28 \text{ for } b, \text{ and 49 for } c.$$

Since the discriminant equals 0, this equation has one real solution.

Quick Check **6** For the given quadratic equation, use the discriminant to determine the number and type of solutions.

a) $x^2 + 3x + 8 = 0$ **b)** $x^2 - 30 = 0$ **c)** $x^2 + 16x + 64 = 0$

Using the Discriminant to Determine whether a Quadratic Expression Is Factorable

Objective 5 **Use the discriminant to determine whether a quadratic expression is factorable.** The discriminant can also tell us whether a quadratic expression is factorable. If the discriminant is equal to 0 or a positive number that is a perfect square (1, 4, 9, etc.), then the expression is factorable. If the expression is not factorable, it is prime.

EXAMPLE ▶ 8 For each expression, use the discriminant to determine whether the expression can be factored.

a) $x^2 - 6x - 40$

Solution

$$(-6)^2 - 4(1)(-40) = 196 \qquad \text{Substitute 1 for } a, -6 \text{ for } b, \text{ and } -40 \text{ for } c.$$

The discriminant is a perfect square ($\sqrt{196} = 14$), so the expression can be factored.

$$x^2 - 6x - 40 = (x - 10)(x + 4)$$

b) $2x^2 + 7x + 2$

Quick Check 7

For the given quadratic expression, use the discriminant to determine if the expression can be factored.

a) $x^2 + 9x + 10$
b) $20x^2 - 33x - 27$

Solution

$$(7)^2 - 4(2)(2) = 33 \qquad \text{Substitute 2 for } a, 7 \text{ for } b, \text{ and 2 for } c.$$

The discriminant is not a perfect square, so the expression is not factorable.

> **Building Your Study Strategy** Time Management, 2 **When to Study** The best time to study new material is as soon as possible after the class period. After you have made a schedule of your current commitments, look for a block of time that is as close to your class period as possible. Try to study each day at the same time. It is ideal to begin each study session by reworking your notes, then move on to attempting the homework exercises.
>
> If possible, establish a second study period during the day that will be used for review purposes. This second study period should take place later in the day, and can be used to review homework or notes, or even to read ahead for the next class period.
>
> You should also try to schedule your math study sessions for the time of day when you feel that you are at your sharpest. If you are most alert in the mornings, then you should reserve as much time as possible in the mornings to study math. Other students will find that it is better for them to study math in the middle of the day, in the evening, or at night. If you constantly get tired while studying late at night, change your study schedule so that you can study math before you get tired.

EXERCISES 9.2

Vocabulary

1. The _____ formula is a formula for calculating the solutions of a quadratic equation.

2. In the quadratic formula, the expression $b^2 - 4ac$ is called the _____.

3. If the discriminant is negative, then the quadratic equation has _____ real solution(s).

4. If the discriminant is zero, then the quadratic equation has _____ real solution(s).

5. If the discriminant is positive, then the quadratic equation has _____ real solution(s).

6. If the discriminant is 0 or a positive perfect square, then the quadratic expression in the equation is _____.

Solve by using the quadratic formula.

7. $x^2 + 7x + 10 = 0$

8. $x^2 + x - 42 = 0$

9. $x^2 - 6x + 4 = 0$

10. $x^2 + 8x - 2 = 0$

11. $x^2 - 6x + 15 = 0$

12. $4x^2 - 8x + 9 = 0$

13. $x^2 - 5x = 0$

14. $x^2 + 14x + 42 = 0$

15. $x^2 + 6x - 27 = 0$

16. $4x^2 + 12x + 19 = 0$

17. $x^2 + 14x + 48 = 0$

18. $x^2 + 9x + 14 = 0$

19. $x^2 - 36 = 0$

20. $4x^2 + 5x - 6 = 0$

21. $x^2 - 20 = -x$

22. $x^2 + 18x = -57$

23. $x^2 - 5x = 21$

24. $x^2 = -3x$

25. $-42 = 2x - 4x^2$

26. $8x - x^2 = 12$

27. $x^2 + 8x + 16 = 0$

28. $x^2 - 14x + 49 = 0$

29. $x^2 - 72 = 0$

30. $x^2 + 81 = 0$

31. $x(x - 2) + 6x = 32$

32. $(3x - 2)(x - 4) = -10$

33. $\frac{1}{3}x^2 + \frac{3}{2}x - 3 = 0$

34. $\frac{1}{6}x^2 + \frac{2}{3}x + \frac{7}{3} = 0$

35. $-x^2 + 11x - 28 = 0$

36. $-2x^2 + 9x = 6$

For the following quadratic equations, use the discriminant to determine the number and type of solutions.

37. $x^2 + 5x - 17 = 0$

38. $x^2 - 10x + 25 = 0$

39. $x^2 - 5x + 8 = 0$

40. $3x^2 + 11x + 9 = 0$

41. $\frac{2}{3}x^2 - 6x + \frac{27}{2} = 0$

42. $3x^2 + 17x + 25 = 0$

43. $7x^2 - 20x + 19 = 0$

44. $9x^2 - 48x + 64 = 0$

Use the discriminant to determine whether each of the following quadratic expressions is factorable or prime.

45. $x^2 - 9x - 20$

46. $x^2 + 5x - 84$

47. $x^2 - 20x + 91$

48. $x^2 - 14x + 32$

49. $2x^2 + x - 45$

50. $3x^2 - 10x - 21$

51. $8x^2 + 46x - 105$

52. $9x^2 - 81x + 152$

(**Mixed Practice, 53–88**)

Solve the following quadratic equations, using the most efficient technique for each (factoring, extracting square roots, completing the square, or quadratic formula).

53. $x^2 - 3x - 17 = 0$

54. $x^2 - 63 = 0$

55. $3x^2 - 17x + 10 = 0$

56. $x^2 - 18x + 72 = 0$

57. $(2x - 7)^2 = 121$

58. $3x(3x - 5) + 6(3x - 5) = 0$

59. $3x^2 - 24x = -49$

60. $x^2 - 121x = 0$

61. $x^2 + 256 = 0$

62. $x^2 + \dfrac{2}{3}x - \dfrac{1}{4} = 0$

63. $6x^2 - 23x - 4 = 0$

64. $x^2 - 2x - 63 = 0$

65. $2(3x - 5)^2 - 19 = 45$

66. $6x^2 + 5x - 8 = 0$

67. $x^2 - 7x - 32 = 0$

68. $9x^2 - 49 = 0$

69. $9x^2 - 42x + 49 = 0$

70. $x^2 - 19x + 90 = 0$

71. $x^2 + 3x + 40 = 0$

72. $x^2 + 10x + 16 = 0$

73. $x^2 + 2x - 9 = 0$

74. $3x^2 + 23x + 14 = 0$

75. $x^2 - 4x + 29 = 0$

76. $(x - 3)^2 - 21 = 43$

77. $\dfrac{2}{3}x^2 + x - 1 = 0$

78. $x^2 + x - 30 = 0$

79. $5x(x - 13) + 4(x - 13) = 0$

80. $(2x + 9)^2 - 12 = 13$

81. $3(x - 4)^2 - 18 = 57$

82. $x^2 + 5x - 36 = 0$

83. $x^2 + x - 6 = 0$

84. $x^2 + 15x + 54 = 0$

85. $x^2 - 14x + 34 = 0$

86. $x^2 - 4x + 9 = 0$

87. $x^2 - 11x + 18 = 0$

88. $3x^2 - 4x - 12 = 0$

89. Derive a formula that would provide the general solution to a linear equation $ax + b = 0$, where a and b are real numbers and $a \neq 0$.

Writing in Mathematics

Answer in complete sentences.

90. Which method of solving quadratic equations do you prefer: solving by factoring or the quadratic formula? Explain your answer.

91. Explain how the discriminant tells us whether an equation has two real solutions, one real solution, or two complex solutions.

92. *Solutions Manual* * Write a solutions manual page for the following problem:

Solve $\dfrac{1}{2}x^2 - \dfrac{5}{6}x + 1 = 0$ *by using the quadratic formula.*

93. *Newsletter* * Write a newsletter explaining how to solve quadratic equations by using the quadratic formula.

*See Appendix B for details and sample answers.

9.3
Applications Using Quadratic Equations

Objectives

1 Solve applied geometric problems.
2 Solve problems using the Pythagorean theorem.
3 Solve applied problems using the Pythagorean theorem.
4 Solve projectile motion problems.

In this section, we will learn how to solve applied problems resulting in quadratic equations.

Solving Applied Geometric Problems

Objective 1   Solve applied geometric problems. Problems involving the area of a geometric figure often lead to quadratic equations, as area is measured in square units. Here are some useful area formulas.

Figure	Square	Rectangle	Triangle	Circle
Dimensions	Side s	Length l, Width w	Base b, Height h	Radius r
Area	$A = s^2$	$A = l \cdot w$	$A = \frac{1}{2}bh$	$A = \pi r^2$

EXAMPLE ▶ 1   The width of a rectangle is 7 centimeters less than its length. The area is 60 cm^2. Find the length and the width of the rectangle.

Solution

In this problem, the unknown quantities are the length and the width, while the known quantity is the area, 60 cm^2. Since the width is given in terms of the length, a wise choice is to represent the length of the rectangle by x. Since the width is 7 cm less than the length, it can be represented by $x - 7$. This information is summarized in the following table.

Unknowns	Known	
Length: x	Area: 60 cm^2	
Width: $x - 7$		

Since the area of a rectangle is equal to its length times its width, the equation we need to solve is $x(x - 7) = 60$.

$$x(x - 7) = 60$$
$$x^2 - 7x = 60 \qquad \text{Distribute.}$$

$$x^2 - 7x - 60 = 0$$
$$(x - 12)(x + 5) = 0$$
$$x - 12 = 0 \quad \text{or} \quad x + 5 = 0$$
$$x = 12 \quad \text{or} \quad x = -5$$

Rewrite in standard form.
Factor.
Set each factor equal to 0.
Solve.

Look back to the table of unknowns. If $x = -5$, then the length is -5 centimeters and the width is -12 centimeters, which is not possible. The solutions derived from $x = -5$ are omitted. If $x = 12$, then the length is 12 centimeters, while the width is $12 - 7$ or 5 centimeters.

Quick Check **1**
The base of a triangle is 3 inches longer than 2 times its height. If the area of the triangle is 22 square inches, find the base and height of the triangle.

Length: $x = 12$
Width: $x - 7 = 12 - 7 = 5$

Now we express the answer in a complete sentence with the proper units: The length of the rectangle is 12 centimeters and the width is 5 centimeters.

Solving Problems Using the Pythagorean Theorem

Objective **2** **Solve problems using the Pythagorean theorem.** Other applications of geometry that lead to quadratic equations involve the Pythagorean theorem, which is an equation that relates the length of the three sides of a right triangle. The side opposite the right angle is called the **hypotenuse**, and is labeled c in the figure at the right. The other two sides are labeled a and b. (It makes no difference which is a and which is b.)

The Pythagorean theorem states that for any right triangle with a hypotenuse measuring length c and two legs of lengths a and b, respectively,

$$a^2 + b^2 = c^2.$$

EXAMPLE **2** A right triangle has a hypotenuse that measures 15 inches, and one of its legs is 7 inches long. Find the length of the other leg, to the nearest tenth of an inch.

Solution

In this problem the length of one of the legs is unknown. We can label the unknown leg as either a or b.

Unknown	Known
a	b: 7 inches
	Hypotenuse (c): 15 inches

$$a^2 + 7^2 = 15^2$$
$$a^2 + 49 = 225$$
$$a^2 = 176$$

Substitute 7 for b and 15 for c in $a^2 + b^2 = c^2$.
Square 7 and 15.
Subtract 49.

$$\sqrt{a^2} = \pm\sqrt{176} \qquad \text{Solve by extracting square roots.}$$
$$a = \pm 4\sqrt{11} \qquad \text{Simplify the square root.}$$

Quick Check 2
One leg of a right triangle measures 3 inches, while the hypotenuse measures 16 inches. Find, to the nearest tenth of an inch, the length of the other leg of the triangle.

Since the length of a leg must be a positive number, we can eliminate the negative solution. The remaining solution, $4\sqrt{11}$, rounds to be 13.3 inches. The length of the other leg is approximately 13.3 inches.

Applications of the Pythagorean Theorem

Objective 3   Solve applied problems using the Pythagorean theorem. Now we turn our attention to solving applied problems using the Pythagorean theorem. In these problems, we begin by drawing a picture of the situation. We must be able to identify a right triangle in our figure in order to apply the Pythagorean theorem.

EXAMPLE 3   A 10-foot ladder is leaning against a wall. If the bottom of the ladder is 6 feet from the base of the wall, how high up the wall is the top of the ladder?

Solution

The ladder, the wall, and the ground form a right triangle, with the ladder as the hypotenuse. In this problem the height of the wall, which is the length of one of the legs in the right triangle, is unknown.

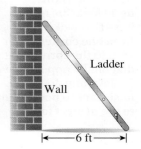

Ladder

Wall

|←——6 ft——→|

Unknown	Known
a	b: 6 feet
	Hypotenuse (c): 10 feet

Quick Check 3
An 8-foot ladder is leaning against a wall. If the bottom of the ladder is 3 feet from the base of the wall, how high up the wall is the top of the ladder? (Round to the nearest tenth of a foot.)

$$a^2 + 6^2 = 10^2 \qquad \text{Substitute 6 for } b \text{ and 10 for } c \text{ in } a^2 + b^2 = c^2.$$
$$a^2 + 36 = 100 \qquad \text{Square 6 and 10.}$$
$$a^2 = 64 \qquad \text{Subtract 36.}$$
$$\sqrt{a^2} = \pm\sqrt{64} \qquad \text{Solve by extracting square roots.}$$
$$a = \pm 8 \qquad \text{Simplify the square root.}$$

Again, the negative solution does not make sense in this problem, so we eliminate it. The ladder is resting at a point on the wall that is 8 feet above the ground.

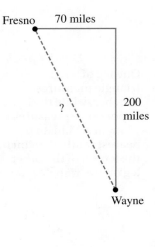

Fresno   70 miles

?   200 miles

Wayne

EXAMPLE 4   Wayne is flying to Fresno, which is located 200 miles north and 70 miles west of his local airport. If Wayne flies directly to Fresno, how many miles is the flight? (Round to the nearest mile.)

Solution

The directions of north and west form a 90-degree angle, so our picture shows us that we have a right triangle whose hypotenuse (the direct distance to Fresno) is unknown.

Unknown	Known
Hypotenuse: c	a: 200 miles
	b: 70 miles

$$200^2 + 70^2 = c^2$$ Substitute 200 for a and 70 for b in $a^2 + b^2 = c^2$.
$$40{,}000 + 4900 = c^2$$ Square 200 and 70.
$$44{,}900 = c^2$$ Simplify.
$$\pm\sqrt{44{,}900} = \sqrt{c^2}$$ Solve by extracting square roots.
$$c \approx \pm 212$$ Approximate the square root using a calculator.

The negative solution does not make sense in this problem. The direct distance to Fresno is approximately 212 miles.

> **Quick Check 4**
>
> **Peggy's backyard is in the shape of a rectangle with dimensions 50 feet by 80 feet. She needs a hose that will extend from one corner of her yard to the corner that is diagonally opposite to it. How long does the hose have to be? (Round to the nearest foot.)**

Projectile Motion Problems

Objective 4   Solve projectile motion problems. Another application that leads to a quadratic equation involves the height after a certain amount of time of an object propelled into the air. Recall that the height of a projectile after t seconds can be found using the function $h(t) = -16t^2 + v_0 t + s$, where v_0 is the initial velocity of the projectile and s is the initial height.

Height of a Projectile

$$h(t) = -16t^2 + v_0 t + s$$

t: Time (seconds)

v_0: Initial Velocity (feet/second)

s: Initial Height (feet)

EXAMPLE 5   A rock is thrown upward from ground level at an initial velocity of 64 feet/second.

a) How long will it take until the rock lands on the ground?

Solution

The initial velocity of the rock is 64 feet per second, so $v_0 = 64$. Since the rock is thrown from ground level, the initial height is 0 feet. The function for the height of the rock after t seconds is $h(t) = -16t^2 + 64t$.

The rock's height when it lands on the ground is 0 feet, so we set the function equal to 0 and solve for the time t in seconds.

$$-16t^2 + 64t = 0 \qquad \text{Set the function equal to 0.}$$
$$0 = 16t^2 - 64t \qquad \text{Collect all terms on the right side of the equation so the leading coefficient is positive.}$$
$$0 = 16t(t - 4) \qquad \text{Factor out the GCF } (16t).$$
$$16t = 0 \quad \text{or} \quad t - 4 = 0 \qquad \text{Set each variable factor equal to 0.}$$
$$t = 0 \quad \text{or} \quad t = 4 \qquad \text{Solve each equation.}$$

The two solutions represent the times when the rock is on the ground. The time of 0 seconds corresponds to the precise moment that the rock was thrown. The time of 4 seconds corresponds to the moment that the rock lands on the ground again.

b) When will the rock be at a height of 48 feet?

Solution

This problem is similar to part (a) with the exception that we will set the function $h(t)$ equal to 48 rather than 0.

Quick Check 5

A rock is thrown upward from ground level at an initial velocity of 96 feet per second.

a) After how many seconds will the rock land on the ground?

b) When will the rock be at a height of 128 feet?

$$-16t^2 + 64t = 48 \qquad \text{Set the function equal to 48.}$$
$$0 = 16t^2 - 64t + 48 \qquad \text{Collect all terms on the right side of the equation.}$$
$$0 = 16(t^2 - 4t + 3) \qquad \text{Factor out the GCF (16).}$$
$$0 = 16(t - 1)(t - 3) \qquad \text{Factor the trinomial.}$$
$$t - 1 = 0 \quad \text{or} \quad t - 3 = 0 \qquad \text{Set each variable factor equal to 0.}$$
$$t = 1 \quad \text{or} \quad t = 3 \qquad \text{Solve each equation.}$$

Do both of these answers make sense? The rock first passes a height of 48 feet after 1 second while it is on its way up. It is at a height of 48 feet again after three seconds while it is on its way down. The rock is 48 feet above the ground after 1 second and again after 3 seconds.

EXAMPLE 6 A marble is launched upward by a slingshot from a platform that is 15 feet high at an initial velocity of 90 feet per second.

a) When will the marble be at a height of 120 feet? (Round to the nearest hundredth of a second.)

Solution

The initial velocity of the marble is 90 feet per second, so $v_0 = 90$. Since the marble was launched from a platform 15 feet high, the initial height $s = 15$. The function for the height of the marble after t seconds is $h(t) = -16t^2 + 90t + 15$.

To find when the marble is at a height of 120 feet, we set the function equal to 120 and solve for the time t in seconds.

$$-16t^2 + 90t + 15 = 120 \qquad \text{Set the function equal to 120.}$$
$$0 = 16t^2 - 90t + 105 \qquad \text{Collect all terms on the right side of the equation.}$$

This trinomial does not factor, so we will use the quadratic formula with $a = 16$, $b = -90$, and $c = 105$ to solve this equation.

$$t = \frac{90 \pm \sqrt{(-90)^2 - 4(16)(105)}}{2(16)}$$ Substitute 16 for a, -90 for b, and 105 for c.

$$t = \frac{90 \pm \sqrt{1380}}{32}$$ Simplify the radicand and the denominator.

We use a calculator to approximate these solutions.

$$\frac{90 + \sqrt{1380}}{32} \approx 3.97$$

$$\frac{90 - \sqrt{1380}}{32} \approx 1.65$$

The marble is at a height of 120 feet after approximately 1.65 seconds and again after approximately 3.97 seconds.

b) Will the marble ever reach a height of 300 feet? If so, when will it be at this height?

Solution

We begin by setting the function equal to 300 and solving for t.

$$-16t^2 + 90t + 15 = 300$$ Set the function equal to 300.
$$0 = 16t^2 - 90t + 285$$ Collect all terms on the right side of the equation.

We will use the quadratic formula with $a = 16$, $b = -90$, and $c = 285$ to solve this equation.

$$t = \frac{90 \pm \sqrt{(-90)^2 - 4(16)(285)}}{2(16)}$$ Substitute 16 for a, -90 for b, and 285 for c.

$$t = \frac{90 \pm \sqrt{-10,140}}{32}$$ Simplify the radicand and the denominator.

Since the radicand is negative, this equation has no real number solutions. The marble never reaches a height of 300 feet.

A Word of Caution   When using the quadratic formula to solve an applied problem, a negative radicand indicates that there are no real solutions to this problem.

Quick Check 6

A projectile is launched upward from the top of a building 20 feet high at an initial velocity of 40 feet per second.

a) When will the projectile be at a height of 35 feet? (Round to the nearest hundredth of a second.)
b) Will the projectile ever reach a height of 50 feet? If so, when will it be at this height?

Building Your Study Strategy Time Management, 3 **Setting Goals**   One way to get the most from a study session is to establish a set of goals to accomplish for each session. For example, during a 90-minute session, you may set a goal of completing the homework assignment and creating a set of note cards for the most recent section in the textbook. Setting goals will encourage you to work quickly and efficiently. Many students set a goal of studying for a certain amount of time, but time alone is not a worthy goal. Create a "to do" list each time you start a study session, and you will find that you will have a much greater chance of reaching your goals.

If you have more than one subject to study, as most students do, then the order in which you study the subjects is important. It's a good idea to arrange your study schedule so that you will be studying the most difficult subject first, while you are most alert. If possible, avoid studying two difficult subjects back to back. If you cannot put some space between these two topics, try to study a less demanding subject between two difficult subjects. Another suggestion to keep your mental energy at its highest while studying is to take short 10-minute study breaks. One study break per hour will help keep you from feeling fatigued.

EXERCISES *9.3*

Vocabulary

1. In a right triangle, the side opposite the right angle is called the _____.

2. In a right triangle, the sides that are adjacent to the right angle are called _____.

3. The _____ theorem states that for any right triangle with a hypotenuse measuring length c and two legs of lengths a and b, respectively, $a^2 + b^2 = c^2$.

4. State the function $h(t)$ for the height in feet of a projectile after t seconds.

For all problems, approximate to the nearest tenth when necessary.

5. The area of a square is 36 square meters. Find the length of a side of the square.

6. The area of a square is 140 square inches. Find the length of a side of the square.

7. The area of a circle is 50π square feet. Find the radius of the circle.

8. The area of a circle is 81π square meters. Find the radius of the circle.

9. The area of a circle is 85 square inches. Find the radius of the circle.

10. The area of a circle is 160 square centimeters. Find the radius of this circle.

11. The width of a rectangle is 6 inches less than its length. If the area of the rectangle is 112 square inches, find the length and width of the rectangle.

12. The length of a rectangle is 1 inch more than twice its width. If the area of the rectangle is 36 square inches, find the length and width of the rectangle.

13. The length of a rectangle is 3 times its width. If the area of the rectangle is 108 square inches, find the length and width of the rectangle.

14. The length of a rectangle is 7 feet less than 3 times its width. If the area of the rectangle is 286 square feet, find the length and width of the rectangle.

15. The length of a rectangle is 3 inches less than 4 times its width. If the area of the rectangle is 80 square inches, find the dimensions of the rectangle.

16. The width of a rectangle is half of its length. If the area of the rectangle is 500 square inches, find the dimensions of the rectangle.

17. The base of a triangle is 3 inches more than its height. If the area of the triangle is 54 square inches, find the base and height of the triangle.

18. The height of a triangle is 3 feet less than 3 times its base. If the area of the triangle is 45 square feet, find the base and height of the triangle.

19. The height of a triangle is 3 inches more than twice its base. If the area of the triangle is 38 square inches, find the base and height of the triangle.

20. The height of a triangle is 6 inches less than its base. If the area of the triangle is 55 square inches, find the base and height of the triangle.

21. A rectangular photograph has an area of 80 square inches. If the height of the photograph is 2 inches more than its width, find the dimensions of the photograph.

22. The area of a rectangular poster is 432 square inches. If the length of the poster is 6 inches more than its width, find the dimensions of the poster.

23. The length of a rectangular quilt is twice its width, and the area of the quilt is 24.5 square feet. Find the dimensions of the quilt.

24. The length of a rectangular room is 5 feet less than twice its width. If the area of the room is 210 square feet, find the dimensions of the room.

25. The length of a rectangular table is 20 inches more than its width. If the area of the table is 570 square inches, find the dimensions of the table.

26. The width of a rectangular lawn is 13 feet less than the length. If the area of the lawn is 5000 square feet, find the dimensions of the lawn.

27. A kite is in the shape of a triangle. The base of the kite is twice its height. If the area of the kite is 225 square inches, find the base and height of the kite.

28. A hang glider is triangular in shape. If the base of the hang glider is 3 feet more than twice its height, and the area of the hang glider is 45 square feet, find the hang glider's base and height.

29. The sail on a sailboat is shaped like a triangle, with an area of 50 square feet. The height of the sail is 7 feet more than the base of the sail. Find the base and height of the sail.

30. A triangular kite is made with 420 square inches of material. The base of the kite is 18 inches longer than the height of the kite. Find the base and the height of the kite.

31. The two legs of a right triangle are 5 inches and 12 inches respectively. Find the hypotenuse of the triangle.

32. The two legs of a right triangle are 9 inches and 16 inches respectively. Find the hypotenuse of the triangle.

33. A right triangle with a hypotenuse of 18 centimeters has a leg that measures 8 centimeters. Find the length of the other leg.

34. A right triangle has a leg that measures 20 inches and a hypotenuse that measures 40 inches. Find the length of the other leg.

35. Patti drove 45 miles to the west, and then drove 60 miles south. How far is she from her starting location?

36. Sheng-yi flew to a city that was 210 miles north and 720 miles east of his starting point. How far did he fly to reach this city?

37. A post is casting a shadow on the ground. If the post is 3 feet taller than the length of the shadow on the ground, and the tip of the shadow is 15 feet from the top of the post, how tall is the post?

38. A 13-foot ladder is leaning against a wall. If the distance between the base of the ladder and the wall is 7 feet less than the height of the top of the ladder on the wall, how high up the wall does the ladder reach?

39. A guy wire 35 feet long runs from the top of a pole to a spot on the ground. If the height of the pole is 6 feet more than the distance from the base of the pole to the spot where the guy wire is anchored, how tall is the pole?

40. A 15-foot ramp leads to a doorway. The height of the ramp at the doorway is 9 feet less than the horizontal distance covered by the ramp. How high above the ground is the doorway?

41. A rectangular computer screen is 15 inches wide and 12.5 inches high. Find the length of its diagonal.

42. The bases on a little league baseball diamond form a square 60 feet on a side. How far is it from home plate to second base?

43. The length of a rectangular quilt is 1 foot more than its width. If the diagonal of the quilt is 8 feet, find the length and width of the quilt.

44. The diagonal of a rectangular table is 11 feet, and the length of the table is 5 feet more than its width. Find the length and width of the table.

For Exercises 45–48, use the fact that the legs of a right triangle represent the base and height of that triangle.

45. The height of a right triangle is 7 feet more than the base of the triangle. The area of the triangle is 30 square feet.

  a) How long are the base and height of this triangle?

  b) How long is the hypotenuse of the triangle?

46. The base of a right triangle is 3 feet more than 3 times the height of the triangle. The area of the triangle is 84 square feet.

  a) How long are the base and height of this triangle?

  b) How long is the hypotenuse of the triangle?

47. A road sign indicating falling rocks is in the shape of an equilateral triangle, with each side measuring 90 centimeters.

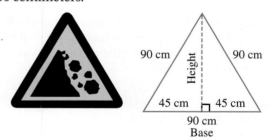

  a) Use the Pythagorean theorem to find the height of the triangle. Round to the nearest tenth of a centimeter.

  b) Find the area of the sign.

48. A preschool installed a sand-box in the shape of an equilateral triangle 20 feet on a side.

  a) Use the Pythagorean theorem to find the height of the triangle. Round to the nearest tenth of a foot.

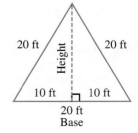

  b) Find the area of the sandbox.

For Exercises 49–58, use the function
$h(t) = -16t^2 + v_0 t + s.$

49. An object is launched upward from the ground with an initial speed of 240 feet per second. How long will it take until the object lands on the ground?

50. An object is launched upward from a building 336 feet above the ground, with an initial speed of 64 feet per second. How long will it take until it lands on the ground?

51. Rob is standing on a cliff 80 feet above a beach. He throws a rock upward at a speed of 60 feet per second. How long will it take until the rock lands on the beach? Round to the nearest tenth of a second.

52. Alejandra is standing on the roof of a building 40 feet high. She launches a water balloon upward at an initial speed of 55 feet per second. How long will it take until it lands on the ground? Round to the nearest tenth of a second.

53. An object is launched upward from ground level with an initial velocity of 96 feet per second. At what time(s) is the object 128 feet above the ground?

54. An object is launched upward from ground level with an initial speed of 90 feet per second. At what time(s) is the object 100 feet above the ground? Round to the nearest tenth of a second.

55. An object is launched upward from the top of a building 36 feet high, with an initial velocity of 132 feet per second. At what time(s) is the object 225 feet above the ground? Round to the nearest tenth of a second.

56. An object is launched upward from a cliff 95 feet above a beach, with an initial velocity of 120 feet per second. At what time(s) is the object 280 feet above the beach? Round to the nearest tenth of a second.

57. A boy throws a rock upward with an initial velocity of 20 feet per second at a streetlight that is 30 feet above the ground. Does the rock ever reach the height of the streetlight? Explain why.

58. A football player kicked a football upward. If it took 3.5 seconds for the ball to land on the ground, find the initial velocity of the football.

Writing in Mathematics

Answer in complete sentences.

59. Write a word problem whose solution is "The length of the rectangle is 20 feet and the width is 12 feet." The problem must lead to a quadratic equation.

60. Write a word problem associated with the equation $45^2 + b^2 = 75^2$. Explain how you created the problem. Solve your problem, explaining each step.

61. Write a word problem associated with the equation $-16t^2 + 64t + 15 = 95$. Explain how you created the problem. Solve your problem, explaining each step.

62. Write a word problem whose solution is "The object will land on the ground after 7 seconds." Explain how you created the problem.

63. **Solutions Manual** * Write a solutions manual page for the following problem:

    The length of a rectangle is 8 centimeters longer than twice its width. If the area of the rectangle is 150 square centimeters, find the length and the width of the rectangle. (Round to the nearest tenth of a centimeter.)

64. **Newsletter** * Write a newsletter explaining how to solve projectile motion problems.

*See Appendix B for details and sample answers.

Objectives

1 Graph quadratic equations.

2 Graph parabolas that open downward.

3 Find the vertex of a parabola by completing the square.

4 Graph quadratic equations of the form $y = a(x - h)^2 + k$.

Graphing Quadratic Equations

Objective 1 **Graph quadratic equations.** The graphs of quadratic equations are not lines like the graphs of linear equations. They are U-shaped and called parabolas. Let's consider the graph of the most basic quadratic equation: $y = x^2$. We will first create a table of ordered pairs that will represent points on our graph.

x	$y = x^2$
-2	$(-2)^2 = 4$
-1	$(-1)^2 = 1$
0	$0^2 = 0$
1	$1^2 = 1$
2	$2^2 = 4$

The figure shows these points and the graph of $y = x^2$. Notice that the shape is not a straight line, but instead is U-shaped.

The point where the graph changes from decreasing to increasing is called the **vertex**. In this example, the vertex is located at the bottom of the parabola. Notice that if we drew a vertical line through the vertex of the parabola, the left and right sides are mirror images. The graph of an equation is said to be **symmetric** if we can fold the graph along a line and the two sides of the graph coincide; in other words, a graph is symmetric if one side of the graph is a mirror image of the other side. A parabola is always symmetric, and we call the vertical line through the vertex the **axis of symmetry.** For this parabola, the equation for the axis of symmetry is $x = 0$.

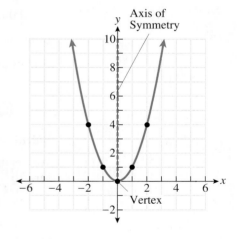

We graph parabolas by plotting points. We can create the graph most easily if we focus on a few important points: the y-intercept, the x-intercept(s) if there are any, and the vertex. We also use the axis of symmetry to help us find "mirror" points that are symmetric to points we have already graphed.

As before, we find the y-intercept by substituting 0 for x and solving for y. The y-intercept of a quadratic equation in standard form ($y = ax^2 + bx + c$) is always the point $(0, c)$. We find the x-intercepts, if there are any, by substituting 0 for y and solving for x. This equation will be quadratic, and we solve it using the techniques we have already learned.

A parabola opens upward if a is positive, with the vertex at the lowest point of the parabola. To learn how to find the coordinates of the vertex, we begin by completing the square for the equation $y = ax^2 + bx + c$.

$$y = ax^2 + bx + c$$

$$y - c = a\left(x^2 + \frac{b}{a}x\right)$$

Factor a from the two terms containing x. Subtract c from both sides.

$$y - c + \frac{b^2}{4a} = a\left(x^2 + \frac{b}{a}x + \frac{b^2}{4a^2}\right)$$

Half of $\frac{b}{a}$ is $\frac{b}{2a}$. Add $\left(\frac{b}{2a}\right)^2$ or $\frac{b^2}{4a^2}$ to the terms inside the parentheses. Since there is a factor in front of the parentheses, we add $a \cdot \frac{b^2}{4a^2}$ or $\frac{b^2}{4a}$ to the left side.

$$y = a\left(x + \frac{b}{2a}\right)^2 + \frac{4ac - b^2}{4a}$$

Factor the trinomial inside the parentheses. Collect all terms on the right side of the equation. $c - \frac{b^2}{4a} = \frac{4ac - b^2}{4a}$.

Since a squared expression cannot be negative, the minimum value of y occurs when $x + \frac{b}{2a} = 0$, or, in other words, when $x = \frac{-b}{2a}$.

If the equation is in standard form, $y = ax^2 + bx + c$, then we can find the x-coordinate of the vertex using the formula $x = \frac{-b}{2a}$. We then find the y-coordinate of the vertex by substituting this value for x in the original equation.

EXAMPLE 1   Graph $y = x^2 - 8x + 12$.

Solution

We begin by noting that $a = 1$, $b = -8$, and $c = 12$, and then setting $x = 0$ and solving for y to find the y-intercept.

$$y = (0)^2 - 8(0) + 12 = 12 \qquad \text{Substitute 0 for } x \text{ in the original equation.}$$

The y-intercept is $(0, 12)$. To find the x-intercepts, we substitute 0 for y and attempt to solve for x.

$$0 = x^2 - 8x + 12 \qquad \text{Substitute 0 for } y \text{ in the original equation.}$$
$$0 = (x - 2)(x - 6) \qquad \text{Factor the trinomial.}$$
$$x = 2 \quad \text{or} \quad x = 6 \qquad \text{Set each factor equal to 0 and solve.}$$

The x-intercepts are $(2, 0)$ and $(6, 0)$. Finally, we find the vertex. We begin by finding the x-coordinate of the vertex.

$$x = \frac{8}{2(1)} = 4 \qquad \text{Substitute 1 for } a \text{ and } -8 \text{ for } b \text{ into } x = \frac{-b}{2a}.$$

Now we substitute this value for x into the original equation and solve for y.

$y = (4)^2 - 8(4) + 12$ Substitute 4 for x.

$y = -4$ Simplify.

The vertex is at $(4, -4)$.

    Here is a sketch showing the locations of the four points we have found, the axis of symmetry, and a point symmetric to the y-intercept. We see that the y-intercept $(0, 12)$ is four units to the left of the axis of symmetry. So there must be a mirror point with the same y-coordinate located 4 units to the right of the axis of symmetry.

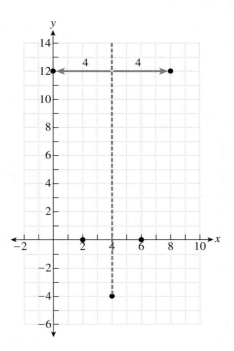

This produces the following graph.

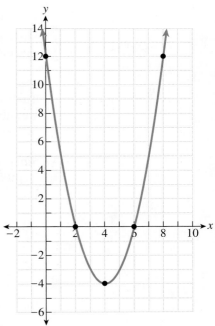

Using Your Calculator   We can use the TI-84 to graph parabolas. To graph $y = x^2 - 8x + 12$, begin by pushing the Y= key and typing $x^2 - 8x + 12$ next to Y_1 as shown.

Quick Check **1**

Graph $y = x^2 + 4x - 5$.

To graph the parabola, press the GRAPH key. Here is the screen that you should see in the standard viewing window.

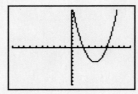

EXAMPLE 2   Graph $y = x^2 - 2x - 6$.

Solution

We begin by noting that $a = 1$, $b = -2$, and $c = -6$. We then substitute 0 for x in the original equation and solve for y to find the y-intercept.

$$y = (0)^2 - 2(0) - 6 = -6 \qquad \text{Substitute 0 for } x.$$

The y-intercept is $(0, -6)$. Again, the y-intercept is the point $(0, c)$. To find the x-intercepts, we substitute 0 for y in the original equation and attempt to solve for x.

$$0 = x^2 - 2x - 6 \qquad \text{Substitute 0 for } y.$$

Since we cannot factor this expression, we will use the quadratic formula to find the x-intercepts.

$$x = \frac{2 \pm \sqrt{(-2)^2 - 4(1)(-6)}}{2(1)} \qquad \text{Substitute 1 for } a, -2 \text{ for } b, \text{ and } -6 \text{ for } c.$$

$$x = \frac{2 \pm \sqrt{28}}{2} \qquad \text{Simplify the radicand and denominator.}$$

$$x = \frac{2 \pm 2\sqrt{7}}{2} \qquad \text{Simplify the square root.}$$

$$x = \frac{\overset{1}{2}(1 \pm \sqrt{7})}{\underset{1}{2}} \qquad \text{Divide out common factors.}$$

$$x = 1 \pm \sqrt{7} \qquad \text{Simplify.}$$

Since $1 + \sqrt{7} \approx 3.6$ and $1 - \sqrt{7} \approx -1.6$, the x-intercepts are approximately $(3.6, 0)$ and $(-1.6, 0)$. Now we find the vertex, using the formula $x = \dfrac{-b}{2a}$.

$$x = \frac{2}{2(1)} = 1 \qquad \text{Substitute 1 for } a \text{ and } -2 \text{ for } b.$$

We substitute 1 for x in the original equation and solve for y to find the y-coordinate of the vertex.

$$y = (1)^2 - 2(1) - 6 \qquad \text{Substitute 1 for } x.$$
$$y = -7 \qquad\qquad\quad \text{Simplify.}$$

The vertex is at $(1, -7)$.

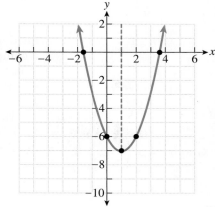

Quick Check **2**

Graph $y = x^2 + 6x - 4$.

Here is a sketch of the parabola showing the vertex, x-intercepts, y-intercept, and the axis of symmetry $(x = 1)$. Also shown is the point $(2, -6)$, which is symmetric to the y-intercept.

Occasionally, a parabola will not have any x-intercepts. In this case the quadratic formula will produce a negative discriminant $(b^2 - 4ac)$. We also can determine that a parabola does not have any x-intercepts by plotting the vertex and y-intercept first, before looking for the x-intercepts.

EXAMPLE 3   Graph $y = x^2 - 5x + 7$.

Solution

Note that $a = 1$, $b = -5$, and $c = 7$. In this example we begin by finding the vertex rather than the intercepts. The x-coordinate of the vertex is found using the formula $x = \dfrac{-b}{2a}$.

$$x = \frac{5}{2(1)} = \frac{5}{2} \qquad \text{Substitute 1 for } a \text{ and } -5 \text{ for } b.$$

Now we substitute $\frac{5}{2}$ for x in the original equation and solve for y.

$$y = \left(\frac{5}{2}\right)^2 - 5\left(\frac{5}{2}\right) + 7 \qquad \text{Substitute } \tfrac{5}{2} \text{ for } x.$$

$$y = \frac{25}{4} - \frac{25}{2} + 7 \qquad\qquad \text{Simplify each term.}$$

$$y = \frac{25}{4} - \frac{50}{4} + \frac{28}{4} \qquad\quad \begin{array}{l}\text{Rewrite each fraction with a}\\ \text{common denominator of 4.}\end{array}$$

$$y = \frac{3}{4} \qquad\qquad\qquad\qquad \text{Simplify.}$$

The vertex is at $\left(\frac{5}{2}, \frac{3}{4}\right)$. To find the y-intercept, we substitute 0 for x in the original equation.

$$y = (0)^2 - 5(0) + 7 = 7 \qquad \text{Substitute 0 for } x.$$

The y-intercept is $(0, 7)$.

If we plot the vertex and the y-intercept on the graph, we see there are no x-intercepts. The vertex is above the x-axis, and the parabola only moves in an upward direction from there. Therefore, this parabola does not have any x-intercepts. If we chose to use the quadratic formula to find the x-intercepts, we would have ended up with $x = \dfrac{5 \pm \sqrt{-3}}{2}$, a complex number. The discriminant would be -3. Since the discriminant is negative, the equation $0 = x^2 - 5x + 7$ does not have any real solutions and the parabola does not have any x-intercepts.

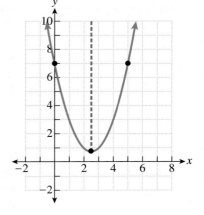

Here is the graph, showing the axis of symmetry

Quick Check 3
Graph $y = x^2 + 4x + 8.$

$\left(x = \frac{5}{2}\right)$ and a third point, $(5, 7)$, that is symmetric to the y-intercept.

A Word of Caution   If a parabola that opens upward has its vertex above the x-axis, then there are no x-intercepts.

Graphing Parabolas That Open Downward

Objective 2   **Graph parabolas that open downward.** Some parabolas open downward rather than upward. The way to determine which way a parabola will open is by writing the equation in standard form: $y = ax^2 + bx + c$. If a is positive, as it was in our previous examples, then the parabola will open upward. If a is negative, then the parabola will open downward. For instance, the graph of $y = -3x^2 + 5x - 7$ would open downward because the leading coefficient a is negative. The following example shows how to graph a parabola that opens downward.

EXAMPLE 4   Graph $y = -x^2 + 6x - 8.$

Solution

In this example, $a = -1$, $b = 6$, and $c = -8$. We begin by finding the x-coordinate of the vertex.

$$x = \frac{-6}{2(-1)} = 3 \qquad \text{Substitute } -1 \text{ for } a \text{ and } 6 \text{ for } b.$$

Now we substitute 3 for x in the original equation and solve for y.

$$y = -(3)^2 + 6(3) - 8 \qquad \text{Substitute 3 for } x.$$
$$y = 1 \qquad\qquad\qquad \text{Simplify.}$$

The vertex is at $(3, 1)$. To find the y-intercept, we substitute 0 for x and solve for y.

$$y = -(0)^2 + 6(0) - 8 = -8 \qquad \text{Substitute 0 for } x.$$

The y-intercept is $(0, -8)$. If we plot the vertex and the y-intercept on the graph, we will see that there must be two x-intercepts. The vertex is above the x-axis, and the parabola opens downward, so the graph must cross the x-axis. To find the x-intercepts, we substitute 0 for y in the original equation and solve for x.

$$0 = -x^2 + 6x - 8 \qquad \text{Substitute 0 for } y.$$
$$x^2 - 6x + 8 = 0 \qquad \text{Collect all terms on the left side of the}$$
equation, so that the coefficient
of the squared term is positive.

$$(x - 2)(x - 4) = 0 \qquad \text{Factor.}$$
$$x = 2 \quad \text{or} \quad x = 4 \qquad \text{Set each factor equal to 0 and solve.}$$

Quick Check 4
Graph $y = -x^2 + 5x + 6$.

The x-intercepts are $(2, 0)$ and $(4, 0)$.

Here is the graph, showing the axis of symmetry $(x = 3)$ and the point, $(6, -8)$, that is symmetric to the y-intercept.

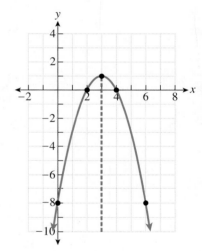

EXAMPLE 5   Graph $y = -\frac{1}{2}x^2 + 2x + 6$.

Solution

In this example, $a = -\frac{1}{2}$, $b = 2$, and $c = 6$. This parabola will open downward because the leading coefficient a is negative. We begin by finding the x-coordinate of the vertex.

$$x = \frac{-2}{2\left(-\dfrac{1}{2}\right)} = 2 \qquad \text{Substitute } -\tfrac{1}{2} \text{ for } a \text{ and 2 for } b.$$

Now we substitute 2 for x in the original equation and solve for y.

$$y = -\frac{1}{2}(2)^2 + 2(2) + 6 \qquad \text{Substitute 2 for } x.$$

$$y = -2 + 4 + 6 \qquad \text{Simplify each term.}$$
$$y = 8 \qquad \text{Simplify.}$$

The vertex is at $(2, 8)$. For an equation of the form $y = ax^2 + bx + c$, the y-intercept is at the point $(0, c)$. The y-intercept of this parabola is $(0, 6)$.

If we plot the vertex and the y-intercept on the graph, we will see that there must be two x-intercepts. The vertex is above the x-axis, and the parabola opens downward, so the graph must cross the x-axis. To find the x-intercepts, we substitute 0 for y in the original equation and solve for x.

$$0 = -\frac{1}{2}x^2 + 2x + 6 \qquad \text{Substitute 0 for } y.$$

$$\frac{1}{2}x^2 - 2x - 6 = 0 \qquad \text{Collect all terms on the left side of the equation.}$$

$$2\left(\frac{1}{2}x^2 - 2x - 6\right) = 2 \cdot 0 \qquad \text{Multiply both sides of the equation by 2 to clear the equation of fractions.}$$

$$x^2 - 4x - 12 = 0 \qquad \text{Distribute and simplify.}$$

$$(x - 6)(x + 2) = 0 \qquad \text{Factor.}$$

$$x = 6 \quad \text{or} \quad x = -2 \qquad \text{Set each factor equal to 0 and solve.}$$

The x-intercepts are $(6, 0)$ and $(-2, 0)$.

Here is the graph, using the axis of symmetry to find a fifth point that is symmetric to the y-intercept.

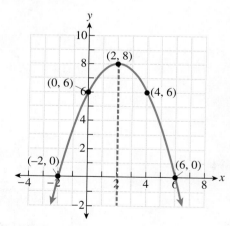

Quick Check 5

Graph $y = -\frac{1}{4}x^2 + 2x - \frac{7}{4}$.

Here is a summary of how to graph parabolas.

Graphing Parabolas

- Determine whether the parabola opens upward or downward. For an equation in standard form $(y = ax^2 + bx + c)$, the parabola will open upward if a is positive. If a is negative, the parabola will open downward.
- Find the vertex of the parabola. Use the formula $x = \dfrac{-b}{2a}$ to find the x-coordinate of the vertex. Substitute this result for x in the original equation to find the y-coordinate.
- Find the y-intercept of the parabola by letting $x = 0$.
- After plotting the vertex and the y-intercept, determine whether there are any x-intercepts. If there are, find them by letting $y = 0$ and solving the resulting quadratic equation for x. If the discriminant of the quadratic formula, $b^2 - 4ac$, is negative, then there are no x-intercepts.

Finding the Vertex of a Parabola by Completing the Square

Objective **3** **Find the vertex of a parabola by completing the square.** As an alternative, we could find the coordinates of the vertex by completing the square.

Finding the Vertex of $y = a(x - h)^2 + k$

The graph of the quadratic equation $y = a(x - h)^2 + k$ is a parabola with axis of symmetry $x = h$ and vertex (h, k). The parabola opens upward if a is positive and opens downward if a is negative.

EXAMPLE 6   Find the vertex and axis of symmetry for the parabola.

a) $y = (x - 2)^2 + 7$

Solution

This equation is in the form $y = a(x - h)^2 + k$. The axis of symmetry is $x = 2$ and the vertex is $(2, 7)$.

b) $y = -(x + 3)^2 - 6$

Solution

This parabola opens downward, but this does not affect how we find the axis of symmetry or the vertex. Note that $x + 3$ is equivalent to $x - (-3)$, so the equation can be rewritten as $y = -(x - (-3))^2 - 6$. The axis of symmetry is $x = -3$ and the vertex is $(-3, -6)$.

Quick Check **6**
Find the vertex and axis of symmetry for the parabola.

a) $y = (x + 9)^2 - 4$
b) $y = -3(x - 1)^2 + 6$

EXAMPLE ▸ **7**   Find the vertex of the parabola $y = x^2 - 12x + 35$ by completing the square.

Solution

We will begin by isolating the terms containing x on the right side of the equation. Then we will complete the square.

$$y = x^2 - 12x + 35$$
$$y - 35 = x^2 - 12x \qquad \text{Subtract 35 to isolate the terms containing } x.$$
$$y - 35 + 36 = x^2 - 12x + 36 \qquad \text{Take half of the coefficient of the first-degree term, square it, and add it to both sides of the equation. } \left(\frac{-12}{2}\right)^2 = (-6)^2 = 36$$
$$y + 1 = (x - 6)^2 \qquad \text{Factor the trinomial as a perfect square.}$$
$$y = (x - 6)^2 - 1 \qquad \text{Subtract 1 to isolate } y.$$

▸ The vertex of the parabola is $(6, -1)$.

EXAMPLE ▸ **8**   Find the vertex of the parabola $y = -2x^2 + 16x - 45$ by completing the square.

Solution

After isolating the terms containing x, we must factor out -2 so that the coefficient of the squared term is 1.

$$y = -2x^2 + 16x - 45$$
$$y + 45 = -2x^2 + 16x \qquad \text{Add 45 to isolate the terms containing } x.$$
$$y + 45 = -2(x^2 - 8x) \qquad \text{Factor out the common factor } -2.$$
$$y + 45 - 32 = -2(x^2 - 8x + 16) \qquad \left(\frac{-8}{2}\right)^2 = (-4)^2 = 16$$

Add 16 in the parentheses. This is equivalent to subtracting 32 on the right side because of the factor -2 that is multiplied by 16. Subtract 32 from the left side.

$$y + 13 = -2(x - 4)^2 \qquad \text{Factor the trinomial as a perfect square.}$$
$$y = -2(x - 4)^2 - 13 \qquad \text{Subtract 13 to isolate } y.$$

The vertex is $(4, -13)$.

Quick Check ▸ **7**

Find the vertex of the parabola by completing the square.

a) $y = x^2 + 10x - 30$
b) $y = -x^2 + 16x - 17$

Graphing Quadratic Equations of the Form $y = a(x - h)^2 + k$

Objective 4 Graph quadratic equations of the form $y = a(x - h)^2 + k$.

EXAMPLE 9   Graph $y = (x + 3)^2 - 4$.

Solution

This parabola opens upward, and the vertex is $(-3, -4)$. Next we find the y-intercept.

$$y = (0 + 3)^2 - 4 \qquad \text{Substitute 0 for } x.$$
$$y = 5 \qquad \text{Simplify.}$$

The y-intercept is $(0, 5)$. Since the parabola opens upward and the vertex is below the x-axis, the parabola has two x-intercepts. We find the coordinates of the x-intercepts by substituting 0 for y and solving for x by extracting square roots.

$$0 = (x + 3)^2 - 4 \qquad \text{Substitute 0 for } y.$$
$$4 = (x + 3)^2 \qquad \text{Add 4 to isolate } (x + 3)^2.$$
$$\pm\sqrt{4} = \sqrt{(x + 3)^2} \qquad \text{Take the square root of each side.}$$
$$\pm 2 = x + 3 \qquad \text{Simplify each square root.}$$
$$-3 \pm 2 = x \qquad \text{Subtract 3 to isolate } x.$$

Quick Check 8

Graph
$y = -(x + 4)^2 + 9.$

The x-coordinates of the x-intercepts are $-3 + 2 = -1$ and $-3 - 2 = -5$. The x-intercepts are $(-1, 0)$ and $(-5, 0)$. The axis of symmetry is $x = -3$, and the point $(-6, 5)$ is symmetric to the y-intercept.

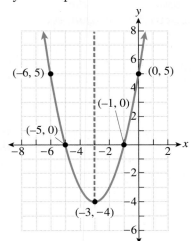

Building Your Study Strategy Time Management, 4 **Studying at Work**   Many students have a job, and this commitment takes a great deal of time away from their studies. If possible, find a job that allows you opportunities to study while you work. For example, if you are working at the checkout counter of the library, your boss may allow you to study while there are no students waiting to check out books. If you are unable to study at work, be sure to take full advantage of any breaks that you receive. A 10-minute work break may not seem like a lot of time, but you can read through a series of note cards or review your class notes during this time.

Vocabulary

1. The graph of a quadratic equation is a U-shaped graph called a _____.

2. A parabola opens upward if the leading coefficient is _____.

3. A parabola opens downward if the leading coefficient is _____.

4. The turning point of a parabola is called its _____.

5. To find the _____ of a parabola, substitute 0 for x and solve for y.

6. To find the _____ of a parabola, substitute 0 for y and solve for x.

Find the vertex of the parabola associated with the following quadratic equations, as well as the equation of the axis of symmetry.

7. $y = x^2 - 8x - 37$

8. $y = x^2 + 6x - 19$

9. $y = x^2 + 16x + 63$

10. $y = x^2 - 2x - 60$

11. $y = -x^2 - 10x + 50$

12. $y = -x^2 + 4x - 11$

13. $y = x^2 - 5x + 21$

14. $y = x^2 + 7x + 15$

15. $y = 3x^2 + 18x - 32$

16. $y = -2x^2 - 10x + 5$

17. $y = x^2 - 10$

18. $y = x^2 - 10x$

19. $y = (x - 6)^2 - 1$

20. $y = (x + 2)^2 + 11$

21. $y = -(x + 9)^2 - 18$

22. $y = -(x - 12)^2 - 3$

Find the x- and y-intercepts of the parabola associated with the following quadratic equations. If necessary, round to the nearest tenth. If the parabola does not have any x-intercepts, state "no x-intercepts."

23. $y = x^2 + 9x + 18$

24. $y = x^2 - 4x - 22$

25. $y = -x^2 + 5x - 9$

26. $y = x^2 - 9x + 20$

27. $y = 2x^2 - 5x - 17$

28. $y = -x^2 + 3x + 40$

29. $y = x^2 + 8x$

30. $y = -x^2 + 8x - 21$

31. $y = x^2 - 12x + 36$

32. $y = -x^2 + 10x - 25$

33. $y = (x + 5)^2 - 9$

34. $y = (x - 2)^2 - 12$

35. $y = -(x - 3)^2 - 4$

36. $y = -(x + 7)^2 + 16$

Graph the following parabolas. Find the vertex and all intercepts.

37. $y = x^2 + 2x - 3$

38. $y = x^2 - 5x + 3$

41. $y = -x^2 + 5x - 9$

39. $y = -x^2 + 6x$

42. $y = -x^2 + 9$

40. $y = x^2 - 4x - 32$

43. $y = 2x^2 + 12x - 15$

44. $y = x^2 - 2x - 7$

47. $y = -x^2 + 8x - 4$

45. $y = x^2 + 4x + 9$

48. $y = x^2 - 2x$

46. $y = x^2 - 10x + 26$

49. $y = x^2 + 6x + 9$

50. $y = x^2 + 5x + 4$

53. $y = \frac{1}{2}x^2 - 3x + \frac{5}{2}$

51. $y = x^2 - 8$

54. $y = \frac{1}{4}x^2 + \frac{1}{2}x - 12$

52. $y = -x^2 + 8x - 16$

55. $y = (x + 1)^2 + 5$

56. $y = (x - 3)^2 - 4$

57. $y = -(x + 2)^2 + 7$

58. $y = -(x - 2)^2 + 6$

Find an equation of a parabola that meets the given conditions.

59. Axis of symmetry: $x = 3$, range: $[-5, \infty)$

60. Axis of symmetry: $x = 7$, range: $(-\infty, 2]$

61. x-intercepts: $(8, 0)$ and $(12, 0)$

62. Axis of symmetry: $x = -4$, no x-intercepts

> **Writing in Mathematics**

Answer in complete sentences.

63. Explain how to determine that a parabola does not have any x-intercepts.

64. Explain how to find the vertex of a parabola by completing the square. Use an example.

65. *Solutions Manual*[*] Write a solutions manual page for the following problem:

    Graph $y = (x - 3)^2 - 4$.

66. *Newsletter*[*] Write a newsletter explaining how to graph an equation of the form $y = ax^2 + bx + c$.

*See Appendix B for details and sample answers.

Section 9.1–Topic	Chapter Review Exercises
Solving Quadratic Equations by Factoring	1–6
Solving Quadratic Equations by Extracting Square Roots	7–12
Solving Quadratic Equations by Completing the Square	13–16
Finding a Quadratic Equation Given Its Solutions	33–38

Section 9.2–Topic	Chapter Review Exercises
Solving Quadratic Equations by Using the Quadratic Formula	17–22
Solving Quadratic Equations by Using the Most Efficient Technique	23–32

Section 9.3–Topic	Chapter Review Exercises
Solving Applied Problems Involving Quadratic Equations	53–58

Section 9.4–Topic	Chapter Review Exercises
Finding the Vertex and Axis of Symmetry of a Parabola	39–42
Finding the Intercepts of a Parabola	43–46
Graphing Parabolas	47–52

Review of Chapter 9 Study Strategies

Poor time management can cause even a bright student to do poorly in a math class. To maximize the time you have available to study mathematics, and to make your use of this time effective, consider the following suggestions:

- Record how you spend your time over the period of one week. Look for how you might be wasting time. Create an efficient schedule to follow.
- Study new material as soon as possible after class. Review work later in the day.
- Arrange your schedule to allow you to study mathematics at the time of day when your mental energy is at its highest.
- Set goals for each study session, and record these goals in a "to do" list.
- Study your most difficult subject first, before you become fatigued.
- If you must work, try to find a job that will allow you to do some studying while you work. At the very least, make efficient use of your work breaks to study.

Solve by factoring. [9.1]

1. $x^2 + 15x + 56 = 0$

2. $x^2 + 4x - 60 = 0$

3. $x^2 + 5x = 0$

4. $3x^2 - 20x + 12 = 0$

5. $x^2 + 5x - 23 = 8x + 31$

6. $x^2 + 4x - 52 = 7(x + 8)$

Solve by extracting square roots. [9.1]

7. $x^2 = 64$

8. $x^2 - 48 = 0$

9. $x^2 + 36 = 0$

10. $(x + 3)^2 = -49$

11. $(x - 4)^2 + 52 = 43$

12. $(2x - 9)^2 - 27 = 142$

Solve by completing the square. [9.1]

13. $x^2 - 3x - 28 = 0$ 14. $x^2 - 8x + 4 = 0$

15. $x^2 + 10x + 65 = 0$ 16. $x^2 + 3x - 14 = 0$

Solve using the quadratic formula. [9.2]

17. $x^2 + 9x + 5 = 0$

18. $x^2 + 4x - 10 = 0$

19. $x^2 - 5x + 8 = 0$

20. $3x^2 - 16x + 19 = 0$

21. $2x^2 - 2x - 17 = x^2 + 2x + 28$

22. $x(x + 5) = 13x - 25$

Solve. [9.1, 9.2]

23. $x^2 - 14x + 48 = 0$

24. $x^2 + 2x - 63 = 0$

25. $(x - 7)^2 + 6 = -18$

26. $x^2 + 6x + 41 = 0$

27. $x^2 + 7x = 0$

28. $x^2 - 8x - 52 = 0$

29. $2x^2 - 9x + 3 = 0$

30. $(x + 10)^2 + 81 = 0$

For the given function $f(x)$, find all values x for which $f(x) = 0$. [9.1, 9.2]

31. $f(x) = (x + 7)^2 - 20$

32. $f(x) = x^2 - 5x - 28$

Find a quadratic equation with integer coefficients that has the given solutions. [9.1]

33. $\{-2, 7\}$

34. $\{8\}$

35. $\left\{-\dfrac{3}{4}, \dfrac{2}{5}\right\}$

36. $\{-5\sqrt{3}, 5\sqrt{3}\}$

37. $\{-i, i\}$

38. $\{-2i\sqrt{7}, 2i\sqrt{7}\}$

Find the vertex and axis of symmetry of the parabola associated with the given quadratic equation. [9.4]

39. $y = x^2 - 10x - 56$

40. $y = x^2 + 3x + 20$

41. $y = -x^2 + 6x + 42$

42. $y = (x - 8)^2 - 15$

Find the x- and y-intercepts of the parabola associated with the given equation. If the parabola does not have any x-intercepts, state this. [9.4]

43. $y = x^2 - 7x + 20$

44. $y = x^2 - 6x - 10$

45. $y = -x^2 - 2x + 35$

46. $y = (x + 8)^2 - 44$

Worked-out solutions to Review Exercises marked with ◯ can be found on page AN–31.

Graph the quadratic equation. Find the vertex, the y-intercept, and any x-intercepts. [9.4]

47. $y = x^2 + 10x + 24$

48. $y = -x^2 - 2x + 8$

49. $y = x^2 - 4x - 6$

50. $y = x^2 - 2x + 6$

51. $y = (x + 2)^2 - 4$

52. $y = -(x - 4)^2 + 8$

55. Jay needs to fly to a city that is 300 miles north and 900 miles east of where he lives. If the airplane flies a direct route, how far will the flight be? Round to the nearest tenth of a mile. [9.3]

56. The diagonal of a rectangular garage is 28 feet long. If the width of the garage is 9 feet less than the length of the garage, find the dimensions of the garage. Round to the nearest tenth of a foot. [9.3]

For Exercises 57 and 58, use the function
$h(t) = -16t^2 + v_0t + s.$

57. A projectile is launched upward from the ground at a speed of 116 feet per second. How long will it take until the projectile lands on the ground? [9.3]

58. A projectile is launched upward from a platform 12 feet high with an initial velocity of 75 feet per second. How long will it take until the projectile lands on the ground? Round to the nearest hundredth of a second. [9.3]

53. A rectangular photo has an area of 560 square centimeters. If the width of the photo is 8 centimeters longer than its height, find the dimensions of the photo. [9.3]

54. A rectangular swimming pool covers an area of 350 square feet. If the length of the pool is 10 feet more than its width, find the dimensions of the pool. Round to the nearest tenth of a foot. [9.3]

For Extra Help | Pass *the* Test | Test solutions are found on the enclosed CD.

1. Solve by factoring. $x^2 + 11x + 24 = 0$

2. Solve by extracting square roots. $(x + 6)^2 - 7 = 113$

3. Solve by completing the square. $x^2 - 10x - 17 = 0$

4. Solve using the quadratic formula. $x^2 + 8x + 20 = 0$

Solve.

5. $(2x - 5)^2 - 19 = 30$

6. $4x^2 - 17x + 18 = 0$

7. $x^2 - 14x + 45 = 0$

8. $x^2 + 6x + 27 = 0$

Find a quadratic equation with integer coefficients that has the given solutions.

9. $\left\{ -2, \dfrac{5}{3} \right\}$

10. $\{ -3i, 3i \}$

11. Find the x- and y-intercepts of the parabola associated with $y = x^2 + 14x + 30$.

12. Graph $y = -x^2 - 6x - 8$. Find the vertex, the y-intercept, and any x-intercepts.

13. Graph $y = (x + 3)^2 - 8$. Find the vertex, the y-intercept, and any x-intercepts.

14. The area of a rectangular parking lot is 3200 square yards. If the width of the parking lot is 45 yards less than the length of the parking lot, find the dimensions of the field. Round to the nearest tenth of a yard.

15. Mark flew to a resort town 1200 miles south and 1000 miles west of where he lives. How far away is the resort town from his home? Round to the nearest tenth of a mile.

Mathematicians in History
Marie-Sophie Germain

Marie-Sophie Germain, frequently referred to as simply Sophie Germain, was a French mathematician who earned the respect of prominent mathematicians at a time when it was difficult for a woman to do so. Among areas of interest for Sophie were number theory and elasticity of surfaces.

Write a one-page summary (*or* make a poster) of the mathematical achievements and life of Sophie Germain.

Interesting issues:

- At the age of 13, Sophie Germain read the account of the death of a prominent mathematician at the hands of a Roman soldier, and this prompted her decision to become a mathematician. Who was the mathematician? Relate the story of his death.
- Sophie obtained lecture notes for many courses at École Polytechnique. Why didn't she attend École Polytechnique?
- What pseudonym did Germain use to submit a paper to the prominent mathematician Joseph Lagrange?
- Germain also correspondended with Carl Gauss. Describe the circumstances by which Gauss discovered that Germain was actually a woman.
- Describe the circumstances by which Germain won a prize from the Institut de France for her work on Chaldini figures.
- One of her significant contributions to mathematics is the concept of the Sophie Germain prime. What are the conditions for a number to be a Sophie Germain prime? Give an example of such a number.
- She also introduced an identity that is commonly known as Sophie Germain's Identity. State this identity.
- Describe the circumstances that led to Germain's death.

Stretch Your Thinking ❭ Chapter 9

The pilot of a small plane planned to fly due east. However there is a powerful wind blowing due north that is causing the plane to fly 7° off its eastern course. The wind has also increased the speed of the plane so that it is flying at 287 miles per hour. The speed of the plane in still air is 250 miles per hour faster than the speed of the wind. Find the speed of the plane in still air and the speed of the wind.

Find all values for which the rational expression is undefined.

1. $\dfrac{x^2 + 11x + 24}{x^2 - 14x + 45}$

Simplify the given rational expression. (Assume that all denominators are nonzero.)

2. $\dfrac{x^2 + 3x - 54}{x^2 + 16x + 63}$

Evaluate the given rational function.

3. $r(x) = \dfrac{x + 12}{x^2 + 7x + 5}, r(3)$

Multiply.

4. $\dfrac{x^2 + 3x - 18}{x^2 + 15x + 44} \cdot \dfrac{x^2 + 2x - 8}{x^2 + 8x + 12}$

Divide.

5. $\dfrac{x^2 + x - 90}{x^2 + 14x + 49} \div \dfrac{x^2 - 8x - 9}{x^2 + 5x - 14}$

Add or subtract.

6. $\dfrac{5x + 7}{x^2 - 2x - 8} - \dfrac{2x + 19}{x^2 - 2x - 8}$

7. $\dfrac{6}{x^2 + x - 2} + \dfrac{2}{x^2 - 6x + 5}$

8. $\dfrac{x - 5}{x^2 - 10x + 21} - \dfrac{3}{x^2 - 9}$

9. $\dfrac{x + 4}{x^2 + 13x + 42} + \dfrac{6}{x^2 + 12x + 35}$

Simplify the complex fraction.

10. $\dfrac{\dfrac{x - 5}{x + 3} - \dfrac{1}{x - 6}}{\dfrac{x^2 + 5x - 24}{x^2 - 3x - 18}}$

11. $\dfrac{1 + \dfrac{1}{x} - \dfrac{20}{x^2}}{1 + \dfrac{13}{x} + \dfrac{40}{x^2}}$

Solve.

12. $6 + \dfrac{4}{x - 3} = \dfrac{3x + 1}{x - 3}$

13. $\dfrac{2x - 3}{x + 4} = \dfrac{x + 3}{x - 4}$

14. $\dfrac{5}{x + 7} + \dfrac{4}{x - 3} = \dfrac{x - 11}{x^2 + 4x - 21}$

Solve for the specified variable.

15. $\dfrac{1}{x} + \dfrac{1}{y} = 2$ for x

16. The sum of the reciprocal of a number and $\dfrac{4}{15}$ is $\dfrac{11}{30}$. Find the number.

17. Two pipes, working together, can fill a tank in 3 hours. Working alone, it would take the smaller pipe 8 hours longer than it would take the larger pipe to fill the tank. How long would it take the smaller pipe alone to fill the tank?

Evaluate the radical function. (Round to the nearest thousandth if necessary.)

18. $f(x) = \sqrt{3x - 19}, f(10)$

Simplify the radical expression. Assume that all variables represent nonnegative values.

19. $\sqrt{60a^5b^{12}}$

20. $\sqrt[3]{48x^9y^{16}z^2}$

Add or subtract.

21. $9\sqrt{18} - 4\sqrt{50}$

22. $4\sqrt{12} + 5\sqrt{48} + 6\sqrt{27}$

Multiply. Assume that all variables represent nonnegative values.

23. $(3\sqrt{2} - 4\sqrt{3})(7\sqrt{2} + 2\sqrt{3})$

24. $(7\sqrt{3} + 6)(7\sqrt{3} - 6)$

Rationalize the denominator and simplify. Assume that all variables represent nonnegative values.

25. $\dfrac{6}{\sqrt{8x}}$

26. $\dfrac{4\sqrt{5} + \sqrt{2}}{3\sqrt{5} - 2\sqrt{2}}$

Solve.

27. $\sqrt{x + 7} - 1 = x$

28. For the function $f(x) = \sqrt{3x + 1} - 5$, find all values x for which $f(x) = 2$.

Express in terms of i.

29. $\sqrt{-108}$

Multiply.

30. $(7 + 4i)(9 - i)$

Rationalize the denominator.

31. $\dfrac{10}{3 - 4i}$

32. $\dfrac{6 + 5i}{3i}$

Rewrite as a radical expression and simplify if possible. (Assume $x \geq 0$.)

33. $(16x^{12})^{5/2}$

Simplify the expression. (Assume $x \geq 0$.)

34. $(x^{2/5})^{9/4}$

Solve by extracting square roots.

35. $(x - 4)^2 + 37 = 29$

Solve by completing the square.

36. $x^2 + 10x - 19 = 0$

Solve using the quadratic formula.

37. $2x^2 - 6x - 15 = 0$

Solve.

38. $x^2 + 12x + 27 = 0$

39. $x(x - 6) = 8x - 15$

40. $2(x - 9)^2 - 15 = 21$

Graph the quadratic equation. Find the vertex, the y-intercept, and any x-intercepts.

41. $y = x^2 - 6x - 7$

42. $y = -x^2 - 3x + 5$

43. $y = (x - 2)^2 + 1$

44. The length of a rectangular lawn is 6 feet more than its width. If the area of the lawn is 750 square feet, find the dimensions of the lawn. Round to the nearest tenth of a foot.

45. A projectile is launched upward from the roof of a building 15 feet high with an initial velocity of 64 feet per second. How long will it take until the projectile lands on the ground? Round to the nearest hundredth of a second. (Use the function $h(t) = -16t^2 + v_0 t + s$.)

Synthetic division is a shorthand alternative to polynomial long division when the divisor is of the form $x - b$. When we use synthetic division, we do not write any variables, only the coefficients.

Suppose that we wanted to divide $\dfrac{2x^2 - x - 25}{x - 4}$. Note that $b = 4$ in this example. Before we begin, we check the dividend to be sure that it is written in descending order and that there are no missing terms. If there are missing terms, we will add placeholders. On the first line, we write down the value for b as well as the coefficients for the dividend.

$$4 \underline{|} \qquad 2 \qquad -1 \qquad -25$$

Notice that we set the value of b apart from the coefficients of the divisor. After we use synthetic division, we will find the quotient and the remainder in the last of our three rows. To begin the actual division, we bring the first coefficient down to the bottom row as follows:

$$
\begin{array}{r|rrr}
4 & 2 & -1 & -25 \\
& \downarrow & & \\
\hline
& 2 & &
\end{array}
$$

This number will be the leading coefficient of our quotient. In the next step, we multiply 4 by this first coefficient and write it under the next coefficient in the dividend (-1). Totaling this column will give us the next coefficient in the quotient.

$$
\begin{array}{r|rrr}
4 & 2 & -1 & -25 \\
& \downarrow & 8 & \\
\hline
& 2 & 7 &
\end{array}
$$

We repeat this step until we total the last column. Multiply 4 by 7 and write the product under -25. Total the last column.

$$
\begin{array}{r|rrr}
4 & 2 & -1 & -25 \\
& \downarrow & 8 & 28 \\
\hline
& 2 & 7 & \boxed{3}
\end{array}
$$

We are now ready to read our quotient and remainder from the bottom row. The remainder is the last number in the bottom row, which in our example is 3. The numbers in front of the remainder are the coefficients of our quotient, written in descending order. In this

example, the quotient is $2x + 7$. So, $\dfrac{2x^2 - x - 25}{x - 4} = 2x + 7 + \dfrac{3}{x - 4}$.

EXAMPLE 1 Use synthetic division to divide $\dfrac{x^3 - 11x^2 - 13x + 6}{x + 2}$.

Solution

The divisor is of the correct form, and $b = -2$. This is because $x + 2$ can be rewritten as $x - (-2)$. The dividend is in descending form with no missing terms, so we may begin.

$$
\begin{array}{r|rrrr}
-2 & 1 & -11 & -13 & 6 \\
   & \downarrow & -2 & 26 & -26 \\
\hline
   & 1 & -13 & 13 & \boxed{-20}
\end{array}
$$

So, $\dfrac{x^3 - 11x^2 - 13x + 6}{x + 2} = x^2 - 13x + 13 - \dfrac{20}{x + 2}$.

Now we turn to an example that has a dividend with missing terms, requiring the use of placeholders.

EXAMPLE 2 Use synthetic division to divide $\dfrac{3x^4 - 8x - 5}{x - 3}$.

Solution

The dividend, written in descending order with placeholders, is $3x^4 + 0x^3 + 0x^2 - 8x - 5$. The value for b is 3.

$$
\begin{array}{r|rrrrr}
3 & 3 & 0 & 0 & -8 & -5 \\
  & \downarrow & 9 & 27 & 81 & 219 \\
\hline
  & 3 & 9 & 27 & 73 & \boxed{214}
\end{array}
$$

So, $\dfrac{3x^4 - 8x - 5}{x - 3} = 3x^3 + 9x^2 + 27x + 73 + \dfrac{214}{x - 3}$.

EXERCISES

Divide using synthetic division.

1. $\dfrac{x^2 - 18x + 77}{x - 7}$

2. $\dfrac{x^2 + 4x - 45}{x - 5}$

3. $(x^2 + 13x - 48) \div (x - 3)$

4. $(2x^2 - 43x + 221) \div (x - 13)$

5. $\dfrac{7x^2 - 11x - 12}{x + 2}$

6. $\dfrac{3x^2 + 50x - 30}{x + 17}$

7. $\dfrac{12x^2 - 179x - 154}{x - 14}$

8. $\dfrac{x^2 + 14x - 13}{x + 1}$

9. $\dfrac{x^3 - 34x + 58}{x - 5}$

10. $\dfrac{x^3 - 59x - 63}{x + 7}$

11. $\dfrac{x^4 + 8x^2 - 33}{x + 8}$

12. $\dfrac{5x^5 - 43x^3 - 8x^2 + 9x - 9}{x - 3}$

SOLUTIONS MANUAL

A **solutions manual page** is an assignment in which you solve a problem as if you were creating a solutions manual for a textbook. Show and explain each step in solving the problem in such a way that a fellow student would be able to read and understand your work.

EXAMPLE   Write a solutions manual page for the following problem:

Simplify $3 - 4(20 \div 4 - 6 \cdot 8) - 5^2$.

Student's Solution

We simplify this expression using the order of operations, beginning with the expression inside the parentheses.

$$3 - 4(20 \div 4 - 6 \cdot 8) - 5^2$$

$= 3 - 4(5 - 6 \cdot 8) - 5^2$	Inside the parentheses, do the division first. $20 \div 4 = 5$.
$= 3 - 4(5 - 48) - 5^2$	Still working inside the parentheses, multiplication comes before subtraction. $6 \cdot 8 = 48$.
$= 3 - 4(-43) - 5^2$	Subtract $5 - 48$ inside the parentheses. $5 - 48 = -43$
$= 3 - 4(-43) - 25$	Now that the work inside parentheses is done, simplify expressions with exponents. $5^2 = 25$. We are squaring 5, not -5.
$= 3 + 172 - 25$	Next in order is multiplication or division, so multiply. $-4(-43) = 172$
$= 150$	Finish with addition and subtraction.

NEWSLETTER

A **newsletter** assignment asks you to explain a certain mathematical topic. Your explanation should be in the form of a short, visually appealing article that might be published in a newsletter read by people who were interested in learning mathematics.

> **EXAMPLE** ▶   Write a newsletter explaining how to solve a system of two linear equations by the substitution method.

Student's Solution

Substitution Method

First, solve one of the two equations for x or y. It's usually easier to solve an equation that has a variable term with a coefficient of 1 or -1, that way you won't have to divide or get involved with fractions. If you solve the first equation for x in terms of y, then you substitute that expression for x in the second equation.

Solve $\begin{aligned} x + 3y &= 1 \\ 4x - 5y &= 30 \end{aligned}$ using the substitution method.

First, solve the first equation for x.

$$\begin{aligned} x + 3y &= -1 \\ x &= -3y - 1 \end{aligned}$$ Subtract $3y$ from both sides.

Next, we can substitute $-3y - 1$ for x in the second equation, $4x - 5y = 30$

$$\begin{aligned} 4x - 5y &= 30 \\ 4(-3y - 1) - 5y &= 30 \end{aligned}$$ Substitute $-3y - 1$ for x.

This gives us an equation with only one variable (y) in it, so we solve this equation for y.

$$\begin{aligned} 4(-3y - 1) - 5 &= 30 \\ -12y - 4 - 5y &= 30 \\ -17y - 4 &= 30 \\ -17 &= 34 \\ \frac{-17y}{-17} &= \frac{34}{-17} \\ y &= -2 \end{aligned}$$ Solve the equation for y.

Now that we have one coordinate of our ordered pair solution, we can substitute for y to find our x-coordinate.

$$\begin{aligned} x &= -3y - 1 \\ x &= -3(-2) - 1 \\ x &= 6 - 1 \\ x &= 5 \end{aligned}$$ Substitute -2 for y and solve.

▶ The solution is $(5, -2)$.

Chapter 1

Quick Check 1.1 **1.** [number line 0–10] **2. a)** $>$ **b)** $<$ **3. a)** 13 **b)** -8 **4. a)** $<$ **b)** $>$ **5.** 9

Section 1.1 **1.** empty set **3.** infinity **5.** [number line 0–10] **7.** [number line 0–10] **9.** [number line 0–10] **11.** $<$ **13.** $>$
15. $>$ **17.** [number line -10 to 0] **19.** [number line 0–10] **21.** [number line -15 to 0] **23.** 7 **25.** -14 **27.** 0 **29.** $>$
31. $<$ **33.** $<$ **35.** $<$ **37.** 15 **39.** 0 **41.** 25 **43.** -7 **45.** -29 **47.** $>$ **49.** $>$ **51.** $<$ **53.** A, B, C, D **55.** B, C, D **57.** C, D
59. $-7, 7$ **61.** 0 **63.** $-6, 6$ **65.** [number line -10 to 10] **67.** [number line -10 to 10, -9 is to the left of -4]
69. No, the absolute value of 0 is not positive. **71.** Negative

Quick Check 1.2 **1.** 8 **2.** -13 **3.** -11 **4.** 18 **5.** -4 **6.** -60 **7.** 63 **8.** -400 **9.** -9

Section 1.2 **1.** the sign of the integer with the larger absolute value **3.** a positive integer **5.** positive **7.** divisor **9.** -2
11. -18 **13.** 1 **15.** 8 **17.** -11 **19.** 2 **21.** -6 **23.** -8 **25.** -21 **27.** 27 **29.** 15 **31.** -6 **33.** 27 **35.** 1 **37.** $\$8$ **39.** 4°C
41. 2150 ft **43.** -48 **45.** -36 **47.** 143 **49.** -82 **51.** 0 **53.** 90 **55.** 180 **57.** -6 **59.** -9 **61.** 7 **63.** -14 **65.** 0
67. Undefined **69.** 132 **71.** -8 **73.** -21 **75.** -144 **77.** -216 **79.** 0 **81.** $\$96$ **83.** $-\$200$ **85.** 26 **87.** -84 **89.** True
91. True **93.** [number line -10 to 10, 3, $+6$] **95.** [number line -10 to 10, $+3$, -7]
97. Answers will vary. **99.** Answers will vary.

Quick Check 1.3 **1.** $\{1, 2, 3, 4, 6, 9, 12, 18, 36\}$ **2. a)** Composite **b)** Prime **c)** Composite **3.** $3 \cdot 3 \cdot 7$ **4. a)** $\frac{3}{14}$ **b)** $\frac{1}{16}$ **5.** $9\frac{4}{13}$ **6.** $\frac{49}{6}$

Section 1.3 **1.** factor set **3.** composite **5.** the number written on the top of the fraction **7.** its numerator and denominator
contain no common factors other than 1 **9.** improper **11.** No **13.** Yes **15.** Yes **17.** $\{1, 3, 9, 27\}$ **19.** $\{1, 2, 4, 5, 10, 20\}$
21. $\{1, 31\}$ **23.** $\{1, 2, 3, 4, 5, 6, 10, 12, 15, 20, 30, 60\}$ **25.** $\{1, 7, 49\}$ **27.** $\{1, 3, 23, 69\}$ **29.** $\{1, 13, 17, 221\}$ **31.** $2 \cdot 3 \cdot 3$
33. $2 \cdot 3 \cdot 7$ **35.** $3 \cdot 11$ **37.** $3 \cdot 3 \cdot 3$ **39.** $5 \cdot 5 \cdot 5$ **41.** Prime **43.** $3 \cdot 3 \cdot 11$ **45.** Prime **47.** $7 \cdot 17$ **49.** $2 \cdot 2 \cdot 2 \cdot 3 \cdot 3 \cdot 5$
51. $2 \cdot 2 \cdot 2 \cdot 13$ **53.** $\frac{5}{8}$ **55.** $\frac{1}{5}$ **57.** $\frac{3}{7}$ **59.** $\frac{27}{64}$ **61.** $\frac{10}{11}$ **63.** $\frac{19}{5}$ **65.** $\frac{50}{17}$ **67.** $\frac{431}{19}$ **69.** $7\frac{4}{5}$ **71.** $14\frac{3}{7}$ **73.** $5\frac{6}{11}$ **75.** Answers will vary; one
possible answer is $\frac{6}{8}, \frac{9}{12}, \frac{12}{16}$, and $\frac{15}{20}$. **77.** Answers will vary; one possible answer is 30, 42, 70, and 105. **79.** Answers will vary.

Quick Check 1.4 **1.** $\frac{5}{56}$ **2.** 23 **3.** $\frac{8}{105}$ **4.** $\frac{5}{14}$ **5.** $\frac{5}{8}$ **6.** $\frac{3}{5}$ **7.** $11\frac{3}{4}$ **8.** 126 **9.** $\frac{26}{45}$ **10.** $\frac{5}{6}$ **11.** $1\frac{11}{30}$

Section 1.4 **1.** multiply the numerators together and multiply the denominators together, dividing out any factors that are common
to a numerator and a denominator **3.** reciprocal **5.** denominator **7.** a **9.** $\frac{3}{20}$ **11.** $-\frac{21}{25}$ **13.** $2\frac{2}{5}$ **15.** $31\frac{1}{2}$ **17.** $\frac{16}{27}$ **19.** $\frac{4}{3}$
21. $-\frac{1}{12}$ **23.** $\frac{17}{20}$ **25.** $\frac{5}{44}$ **27.** $-\frac{21}{50}$ **29.** $\frac{11}{15}$ **31.** 1 **33.** $-\frac{2}{5}$ **35.** $\frac{3}{8}$ **37.** 36 **39.** 70 **41.** 36 **43.** 36 **45.** $\frac{5}{6}$ **47.** $\frac{31}{20}$ **49.** $\frac{53}{40}$ **51.** $\frac{8}{15}$
53. $11\frac{1}{5}$ **55.** $11\frac{5}{6}$ **57.** $\frac{1}{2}$ **59.** $\frac{1}{5}$ **61.** $\frac{13}{28}$ **63.** $\frac{2}{15}$ **65.** $4\frac{1}{5}$ **67.** $3\frac{3}{10}$ **69.** $-\frac{41}{36}$ **71.** $-\frac{17}{48}$ **73.** $\frac{146}{105}$ **75.** $-\frac{89}{90}$ **77.** $-\frac{67}{240}$ **79.** $\frac{8}{15}$ **81.** $\frac{4}{3}$
83. $\frac{8}{25}$ **85.** $-\frac{17}{24}$ **87.** $\frac{56}{45}$ **89.** 104 **91.** $2\frac{6}{35}$ **93.** $-\frac{83}{168}$ **95.** $-\frac{77}{468}$ **97.** 8 **99.** 14 **101.** $1\frac{7}{12}$ cups **103.** $\frac{9}{20}$ fluid ounce **105.** $\frac{7}{10}$ of a
foot **107.** No, the craftsman needs 3 feet of wood. **109.** $\frac{2}{3}$ cups **111.** [bar model figure]
113. [bar model figure] **115.** Answers will vary.

Quick Check 1.5 **1.** 0.75 **2.** $0.2\overline{7}$ **3.** $\frac{1}{25}$ **4.** $\frac{17}{40}$ **5.** 70% **6.** $52\frac{1}{2}\%$ **7.** 42% **8.** $\frac{7}{20}$ **9.** $\frac{7}{60}$ **10.** 0.08 **11.** 2.4

Section 1.5 **1.** tenths **3.** divide the numerator by the denominator **5.** divide it by 100 and omit the percent sign **7.** 7.85
9. -36.4 **11.** 4.7 **13.** 33.92 **15.** -8.03 **17.** 35.032 **19.** 55.2825 **21.** 94 **23.** 0.6 **25.** -0.875 **27.** 0.64 **29.** 24.58 **31.** $\frac{1}{5}$
33. $\frac{11}{25}$ **35.** $-\frac{3}{4}$ **37.** $\frac{3}{8}$ **39.** 101.4°F **41.** $\$245.95$ **43.** $\$257.64$ **45.** 50% **47.** 75% **49.** 80% **51.** $87\frac{1}{2}\%$ **53.** 675% **55.** 40%
57. 15% **59.** 9% **61.** 320% **63.** $\frac{53}{100}$ **65.** $\frac{21}{25}$ **67.** $\frac{7}{100}$ **69.** $\frac{1}{9}$ **71.** $5\frac{1}{5}$ **73.** 0.32 **75.** 0.16 **77.** 0.04 **79.** 0.003 **81.** 4
83. Answers will vary, possible answers include $\frac{3}{8}, \frac{6}{16}, \frac{9}{24}$ **85.** $\frac{1}{5}$, 20% **87.** $\frac{8}{25}$, 0.32 **89.** 1.625, 162.5% **91.** Answers will vary.

Quick Check 1.6

1.

1. Is There Adequate Parking on Campus?

Unsure 20% Yes 22% No 58%

2. Unemployment Rates

7.6, 4.7, 4.2, 3.3, 2.3 — Less than high school, High school, no college, Some college, Associate's degree, Bachelor's degree or higher; Level of Education

3. Money Raised for Charity by Runners (2002–2005)

Section 1.6 **1.** pie chart **3.** compare categories and their frequencies in a set of data **5.**

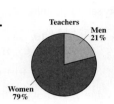

Teachers — Men 21%, Women 79%

7. Americans with College Degrees — College Degree 36%, No College Degree 64%

9. Checking In on Vacation — Never 30%, Occasionally 49%, Constantly 6%, Daily 15%

11. Should Moviegoers Pay Less? — Unsure 19%, No 10%, Yes 71%

13. When Freshmen Planned on Attending College — Unsure 7%, Never 17%, Right after Graduation 64%, After Working at Least 2 Years 12%

15. College Freshmen — Male 45%, Female 55%; College Faculty — Female 43%, Male 57%

17. Hardship from Gasoline Prices — 121 Severe Hardship, 414 Moderate Hardship, 464 No Hardship, 10 Unsure; Level of Hardship

19. Eighth-Graders Who Scored at or above "Proficient" on National Assessments of Reading Skills — 11 Atlanta, 22 Boston, 15 Chicago, 20 San Diego, 10 Washington, DC; City

21. 2005 Patents — 2941 IBM, 1828 Canon, 1797 Hewlett-Packard, 1688 Matsushita Electric, 1641 Samsung; Company

23. Hours Spent Listening to Recorded Music (1999–2004)

25. Worldwide Spam Messages Sent Daily (1999–2004)

27. Average Size of New Home (1940–2010)

29. Gastric Bypass Surgeries (1997–2005)

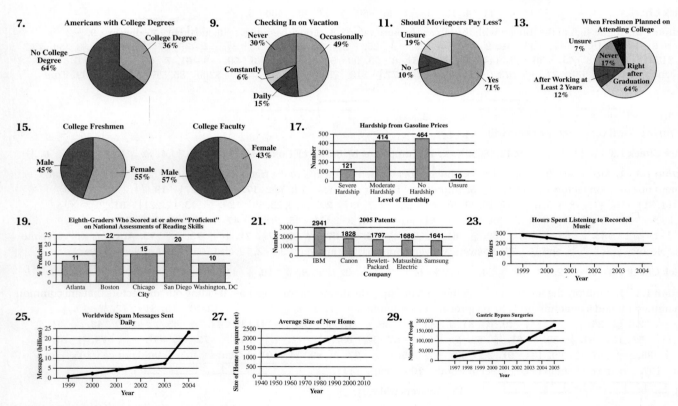

31. Answers will vary **33.** Answers will vary.

Quick Review **1.** -7 **2.** -120 **3.** -20 **4.** $-\frac{110}{21}$

Quick Check 1.7 **1.** 64 **2.** $\frac{1}{4096}$ **3.** -84 **4.** 1 **5.** 360 **6.** -84

Section 1.7 **1.** base **3.** squared **5.** grouping **7.** 2^3 **9.** $\left(\frac{2}{9}\right)^6$ **11.** $(-2)^4$ **13.** -3^5 **15.** 5^3 **17.** 32 **19.** 2401 **21.** 1000 **23.** 1 **25.** $\frac{27}{64}$ **27.** 0.00032 **29.** 81 **31.** -128 **33.** 72 **35.** 576 **37.** 7 **39.** 10 **41.** 16 **43.** -14.44 **45.** 14.9 **47.** -71 **49.** 49 **51.** 51 **53.** $\frac{11}{14}$ **55.** $\frac{11}{125}$ **57.** $\frac{1}{21}$ **59.** $-\frac{1}{4}$ **61.** $\frac{10}{3}$ **63.** 14 **65.** 30 **67.** -64 **69.** -1995 **71.** Answers will vary. **73.** Answers will vary. **75.** 1024 **77.** 1,073,741,824 **79.** 6 **81.** 6 **83.** 5 **85.** $3 \cdot 5 - 9 + 8 = 14$ **87.** $(3 + 7 \cdot 9) \div 2 = 33$ **89.** Answers will vary. $(-2)^6 = 64, -2^6 = -64$

Quick Check 1.8   **1.** $x + 9$ **2.** $x - 25$ **3.** $2x$ **4.** $\frac{x}{20}$ **5.** 57   **6.** 128   **7.** 97   **8.** $24a$ **9.** $9x + 34$ **10.** $35x - 28$ **11.** $12x - 24y + 36z$ **12.** $-24x - 66$ **13.** Four terms: $x^3, -x^2, 23x,$ and -59, Coefficients: $1, -1, 23, -59$ **14.** $-2x + 8y$ **15.** $19x + 7$ **16.** $-2x + 51$

Section 1.8   **1.** variable   **3.** evaluate   **5.** associative   **7.** term   **9.** like terms   **11.** $x + 11$ **13.** $5x$ **15.** $13 - x$ **17.** $3x - 4$ **19.** $x + y$ **21.** $7(x - y)$ **23.** $325c$ **25.** $25{,}000 + 22a$ **27.** -1 **29.** 20   **31.** 6   **33.** 88   **35.** -20 **37.** 0   **39.** -754 **41.** 49   **43.** 44   **45.** -31.95 **47.** 8.91   **49. a)** Answers will vary. **b)** Yes, a and b must be equal.   **51.** $3x - 27$ **53.** $15 - 35x$ **55.** $16x$ **57.** $-5x$ **59.** $7x + 4$ **61.** $38a - 63$ **63.** $4a + 5b$ **65.** $-2x - 22y$ **67.** $10x - 8$ **69.** $-8x + 7$ **71.** $-9y + 85$ **73.** $30z + 6$ **75.** $-10a - 5b + 26c$ **77. a)** 4 **b)** $5x^3, 3x^2, -7x, -15$ **c)** $5, 3, -7, -15$ **79. a)** 2 **b)** $3x, -17$ **c)** $3, -17$ **81. a)** 3 **b)** $15x^2, -41x, 55$ **c)** $15, -41, 55$ **83. a)** 4 **b)** $-35a, -33b, 52c, 69$ **c)** $-35, -33, 52, 69$ **85.** Answers will vary; some examples are $3(4x - 5)$, $5x + 7x - 15, 3x + 7 + 9x - 22,$ and $16x - 5 - 4x - 10$.   **87.** Answers will vary.

Chapter 1 Review   **1.** $<$ **2.** $>$ **3.** 8   **4.** 13   **5.** 6   **6.** -12 **7.** -7 **8.** -3 **9.** -41 **10.** 27   **11.** -37 **12.** 20   **13.** -54 **14.** 16   **15.** $\{1, 2, 3, 6, 7, 14, 21, 42\}$ **16.** $\{1, 2, 3, 4, 6, 9, 12, 18, 27, 36, 54, 108\}$ **17.** $2 \cdot 2 \cdot 2 \cdot 2 \cdot 2$ **18.** $2 \cdot 2 \cdot 3 \cdot 5$ **19.** $\frac{4}{7}$ **20.** $\frac{1}{8}$ **21.** $\frac{20}{117}$ **22.** 14   **23.** $\frac{17}{3}$ **24.** $\frac{307}{25}$ **25.** $3\frac{4}{5}$ **26.** $9\frac{1}{6}$ **27.** $\frac{21}{100}$ **28.** $2\frac{1}{3}$ **29.** $\frac{14}{25}$ **30.** $\frac{32}{75}$ **31.** $\frac{29}{34}$ **32.** $\frac{89}{60}$ **33.** $\frac{1}{6}$ **34.** $-\frac{29}{252}$ **35.** $\frac{25}{77}$ **36.** $\frac{5}{24}$ **37.** $\frac{21}{104}$ **38.** $\frac{5}{12}$ **39.** 12.62   **40.** 8.818   **41.** 30.24   **42.** 8.8   **43.** 0.32   **44.** 0.9375   **45.** $\frac{3}{4}$ **46.** $\frac{7}{25}$ **47.** 40%   **48.** 28%   **49.** 90%   **50.** 45%   **51.** $\frac{3}{10}$ **52.** $\frac{11}{20}$ **53.** 0.9   **54.** 0.04   **55.** \$17   **56.** $-\$47$ **57.** \$12,600   **58.** $4\frac{1}{6}$ cups

59.
Freshmen at Two-Year Colleges

60.

61.

62. 64   **63.** $\frac{8}{125}$ **64.** -64 **65.** 6075   **66.** 37   **67.** 22   **68.** 48   **69.** 60   **70.** 51   **71.** $\frac{87}{100}$ **72.** $n + 14$ **73.** $n - 20$ **74.** $2n - 8$ **75.** $6n + 9$ **76.** $3.55c$ **77.** $20 + 0.15m$ **78.** 44   **79.** 25   **80.** 56   **81.** -69 **82.** 0   **83.** 84   **84.** $5x + 35$ **85.** $27x$ **86.** $5x - 8$ **87.** $-164 + 1$ **88.** $5k - 48$ **89.** $3x - 303$ **90. a)** 4 **b)** $x^3, -4x^2, -10x, 41$ **c)** $1, -4, -10, 41$ **91. a)** 3 **b)** $-x^2, 5x, -30$ **c)** $-1, 5, -30$

Chapter 1 Review Exercises: Worked-Out Solutions

12. $8 - (-19) - 7$
$= 8 + 19 - 7$
$= 27 - 7$
$= 20$

15.
$\begin{array}{r} 42 \\ \hline 1 \cdot 42 \\ 2 \cdot 21 \\ 3 \cdot 14 \\ 6 \cdot 7 \end{array}$
$\{1, 2, 3, 6, 7, 14, 21, 42\}$

18.
60
$2 \quad 30$
$2 \quad 15$
$3 \quad 5$
$60 = 2 \cdot 2 \cdot 3 \cdot 5$

19. $\dfrac{24}{42} = \dfrac{\cancel{2} \cdot 2 \cdot 2 \cdot \cancel{3}}{\cancel{2} \cdot \cancel{3} \cdot 7} = \dfrac{4}{7}$

24. $12 \cdot 25 + 7 = 307$
$12\frac{7}{25} = \frac{307}{25}$

26. $\begin{array}{r} 9 \\ 6\overline{)\,55} \\ -54 \\ \hline 1 \end{array}$
$\frac{55}{6} = 9\frac{1}{6}$

27. $\dfrac{3}{\cancel{9}} \cdot \dfrac{7}{\cancel{26}} = \dfrac{21}{100}$
$\dfrac{\cancel{9}}{\cancel{16}} \cdot \dfrac{\cancel{28}}{\cancel{75}}$

31. $\frac{5}{8} + \frac{7}{12}$
$= \frac{15}{24} + \frac{14}{24}$
$= \frac{29}{24}$

43. $\begin{array}{r} 0.32 \\ 25\overline{)\,8.00} \\ -7\,5 \\ \hline 50 \\ -50 \\ \hline 0 \end{array}$
$\frac{8}{25} = 0.32$

45. $0.75 = \frac{75}{100} = \frac{3}{4}$

47. $\dfrac{2}{\cancel{5}} \cdot \overset{20}{\cancel{100}}\% = 40\%$

51. $30\% = \frac{30}{100} = \frac{3}{10}$

70. $54 - 27 \div 3^2$
$= 54 - 27 \div 9$
$= 54 - 3$
$= 51$

82. $x^2 - 7x - 30$
$(-3)^2 - 7(-3) - 30$
$= 9 + 21 - 30$
$= 0$

87. $8y - 6(4y - 21)$
$= 8y - 24y + 126$
$= -16y + 126$

Chapter 1 Test   **1.** $>$ **2.** 17   **3.** -6 **4.** 63   **5.** $\{1, 3, 5, 9, 15, 45\}$ **6.** $2 \cdot 2 \cdot 3 \cdot 3 \cdot 3$ **7.** $\frac{5}{7}$ **8.** $3\frac{13}{18}$ **9.** $\frac{5}{84}$ **10.** $\frac{7}{12}$ **11.** $\frac{91}{60}$ **12.** $-\frac{17}{72}$ **13.** 18.2735   **14.** $\frac{9}{25}$ **15.** \$6487   **16.** \$795.51   **17.** $\frac{18}{25}$ **18.** 0.06   **19.**
College Students

20. -24 **21.** 25   **22.** $\frac{5}{11}$ **23.** $4n - 7$ **24.** $50 + 20h$ **25.** 61   **26.** -1 **27.** $10x - 65$ **28.** $-9y + 240$

Chapter 2

Quick Check 2.1 **1.** Yes **2.** $\{-16\}$ **3.** $\{14\}$ **4.** $\{-16\}$ **5.** $\{\frac{14}{3}\}$ **6.** $\{-30\}$ **7.** $\{15\}$ **8.** $\{\frac{31}{12}\}$ **9.** $65

Section 2.1 **1.** equation **3.** The solution set of an equation is the set of all solutions to that equation. **5.** For any algebraic expressions A and B, and any number n, if $A = B$ then $A + n = B + n$ and $A - n = B - n$. **7.** No, there is a term containing x^2. **9.** Yes **11.** Yes **13.** Yes **15.** No, there is a variable in a denominator. **17.** Yes **19.** No **21.** Yes **23.** No **25.** $\{18\}$ **27.** $\{-10\}$ **29.** $\{\frac{5}{3}\}$ **31.** $\{0\}$ **33.** $\{-7\}$ **35.** $\{14\}$ **37.** $\{-45\}$ **39.** $\{21\}$ **41.** $\{-96\}$ **43.** $\{8\}$ **45.** $\{-10\}$ **47.** $\{2.17\}$ **49.** $\{3\}$ **51.** $\{-4\}$ **53.** $\{17\}$ **55.** $\{2.5\}$ **57.** $\{9\}$ **59.** $\{20\}$ **61.** $\{3\}$ **63.** $\{23\}$ **65.** $\{-9\}$ **67.** $\{\frac{5}{36}\}$ **69.** $\{-4\}$ **71.** $\{12\}$ **73.** $\{16\}$ **75.** $\{30\}$ **77.** $\{-41\}$ **79.** $\{-77\}$ **81.** $\{-\frac{29}{30}\}$ **83.** $\{-56\}$ **85.** $\{\frac{3}{8}\}$ **87.** $\{-\frac{21}{10}\}$ **89.** $\{0\}$ **91.** Answers will vary; example $2x = 14$. **93.** Answers will vary; example $2n = 5$. **95.** Answers will vary; example $x + 2 = -4$. **97.** Answers will vary; example $b + \frac{5}{6} = 1$. **99.** 27 nickels **101.** 159 people **103.** 166 employees **105.** 56° F **107.** Answers will vary. **109.** Answers will vary.

Quick Check 2.2 **1.** $\{7\}$ **2.** $\{-8\}$ **3.** $\{\frac{35}{12}\}$ **4.** $\{-9\}$ **5.** $\{2\}$ **6.** $\{-3\}$ **7.** $\{4\}$ **8.** $\{-6\}$ **9.** $\emptyset$ **10.** $\mathbb{R}$ **11.** $y = \frac{5-x}{2}$ **12.** $x = \frac{5z}{y}$

Section 2.2 **1.** the LCM of the denominators **3.** the empty set **5.** b **7.** $\{-3\}$ **9.** $\{\frac{1}{2}\}$ **11.** $\{-\frac{5}{2}\}$ **13.** $\{0\}$ **15.** $\{5.7\}$ **17.** $\{3\}$ **19.** $\{37\}$ **21.** $\{-10\}$ **23.** $\{-\frac{15}{2}\}$ **25.** $\mathbb{R}$ **27.** $\{\frac{24}{5}\}$ **29.** $\{\frac{17}{3}\}$ **31.** $\{6\}$ **33.** $\{\frac{104}{25}\}$ **35.** $\{-2\}$ **37.** $\mathbb{R}$ **39.** $\emptyset$ **41.** $\{\frac{5}{3}\}$ **43.** $\emptyset$ **45.** $\{-5\}$ **47.** Answers will vary; example $x - 5 = 4$. **49.** Answers will vary; example $9a - 1 = 0$. **51.** Answers will vary; example $5x - 6 = 5x - 7$. **53.** Answers will vary; example $2(x - 4) - 3 = 2x - 11$. **55.** Answers will vary; example $x + 3 = -8$. **57.** $y = -5x - 2$ **59.** $y = \frac{-7x + 4}{2}$ **61.** $y = \frac{-6x + 13}{-3}$ **63.** $b = P - a - c$ **65.** $t = \frac{d}{r}$ **67.** $r = \frac{C}{2\pi}$ **69.** 20° C **71.** 354 **73. a)** 80 kites **b)** $6.50 **c)** $12.75 **75.** 4 **77.** 15 **79.** Answers will vary.

Quick Check 2.3 **1.** 8 **2.** Length: 24 feet, Width: 16 feet **3.** 14 inches, 48 inches, 50 inches **4.** 70, 71, 72 **5.** 7 hours **6.** 17 and 86 **7.** 22 nickels and 9 quarters

Section 2.3 **1.** b **3.** b **5.** $x + 1, x + 2$ **7.** $x + 2, x + 4$ **9.** 46 **11.** 145 **13.** 8.9 **15.** Length: 9 m, Width: 4 m **17.** Length: 17 feet, Width: 10 feet **19.** Length: 52 cm, Width: 13 cm **21.** 55 ft **23.** 32 cm **25.** 5 inches, 9 inches, 12 inches **27.** 11 inches **29.** A: 60°, B: 30° **31.** 20°, 70° **33.** 42°, 138° **35.** 50°, 60°, 70° **37.** 667 square inches **39.** 52 inches **41.** 152, 153 **43.** 157 **45.** 73, 75, 77 **47.** 93 **49.** 138, 139 **51.** 15, 16, 17 **53.** 500 miles **55.** $5\frac{1}{2}$ hours **57.** $1\frac{11}{14}$ meters/second **59. a)** 15 hours **b)** 560 miles **61.** 30 hours **63.** 27, 44 **65.** 44, 52 **67.** 13, 18 **69.** 55 **71.** 18 **73.** 90 **75.** Answers will vary. **77.** Answers will vary.

Quick Check 2.4 **1.** 11 **2.** 65 **3.** 70% **4.** 140 **5.** 54% **6.** $60.90 **7.** $5000 at 3%, $7000 at 5% **8.** $850 at 4%, $1050 at 5% **9.** 20 ml of 30% and 60 ml of 42% **10.** $\{33\}$ **11.** 68

Section 2.4 **1.** Amount = Percent · Base **3.** original value **5.** $I = P \cdot r \cdot t$ **7.** proportion **9.** 7 **11.** 140 **13.** 70% **15.** 39 **17.** 163.2 **19.** 44 **21.** 84 **23.** 1025 **25.** 300 ml **27.** $18.40 **29.** $3.25 **31.** $21.21 **33.** 7.5% **35.** 25% **37.** $700 at 2%, $1400 at 5% **39.** $2500 at 4%, $4000 at 5% **41.** 24 gal of 70%, 36 gal of 40% **43.** 400 gal of 5%, 600 gal of 2% **45.** 240 ml **47.** $\{32\}$ **49.** $\{36\}$ **51.** $\{50.4\}$ **53.** $\{87.5\}$ **55.** $\{9\}$ **57.** $\{3\}$ **59.** 5520 **61.** 22.5 **63.** 144 **65.** 144 **67.** Answers will vary.

Quick Review **1.** $\{6\}$ **2.** $\{-16\}$ **3.** $\{13\}$ **4.** $\{16\}$

Quick Check 2.5 **1.** **2.** $x < 7$, , $(-\infty, 7)$ **3.** $x \geq 5$, , $[5, \infty)$ **4.** $x < -3$, , $(-\infty, -3)$ **5.** $x \geq -6$, , $[-6, \infty)$ **6.** $x \leq 4$ or $x \geq 7$, , $(-\infty, 4] \cup [7, \infty)$ **7.** $-3 < x < \frac{3}{2}$, , $(-3, \frac{3}{2})$ **8.** $x < 130$ **9.** 48 or lower

Section 2.5 **1.** linear inequality **3.** interval **5.** compound **7.** , $(-\infty, 3)$ **9.** , $[-1, \infty)$ **11.** , $(-2, 8)$ **13.** , $(\frac{9}{2}, \infty)$ **15.** , $(-\infty, 2.4]$ **17.** , $(-\infty, 2] \cup (8, \infty)$ **19.** , $(-\infty, \frac{3}{2}) \cup (\frac{13}{4}, \infty)$ **21.** $x > 5$ **23.** $-4 \leq x \leq 7$ **25.** $x < 2$ or $x > 9$ **27.** $x < -4$ **29.** $x < -4$, , $(-\infty, -4)$ **31.** $x > 6$, , $(6, \infty)$ **33.** $x < -3$, , $(-\infty, -3)$ **35.** $x \geq -3.5$, , $[-3.5, \infty)$ **37.** $x < 4$, , $(-\infty, 4)$ **39.** $x \leq 2$, , $(-\infty, 2]$ **41.** $x \geq -10$, , $[-10, \infty)$ **43.** $x > 7$, , $(7, \infty)$ **45.** $x > \frac{5}{3}$, , $(\frac{5}{3}, \infty)$ **47.** $x \leq -7$, , $(-\infty, -7]$ **49.** $x < 7$, , $(-\infty, 7)$

51. $x > -8$, $(-8, \infty)$ **53.** $x > -\frac{11}{2}$, $(-\frac{11}{2}, \infty)$ **55.** $x < -4$ or $x > -2$, $(-\infty, -4) \cup (-2, \infty)$ **57.** $x < -2$ or $x > \frac{14}{3}$, $(-\infty, -2) \cup (\frac{14}{3}, \infty)$ **59.** $x > 6$ or $x < 3$, $(-\infty, 3) \cup (6, \infty)$ **61.** $x \leq \frac{3}{2}$ or $x \geq 3$, $(-\infty, \frac{3}{2}] \cup [3, \infty)$ **63.** $-1 < x < 4$, $(-1, 4)$ **65.** $-\frac{3}{2} \leq x \leq \frac{9}{4}$, $[-\frac{3}{2}, \frac{9}{4}]$ **67.** $1 < x < 7$, $(1, 7)$

69. $-3 < x \leq 6$, $(-3, 6]$ **71.** $\frac{16}{15} \leq x \leq \frac{25}{6}$, $[\frac{16}{15}, \frac{25}{6}]$ **73.** Answers will vary. **75.** $x \geq 2$ **77.** $x < 30$ **79.** $x \geq 12,500$ **81.** At least 300 **83.** At most 929 **85.** At least \$5625 **87.** 34 to 100 **89.** Answers will vary.

Chapter 2 Review 1. No **2.** Yes **3.** $\{16\}$ **4.** $\{-6\}$ **5.** $\{-6\}$ **6.** $\{-6\}$ **7.** 118 **8.** 212 **9.** $\{7\}$ **10.** $\{-9\}$ **11.** $\{2\}$ **12.** $\{36\}$ **13.** $\{10\}$ **14.** $\{\frac{35}{4}\}$ **15.** $\{-33\}$ **16.** $\{-\frac{3}{5}\}$ **17.** $\{16\}$ **18.** $\{-7\}$ **19.** $y = 8 - 7x$ **20.** $y = \frac{30 - 15x}{7}$ **21.** $d = \frac{C}{\pi}$ **22.** $a = \frac{P - b - 2c}{3}$ **23.** 36, 62 **24.** Length: 16 feet, Width: 9 feet **25.** 66, 68, 70 **26.** $4\frac{3}{4}$ hours **27.** 41 **28.** 19 **29.** $37\frac{1}{2}\%$ **30.** 4250 **31.** 5786 **32.** $63\frac{3}{4}\%$ **33.** \$12,300 **34.** \$47.96 **35.** \$1700 at 6\%, \$300 at 5\% **36.** $\{4\}$ **37.** $\{32\}$ **38.** $\{78\}$ **39.** $\{50\}$ **40.** 6408 **41.** 15

42. $x < -3$ $(-\infty, -3)$ **43.** $x > -5$ $(-5, \infty)$ **44.** $x \geq 5$ $[5, \infty)$ **45.** $x \leq -4$ $(-\infty, -4]$ **46.** $x < -6$ or $x > 7$ $(-\infty, -6) \cup (7, \infty)$ **47.** $x \leq -8$ or $x \geq -3$ $(-\infty, -8] \cup [-3, \infty)$ **48.** $-5 \leq x \leq 10$ $[-5, 10]$ **49.** $-1 < x < 10$ $(-1, 10)$ **50.** $x \geq 7$ **51.** At least 67 **52.** 79 to 100

Chapter 2 Review Exercises: Worked-Out Solutions

1.
$$4x + 9 = 19$$
$$4(7) + 9 = 19$$
$$28 + 9 = 19$$
$$37 = 19$$
False
$x = 7$ is not a solution.

4.
$$-9x = 54$$
$$\frac{-9x}{-9} = \frac{54}{-9}$$
$$x = -6$$
$$\{-6\}$$

6.
$$a + 14 = 8$$
$$a + 14 - 14 = 8 - 14$$
$$a = -6$$
$$\{-6\}$$

9.
$$5x - 8 = 27$$
$$5x - 8 + 8 = 27 + 8$$
$$5x = 35$$
$$\frac{5x}{5} = \frac{35}{5}$$
$$x = 7$$
$$\{7\}$$

13.
$$\frac{x}{6} - \frac{5}{12} = \frac{5}{4}$$
$$12 \cdot \frac{x}{6} - 12 \cdot \frac{5}{12} = 12 \cdot \frac{5}{4}$$
$$2x - 5 = 15$$
$$2x - 5 + 5 = 15 + 5$$
$$2x = 20$$
$$\frac{2x}{2} = \frac{20}{2}$$
$$x = 10$$
$$\{10\}$$

16.
$$2m + 23 = 17 - 8m$$
$$2m + 23 + 8m = 17 - 8m + 8m$$
$$10m + 23 = 17$$
$$10m + 23 - 23 = 17 - 23$$
$$10m = -6$$
$$\frac{10m}{10} = \frac{-6}{10}$$
$$m = -\frac{3}{5}$$
$$\{-\frac{3}{5}\}$$

22.
$$P = 3a + b + 2c$$
$$P - b - 2c = 3a$$
$$\frac{P - b - 2c}{3} = \frac{3a}{3}$$
$$a = \frac{P - b - 2c}{3}$$

24. Length: x
Width: $x - 7$
$$2x + 2(x - 7) = 50$$
$$2x + 2x - 14 = 50$$
$$4x - 14 = 50$$
$$4x = 64$$
$$x = 16$$
Length: $x = 16$
Width: $x - 7 = 16 - 7 = 9$

30.
$$1360 = 0.32x$$
$$\frac{1360}{0.32} = \frac{0.32x}{0.32}$$
$$4250 = x$$

31.
$$x = 0.55 \cdot 10,520$$
$$x = 5786$$

36.
$$\frac{n}{6} = \frac{18}{27}$$
$$n \cdot 27 = 6 \cdot 18$$
$$27n = 108$$
$$n = 4$$

40.
$$\frac{3}{7} = \frac{n}{14,952}$$
$$3 \cdot 14,952 = 7 \cdot n$$
$$44,856 = 7n$$
$$6408 = n$$

45.
$$3x + 4 \leq -8$$
$$3x \leq -12$$
$$x \leq -4$$
$$(-\infty, -4]$$

48.
$$-21 \leq 3x - 6 \leq 24$$
$$-15 \leq 3x \leq 30$$
$$-5 \leq x \leq 10$$
$$[-5, 10]$$

51.
$$45x \geq 3000$$
$$x \geq \frac{200}{3}$$
$$x \geq 66\frac{2}{3}$$
They need to sell at least 67.

Chapter 2 Test 1. $\{-7\}$ **2.** $\{-39\}$ **3.** $\{8\}$ **4.** $\{-\frac{21}{4}\}$ **5.** $\{\frac{55}{2}\}$ **6.** $\{-3\}$ **7.** $\{12\}$ **8.** $\{55\}$ **9.** $y = \frac{28 - 4x}{3}$ **10.** 28 **11.** 15\% **12.** $(-\infty, -\frac{5}{2}]$ **13.** $(-\infty, -9] \cup [-\frac{3}{2}, \infty)$ **14.** $(-10, \frac{16}{3})$ **15.** Length: 23 feet, Width: 13 feet **16.** 17 **17.** 88 **18.** At least 34

Chapter 3

Quick Check 3.1   **1.** Yes   **2.**

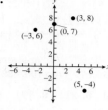

3. $A(-6, 2), B(-2, 6), C(0, -5), D(-4, -2), E(8, 1)$ **4.** $y = -4$

5.

x	4	9	-1
y	-2	3	-7

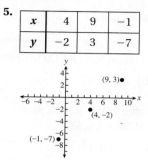

6.

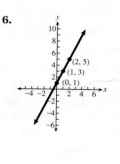

Section 3.1   **1.** $Ax + By = C$ **3.** x-axis   **5.** origin   **7.** Yes   **9.** No   **11.** Yes   **13.** Yes   **15.** Yes

17.

19.

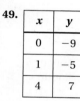

21.

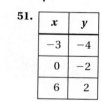

23. $A(-3, 1), B(2, -5), C(-6, -4), D(0, 3)$
25. $A(30, -15), B(-40, 0), C(-10, -30),$
$D(-15, 15)$ **27.** IV   **29.** I   **31.** I   **33.** IV
35. -1 **37.** -7 **39.** -2 **41.** 2   **43.** -2

45.

x	y
1	5
3	1
0	7

47.

x	y
-3	-1
0	4
3	9

49.

x	y
0	-9
1	-5
4	7

51.

x	y
-3	-4
0	-2
6	2

53. Answers will vary; an example is
$x + y = 7$.   **55.** Answers will vary; an
example is $5x + 4y = 24$.

57.

59.

61.

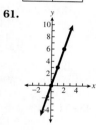

63.

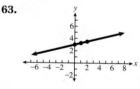

65.

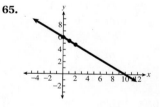

67.

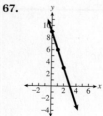

69. Answers will vary.

Quick Check 3.2 **1.** $(-4, 0), (0, 5)$ **2.** $(4, 0), (0, -3)$ **3.** $(\frac{9}{2}, 0), (0, -9)$

4.
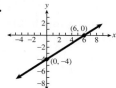

5.
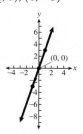

6.
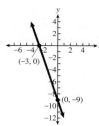

7.
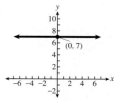

8.
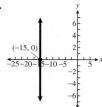

9. a) Nancy was initially $10,000 in debt. **b)** It will take five months before Nancy breaks even.

Section 3.2 **1.** x-intercept **3.** vertical **5.** Substitute 0 for y and solve for x. **7.** $(-3, 0), (0, 6)$ **9.** $(1, 0), (0, 5)$
11. No x-intercept, $(0, 2)$ **13.** $(6, 0), (0, 6)$ **15.** $(-8, 0), (0, 8)$ **17.** $(3, 0), (0, 9)$ **19.** $(-15, 0), (0, 9)$ **21.** $(-\frac{7}{2}, 0), (0, \frac{7}{3})$
23. $(\frac{25}{3}, 0), (0, \frac{100}{3})$ **25.** $(5, 0), (0, 6)$ **27.** $(8, 0), (0, 8)$ **29.** $(4, 0), (0, -12)$ **31.** $(0, 0), (0, 0)$ **33.** $(\frac{10}{3}, 0), (0, -2)$
35. No x-intercept, $(0, 2)$ **37.** Answers will vary; an example is $x + y = 4$. **39.** Answers will vary; an example is $-2x + 5y = -50$.
41. $(4, 0), (0, 4)$ **43.** $(5, 0), (0, -2)$ **45.** $(8, 0), (0, -6)$ **47.** $(3, 0), (0, 9)$ **49.** $(-\frac{7}{2}, 0), (0, 2)$

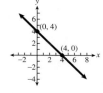 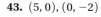 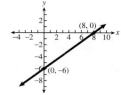 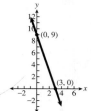 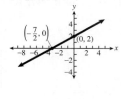

51. $(4, 0), (0, -4)$ **53.** $(0, 0), (0, 0)$ **55.** $(\frac{7}{2}, 0), (0, 7)$ **57.** $(2, 0), (0, -7)$ **59.** No x-intercept, $(0, 4)$

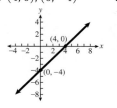 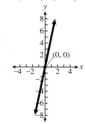 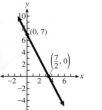  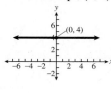

61. $(3, 0)$, No y-intercept **63. a)** $(0, 12{,}000)$ The original value of the copy machine is $12,000. **b)** $(8, 0)$ The copy machine has no value after 8 years. **c)** $7500 **65. a)** $(0, 5000)$ The cost of membership without playing any rounds of golf is $5000. **b)** Since the minimum fee is $5000, the cost to a member can never be $0. **c)** $8900

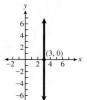

Quick Review   **1.** $y = -2x + 8$ **2.** $y = 4x - 6$ **3.** $y = 4x + 5$ **4.** $y = -\frac{2}{3}x + 2$

Quick Check 3.3   **1.** -2 **2.** 3   **3.** $-\frac{5}{2}$ **4.** Slope is undefined.   **5.** $m = 0$ **6.** $m = \frac{5}{2}, (0, -6)$ **7.** $m = -3, (0, 5)$ **8.** $y = \frac{1}{2}x - \frac{4}{7}$

9. 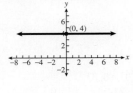   **10.**    **11.** Slope is 218: Number of women accepted to medical school increases by 218 per year. y-intercept is (0, 7485): Approximately 7485 women were accepted to medical school in 1997.

Section 3.3   **1.** slope   **3.** negative   **5.** horizontal   **7.** slope–intercept   **9.** Negative   **11.** Positive   **13.** $-\frac{1}{2}$ **15.** $-\frac{3}{4}$ **17.** 0
19. 2   **21.** 3   **23.** 0   **25.** Undefined
27. y-intercept: $(0, 4)$, m: 0   **29.** x-intercept: $(0, 0)$, m: undefined   **31.** x-intercept: $(8, 0)$, m: undefined   **33.** $x = 6$, m: undefined

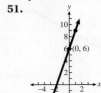   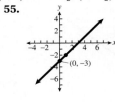   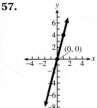

35. $y = 2, m$: 0   **37.** 6, $(0, -7)$ **39.** $-2, (0, 3)$ **41.** $-\frac{3}{2}, (0, -\frac{5}{2})$ **43.** $\frac{1}{5}, (0, -\frac{8}{5})$ **45.** $y = -2x + 5$ **47.** $y = 3x - 6$ **49.** $y = -4$
51. 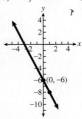   **53.**   **55.**   **57.** 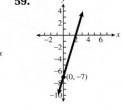   **59.**

61.   **63.**   **65.**   **67.**   **69.**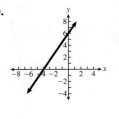

71. 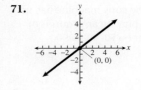   **73.** 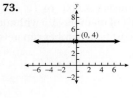   **75.**    **77.** 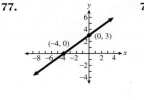   **79.**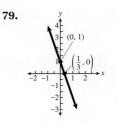

81.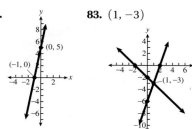

83. $(1, -3)$

85. a) 4. The value increased by \$4/year. **b)** $(0, 50)$. The original value of the card was \$50. **c)** \$150 **87.** $\frac{1}{20}$ **89.** Negative. The heart rate (y) decreases as the animal's weight (x) increases. **91.** For the first equation, using the slope and y-intercept is more efficient. For the second equation, finding the x- and y-intercepts is more efficient.

Quick Check 3.4 **1.** Cost $= 19.95 + 0.15x$. Domain: Set of all possible miles. Range: Set of all possible costs. **2.** -3 **3.** -23
4. $7a + 44$ **5.** 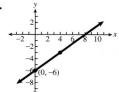 **6.**   **7. a)** $(4, 0), (0, -8)$ **b)** 4 **c)** 7

Section 3.4 **1.** function **3.** range **5.** linear **7.** Yes, each player is listed with only one team. **9.** Yes, each person has only one mother. **11. a)** Yes **b)** No **13. a)** Yes **b)** No **15.** Yes **17.** No, some x-coordinates are associated with more than one y-coordinate. **19.** No, some x-coordinates are associated with more than one y-coordinate. **21. a)** $F(x) = \frac{9}{5}x + 32$
b) $32°F, 212°F, 86°F, 14°F, -40°F$ **23. a)** $f(x) = 7x + 36$ **b)** \$120 **25.** $f(x) = 4x + 3$ **27.** $f(x) = -3x - 4$
29. $f(x) = 6$ **31.** -3 **33.** 7 **35.** 32 **37.** -25 **39.** 1 **41.** $3a + 4$ **43.** $7a + 19$ **45.** $-6a + 31$ **47.** $6x + 6h + 4$
49. 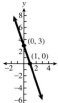 **51.** **53.**  **55.** 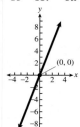 **57.** 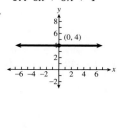 **59.**

61. a) $(3, 0)$ **b)** $(0, -6)$ **c)** 4 **d)** -1 **63. a)** $(4, 0)$ **b)** $(0, 6)$ **c)** -3 **d)** -2 **65.** Domain: $(-\infty, \infty)$, Range: $(-\infty, \infty)$
67. Domain: $(-\infty, \infty)$, Range: $\{5\}$ **69.** Answers will vary.

Quick Review **1.** $m = -3$ **2.** $m = \frac{5}{4}$ **3.** $m = \frac{1}{5}$ **4.** $m = -\frac{7}{5}$

Quick Check 3.5 **1.** No **2.** Yes **3.** $\frac{2}{9}$ **4.** Yes **5.** Yes **6.** $-\frac{3}{7}$ **7.** Neither **8.** Parallel **9.** Perpendicular
10. $y = -\frac{3}{2}x - 7$

Section 3.5 **1.** parallel **3.** vertical **5.** parallel **7.** Yes **9.** No **11.** Yes **13.** No **15.** Yes **17.** No **19.** Neither
21. Perpendicular **23.** Parallel **25.** Parallel **27.** Neither **29.** Perpendicular **31.** Parallel **33.** Neither **35.** Parallel **37.** -6
39. 5 **41.** -3 **43.** -4 **45.** $-\frac{5}{3}$ **47.** $\frac{8}{5}$ **49.** $-\frac{4}{B}$ **51.** Yes **53.** Yes

55. $y = 2x - 6$ **57.** $y = -4$ **59.** $y = -\frac{3}{5}x - 3$ **61.** Answers will vary. **63.** Answers will vary.

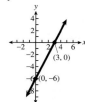 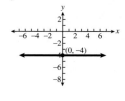 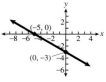

Quick Review   **1.** $m = -2$ **2.** $m = \frac{4}{3}$ **3.** $m = 5$ **4.** $m = \frac{3}{4}$

Quick Check 3.6   **1.** $y = 2x + 3$ **2.** $y = -3x + 6$ **3.** $f(x) = -4x + 5$ **4.** $y = -\frac{3}{2}x + 4$ **5.** $y = 4$ **6.** $y = 3.9x + 84.6$
7. $y = \frac{2}{5}x + 2$ **8.** $y = -\frac{4}{3}x + 6$

Section 3.6   **1.** $y - y_1 = m(x - x_1)$ **3.** $y = -2x + 4$ **5.** $y = \frac{1}{5}x - 2$ **7.** $y = 7x - 3$ **9.** $y = -3x + 5$ **11.** $y = \frac{2}{3}x - 3$
13. $y = 5x + 8$ **15.** $y = x - 1$ **17.** $y = -3x + 3$ **19.** $y = \frac{3}{2}x$ **21.** $y = 5$ **23.** $f(x) = -2x + 15$

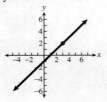
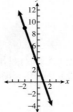

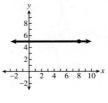

25. $f(x) = \frac{3}{5}x - 8$ **27.** $y = x - 5$ **29.** $y = 3x + 6$ **31.** $y = 2x + 8$ **33.** $x = -2$ **35.** $y = -\frac{1}{3}x + 2$

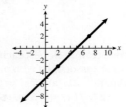
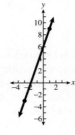
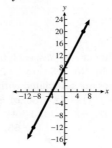

37. $y = x - 4$ **39.** $y = x - 6$ **41. a)** $y = 10x + 36$ **b)** \$36   **43. a)** $y = 70x + 4200$ **b)** \$4200 **c)** \$70   **45.** $y = -2x - 15$
47. $y = 3x - 18$ **49.** $y = 7$ **51.** $y = -\frac{1}{3}x + 7$ **53.** $y = \frac{5}{4}x + 7$ **55.** $x = -5$ **57.** $y = x + 3$ **59.** $y = -\frac{1}{4}x + 2$ **61.** Answers will vary.

Quick Review   **1.**
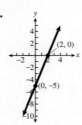
2.

3.
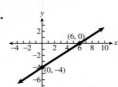
4.
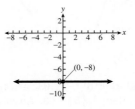

Quick Check 3.7   **1.**
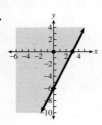
2.
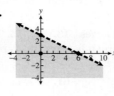
3.
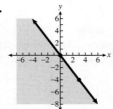
4. $x + y \geq 60$
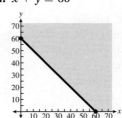

Section 3.7 **1.** solution **3.** dashed **5. a)** Yes **b)** No **c)** Yes **d)** Yes **7. a)** Yes **b)** No **c)** No **d)** No **9. a)** Yes **b)** Yes **c)** No **d)** No **11.**

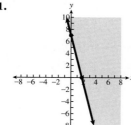

13.

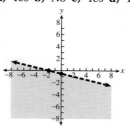

15. A **17.** A **19.** ≤ **21.** < **23.**

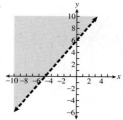

25.

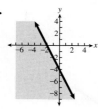

27.

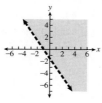

29.

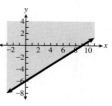

31.

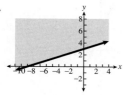

33.

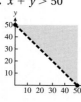

35.

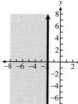

37.

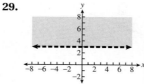

39.

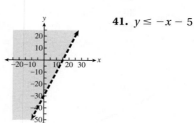

41. $y \leq -x - 5$

43. $y > 8x$ **45.** $x + y > 50$ **47.**

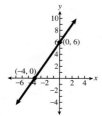

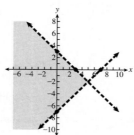

49. False. For some ordered pairs $3x - 8y = 24$.

Chapter 3 Review **1.** $A(7, -2)$, $B(-5, -1)$, $C(-1, -5)$, $D(0, 8)$ **2.** III **3.** II **4.** No **5.** Yes
6. Yes **7.** $(-20, 0)$, $(0, 5)$ **8.** $(-3, 0)$, $(0, 15)$ **9.** $(0, 0)$ **10.** $(\frac{7}{2}, 0)$, $(0, -2)$
11. $(-4, 0)$, $(0, 6)$ **12.** $(-8, 0)$, $(0, 2)$ **13.** $(0, 0)$, $(0, 0)$ **14.** $(-\frac{3}{2}, 0)$, $(0, -2)$

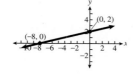
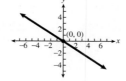
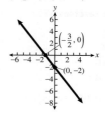

15. $(4,0),(0,6)$ **16.** $(2,0),(0,-4)$ **17.** 3 **18.** -4 **19.** 1 **20.** 0 **21.** $-2,(0,7)$ **22.** $4,(0,-6)$
23. $-\frac{2}{3},(0,0)$ **24.** $2,(0,-\frac{9}{2})$ **25.** $-\frac{5}{3},(0,6)$ **26.** $0,(0,-7)$

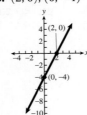
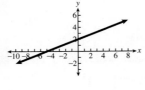

27.  **28.**  **29.** 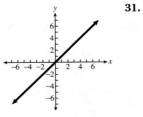 **30.**  **31.**

32. 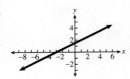   **33.** Neither **34.** Perpendicular **35.** Parallel **36.** $y=-4x-2$ **37.** $y=\frac{2}{5}x-6$
38. a) $f(x)=720+50x$ **b)** \$1270 **c)** 41 months **39.** -11 **40.** 43 **41.** $15a+4$
42. a) -7 **b)** 6 **c)** $(-\infty,\infty)$ **d)** $(-\infty,\infty)$ **43.** $y=x-6$ **44.** $y=-5x+8$ **45.** $y=-\frac{3}{2}x+3$
46. $y=-2x-3$ **47.** $y=\frac{3}{2}x+4$ **48.** $y=3$ **49.** $y=\frac{3}{2}x+6$ **50.** $y=3x$

51.  **52.** 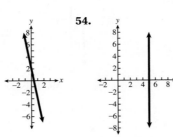 **53.** **54.** **55.**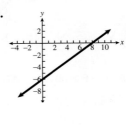

56. 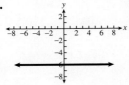 **57.** **58.** 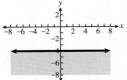 **59.** 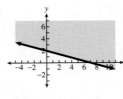 **60.**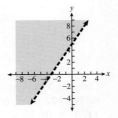

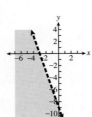
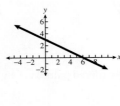

Chapter 3 Review Exercises: Worked-Out Solutions

4. $4x-2y=16$
  $4(5)-2(-2)=16$
  $20+4=16$
  $24=16$
  False, so $(5,-2)$ is not a solution

7. <u>x-intercept</u>   <u>y-intercept</u>
  $2x-8(0)=-40$ $2(0)-8y=-40$
  $2x=-40$ $-8y=-40$
  $\frac{2x}{2}=\frac{-40}{2}$ $\frac{-8y}{-8}=\frac{-40}{-8}$
  $x=-20$ $y=5$
  $(-20,0)$ $(0,5)$

11. <u>x-intercept</u>   <u>y-intercept</u>
$-3x + 2(0) = 12$ $-3(0) + 2y = 12$
$-3x = 12$ $2y = 12$
$x = -4$ $y = 6$
$(-4, 0)$ $(0, 6)$

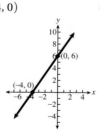

17. $(x_1, y_1): (-5, 3)(x_2, y_2): (-3, 9)$
$m = \frac{y_2 - y_1}{x_2 - x_1}$
$m = \frac{9 - 3}{-3 - (-5)}$
$m = \frac{6}{2}$
$m = 3$

24. $4x - 2y = 9$
$-2y = -4x + 9$
$\frac{-2y}{-2} = \frac{-4x}{-2} + \frac{9}{-2}$
$y = 2x - \frac{9}{2}$
$m = 2$, y-intercept: $(0, -\frac{9}{2})$

27. Put the y-intercept $(0, -6)$ on the graph
The slope is 3.
Beginning at $(0, -6)$, move up 3 units
and 1 unit to the right.
Plot a point at the location $(1, -3)$.
Graph the line that goes through these
two points.

33. The slope of the line $y = 2x - 9$ is 2.
The slope of the line $y = -2x - 9$ is -2.
Since the two slopes are not equal, the lines are not parallel.
Since the two slopes are not negative reciprocals, the lines are not
perpendicular.
So, the two lines are neither parallel nor perpendicular.

38. a) The fund started with $720 and added $50 for each
month (x), so the function is $f(x) = 720 + 50x$
b) Substitue 11 for the number of months.
$f(x) = 720 + 50(11) = 1270$
There will be $1270 after 11 months.
c) Set the function equal to $2750 and solve for x.
$720 + 50x = 2750$
$50x = 2030$
$x = 40.6$
Rounding up to the next month, it will take
41 months.

39. $f(x) = 9x + 7$
$f(-2) = 9(-2) + 7$
$f(-2) = -18 + 7$
$f(-2) = -11$

44. $y - y_1 = m(x - x_1)$
$y - 3 = -5(x - 1)$
$y - 3 = -5x + 5$
$y = -5x + 8$

46. $(x_1, y_1): (-2, 1), (x_2, y_2): (2, -7)$
First, find the slope of the line.
$m = \frac{y_2 - y_1}{x_2 - x_1}$
$m = \frac{-7 - 1}{2 - (-2)}$
$m = \frac{-8}{4}$
$m = -2$
Then find the equation of the line.
$y - y_1 = m(x - x_1)$
$y - 1 = -2(x - (-2))$
$y - 1 = -2(x + 2)$
$y - 1 = -2x - 4$
$y = -2x - 3$

49. The line passes through $(-2, 3)$
and $(2, 9)$.
$m = \frac{9 - 3}{2 - (-2)} = \frac{6}{4} = \frac{3}{2}$
$y - 3 = \frac{3}{2}(x - (-2))$
$y - 3 = \frac{3}{2}x + 3$
$y = \frac{3}{2}x + 6$

57. Graph the dashed line $3x + y = -9$
The x-intercept of the line is $(-3, 0)$
The y-intercept of the line is $(0, -9)$.
Select $(0, 0)$ as a test point.
$3(0) + (0) < -9$
$0 < -9$
Since this inequality is false, the test point
$(0,0)$ is not a solution.
Shade the half-plane that does not
contain $(0,0)$

Chapter 3 Test   **1.** II   **2.** Yes   **3.** $(-3, 0), (0, \frac{5}{2})$ **4.** $(6, 0), (0, 21)$

5. $(\frac{3}{2}, 0), (0, -6)$ **6.** $(6, 0), (0, -4)$ **7.** $-\frac{1}{5}$ **8.** $7, (0, -8)$ **9.** $-\frac{3}{5}, (0, 9)$ **10.**

11.

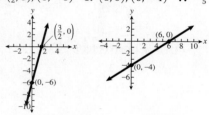

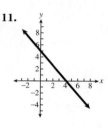

12. Parallel   **13.** $y = -6x + 5$ **14.** 13   **15.** $y = -\frac{5}{2}x + 4$ **16.** $y = -2x - 8$

17.   **18.**   **19.**   **20.**

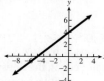

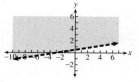

Cumulative Review Chapters 1–3   **1.** $>$ **2.** -34 **3.** -165 **4.** 13   **5.** $2^3 \cdot 3^2$ **6.** $\frac{2}{49}$ **7.** $\frac{33}{20}$ **8.** 6.08   **9.** 17.86   **10.** $448

11. $4\frac{1}{12}$ cups   **12.** -43 **13.** 15   **14.** -28 **15.** $5x - 3$ **16.** $-11x - 69$ **17.** $\{-9\}$ **18.** $\{\frac{7}{2}\}$ **19.** $\{14\}$ **20.** $\{-14\}$

21. Length: 13 ft, Width: 22 ft   **22.** 19   **23.** 25   **24.** 24%   **25.** $248,600   **26.** $\{6\}$ **27.** 4615   **28.** 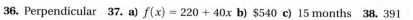 $(-\infty, 3]$

29. 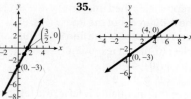 $[-5, 1]$ **30.** $(3, 0), (0, -6)$ **31.** $(8, 0), (0, -6)$ **32.** 2   **33.** $m = -\frac{2}{3}, (0, 7)$

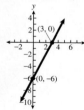
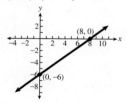

34.   **35.**   **36.** Perpendicular   **37. a)** $f(x) = 220 + 40x$ **b)** $540   **c)** 15 months   **38.** 391

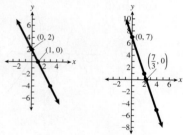
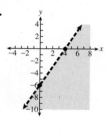

39. 1129   **40.** $y = -2x + 2$ **41.** $y = -3x + 7$ **42.**

Chapter 4

Quick Check 4.1   **1.** Yes   **2.** $(5, 0)$ **3.** $(2, 4)$ **4.** $(-5, 2)$ **5.** No solution   **6.** $(x, -\frac{2}{3}x + 2)$

Section 4.1   **1.** system of linear equations   **3.** it is a solution to each equation in the system   **5.** inconsistent   **7.** Yes   **9.** No   **11.** Yes   **13.** $(2, 4)$ **15.** $(-3, 6)$ **17.** $(2, 9)$ **19.** $(2, 7)$ **21.** $(-6, 4)$ **23.** $(-3, -10)$ **25.** $(0, 4)$ **27.** Inconsistent, $\varnothing$ **29.** $(-4, -9)$ **31.** Answers will vary.   **33.** Answers will vary.   **35.** Answers will vary.   **37.** Answers will vary.

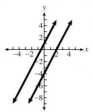
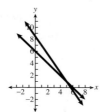
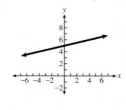

Quick Review   **1.** $\{3\}$ **2.** $\{-4\}$ **3.** $\{-5\}$ **4.** $\{-7\}$

Quick Check 4.2   **1.** $(5, 2)$ **2.** $(2, 6)$ **3.** $\varnothing$ **4.** $(x, \frac{1}{6}x - \frac{2}{3})$ **5.** $(1, -4)$ **6.** 11 desktops and 4 laptops

Section 4.2   **1.** one of the equations for one of the variables   **3.** dependent   **5.** $(3, 4)$ **7.** $(17, -3)$ **9.** Inconsistent, $\varnothing$ **11.** Dependent, $(x, -3x - 2)$ **13.** $(4, 2)$ **15.** $(2, 9)$ **17.** $(\frac{1}{2}, \frac{3}{4})$ **19.** $(13, -2)$ **21.** $(4, 1)$ **23.** $(4, 2)$ **25.** Dependent, $(x, -6x + 21)$ **27.** $(-5, 3)$ **29.** $(-8, 5)$ **31.** $(4, 3)$ **33.** $(2, 1)$ **35.** $(-2, 4)$ **37.** Example of correct solution: $y = 2x - 6$ **39.** Example of correct solution: $x + y = -10$ **41.** 250 students, 950 nonstudents   **43.** 47 chicken, 95 steak   **45.** 13 regular coffees, 5 café mochas   **47.** 2090   **49.** Answers will vary.

Quick Check 4.3   **1.** $(7, 1)$ **2.** $(3, -3)$ **3.** $(-8, -1)$ **4.** $(x, \frac{2}{3}x - 4)$ **5.** $\varnothing$ **6.** $(-6, 4)$ **7.** $(275, 425)$ **8.** 43 nickels, 32 dimes

Section 4.3   **1.** the LCM of the denominators   **3.** dependent   **5.** $(-2, -3)$ **7.** $(4, 1)$ **9.** $(6, 1)$ **11.** Dependent, $(x, \frac{1}{2}x - \frac{11}{4})$ **13.** $(7, 1)$ **15.** $(-2, -5)$ **17.** $(2, -3)$ **19.** $(6, 2)$ **21.** $(1, -4)$ **23.** $(\frac{3}{2}, 4)$ **25.** Inconsistent, $\varnothing$ **27.** $(\frac{5}{2}, \frac{1}{2})$ **29.** $(7, 2)$ **31.** $(3, -2)$ **33.** $(2, 1)$ **35.** $(3, 12)$ **37.** $(4, 2)$ **39.** $(-2, -3)$ **41.** 4 roses, 11 carnations   **43.** 13 5-gallon trees, 7 15-gallon trees   **45.** 49 dimes, 32 nickels   **47.** 29 nickels, 22 quarters   **49.** Example of correct solution: $6x - 8y = 36$ **51.** Example of correct solution: $x + 3y = -1$ **53.** $(9, -1)$ **55.** $(-7, 6)$ **57.** $(12, 10)$ **59.** Answers will vary.

Quick Check 4.4   **1.** 40   **2.** 75 feet by 40 feet   **3.** $3200 at 5%, $1000 at 4%   **4.** 32 pounds of almonds and 16 pounds of cashews   **5.** 160 ml of 60% solution and 240 ml of 50% solution   **6.** $\frac{1}{2}$ hour   **7.** plane 475 mph, wind 25 mph

Section 4.4   **1.** $P = 2L + 2W$ **3.** $d = rt$ **5.** algebra: 38, statistics: 55   **7.** Andre: 35, father: 63   **9.** 24, 55   **11.** 21 multiple choice, 7 true–false   **13.** 5 free throws, 16 field goals   **15.** 22   **17.** pizza: $11, soda: $4   **19.** 34°, 56°   **21.** 49°, 131°   **23.** length: 27 in., width: 19 in.   **25.** length: 135 ft, width: 30 ft   **27.** length: 175 cm, width: 75 cm   **29.** length: 40 ft, width: 25 ft   **31.** $800 at 6%, $1200 at 3%   **33.** $20,000 at 4.25%, $5000 at 3.5%   **35.** $6000 at 4% profit, $1500 at 13% loss   **37.** 80 ml of 80%, 240 ml of 44%   **39.** 12 oz of 20%, 6 oz of 50%   **41.** 2.2 L of vodka, 1.8 L of tonic water   **43.** George: 2 hours, Tina: 4 hours   **45.** after 1.5 hours   **47.** kayak: 4 mph, current: 2 mph   **49.** plane: 550 mph, wind: 50 mph   **51.** Answers will vary.   **53.** Answers will vary.

Quick Review   **1.**

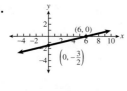
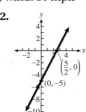
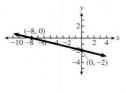
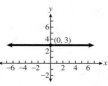

2.   **3.**   **4.**

Quick Check 4.5   **1.**   **2.**   **3.**

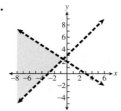
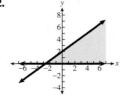
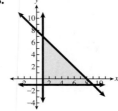

Section 4.5   **1.** system of linear inequalities   **3.** solid

5. 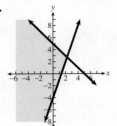   **7.** 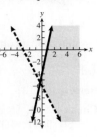   **9.** 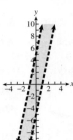   **11.**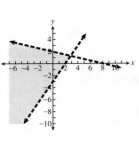

13. 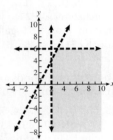   **15.** 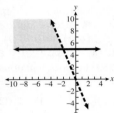   **17.** 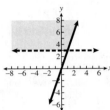   **19.**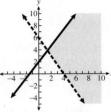

21. 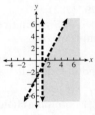   **23.** 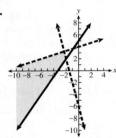   **25.** 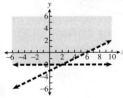

27. $x > 0$ **29.** $x + y > 4$ (Answers will vary.)
$\quad\quad\; y > 0$ $\quad\quad\; x + y < -2$
31. 18 square units

Chapter 4 Review   **1.** $(-4, -5)$ **2.** $(6, -1)$ **3.** $(-3, 6)$ **4.** $(-7, 0)$ **5.** $(8, 3)$ **6.** $(3, -3)$ **7.** $(2, 4)$ **8.** $(-6, -4)$ **9.** $(-3, -2)$
10. $(-7, 4)$ **11.** $(0, -5)$ **12.** $\varnothing$ **13.** $(9, 1)$ **14.** $\left(4, \frac{3}{2}\right)$ **15.** $\left(-\frac{5}{3}, -8\right)$ **16.** $(x, 2x - 4)$ **17.** $(-2, -3)$ **18.** $(-4, 6)$ **19.** $(2, -4)$
20. $(-3, 7)$ **21.** $(-2, -5)$ **22.** $(5, -3)$ **23.** $(x, 3x - 2)$ **24.** $\left(-\frac{5}{2}, -\frac{3}{2}\right)$ **25.** $(0, -6)$ **26.** $(27, -16)$ **27.** $\varnothing$ **28.** $(-3, 5)$
29. $(5, 3.9)$ **30.** $(6000, 2000)$ **31.** $23, 64$ **32.** 1077 **33.** 834 **34.** 46 **35.** 400 ft by 260 ft   **36.** 76 in.   **37.** $1200 at 5%, $2800 at
4.25%   **38.** $1000 at 20%, $1500 at 8%   **39.** 15 lb 4%, 30 lb 13%   **40.** 5 mph

41. 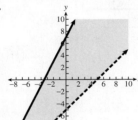   **42.** 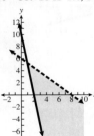   **43.** 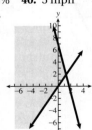   **44.**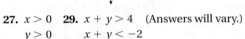

Chapter 4 Review Exercises: Worked-Out Solutions **1.** Look for the point where the two lines intersect. The solution to the system is $(-4, -5)$

5. 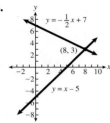 $(8, 3)$

9. $x = 2y + 1$
$2x - 5y = 4$
$2(2y + 1) - 5y = 4$
$4y + 2 - 5y = 4$
$-y + 2 = 4$
$-y = 2$
$y = -2$
$x = 2(-2) + 1$
$x = -4 + 1$
$x = -3$
$(-3, -2)$

15. $6x + 7y = -66$
$-3x + 4y = -27$ $\xrightarrow{\text{Multiply by 2}}$

$6x + 7y = -66$
$-6x + 8y = -54$

$6x + 7y = -66$
$\underline{-6x + 8y = -54}$
$15y = -120$
$y = -8$

$6x + 7(-8) = -66$
$6x - 56 = -66$
$6x = -10$
$x = -\dfrac{10}{6}$
$x = -\dfrac{5}{3}$

$\left(-\tfrac{5}{3}, -8\right)$

33. Women: x
Men: y

$x + y = 1764$ $\xrightarrow{\text{Multiply by } -5}$ $-5x - 5y = -8820$
$5x + 10y = 13470$ $\qquad\qquad\qquad 5x + 10y = 13470$

$-5x - 5y = -8820$
$\underline{5x + 10y = 13470}$
$5y = 4650$
$y = 930$

$x + 930 = 1764$
$x = 834$

There were 834 ladies.

35. Length: l
Width: w

$2l + 2w = 1320$
$l = w + 140$

$2l + 2w = 1320$
$2(w + 140) + 2w = 1320$
$2w + 280 + 2w = 1320$
$4w + 280 = 1320$
$4w = 1040$
$w = 260$

$l = 260 + 140$
$l = 400$

The length is 400 feet and the width is 260 feet.

37. First account (5%): x
Second account (4.25%): y

$x + y = 4000$
$0.05x + 0.0425y = 179$ $\xrightarrow{\text{Multiply by 10,000}}$

$x + y = 4000$
$500x + 425y = 1{,}790{,}000$

$x + y = 4000$ $\xrightarrow{\text{Multiply by } -425}$ $-425x - 425y = -1{,}700{,}000$
$500x + 425y = 1{,}790{,}000$ $\qquad\qquad 500x + 425y = 1{,}790{,}000$

$-425x - 425y = -1{,}700{,}000$
$\underline{500x + 425y = 1{,}790{,}000}$
$75x = 90{,}000$

$x = 1200$

$1200 + y = 4000$
$y = 2800$

He invested $1200 at 5% interest and $2800 at 4.25% interest.

41.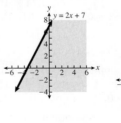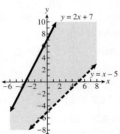

Chapter 4 Test **1.** $(-5, 2)$ **2.** $(4, -5)$ **3.** $(3, -3)$ **4.** $(-4, -1)$ **5.** $(-3, 2)$ **6.** $\varnothing$ **7.** $(5, 6)$ **8.** $(9, 10)$ **9.** $(4, 0)$
10. $(x, \frac{2}{3}x-3)$ **11.** 180 **12.** 5400 **13.** Length: 25 ft, Width: 8 ft **14.** \$20,000 at 7%, \$12,000 at 3%
15.

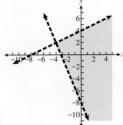

Chapter 5

Quick Check 5.1 **1. a)** x^9 **b)** 729 **2.** $(a-6)^{27}$ **3.** $a^{15}b^{10}$ **4.** x^{63} **5.** x^{34} **6.** $a^{40}b^{24}$ **7.** $64a^{26}b^{19}c^{35}$ **8. a)** x^{16} **b)** x^{16} **9.** $4a^8b^6c^6$

10. a) 1 **b)** 1 **c)** 9 **11.** $\frac{a^5b^{35}}{c^{15}d^{20}}$ **12.** 4096 **13.** a^{56}

Section 5.1 **1.** $x^m \cdot x^n = x^{m+n}$ **3.** $(xy)^n = x^n y^n$ **5.** $x^0 = 1$ **7.** 32 **9.** x^{16} **11.** m^{35} **13.** b^{14} **15.** $(2x-3)^{14}$ **17.** x^8y^{13} **19.** x^5
21. b **23.** x^{10} **25.** a^{56} **27.** 64 **29.** x^{29} **31.** 7 **33.** 5 **35.** $125x^3$ **37.** $-64y^9$ **39.** $x^{21}y^{28}$ **41.** $81x^{16}y^4$ **43.** $x^{14}y^{67}z^{61}$ **45.** 7
47. $6, 13$ **49.** x^2 **51.** d^{30} **53.** $4x^5$ **55.** $(a+5b)^2$ **57.** a^5b^5 **59.** 12 **61.** 30 **63.** 1 **65.** 1 **67.** 3 **69.** 1 **71.** $\frac{16}{625}$ **73.** $\frac{a^7}{b^7}$
75. $\frac{x^{14}}{y^{21}}$ **77.** $\frac{8a^{18}}{b^{21}}$ **79.** $\frac{a^{45}b^{18}}{c^{63}}$ **81.** 4 **83.** 12 **85.** 16 **87.** 16 **89.** a^{24} **91.** a^{42} **93.** $\frac{125x^{12}}{8y^{21}}$ **95.** x^{12}
97. $\frac{a^{40}b^{45}}{32c^{10}}$ **99.** b^{20} **101.** x^{90} **103.** $\frac{81a^8b^{36}}{2401c^4d^{16}}$ **105.** 144 feet **107.** 8100 square feet **109.** 615.44 square feet
111. $x^5 \cdot x^4$; explanations will vary. **113.** Answers will vary.

Quick Check 5.2 **1.** $\frac{1}{64}$ **2.** $\frac{a^5}{b^4}$ **3. a)** $x^{12}y^7$ **b)** $\frac{y^4zw^9}{x^2}$ **4.** $\frac{1}{x^7}$ **5.** x^{42} **6.** $\frac{b^6}{81a^2c^4}$ **7.** $\frac{1}{x^6}$ **8.** $\frac{1}{x^{20}y^{16}}$ **9. a)** 4.6×10^{-3}
b) 3.57×10^6 **10. a)** $3,200,000$ **b)** 0.000721 **11.** 6.96×10^{13} **12.** 3.0×10^{12} **13.** 8000 seconds

Section 5.2 **1.** $\frac{1}{x^n}$ **3.** $\frac{1}{16}$ **5.** $\frac{1}{216}$ **7.** $-\frac{1}{343}$ **9.** $\frac{1}{a^{10}}$ **11.** $\frac{4}{x^3}$ **13.** $-\frac{5}{m^{19}}$ **15.** x^5 **17.** $3y^4$ **19.** $\frac{y^2}{x^{11}}$ **21.** $-\frac{8b^5}{a^6c^7}$ **23.** $\frac{5c^4}{a^6b^9d^5}$
25. x^3 **27.** $\frac{1}{a^{16}}$ **29.** $\frac{1}{m^4}$ **31.** $\frac{1}{x^{20}}$ **33.** x^{42} **35.** $\frac{z^{12}}{x^{10}y^8}$ **37.** $\frac{a^{15}b^{24}}{64z^6}$ **39.** $\frac{1}{x^{13}}$ **41.** a^{12} **43.** $\frac{1}{y^2}$ **45.** $\frac{1}{x^{11}}$ **47.** 1 **49.** $\frac{y^{20}}{x^{15}}$
51. $\frac{9c^{10}}{a^8b^{14}d^{16}}$ **53.** 0.000000307 **55.** $8,935,000,000$ **57.** $90,210$ **59.** 4.5×10^{-6} **61.** 4.7×10^7 **63.** 6.0×10^{-9} **65.** 9.28×10^{18}
67. 3.72×10^{14} **69.** 4.1496×10^{-11} **71.** 1.74×10^5 **73.** 3.36×10^4 seconds (33,600 seconds) **75.** 1.116×10^8 miles
77. 1.1×10^{45} grams **79.** \$63,590,000,000 **81.** $151,200,000,000$ calculations **83.** \$6,438,000,000 **85.** Answers will vary.
87. Answers will vary.

Quick Review **1.** $11x$ **2.** $-7a$ **3.** $-6x^2 + 11x$ **4.** $12y^3 + 5y^2 - 13y - 37$

Quick Check 5.3 **1. a)** Trinomial; 2, 1, 0 **b)** Binomial; 2, 0 **c)** Monomial; 7 **2.** $1, -6, -11, 32$ **3.** $-x^3 + 4x^2 + x + 9$, leading
term: $-x^3$, leading coefficient: -1, degree: 3 **4.** 111 **5.** 280 **6.** $x^3 + 3x + 169$ **7.** $-3x^3 - x^2 + 15x + 5$ **8.** $6x^2 + 13x - 97$
9. 289 **10.** Each term: 5, 8, 10, Polynomial: 10 **11.** $13x^4y^2 - 4x^3y^3$

Section 5.3 **1.** polynomial **3.** monomial **5.** trinomial **7.** descending order **9.** 4, 2, 1, 0 **11.** 1, 7, 4 **13.** 7, 1, -15
15. $10, -17, 6, -1, 2$ **17.** trinomial **19.** binomial **21.** monomial **23.** $3x^2 + 8x - 7, 3x^2, 3, 2$ **25.** $2x^4 + 6x^2 - 11x + 10, 2x^4, 2, 4$
27. 22 **29.** -38 **31.** -252 **33.** -2 **35.** -591 **37.** $8x^2 - 6x + 3$ **39.** $2x^2 - 17x + 80$ **41.** $2x^4 + x^3 + 2x^2 - 27x - 20$
43. $x^3 - 9x^2 + 20x - 16$ **45.** $2x^9 + 7x^6 - 4x^5 - 5x^4 + 7x^2 + 12$ **47.** $4x^2 - 3x - 13$ **49.** $2x^4 - 3x^3 + x^2 + 11x + 11$
51. $12x^2 + x + 9, 2x^2 + 19x - 3$ **53.** $5x^3 + x^2 - 5x - 6, -3x^3 - x^2 - 11x - 56$ **55.** 184 **57.** 148 **59.** $75,680$
61. 10, 6, 3, Polynomial: 10 **63.** 8, 9, 11, Polynomial: 11 **65.** $7x^2y + 11xy^2 - 3x^2y^3$ **67.** $-14a^3b^2 + 13a^5b + 5a^2b^3$
69. $2x^3yz^2 - 15xy^4z^3 - 7x^2y^2z^5 - 4x^2yz^3$ **71.** \$1800 **73.** 1486 **75.** Answers will vary.

Quick Check 5.4 **1.** $56x^2$ **2.** $60x^{14}yz^{13}$ **3.** $28x^4 - 24x^3 + 20x^2 - 32x$ **4.** $6x^9 - 3x^8 + 3x^7 + 45x^6 - 63x^5$ **5.** $x^2 + 20x + 99$
6. $10x^2 - 49x + 18$ **7.** $3x^3 + 30x^2 + 46x - 16$ **8.** $x^2 - 100$ **9.** $4x^2 - 49$ **10.** $25x^2 - 80x + 64$ **11.** $x^2 + 12x + 36$

Section 5.4 **1.** add **3.** multiply **5.** $35x^4$ **7.** $-18x^9$ **9.** $96x^8y^6$ **11.** $-70x^4y^2z^9w^4$ **13.** $84x^{10}$ **15.** $5x^3$ **17.** $-12x^9$ **19.** $10x - 16$ **21.** $-8x + 36$ **23.** $18x^2 + 30x$ **25.** $3x^5 - 4x^4 + 7x^3$ **27.** $6x^3y^2 - 12x^2y^3 + 14xy^4$ **29.** $x^2 + 5x + 6$ **31.** $x^2 - 19x + 90$ **33.** $2x^2 - 9x - 5$ **35.** $3x^3 - 13x^2 - 37x - 18$ **37.** $x^4 - 4x^3 - 51x^2 + 310x - 400$ **39.** $x^2 - 2xy - 8y^2$ **41.** $3x^3$ **43.** $3x^3 - 5x^2 - 12$ **45.** 3 **47.** 3, 6 **49.** $x^2 - 7x - 18$ **51.** $-18x^9$ **53.** $-4x^8 + 20x^7 - 32x^6 - 12x^5$ **55.** $x^2 - 64$ **57.** $4x^2 - 25$ **59.** $x^2 + 12x + 36$ **61.** $4x^2 - 12x + 9$ **63.** $x - 9$ **65.** $x - 6$ **67.** $x^2 + 6x - 7$ **69.** $5x^3 - 40x^2 - 45x$ **71.** $-28x^6y^9$ **73.** $20x^{14}$ **75.** $25x^2 - 9$ **77.** $x^2 + 26x + 169$ **79.** $6x^4y - 2x^6y^4 - 14x^2y^2$ **81.** $16x^2 - 72x + 81$ **83.** $x^4 - 2x^3 - 7x^2 - 8x + 16$ **85.** Answers will vary. **87.** Answers will vary.

Quick Check 5.5 **1.** $8x^5$ **2.** $5x^2y^9$ **3.** $x^2 - 3x - 7$ **4.** $8x^8 + 2x^5 + 5x^3$ **5.** $x + 9$ **6.** $x - 10 + \frac{62}{x+7}$ **7.** $4x + 3 + \frac{16}{3x-4}$ **8.** $x^2 - x - 2 - \frac{13}{x-2}$

Section 5.5 **1.** term **3.** factor **5.** $5x^2$ **7.** $-11x^{10}$ **9.** $17x^4y$ **11.** $5x^4$ **13.** $\frac{3x^2}{2}$ **15.** $\frac{x^8}{3}$ **17.** $10x^{11}$ **19.** $-4x^5$ **21.** $6x^2 + 7x - 4$ **23.** $x^3 + 7x^2 - 9x + 3$ **25.** $2x^6 + 3x^5 + 5x^4$ **27.** $-2x^4 + 3x^2$ **29.** $x^5y^4 - x^3y^3 + xy^2$ **31.** $6x$ **33.** $6x^7 - 8x^5 - 18x^3$ **35.** $x + 3$ **37.** $x + 3$ **39.** $x - 8$ **41.** $x + 4 + \frac{3}{x-8}$ **43.** $x^2 - 12x - 25 + \frac{39}{x+1}$ **45.** $2x - 7 + \frac{3}{x+5}$ **47.** $4x + 3 + \frac{3}{3x-5}$ **49.** $x + 10$ **51.** $x^3 + 3x^2 + 11x + 18 + \frac{86}{x-3}$ **53.** $x^2 + 5x + 25$ **55.** $x^2 + 5x - 24$ **57.** $x + 13$ **59.** Yes **61.** No **63.** $x - 5 - \frac{6}{x-3}$ **65.** $x^2 - 4x + 24 - \frac{115}{x+4}$ **67.** $3x^4 - 2x^2 + 5x - 1$ **69.** $8x - 6 + \frac{11}{x+6}$ **71.** $5x - 9 - \frac{33}{4x-3}$ **73.** $-9x^7 + 6x^5 + 2x^4 + x$ **75.** $2x^2 - 5x - 9 + \frac{27}{4x+10}$ **77.** $3x^2 - 5x - 17$ **79.** $-3x^2yz^5$

Chapter 5 Review **1.** x^6 **2.** $64x^{21}$ **3.** 7 **4.** $\frac{625x^{36}}{16y^8}$ **5.** $28x^{19}$ **6.** $24x^{16}y^8$ **7.** $36a^{10}b^6c^{12}$ **8.** -2 **9.** $x^{40}y^{52}z^{16}$ **10.** x^{10} **11.** $\frac{1}{25}$ **12.** $\frac{1}{1000}$ **13.** $\frac{7}{x^4}$ **14.** $-\frac{6}{x^6}$ **15.** $-8y^5$ **16.** $\frac{27}{x^{12}}$ **17.** $\frac{x^{10}}{4}$ **18.** x^{15} **19.** $\frac{b^{21}}{a^{12}}$ **20.** $\frac{1}{m^6}$ **21.** $\frac{x^{15}z^{21}}{64y^{12}}$ **22.** $\frac{y^{36}}{x^{66}}$ **23.** $\frac{1}{a^{20}}$ **24.** $\frac{1}{x^{11}}$ **25.** 1.4×10^9 **26.** 2.1×10^{-12} **27.** 5.002×10^{-6} **28.** 0.000123 **29.** 40,750,000 **30.** 6,127,500,000,000 **31.** 1.5×10^{14} **32.** 2.5×10^{-18} **33.** 2.49×10^{-12} grams **34.** 15,000 seconds **35.** 80 **36.** -47 **37.** 30 **38.** -112 **39.** 27 **40.** 371 **41.** $2x^2 - 6x - 9$ **42.** $8x^2 - 19x - 50$ **43.** $8x - 21$ **44.** $x^2 + 9x - 22$ **45.** $3x^3 - 5x^2 + 13x - 30$ **46.** $-4x^3 + 4x^2 - 6x - 115$ **47.** -9 **48.** 37 **49.** $a^8 - 7a^4 + 10$ **50.** $4a^6 + 6a^3 - 30$ **51.** $f(x) + g(x) = 10x^2 - 2x + 31, f(x) - g(x) = -4x^2 - 14x - 49$ **52.** $f(x) + g(x) = -7x^2 - 10x - 24, f(x) - g(x) = 9x^2 + 20x - 36$ **53.** $x^2 - 10x + 25$ **54.** $x^2 + 9x - 52$ **55.** $12x^3 - 28x^2 - 64x$ **56.** $4x^2 - 28x + 49$ **57.** $15x^{13}$ **58.** $x^2 - 64$ **59.** $6x^2 - 19x - 130$ **60.** $-18x^6$ **61.** $36x^2 - 25$ **62.** $x^2 + 20x + 100$ **63.** $16x^2 + 56x + 49$ **64.** $-24x^7 + 56x^6 + 48x^5$ **65.** $x^3 + 3x^2 - 42x + 108$ **66.** $3x^3 - 11x^2 + x + 18$ **67.** $-120x^{18}$ **68.** $10x^6 + 22x^5 - 24x^4$ **69.** $f(x) \cdot g(x) = x^2 + 3x - 70$ **70.** $f(x) \cdot g(x) = 24x^8 + 54x^7 - 120x^6$ **71.** $3x^2 - 4x + 10$ **72.** $3x + 4 - \frac{9}{x-4}$ **73.** $-2y^6$ **74.** $x^2 - 2x + 10 - \frac{65}{x+5}$ **75.** $x^3 - 3x^2 + 9x - 27 + \frac{162}{x+3}$ **76.** $6yz^5$ **77.** $8x + 15 + \frac{36}{x+9}$ **78.** $6x + 12$ **79.** $3x - 17$ **80.** $2x + 1 + \frac{23}{8x-3}$

Chapter 5 Review Exercises: Worked-Out Solutions

1. $\frac{x^9}{x^3} = x^{9-3}$ **6.** $3x^8y^5 \cdot 8x^8y^3 = 24x^{8+8}y^{5+3}$ **7.** $(-6a^5b^3c^6)^2 = (-6)^2a^{5\cdot2}b^{3\cdot2}c^{6\cdot2}$ **18.** $\frac{x^{10}}{x^{-5}} = x^{10-(-5)}$
$\quad = x^6$ $\quad\quad = 24x^{16}y^8$ $\quad\quad = 36a^{10}b^6c^{12}$ $\quad\quad = x^{15}$

21. $(4x^{-5}y^4z^{-7})^{-3} = 4^{-3}x^{15}y^{-12}z^{21}$ **23.** $a^{-15} \cdot a^{-5} = a^{-15+(-5)}$ **32.** $(3.0 \times 10^{-13}) \div (1.2 \times 10^5) = \frac{3.0}{1.2} \times \frac{10^{-13}}{10^5}$
$\quad = \frac{x^{15}z^{21}}{4^3y^{12}}$ $\quad\quad = a^{-20}$ $\quad\quad = 2.5 \times 10^{-13-5}$
$\quad = \frac{x^{15}z^{21}}{64y^{12}}$ $\quad\quad = \frac{1}{a^{20}}$ $\quad\quad = 2.5 \times 10^{-18}$

33. $1.66 \times 10^{-24} \cdot 1.5 \times 10^{12} = 1.66 \cdot 1.5 \times 10^{-24} \cdot 10^{12}$ **39.** $5x^2 + 10x + 12$
$\quad\quad = 2.49 \times 10^{-24+12}$ $\quad\quad 5(-3)^2 + 10(-3) + 12$
$\quad\quad = 2.49 \times 10^{-12}$ $\quad\quad = 5(9) + 10(-3) + 12$
$\quad\quad 2.49 \times 10^{-12}$ grams $\quad\quad = 45 - 30 + 12$
$\quad\quad = 27$

41. $(x^2 + 3x - 15) + (x^2 - 9x + 6) = x^2 + 3x - 15 + x^2 - 9x + 6$
$\quad\quad = 2x^2 - 6x - 9$

43. $(x^2 - 6x - 13) - (x^2 - 14x + 8) = x^2 - 6x - 13 - x^2 + 14x - 8$
$\quad\quad = 8x - 21$

47. $f(x) = x^2 - 25$

$f(-4) = (-4)^2 - 25$

$= 16 - 25$

$= -9$

51. $f(x) + g(x) = (3x^2 - 8x - 9) + (7x^2 + 6x + 40)$

$= 3x^2 - 8x - 9 + 7x^2 + 6x + 40$

$= 10x^2 - 2x + 31$

$f(x) - g(x) = (3x^2 - 8x - 9) - (7x^2 + 6x + 40)$

$= 3x^2 - 8x - 9 - 7x^2 - 6x - 40$

$= -4x^2 - 14x - 49$

54. $(x - 4)(x + 13) = x^2 + 13x - 4x - 52$

$= x^2 + 9x - 52$

55. $4x(3x^2 - 7x - 16) = 4x \cdot 3x^2 - 4x \cdot 7x - 4x \cdot 16$

$= 12x^3 - 28x^2 - 64x$

56. $(2x - 7)^2 = (2x - 7)(2x - 7)$

$= 4x^2 - 14x - 14x + 49$

$= 4x^2 - 28x + 49$

57. $5x^7 \cdot 3x^6 = 15x^{7+6}$

$= 15x^{13}$

69. $f(x) \cdot g(x) = (x + 10)(x - 7)$

$= x^2 - 7x + 10x - 70$

$= x^2 + 3x - 70$

71. $\frac{6x^4 - 8x^3 + 20x^2}{2x^2} = \frac{6x^4}{2x^2} - \frac{8x^3}{2x^2} + \frac{20x^2}{2x^2}$

$= 3x^2 - 4x + 10$

72.

$$
\begin{array}{r}
3x + 4 \\
x - 4 \overline{)3x^2 - 8x - 25} \\
\underline{3x^2 \not{+} 12x} \downarrow \\
4x - 25 \\
\underline{4x \not{+} 16} \\
-9
\end{array}
$$

$\frac{3x^2 - 8x - 25}{x - 4} = 3x + 4 - \frac{9}{x - 4}$

74.

$$
\begin{array}{r}
x^2 - 2x + 10 \\
x + 5 \overline{)x^3 + 3x^2 + 0x - 15} \\
\underline{x^3 \not{+} 5x^2} \downarrow \downarrow \\
-2x^2 + 0x - 15 \\
\underline{\not{+}2x^2 \not{+} 10x} \downarrow \\
10x - 15 \\
\underline{10x \not{+} 50} \\
-65
\end{array}
$$

$\frac{x^3 + 3x^2 - 15}{x + 5} = x^2 - 2x + 10 - \frac{65}{x + 5}$

Chapter 5 Test 1. x^8 **2.** $\frac{x^{30}}{729y^{24}}$ **3.** $x^{72}y^{40}z^{64}$ **4.** $\frac{1}{256}$ **5.** $\frac{32}{x^{35}}$ **6.** x^{17} **7.** x^7 **8.** $\frac{y^{12}}{9x^2z^{10}}$ **9.** 2.35×10^7 **10.** 0.000000047
11. 7.74×10^4 **12.** -70 **13.** $-4x^2 - 14x + 1$ **14.** $2x^2 + 14x - 64$ **15.** 14 **16.** $30x^5 + 40x^4 - 85x^3$ **17.** $x^2 - 36$
18. $20x^2 - x - 63$ **19.** $7x^5 + 11x^4 - 5x^3$ **20.** $6x - 29 + \frac{125}{x + 3}$

Chapter 6

Quick Check 6.1 1. 4 **2.** 12 **3. a)** $4x^3$ **b)** x^3z^4 **4.** $6x^2(x^2 - 7x - 15)$ **5.** $3x^4(5x^3 - 10x + 1)$ **6. a)** $(x - 9)(5x + 14)$
b) $(x - 8)(7x - 6)$ **7.** $(x + 4)(x^2 + 7)$ **8.** $(x - 5)(2x - 9)$ **9.** $(x - 9)(4x^2 + 1)$ **10.** $3(x + 8)(x - 4)$

Section 6.1 1. factored **3.** smallest **5.** 2 **7.** 6 **9.** 4 **11.** x^3 **13.** a^2b^2 **15.** $2x^3$ **17.** $5a^2c$ **19.** $7(x - 2)$ **21.** $x(5x + 4)$
23. $4x(2x^2 + 5)$ **25.** $7(x^2 + 8x + 10)$ **27.** $2x^3(9x^3 - 7x - 4)$ **29.** $m^2n(n^2 - m^3n + m)$ **31.** $8a^3b^5(2a^6b^2 - 9a^2 - 10b)$
33. $4x(-2x^2 + 3x - 4)$ **35.** $(2x - 7)(5x + 8)$ **37.** $(3x + 11)(x - 7)$ **39.** $(x + 10)(2x + 1)$ **41.** $(x + 8)(x + 7)$
43. $(x - 5)(x + 4)$ **45.** $(x + 2)(x - 9)$ **47.** $(x + 5)(3x + 4)$ **49.** $(x + 4)(7x - 6)$ **51.** $(x + 7)(3x + 1)$ **53.** $(x - 5)(2x - 1)$
55. $(x + 8)(x^2 + 6)$ **57.** $(x^2 + 3)(4x + 3)$ **59.** $2(x + 5)(x + 3)$ **61.** $x^2 + 3x - 40$ **63.** $2x^2 + x - 36$ **65.** Answers will vary.
67. Answers will vary.

Quick Review 1. $x^2 + 11x + 28$ **2.** $x^2 - 20x + 99$ **3.** $x^2 - 3x - 40$ **4.** $x^2 - 4x - 60$

Quick Check 6.2 1. a) $(x + 2)(x + 5)$ **b)** $(x - 5)(x - 6)$ **2. a)** $(x + 12)(x - 3)$ **b)** $(x + 6)(x - 7)$ **3. a)** $(x + 6)(x - 1)$
b) $(x + 2)(x + 3)$ **c)** $(x - 2)(x - 3)$ **d)** $(x - 6)(x + 1)$ **4.** $(x + 5)^2$ **5.** Prime **6.** $5(x + 2)(x + 6)$ **7.** $-(x - 6)(x - 8)$
8. $(x - 8y)(x + 4y)$ **9.** $(xy + 12)(xy - 2)$

Section 6.2 **1.** leading coefficient **3.** prime **5.** $(x - 10)(x + 2)$ **7.** $(x + 9)(x - 4)$ **9.** Prime **11.** $(x + 6)(x + 8)$ **13.** $(x + 15)(x - 2)$ **15.** Prime **17.** $(x - 20)(x + 3)$ **19.** $(x + 12)(x - 1)$ **21.** $(x + 7)^2$ **23.** $(x + 7)(x + 13)$ **25.** $(x - 12)^2$ **27.** $(x - 5)(x - 8)$ **29.** $(x - 15)(x + 2)$ **31.** $(x - 4)(x - 5)$ **33.** $(x - 6)(x - 10)$ **35.** $5(x^2 + 6x - 11)$ **37.** $21(x + 2)(x + 5)$ **39.** $-3(x - 8)^2$ **41.** $x(x + 4)^2$ **43.** $-3x^3(x + 10)(x - 8)$ **45.** $10x^6(x + 11)(x - 9)$ **47.** $(x + 7y)(x - 6y)$ **49.** $(xy - 3)(xy - 4)$ **51.** $(x - 13y)(x + 3y)$ **53.** $5x^3(x + 3y)(x + 7y)$ **55.** $(x + 9)(x^3 + 5)$ **57.** Prime **59.** $(x + 2)(x + 20)$ **61.** $(x + 20)^2$ **63.** $4(x + 6)(x - 4)$ **65.** $x^3(x - 15)$ **67.** $(x - 7)(x - 9)$ **69.** Prime **71.** $(x - 25)(x + 4)$ **73.** 16 **75.** 60 **77.** Answers will vary.

Quick Review **1.** $2x^2 + 13x + 20$ **2.** $6x^2 - 37x + 56$ **3.** $3x^2 + 14x - 24$ **4.** $8x^2 + 30x - 27$

Quick Check 6.3 **1.** $(3x + 4)(x + 3)$ **2.** $(4x - 9)(x - 4)$ **3.** $(2x - 3)(8x - 7)$ **4.** $(x + 4)(2x - 7)$ **5.** $(4x - 3)(5x + 8)$ **6.** $9(x + 6)(x - 3)$

Section 6.3 **1.** $ax^2 + bx + c$, where $a \ne 1$ **3.** $(5x + 2)(x - 4)$ **5.** $(2x + 1)(x + 4)$ **7.** $(3x - 4)(x + 1)$ **9.** $(2x + 3)(2x - 1)$ **11.** $(2x + 1)(3x + 2)$ **13.** $(3x - 5)(3x + 4)$ **15.** $(x + 8)(20x + 1)$ **17.** $(3x - 2)(x + 5)$ **19.** $(3x + 8)(4x - 7)$ **21.** $(4x - 3)^2$ **23.** $3(5x + 2)(x - 2)$ **25.** $8(2x + 5)(x - 1)$ **27.** $16(x - 2)^2$ **29.** $5(3x + 16)(x - 7)$ **31.** $(x + 1)(x + 13)$ **33.** $9(2x - 1)$ **35.** $(2x - 3)(x - 2)$ **37.** $(x + 20)(x - 1)$ **39.** $x^4(x - 6)$ **41.** $4x(x - 5)$ **43.** $-(3x + 2)(4x + 5)$ **45.** Prime **47.** $(x^2 + 3)(x^4 + 8)$ **49.** $(x + 2)(x^2 - 28)$ **51.** $(x - 12)^2$ **53.** $9(x - 6)(x + 1)$ **55.** $-(5x + 2)(x - 6)$ **57.** $(x - 9)(2x + 3)$ **59.** $(x + 15)(x - 3)$ **61.** $-3(x + 8)(x - 4)$ **63.** $(5x - 3)(3x + 5)$ **65.** $(2x + 1)(x + 4)$ **67.** Answers will vary.

Quick Review **1.** $x^2 - 64$ **2.** $16x^2 - 9$ **3.** $x^3 + 512$ **4.** $8x^3 - 125$

Quick Check 6.4 **1.** $(x + 5)(x - 5)$ **2.** $(9x^6 + 10y^8)(9x^6 - 10y^8)$ **3.** $7(x + 5)(x - 5)$ **4.** $(x^2 + y^2)(x + y)(x - y)$ **5.** $(x - 5)(x^2 + 5x + 25)$ **6.** $(x^4 - 2)(x^8 + 2x^4 + 4)$ **7.** $(x + 6)(x^2 - 6x + 36)$

Section 6.4 **1.** difference of squares **3.** multiples of 2 **5.** difference of cubes **7.** $(x + 2)(x - 2)$ **9.** $(x + 6)(x - 6)$ **11.** $(10x + 9)(10x - 9)$ **13.** Prime **15.** $(a + 4b)(a - 4b)$ **17.** $(5x + 8y)(5x - 8y)$ **19.** $(4 + x)(4 - x)$ **21.** $3(x + 5)(x - 5)$ **23.** Prime **25.** $5(x^2 + 4)$ **27.** $(x^2 + 4)(x + 2)(x - 2)$ **29.** $(x^2 + 9y)(x^2 - 9y)$ **31.** $(x - 1)(x^2 + x + 1)$ **33.** $(x - 2)(x^2 + 2x + 4)$ **35.** $(10 - y)(100 + 10y + y^2)$ **37.** $(x - 2y)(x^2 + 2xy + 4y^2)$ **39.** $(2a - 9b)(4a^2 + 18ab + 81b^2)$ **41.** $(x + y)(x^2 - xy + y^2)$ **43.** $(x + 2)(x^2 - 2x + 4)$ **45.** $(x^2 + 3)(x^4 - 3x^2 + 9)$ **47.** $(4m + n)(16m^2 - 4mn + n^2)$ **49.** $6(x - 3)(x^2 + 3x + 9)$ **51.** $2x(2x + 5y)(4x^2 - 10xy + 25y^2)$ **53.** $(x - 8)^2$ **55.** $(x + 2)(x + 14)$ **57.** $2(7x - 3)$ **59.** $(2x - 5)(3x + 5)$ **61.** $x^2(x - 9)(x + 7)$ **63.** $-8(x^3 + 9)$ **65.** $-6(x - 4)(x - 6)$ **67.** $(x - 13)(x + 3)$ **69.** $(5x + 6y^4)(5x - 6y^4)$ **71.** Prime **73.** $(x + 8y^3)(x - 8y^3)$ **75.** $3x^2(3x^3 - 8x^2 - 15)$ **77.** $4(x^2 + 49)$ **79.** $(x + 5)(x + 7)$ **81.** $(x + 2)(3x - 19)$ **83.** $(x - 10)(x^2 - 3)$ **85.** $(x + 5y)(x^2 - 5xy + 25y^2)$ **87.** $x^5(3x + 5)^2$ **89.** d **91.** f **93.** g **95.** b **97.** Answers will vary. **99.** Answers will vary.

Quick Check 6.5 **1.** $-6(x - 10)(x + 1)$ **2.** $(6x^4y^7 + 5)(36x^8y^{14} - 30x^4y^7 + 25)$ **3.** $(x + 5)(x - 2)(x^2 + 2x + 4)$ **4.** $(x^2 + y)(x^4 - x^2y + y^2)(x^2 - y)(x^4 + x^2y + y^2)$ **5.** $(x^2 + 4)(x^4 - 4x^2 + 16)$ **6.** $2(4x - 3)(x + 6)$

Section 6.5 **1.** common factors **3.** sum of squares **5.** sum of cubes **7.** $(3x - 2y)(9x^2 + 6xy + 4y^2)$ **9.** $(x - 7)(2x + 1)(2x - 1)$ **11.** $(x + 1)(x^2 - x + 1)(x - 1)(x^2 + x + 1)$ **13.** $(x - 3)(x - 15)$ **15.** $(x^2 + 4y^4)(x^4 - 4x^2y^4 + 16y^8)$ **17.** $4(x + 6)(x + 8)$ **19.** $(3x - 8)(x + 6)$ **21.** $(x + 6)(x + 7)$ **23.** $(x + 16)(x + 1)(x^2 - x + 1)$ **25.** $(3x - 5)^2$ **27.** $(x - 7y)^2$ **29.** $(x + 15)(x - 11)$ **31.** $4(3x + 5)$ **33.** $(x + 14)(x - 1)$ **35.** $(x - 5)(6x + 1)$ **37.** $4(21x^2 + x - 5)$ **39.** $(x + 4)(7x + 1)$ **41.** $(x + 2)(3x + 2)(3x - 2)$ **43.** $(x - 20)(x + 14)$ **45.** $4(x + 8)(x - 7)$ **47.** $(x + y)(x^2 - xy + y^2)(x - y)(x^2 + xy + y^2)$ **49.** $-(x - 3)(x - 4)$ **51.** $(3x - 2)(2x + 5)$ **53.** $(x^2 + 9)(x^4 - 9x^2 + 81)$ **55.** $(3x^3 + 5y^5)(9x^6 - 15x^3y^5 + 25y^{10})$ **57.** $(x - 14)(x + 11)$ **59.** $(x + 10)^2$ **61.** $3(x + 5)(2x + 15)$ **63.** 30 **65.** 38, 38 **67.** $7x - 5z^2$

Quick Review **1.** $\{7\}$ **2.** $\{-3\}$ **3.** $\left\{\frac{12}{5}\right\}$ **4.** $\left\{-\frac{29}{2}\right\}$

Quick Check 6.6 **1.** a) $\{2, -8\}$ b) $\left\{0, -\frac{9}{4}\right\}$ **2.** $\{5, -4\}$ **3.** $\{-7, -8\}$ **4.** $\left\{\frac{7}{2}, \frac{1}{3}\right\}$ **5.** $\{-9, 5\}$ **6.** $\{-7, 10\}$ **7.** $\{3, -9\}$ **8.** $x^2 - 2x - 48 = 0$

Section 6.6 **1.** quadratic equation **3.** zero-factor **5.** $\{-7, 100\}$ **7.** $\{0, 12\}$ **9.** $\{4, 8\}$ **11.** $\left\{-\frac{7}{2}, 5\right\}$ **13.** $\{1, 4\}$ **15.** $\{-5, -6\}$ **17.** $\{-10, 4\}$ **19.** $\{15, -3\}$ **21.** $\{8\}$ **23.** $\{9, -9\}$ **25.** $\{0, -7\}$ **27.** $\left\{0, \frac{22}{5}\right\}$ **29.** $\{-11, 7\}$ **31.** $\{1, 13\}$ **33.** $\{-14, 3\}$ **35.** $\left\{-\frac{2}{9}, \frac{1}{3}\right\}$ **37.** $\{7, -5\}$ **39.** $\{4\}$ **41.** $\{-2, -5\}$ **43.** $\{14, -2\}$ **45.** $\{3, -3\}$ **47.** $\{6, -6\}$ **49.** $\{8, -8\}$ **51.** $\{-5, -8\}$ **53.** $\{10, -3\}$ **55.** $\{-17\}$ **57.** $\left\{-\frac{3}{2}, 9\right\}$ **59.** $\left\{\frac{5}{4}, 4\right\}$ **61.** $x^2 - 9x + 20 = 0$ **63.** $x^2 - 6x = 0$ **65.** $x^2 - 64 = 0$ **67.** $21x^2 - 40x - 21 = 0$ **69.** $x = -\frac{14}{3}$ **71.** $x = -29$ **73.** $x = \frac{28}{5}$ **75.** Answers will vary. **77.** Answers will vary.

Quick Review **1.** 23 **2.** 101 **3.** $-14b - 40$ **4.** 58

Quick Check 6.7 **1.** 495 **2.** 8, 9 **3.** −6, 12 **4. a)** 48 feet **b)** 3 seconds **c)** 1 second **d)** 64 feet

Section 6.7 **1.** quadratic function **3.** 37 **5.** −7 **7.** 0 **9.** 0 **11.** 147 **13.** 17 **15.** 160 **17.** $9a^2 + 39a + 30$ **19.** $a^2 − 5a − 35$ **21.** −10, 6 **23.** −6, −7 **25.** 9, 14 **27.** 7, −7 **29.** −8, 2 **31.** −5, −8 **33. a)** 5 seconds **b)** 336 feet **35. a)** 192 feet **b)** 5 seconds **c)** 1 second **d)** 256 feet **37. a)** $70 **b)** 5 mugs **c)** 20 mugs **d)** $80 **39.** Answers will vary.

Quick Check 6.8 **1.** 11 and 12 **2.** 9 and 12 **3.** Length: 15 feet, Width: 7 feet **4.** 11 feet by 14 feet **5. a)** $h(t) = −16t^2 + 144t$ **b)** 9 seconds **c)** $[0, 9]$ **6. a)** 5050 **b)** 9

Section 6.8 **1.** $x, x + 1, x + 2$ **3.** Area = Length · Width **5.** 13, 14 **7.** 16, 18 **9.** −9, −11 **11.** 8, 9 **13.** 20, 22 **15.** 6, 11 **17.** 6, 14 **19.** 12, 20 **21.** 3, 14 **23.** 8, 11 **25.** 14 **27.** 21 **29.** Length: 8 feet, Width: 5 feet **31.** Length: 36 feet, Width: 12 feet **33.** Length: 15 m, Width: 7 m **35.** Length: 40 feet, Width: 24 feet **37.** Length: 8 inches, Width: 5 inches **39.** Length: 9 feet, Width: 5 feet **41.** 8 meters **43.** Base: 8 feet, Height: 21 feet **45.** 10 seconds **47.** 5 seconds **49.** 1 second, 3 seconds **51.** 10 seconds **53.** 128 feet/second **55.** 3 seconds **57.** 8 **59.** 20 **61.** Answers will vary. **63.** Answers will vary.

Chapter 6 Review **1.** $3(x − 7)$ **2.** $x(x − 20)$ **3.** $5x^3(x^4 + 3x − 4)$ **4.** $a^3b(6a^2b − 10b^2 + 15a)$ **5.** $(x + 6)(x^2 + 4)$ **6.** $(x + 9)(x^2 − 7)$ **7.** $(x − 3)(x^2 − 11)$ **8.** $(x − 3)^2(x + 3)$ **9.** $(x − 4)^2$ **10.** $(x + 7)(x − 4)$ **11.** $(x − 3)(x − 8)$ **12.** $−5(x + 8)(x − 4)$ **13.** $(x + 5y)(x + 9y)$ **14.** Prime **15.** $(x + 2)(x + 15)$ **16.** $(x − 7)(x + 6)$ **17.** $(x − 8)(10x − 3)$ **18.** $(4x + 5)(x − 2)$ **19.** $6(3x − 1)(x + 1)$ **20.** $(2x − 3)^2$ **21.** $3(x + 5)(x − 5)$ **22.** $(2x + 5y)(2x − 5y)$ **23.** $(x + 2y)(x^2 − 2xy + 4y^2)$ **24.** Prime **25.** $(x − 6)(x^2 + 6x + 36)$ **26.** $7(x − 3)(x^2 + 3x + 9)$ **27.** $\{5, −7\}$ **28.** $\{\frac{8}{3}, −\frac{1}{2}\}$ **29.** $\{5, −5\}$ **30.** $\{4, 10\}$ **31.** $\{−8\}$ **32.** $\{−12, 9\}$ **33.** $\{14, −5\}$ **34.** $\{0, 16\}$ **35.** $\{−4, −15\}$ **36.** $\{8, 10\}$ **37.** $\{16, −1\}$ **38.** $\{\frac{15}{2}, −4\}$ **39.** $\{6, −6\}$ **40.** $\{3, 12\}$ **41.** $\{−3, −7\}$ **42.** $\{−13, 4\}$ **43.** $x^2 − 13x + 42 = 0$ **44.** $x^2 − 16 = 0$ **45.** $5x^2 − 7x − 6 = 0$ **46.** $28x^2 + 27x − 10 = 0$ **47.** 9 **48.** −16 **49.** 8 **50.** 96 **51.** 6, −6 **52.** −9, 6 **53.** −2, −10 **54.** 11, −4 **55. a)** 80 feet **b)** 4 seconds **c)** 1 second **d)** 144 feet **56. a)** $20.85 **b)** 400 **c)** $4.85 **57.** 16, 18 **58.** Rosa: 7, Dale: 15 **59.** Length: 13 m, Width: 8 m **60. a)** 8 seconds **b)** 2 seconds, 3 seconds

Chapter 6 Review Exercises: Worked-Out Solutions

3. GCF: $5x^3$
$5x^7 + 15x^4 − 20x^3 = 5x^3(x^4 + 3x − 4)$

5. $x^3 + 6x^2 + 4x + 24 = x^2(x + 6) + 4(x + 6)$
  $= (x + 6)(x^2 + 4)$

10. $m \cdot n = −28, m + n = 3$
$x^2 + 3x − 28 = (x + 7)(x − 4)$

11. $m \cdot n = 24, m + n = −11$
$x^2 − 11x + 24 = (x − 3)(x − 8)$

15. $m \cdot n = 30, m + n = 17$
$x^2 + 17x + 30 = (x + 2)(x + 15)$

17. By grouping:
$10x^2 − 83x + 24 = 10x^2 − 80x − 3x + 24$
$= 10x(x − 8) − 3(x − 8)$
$= (x − 8)(10x − 3)$

21. $3x^2 − 75 = 3(x^2 − 25)$
  $= 3(x + 5)(x − 5)$

23. $x^3 + 8y^3 = (x)^3 + (2y)^3$
  $= (x + 2y)(x^2 − 2xy + 4y^2)$

25. $x^3 − 216 = (x)^3 − (6)^3$
  $= (x − 6)(x^2 + 6x + 36)$

30. $x^2 − 14x + 40 = 0$
$(x − 4)(x − 10) = 0$
$x − 4 = 0$ or $x − 10 = 0$
$x = 4$ or $x = 10$
$\{4, 10\}$

41. $x(x + 10) = −21$
$x^2 + 10x = −21$
$x^2 + 10x + 21 = 0$
$(x + 3)(x + 7) = 0$
$x + 3 = 0$ or $x + 7 = 0$
$x = −3$ or $x = −7$
$\{−3, −7\}$

43. $\{6, 7\}$
$x = 6$ or $x = 7$
$x − 6 = 0$ or $x − 7 = 0$
$(x − 6)(x − 7) = 0$
$x^2 − 13x + 42 = 0$

49. $f(−8) = (−8)^2 + 10(−8) + 24$
  $= 64 − 80 + 24$
  $= 8$

51. $f(x) = 0$
$x^2 − 36 = 0$
$(x + 6)(x − 6) = 0$
$x + 6 = 0$ or $x − 6 = 0$
$x = −6$ or $x = 6$

56. a) $f(800) = 0.0001(800)^2 − 0.08(800) + 20.85$
  $= 0.0001(640,000) − 0.08(800) + 20.85$
  $= 64 − 64 + 20.85$
  $= 20.85$

b) The lowest point on the graph occurs when the number of toys is 400.

c) $f(400) = 0.0001(400)^2 − 0.08(400) + 20.85$
  $= 0.0001(160,000) − 0.08(400) + 20.85$
  $= 16 − 32 + 20.85$
  $= 4.85$

57. First: x
Second: $x + 2$
$$x(x + 2) = 288$$
$$x^2 + 2x = 288$$
$$x^2 + 2x - 288 = 0$$
$$(x + 18)(x - 16) = 0$$
$$x + 18 = 0 \quad \text{or} \quad x - 16 = 0$$
$$x = -18 \quad \text{or} \quad x = 16$$
Omit the negative solution.
First: $x = 16$
Second: $x + 2 = 16 + 2 = 18$
The two numbers are 16 and 18.

59. Length: $2x - 3$
Width: x
$$x(2x - 3) = 104$$
$$2x^2 - 3x = 104$$
$$2x^2 - 3x - 104 = 0$$
$$(2x + 13)(x - 8) = 0$$
$$2x + 13 = 0 \quad \text{or} \quad x - 8 = 0$$
$$x = -\tfrac{13}{2} \quad \text{or} \quad x = 8$$
Omit the negative solution.
Length: $2x - 3 = 2(8) - 3 = 13$
Width: $x = 8$
The length is 13 meters and the width is 8 meters.

Chapter 6 Test **1.** $x(x - 25)$ [6.1] **2.** $(x - 3)(x^2 - 5)$ [6.1] **3.** $(x - 4)(x - 5)$ [6.2] **4.** $(x - 18)(x + 4)$ [6.2]
5. $3(x + 1)(x + 19)$ [6.2] **6.** $(6x + 5)(x - 1)$ [6.3] **7.** $(x + 5y)(x - 5y)$ [6.4] **8.** $(x - 6y)(x^2 + 6xy + 36y^2)$ [6.4]
9. $\{10, -10\}$ [6.6] **10.** $\{4, 9\}$ [6.6] **11.** $\{-12, 5\}$ [6.6] **12.** $\{-6, 13\}$ [6.6] **13.** $x^2 - 4 = 0$ [6.6] **14.** $5x^2 + 21x - 54 = 0$ [6.6]
15. 120 [6.7] **16.** -18 [6.7] **17.** $-6, -8$ [6.7] **18. a)** \$2.245 **b)** 8000 **c)** \$2.11 [6.7] **19.** Length: 20 feet, Width: 5 feet [6.8]
20. a) 5 seconds **b)** 3 seconds [6.8]

Cumulative Review Chapters 4–6 **1.** $(5, 2)$ **2.** $(3, 7)$ **3.** $(1, 2)$ **4.** $(-1, 10)$ **5.** 250 children **6.** Length: 40 feet, Width: 25 feet
7. \$11,000 at 3%, \$4000 at 2.5% **8.**   **9.** x^7 **10.** $108m^{16}n^{21}$ **11.** $\frac{625a^{24}b^{44}}{c^{28}}$ **12.** t^{21} **13.** $\frac{a^{48}}{b^{56}}$ **14.** $\frac{1}{x^{35}}$

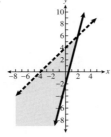

15. 3.3×10^{11} **16.** 7.14×10^{-13} **17.** 0.001 second **18.** -44 **19.** $2x^2 - 7x - 59$ **20.** $-x^2 - 12x + 22$ **21.** 67
22. $15x^3 - 24x^2 + 36x$ **23.** $12x^2 - 53x + 56$ **24.** $5x^7 + 12x^4 - 9x$ **25.** $x + 7 - \frac{3}{x + 6}$ **26.** $(x - 4)(x^2 + 6)$ **27.** $2(x - 5)(x - 8)$
28. $(x + 9)(x - 4)$ **29.** $(2x - 5)(x + 2)$ **30.** $(x - 5)(x^2 + 5x + 25)$ **31.** $(x + 9)(x - 9)$ **32.** $\{-7, 7\}$ **33.** $\{3, 8\}$ **34.** $\{-7\}$
35. $\{-10, 2\}$ **36.** $x^2 - 5x - 14 = 0$ **37.** $-10, 10$ **38.** $-7, 5$ **39. a)** 240 feet **b)** 6 seconds **c)** 2 seconds **d)** 256 feet
40. 12, 14 **41.** Length: 19 meters, Width: 7 meters

Chapter 7

Quick Check 7.1 **1.** $\frac{1}{3}$ **2.** $\frac{7}{2}$ **3.** $-1, 9$ **4.** $\frac{5x}{7}$ **5.** $\frac{x + 9}{x + 3}$ **6.** $\frac{x + 4}{x - 8}$ **7.** $\frac{3x + 1}{x + 3}$ **8.** $-\frac{x + 9}{1 + x}$ **9.** -2 **10.** All real numbers except 9 and -5.
$(-\infty, -5) \cup (-5, 9) \cup (9, \infty)$

Section 7.1 **1.** rational expression **3.** lowest terms **5.** rational function **7.** $\frac{3}{4}$ **9.** $\frac{2}{3}$ **11.** $\frac{26}{7}$ **13.** $-\frac{75}{26}$ **15.** 3 **17.** 0, 8
19. $-2, 8$ **21.** $-7, 7$ **23.** $\frac{1}{3x^3}$ **25.** $\frac{1}{x + 6}$ **27.** $\frac{x + 2}{x - 2}$ **29.** $\frac{x + 4}{x - 7}$ **31.** $\frac{x - 9}{x + 6}$ **33.** $\frac{x + 8}{x - 4}$ **35.** $\frac{3x + 5}{x + 7}$ **37.** $\frac{3(x + 5)}{x - 4}$ **39.** Not opposites
41. Opposites **43.** Opposites **45.** $-\frac{3}{5 + x}$ **47.** $-\frac{x + 6}{9 + x}$ **49.** $\frac{3 - x}{x + 9}$ **51.** $\frac{1}{3}$ **53.** $-\frac{8}{7}$ **55.** $\frac{21}{16}$ **57.** All real numbers except 0 and -10.
59. All real numbers except 5 and -3. **61.** All real numbers except 3 and -6. **63.** Quadratic **65.** Rational **67.** Linear **69. a)** 4
b) -1 **71. a)** 6 **b)** 0, 5 **73.** Answers will vary. **75.** Answers will vary

Quick Review 7.1 **1.** $\frac{41}{60}$ **2.** $\frac{49}{120}$ **3.** $\frac{1}{6}$ **4.** $\frac{68}{231}$

Quick Check 7.2 **1.** $\frac{(x - 6)(x + 2)}{(x - 4)(x + 10)}$ **2.** $-\frac{6 + x}{x + 7}$ **3.** $-\frac{1}{7 + x}$ **4.** $\frac{(x - 2)(x + 7)}{(x - 1)(x - 8)}$ **5.** $-(x + 5)$ **6.** $\frac{x - 6}{(x - 1)(x - 10)(x - 2)}$

Section 7.2 **1.** factoring **3.** factored **5.** $\frac{x + 5}{(x - 2)(x - 3)}$ **7.** $\frac{(x + 6)(x + 1)}{(x + 9)(x - 2)}$ **9.** $\frac{x(x + 2)}{(x - 9)(x + 10)}$ **11.** $\frac{(x - 11)(2x + 1)}{(x - 4)(x + 2)}$ **13.** $\frac{(x - 11)(x + 9)}{(x + 4)(x + 6)}$

15. $\frac{2(x+8)(x+9)}{3(x-1)(x+2)}$ **17.** $-\frac{(2+x)(x+4)}{(x-6)(x+9)}$ **19.** 1 **21.** $-\frac{x(2x-9)}{(x+10)(x-6)}$ **23.** $\frac{2}{x+4}$ **25.** $\frac{(x-11)(x+5)}{(x+12)(x-4)}$ **27.** $\frac{(x-2)(x+11)}{(x+4)(x+9)}$ **29.** $\frac{x-7}{x-2}$

31. $\frac{(x-3)(x+11)}{(2-x)(x+2)}$ **33.** $\frac{x-6}{3x+1}$ **35.** $\frac{(x+5)(x-3)}{(x-7)(x+10)}$ **37.** $\frac{x-6}{x+10}$ **39.** $\frac{12(x-2)}{x+9}$ **41.** $\frac{(x-1)(x-4)}{(x+5)(x+6)}$ **43.** 1 **45.** $-\frac{(x+7)(x-10)}{(x-7)(9+x)}$ **47.** $\frac{x+4}{4x+1}$

49. $-\frac{(x+7)(x-3)}{x(x-8)}$ **51.** $\frac{x-1}{(x-5)^2}$ **53.** $\frac{(x-6)(x-8)}{x(x-3)}$ **55.** $\frac{x(x-9)}{(x-1)(x+3)}$ **57.** $\frac{x-6}{x-3}$ **59.** $\frac{(x-9)(x+4)}{x(x-12)}$ **61.** Answers will vary.

Quick Review 7.2 **1.** $9x+3$ **2.** $2x^2-6x-29$ **3.** $3x-42$ **4.** $-2x^2-2x-29$

Quick Check 7.3 **1.** $\frac{21}{2x+7}$ **2.** $\frac{5}{x-4}$ **3.** $\frac{x+4}{x-3}$ **4.** $\frac{4}{3(x+6)}$ **5.** 3 **6.** $-\frac{x-3}{8+x}$ **7.** $\frac{1}{5}$ **8.** $\frac{x+9}{x+4}$

Section 7.3 **1.** numerators **3.** first term **5.** $\frac{13}{x+3}$ **7.** $\frac{x+10}{x-5}$ **9.** 2 **11.** $\frac{1}{x-2}$ **13.** $\frac{3}{x+11}$ **15.** $\frac{x+9}{x-8}$ **17.** $\frac{2(x-12)}{x-6}$ **19.** $\frac{8}{x}$ **21.** $\frac{x+2}{x-9}$

23. $\frac{4}{x+1}$ **25.** $\frac{x-7}{x+7}$ **27.** 4 **29.** 2 **31.** $\frac{1}{x-10}$ **33.** $\frac{2}{x-4}$ **35.** $\frac{x-10}{x+6}$ **37.** $\frac{x-8}{x+8}$ **39.** $\frac{(x-3)(x-9)}{(x-5)(x+3)}$ **41.** $-\frac{11}{x+8}$ **43.** $\frac{(x-7)(x+2)}{(x-2)(x-9)}$ **45.** 2

47. $x+11$ **49.** 4 **51.** $\frac{x+3}{2}$ **53.** $\frac{x-4}{x-3}$ **55.** $\frac{x+8}{x}$ **57.** $\frac{2(x+4)}{x-9}$ **59.** $\frac{x+2}{x-4}$ **61.** $\frac{3(x-10)}{x+4}$ **63.** $\frac{2(3x+2)}{3x-2}$ **65.** $\frac{x+9}{x-9}$ **67.** $\frac{x-3}{x+2}$ **69.** $3x$

71. $4x+7$ **73.** Answers will vary.

Quick Check 7.4 **1.** $36r^2s^6$ **2.** $(x-8)(x-5)(x+4)$ **3.** $(x+9)^2(x-9)$ **4.** $\frac{11x+18}{(x-2)(x+6)}$ **5.** $\frac{x+5}{(x-5)(x+3)}$ **6.** $\frac{2(x-1)}{(x+2)(x-6)}$

Section 7.4 **1.** LCD **3.** $6a$ **5.** $(x-9)(x+5)$ **7.** $(x+3)(x-3)(x-6)$ **9.** $(x+2)^2(x+5)$ **11.** $\frac{11x}{12}$ **13.** $\frac{14}{15a}$ **15.** $\frac{4m^2+5n^2}{m^4n^3}$

17. $\frac{2(4x+13)}{(x+2)(x+4)}$ **19.** $\frac{6(x-7)}{(x+3)(x-3)}$ **21.** $\frac{9x}{(x+4)(x-2)}$ **23.** $\frac{7}{(x+5)(x-2)}$ **25.** $\frac{4}{(x+1)(x-3)}$ **27.** $\frac{2}{(x+3)(x-3)}$ **29.** $\frac{14}{(x-10)(x+4)}$

31. $-\frac{1}{(x-5)(x-4)}$ **33.** $\frac{6}{(x+2)(x-4)}$ **35.** $\frac{x^2+9x+27}{(x+3)(x-3)(x+4)}$ **37.** $\frac{x^2-3x-12}{(x-5)(x+4)(x-6)}$ **39.** $\frac{x^2-x+3}{(x+8)(x-3)(x+3)}$ **41.** $\frac{x^2+11x+110}{(x+10)(x-5)(x+5)}$

43. $\frac{x^2+7x+70}{x(x+7)(x+10)}$ **45.** $\frac{x^2+12x+132}{(x+11)(x-6)(x+9)}$ **47.** 3 **49.** 4 **51.** Answers will vary.

Quick Check 7.5 **1.** $\frac{31}{38}$ **2.** $\frac{x+5}{x}$ **3.** $\frac{x+8}{8x}$ **4.** $\frac{11}{x+7}$ **5.** $\frac{(x-8)(x-2)}{(x+10)(x+1)}$

Section 7.5 **1.** complex fraction **3.** $\frac{3}{68}$ **5.** $\frac{39}{38}$ **7.** $\frac{7(5x+3)}{5(7x+4)}$ **9.** $\frac{6}{x}$ **11.** $\frac{2}{3}$ **13.** $\frac{2x}{x-4}$ **15.** $\frac{4}{x+5}$ **17.** $\frac{12}{x}$ **19.** $\frac{x-4}{x+5}$ **21.** $\frac{2(x-2)}{x+2}$

23. $\frac{x-6}{x-5}$ **25.** $\frac{x(x-4)}{(x-8)(x+9)}$ **27.** $\frac{x-7}{x+10}$ **29.** $\frac{x-1}{x-7}$ **31.** $\frac{x^2+6x+14}{(x-6)(x+4)(x-2)}$ **33.** $\frac{x(x+2)}{(3x-4)(x+5)}$ **35.** $\frac{x+6}{x(x+7)}$ **37.** 5 **39.** $\frac{5}{x-3}$

41. $\frac{2x-19}{(x+10)(x-8)(x-3)}$ **43.** $\frac{8x}{x-8}$ **45.** $\frac{12(x+5)}{(x+3)(x+4)(x+6)}$ **47.** $\frac{x+11}{x-8}$ **49.** $\frac{x+6}{x+1}$ **51.** $\frac{x(x+5)}{(x+11)(3x-7)}$ **53.** Answers will vary.

Quick Check 7.6 **1.** $\{20\}$ **2.** $\{-3,8\}$ **3.** $\varnothing$ **4.** $\{-3,6\}$ **5.** $\{4\}$ **6.** $x=\frac{2yz}{4y-3z}$ **7.** $x=\frac{16(y-1)}{3y-4}$

Section 7.6 **1.** rational equation **3.** extraneous solution **5.** $\{\frac{75}{4}\}$ **7.** $\{\frac{1}{16}\}$ **9.** $\{\frac{72}{13}\}$ **11.** $\{\frac{80}{21}\}$ **13.** $\{3,4\}$ **15.** $\{2,6\}$ **17.** $\{3\}$

19. $\varnothing$ **21.** $\{2\}$ **23.** $\varnothing$ **25.** $\{7\}$ **27.** $\{16\}$ **29.** $\{4\}$ **31.** $\{-7,5\}$ **33.** $\{0,9\}$ **35.** $\{2\}$ **37.** $\varnothing$ **39.** $\{-2,5\}$ **41.** $\{\frac{7}{2},10\}$

43. $\{5,18\}$ **45.** $\{1\}$ **47.** $\{-1,15\}$ **49.** $W=\frac{A}{L}$ **51.** $x=-\frac{5y}{2y-1}$ **53.** $r=\frac{2x+y}{2}$ **55.** $b=\frac{ac}{a-c}$ **57.** $x=\frac{y-y_1+mx_1}{m}$

59. $n_1=\frac{n_2}{nn_2-1}$ **61. a)** 116 **b)** $x=-15$ **63.** Answers will vary. **65.** Answers will vary.

Quick Check 7.7 **1.** 10 **2.** 3 and 12 **3.** $12\frac{8}{11}$ minutes **4.** 24 hours **5.** 75 mph **6.** 6 mph **7.** $360 **8.** $163\frac{3}{4}$ minutes **9.** 864

Section 7.7 **1.** reciprocal **3.** divided **5.** vary directly **7.** 5 **9.** 11 **11.** 10 **13.** 6, 9 **15.** 10, 20 **17.** 6 minutes

19. $51\frac{3}{7}$ minutes **21.** $\frac{30}{31}$ hours or $58\frac{2}{31}$ minutes **23.** 28 hours **25.** 18 hours **27.** 6 hours **29.** 6 hours **31.** 10 mph **33.** 6 mph

35. 40 kilometers/hour **37.** 150 mph **39.** 80 **41.** 90 **43.** 14 **45.** $\frac{16}{3}$ **47.** 294 **49.** 108 **51.** $272 **53.** 3 amperes **55.** $15

57. 18 amperes **59.** $7\frac{1}{2}$ foot-candles **61.** Answers will vary. **63.** Answers will vary.

Chapter 7 Review **1.** $-\frac{1}{5}$ **2.** $\frac{3}{8}$ **3.** $\frac{5}{2}$ **4.** $\frac{3}{2}$ **5.** -9 **6.** $-6,9$ **7.** $\frac{1}{x-3}$ **8.** $\frac{x+10}{x-5}$ **9.** $-\frac{6+x}{x-1}$ **10.** $\frac{x+3}{4x+3}$ **11.** $\frac{5}{9}$ **12.** $\frac{17}{22}$ **13.** All real

numbers except -6 and 0. **14.** All real numbers except 1 and 4. **15.** $\frac{x}{x+5}$ **16.** $\frac{(x+9)(x+8)}{(x-7)(x-2)}$ **17.** $\frac{x-4}{x-6}$ **18.** $-\frac{(4+x)(7x-1)}{x+6}$

19. $\frac{(x+1)(x+5)}{(x-1)(x+3)}$ **20.** $\frac{(x+8)(7-x)}{(x-6)(x+3)}$ **21.** $\frac{x+6}{x-4}$ **22.** $\frac{(x-3)(x+12)}{(x-11)(x+8)}$ **23.** $\frac{(x+1)(x-1)}{x(3x+2)}$ **24.** $-\frac{(3+x)(x+2)}{x-5}$ **25.** $\frac{x(x-6)}{(x-9)(x+5)}$

26. $\frac{(x+2)(x+3)}{(x+1)(x-10)}$ **27.** $\frac{12}{x+6}$ **28.** $\frac{x-3}{x+7}$ **29.** $\frac{x+9}{x+4}$ **30.** 6 **31.** $x+6$ **32.** $\frac{x-6}{x-5}$ **33.** $\frac{7}{(x+4)(x-3)}$ **34.** $\frac{5}{(x-2)(x-7)}$ **35.** $\frac{x+2}{(x+5)(x-4)}$

36. $\frac{x-3}{(x+3)(x+9)}$ **37.** $\frac{2(x+4)}{(x+6)(x-2)}$ **38.** $\frac{x+5}{(x+1)(x-5)}$ **39.** $\frac{x+3}{(x-1)(x+7)}$ **40.** $\frac{x}{x+9}$ **41.** $\frac{x-11}{x+11}$ **42.** $\frac{x+7}{x-2}$ **43.** $\frac{(x-3)(x+3)}{(x-2)(x+6)}$ **44.** $\{\frac{72}{13}\}$

45. $\{3,4\}$ **46.** $\{16\}$ **47.** $\{-8,5\}$ **48.** $\{2\}$ **49.** $\{-3\}$ **50.** $\{-2,5\}$ **51.** $\{-2,-1\}$ **52.** $h=\frac{2A}{b}$ **53.** $x=\frac{7y}{4y-3}$ **54.** $r=\frac{15x-10y}{6}$

55. 8 **56.** 5, 10 **57.** 20 minutes **58.** 84 minutes **59.** 60 mph **60.** 6 mph **61.** 130 calories **62.** 225 feet **63.** 675 pounds **64.** 80 foot-candles

Chapter 7 Review Exercises: Worked-Out Solutions

4. $\dfrac{(-4)^2 - 5(-4) - 15}{(-4)^2 + 17(-4) + 66} = \dfrac{16 + 20 - 15}{16 - 68 + 66}$

$= \dfrac{21}{14}$

$= \dfrac{3}{2}$

6. $x^2 - 3x - 54 = 0$

$(x + 6)(x - 9) = 0$

$x = -6 \quad \text{or} \quad x = 9$

8. $\dfrac{x^2 + 6x - 40}{x^2 - 9x + 20} = \dfrac{(x + 10)(x - 4)}{(x - 4)(x - 5)}$

$= \dfrac{x + 10}{x - 5}$

11. $r(3) = \dfrac{(3) + 7}{(3)^2 + 8(3) - 15}$

$= \dfrac{3 + 7}{9 + 24 - 15}$

$= \dfrac{10}{18}$

$= \dfrac{5}{9}$

13. $x^2 + 6x = 0$

$x(x + 6) = 0$

$x = 0 \quad \text{or} \quad x = -6$

All real numbers except -6 and 0.

16. $\dfrac{x^2 + 14x + 45}{x^2 - 6x - 7} \cdot \dfrac{x^2 + 9x + 8}{x^2 + 3x - 10} = \dfrac{(x + 5)(x + 9)}{(x - 7)(x + 1)} \cdot \dfrac{(x + 1)(x + 8)}{(x + 5)(x - 2)}$

$= \dfrac{(x + 9)(x + 8)}{(x - 7)(x - 2)}$

19. $f(x) \cdot g(x) = \dfrac{x - 6}{x - 1} \cdot \dfrac{x^2 + 6x + 5}{x^2 - 3x - 18}$

$= \dfrac{x - 6}{x - 1} \cdot \dfrac{(x + 1)(x + 5)}{(x - 6)(x + 3)}$

$= \dfrac{(x + 1)(x + 5)}{(x - 1)(x + 3)}$

22. $\dfrac{x^2 - x - 6}{x^2 - 19x + 88} \div \dfrac{x^2 + 10x + 16}{x^2 + 4x - 96} = \dfrac{(x - 3)(x + 2)}{(x - 11)(x - 8)} \cdot \dfrac{(x + 12)(x - 8)}{(x + 8)(x + 2)}$

$= \dfrac{(x - 3)(x + 12)}{(x - 11)(x + 8)}$

28. $\dfrac{x^2 + 10x}{x^2 + 15x + 56} - \dfrac{5x + 24}{x^2 + 15x + 56} = \dfrac{x^2 + 10x - 5x - 24}{x^2 + 15x + 56}$

$= \dfrac{x^2 + 5x - 24}{x^2 + 15x + 56}$

$= \dfrac{(x + 8)(x - 3)}{(x + 7)(x + 8)}$

$= \dfrac{x - 3}{x + 7}$

32. $\dfrac{x^2 - 3x - 12}{x^2 - 25} - \dfrac{2x - 18}{25 - x^2} = \dfrac{x^2 - 3x - 12}{x^2 - 25} + \dfrac{2x - 18}{x^2 - 25}$

$= \dfrac{x^2 - 3x - 12 + 2x - 18}{x^2 - 25}$

$= \dfrac{x^2 - x - 30}{x^2 - 25}$

$= \dfrac{(x - 6)(x + 5)}{(x + 5)(x - 5)}$

$= \dfrac{x - 6}{x - 5}$

33. $\dfrac{5}{x^2 + 3x - 4} + \dfrac{2}{x^2 - 4x + 3} = \dfrac{5}{(x + 4)(x - 1)} + \dfrac{2}{(x - 3)(x - 1)}$

$= \dfrac{5}{(x + 4)(x - 1)} \cdot \dfrac{x - 3}{x - 3} + \dfrac{2}{(x - 3)(x - 1)} \cdot \dfrac{x + 4}{x + 4}$

$= \dfrac{5x - 15 + 2x + 8}{(x + 4)(x - 1)(x - 3)}$

$= \dfrac{7x - 7}{(x + 4)(x - 1)(x - 3)}$

$= \dfrac{7(x - 1)}{(x + 4)(x - 1)(x - 3)}$

$= \dfrac{7}{(x + 4)(x - 3)}$

39. $f(x) - g(x) = \dfrac{x + 2}{x^2 + 4x - 5} - \dfrac{1}{x^2 + 12x + 35}$

$= \dfrac{x + 2}{(x + 5)(x - 1)} - \dfrac{1}{(x + 5)(x + 7)}$

$= \dfrac{x + 2}{(x + 5)(x - 1)} \cdot \dfrac{x + 7}{x + 7} - \dfrac{1}{(x + 5)(x + 7)} \cdot \dfrac{x - 1}{x - 1}$

$= \dfrac{(x^2 + 9x + 14) - (x - 1)}{(x + 5)(x - 1)(x + 7)}$

$= \dfrac{x^2 + 9x + 14 - x + 1}{(x + 5)(x - 1)(x + 7)}$

$= \dfrac{x^2 + 8x + 15}{(x + 5)(x - 1)(x + 7)}$

$= \dfrac{(x + 3)(x + 5)}{(x + 5)(x - 1)(x + 7)}$

$= \dfrac{(x + 3)}{(x - 1)(x + 7)}$

42. $\dfrac{1 + \frac{2}{x} - \frac{35}{x^2}}{1 - \frac{7}{x} + \frac{10}{x^2}} = \dfrac{1 + \frac{2}{x} - \frac{35}{x^2}}{1 - \frac{7}{x} + \frac{10}{x^2}} \cdot \dfrac{x^2}{x^2}$

$= \dfrac{1 \cdot x^2 + \frac{2}{x} \cdot x^2 - \frac{35}{x^2} \cdot x^2}{1 \cdot x^2 - \frac{7}{x} \cdot x^2 + \frac{10}{x^2} \cdot x^2}$

$= \dfrac{x^2 + 2x - 35}{x^2 - 7x + 10}$

$= \dfrac{(x + 7)(x - 5)}{(x - 5)(x - 2)}$

$= \dfrac{x + 7}{x - 2}$

45. $x - 5 + \dfrac{12}{x} = 2$

$x \cdot x - 5 \cdot x + \dfrac{12}{x} \cdot x = 2 \cdot x$

$x^2 - 5x + 12 = 2x$

$x^2 - 7x + 12 = 0$

$(x - 3)(x - 4) = 0$

$x = 3 \quad \text{or} \quad x = 4$

$\{3, 4\}$

49. $\dfrac{x}{x + 5} + \dfrac{3}{x - 7} = \dfrac{36}{x^2 - 2x - 35}$

$\dfrac{x}{x + 5} + \dfrac{3}{x - 7} = \dfrac{36}{(x + 5)(x - 7)}$

$(x + 5)(x - 7) \cdot \dfrac{x}{x + 5} + (x + 5)(x - 7) \cdot \dfrac{3}{x - 7} = (x + 5)(x - 7) \cdot \dfrac{36}{(x + 5)(x - 7)}$

$x(x - 7) + 3(x + 5) = 36$

$x^2 - 7x + 3x + 15 = 36$

$x^2 - 4x + 15 = 36$

$x^2 - 4x - 21 = 0$

$(x + 3)(x - 7) = 0$

$x = -3 \quad \text{or} \quad x = 7$

$x = 7$ is an extraneous solution.

$\{-3\}$

53. $y = \dfrac{3x}{4x - 7}$

$y(4x - 7) = 3x$

$4xy - 7y = 3x$

$4xy - 3x = 7y$

$x(4y - 3) = 7y$

$x = \dfrac{7y}{4y - 3}$

58.

Pipe	Time alone	Work-rate	Time Working	Portion Completed
Smaller	$t + 42$ minutes	$\frac{1}{t+42}$	28 minutes	$\frac{28}{t+42}$
Larger	t minutes	$\frac{1}{t}$	28 minutes	$\frac{28}{t}$

$$\frac{28}{t+42} + \frac{28}{t} = 1$$
$$t(t+42) \cdot \frac{28}{t+42} + t(t+42) \cdot \frac{28}{t} = 1 \cdot t(t+42)$$
$$28t + 28(t+42) = t(t+42)$$
$$28t + 28t + 1176 = t^2 + 42t$$
$$56t + 1176 = t^2 + 42t$$
$$0 = t^2 - 14t - 1176$$
$$0 = (t-42)(t+28)$$

$t = 42$ or $t = -28$ (Omit negative solution.)
It would take the smaller pipe $42 + 42 = 84$ minutes.

60. Rate upstream: $r - 2$, Rate downstream: $r + 2$
$$\frac{10}{r+2} = \frac{5}{r-2}$$
$$(r-2)(r+2) \cdot \frac{10}{r+2} = (r-2)(r+2) \cdot \frac{5}{r-2}$$
$$10(r-2) = 5(r+2)$$
$$10r - 20 = 5r + 10$$
$$5r = 30$$
$$r = 6$$

Dominique can paddle 6 mph in still water.

Chapter 7 Test **1.** $-\frac{5}{6}$ **2.** $-3, -7$ **3.** $\frac{x-5}{x+10}$ **4.** -13 **5.** $\frac{x-1}{x-3}$ **6.** $-\frac{(x-5)^2}{(x+3)(5+x)}$ **7.** 3 **8.** $x + 8$ **9.** $\frac{11}{(x+7)(x-4)}$
10. $\frac{x-6}{(x-5)(x-2)}$ **11.** $\frac{x+9}{x+7}$ **12.** $\frac{x-5}{(x+9)(x+11)}$ **13.** $\frac{(x+5)(x+9)}{x(x+6)}$ **14.** $\frac{x+6}{x+2}$ **15.** $\frac{(x-10)(x+3)}{(x-1)(x+8)(x+2)}$ **16.** $\{36\}$
17. $\{-2, 9\}$ **18.** $y = \frac{5x}{7x-2}$ **19.** 90 minutes **20.** 20 mph

Chapter 8

Quick Check 8.1 **1. a)** 5 **b)** $\frac{1}{9}$ **c)** -7 **2. a)** $|b^{13}|$ **b)** $7|a^3|$ **3.** $4x^{10}$ **4.** 13.229 **5. a)** 10 **b)** 6 **c)** -8 **6.** x^7 **7.** $-4x^4y^{10}$
8. $14b^5$ **9.** 198 **10.** 6 **11.** 16 **12.** $[-3, \infty)$

Section 8.1 **1.** square root **3.** nth root **5.** radicand **7.** radical function **9.** 4 **11.** 3 **13.** $\frac{1}{10}$ **15.** $\frac{7}{8}$ **17.** -5
19. Not a real number **21.** a^{10} **23.** $|x^{11}|$ **25.** $3|x^5|$ **27.** $\frac{3}{8}x^4$ **29.** $|x^{13}y^7|$ **31.** $a^8b^4c^6$ **33.** 49 **35.** $25a^2$ **37.** 8.367 **39.** 15.716
41. 0.424 **43.** 4 **45.** 3 **47.** x^5 **49.** $-5x^{11}$ **51.** $4m^5n^6$ **53.** 4 **55.** 20 **57.** 6 **59.** 9 **61.** 4 **63.** 15 **65.** 2 **67.** 42 **69.** 84
71. 72 **73.** 1350 **75.** 3 **77.** 0.083 **79.** 9 **81.** $[5, \infty)$ **83.** $[\frac{13}{4}, \infty)$ **85.** $[0, \infty)$ **87.** $(-\infty, \infty)$ **89.** Real: $-\sqrt{64}, -\sqrt[3]{64}, \sqrt[3]{-64}$;
Not real: $\sqrt{-64}$; Explanations will vary. **91.** Answers will vary.

Quick Check 8.2 **1.** $x^4\sqrt[5]{x^3}$ **2.** $x^2z^5w^6\sqrt{xyw}$ **3.** $5\sqrt{3}$ **4.** $3\sqrt[3]{22}$ **5.** $x^7y^2z^2\sqrt[6]{xy^5}$ **6.** $27\sqrt{2xy}$ **7.** $-10\sqrt{3} - 19\sqrt{5}$
8. $3\sqrt{7}$ **9.** $-2\sqrt{3}$

Section 8.2 **1.** like radicals **3.** $5\sqrt{3}$ **5.** $7\sqrt{2}$ **7.** $12\sqrt{5}$ **9.** $3\sqrt[3]{2}$ **11.** $4\sqrt[3]{6}$ **13.** $3\sqrt[4]{21}$ **15.** $x^3\sqrt{x}$ **17.** $a^7\sqrt[3]{a^2}$
19. $m^8\sqrt[4]{m^2}$ **21.** $x^6y^{10}\sqrt{x}$ **23.** $b^5\sqrt{a}$ **25.** $x^8y^3z^{15}\sqrt{xz}$ **27.** $a^5b^3c\sqrt[3]{bc^2}$ **29.** $x^4y^6z^{11}\sqrt[5]{x^3y^2}$ **31.** $3x^3y^3\sqrt{2x}$ **33.** $2a^5b^3\sqrt[3]{6a^2b}$
35. $17\sqrt{2}$ **37.** $14\sqrt[3]{6}$ **39.** $5\sqrt{10}$ **41.** $6\sqrt{14} - 15\sqrt{21}$ **43.** $5\sqrt{5} + 11\sqrt{2}$ **45.** $18\sqrt{a}$ **47.** $-11\sqrt[5]{x}$ **49.** $-6\sqrt{a} + 7\sqrt{b}$
51. $17\sqrt{3}$ **53.** $-22\sqrt{2}$ **55.** $29\sqrt[3]{2}$ **57.** $-18\sqrt{2} + 12\sqrt{6}$ **59.** $8\sqrt{5} + 4\sqrt{2}$ **61.** $-4\sqrt{3} - 13\sqrt{2}$ **63.** Answers will vary; one
example is $2\sqrt{3} + 4\sqrt{3}$. **65.** Answers will vary; one example is $8\sqrt{5} + \sqrt{45} + \sqrt{3} - 3\sqrt{3}$. **67.** Answers will vary.

Quick Check 8.3 **1. a)** 10 **b)** $a^3b^5c^5\sqrt[3]{bc}$ **c)** $72\sqrt{14}$ **2.** $12\sqrt{5} + 6\sqrt{3}$ **3.** $x^5y^6\sqrt[3]{y} - x^7y^8\sqrt[3]{x^2}$ **4. a)** $128 + 35\sqrt{10}$
b) $13 + 2\sqrt{42}$ **5.** 529 **6.** $\frac{\sqrt{42}}{2}$ **7.** $\frac{2\sqrt{5}}{3}$ **8.** $\frac{\sqrt{6ab}}{3a^3b^2}$ **9.** $\frac{7\sqrt{6} + 3\sqrt{14}}{4}$ **10.** $\frac{10 + \sqrt{2}}{7}$

Section 8.3 **1.** x **3.** rationalizing **5.** 30 **7.** 10 **9.** 6 **11.** 320 **13.** x^{11} **15.** $m^9 n^{11}\sqrt{n}$ **17.** $x^5 y^7 \sqrt[4]{x^2 y^3}$ **19.** $5\sqrt{3} + 5\sqrt{2}$
21. $15\sqrt{6} + 6\sqrt{3}$ **23.** $14\sqrt{77} - 6\sqrt{33}$ **25.** $m^5 n^5 \sqrt{n} + m^5 n^{10}\sqrt{m}$ **27.** $a^4 b^4 c^9 \sqrt[3]{a} + a^8 b^8 c^9 \sqrt[3]{c}$ **29.** $3 - \sqrt{6} + \sqrt{21} - \sqrt{14}$
31. 4 **33.** $23\sqrt{6} - 3$ **35.** $13 + 67\sqrt{3}$ **37.** $210 - 56\sqrt{14}$ **39.** 9 **41.** 97 **43.** -471 **45.** $\frac{9}{5}$ **47.** $\frac{3\sqrt{2}}{x^5 y^4}$ **49.** $2\sqrt{3}$ **51.** $\frac{2\sqrt{2}}{3}$
53. $\frac{3\sqrt{10}}{2}$ **55.** $\frac{\sqrt{6}}{6}$ **57.** $\frac{2\sqrt{10a}}{5a^3}$ **59.** $\frac{s^4 \sqrt[3]{9r^2 st}}{3r^2 t^2}$ **61.** $\frac{3\sqrt{5}}{5}$ **63.** $\frac{b^2 \sqrt{ac}}{a^3 c^5}$ **65.** $\frac{3\sqrt{15} + 3\sqrt{11}}{2}$ **67.** $\frac{-7 - 7\sqrt{3}}{2}$ **69.** $\frac{15\sqrt{77} + 30\sqrt{7}}{7}$ **71.** $\frac{66 + 19\sqrt{6}}{73}$
73. -2 **75.** $\frac{31 + 4\sqrt{15}}{14}$ **77.** $\frac{2\sqrt{2}}{7}$ **79.** $85 + 10\sqrt{30}$ **81.** $3a^5 b^3 c^8 \sqrt[3]{3a^2}$ **83.** $\frac{3 - \sqrt{35}}{13}$ **85.** $280 + 168\sqrt{2} - 30\sqrt{7} - 18\sqrt{14}$
87. $30x^7 \sqrt{3x} - 5x^{10}\sqrt{6x}$ **89.** $\frac{a^6 b^5 \sqrt{7bc}}{21c^4}$ **91.** $3x^3 \sqrt[3]{2x^2}$ **93.** 712 **95.** $a + 2\sqrt{ab} + b$ **97.** Answers will vary; one example is
$\sqrt{21} + \sqrt{6}$ and $\sqrt{21} - \sqrt{6}$. **99.** Answers will vary; one example is $3\sqrt{2} + 4\sqrt{3}$. **101.** Answers will vary.

Quick Review Exercises 8.3 **1.** $2^4 \cdot 3^2$ **2.** $2^3 \cdot 5^2 \cdot 7$ **3.** 2^{10} **4.** $2 \cdot 3^3 \cdot 7^2 \cdot 11$

Quick Check 8.4 **1. a)** $\{41\}$ **b)** $\{-127\}$ **2.** $\varnothing$ **3.** 37 **4.** $\{3, 4\}$ **5.** $\{12\}$ **6.** $\{16\}$ **7.** $\{9\}$

Section 8.4 **1.** radical **3.** isolate **5.** $\{46\}$ **7.** $\{14\}$ **9.** $\{131\}$ **11.** $\varnothing$ **13.** $\{96\}$ **15.** $\{22\}$ **17.** $\{80\}$ **19.** $\varnothing$ **21.** $\{-11, 2\}$
23. 17 **25.** 42 **27.** 2, 15 **29.** $\{2, 5\}$ **31.** $\{3\}$ **33.** $\{12\}$ **35.** $\varnothing$ **37.** $\{\frac{7}{2}\}$ **39.** $\{4\}$ **41.** $\{11\}$ **43.** $\{7\}$ **45.** $\{-6, \frac{1}{2}\}$
47. $\{7\}$ **49.** $\{10\}$ **51.** $\{263\}$ **53.** Answers will vary; one example is $\sqrt{2x + 13} = 5$. **55.** Answers will vary.

Quick Check 8.5 **1.** 2.48 seconds **2.** 12.97 feet **3.** 82 miles per hour

Section 8.5 **1.** pendulum **3.** 1.92 seconds **5.** 1.36 seconds **7.** 0.8 ft **9.** 1.60 seconds **11.** 4.01 seconds **13.** 56 mph
15. 76 mph **17.** 250 feet **19.** 158.4 gallons/minute **21.** 125.2 gallons/minute **23.** 6.4 feet **25.** 7.3 knots **27.** 35.6 feet
29. 689 mph **31.** 2.6 miles **33.** 19.8 feet **35.** Answers will vary.

Quick Check 8.6 **1. a)** $9i$ **b)** $2i\sqrt{11}$ **2. a)** $3 + 14i$ **b)** $-4 - 11i$ **3.** -56 **4.** $-10\sqrt{7}$ **5.** $-14 - 12i$ **6.** $34 - 19i$ **7.** $91 - 60i$
8. 85 **9.** $\frac{15 - 9i}{17}$ or $\frac{15}{17} - \frac{9}{17}i$ **10.** $\frac{11 + 13i}{10}$ or $\frac{11}{10} + \frac{13}{10}i$ **11.** $\frac{7 - 6i}{5}$ or $\frac{7}{5} - \frac{6}{5}i$

Section 8.6 **1.** imaginary **3.** -1 **5.** real **7.** $3i$ **9.** $7i$ **11.** $-13i$ **13.** $4i\sqrt{3}$ **15.** $9i\sqrt{3}$ **17.** $14 + 7i$ **19.** $-17 + 2i$
21. $5 + i$ **23.** $-8i$ **25.** $17 - 20i$ **27.** $7 - 3i$ **29.** $-9 + 9i$ **31.** -18 **33.** 72 **35.** -15 **37.** $-15\sqrt{2}$ **39.** -70 **41.** $6i \cdot 9i$;
Answers may vary. **43.** $-2i \cdot 16i$; Answers may vary. **45.** $20 + 12i$ **47.** $6 + 16i$ **49.** $17 - 11i$ **51.** $21 - 20i$ **53.** 74
55. $19 + 180i$ **57.** 25 **59.** 221 **61.** $\frac{18 - 6i}{5}$ or $\frac{18}{5} - \frac{6}{5}i$ **63.** $\frac{-21 + 6i}{53}$ or $-\frac{21}{53} + \frac{6}{53}i$ **65.** $\frac{23 - 2i}{41}$ or $\frac{23}{41} - \frac{2}{41}i$ **67.** $\frac{23 - 11i}{25}$ or $\frac{23}{25} - \frac{11}{25}i$
69. $-5i$ **71.** $\frac{-9 - 4i}{6}$ or $-\frac{3}{2} - \frac{2}{3}i$ **73.** $\frac{-1 + 5i}{4}$ or $-\frac{1}{4} + \frac{5}{4}i$ **75.** $\frac{-5 + 7i}{37}$ or $-\frac{5}{37} + \frac{7}{37}i$ **77.** $\frac{1 - 13i}{5}$ or $\frac{1}{5} - \frac{13}{5}i$ **79.** $2 - 5i$ **81.** $14 - 8i$
83. $62 + 5i$ **85.** $\frac{2 - 11i}{3}$ or $\frac{2}{3} - \frac{11}{3}i$ **87.** $\frac{33 + 12i}{137}$ or $\frac{33}{137} + \frac{12}{137}i$ **89.** 106 **91.** $-20\sqrt{3}$ **93.** $\frac{-30 + 54i}{53}$ or $-\frac{30}{53} + \frac{54}{53}i$ **95.** $5i\sqrt{6}$
97. $39 + 80i$ **99.** $11 - 6i$ **101.** 56 **103.** $-39 + 13i$ **105.** Answers will vary.

Quick Check 8.7 **1. a)** $\sqrt{81} = 9$ **b)** $\sqrt[3]{-27x^{15}} = -3x^5$ **2.** $\sqrt[4]{a^{40} b^{64} c^{24}} = a^{10} b^{16} c^6$ **3.** $a^{1/8}$ **4.** $(\sqrt[6]{64x^{30}})^5 = 32x^{25}$ **5.** $x^{5/14}$
6. $x^{11/9}$ **7.** $x^{6/5}$ **8.** $\frac{x^6}{y^2 z^4}$ **9.** $\frac{1}{3125}$

Section 8.7 **1.** $\sqrt[n]{a}$ **3.** $x^{1/6}$ **5.** $10^{1/2}$ **7.** $4x^{1/5}$ **9.** 6 **11.** 2 **13.** -6 **15.** x^7 **17.** $a^6 b^5$ **19.** $3x^4 y^6$ **21.** $-3x^3 y^4 z^5$ **23.** $x^{4/7}$
25. $a^{9/10}$ **27.** $x^{3/8}$ **29.** $(5x^2)^{8/3}$ **31.** 27 **33.** 100,000 **35.** x^{12} **37.** $125a^9 b^{12}$ **39.** $81x^{20} y^{28} z^4$ **41.** $x^{1/2}$ **43.** $x^{11/12}$ **45.** $x^{43/20}$
47. $a^{13/12}$ **49.** $x^{13/14} y^{23/20}$ **51.** $x^{1/6}$ **53.** $m^{9/25}$ **55.** $x^{21/4}$ **57.** $x^{1/2}$ **59.** $x^{11/24}$ **61.** 1 **63.** 1 **65.** $\frac{1}{2}$ **67.** $\frac{1}{4}$ **69.** $\frac{1}{2187}$ **71.** $\frac{1}{12}$ **73.** $\frac{1}{625}$
75. $\frac{1}{256}$ **77.** Answers will vary; one example is $9^{1/2}$. **79.** Answers will vary; one example is $(49x^8)^{1/2}$. **81.** Yes; explanations will vary.

Chapter 8 Review **1.** 9 **2.** 14 **3.** -5 **4.** x^5 **5.** $5x^9$ **6.** $4x^7 y^6 z^{10}$ **7.** 7.746 **8.** 15.460 **9.** 8 **10.** 14.832 **11.** $[3, \infty)$
12. $[-\frac{7}{3}, \infty)$ **13.** $3\sqrt{6}$ **14.** $x^5 \sqrt[3]{x^2}$ **15.** $a^5 b^4 c^3 \sqrt[4]{a^3 b^2}$ **16.** $2x^8 y^{11} \sqrt[3]{55xy^2}$ **17.** $5x^{16} y^7 \sqrt{6xz}$ **18.** $3r^7 s^{10} t^{19}\sqrt{7rt}$ **19.** $16\sqrt{5}$
20. $8\sqrt[3]{x}$ **21.** $67\sqrt{3}$ **22.** $58\sqrt{2} - 40\sqrt{3}$ **23.** $3x^3 \sqrt[3]{10x}$ **24.** $126\sqrt{3} - 56\sqrt{5}$ **25.** $48 + 7\sqrt{21}$ **26.** $-33 + 3\sqrt{10}$
27. $290 + 88\sqrt{6}$ **28.** -3 **29.** $2x^4 \sqrt{21}$ **30.** $\frac{3a^3 b^3 \sqrt{b}}{2c^7}$ **31.** $\frac{2\sqrt{15}}{5}$ **32.** $\frac{5\sqrt{6}}{3}$ **33.** $\frac{\sqrt{35}}{10}$ **34.** $x^6 y^2 z^5 \sqrt{yz}$ **35.** $\frac{3(\sqrt{13} - \sqrt{5})}{4}$ **36.** $\frac{18\sqrt{3} + 9\sqrt{7}}{5}$
37. $\frac{83 + 23\sqrt{10}}{123}$ **38.** $\frac{50 + 28\sqrt{3}}{37}$ **39.** $\{7\}$ **40.** $\{\frac{73}{8}\}$ **41.** $\{4, 6\}$ **42.** $\{9\}$ **43.** $\{11\}$ **44.** $\{8\}$ **45.** 12 **46.** $-3, 10$ **47.** 2.36 seconds
48. 85 mph **49.** $4i$ **50.** $2i\sqrt{14}$ **51.** $21i\sqrt{2}$ **52.** $-5i\sqrt{13}$ **53.** $12 + 14i$ **54.** $4 - 3i$ **55.** $-4 - 11i$ **56.** $1 + 10i$ **57.** -36
58. $-6\sqrt{7}$ **59.** $-15 + 40i$ **60.** $53 + 53i$ **61.** $9 + 40i$ **62.** 73 **63.** $\frac{10 - 6i}{17}$ or $\frac{10}{17} - \frac{6}{17}i$ **64.** $\frac{-8 + 6i}{5}$ or $-\frac{8}{5} + \frac{6}{5}i$ **65.** $\frac{31 - 27i}{65}$ or
$\frac{31}{65} - \frac{27}{65}i$ **66.** $-8i$ **67.** $x^{1/3}$ **68.** $a^{5/8}$ **69.** $m^{11/2}$ **70.** $x^{9/4}$ **71.** 3 **72.** $8x^{12}$ **73.** $x^{5/2}$ **74.** $x^{3/5}$ **75.** $x^{1/5}$ **76.** $\frac{1}{6}$ **77.** $\frac{1}{81}$ **78.** $\frac{1}{64}$

Chapter 8 Review Exercises: Worked-Out Solutions

4. $\sqrt[4]{x^{20}} = \sqrt[4]{(x^5)^4}$ **9.** $f(x) = \sqrt{3x + 4}$ **11.** $5x - 15 \geq 0$ **15.** $\sqrt[4]{a^{23} b^{18} c^{12}} = \sqrt[4]{a^{20} b^{16} c^{12}} \cdot \sqrt[4]{a^3 b^2}$
$\qquad = x^5$ $\qquad f(20) = \sqrt{3(20) + 4}$ $\qquad 5x \geq 15$ $\qquad = a^5 b^4 c^3 \sqrt[4]{a^3 b^2}$
$\qquad\qquad\qquad = \sqrt{64}$ $\qquad x \geq 3$
$\qquad\qquad\qquad = 8$ $\qquad [3, \infty)$

21. $9\sqrt{12} + 7\sqrt{147} = 9\sqrt{4 \cdot 3} + 7\sqrt{49 \cdot 3}$
$\qquad\qquad = 9 \cdot 2\sqrt{3} + 7 \cdot 7\sqrt{3}$
$\qquad\qquad = 18\sqrt{3} + 49\sqrt{3}$
$\qquad\qquad = 67\sqrt{3}$

25. $(2\sqrt{7} + \sqrt{3})(3\sqrt{7} + 2\sqrt{3}) = 2\sqrt{7} \cdot 3\sqrt{7} + 2\sqrt{7} \cdot 2\sqrt{3} + \sqrt{3} \cdot 3\sqrt{7} + \sqrt{3} \cdot 2\sqrt{3}$
$\qquad\qquad\qquad\qquad\qquad\quad = 6 \cdot 7 + 4\sqrt{21} + 3\sqrt{21} + 2 \cdot 3$
$\qquad\qquad\qquad\qquad\qquad\quad = 42 + 7\sqrt{21} + 6$
$\qquad\qquad\qquad\qquad\qquad\quad = 48 + 7\sqrt{21}$

33. $\sqrt{\frac{7}{20}} = \frac{\sqrt{7}}{\sqrt{20}} \cdot \frac{\sqrt{5}}{\sqrt{5}}$
$\qquad = \frac{\sqrt{35}}{\sqrt{100}}$
$\qquad = \frac{\sqrt{35}}{10}$

36. $\frac{9\sqrt{2}}{2\sqrt{6} - \sqrt{14}} = \frac{9\sqrt{2}}{2\sqrt{6} - \sqrt{14}} \cdot \frac{2\sqrt{6} + \sqrt{14}}{2\sqrt{6} + \sqrt{14}}$
$\qquad\qquad = \frac{18\sqrt{12} + 9\sqrt{28}}{4 \cdot 6 - 14}$
$\qquad\qquad = \frac{18 \cdot 2\sqrt{3} + 9 \cdot 2\sqrt{7}}{24 - 14}$
$\qquad\qquad = \frac{36\sqrt{3} + 18\sqrt{7}}{10}$
$\qquad\qquad = \frac{18\sqrt{3} + 9\sqrt{7}}{5}$

39. $\sqrt{2x + 11} = 5$
$(\sqrt{2x + 11})^2 = 5^2$
$2x + 11 = 25$
$2x = 14$
$x = 7$
$\{7\}$

42. $\sqrt{2x + 31} = x - 2$
$(\sqrt{2x + 31})^2 = (x - 2)^2$
$2x + 31 = (x - 2)(x - 2)$
$2x + 31 = x^2 - 4x + 4$
$0 = x^2 - 6x - 27$
$0 = (x - 9)(x + 3)$
$x = 9$ or $x = -3$
$x = -3$ is an extraneous solution.
$\{9\}$

48. $s = \sqrt{30df}$
$s = \sqrt{30(320)(0.75)}$
$s = \sqrt{7200}$
$s \approx 85$

50. $\sqrt{-56} = \sqrt{-1} \cdot \sqrt{56}$
$\qquad\quad = i \cdot \sqrt{4 \cdot 14}$
$\qquad\quad = i \cdot 2\sqrt{14}$
$\qquad\quad = 2i\sqrt{14}$

53. $(9 + 4i) + (3 + 10i) = 9 + 3 + 4i + 10i$
$\qquad\qquad\qquad\qquad = 12 + 14i$

60. $(9 - 5i)(2 + 7i) = 18 + 63i - 10i - 35i^2$
$\qquad\qquad\qquad\quad = 18 + 53i + 35$
$\qquad\qquad\qquad\quad = 53 + 53i$

63. $\frac{4}{5 + 3i} = \frac{4}{5 + 3i} \cdot \frac{5 - 3i}{5 - 3i}$
$\qquad\quad = \frac{20 - 12i}{25 - 9i^2}$
$\qquad\quad = \frac{20 - 12i}{25 + 9}$
$\qquad\quad = \frac{20 - 12i}{34}$
$\qquad\quad = \frac{10 - 6i}{17}$ or $\frac{10}{17} - \frac{6}{17}i$

66. $\frac{8}{i} = \frac{8}{i} \cdot \frac{i}{i}$
$\qquad = \frac{8i}{i^2}$
$\qquad = \frac{8i}{-1}$
$\qquad = -8i$

72. $(32x^{20})^{3/5} = (\sqrt[5]{32x^{20}})^3$
$\qquad\qquad\quad = (2x^4)^3$
$\qquad\qquad\quad = 8x^{12}$

73. $x^{7/3} \cdot x^{1/6} = x^{7/3 + 1/6}$
$\qquad\qquad = x^{14/6 + 1/6}$
$\qquad\qquad = x^{15/6}$
$\qquad\qquad = x^{5/2}$

Chapter 8 Test **1.** $6a^7b^6$ **2.** 8.775 **3.** $2\sqrt{22}$ **4.** $x^5y^3z^8\sqrt[5]{x^4y^2}$ **5.** $29\sqrt{2}$ **6.** $84\sqrt{2} - 63\sqrt{5}$ **7.** $81 - 34\sqrt{6}$ **8.** $a^3b^2\sqrt{a}$
9. $-\frac{84 + 29\sqrt{30}}{39}$ **10.** $\{171\}$ **11.** $\{7\}$ **12.** 8.3 feet **13.** $2i\sqrt{17}$ **14.** $-19 - 13i$ **15.** 109 **16.** $\frac{3 + 5i}{2}$ or $\frac{3}{2} + \frac{5}{2}i$ **17.** $x^{4/5}$ **18.** $n^{11/24}$
19. $\frac{1}{36}$

Chapter 9

Quick Check 9.1 **1. a)** $\{3, 6\}$ **b)** $\{-10, 10\}$ **2. a)** $\{-3\sqrt{2}, 3\sqrt{2}\}$ **b)** $\{-2i\sqrt{6}, 2i\sqrt{6}\}$ **3. a)** $\{-\frac{1}{3}, 5\}$ **b)** $\{\frac{1 - 2\sqrt{3}}{2}, \frac{1 + 2\sqrt{3}}{2}\}$
4. a) $\{-5, -3\}$ **b)** $\{3 - i\sqrt{11}, 3 + i\sqrt{11}\}$ **5.** $\{\frac{-5 - 3i\sqrt{3}}{2}, \frac{-5 + 3i\sqrt{3}}{2}\}$ **6.** $\{4 - 2\sqrt{2}, 4 + 2\sqrt{2}\}$ **7.** $4x^2 + 25x - 21 = 0$
8. $x^2 + 4 = 0$

Section 9.1 **1.** extracting square roots **3.** completing the square **5.** $\{-2, 9\}$ **7.** $\{2, 7\}$ **9.** $\{-7, -1\}$ **11.** $\{-5, 10\}$
13. $\{-7, 7\}$ **15.** $\{5, 11\}$ **17.** $\{-5, 5\}$ **19.** $\{\pm 4\sqrt{5}\}$ **21.** $\{\pm 6i\}$ **23.** $\{\pm 4\sqrt{2}\}$ **25.** $\{-5, 13\}$ **27.** $\{-5 \pm 3i\sqrt{2}\}$
29. $\{-2 \pm 4i\sqrt{3}\}$ **31.** $\{5, 15\}$ **33.** $\{-1, \frac{3}{2}\}$ **35.** $x^2 + 10x + 25 = (x + 5)^2$ **37.** $x^2 - 9x + \frac{81}{4} = (x - \frac{9}{2})^2$
39. $x^2 - 12x + 36 = (x - 6)^2$ **41.** $x^2 + \frac{1}{6}x + \frac{1}{144} = (x + \frac{1}{12})^2$ **43.** $\{-4, 10\}$ **45.** $\{-1 \pm 2i\sqrt{3}\}$ **47.** $\{-8, 2\}$ **49.** $\{6, 8\}$
51. $\{-6, -3\}$ **53.** $\{\frac{-5 \pm \sqrt{73}}{2}\}$ **55.** $\{-2, \frac{3}{2}\}$ **57.** $\{4, 8\}$ **59.** $\{3 \pm 2i\}$ **61.** $x^2 - 2x - 8 = 0$ **63.** $x^2 - 13x + 30 = 0$
65. $4x^2 - 13x + 3 = 0$ **67.** $x^2 - 25 = 0$ **69.** $x^2 + 49 = 0$ **71.** $\{-2, 15\}$ **73.** $\{-1 \pm 2\sqrt{10}\}$ **75.** $\{-12, 1\}$ **77.** $\{2 \pm 2\sqrt{2}\}$
79. $\{2 \pm 4\sqrt{2}\}$ **81.** $\{-7, 5\}$ **83.** $\{-5 \pm i\sqrt{15}\}$ **85.** $\{-5 \pm 3i\sqrt{3}\}$ **87.** $\{\frac{-5 \pm \sqrt{77}}{2}\}$ **89.** $\{-7, \frac{3}{2}\}$ **91.** $\{-1, 6\}$ **93.** $\{-6, 4\}$
95. $-4, 9$ **97.** $\pm 2\sqrt{6}$ **99.** $-2, 8$ **101. a)** 89 **b)** 33 **103. a)** Answers will vary; any number greater than -47 will work.
b) Answers will vary; any number less than -47 will work. **105.** Answers will vary.

Quick Check 9.2 **1. a)** $a = 1$, $b = 10$, and $c = -24$ **b)** $a = 4$, $b = -5$, and $c = -11$ **c)** $a = 1$, $b = 0$, and $c = 25$ **2. a)** $\{3, 5\}$
b) $\{\frac{-13 - \sqrt{41}}{4}, \frac{-13 + \sqrt{41}}{4}\}$ **3.** $\{\frac{-5 - 3i\sqrt{3}}{2}, \frac{-5 + 3i\sqrt{3}}{2}\}$ **4.** $\{-4 - \sqrt{7}, -4 + \sqrt{7}\}$ **5.** $\{-\frac{1}{5}, \frac{4}{3}\}$ **6. a)** Two complex solutions
b) Two real solutions **c)** One real solution **7. a)** Not factorable **b)** Factorable

Section 9.2 **1.** quadratic **3.** zero **5.** two **7.** $\{-5, -2\}$ **9.** $\{3 \pm \sqrt{5}\}$ **11.** $\{3 \pm i\sqrt{6}\}$ **13.** $\{0, 5\}$ **15.** $\{-9, 3\}$ **17.** $\{-8, -6\}$ **19.** $\{-6, 6\}$ **21.** $\{-5, 4\}$ **23.** $\{\frac{5 \pm \sqrt{109}}{2}\}$ **25.** $\{-3, \frac{7}{2}\}$ **27.** $\{-4\}$ **29.** $\{\pm 6\sqrt{2}\}$ **31.** $\{-8, 4\}$ **33.** $\{-6, \frac{3}{2}\}$ **35.** $\{4, 7\}$ **37.** Two real **39.** Two complex **41.** One real **43.** Two complex **45.** Prime **47.** Factorable **49.** Factorable **51.** Factorable **53.** $\{\frac{3 \pm \sqrt{77}}{2}\}$ **55.** $\{\frac{2}{3}, 5\}$ **57.** $\{-2, 9\}$ **59.** $\{\frac{12 \pm i\sqrt{3}}{3}\}$ **61.** $\{\pm 16i\}$ **63.** $\{-\frac{1}{6}, 4\}$ **65.** $\{\frac{5 \pm 4\sqrt{2}}{3}\}$ **67.** $\{\frac{7 \pm \sqrt{177}}{2}\}$ **69.** $\{\frac{7}{3}\}$ **71.** $\{\frac{-3 \pm i\sqrt{151}}{2}\}$ **73.** $\{-1 \pm \sqrt{10}\}$ **75.** $\{2 \pm 5i\}$ **77.** $\{\frac{-3 \pm \sqrt{33}}{4}\}$ **79.** $\{-\frac{4}{5}, 13\}$ **81.** $\{-1, 9\}$ **83.** $\{-3, 2\}$ **85.** $\{7 \pm \sqrt{15}\}$ **87.** $\{2, 9\}$ **89.** $x = -\frac{b}{a}$ **91.** Answers will vary.

Quick Check 9.3 **1.** Base: 11 inches, Height: 4 inches **2.** 15.7 inches **3.** 7.4 feet **4.** 94 feet **5. a)** 6 seconds **b)** 2 seconds, 4 seconds **6. a)** 0.46 seconds, 2.04 seconds **b)** No

Section 9.3 **1.** hypotenuse **3.** Pythagorean **5.** 6 meters **7.** 7.1 feet **9.** 5.2 inches **11.** Length: 14 inches, Width: 8 inches **13.** Length: 18 inches, Width: 6 inches **15.** Length: 16.6 inches, Width: 4.9 inches **17.** Base: 12 inches, Height: 9 inches **19.** Base: 5.5 inches, Height: 14.0 inches **21.** Height: 10 inches, Width: 8 inches **23.** Length: 7 feet, Width: 3.5 feet **25.** Length: 35.9 inches, Width: 15.9 inches **27.** Base: 30 inches, Height: 15 inches **29.** Base: 7.1 feet, Height: 14.1 feet **31.** 13 inches **33.** 16.1 centimeters **35.** 75 miles **37.** 12 feet **39.** 27.6 feet **41.** 19.5 inches **43.** Length: 6.1 feet, Width: 5.1 feet **45. a)** Base: 5 feet, Height: 12 feet **b)** 13 feet **47. a)** 77.9 centimeters **b)** 3505.5 square centimeters **49.** 15 seconds **51.** 4.8 seconds **53.** 2 seconds and 4 seconds **55.** 1.8 seconds and 6.4 seconds **57.** No, the equation $-16t^2 + 20t = 30$ has no solution. **59.** Answers will vary. **61.** Answers will vary.

Quick Check 9.4 **1.** 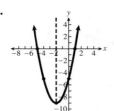 **2.** 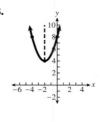 **3.** **4.**

5. 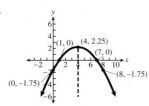 **6. a)** $(-9, -4)$, $x = -9$ **b)** $(1, 6)$, $x = 1$ **7. a)** $y = (x + 5)^2 - 55$, Vertex: $(-5, -55)$ **b)** $y = -(x - 8)^2 + 47$, Vertex: $(8, 47)$ **8.**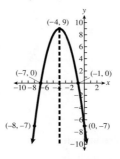

Section 9.4 **1.** parabola **3.** negative **5.** y-intercept **7.** $(4, -53)$, $x = 4$ **9.** $(-8, -1)$, $x = -8$ **11.** $(-5, 75)$, $x = -5$ **13.** $(\frac{5}{2}, \frac{59}{4})$, $x = \frac{5}{2}$ **15.** $(-3, -59)$, $x = -3$ **17.** $(0, -10)$, $x = 0$ **19.** $(6, -1)$, $x = 6$ **21.** $(-9, -18)$, $x = -9$ **23.** $(-6, 0), (-3, 0), (0, 18)$ **25.** No x-intercepts, $(0, -9)$ **27.** $(-1.9, 0), (4.4, 0), (0, -17)$ **29.** $(-8, 0), (0, 0)$ **31.** $(6, 0), (0, 36)$ **33.** $(-2, 0), (-8, 0), (0, 16)$ **35.** No x-intercepts, $(0, -13)$

37. Vertex: $(-1, -4)$,
y-intercept: $(0, -3)$,
x-intercepts: $(-3, 0), (1, 0)$

39. Vertex: $(3, 9)$,
y-intercept: $(0, 0)$,
x-intercepts: $(0, 0), (6, 0)$

41. Vertex: $\left(\frac{5}{2}, -\frac{11}{4}\right)$,
y-intercept: $(0, -9)$,
x-intercept: None

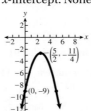

43. Vertex: $(-3, -33)$,
y-intercept: $(0, -15)$,
x-intercepts: $\left(\frac{-6 \pm \sqrt{66}}{2}, 0\right)$

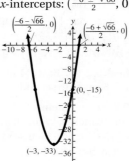

45. Vertex: $(-2, 5)$,
y-intercept: $(0, 9)$,
x-intercept: None

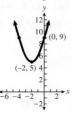

47. Vertex: $(4, 12)$,
y-intercept: $(0, -4)$,
x-intercepts: $(4 \pm 2\sqrt{3}, 0)$

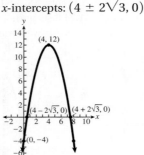

49. Vertex: $(-3, 0)$,
y-intercept: $(0, 9)$,
x-intercept: $(-3, 0)$

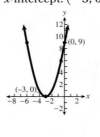

51. Vertex: $(0, -8)$,
y-intercept: $(0, -8)$,
x-intercepts: $(\pm 2\sqrt{2}, 0)$

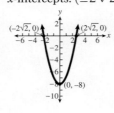

53. Vertex: $(3, -2)$,
y-intercept: $(0, \frac{5}{2})$,
x-intercepts: $(1, 0), (5, 0)$

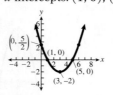

55. Vertex: $(-1, 5)$,
y-intercept: $(0, 6)$,
x-intercept: None

57. Vertex: $(-2, 7)$,
y-intercept: $(0, 3)$,
x-intercepts: $(-2 \pm \sqrt{7}, 0)$

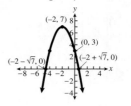

59. Answers will vary; examples include $y = (x - 3)^2 - 5$ or $y = x^2 - 6x + 4$.

61. Answers will vary; examples include $y = x^2 - 20x + 96$ or $y = (x - 10)^2 - 4$.

63. Answers will vary.

Chapter 9 Review   **1.** $\{-8, -7\}$ **2.** $\{-10, 6\}$ **3.** $\{-5, 0\}$ **4.** $\{\frac{2}{3}, 6\}$ **5.** $\{-6, 9\}$ **6.** $\{-9, 12\}$ **7.** $\{-8, 8\}$ **8.** $\{\pm 4\sqrt{3}\}$
9. $\{\pm 6i\}$ **10.** $\{-3 \pm 7i\}$ **11.** $\{4 \pm 3i\}$ **12.** $\{-2, 11\}$ **13.** $\{-4, 7\}$ **14.** $\{4 \pm 2\sqrt{3}\}$ **15.** $\{-5 \pm 2i\sqrt{10}\}$ **16.** $\{\frac{-3 \pm \sqrt{65}}{2}\}$
17. $\{\frac{-9 \pm \sqrt{61}}{2}\}$ **18.** $\{-2 \pm \sqrt{14}\}$ **19.** $\{\frac{5 \pm i\sqrt{7}}{2}\}$ **20.** $\frac{8 \pm \sqrt{7}}{3}$ **21.** $\{-5, 9\}$ **22.** $\{4 \pm 3i\}$ **23.** $\{6, 8\}$ **24.** $\{-9, 7\}$
25. $\{7 \pm 2i\sqrt{6}\}$ **26.** $\{-3 \pm 4i\sqrt{2}\}$ **27.** $\{-7, 0\}$ **28.** $\{4 \pm 2\sqrt{17}\}$ **29.** $\{\frac{9 \pm \sqrt{57}}{4}\}$ **30.** $\{-10 \pm 9i\}$ **31.** $\{-7 \pm 2\sqrt{5}\}$
32. $\{\frac{5 \pm \sqrt{137}}{2}\}$ **33.** $x^2 - 5x - 14 = 0$ **34.** $x^2 - 16x + 64 = 0$ **35.** $20x^2 + 7x - 6 = 0$ **36.** $x^2 - 75 = 0$ **37.** $x^2 + 1 = 0$
38. $x^2 + 28 = 0$ **39.** $(5, -81), x = 5$ **40.** $(-\frac{3}{2}, \frac{71}{4}), x = -\frac{3}{2}$ **41.** $(3, 51), x = 3$ **42.** $(8, -15), x = 8$ **43.** x-intercept: None,
y-intercept: $(0, 20)$ **44.** x-intercepts: $(3 \pm \sqrt{19}, 0)$, y-intercept: $(0, -10)$ **45.** x-intercepts: $(-7, 0), (5, 0)$, y-intercept: $(0, 35)$
46. x-intercepts: $(-8 \pm 2\sqrt{11}, 0)$, y-intercept: $(0, 20)$

47. Vertex: $(-5, -1)$, y-intercept: $(0, 24)$,
  x-intercepts: $(-6, 0)$, $(-4, 0)$

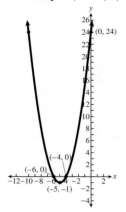

48. Vertex: $(-1, 9)$, y-intercept: $(0, 8)$,
  x-intercepts: $(-4, 0)$, $(2, 0)$

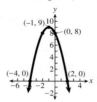

49. Vertex: $(2, -10)$, y-intercept: $(0, -6)$,
  x-intercept: $(2 \pm \sqrt{10}, 0)$

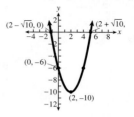

50. Vertex: $(1, 5)$, y-intercept: $(0, 6)$,
  x-intercept: None

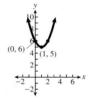

51. Vertex: $(-2, -4)$, y-intercept: $(0, 0)$,
  x-intercepts: $(-4, 0)$, $(0, 0)$

52. Vertex: $(4, 8)$, y-intercept: $(0, -8)$,
  x-intercept: $(4 \pm 2\sqrt{2}, 0)$

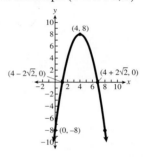

53. Height: 20 centimeters, Width: 28 centimeters   **54.** Length: 24.4 feet, Width: 14.4 feet   **55.** 948.7 miles
56. Length: 23.8 feet, Width: 14.8 feet   **57.** $7\frac{1}{4}$ seconds   **58.** 4.84 seconds

Chapter 9 Review Exercises: Worked-Out Solutions

2. $x^2 + 4x - 60 = 0$
  $(x + 10)(x - 6) = 0$
  $x = -10$ or $x = 6$
  $\{-10, 6\}$

8. $x^2 - 48 = 0$
  $x^2 = 48$
  $\sqrt{x^2} = \pm\sqrt{48}$
  $x = \pm 4\sqrt{3}$
  $\{\pm 4\sqrt{3}\}$

11. $(x - 4)^2 + 52 = 43$
  $(x - 4)^2 = -9$
  $\sqrt{(x - 4)^2} = \pm\sqrt{-9}$
  $x - 4 = \pm 3i$
  $x = 4 \pm 3i$
  $\{4 \pm 3i\}$

14. $x^2 - 8x + 4 = 0$
  $x^2 - 8x = -4$
  $x^2 - 8x + 16 = -4 + 16$
  $(x - 4)^2 = 12$
  $\sqrt{(x - 4)^2} = \pm\sqrt{12}$
  $x - 4 = \pm 2\sqrt{3}$
  $x = 4 \pm 2\sqrt{3}$
  $\{4 \pm 2\sqrt{3}\}$

18. $x = \dfrac{-4 \pm \sqrt{4^2 - 4(1)(-10)}}{2(1)}$
  $x = \dfrac{-4 \pm \sqrt{56}}{2}$
  $x = \dfrac{-4 \pm 2\sqrt{14}}{2}$
  $x = -2 \pm \sqrt{14}$
  $\{-2 \pm \sqrt{14}\}$

19. $x = \dfrac{5 \pm \sqrt{(-5)^2 - 4(1)(8)}}{2(1)}$
  $x = \dfrac{5 \pm \sqrt{-7}}{2}$
  $x = \dfrac{5 \pm i\sqrt{7}}{2}$
  $\{\frac{5 \pm i\sqrt{7}}{2}\}$

31. $f(x) = 0$
  $(x + 7)^2 - 20 = 0$
  $(x + 7)^2 = 20$
  $\sqrt{(x + 7)^2} = \pm\sqrt{20}$
  $x + 7 = \pm 2\sqrt{5}$
  $x = -7 \pm 2\sqrt{5}$
  $\{-7 \pm 2\sqrt{5}\}$

33. $x = -2$ or $x = 7$
  $x + 2 = 0$ or $x - 7 = 0$
  $(x + 2)(x - 7) = 0$
  $x^2 - 5x - 14 = 0$

39. $x = \dfrac{-(-10)}{2(1)} = \dfrac{10}{2} = 5$
  $y = 5^2 - 10(5) - 56$
  $= 25 - 50 - 56$
  $= -81$
  Vertex: $(5, -81)$
  Axis of symmetry: $x = 5$

44. x-intercepts

$x^2 - 6x - 10 = 0$

$x = \dfrac{6 \pm \sqrt{(-6)^2 - 4(1)(-10)}}{2(1)}$

$x = \dfrac{6 \pm \sqrt{76}}{2}$

$x = \dfrac{6 \pm 2\sqrt{19}}{2}$

$x = 3 \pm \sqrt{19}$

$(3 \pm \sqrt{19}, 0)$

y-intercept

$y = (0)^2 - 6(0) - 10 = -10$

$(0, -10)$

47. Vertex

$x = \dfrac{-(10)}{2(1)} = \dfrac{-10}{2} = -5$

$y = (-5)^2 + 10(-5) + 24$

$\quad = 25 - 50 + 24$

$\quad = -1$

$(-5, -1)$

y-intercept

$y = (0)^2 + 10(0) + 24 = 24$

$(0, 24)$

x-intercepts

$x^2 + 10x + 24 = 0$

$(x + 4)(x + 6) = 0$

$x = -4 \quad \text{or} \quad x = -6$

$(-4, 0), (-6, 0)$

48. Vertex

$x = \dfrac{-(-2)}{2(-1)} = \dfrac{2}{-2} = -1$

$y = -(-1)^2 - 2(-1) + 8$

$\quad = -1 + 2 + 8$

$\quad = 9$

$(-1, 9)$

y-intercept

$y = -(0)^2 - 2(0) + 8 = 8$

$(0, 8)$

x-intercepts

$-x^2 - 2x + 8 = 0$

$0 = x^2 + 2x - 8$

$0 = (x + 4)(x - 2)$

$x = -4 \quad \text{or} \quad x = 2$

$(-4, 0), (2, 0)$

54. Length: $x + 10$, Width: x

$x(x + 10) = 350$

$x^2 + 10x - 350 = 0$

$x = \dfrac{-10 \pm \sqrt{(10)^2 - 4(1)(-350)}}{2(1)}$

$x = \dfrac{-10 \pm \sqrt{1500}}{2}$

$\dfrac{-10 + \sqrt{1500}}{2} \approx 14.4$

$\dfrac{-10 - \sqrt{1500}}{2} \approx -24.4$

Omit the negative solution. The length is approximately $14.4 + 10 = 24.4$ feet and the width is approximately 14.4 feet.

55. $300^2 + 900^2 = c^2$

$90{,}000 + 810{,}000 = c^2$

$900{,}000 = c^2$

$\sqrt{900{,}000} = \sqrt{c^2}$

$c \approx 948.7$

The flight is approximately 948.7 miles.

Chapter 9 Test   **1.** $\{-8, -3\}$ **2.** $\{-6 \pm 2\sqrt{30}\}$ **3.** $\{5 \pm \sqrt{42}\}$ **4.** $\{-4 \pm 2i\}$ **5.** $\{-1, 6\}$ **6.** $\{2, \frac{9}{4}\}$ **7.** $\{5, 9\}$

8. $\{-3 \pm 3i\sqrt{2}\}$ **9.** $3x^2 + x - 10 = 0$ **10.** $x^2 + 9 = 0$ **11.** x-intercepts: $(-7 \pm \sqrt{19}, 0)$, y-intercept: $(0, 30)$

12. Vertex: $(-3, 1)$, y-intercept: $(0, -8)$, x-intercepts: $(-4, 0), (-2, 0)$

13. Vertex: $(-3, -8)$, y-intercept: $(0, 1)$, x-intercepts: $(-3 \pm 2\sqrt{2}, 0)$

14. Length: 83.4 yards, Width: 38.4 yards

15. 1562.0 miles

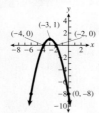

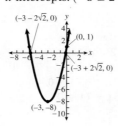

Cumulative Review Chapters 7–9   **1.** $5, 9$ **2.** $\frac{x-6}{x+7}$ **3.** $\frac{3}{7}$ **4.** $\frac{(x-3)(x-2)}{(x+11)(x+2)}$ **5.** $\frac{(x+10)(x-2)}{(x+7)(x+1)}$ **6.** $\frac{3}{x+2}$ **7.** $\frac{2(4x-13)}{(x+2)(x-1)(x-5)}$
8. $\frac{x-2}{(x-7)(x+3)}$ **9.** $\frac{x+8}{(x+6)(x+5)}$ **10.** $\frac{x-9}{x+8}$ **11.** $\frac{x-4}{x+8}$ **12.** $\{5\}$ **13.** $\{0, 18\}$ **14.** $\{-3\}$ **15.** $x = \frac{y}{2y-1}$ **16.** 10 **17.** 12 hours
18. 3.317 **19.** $2a^2b^6\sqrt{15a}$ **20.** $2x^3y^5\sqrt[3]{6yz^2}$ **21.** $7\sqrt{2}$ **22.** $46\sqrt{3}$ **23.** $18 - 22\sqrt{6}$ **24.** 111 **25.** $\frac{3\sqrt{2x}}{2x}$ **26.** $\frac{64 + 11\sqrt{10}}{37}$ **27.** $\{2\}$
28. 16 **29.** $6i\sqrt{3}$ **30.** $67 + 29i$ **31.** $\frac{6+8i}{5}$ or $\frac{6}{5} + \frac{8}{5}i$ **32.** $\frac{5-6i}{3}$ or $\frac{5}{3} - 2i$ **33.** $1024x^{30}$ **34.** $x^{9/10}$ **35.** $\{4 \pm 2i\sqrt{2}\}$
36. $\{-5 \pm 2\sqrt{11}\}$ **37.** $\{\frac{3 \pm \sqrt{39}}{2}\}$ **38.** $\{-9, -3\}$ **39.** $\{7 \pm \sqrt{34}\}$ **40.** $\{9 \pm 3\sqrt{2}\}$

41. Vertex: $(3, -16)$, y-intercept: $(0, -7)$,   **42.** Vertex: $(-\frac{3}{2}, \frac{29}{4})$, y-intercept: $(0, 5)$,   **43.** Vertex: $(2, 1)$, y-intercept: $(0, 5)$,
x-intercepts: $(-1, 0), (7, 0)$ x-intercepts: $(\frac{-3 \pm \sqrt{29}}{2}, 0)$ x-intercept: None

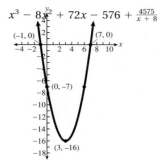

$x^3 - 8x^2 + 72x - 576 + \frac{4575}{x+8}$

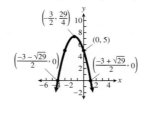

44. Length: 30.5 feet, Width: 24.5 feet   **45.** 4.22 seconds

Appendix A

1. $x - 11$ **3.** $x + 16$ **5.** $7x - 25 + \frac{38}{x+2}$ **7.** $12x - 11 - \frac{308}{x-14}$ **9.** $x^2 + 5x - 9 + \frac{13}{x-5}$
11. $x^3 - 8x^2 + 72x - 576 + \frac{4575}{x+8}$

Photo Credits

Agriculture

Feed, 284, 297, 482
Fencing, 108, 283, 297, 420
Gardening, 282, 297, 412, 481
Goat pen, 413
Peach farmer, 404
Plant food, 121
Weed spray, 32, 137

Automotive/Transportation

Airplane travel, 284, 483, 585, 606, 607
Antifreeze solution, 121, 284
Auto plant production, 120
Auto sales, 133
Car travel, 482, 488, 585
Miles per hour, 108, 284
Race car, 108, 284
Rental car, 134
Road sign, 585
Skid marks, 532, 533, 551
Train speed, 108, 109
Truck driver, 284

Business

Bookstore, 120
Booth rental, 96
Business executives, 47
Coffeehouse, 71, 261
Company losses, 16
Copy machine, 166, 481
Day care center, 121
Discounting prices, 136
Finding average cost, 330
Gross pay, 483
Insurance company, 85
Internet company, 16
New company, 69
Online retail sales, 51
Rental company, 71
Sales, 84, 120, 137, 138, 194, 214, 321, 404
Stuffing envelopes, 482
Toy production, 417
Winery, 134

Construction

Concrete slab, 283
Doors, 297, 412
Electrical work, 283, 482
Guy wire, 585

Highway, 184
Landscaping, 72, 133, 214
Lumber, 32
Painting, 481
Picture frame, 32
Sandbox, 107, 585
Support beam, 483, 488
Swimming pool, 283, 412, 481, 606
Water pipes, 482, 488, 489, 459
Water tank, 532, 609

Consumer Applications

Admission to amusement park, 297
Cell phone use, 50
Cost of a gift, 39
Cost of a hot dog, 39
Cost of a movie, 15, 47, 110, 272, 297
Cost of clothing, 282
Cost of paper, 39
Holiday debt, 47
Holiday shopping, 49
Price of flowers, 272
Price of gasoline, 48
Price of milk, 120
Price of pizza, 83
Purchase of a formal dress, 134

Economics

Accumulating interest, 136
Cash, 106, 138, 239, 261
Cash box, 272
Certificates of deposit, 120, 283, 297, 420
Checking account, 15, 39, 69
Coins, 84, 109, 282
Fiscal year, 321
Mutual funds, 50, 120, 283, 297
Piggy bank, 136, 272
Postage stamps, 272
Savings account, 109, 120, 136
Shares of stock, 120
Stock market, 39, 40
Stock values, 16
Vacation fund, 233

Education

Algebra class, 283
Attending college, 47
College degrees, 46, 136, 138, 261, 330
College tuition, 65

Community college, 70, 120, 121, 134, 137, 167, 239
School fund-raiser, 110, 134, 194
Exam, 85, 120, 134, 282
Financial aid, 321
Math book, 283
Math department, 47
National assessment, 46, 48, 49
Nursing students, 184
Reading tutor, 166
School teachers, 46
Size of a classroom, 412
Sorority, 261
Teacher qualifications, 121
Test score, 120, 134, 137

Entertainment/Sports

Baseball, 96, 107, 239, 272, 282, 418, 420
Basketball game, 229, 260, 261
Basketball players, 65, 192, 283
Bicycling, 489
Blackjack, 261
Canoeing, 481
Country club, 167, 214
Dance recital, 108
Dartboard, 413
Football, 193
Golfing, 15
Hang gliding, 584
Hockey team, 297
Kayaking, 284, 297, 483, 488
Little league baseball, 585
Music, 49
Rowboat, 284
Running, 482
Sailboat, 532, 533, 584
Scoring a goal, 133, 282
Skydiving, 311
Soccer, 70, 282, 412
Softball, 282
Swimming, 109, 483
Television, 65
Wine-tasting festival, 134
Winery tour, 134

Environment

Cold front, 85
Elevation change, 15
Salmon swimming upstream, 482
Temperature, 15, 96, 193
Tsunami, 533

Index

A

Absolute values, 5
Addition
  associative property in, 61
  commutative property in, 60–61
  of complex numbers, 535–36
  of decimals, 33
  of fractions
    with different denominators, 28
    with same denominator, 26
  of integers, 8–11
  in order of operations, 53
  of polynomials, 325
  of radical expressions containing like
    radicals, 505–7
  of rational expressions
    that have different denominators, 450–53
    that have opposite denominators, 444–45
    that have same denominator, 441–42
Addition method
  solving applied problems involving systems
    of equations by using, 273–74
  solving systems of equations with, 262–66
    coefficients that are fractions or decimals
      using, 266–68
Addition property of equality, 80–81, 86–87
Algebraic fractions, 424. *See also* Rational
  expressions
Amount, 111
Area
  of circle, 101
  of rectangle, 101
  of square, 100
  of triangle, 100
Arithmetic expressions, order of operations in
  simplifying, 53–55
Arithmetic operations, performing, using
  numbers in scientific notation, 317–18
Associative property
  in addition, 61
  in multiplication, 61
Axis, 143
  horizontal, 143
  of symmetry, 587
  vertical, 143

B

Bar graphs, 42–44
Base
  of exponents, 52, 302
  in percent equation, 111
Binomial factors, factoring common, out of
  polynomial, 356

Binomials, 324
  factoring, 381
  factoring special, 377–81
Braces ({ }), 2

C

Calculator
  approximating square roots using, 495–96
  in checking solutions to linear equations, 89
  for exponents, 52
  in graphing lines, 157
  in graphing parabolas, 590
  in multiplication of integers, 13
  in rewriting decimals as fractions, 35
  scientific notation on, 318
  in simplifying expressions, 11, 54
  in solving quadratic equations, 570
  in solving system of linear equations, 245–46
Cartesian plane, 143
Charts, pie, 41–42
Circle
  area of, 101
  circumference of, 100
Circle graph. *See* Pie charts
Circumference of a circle, 100
Coefficient(s), 62
  identifying, to be used in the quadratic
    formula, 568
  of term, 324
Commutative property
  in addition, 60–61
  in multiplication, 60–61
Completing the square
  finding vertex of parabola by, 595–96
  solving quadratic equations by, 559–62
Complex fractions, 456
  multiplying, 537–38
  simplifying, 456–57
    containing variables, 457–59
Complex numbers, 535
  addition of, 535–36
  division by, 538–39
  multiplication of, 537–38
  subtraction of, 535–36
Composite numbers, 18
Compound inequalities, solving, 127–29
  involving intersection of linear inequalities,
    128–29
  involving union of linear inequalities, 127–28
Conjugates, 513
  multiplying, 513–14
Consecutive integers, 102–3
  solving problems involving, 102–3, 405–6
Constant functions, 188

Constant of variation, 478
Constant term, 323
Contradiction, 92
Coordinates, 142
Cross-canceling, 24
Cross multiplying, 117–18
Cross products of proportion, 117
Cubed, 52
Cube roots, 496

D

Data, 41
  displaying over time, using line graph,
    44–45
  finding linear equation to describe linear,
    209–10
  representing graphically, 41–45
Decimal notation, 33
Decimals, 33
  addition of, 33
  division of, 34
  multiplication of, 33–34
  rewriting
    as fractions, 34–35
    as percents, 36
  rewriting percents as, 36–37
  solving systems of equations with
    coefficients that are, 266–68
  subtraction of, 33
Degree of each term, 323
Degree of polynomial, 324–25
Denominators, 20
  addition of fractions
    with different, 28
    with same, 26
  addition of rational expressions
    that have different, 450–53
    that have opposite, 444–45
    that have same, 441–42
  least common, 449–50
  rationalizing, 514–16
    that has two terms, 516–18
  subtraction of fractions
    with different, 28
    with same, 26
  subtraction of rational expressions
    that have different, 450–53
    that have opposite, 444–45
    that have same, 442–43
Dependent systems of equations, 247
  addition method in identifying, 265–66
  substitution method in identifying, 255–56
Descartes, René, 143, 227, 237
Descending order, 324

Tab Your Way to Success

Use these tabs to mark important pages of your textbook for quick reference and review.

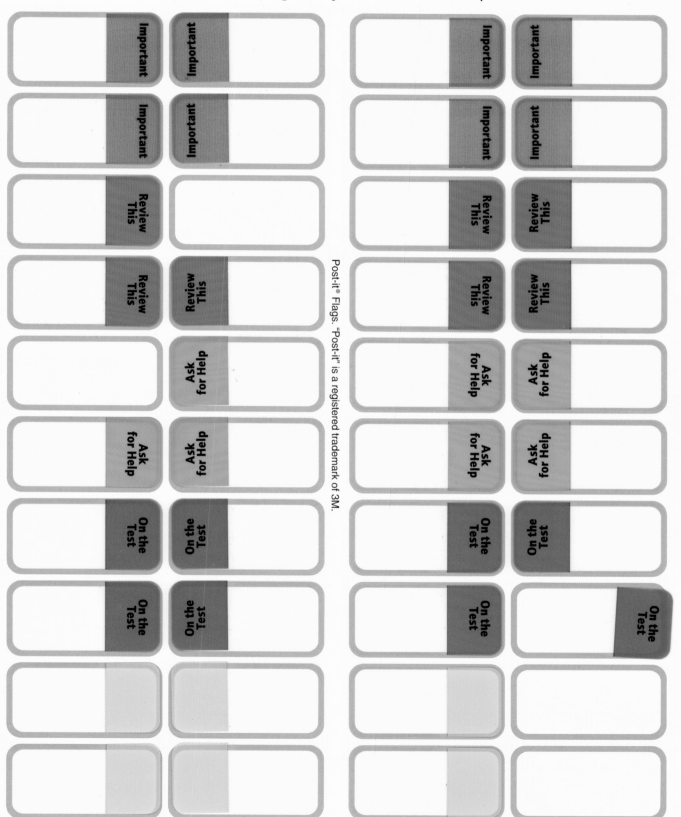

Important

Important

Important

Important

Important

Important

Important

Important

Review This

Review This

Review This

Review This

Review This

Ask for Help

Ask for Help

Ask for Help

Ask for Help

Ask for Help

Ask for Help

On the Test

On the Test

On the Test

On the Test

On the Test

On the Test

On the Test

Post-it® Flags. "Post-it" is a registered trademark of 3M.

Using the Tabs

Customize Your Textbook and Make It Work for You!

These removable and reusable tabs offer you five ways to be successful in your math course by letting you bookmark pages with helpful reminders.

 Use these tabs to flag anything your instructor indicates is important.

 Mark important definitions, procedures, or key terms to review later.

 Not sure of something? Need more instruction? Place these tabs in your textbook to address any questions with your instructor during your next class meeting or with your tutor during your next tutoring session.

 If your instructor alerts you that something will be covered on a test, use these tabs to bookmark it.

 Write your own notes or create more of the preceding tabs to help you succeed in your math course.

ISBN-13: 978-0-321-53637-2
ISBN-10: 0-321-53637-1

EAN

90000

9 780321 536372

Geometry Formulas and Definitions

- Perimeter of a triangle with sides s_1, s_2, and s_3:
  $$P = s_1 + s_2 + s_3$$

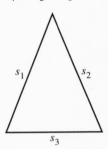

- **Circumference** (distance around the outside) of a circle with radius r: $C = 2\pi r$

- Area of a rectangle with length L and width W: $A = LW$

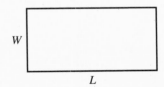

- An **equilateral triangle** is a triangle that has three equal sides.

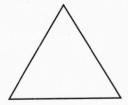

- Perimeter of a square with side s: $P = 4s$

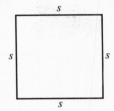

- Area of a triangle with base b and height h:
  $$A = \frac{1}{2}bh$$

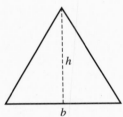

- Area of a circle with radius r:
  $$A = \pi r^2$$

- An **isosceles triangle** is a triangle that has at least two equal sides.

- Perimeter of a rectangle with length L and width W:
  $$P = 2L + 2W$$

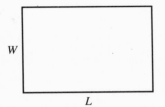

- Area of a square with side s: $A = s^2$

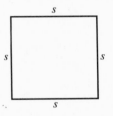

Formulas and Equations

Distance Formula (see page 103) $d = r \cdot t$

Basic Percent Equation (see page 111)

Amount = Percent · Base

Simple Interest (see page 114) $I = P \cdot r \cdot t$

Horizontal Line (see page 159) $y = b$

Vertical Line (see page 160) $x = a$

Slope Formula (see page 171) $m = \dfrac{y_2 - y_1}{x_2 - x_1}$

Slope–Intercept Form of a Line (see page 174)
$y = mx + b$

Point–Slope Form of a Line (see page 206)
$y - y_1 = m(x - x_1)$

Rules for Exponents (see pages 308, 312)

$x^m \cdot x^n = x^{m+n}$

$(x^m)^n = x^{m \cdot n}$

$(xy)^n = x^n y^n$

$\dfrac{x^m}{x^n} = x^{m-n} \ (x \neq 0)$

$x^0 = 1 \ (x \neq 0)$

$\left(\dfrac{x}{y}\right)^n = \dfrac{x^n}{y^n} \ (y \neq 0)$

$x^{-n} = \dfrac{1}{x^n} \ (x \neq 0)$

Special Products (see page 334)

$(a + b)(a - b) = a^2 - b^2$

$(a + b)^2 = a^2 + 2ab + b^2$

$(a - b)^2 = a^2 - 2ab + b^2$

Factoring Formulas

Difference of Squares:
$a^2 - b^2 = (a + b)(a - b)$ (see page 377)

Difference of Cubes:
$a^3 - b^3 = (a - b)(a^2 + ab + b^2)$ (see page 379)

Sum of Cubes:
$a^3 + b^3 = (a + b)(a^2 - ab + b^2)$ (see page 380)

Height, in feet, of a projectile after t seconds (see page 409)
$h(t) = -16t^2 + v_0 t + s$

Direct Variation (see page 478) $y = kx$

Inverse Variation (see page 479) $y = \dfrac{k}{x}$

Properties of Radicals (see page 498)

$\sqrt[n]{a} \cdot \sqrt[n]{b} = \sqrt[n]{ab}$

$\dfrac{\sqrt[n]{a}}{\sqrt[n]{b}} = \sqrt[n]{\dfrac{a}{b}}$

Fractional Exponents (see page 544) $a^{m/n} = (\sqrt[n]{a})^m$

Period of a Pendulum (see page 529) $T = 2\pi\sqrt{\dfrac{L}{32}}$

Imaginary Numbers (see page 534) $i = \sqrt{-1}, \ i^2 = -1$

Quadratic Formula (see page 568) $x = \dfrac{-b \pm \sqrt{b^2 - 4ac}}{2a}$

Pythagorean Theorem (see page 578) $a^2 + b^2 = c^2$